AF762822

Ice Caves

Ice Caves

Edited by

Aurel Perşoiu

Stein-Erik Lauritzen

ELSEVIER

Elsevier
Radarweg 29, PO Box 211, 1000 AE Amsterdam, Netherlands
The Boulevard, Langford Lane, Kidlington, Oxford OX5 1GB, United Kingdom
50 Hampshire Street, 5th Floor, Cambridge, MA 02139, United States

Library of Congress Cataloging-in-Publication Data
A catalog record for this book is available from the Library of Congress

British Library Cataloguing-in-Publication Data
A catalogue record for this book is available from the British Library

ISBN: 978-0-12-811739-2

Publisher: Candice Janco
Acquisition Editor: Laura S. Kelleher
Editorial Project Manager: Tasha Frank
Production Project Manager: Anitha Sivaraj
Cover Designer: Christian J. Bilbow

Typeset by SPi Global, India

The contributions presented in this book are the result of studies conducted by both amateur cavers and scientists; however, most of the work behind exploring the caves and gathering of data was done by anonymous cavers pursuing their curiosity and yearning for exploration—the children deep inside us.
This book is dedicated to them.

Contents

Contributors

Lene Baastad
University of Bergen, Bergen, Norway

Carmen-Andreea Bădăluță
Ștefan cel Mare University, Suceava, Romania

Darko Bakšić
University of Zagreb, Zagreb, Croatia

Goran Barović
University of Montenegro, Podgorica, Montenegro

Miguel Bartolomé
Instituto Pirenaico de Ecología, CSIC, Zaragoza, Spain

Pavel Bella
State Nature Conservancy of the Slovak Republic, Slovak Caves Administration, Liptovský Mikuláš; Catholic University in Ružomberok, Ružomberok, Slovakia

Ánchel Belmonte
Universidad de Zaragoza, Zaragoza, Spain

Neven Bočić
University of Zagreb, Zagreb, Croatia

Traian Brad
Emil Racoviță Institute of Speleology, Cluj-Napoca, Romania

Nenad Buzjak
University of Zagreb, Zagreb, Croatia

Jelena Ćalić
Geographical Institute "Jovan Cvijić" of the Serbian Academy of Sciences and Arts, Belgrade, Serbia

Renato R. Colucci
ISMAR-CNR; Commissione Grotte E. Boegan, Società Alpina delle Giulie, Trieste, Italy

Yuri Dublyansky
University of Innsbruck, Innsbruck, Austria

Vinka Dubovečak
Speleological Association Kraševski zviri, Ivanec, Croatia

Julie Engelien
University of Bergen, Bergen, Norway

Michal Filippi
Institute of Geology of the Czech Academy of Sciences, Prague, Czech Republic

Derek Ford
McMaster University, Hamilton, ON, Canada

Gaetano Giudice
Istituto Nazionale di Geofisica e Vulcanologia sezione di Palermo – Geochimica (INGV), Palermo; Centro Speleologico Etneo (CSE), Catania, Italy

Manue Gómez-Lende
Universidad de Valladolid, Valladolid, Spain

Michał Gradziński
Jagiellonian University, Kraków, Poland

Stephen R. Higham
Consultant, Georgetown, MA, United States

Bernard Hivert
AS Charentaise, Angouleme, France

Greg Horne
Jasper National Park of Canada, Jasper, AB, Canada

Sanda Iepure
IMDEA Water Institute, Alcala de Henares, Spain

Pierre-Yves Jeannin
Swiss Institute of Speleology and Karst Studies (SISKA), La Chaux-de-Fonds, Switzerland

Olga I. Kadebskaya
Mining Institute of the Ural Branch of the Russian Academy of Sciences, Perm, Russia

Zoltán Kern
Institute for Geological and Geochemical Research, Research Centre for Astronomy and Earth Sciences, MTA, Budapest, Hungary

Ditta Kicińska
Adam Mickiewicz University, Poznań, Poland

Bernard Lauriol
Université d'Ottawa, Ottawa, ON Canada

Stein-Erik Lauritzen
University of Bergen, Bergen, Norway

María Leunda
Instituto Pirenaico de Ecología, CSIC, Zaragoza, Spain

Marc Luetscher
Swiss Institute of Speleology and Karst Studies (SISKA), La Chaux-de-Fonds, Switzerland; Austrian Academy of Sciences, Innsbruck, Austria; University of Innsbruck, Innsbruck, Austria

Valter Maggi
University of Milano-Bicocca, Milano; Institute of Geosciences and Earth Resource (CNR), Pisa, Italy

Mihajlo Mandić
Geological Survey of Serbia, Belgrade, Serbia

Bulat R. Mavlyudov
Institute of Geography of the Russian Academy of Sciences, Moscow, Russia

Christiane Meyer
Ruhr-University Bochum, Bochum, Germany

Andrej Mihevc
Karst Research Institute, Postojna, Slovenia

Ana Moreno
Instituto Pirenaico de Ecología, CSIC, Zaragoza, Spain

Jasminko Mulaomerović
Center for Karst and Speleology, Sarajevo, Bosnia and Herzegovina

Alireza Nadimi
University of Isfahan, Isfahan, Iran

Mohammad Nakhaei
Kharazmi University, Tehran, Iran

Dragan Nešić
Institute for Nature Conservation of Serbia, Niš, Serbia

Jakub Nowak
Cracow Caving Club, Kraków, Poland

Friedrich Oedl
Eisriesenwelt Cave, Salzburg, Austria; International Show Caves Association (ISCA), Ancona, Italy; Austrian Caving Association VÖH, Vienna, Austria

Bogdan P. Onac
University of South Florida, Tampa, FL, United States; Emil Racoviță Institute of Speleology, Cluj-Napoca, Romania

Dalibor Paar
University of Zagreb, Zagreb, Croatia

Arthur N. Palmer
State University of New York, Oneonta, NY, United States

Rudolf Pavuza
Natural History Museum Vienna, Vienna, Austria

Christos Pennos
Aristotle University of Thessaloniki, Thessaloniki, Greece

Aurel Perșoiu
Emil Racoviță Institute of Speleology, Cluj-Napoca, Romania

Andreas Pflitsch
Ruhr-University Bochum, Bochum, Germany

Lukas Plan
Natural History Museum Vienna, Vienna, Austria

Cristina Purcarea
Institute of Biology, Bucharest, Romania

Grzegorz Rachlewicz
Adam Mickiewicz University in Poznan, Poznań, Poland

Luca Randazzo
Centro Speleologico Etneo (CSE), Catania, Italy

Javier Sánchez-Benítez
CES Alfa, Madrid, Spain

Carlos Sancho
Universidad de Zaragoza, Zaragoza, Spain

Jacques Schroeder
Université du Québec à Montréal, Montréal, QC, Canada

Federico Scoto
Ca' Foscari University of Venice, Venezia Mestre; Fondazione CMCC Centro euro-Mediterraneo sui Cambiamenti Climatici, Bologna, Italy

Enrique Serrano
Universidad de Valladolid, Valladolid, Spain

Wiesław Siarzewki
Tatra National Park, Zakopane, Poland

Mojgan Soleymani
Kharazmi University, Tehran, Iran

Yorgos Sotiriadis
Technological Institute of Eastern Macedonia and Thrace, Kavala, Greece

Christoph Spötl
University of Innsbruck, Innsbruck, Austria

Krzysztof Strug
The Institute of Meteorology and Water Management—National Research Institute, Wrocław, Poland

Michael Styllas
Geoservice LTD, Thessaloniki, Greece

Witold Szczuciński
Adam Mickiewicz University in Poznan, Poznań, Poland

Marjan Temovski
Institute for Nuclear Research, Hungarian Academy of Sciences, Debrecen, Hungary

Markos Vaxevanopoulos
Aristotle University of Thessaloniki, Thessaloniki; Natural History Museum of Volos, Volos, Greece

Maximilian Wimmer
Landesverein für Höhlenkunde in Oberösterreich, Linz, Austria

Charles J. Yonge
Yonge Cave and Karst Consulting Inc., Canmore, AB, Canada

Karel Žák
Institute of Geology of the Czech Academy of Sciences, Prague, Czech Republic

Ján Zelinka
State Nature Conservancy of the Slovak Republic, Slovak Caves Administration, Liptovský Mikuláš, Slovakia

PART

I

PERENNIAL ICE ACCUMULATIONS IN CAVES: OVERVIEW

CHAPTER

INTRODUCTION

1

Aurel Perşoiu*, Stein-Erik Lauritzen†

Emil Racoviţă Institute of Speleology, Cluj-Napoca, Romania University of Bergen, Bergen, Norway†*

Ice caves (i.e., caves hosting perennial ice accumulations) are arguably the least well-known part of the global cryosphere. They occur in places where peculiar cave morphology and climatic conditions combine to allow for ice to form and persist in otherwise adverse conditions. By their nature, ice caves are both sensitive indicators of present-day climatic conditions, as well as archives of past climates. Recent climatic change is in fact jeopardizing their existence (Kern and Perşoiu, 2013). While ice caves became an increasingly important target for the scientific community in the past decade, they have been the subjects of scientific studies for more than a century. Nevertheless, the vast majority of the results of these studies were published in various languages outside the mainstream scientific languages of the past century (French, German, and English).

Since the monographs of Balch (1900) and Kyrle (1923), no comprehensive overview of the ice caves phenomenon has appeared. The present volume attempts to present an up-to-date review on the physical processes and physiography of ice caves, and their global distribution. The book is divided into two sections. First, a series of topical chapters describes theoretical and general aspects of the physical processes and formation mechanisms for cave ice.

Meyer reviews the history of ice cave observations and research (which actually commenced in the 12th century) and takes the development of ideas up to the present. Perşoiu discusses the climatic conditions prevailing in ice caves, including enthalpy dynamics in single and multi-entrance caves, as well the palaeoclimatic significance of perennial ice accumulations in caves. Bulat, Bella, and Perşoiu present the physical processes of ice formation and the various morphological forms resulting from cooling. Kern summarizes the methods available for dating cave ice. Žák discusses cryogenic and periglacial processes taking place on and in the vicinity of cave ice, like mineralization and mobilization of sediments. Iepure and Purcarea consider the ice cave habitat and present the occurrence of life forms (fauna and microbiota) on and in cave ice. This knowledge has wide consequences, and is in fact crucial for the possibility of finding extraterrestrial life, and for sustaining human occupation, e.g., on Mars (e.g., Williams et al., 2010). How to manage our ice caves, of which some are tourist attractions? Oedl presents his experiences from Eisriesenwelt in, Austria and other caves. The presence of ice in caves is a widespread global phenomenon with obvious latitudinal constraints (Fig. 1.1). Bulat presents the final chapter on the geography of cave glaciation. In the next 20 chapters, national experts present the most important ice caves in their respective countries (Fig. 1.1) and document existing micrometeorology, stratigraphy, and their implications for climate history.

Ice Caves. https://doi.org/10.1016/B978-0-12-811739-2.00001-2

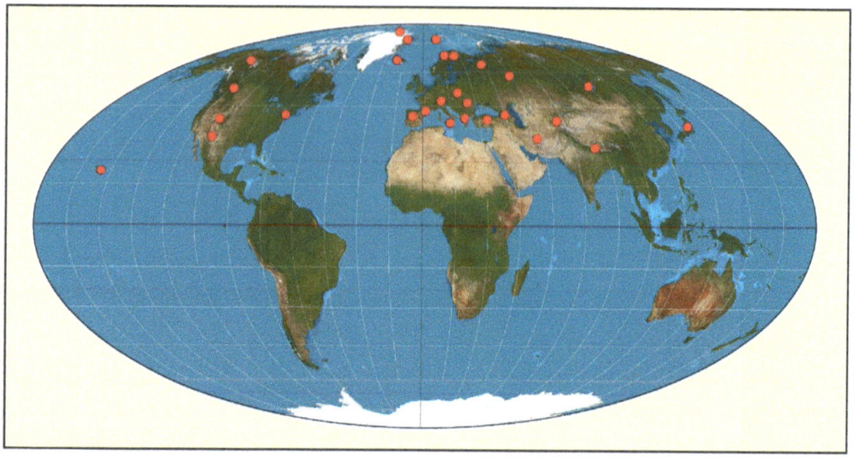

FIG. 1.1

Global distribution of ice cave sites (dots represent regions, rather than single caves) presented in the book.

We hope that the present book will serve as a trigger for the mind and a platform for new ideas, and that it will further research on a fragile resource that is literally melting away in our hands (i.e., rapidly ablating in our time).

ACKNOWLEDGMENTS

This work was partly supported by the Research Council of Norway through its Centre of Excellence's funding scheme, SFF Centre for Early Sapiens Behaviour (SapienCE), project number 262618, and by The Bjerknes Centre for Climate Research. We are also indebted to grants numbers PNII-RU-TE-2014-4-1993, financed by UEFISCDI, Romania, and ELAC2014/DCC-0178 (contract no. 24/2016), financed by the EU ERA-NET LAC program. Arthur Palmer, Megan Palmer, Charles Yonge, John Mylroie, Bogdan P. Onac, Dylan Parmenter, Norbert Schorghofer, Ethan Upton, and Vasile Ersek all have our thanks for crucial language corrections.

REFERENCES

Balch, E.S., 1900. Glacieres, or Freezing Caverns. Allen Lane & Scott, Philadelphia, p. 406.

Kern, Z., Perşoiu, A., 2013. Cave ice - the imminent loss of untapped mid-latitude cryospheric palaeoenvironmental archives. Quaternary Science Reviews 67, 1–7.

Kyrle, G., 1923. Grundriß der theoretischen Speläologie. Österreichische Staatsdruckerei, Vienna.

Williams, K.E., McKay, C.P., Toon, O.B., Head, J.W., 2010. Do ice caves exist on Mars? Icarus 209 (2), 358–368.

CHAPTER

HISTORY OF ICE CAVES RESEARCH

2

Christiane Meyer
Ruhr-University Bochum, Bochum, Germany

CHAPTER OUTLINE

2.1 INTRODUCTION

The natural phenomenon ice cave has been known for centuries. Today ice cave sites are reported in many regions around the world, and the research activity is very high in many countries. But in the beginning, ice caves had been recognized only as a single phenomenon. Over the course of time the increase of knowledge was reduced and retarded by several factors: first, the inaccessibility at remote places in High Mountain regions, and second, the inconsistency of the related nomenclature. As a result the history of ice cave research didn't proceed consistently in one direction of development, and numerous theories evolved concurrently. But apart from that a substantial, mainly descriptive, literature developed with descriptions of ice caves worldwide, a related nomenclature, geographical setting, ice morphology, size and the dynamics of the ice body, and the cave climate.

In this chapter, we will give an overview of the most important aspects of ice cave history in general. The course of the ice cave history on a regional level may differ, since the starting point of the ice cave research, or the recognition of the phenomenon, is very different. In some countries (e.g., France, Slovenia, Russia), the phenomenon ice cave has been known of for centuries, while in some countries like Croatia (Bočić et al., 2014), ice cave research just recently started. Instead of a common presentation of the historical chronology, we will focus on specific aspects of ice cave history and describe their development over the centuries. Beginning with an overview of the first historical reports on ice caves worldwide, we will proceed with the development of the nomenclature and definitions related to ice caves, the history of ice caves theories, and finally come to the first systematical investigations in the past.

Ice Caves. https://doi.org/10.1016/B978-0-12-811739-2.00002-4

2.2 FIRST HISTORICAL REPORTS ON ICE CAVES WORLDWIDE

One of the first questions arising when talking about the history of ice cave research is where and when ice caves have been reported in the past in the different regions of the world. For this purpose, we focus on the earliest documents, in which the phenomenon ice cave was first mentioned. Different kinds of resources have been used, e.g., chronicles, historical travel reports, fiction books, old scientific articles and monographs, religious texts, and many more to gather as much information as possible. The results deliver information about the facts, where ice caves were known in the different centuries. But this can not be equated with the geographical distribution of ice caves, since we can assume that surely not all ice caves have been discovered or documented in the past. Due to the age of some publications, it was not always possible to get access to all known literature and to review them. But whenever possible the original texts, some of them over 500 years old, were digitized and reviewed. In the case where it was not possible to get access to the literature, multiple entries in the secondary literature were used as proof. Additionally, the access to original sources is also complicated due to the diverse languages in which they were published. Modern publications to find an overview about the ice cave history are Mavlyudov (2008a), Turri et al. (2009), and Meyer et al. (2016).

The oldest document we found, which refers to the existence of an ice cave, dates back to the 12th century. In his chronicle of the kings of Kashmir, Kalhaṇa describes the pilgrimage to Amarnath cave, as it is also written in ancient Sanskrit texts and still visited by pilgrims today (Stein, 1961). Today this cave is located in India. Kalhana wrote in his chronicles around AD 1149 (Stein, 1961, p. 6) "The lake of dazzling whiteness [resembling] a sea of milk, which he created [for himself as residence] on a far-off mountain, is to the present day seen by the people on the pilgrimage to Amarésvara." (Stein, 1961, p. 40) and "And the most delightful Kasmir summer which is not to be found [elsewhere] in the whole world, was used to good purpose over the worship of Lingas formed of snow in the regions above the forests." (Stein, 1961, p. 68) refer to this pilgrimage. Stein (1961, p. 41) explained that "the Amarésvara-yãtrã is directed to the famous cave of Amarnath. … In it S'iva Amarésvara is believed to have manifested himself to the gods who entreated him for protection against death. The god is worshipped in a linga-shaped ice-block."

Around 1490 Petrus Ranzanus mentioned, in his *Epithoma rerum Hungararum*, an ice cave in the Carpathian region (Schönviszky, 1968): "[…] near Scepusium there are cliffs, where […] water […] in summer is frozen" (Ransanus, 1977, p. 71, own translation). At this time Da Vinci also reports in his notebooks about the ice cave of Moncodeno: "These excursions are to be made in the month of May. And the largest bare rocks that are to be found in this part of the country are the mountains of Mandello near to those of Lecco, and of Gravidona towards Bellinzona, 30 miles from Lecco, and those of the valley of Chiavenna; but the greatest of all is that of Mandello, which has at its base an opening towards the lake, which goes down 200 steps, and there at all times is ice and wind." (Richter, 1883, p. 238f; Balch, 1900, p.211). Some of the best known descriptions of a European ice caves date back to the 16th century. Fugger (1893) mentioned Poissenot (1586, pp. 436–453) and (Gollut (1592, p. 88), who reported about Chaux-les-Passavant (Fig. 2.1).

Poissenot writes in a letter: "[…] having come to the cave, which we found of the length and width of a large hall, all surfaced with ice in the lower part […] After having searched in my mind the cause for this antiperistase, I did not find another but this: namely it is that as the heat dominates in summer, the cold retreats to places low and subterranean like this to which the rays of the sun cannot reach" (Poissenot, 1586, p. 265).

FIG. 2.1

Chaux-les-Passavent (Balch, 1900).

Gollut notes: "[...] since at the bottom of a mountain of Leugne ice is found in summer, for the pleasure of those who wish to drink cool. Nevertheless at this time, this is disappearing, for no other reason (as I think) except, that they have despoiled the top of the mountain, of a thick and high mass of woods, which did not permit that the rays of the sun came to warm the earth, and dry up the distillations, which slipped down to the lowest and coldest part of the mountain where (by antiperistase) the cold got thicker, and contracted itself against the heats surrounding and in the neighborhood during the whole summer, all the external circumference of the mountain" (Gollut, 1592, p. 88 [translated by Balch, 1900, p. 202]). Schwalbe (1887) explained that Chaux les-Passavant and Baume or Grâce de Dieu near Besançon in the Jura Mountains were often mixed up, although it is the same cave. Over the following centuries, Chaux-les-Passavant has continuous publications (e.g., Girardot and Trouillet, 1885) during each century since 1586 (Billerez, 1712; Boz, 1726; Prevost, 1789; Pictet, 1822; Fugger, 1891; Balch, 1900 and others), and thus is, apparently, the best documented ice cave in Europe.

In the 17th century, Valvasor's descriptions of ice caves in the former archduchy Carniola (1689), today Slovenia, can be assumed as the earliest descriptions of ice caves in this region. Later, in 1730, Matthias Bel then published a Latin article about the ice cave of Szilicze (today Slovakia) (Fugger, 1893). Mavlyudov (2008a), as well as Turri et al. (2009), refer consistently to the first reference of ice caves in the Volga valley in Russia from 1690, also for the 18th century numerous publications about Russian ice caves are cited. Until this point of time we can assume that only individual sites were known, and an ice cave was assumed to be a unique phenomenon. For example, the first historical report about an ice cave in Germany dates back to 1703, when ice caves are mentioned in the Harz mountains near Questenberg by Bel and by Behrens. Until the 19th century, no other report on ice caves in Germany could be found. Also, for Russia, Mavlyudov (2008a) describes ice caves reported are at different places in Russia and presents a comprehensive picture of the ice cave research in Russia. According to Mavlyudov (2008a), most of the publications till the mid-19th century show a descriptive character. Furthermore, he drew the following conclusions from this descriptive period:

- Ice can be built in caves and exists in special caves.
- Caves with ice are to be found at different places.
- It is possible to find different forms of ice in caves.

We can adapt these conclusions for Russia to include in the general ice cave history of that time. In the 18th and 19th century more and more descriptions of ice caves were published in diverse regions of the world. We find numerous reviews of historical ice cave sources and compilations of ice cave locations in the works of Schwalbe (1887), Fugger (1891, 1893), Balch (1900), and others. The listed reports about ice caves show clearly that the knowledge about ice caves before the 19th century was not only numerically small, but also narrowly defined geographically. Beginning in Middle Europe and Russia,then in the other parts of Europe, and on other continents, ice caves became known at some remote period. Holmgren et al. (2016) present a comprehensive description of the ice cave research history in the United States. In the 19th century, a wide range of investigations, first descriptions, and new discoveries were done. This set the phenomenon ice cave, and their geographical distribution, in the focus of interest. The earliest notifications about some of today's best known ice caves in Slovakia and Romania date back to the 19th century. Among these are, for example, Demänovská ľadová jaskyňa (Berghaus, 1836), Dobšinská ľadová jaskyňa (Krenner, 1874), and Peștera Scărișoara (e.g., Vass, 1857).

As mentioned before, ice caves in the Jura and Western Alps (today France and Switzerland) have been known for a long time, but there was not much information for other ice caves except for Chaux-les-Passavant. All other reports are almost exclusively from the 19th century, among them St. George (esp. Pictet, 1822), the Upper and Lower ice cave of Pré de St. Livres (esp. Thury, 1861), the ice cave of Monthezy (esp. Ebel, 1818), the ice cave of Arc-sous-Cicon (Browne, 1865), the ice cave of Grand Anu (Thury, 1861), the ice cave of Reposoir (Saussure, 1796), the ice cave of Brezon (Pictet, 1822), the ice cave of Fondeurle (esp. Gilbert, 1815), and many more.

In the Northern Calcarious and Austrian Alps the first notifications also date back to the 19th century. Schwalbe (1887) mentions caves at Untersberg like Schellenberger Eishöhle (Fig. 2.2), which were studied by Fugger (1888), but also by Posselt-Czorich (1880). At Tennengebirge, Seeofen (Posselt 1880) and the Posselt Cave are described. The ice cave at Dachstein (Kraus, 1894) and "a new discovered ice cave" at Steinernes Meer are mentioned in a private letter from 1886 (Schwalbe, 1887). Furthermore, Schwalbe (1887) mentions, for example, the ice cave at Roten Kogel, Klimsteinhöhle at

FIG. 2.2

Schellenberger Eishöhle (Fugger, 1888).

Gmunden, the ice cave at Kasberg, Beilsteinhöhle near Gams (Mandl, 1838), Frauenmauerhöhle near Eisenerz, Bärenloch, Geldloch at Oetscher, Taberloch near Wien, Eiskapelle at Raxalpe were reported. In Middle Germany ice caves were also described at that time. Among these are the ice cave near Roth at Eifel (Scrope, 1826), the ice cave near Dürrberg, ice caves near Rosendorf, and ice formations at Sauberge (Reich, 1834).

Balch mentioned, in 1900, that ice caves are generally found in different parts of Europe, Asia and America, mostly at smaller mountain ranges and their foothills, mainly in limestone and sometimes in Basalt: Jura Mountains, Swiss and Italian Alps, Eastern Alps at Tyrol and Carinthia, Hungary, Russia, Iceland (Surtshellir, Fugger (1891), Balch (1900)), Tenerife (Cueva de la Nieve, e.g., Humboldt (1814), Fugger (1891)), Siberia, Central Asia, Himalaya, Japan (Glacière Lava cave near Shoji, Balch (1900)), Korea (Glacière cave on the Han Gang, Balch (1900)), United States (compare Holmgren et al. (2016)), and altogether about 300 worldwide.

In the 20th century, the evaluations of the phenomenon ice cave changed. The scientific investigation of ice caves became the focus of many publications, although still in some countries the simple notification continued. Scientists started to collect historical data to clarify the question regarding where ice caves were known in the past and at what time. Even today this question is still discussed, e.g., Mavlyudov (2008b). Only now have ice caves become a well-known phenomenon. In 1956 Saar wrote: "Apart from the European limestone high mountain regions [...] they are found in the Jura Mountains, the Alpine foothills [...] Several hundred ice caves are known in the United States and in Canada [...] a very large number of ice caves can also be expected to exist in other continents of the northern and southern hemisphere. Provided that karstic rock is present, their occurrence above the 35th degree of latitude north or south seems to be ensured, with their height above sea level decreasing with increasing latitude" (Saar, 1956, p. 58, own translation). In the United States, first Merriam (1950), then Halliday (1954), compiled ice cave locations in several different states.

Today there is an even greater interest in the geographical distribution of ice caves as shown by Mavlyudov (2008b) in the paper "Geography of caves glaciations" and also on historical data in "Cave glaciation in the past" (Mavlyudov, 2010), based on historical data. First descriptions are still published, but nowadays in connection with scientific investigations, e.g., Belmonte and Marcén (2010), Buzjak et al. (2014), Bočić et al. (2014), Colucci et al. (2014), Pflitsch et al (2016), Gómez Lende et al. (2016), or like in Macedonia ice cave research just recently started (Temovski, 2016). The publications of caving organizations and local caving clubs in many parts of the world have not been evaluated yet. Therefore, we can assume that these publications probably hold information about numerous ice cave locations.

2.3 DEVELOPMENT OF THE TERMINOLOGY RELATED TO ICE CAVES

As already mentioned, untill the 19th century, the existing literature had a mainly descriptive character. Nevertheless, over time different definitions and a related terminology evolved depending on the country and the language. In this part of the chapter the main steps of this development are presented. The terminology is closely connected to the existing knowledge about ice caves at a given time.

Still in 1975 Racovitza criticized that a heterogenic nomenclature exists, in which numerous concepts are in use, whose exact meaning or scope are not sufficiently specified, and therefore precision is needed. The vocabulary that developed showed a descriptive character. This is reflected in the cave names that, for example, always contain words like *glacière, Eisloch, Eisgrotte, Eiswinkel, Eiskeller, Eisschacht or Snow-Hole* (Fugger, 1891). Balch (1900) discusses, in his book "Glacieres Or Freezing Caverns" the term ice cave, what he regards as a mistake that the content is mentioned before the geologic formation. Furthermore, he proposes to use the terms "glacière naturelle" and "artificielle," because the presence of ice seems to be the main criteria to describe a cave as an ice cave. Three years earlier he noted already in a paper (Balch, 1897, p. 162): "The term glacière seems to me the most accurate in use in any language, and, if it were not too late to do so, it would be an advantage to use it, especially as we have adopted the term glacier from the same part of the world. In my opinion, the term 'ice cave' should especially apply to the hollows in the ice at the lower end of glaciers, whence the glacier waters make their exit."

Thury (1861) proposed defining ice caves by the primary processes that lead to the existence of underground ice, namely the airflow. He first introduces the differentiation between static and dynamic ice caves (36 f own translation): "Mr. Thury, in his memory, distinguishes two kinds of ice caves, the static ice cave, where the air rests immobile in summer, and the dynamic ice caves, where the airflow normally plays a certain role." Furthermore, on page 41 (own translation): "There is reason to believe, until new and more complete observations are conducted during all seasons of the year, that the ice cave of Vergy illustrates the coexistence of two classifications of phenomena distinguished by Pictet and Deluc. Even if the situation with the opening higher than the ground, and the fact that in August the ice doesn't form, give rise to believe that static theory have to be partly applicable. Whereas the well stated existence of airflow shows that the dynamic theory could have a share in the explication of the phenomena."

On the basis of his studies in the ice cave of Vergy, he found the proof that this ice cave shows the coexistence of the static and the dynamic type, which Bögli (1978) described as statodynamic over a hundred years later. Thus the differentiation between the static and the dynamic type by Thury (1861) was determined by the airflow system. For the first type the air remains immobile all summer, and in the second, air stream played a certain role for him, but he didn't define it precisely enough.

Girardot and Trouillet (1885), who first described the relation between outside and cave air temperature, introduced the terms open and closed period. They regard the open period as the time in the annual cycle, in which the outside air temperature is below the cave temperature. This leads to the fact that colder, and specifically heavier, air sinks from outside into the ice cave, and air exchange with the outside atmosphere takes place (Fig. 2.3A).

On the contrary, the closed period is the time in the annual cycle, in which the outside air temperature is above the cave temperature. This leads to the fact that the warmer, and thereby specific lighter, air does not sink from outside in the ice cave, so air exchange with the outside atmosphere is prevented (Fig. 2.3B).

Balch (1900) had the same idea to classify ice caves in relation to their air movement. But the presence of air streams was not the only precondition for him for a definition of ice caves. For a complete definition, the occurrence of ice in a cave for at least a part of the year was determinative. However, for him ice is only built in a cave, if the airflow brings in the cold air from outside and a water supply exists at the same time, and therefore the air stream is the main aspect after all. But Balch (1900) also doubted this clear classification, as he was not sure if the air in static ice caves was indeed immobile in summer. His solution was to introduce the term "apparently static ice caves" in contrast to dynamic ice caves and windholes. Apparently, static ice caves are characterized by one or more openings, which are close together above the cave floor, according to him. In summer the air movement is nearly immobile, while in

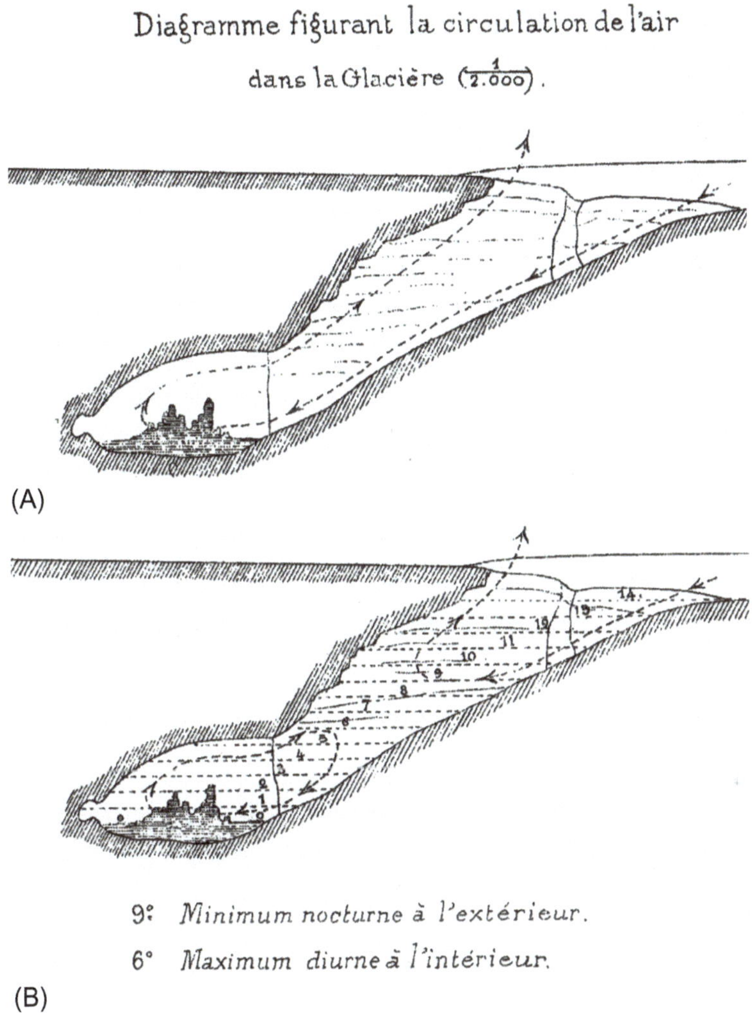

FIG. 2.3

Scheme of air circulation after (Girardot and Trouillet, 1885) during the open (A) and closed (B) period.

winter rotating air movements are initiated, as soon as the outside air temperature drops under the cave air temperature. The last aspect is attributed to most of the caves, which Balch knew during his time. Dynamic ice caves are defined for him by cave openings at different altitudes and at different parts of the cave, whereby airflow occurs. Balch (1900) refers to Thury's remarks from 1861, which were in use since then. Furthermore, he adds that airflow occurs only in some caves in summer, whereas in pit caves and so-called cliff caves it is windless.

Strong air movement generally characterizes the winter circulation. As soon as the outside temperature drops under the cave temperature (Balch, 1900), the outside air begins to sink into the cave.

At the upper entrance the warm air is sucked in during the summer and streamed out in winter; but at the lower entrance, the cold air is sucked in during winter and streams out in summer. Daily variations are subject to changes in the outside temperature. But the general cause for air movement is the cooling down of the air inside the cave and its descent due to the law of gravitation. In order that no vacuum arises, warm air flows down from the upper part. In winter the air warms up in contact with the rock, becomes lighter than the outside air, ascends, streams out at the upper entrance, and the vacuum is substituted by heavy cold air from the lower entrance. The velocity is dependent on the difference in temperature.

The third term Balch (1900) introduces, which is differentiated by the airflow, is the so-called windhole, an underground cavity with at least two openings and distinct airflow, whereby it can, but doesn't have to, contain ice to fulfill the definition. Already in Balch's paper from 1897, "Ice caves and the causes of subterranean ice," a classification according to position in the surface terrain, shape, and size was made (p. 164): "(1) Those at or near the base of cliffs, entering directly into the mountain with a down slope. This class is found in limestone and in volcanic rocks. Examples: The Kolowratshöhle, Dobsina, Roth in the Eifel. (2) Those at or near the base of cliffs where a long passageway exists before the ice cave proper is reached. All instances I know of in this class are in limestone rock. Examples: Demenyfálva, the Frauenmauer. (3) Those where a large pit opens into the ground, and the ice cave is found at the bottom opening into the pit. These are in limestone. Examples: Chaux-les-Passavant and la Genollière."

An additional aspect, which suggested another classification to the author, is the cave temperature. Thereby caves with ice (1), cold caves (2), normal caves (3), and hot caves (4) are distinguished independent from the air movement (Balch, 1900, p.112f). However, the differentiation of underground temperatures still seemed to contain several uncertain factors for Balch, since he found it to be difficult to categorize the different forms of natural refrigerators. For that reason, he applied, additionally, the classification of ice caves according to the rock formation (p. 114): "1. Gullies, gorges, and troughs where ice and snow remain. 2. Soil or rocks overlaying ice sheets. 3. Taluses and boulder heaps retaining ice. 4. Wells, mines and tunnels in which ice sometimes forms. 5. Caves with abnormally low temperatures and often containing ice." The fifth subgroup, "Caves with abnormally low temperatures", which is especially interesting for us, is furthermore subdivided by Balch between cold caves without ice and frozen caves. He even further distinguished between the underground ice formations in permanent and periodical ice caves. But most of the authors of that time don't note a difference whether the ice occurs perennially or periodically in a cave. The permanence of glaciation is mentioned in many descriptive reports (e.g., Thury, 1861; Fugger, 1891, 1892, 1893; Browne, 1865 etc.), but it doesn't play a role really, the underground ice by itself is the phenomenon for them.

Another point of discussion is the difference between static and dynamic ice caves, whose importance was for Kraus (1894) inexplicable. He is aware of the fact that in static ice caves air movement, like in dynamic ice caves, doesn't exist, but for him, both terms seem to be the outermost borders of the same thing (Kraus, 1894). A sharp differentiation appears unachievable. Ice caves resemble windholes only very little, but they are very difficult to distinguish. And the stagnation of the air is a fact no one would dare confirm. The air circulation in some caves only takes place in the entrance zone and depends on the shape of the entrance portal relative to the rest of the cave (Kraus, 1894). In other caves, air circulation occurs through wider fissures and erosion chimneys that at all times go along with various crossings. Kraus (1894) goes even so far as to question if static ice caves do really exist. For him the conditions for an absolute static state of an ice cave are non-existent, each cave has fissures to deeper parts, etc. No cave shows a hermetic seal of impermeable rock.

For Schwalbe (1886) the differentiation is completely dispensable (p. 19, translation): "If one associates with the term dynamic and tube caves the idea of regular air exchange at a certain time, then such caves are extremely rare, while one can not deny that chimneys can facilitate during winter, and at night, the air exchange and infiltration of cold air. But in most cases, they are absent and don't belong to the conditional factors of the phenomenon." In addition, he also adds further terms to the nomenclature. All phenomena of the moderate climates are distinguished in three main classes: ice caves (Kryoantren), ice holes (Kryotrymen) and "Ventarolen" (Psychroauren) (Schwalbe, 1886). He summarizes: "The real ice caves consist of more or less extended caves of different shape. [...] Lively airflow is nowhere. on the contrary the air is calm near by ice formations, while, of course, at the entrance air exchange occurs, a regular air circulation is in outlines noticeable in single cases. [...] The entrance is usually protected [...] precipitous [...], the opening itself is North or North-East directed [...], the width is also varying [...]. The ice formations are normally located (after summer observations) some distance to the entrance, but in some cases ice formations were observed directly at the entrance [...]." (Schwalbe, 1886, p.10f, own translation).

The mentioned quotations of some of the best known publications of the 19th century show that, until this point of time, the nomenclature was as inconsistent as the theory of construction for ice caves. More and more terms were introduced to attempt to classify the phenomenon ice cave. Since no one agreed upon the main characterizing elements and factors of ice caves by then, no clear definitions were established. Too multifarious seemed to be the forms of appearance and the observations that had been made by scientists and layman. No universal and comprising definition was formulated, instead numerous individual aspects were described. The combined classification of airflow, temperature, position, shape and size was not carried out by the authors.

In the 20th century Saar summarized, in the middle of the 1950s, the typecast of ice caves like this (Saar, 1956, p. 1, own translation): "One distinguishes two types of ice caves: static and dynamic; the first are in-the-ground sagging cavities with only one entrance, the latter drawing through the mountain and tunnel-shaped cavity systems with at least two entrances in different altitudes with connection to the surface. The air of both cave systems is on one side under the primary influence of the meteorological elements of the surface climate and on the other hand under the influence of the specific soil heat (geothermal gradient)."

Finally, at that time, the differentiation of static and dynamic ice caves was, thanks to Saar's long-term research activity in the Austrian Alps, acknowledged. In fact, the discovery of Dachstein-Rieseneishöhle and Eisriesenwelt demonstrated to the scientific world that the classification by Thury, done over 100 years before, was valid, and that dynamic ice caves were no margin phenomenon. However, the terminology in this century was by no means unambiguous.

Only in 1975, when Racovitza presented his paper "La classification topoclimatique des cavités souterraines," were comprehensively exact definitions of the different climatic zones in ice caves established. Racovitza (1975, own translation) refers to the newly introduced term "topoclimate," and defined it as the "sum of all phenomena, which describe the physical state of the atmosphere in an underground chamber, with specific topographic boundaries and shape." This represents the uppermost climatic unit with regard to the whole cave, for example, the topoclimate of a horizontal, ascending, descending cave, or a cave with several openings (Racovitza, 1975). The term topoclimate is very complex, for that reason he introduced the term meroclimate (derived from Greek, meros = part of the whole): "Sum of all physical phenomena, which occur in the cave atmosphere in distinctly different parts of the cave, individualized by topographic elements, but always by specific climate elements" (Racovitza, 1975, own translation). For him this term was indispensable because of the extensive meaning of the term topoclimate and the limited meaning of the term microclimate. Meroclimate fills a gap because of the absence of a climatical unit, which characterizes the phenomena between different zones of a cave. Therefore, the zone next to the entrance is the meroclimate of the transition zone between the surface climate and underground topoclimate (Racovitza, 1975).

The phenomena directly in context with diverse cave substrates lead to the smallest unit – the microclimate, the "sum of all thermodynamic and aerodynamic phenomena, which describe the mechanism of heat exchange and of masses on the surface of rocks and the substrates" (Racovitza, 1975, own translation). Characteristics of the microclimate are extremely reduced in amplitude of variations, as well as typical phenomena like condensation, evaporation, and the general energy transport, whose peculiarity depends not on topoclimatic or meroclimatic variations. Indeed, the topoclimate can cause a bigger perturbation as the microclimate, but these variations also occur in a constant topoclimate (Racovitza, 1975).

He further explained that all three climatic units of the general climate of the karst massif are subjected to the regional climate, whereby the topoclimatical factors have a certain value in each case, though their interrelation and separate functioning is difficult to determine. Primary topoclimatical factors are air temperature, relative humidity, and air pressure. They result in the specific cave climate. They are primarily not influenced by external variations but by geographical position and topography (Racovitza, 1975). This indicates that the characterization of a cave can be especially conducted by the temperature, as the middle absolute value can be used for a differentiation between caves of various regions or between various topographical types of caves. Deduced topoclimatical factors include the airflow regime, the evaporation, the condensation and the chill effect, processes which are, in every single moment, determined by the primary factors (Racovitza, 1975). For the description of the topoclimate the airflow regime is mentioned as the most important physical factor. In contradistinction to historical publications the existence of airflow today is not considered an exception anymore, but as a fact that all caves have mass exchange with the surface atmosphere (Racovitza, 1975). Finally, he distinguishes two types of topoclimate: the unidirectional airflow, typical for caves with several openings, and the bidirectional airflow, typical for caves with a single opening. Although there is a clear correlation between the topography and the topoclimate of a cave, the classification in terms of the topoclimate have to be done without regard to the airflow regime. Racovitza's achievement is to introduce definite and clear definitions in the terminology of ice caves, which became more exact in modern times.

2.4 HISTORY OF ICE CAVES THEORIES

Since the phenomenon ice cave was discovered, diverse ice cave theories have been developed over the centuries. Like the development of the nomenclature, the ice cave theories are numerous and contradictory, nevertheless each aimed to find a universal explanation for the existence of subterranean ice (Grebe, 2010). Some ideas were only discussed for a relatively short period of time, while others are still in use today. In Fig. 2.4 we show the development of the ice cave theories based on evaluation of the literature.

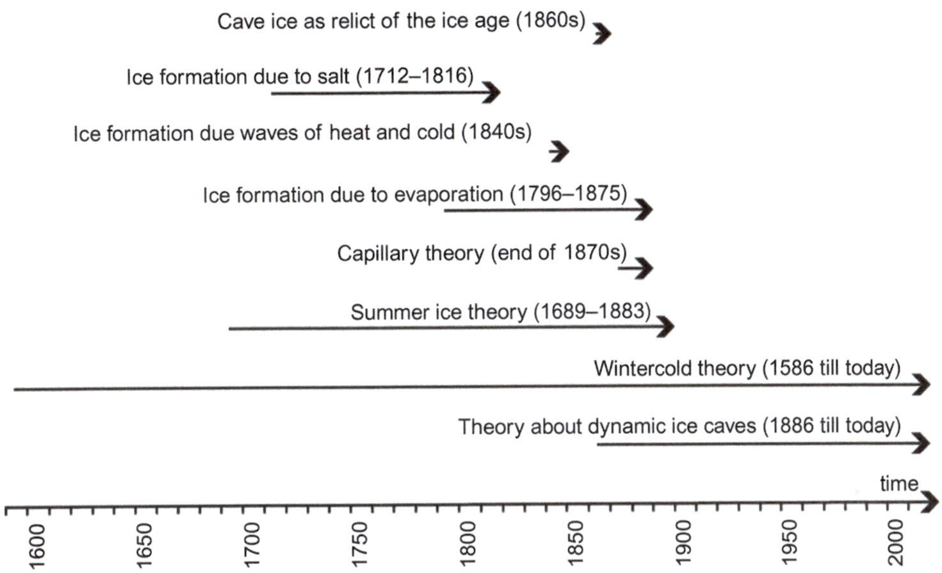

FIG. 2.4

Chronology of the ice cave theories.

From 1689 to 1883 many authors, including Valvasor, Billerez, Behrens, and Scope agreed that ice in caves only formed during the summer months (Fugger, 1893). However, they gave different explanations for this phenomenon. The most famous summer-ice theory was developed by Pictet in 1822 (reported by Fugger, 1893), who stated that ice is only formed due to evaporation driven by air currents, which were stronger in the summer than in the winter (Fugger, 1893). Diverse measurements and observations, however, disproved the summer-ice theory.

Another ice cave theory, developed by Billerez and discussed from 1712 to 1816, held that subterranean ice formed by salt (Fugger, 1893). The soil above the cave contained saltpeter and other salts, which dissolved in water and flowed into the cave, where they produced cold through solution. Consequently, the water inside the cave froze. According to Fugger (1893), a chemical analysis by Cossigny in 1743, however, did not find any saltpeter or any other salts needed for such cooling (Fugger, 1893). According to Schwalbe (1886), this theory is only of historic interest, but not scientifically tenable due to the absence of salts in soils.

De Saussure (1796) published his observations of cold-current caves in the Alps, in which the temperature was reduced by air currents flowing along the wet walls of the cave (cited by Balch, 1900). According to Fugger (1893), Parrot (1815) stated that dry and sufficiently deep caves showed steady temperatures of 10–12°C. Evaporative cooling, however, could lower the temperature until the air was saturated. A steady inflow of warm and dry air led to maximum cooling on hot days. After observations in Saint-Georges and Grand Cave de Matarquis, Thury (1861) disproved the theory that evaporation caused the formation of ice in caves (Balch, 1900).

The theory of ice formation through waves of heat and cold was only discussed in the 1840s, when Hope and Herschel explained that ice in caves formed during the summer months, when cavern water froze due to the penetration of cold winter waves. In contrast, an advancing warm summer wave led to warmer temperatures and therefore melting in winter (Schwalbe, 1886). This concept, however, is contradictory to the distribution of soil temperature.

Fugger (1893) and Turri et al. (2009) report that Hitchcook (1861) and Dawkins (1874) explained ice in caves as a relic from the Pleistocene. According to them the ice formed during the ice ages and persisted under the surface (Fugger, 1893; Turri et al., 2009). According to Schwalbe (1886), the ice-age theory only has a minor significance, as most caves have been intermittently ice-free and show a steady formation of new ice.

Lowe first formulated the theory that subterranean ice formed by capillary forces (Balch, 1900). Bubbles of air in water, which flow down through fissures in rocks, are liberated at the bottom of the cave. The air has lost its heat due to its compression and therefore absorbs the heat from the air and water in the cave, leading to a decrease in temperature (Balch, 1900). However, while almost all caves contain dripping water, not all caves contain ice, which contradicts the theory. In addition, no ice caves are found in hot climates.

Unlike all the theories about the formation of subterranean ice mentioned above, the winter-cold theory as well as the theory of the dynamic ice cave are still viable today. The former was first mentioned by Poissenot (1586) and therefore is considered the oldest ice cave theory (Fugger, 1893). Fugger (1893) cites Prévost as writing in 1789 that caves serve as reservoirs for ice that forms during the winter and does not completely melt during the summer (Fugger, 1893). Also, Balch (1900) describes caves as "iceboxes" preserving ice and snow from the winter months, because in the summer warm air cannot enter the cave.

Thury made a distinction between static and dynamic ice caves in 1861, the latter were not accepted as an independent type until the end of the 19th century. Bock (1913) explained the formation of subterranean ice by a temperature decrease resulting from uneven air currents during the summer and winter months. This theory was supported by long-term measurements in ice caves in Austria (Saar, 1954).

2.5 FIRST SYSTEMATICAL INVESTIGATIONS

While the phenomenon ice cave had been known for centuries, we only find evidence for the beginnings of systematical investigations in the 18th century. Cossigny (1750) made observations in ice caves throughout the seasons and thereby made out that the cave conditions are not independent from the outside temperatures (Balch, 1900). He is followed by Girod-Chantrans (1783), from whom we have the first proof of ice level measurements in Chaux-les-Passavant (Balch, 1900). He leaned his analysis on the measurements by Oudot in 1779–80, who installed wooden sticks on the upper end of ice columns and in this way observed an ice increase of about 30cm from Jan. 1779 until Feb. 1780 (Balch, 1900). Shortly after, Hablizl (1788) detected that there is less ice in the cave in autumn, which could be explained by an ice decrease during the months of July and August (Balch, 1900). In 1797 possibly the first English publication about ice caves by Townson was published. He conducted temperature measurements during summer time and proved that the cave atmosphere showed temperatures around 0°C and thus in thawing status (Balch, 1900). After an apparent break of several decades the next investigations are reported in the second half of the 19th century. Thury (1861) published a series of 25–30 temperature measurements, in which he observed the daily changes in the cave air temperature. Furthermore, Browne (1865) describes in his book "Ice Caves of France and Switzerland," temperature measurements in the diverse ice caves he visited(Fugger, 1893). As already mentioned before, it is Girardot and Trouillet (1885) who describe the relationship between outside air and cave air temperature, as well as the terms open and closed period (compare Fig. 2.3). In Fig. 2.5, we show the temperature measurements they conducted for their investigations. In their results, they could detect numerous open and also closed periods in Chaux-les-Passavant with strongly varying lengths.

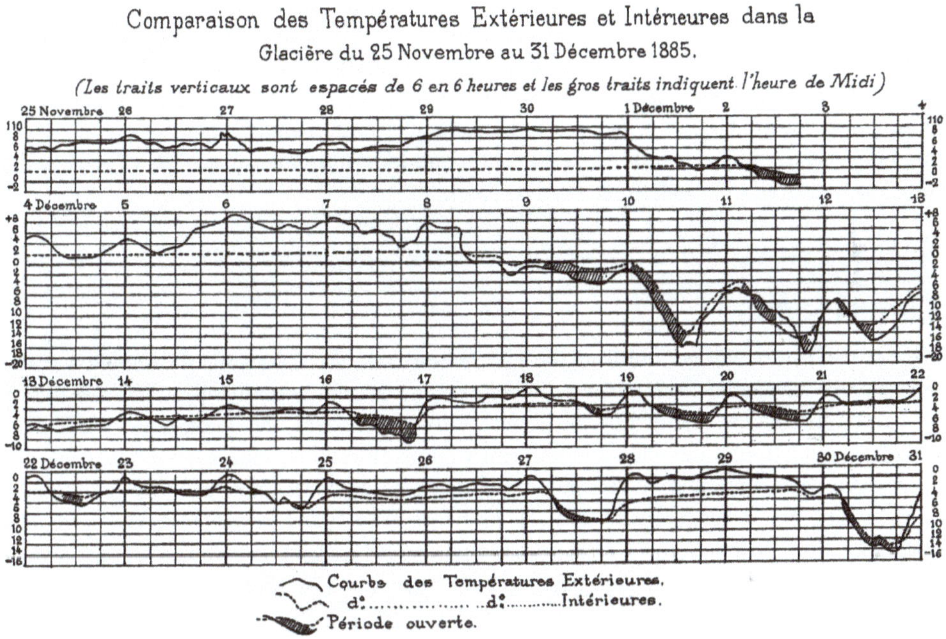

FIG. 2.5

Comparison between outside and cave air temperature for the definition of the open and closed period in Chaux-les-Passavant (Girardot and Trouillet, 1885).

Fugger (1888) published a comprehensive study about the ice caves at Untersberg (Germany), in which he reported temperature measurements, ice level measurements, and morphological observations. Beside this, he published a two-part monography in 1891 and 1892 about ice caves, in which numerous bibliographical references are compiled. Following Mavlyudov (2008a), it is Balch, at the end of the 18th century, who observed first the movement of the ice body, and recognized that in vertical caves ice and snow are moving gradually downwards following the gravitation. Furthermore, he added that the study of the interior structure of the cave ice only began in the beginning of the 20th century. Kraus (1894) complained that observations, which are temporally far apart, are not meaningful, since the meteorological conditions during each season are too different from each other. He further stated that this fact is also valid for the dry and wet periods in the seasons, as well as for each season in consecutive years. Continuous temperature observations are rather needed in all types of ice caves, periodic and permanent. Long-term measurements and comprehensive analysis like we know today could only be conducted since the beginning of the 20th century (e.g., Bock, 1913; Saar, 1956; Racovitza, 1972).

Saar (1956) published continuous temperature measurements from Eisriesenwelt in Austria and described how the cave ice is influenced by annual, periodical, and aperiodical fluctuations of regeneration and degeneration, which are also connected to longer regional climate variations. Among others, he could observe that a dynamic ice cave is not only influenced by the external atmosphere in front of the entrance, but also by the regional climate and the macroclimate. He could prove the principles of the theory of dynamic ice caves, like the seasonal changes in airflow direction, but also the temperature distribution inside the cave with air temperatures in the ice part below 0°C for 8 months a year, while in non-ice parts the air temperature never decreased below 0°C (Saar, 1956). But the air temperature in the non-ice parts also never increases above the 4.5°C border, the mean annual temperature at 1450 m a.s.l. These are only some selected results from the important long-term measurements of Saar, which can be regarded as milestones in modern ice cave research.

2.6 CONCLUSIONS

The review of the historical literature from the 12th century onward shows that the knowledge about ice caves very slowly developed from a rare to widely known natural phenomenon. The research history of ice caves was surely slowed down by the accessibility of the caves, the inconsistency in the description, and also explanation of the phenomenon. But what is more important, even today, centuries after the first ice cave was mentioned, is that modern ice cave research has not reach its zenith and is still developing. Today, first descriptions of new ice caves are often made by speleologists discovering ice caves in remote alpine karst regions. Therefore the potential for gaining new knowledge is very high. More detailed information about the specific ice cave history can be found in the national chapters in the second part of this book.

REFERENCES

Balch, E.S., 1897. Ice caves and the causes of subterranean ice. J. Franklin Inst. CXLIII (3), 161–178. Philadelphia.

Balch, E.S., 1900. Glacières or Freezing Cavernes. Allen Lane and Scot, Philadelphia, PA.

Belmonte, R., Marcén, C.S., 2010. First data from a Pyrenean ice cave (A294 cave, Cotiella massif, Spain). In: 4th International Workshop on Ice Caves, Volume of Abstracts. 7.

Berghaus, H.K.W., 1836. Annalen der Erd-, Völker- u. Staatenbund, Berlin.

Billerez, M. de, 1712. Histoire de l'Académie Royale des Sciences.

Bočić, N., Buzjak, N., Kern, Z., 2014. Some new potential subterranean glaciation research sites From Velebit Mt. (Croatia). In: Land, L., Kern, Z., Maggi, V., Turri, S. (Eds.), Proceedings of the Sixth International Workshop on Ice Caves, August 17–22, Idaho Falls, ID: NCKRI Symposium 4. National Cave and Karst Research Institute, Carlsbad, NM.

Bock, H., 1913. Mathematisch-physikalische Untersuchung der Eishöhlen und Windröhren. In: Die Höhlen im Dachstein. Verein für Höhlenkunde in Österreich, Graz, pp. 102–144.

Bögli, A., 1978. Karsthydrographie und physische Speläologie. Springer Verlag, Berlin-Heidelberg-New York.

de Boz Ingénieur du Roy, M., 1726. Histoire de l'Académie Royale des Sciences.

Browne, G.F., 1865. Ice caves of France and Switzerland. London.

Buzjak, N., Vinka, D., Paar, D., Bočić, N., 2014. The influence of karst topography to ice cave occurrence—example of Ledena jama in Lomska duliba (Croatia). In: Land, L., Kern, Z., Maggi, V., Turri, S. (Eds.), Proceedings of the Sixth International Workshop on Ice Caves, August 17–22, Idaho Falls, ID: NCKRI Symposium 4. National Cave and Karst Research Institute, Carlsbad, NM.

Colucci, R.R., Fontana, D., Forte, E., 2014. Characterization of two permanent ice cave deposits in the southeastern Alps (Italy) by means of ground penetrating radar (Gpr). In: Land, L., Kern, Z., Maggi, V., Turri, S. (Eds.), Proceedings of the Sixth International Workshop on Ice Caves, August 17-22, Idaho Falls, ID: NCKRI Symposium 4. National Cave and Karst Research Institute, Carlsbad, NM.

de Cossigny, Ingénieur en chef de Besançon, M., 1750. Mémoires de Mathématique et de physique présentés a l'Académie Royale des Sciences, vol. I, Paris.

Dawkins, W.B., 1874. Cave Hunting. London.

De Saussure, H.B., 1796. Voyages dans les Alpes, Tome III., sections 1404-1416.

Ebel, G., 1818. *Manual* du voyageur en Suisse. Zürich.

Fugger, E., 1888. Beobachtungen in den Eishöhlen des Untersberges bei Salzburg. Salzburg.

Fugger, E., 1891. Eishöhlen und Windröhren. Separat-Abdruck XXIV. Jahresberichte der K.K. Ober-Realschule in Salzburg. Salzburg.

Fugger, E., 1892. Eishöhlen und Windröhren. Zweiter Theil, Separat-Abdruck XXV. Jahresberichte der K.K. Ober-Realschule in Salzburg. Salzburg.

Fugger, E., 1893. Eishöhlen und Windröhren. Dritter Theil (Schluss). In: Sechsundzwanzigster Jahres-Bericht der K.K. Ober-Realschule in Salzburg, pp. 5–88. Salzburg.

Gilbert, L.W., 1815. Ann. Phys. XIX.

Girardot, A., Trouillet, L., 1885. La Glacière de Chaux-les-Passavant.—Mémoires de la Société d'Emulation du Doubs. Cinquième série neuvième volume, 449–524.

Girod-Chantrans. L.C., 1783. Journal des Mines, Prairial, An. IV., 65–72.

Gollut, L., 1592. Les Mémoires Historiques de la Repub. Sequanoise, Dole.

Gómez Lende, M., Serrano, E., Bordehore, L.J., Sandoval, S., 2016. The role of GPR techniques in determining ice cave properties: Peña Castil ice cave, Picos de Europa. Earth Surf. Process. Landforms 41, 2177–2190. https://doi.org/10.1002/esp.3976.

Grebe, C., 2010. Eishöhlenforschung vom 16. Jahrhundert bis in die Modern - Vom Phänomen zur aktuellen Forschung (Master's thesis). Bochum.

Hablizl, 1788. Description physique de la contrée de la Tauride, La Haye.

Halliday, W.R., 1954. Ice caves of the United States. NSS Bull. 16, 3–28.

Hitchcook, E., 1861. Geology of Vermont, vol. 1.

Holmgren, D., Pflitsch, A., Ringeis, J., Meyer, C., 2016. Ice cave research of the United States of America. J. Cave Karst Stud.

Humboldt, A., 1814. Personal Narrative of Travels to the Equinoctial Regions. vol. I. London.

Kraus, F., 1894. Höhlenkunde. Wien.

Krenner, J.A., 1874. Die Eishöhle von Dobschau. Budapest.

Mandl, A., 1838. Steiermärkische Zeitschrift, neue Folge, V Jahrg, H. 2, p. 151.

Mavlyudov, B.R., 2008. History of researches of caves with ice; non published translation of the chapter 1.1. of the author from Mavlyudov B. R. (2008). Cave glaciation, Moskow.

Mavlyudov, B.R., 2008b. Geography of cave glaciation. In: 3rd International Workshop on Ice Caves. Proceedings. Kungur Ice Cave Russia, May 12–17, 2008, Kungur, pp. 38–44.

Mavlyudov, B.R., 2010. Cave glaciation in the past. In: 4th International Workshop on Ice Caves, June 5–11, 2010, Obertraun, Austria, pp. 21. Abstract Volume.

Merriam, P., 1950. Ice caves. NSS Bull. 12, 32–37.

Meyer, C., Pflitsch, A., Ringeis, J., Maggi, V., 2016. Reports on ice caves in literature from the twelfth to the middle of the twentieth century. J. Cave Karst Stud.

Parrot, G.F., 1815. Grundriss der Physik der Erde und Geologie. Riga and Leipzig.
Pflitsch, A., Schörghofer, N., Smith, S.M., Holmgren, D., 2016. Massive Ice Loss from the Mauna Loa Ice cave, Hawaii. Arct. Antarct. Alp. Res. 48 (1), 33–43. https://doi.org/10.1657/AAAR0014-095.
Pictet, M.A., 1822. Memoire sur les glacières naturelles du Jura et des Alpes. Bibliothèque Universelle de Genève XX, 261. et seq.
Poissenot, B., 1586. Nouvelles histoires tragiques: (1586) / Bénigne Poissenot; édition établie et annotée par Jean-Claude Arnould et Richard A. Carr., Genève.
Posselt-Czorich, A., 1880. Zeitschrift des Deutschen und Oesterreichischen Alpen Verein, XI. 261.
Prevost, P., 1789. Journal de Genève, 21 March.
Racovitza, G., 1972. Sur la corrélation entre l'évolution du climat et la dynamique des dépôts de glace de la grotte de Scărişoara. Trav. Inst. Spéol., E. Racovitza XI, 373–392.
Racovitza, G., 1975. La classification topoclimatique des cavités souterraines. Trav. Inst. Spéol., E. Racovitza 14, 197–216.
Ransanus, P., 1977. Epithoma rerum Hungararum, id est annalium omnium temporum liber primus et sexagesimus/ Petrus Ransanus; curam gerebat Petrus Kulcsár, Budapest.
Reich, F., 1834. Beobachtungen über die Temperatur des Gesteines, Freiberg.
Richter, J.P., 1883. The literary works of Leonardo Da Vinci-compiled and edited from the Original Manuscripts. London.
Saar, R., 1954. Meteorologisch-physikalische Beobachtungen in der Dachstein-Rieseneishöhle. In: Die Höhle 5 (¾), pp. 49–62. Wien.
Saar, R., 1956. Eishöhlen, ein meteorologisch-geophysikalisches Phänomen. Untersuchungen an der Rieseneishöhle (R.E.H.) im Dachstein, Oberösterreich. Geogr. Ann. 38 (1), 1–63.
Schönviszky, L., 1968. The earliest known ice cave of the Carpahtians. In: Karszt s barlang. vols. I–II, pp. 11–16.
Schwalbe, B., 1886. Über Eishöhlen und Eislöcher. Berlin.
Schwalbe, B., 1887. Uebersichtliche Zusammenstellung litterarischer Notizen über Eishöhlen und Eislöcher nebst einigen Zusätzen. In: Mitth. d. Section f. Höhlenkunde d. Oesterreichischen Touristen-Club, Jg. VI. vols. 2–3, pp. 13–39. Wien.
Scrope, G.P., 1826. Edinburgh J. Sci. V, 154.
Stein, M.A., 1961. Kalhaṇa's Rājataraṅgiṇī : a chronicle of the kings of Kaśmīr/translated with an introd., commentary and appendices by M.A. Stein, London.
Temovski, M., 2016. Ice Caves in Macedonia. In: 7th International Workshop on Ice Caves, Programm Guide and Abstracts, Postojna, May 16–22, 2016.
Thury, M., 1861. Etudes sur les glacières naturelles par M. Thury. Tiré des archives des sciences de la bibliothèque universelle. Genève.
Townson, R., 1797. Travels in Hungary. London.
Turri, S., Trofimova, E., Bini, A., et al., 2009. Ice caves scientific research history: from XV to XIX centuries. Materialy glaciologicheskikh Issledovanij (Data Glaciological Studies)107. Moscow.
Valvasor, J.W., 1689. Die Ehre des Herzogthumes Crain. Laybach.
Vass, I., 1857. Eine Wanderung nach der Eishöhle bei Skerisora. Verh. u. Mitt. Des sieb. Für Naturwiss. zu Hermannstadt VIII, 162–170.

FURTHER READING

Behrens, G.H., 1703. Hercynia Curiosa. Nordhausen.
Bel, M., 1739. Philos. Trans. XLI.
Lohmann, H., 1895. Das Höhleneis unter besonderer Berücksichtigung einiger Eishöhlen des Erzgebirges. Diss, Univ. Leipzig.

CHAPTER

ICE CAVES CLIMATE

3

Aurel Perșoiu

Emil Racoviță Institute of Speleology, Cluj-Napoca, Romania

CHAPTER OUTLINE

The study of (ice) caves climate implies actually the study of cave meteorology—the values, distribution and dynamics of temperature (air and substrata), humidity and air movement in caves. The interplay of these, on a background of peculiar cave morphology, is responsible for the genesis, accumulation, and dynamics of ice in caves. With few exceptions, the vast majority of ice caves occur in regions where the mean annual air temperature (MAAT) is well above 0°C, the peculiar cave climate responsible for cave glaciations being the result of specific types of air circulation. These in turn lead to cave undercooling, which further reinforces air circulation types favoring glaciations, in a positive feedback loop that could lead to large (>100,000 m^3) ice masses to accumulate in caves over extended periods of time (>10,000 years). Positive MAATs outside ice caves result in the underground glaciers to be in a sensitive equilibrium with external climatic conditions, continuously subjected to the risk of melt (Kern and Perșoiu, 2013).

As air circulation is the leading (cave-specific) factor responsible for cave ice genesis, it will be treated first, followed by air and substratum (rock and ice) temperature and relative humidity changes.

3.1 AIR CIRCULATION

Due to their location in areas with MAAT>0°C, most caves hosting ice require (1) undercooling during winter months and (2) a mechanism for the preservation of negative temperatures during summer. Undercooling can be achieved either by conductive heat transfer (from the caves outwards) or by forced advection of cold air, driven by temperature (and hence density) differences between the cave and outside environment, pressure fluctuations, gravitational settling, and diphasic flow due to water circulation (Wigley and Brown 1976).

Ice Caves. https://doi.org/10.1016/B978-0-12-811739-2.00003-6

Changes in external air pressure next to cave entrances will necessarily drive flow of air into or outside that caves, depending on the direction of pressure gradient. However, the limited amplitude and rapid reversal of direction of pressure changes result in low speeds of air movement and hence reduced volumes of cold air advected to caves. Both of these (speed and volume) increase with the volume of the cave, but the cold air's penetration depth is limited by geothermal heat transferred through the rock walls and latent heat released during the phase-changes of water (Perşoiu et al., 2011). Similarly, low volumes of air can be dragged in caves by inflowing streams; but undercooling is prevented by the heat transported by the water itself; therefore, this mechanism does not result in ice formation in caves.

The most important cooling mechanisms responsible for cave ice formation (and preservation) are cold air advection through *gravitational settling* (Fig. 3.1A) and *chimney effect* (Fig. 3.1B).

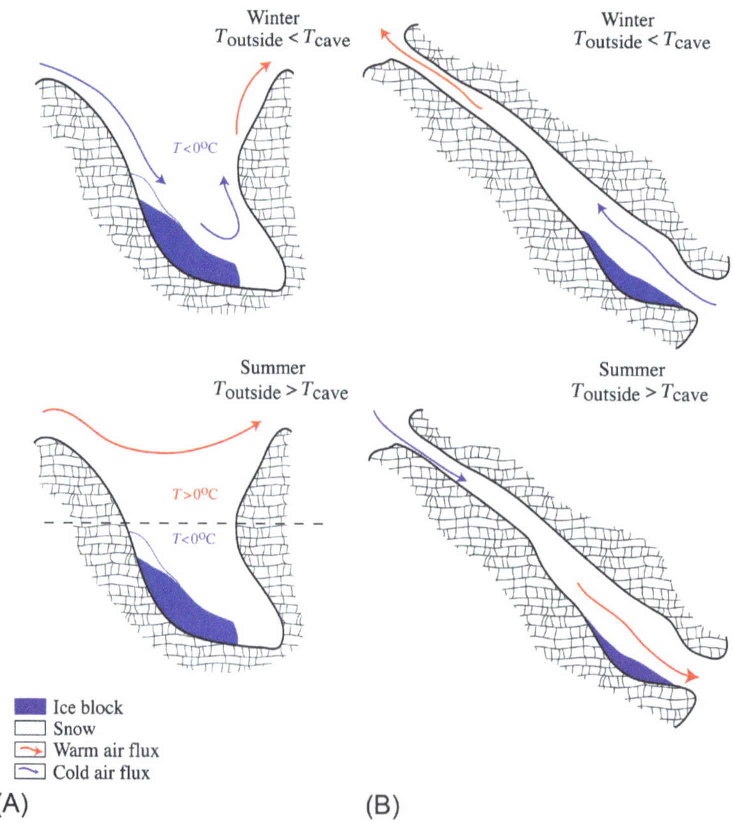

FIG. 3.1

Types of air circulation in caves: (A) gravitational settling (in winter) and in-cave circulation in summer; (B) unidirectional circulation in caves with multiple entrances.

From Perşoiu, A., Onac, B.P., 2012. Ice in caves. In: White, W., Culver, D.C. (Eds.), Encyclopedia of Caves, Elsevier, Amsterdam, pp. 399–404.

Gravitational settling of cold air is the main cooling mechanism behind cave ice accumulation. It occurs in single entrance, descending caves (or in caves with multiple entrances situated at roughly similar elevations) during winter months, when thermal differences between the external and internal air translate into mass differences, resulting in cold-air avalanches (Perrier et al., 2005) cascading down into caves (Fig. 3.1). Depending on the diameter of the entrance and the thermal gradient, the displaced volumes of air could amount to tens of m^3/s, at speeds exceeding 1 m/s. This inflow of cold air will necessarily push out a similar volume of warm air, leading to a steady undercooling of the cave atmosphere (Racoviță, 1994, Perșoiu et al., 2011) and walls, freezing of water, and formation of congelation ice (Fig. 3.2).

FIG. 3.2

Formation of congelation ice by the freezing of standing pond of water in Scărișoara Ice Cave (Romania). Cold air is flowing in the cave from through a ~46 m deep shaft, located on the left of the image.

Cooling of the cave's atmosphere is further accompanied by evaporative cooling, induced by evaporation of moisture from the walls in the inflowing stream of cold and dry air. The undercooled walls further reduce cave temperature by absorbing the geothermal heat transferred through the rock walls, thus helping maintain a "cold-path" for the inflowing air and enhancing water freezing and ice formation.

Inflow of air into caves due to the chimney effect occurs in caves with multiple (at least two) entrances located at different altitudes, as a direct consequence of temperature contrast between the cave's atmosphere and the external environment. These differences result in a pressure gradient between the cave entrances, which in turn triggers airflow through the cave. In winter, when the temperature inside the cave at the lower entrance is higher than outside (Fig. 3.2), cold air will be forced into the cave and warm air will be pushed out through the upper entrance. In summer the cave air temperature is lower than the outside temperature, and as a consequence, warm air will flow out from the cave through the lower entrance, and cold air will be pushed inside the cave by the reversed pressure gradient through the upper entrance. The speed of airflow is controlled by the interplay between the magnitude of pressure imbalance between the cave and the exterior (Wigley and Brown, 1976) and the frictional resistance of the cave itself. Thus, increasing pressure differences will trigger higher velocities, which in turn, will lead to stronger friction, the end result being a limitation of the maximum speeds attainable in (ice) caves (e.g., 5 m/s in Dachstein Rieseneishohle, Saar, 1956). The directions of air movement can change daily, especially in winter, in relation to external synoptic conditions (Obleitner and Spötl, 2011).

Inside caves, air circulation is a continuation of the exchanges with the exterior, except for cold air trap caves, in which colder than external (and thus heavier) air prevents flow. An example of such a system is Scărișoara Ice Cave, in which two seasonal types of circulation can be distinguished, one occurring in summer and the other in winter (Fig. 3.3).

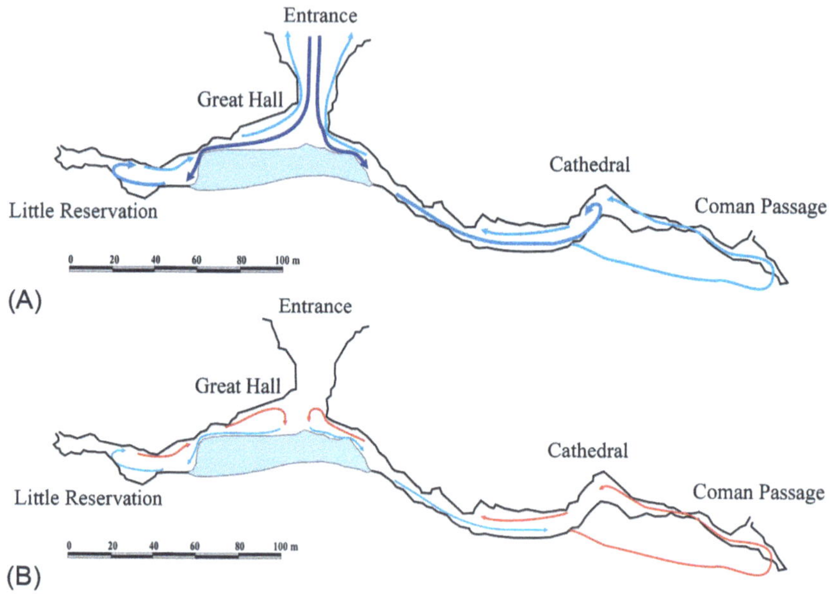

FIG. 3.3

Air circulation in Scărișoara Ice Cave (cold air trap, with a single entrance at the upper part of the cave). (A) Winter circulation, (B) summer circulation.

In Scărișoara Ice Cave, between November and April, airflow triggered by the higher density of external cold air is directed from the surface downward into the cave. The inflowing cold ($T < -15°C$) and dry ($RH < 75\%$) air first reaches the Great Hall, from where it descends along the flanks of the ice block into the Little and Great Reservations, replacing the warmer air, which is pushed out along the ceilings. Between May and October, no air mass exchange occurs between the cave and the outside, as the cold air in the cave is denser than the external air. Meanwhile, the overcooled walls of the cave and ice block account for the persistence of cold air masses in the Great Hall, while the inner parts of the two reservations warm slowly due to the effect of geothermal heat. Thus, this difference in air temperature triggers a slow air movement between the Great Hall and the two reservations, the airflow following almost the same path as in winter. The difference is that the uprising warm air is progressively cooled as it reaches the walls of the Great Hall and sinks back to the bottom of it, closing the convective cell. In April and October, rapid changes between summer and winter types of circulation occur as temperature varies

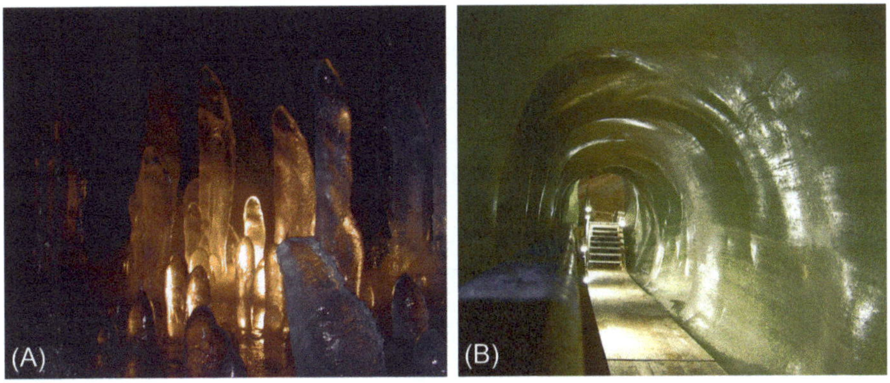

FIG. 3.4

Sublimation of cave ice under the impact of inflowing cold air: (A) ice stalagmites in Scărișoara Ice Cave (Romania) with bilateral symmetry (cold air is flowing from the left); (B) scallops on the walls of an artificial tunnel in Dobsinska Ice Cave, Slovakia (cold air is flowing towards the camera).

around 0°C (Perșoiu and Onac, 2012). Similar patterns of air circulation are present in other caves with large ice accumulation – Dobsinska Ice Cave, Slovakia (Piasecki et al., 2008), Schellenberge Ice Cave, Germany (Meyer et al., 2014) etc. Flow of cold and dry air on top of cave ice accumulations has a strong impact on ice dynamics, leading to ice sublimation and formation of scallops (very much similar to those formed in cave streams) and ice stalagmites with peculiar shapes (Fig. 3.4). As a result of these ablation processes, the airflow(s) warms-up and becomes enriched in moisture and once reaching the still undercooled walls of the caves deposit hoarfrost as large ice crystals (up to 0.5 m in length).

Regardless of its leading cause, speed and volume, air movement is the most important factor for cave climatology, being the main mechanisms for cave cooling and thus ice development.

3.2 AIR TEMPERATURE AND HUMIDITY

A direct consequence of air circulation between caves and the outside environment are changes in temperature and humidity in caves atmosphere and substrata. A common conception is that caves have constant temperature and humidity condition. While this assumption is generally valid for inner, isolated, sections of caves, their entrances, where seasonal and permanent ice deposits develop, are in constant state of changing meteorological conditions.

In studying ice caves, and especially the climatic conditions leading to ice genesis, accumulation and preservation, one of the basic and most frequent assumption is that of "*cold air accumulation in descending caves (cold traps),*", i.e., gravitational settling in winter (or, better, when $T_{outside} < T_{inside}$) of heavier and colder air at the bottom of descending shafts. While cold air does get into descending caves in winters (see above), it *does not accumulate per se* there over long-term periods, although *negative temperatures do persist* in such caves throughout the year (e.g., Viehmann et al., 1965). The confusion arises from the interplay of different processes leading to ice occurrence in caves: freezing and melting of ice. During cold air inflow into caves, liquid water freezes to form ice, a process that leads to rapid release of the latent heat stored in the water and subsequent warming of the inflowing air (Perşoiu et al., 2011). In the absence of a continuous inflow of cold air, this process will rapidly lead to the warming of the cave atmosphere resulting in the loss of ice and prevention of cave glaciation to occur. Further, constant supply of geothermal heat through the cave walls adds energy to the air inside, leading to its continuous warming. However, once ice has formed inside caves, it will be in metastable equilibrium, continuously melting as it "absorbs" heat (advected through walls, the air column(s) in the entrance(s) and incoming warm air and water). As long as ice is present, the enormous amount of enthalpy of fusion of water (334 kj/kg) needed to melt it will prevent warming of the caves' atmospheres above 0°C, so that even in the absence of external input of cold air, such caves will be in a continuous state of negative thermal anomalies. However, "trapping" of cold air inside caves does occur, given the peculiar morphology of these cavities (Fig. 3.1A), but it is not winter cold air that is being trapped, but rather the cold air generated by the very processes of ice melting occurring during summer months.

How far inside caves penetrate the external climatic conditions depends on the thermal difference between the two environments and the morphology of the passages. In Eisriesenwelt and Dachstein Rieseneishohle (Austria), two dynamically ventilated caves (with entrances situated at different elevations, the upper ones well above 2000 m asl), negative temperature anomalies and ice can be found as far as 1 km inside the mountains, while in shafts in the Alps, Velebit Mountains (Croatia), or Central Asia, ice and snow can be found at depths exceeding 500 m. The attenuation of external cold air influence in caves was described by Racoviță (1984), who distinguished four climatic zones in Scărişoara Ice Cave (Fig. 3.3): a transitional zone in the entrance shaft, a glacial zone comprising the area occupied by the permanent ice block (Great Hall), a periglacial zone (Little and Great Reservation) with annual or multiannual ice stalagmites, and a warm, non-glaciated, climatic zone in the inner parts of the cave (Coman Passage). An illustration of this type of temperature distribution is shown for Scărişoara Ice Cave, Romania (Fig. 3.5).

In winter, cold air inflow through the entrance shafts cools down the Great Hall (Sala Mare in Fig. 3.5) and also flows down towards the Great Reservation (Rezervația Mare) along the Maxim Pop Gallery (see also Fig. 3.3). In summer, inflow of cold air ceases, but the melting of ice in the Great Hall keeps temperatures below 0°C, and the thermal differences between this sector of the cave and the lower, warmer one leads to a weak (~10 cm/s) thermal circulation (see also Fig. 3.3) along the same path as in winter.

The dynamics of air temperature (and to a lesser extent of relative humidity) inside caves have been studied in detail in numerous caves in Europe, North America, and Asia, with the results (detailed in the chapters of this book) pointing towards similar findings (Fig. 3.6):

(1) Fluctuations of cave air temperatures follow external ones in both dynamically ventilated caves (caves with two or more entrances) and cold-air traps, the amplitudes decreasing with increasing distance from the entrance.

(2) In cold-air traps during summer months, the external and cave environment are not connected via air circulation; conductive transfer through the air column in the entrance and the rock walls, and dripping water are the main heat sources for the cave atmosphere, while the latent heat consumed in thawing the ice and sensible heat responsible for warming ice and rock are the main heat sinks.

(3) Daily temperature oscillations outside ice caves are transferred relatively fast inside caves in winter (with a delay in the order of 2–3 h, Perşoiu et al., 2011), and are virtually unfelt in summer.

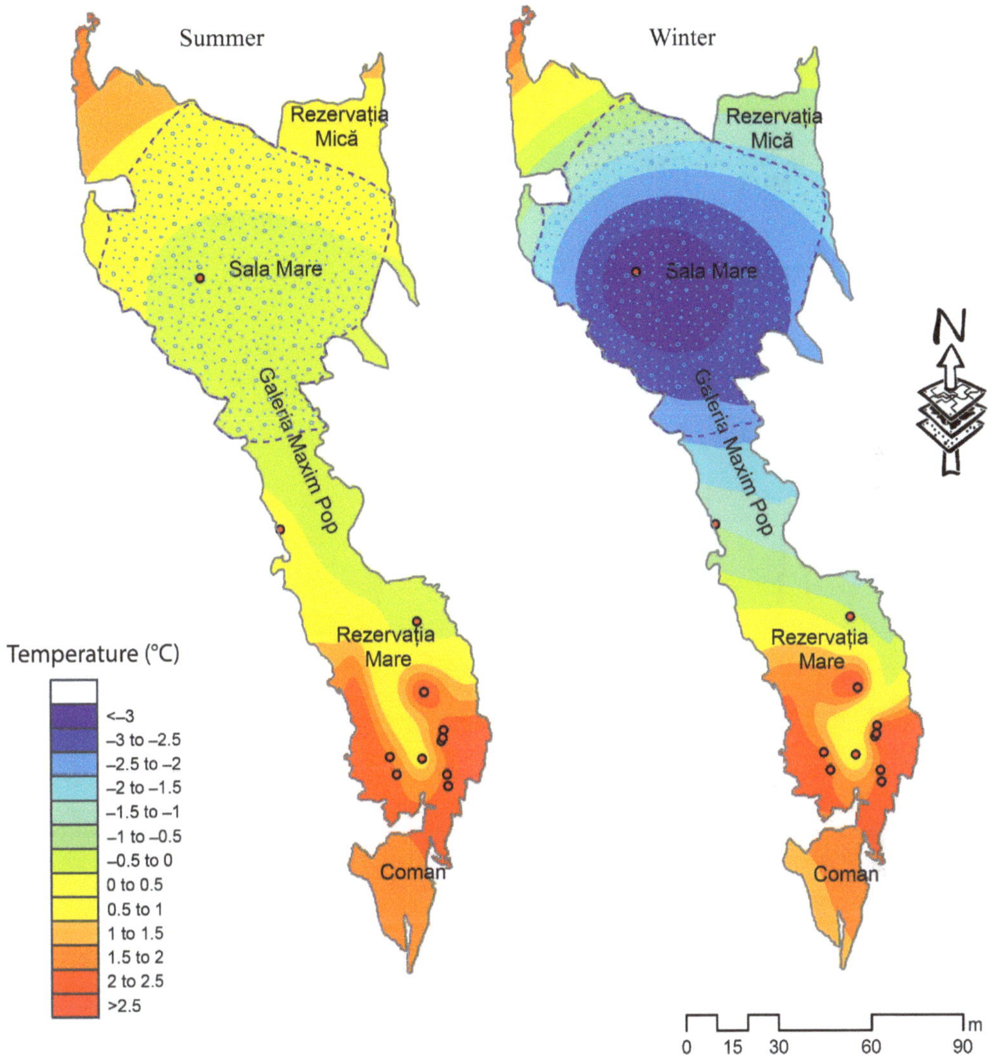

FIG. 3.5

Distribution of air temperature in Scărișoara Ice Cave, Romania. *Red dots* indicate the position of dataloggers, where hourly air temperatures were measured for one year.

Map courtesy of Carmen Bădăluță.

(4) Humidity content of inflowing air plays an important role in modulating the thermal behavior of caves; in summer, condensation and sublimation releasing heat that temporarily could overcome the heat loss to the ice and walls, while in winter, warming of air in contact with walls is compensated by cooling through evaporation of moisture from the walls and sublimation of snow/ice.

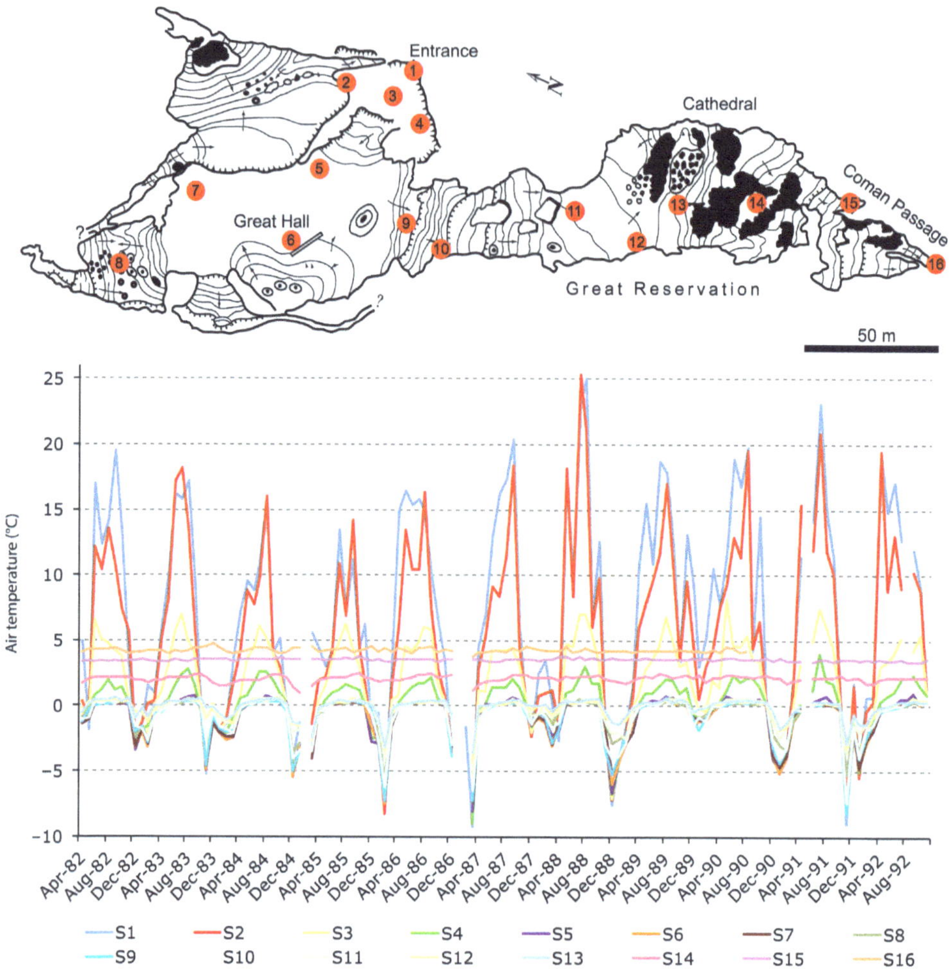

FIG. 3.6

Results of long-term monitoring of air temperature in Scărişoara Ice Cave, Romania. The map shows the location of monitoring station 1–16. The amplitude of annual variations decreases with increased distance form the entrance.

On a yearly basis, the dynamics of cave air temperature go through three different phases (Fig. 3.7): (1) summer phase, when the cave air temperature rises above 0°C and stays constant at this value, (2) cooling phase, occurring between the first inflow of cold air and the moment when air temperatures inside the cave stabilize below 0°C, and (3) winter phase, with air temperatures below 0°C. The separation of cooling and winter phases was made on the basis of the role played by cold air inflow on ice genesis. We have considered the moment when all liquid water available in the cave is frozen, hence no latent heat of freezing is "used" to warm the inflowing cold air, and temperatures drop and stay below 0°C (Perşoiu et al., 2011).

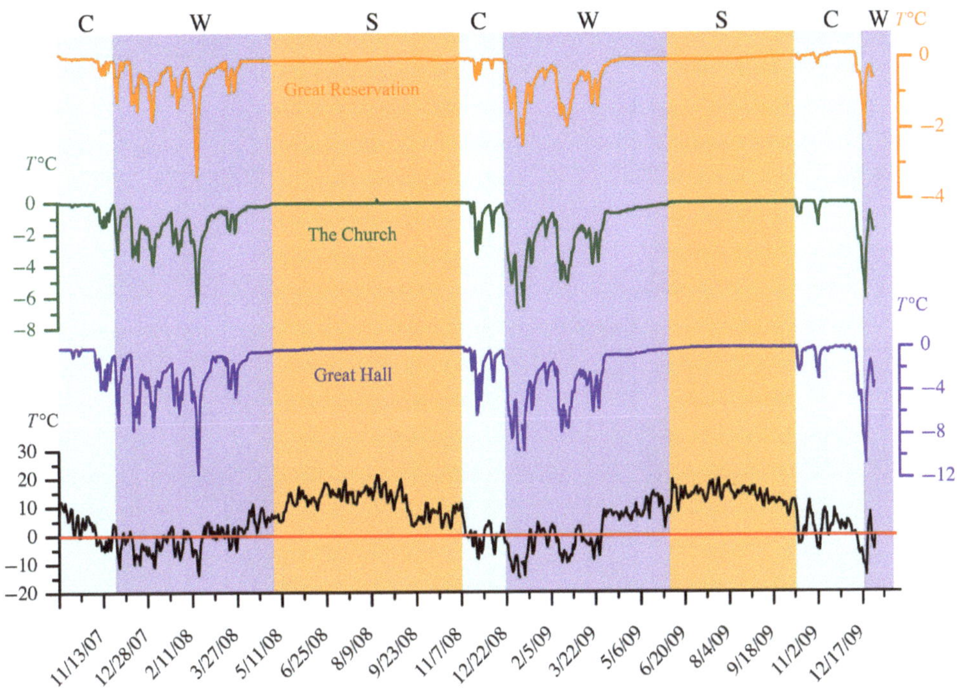

FIG. 3.7

Annual variations of air temperatures in Scărişoara Ice Cave, Romania, with indications of the three main "phases": C—cooling phase, W—winter phase, S—summer phase.

Modified from Perşoiu, A., Onac, B.P., Perşoiu, I., 2011. The interplay between air temperature and ice dynamics in Scărişoara Ice Cave, Romania. Acta Carsologica 40 (3), 445–456.

Air temperature variations are transmitted, by conduction, in both the rock and ice walls of caves. Previous studies (Racoviţă et al., 1991) in Scărişoara Ice Cave have shown that air temperature variations play a major role in the changes of ice temperature that are felt to a depth of ca. 6 m. Luetscher (2005) has shown the presence of annual temperature oscillation at 8 m below surface, and also that the cooling of the ice block in winter reduces the duration of melting period.

3.3 CONCLUSIONS – A CONCEPTUAL MODEL OF ICE CAVES CLIMATE

Cave climate was monitored for 60 years in Scărişoara Ice Cave, Romania, with two periods of monthly temperature measurements (1963–68, 1982, and 1992) and one of hourly measurements (since 2007), while ice dynamics and mass balance changes have been recorded since 1947 (Perşoiu and Pazdur, 2011). The results of these measurements allowed the construction of a conceptual model of cave ice climate (Perşoiu et al., 2011), summarized below from the above reference.

The thermal balance of ice caves' air is controlled by the heat fluxes conducted through the rock walls, those advected through the cave entrance(s) by incoming air and water fluxes, heat absorbed and released during the water phase changes and by the presence of ice and its phase changes with the

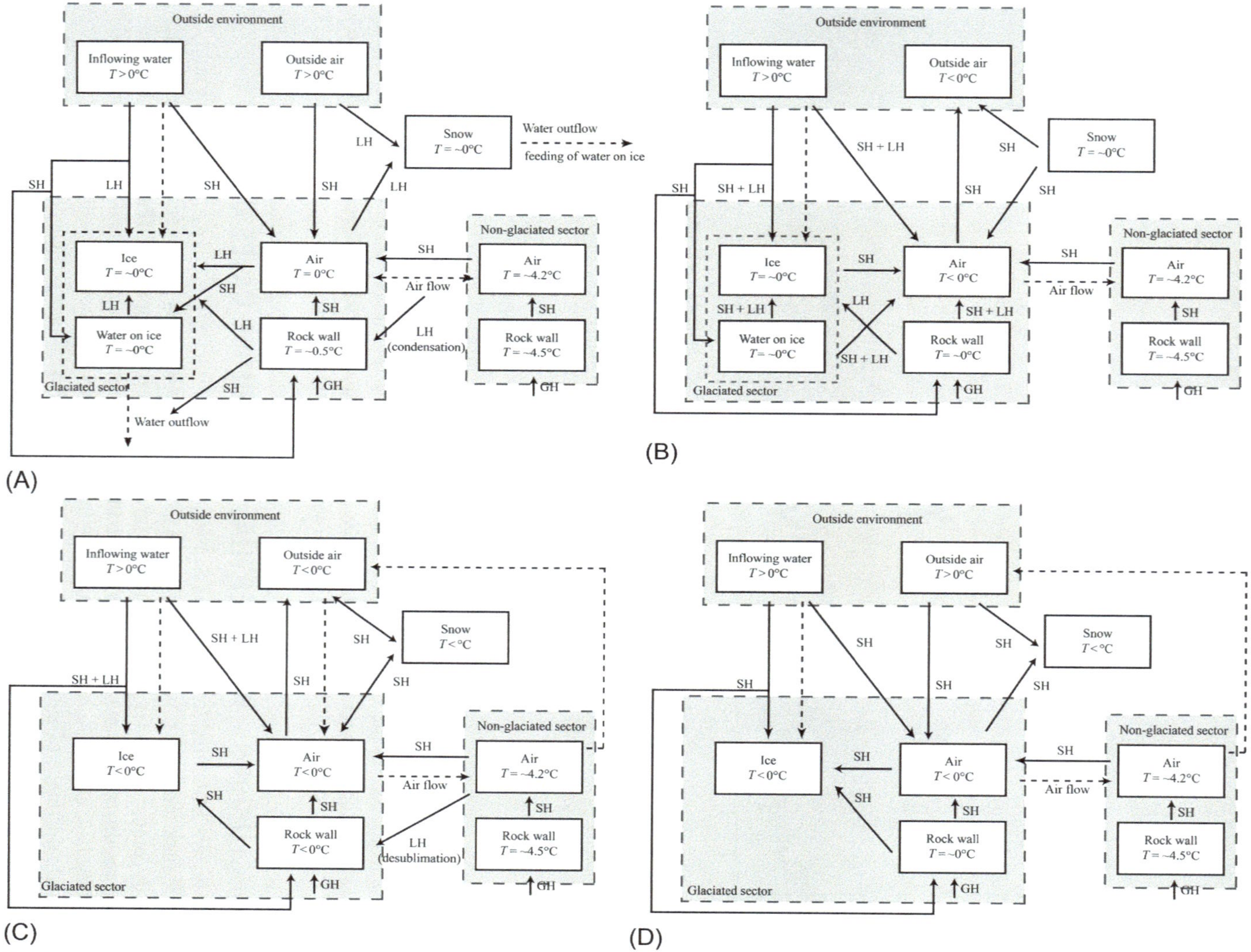

FIG. 3.8

Conceptual model of heat and matter fluxes in Scărișoara Ice Cave, Romania.

From Perșoiu, A., Onac, B.P., Perșoiu, I., 2011. The interplay between air temperature and ice dynamics in Scărișoara Ice Cave, Romania. Acta Carsologica 40 (3), 445–456.

associated heat (and matter) fluxes. Thee fluxes are discussed separately for the summer, cooling, and winter phases (states) describe above.

In the summer state, all heat fluxes are directed towards the ice block and the air in the glaciated part of the cave (Fig. 3.8A). The inner, non-glaciated part of the cave and the warm outside air act as heat sources, while the melting ice (and snow at the bottom of the entrance shaft) is a strong heat sink, thus keeping the air temperature in the glaciated part of the cave at 0°C. The sensible and latent heat absorbed by the melting ice is carried away by outflowing water. A side, but notable effect of the strong heat fluxes towards the melting ice block is the lowering of the mean annual air temperature (MAT) in the non-glaciated sector of the cave by ~1°C below the external MAT (Fig. 3.5).

The cooling state is a transitional one, the major role in the dynamics of the heat fluxes being the inflow of cold air, which acts as a heat sink (Fig. 3.8B). The heat fluxes between the ice (and snow) and air reverse, leading to cave cooling and the freezing of water and genesis of ice. The latent heat of freezing in the lake water and the sensible heat from the inner parts of the cave are the main sources of heat in this state, with the latter acting continuously, and the former only for c. 1–2 months, depending on the strength and temperature of inflowing cold air.

Once the water inside the cave is frozen (except for any occasional inflow), the winter state (a simplified version of the cooling phase) establishes inside the cave (Fig. 3.8C). The heat advected from the inner parts of the cave and the rock walls is consumed in warming the down flowing cold air, which in this case is the main heat sink.

In all three states above, supplemental heat originates from inflowing water, and is generally quickly absorbed by the cold air and the ice, ultimately leading to ice speleothem formation.

However, during winter state, if external air temperatures are above internal ones, the heat fluxes quickly reverse, being directed towards the cave air and ice, slightly warming both of them (Fig. 3.8D). External cooling and inflow of cold air once again reverses the heat fluxes back to the situation described in Fig. 3.8C. At the end of the winter, heat originating in the external warm air and increased inflow of dripwater causes another reversal of heat fluxes, which are now directed towards both air and ice formations. Heat (both sensible and latent) is rapidly absorbed by the (initially warming and later melting) ice, keeping air temperature close to 0°C (Fig. 3.8D). Additional sources of heat could be supplied by the touristic use of the caves, both through their lighting equipment and the body heat of the tourists themselves; however, these play a reduced role in the overall dynamics of temperatures inside caves.

REFERENCES

Kern, Z., Perşoiu, A., 2013. Cave ice - the imminent loss of untapped mid-latitude cryospheric palaeoenvironmental archives. Quat. Sci. Rev. 67, 1–7.

Luetscher, M., 2005. Processes in Ice Caves and their Significance for Paleoenviromental Reconstructions. Swiss Institute for Speleology and Karst Studies (SISKA), La Chaux-de-Fonds, p. 154.

Meyer, C., Pflitsch, A., Holmgren, D., Maggi, V., 2014. In: Land, L., Kern, Z., Maggi, M., Turri, S. (Eds.), Schellenberger Ice Cave (Germany): a conceptual model of temperature and airflow. Proceedings of the Sixth International Workshop on Ice Caves, NCKRI Symposium 4. Carlsbad (NM): National Cave and Karst Research Institute, August 17–22. Idaho Falls, USA, pp. 82–86.

Obleitner, F., Spötl, C., 2011. The mass and energy balance of ice within the Eisriesenwelt cave, Austria. Cryosphere 5, 245–257.

Perrier, F., Le Mouel, J.L., Kossobokov, V., Crouzeix, C., Morat, P., Richon, P., 2005. Properties of turbulent air avalanches in a vertical pit. Eur. Phys. J. B 46 (4), 563–579.

Perşoiu, A., Onac, B.P., Perşoiu, I., 2011. The interplay between air temperature and ice dynamics in Scărişoara Ice Cave, Romania. Acta Carsologica 40 (3), 445–456.

Perşoiu, A., Onac, B.P., 2012. Ice in caves. In: White, W., Culver, D.C. (Eds.), Encyclopedia of Caves. Elsevier, Amsterdam, pp. 399–404.

Perşoiu, A., Pazdur, A., 2011. Ice genesis and its long-term mass balance and dynamics in Scărişoara Ice Cave, Romania. Cryosphere 5 (1), 45–53.

Piasecki, J., Pflitsch, A., Sawiński, T., Zelinka, J., 2008. In: Kadebskaya, O., Mavludov, B., Pyatunin, M. (Eds.), Dobsinska Ice Cave thermal-circulation system in the light of the newest studies. Proceedings of the 3rd International Workshop on Ice Caves, Kungur, Russia, May 12–17. Perm State University, Perm, pp. 23–25.

Racoviţă, G., 1984. Sur la structure meeroclimatique des cavites souterraines. Theor. Appl. Karstol. 1, 123–130.

Racoviţă, G., 1994. Bilan climatique de la grotte glacière de Scărişoara, dressé sur dix années d'observations. Travaux de l'Institute de Spéologie "Emil Racovitza", 33, pp. 107–158.

Racoviţă, G., Boghean, V., Silvestru, E., Petrescu, M., 1991. Etude thermometrique des substratums de la Grotte de Scărişoara. Travaux de l'Institute de Speleologie "Emil Racovitza", XXX. pp. 159–184.

Saar, R., 1956. Eishohlen, ein meteorologisch-geophysikalisches Phanomen, Untersuchungen an der Rieseneishohle (R.E.H.) im Dachstein. Oberosterreich. Geogr. Ann. A. 38, 1–63.

Viehmann, I., Racoviţă, G., Şerban, M., 1965. Observaţii asupra microclimei Gheţarului de la Scărişoara. Lucrările Institutului de Speologie "Emil Racoviţă," tom IV, Bucureşti.

Wigley, T.M.L., Brown, M.C., 1976. The Physics of Caves. In: Ford, T.D., Cullingford, C.H.D. (Eds.), The Science of Speleology. Elsevier, New York, pp. 329–358.

FURTHER READING

Luetscher, M., 2013. Glacial processes in caves. In: Shroder, J., Frumkin, A. (Eds.), Treatise on Geomorphology. vol. 6. Academic Press, San Diego, CA, pp. 258–266. Karst Geomorphology.

CHAPTER

ICE GENESIS, MORPHOLOGY AND DYNAMICS

4

CHAPTER

4.1 ICE GENESIS AND TYPES OF ICE CAVES

Bulat R. Mavlyudov
Institute of Geography of the Russian Academy of Sciences, Moscow, Russia

CHAPTER OUTLINE

To understand ice genesis and types of ice caves, it is necessary to define cave glaciation and explain its properties.

4.1.1 CAVE GLACIATION

Dmitriev (1980a,b) was the first to have actually considered cave glaciation. He considered, that "all set of various ice of karst cavities, as an unit, united by the environment in which they develop—the cave, it is necessary to consider as independent group of a spectrum of a modern glaciation of the Earth." So Dmitriev V.E. understands all sets of ice in cavities as cave glaciation.

Cave glaciation is shown to exist by observing various forms of snow and ice in cavities, their form, and dynamics.

Ice Caves. https://doi.org/10.1016/B978-0-12-811739-2.00032-2

4.1.2 REASONS AND CONDITIONS OF CAVE GLACIATION

For cave glaciation to occur, it is first necessary that a cave exist. Before we consider reasons and conditions of cave glaciation, we will shortly analyze the origin of cavities in the upper part of the earth's crust.

4.1.2.1 ORIGIN OF NATURAL CAVITIES

Natural cavities can have both an artificial and a natural origin. Glaciation can develop in both. However, the position of artificial cavities does not depend on the geological situation around them, that is, they can occur almost any place both above and below the earth's surface. So we can omit geologic constraints. Natural cavities can have a karst, tectonic, piping, abrasion, thermokarst, glacial or volcanic origin.

Karst caves can form as a result of 4 basic conditions: the presence of soluble rocks, systems of cracks and fissures, aggressive water (in relation to rocks), and moving water (Ford and Williams, 1989; Kruber, 1915; Sokolov, 1962; Thornbury, 1954). Soluble rocks include: limestone, dolomite, marble, marl, chalk, gypsum, anhydrite and salt. There are some poorly soluble rocks: calcareous sandstones, conglomerates and breccia with calcareous cement, in rare cases igneous rocks. According to Maksimovich (1963) karst rocks are present in approximately 40% of the former USSR (Fig. II.17.1 from Part II). Conditions of karst formation are observed in almost all areas of the former USSR (Gvozdetskij, 1954, 1972). As these cavities develop non-uniformly, in different conditions, and have different ages, it is possible even within one area for caves to exist at different stages of development. The first one—phreatic (cavities are completely filled by water), the second—vadose (cavities are often drained, i.e., they are situated in the aeration zone of the karst upland), the third epoch—dry (cavities are completely drained). Glaciation can develop only in the cavities, which are in vadose and dry stages of development. Clearly in those cavities with high discharge, glaciation will occur with more difficulty than in drier cavities.

Karst cavities are genetically subdivided into corrosive-gravitational, erosive-corrosion, and nival-corrosion (Dublyanskij, 1977). In the first two types, only some cavities are subject to glaciation, but nival-corrosion cavities are obliged to undergo glaciation because of their origin. These cavities are formed as a result of rock dissolution by meltwater of snow and ice in caves on contact with rocks (Dmitriev, 1977a,b, 1980a; Dublyanskij, 1963; Shutov and Sevast'yanov, 1983). Basically they are vertical cavities. Deepening of these cavities in Crimea is estimated to be 75 micron/year (Dublyanskij and Shutov, 1967). Such cavities are known also in other mountain areas.

By form karst cavities are subdivided into horizontal, inclined, and vertical cavities (Kruber, 1915). Inclined cavities are divided into ascending and descending ones, and vertical into pits, shafts and vertical cave systems. The following cavities are subjected to glaciation: "horizontal" (with entrances at different elevations), inclined descending, and vertical cavities.

Glacial cavities are formed in ice, firn, and the snow of glaciers and snowfields (Ezhov, 1990; Mavlyudov, 1992, 2006). Conditions for cavity formation as a whole are the same as for karst rocks, but instead of chemical dissolution of rocks there is ice thermoerosion. Features forming glacier cavities include ice plasticity (which can close cracks and cavities), and glacier movement (which can move cavities to new structural and tectonic conditions). Cavities of this type are developed in ice of temperate, polythermal, and cold glaciers. In glaciers there are mainly vertical and horizontal cavities, inclined cavities are possible to find, but are rarer. There are also cavities, which are formed on a glacial bed behind rocky ledges.

Thermokarst cavities are formed in rocks containing ice (Kotlyakov, 1984) (the same phenomena in ice we suggest naming glacial karst; Kotlyakov, 1984). Formation of thermokarst cavities occurs the same way as glacial caves, but because of the small quantity of ice in frozen rocks, it is thus necessary to actively carry out friable material by flowing water. Cavities of this type are developed in frozen rocks. The form of cavities depends on the form of ice bodies in the frozen rocks. Inclined and horizontal cavities prevail among thermokarst cavities.

Tectonic cavities are formed mainly in hard rocks as a result of blocks shifting along cracks and faults. The cavities representing empty places among big blocks of rocks and cavities of crevasses in glaciers belong to the same group. Distribution of these cavities is subject to fewer laws than karst cavities, but more often they are located in mountainous or high relief areas. Quite often such cavities were formed by karst processes. Among them there can be horizontal, inclined, and vertical cavities.

Volcanic cavities are formed in volcanic areas as volcanic tubes. As they form not far from the surface, with time the roof of lava tubes can collapse and open up an entrance. Volcanic caves are mainly horizontal or inclined.

4.1.2.2 COMMON PRECONDITIONS OF CAVES GLACIATION

Cave glaciation depends on certain conditions in the outside climate, cave climate, and hydrology.

Outside climate preconditions. For cave glaciation to occur it is necessary that the air temperature outside the caves be negative either for all or part of a year; such conditions exist in many regions of the world.

Cave climate preconditions. For cave glaciation to occur it is necessary that the temperature of the air and rocks in a cavity be below 0°C. Thus in the cavities located within permafrost areas, subzero air temperatures are preserved all year-round and are supported by the frozen condition of the country rocks. Outside of permafrost areas where rock temperature is above freezing, cave glaciation can be supported only by the existence of cooling sources located outside of the caves. This source can be frosty air penetrating into cavities, and/or snow and ice accumulation in them. In the majority of caves, frosty air easily penetrates into their entrances (because of their openness to external influences), in a zone of seasonal temperature fluctuations. The cooling of internal parts of caves occurs if cavities that have certain forms (Fig. 4.1.1), which produce an air circulation system in a cave and a distribution of cooling zones. Some authors propose seven different forms of caves favorable for cave glaciation (Gvozdetskij, 1968, 1972), but it is possible to combine them all into three main types.

Hydrological preconditions. For cave glaciation to occur it is necessary that the cavity receive water in a gaseous, liquid or solid condition.

4.1.2.3 CONDITIONS OF CAVE GLACIATION

It is possible to separate conditions controlling caves glaciation into two groups: the basic and the minor. The first group must have: (1) a difference in temperature between the rock and the outside air; (2) defined forms of cavities; (3) cold reserve in cavities; (4) water inflow into cavities. Groups of minor

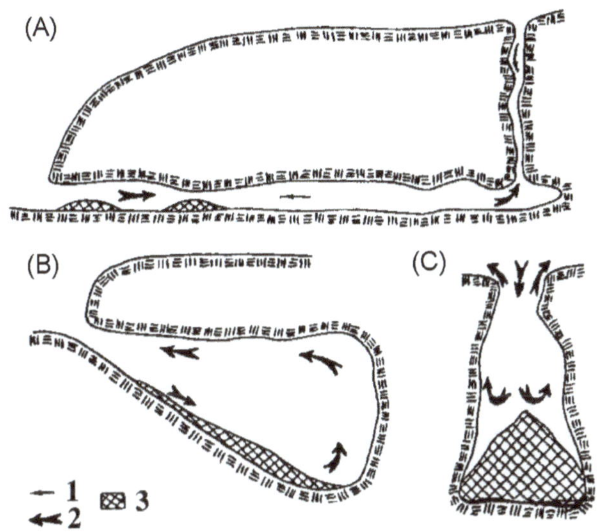

FIG. 4.1.1

The basic forms of cavities favorable for glaciation. (A) "horizontal" cave with entrances situated at different elevations, (B) inclined descending cave (cold trap), (C) vertical cavity (karst pit); 1—direction of air movement in cavities in summer, 2—direction of air movement in cavities in winter, and 3—snow and ice accumulations.

conditions for cave glaciation include: influence of atmospheric phenomena (pressure, air humidity, solar radiation, winds), cavity structure, thermalphysical rock properties, geomorphological conditions around caves, presence of vegetation.

4.1.2.3.1 Main reasons for cave glaciation

Temperature ratio between the rock and outside air

Cave glaciation at above freezing temperatures in the bedrock is potentially possible only at certain winter temperatures where the external air can cool the cave walls to subzero temperatures (Mavlyudov, 1989d, 2008). It is possible to achieve a ratio of external winter air temperatures and rock temperature favorable for cave glaciation on the basis of the analysis of mathematical models of thermal conditions in caves (Golod and Golod, 1974; Mavlyudov, 1985). These models strongly simplify cave conditions and demand many assumptions, but despite that they give a picture of the interaction of external air temperature and cave temperature fields that are close enough to be real. However good models are developed only for horizontal caves and the calculations are laborious.

An approximation of the interaction between external temperatures and cave temperature fields is possible based on the analysis of the ratio of average minimum winter air temperatures outside of caves to the bedrock temperature. It can be made with the use of the empirical temperature index of cave glaciation (Mavlyudov, 1988b), which consists in data analysis of modeling calculations and real situations in ice caves:

$$K = -T_{Jan} / (T_M - T_{Jan}), \quad (4.1.1)$$

where *K*—temperature index of cave glaciation; T_{Jan}—mean air temperature of the coldest month of the year (more often January for the northern hemisphere and July for the southern one), T_m—temperature of the rock containing the cavity that is equal $T_m = T_0 + \boldsymbol{a}$, where T_0—MAAT (mean annual air temperature) of the area where the cavity is situated, ***a***—the additional number which can vary for different areas from 0 to 6°C (for example, in Europe $\boldsymbol{a} \approx 0$°C but for the former USSR it varies from 2 to 6°C (Fig. 4.1.2) (Frolov, 1976)). For simplicity we use an average of $\boldsymbol{a} = 3$°C for all areas of the former USSR and subsequently for eastern Eurasia (Mavlyudov, 1985). For Canada ***a*** can change from 1 to 5°C (Fig. 4.1.3) (Williams and Gold, 1976). This means that for Canada and for all continental North America, ***a*** can be also used as equal 3°C.

Thus ***K*** represents the degree of possible cooling of the bedrock by winter air temperatures: the greater the value of ***K***, the greater the degree of possible cavity cooling. Cave glaciation is possible in those areas where ***K*** has values from 0 to 1. As ***K*** approaches 1, it crosses the boundary of permafrost (where $T_m = 0$) and zero is a zero isotherm of mean January air temperature. The analysis of available data about ice caves shows that in areas where ***K*** has values: (1) from 0 to 0.25 seasonal cave glaciation is possible, (2) from 0.25 to 1.0—permanent glaciation of some caves is possible, (3) >1—permanent glaciation in the majority of caves can develop (Mavlyudov, 2008).

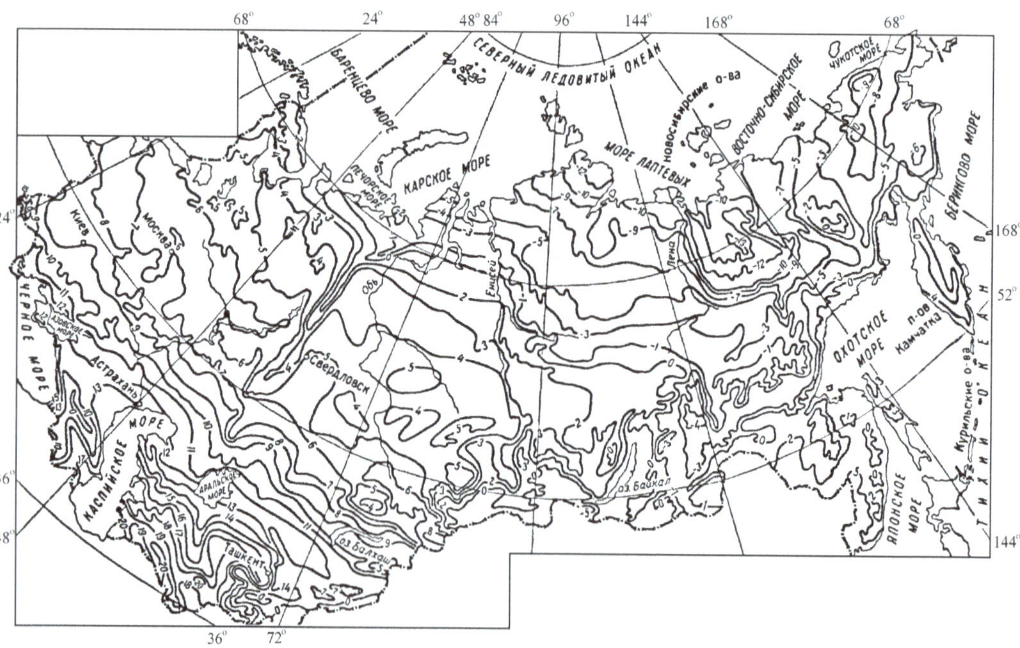

FIG. 4.1.2

Map of rock temperatures at "neutral layer" on the area of the former USSR. Isolines show rock temperature at a depth about 25 m below the surface (Frolov, 1976).

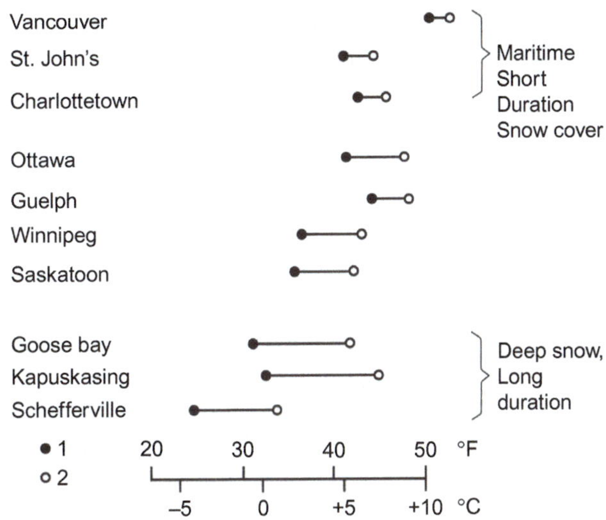

FIG. 4.1.3

Examples of the difference between MAT of soil (temperature of a neutral layer) (1) and MAAT (2) in Canada (Williams and Gold, 1976).

The favorable form of cavities

Other than in permafrost areas, the following cavities are subject to glaciation: "horizontal" cavities with entrances at different elevations, as well as inclined, descending, and vertical cavities (Fig. 4.1.1). The cavity form defines the air circulation system in it and the distribution of cooling zones (Mavlyudov, 1994). Thus in cooling areas in these cavities there are zones with MAAT and rock temperatures lower than the temperature of the country rock, that is, zones of negative temperature anomalies (NTAs) are created. Those NTA zones, where negative air and rock temperatures exist, are favorable for glaciation. Permanent glaciation is developed in those parts of cavities where MAAT and rock temperatures are negative or close to zero.

"Horizontal" caves. Caves of this type represent subhorizontal tunnels with two or more entrances at different elevations. Only lower entrances, in most cases, are accessible to people. The upper entrance is usually a system of impassable cracks and fissures. In such caves, a natural (chimney) air draught is observed, which during the year changes its direction (in the summer—downwards, in the winter—upwards). Changing cave wind directions occur when the outside air temperature reaches rock temperatures (the draught is directed upwards at $T_{sur} < T_m$, draught is directed downwards at $T_{sur} > T_m$). The distribution of air temperature with cave length is shown in Fig. 4.1.4.

Close to the lower cave entrance, is an NTA zone of air and rocks where their MAT (mean annual temperature) is considerably below the temperature of the regional bedrock, but at the upper entrances there is a positive temperature anomaly (PTA) where MAAT and rock temperatures are above the temperature of the regional bedrock (Listov, 1885; Lukin, 1965, 1969). Usually the NTA area is easily found because of the accessibility of the lower cave entrances, but upper entrances into such caves are quite often unknown. For example, the PTA area in the Kungurskaya Ice Cave in the Ural Mountains (Russia) was, for a long time, only assumed. Our measurements in October 1985 showed that the temperature of air flowing out through the upper passages reached 13°C, with the temperature of the bedrock about 5°C, when the air temperature outside the cave was below zero. These observations indicate the presence of a PTA zone in Kungurskaya Ice Cave (Mavlyudov, 2008).

An NTA zone in "horizontal" caves is favored where: (1) areas of NTA and PTA occur in different parts of the cave passage and are formed at various times of the year (Fig. 4.1.4) so they do not compensate each other during an annual cycle (Golod and Golod, 1974); (2) the temperature of the rock containing the cave is usually some degrees above the MAAT of the given district near the cave. As a result an NTA area in "horizontal" caves will be a little longer than the PTA area, because the period with ascending air draught will be longer than with the descending one. For example, in the Kungurskaya Ice Cave (Ural Mountains) the period with ascending draught is about 180 days, and with descending it is about 140 days per year averaged over 20 years (Dorofeev and Mavlyudov, 1993; Mavlyudov, 1997).

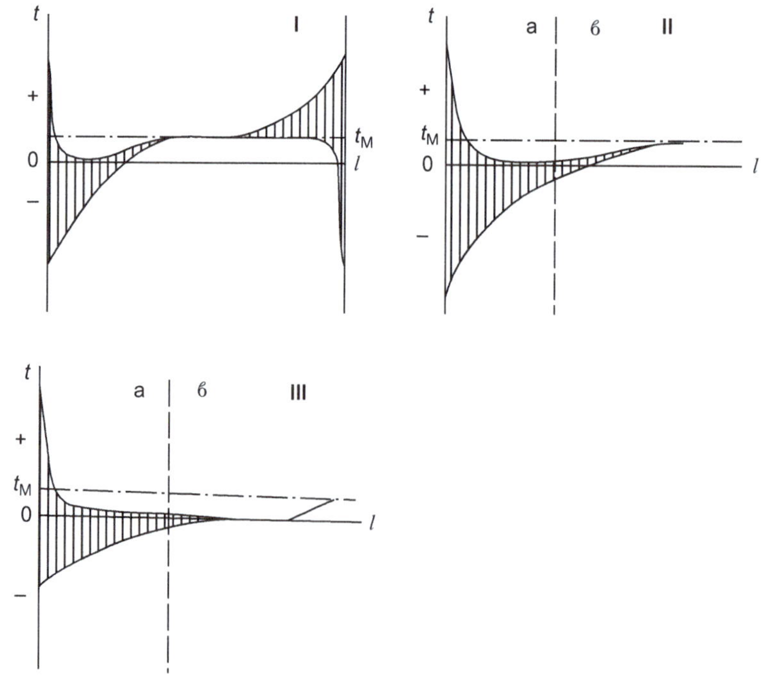

FIG. 4.1.4

Distribution of air temperature with distance into several types of caves: I—"horizontal" caves with entrances at different elevations; II—inclined descending caves; III—vertical cavities. t—air temperature in the cave passage (+: positive, –: negative), t_M—regional rock temperature, l—cavity length, a—short cavities, b—long cavities.

Inclined caves. Caves of this type represent cavities, which descend from the entrance inside a body of rock. As a rule, the width of the cavity increases down from the cave entrance while the ceiling rises forming a large room. Such cavities are often short (a first hundreds meters), and begin at the base of a hill, or at the bottom of dolines and depressions (Fig. 4.1.4). Cavities should have enough high entrances to permit heavy cold frost air to flow into the cavities. Glaciation in such cavities is caused by an accumulation of winter cold air, which is carried out by the exchange of cold and heavy external air with warmer air from the cave.

Intensive air exchange in such cavities occurs only during the winter and autumn, when the air temperature inside is above the outside temperature of the cavity. The most intensive air exchange is observed at sharp outside cold snaps. The cooling of the cavity by a descending stream of cold air begins after the external air temperature becomes lower than the air temperature in the cave. If the air temperature outside is higher than in a distant part of the cave, but lower than at its entrance, descending air movement and cooling will affect only the entrance part of the cave. If the air temperature outside the cave begins to gradually decrease, air will start to flow into deeper and deeper parts of a cave and, eventually, air exchange will involve all active parts of the cavity although the air temperature in the cave interior will not approach the external air temperature.

The velocity of descending wind is proportional to the difference in temperature outside of the cave and in its deepest part. In Fig. 4.1.5 the dependence of wind velocity in an entrance part of Skhvava Caves (Caucasus) on the difference of external and internal air temperatures in December, 1984 is shown. Observations have shown that there is also a daily rhythm of wind velocity change in the caves (Fig. 4.1.6).

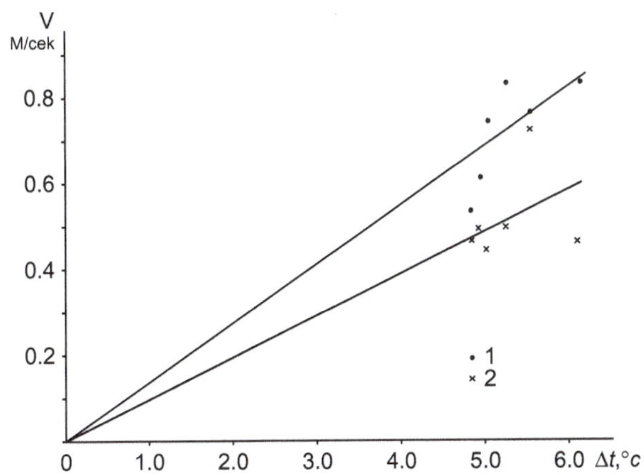

FIG. 4.1.5

Shows the correlation of wind velocity (V) in Skhvava Cave (Caucasus) and differences of external and internal air temperatures (Δt) from observations in December, 1984. 1—at a depth of about 42 m from the surface, 2—at a depth about 50 m from the surface.

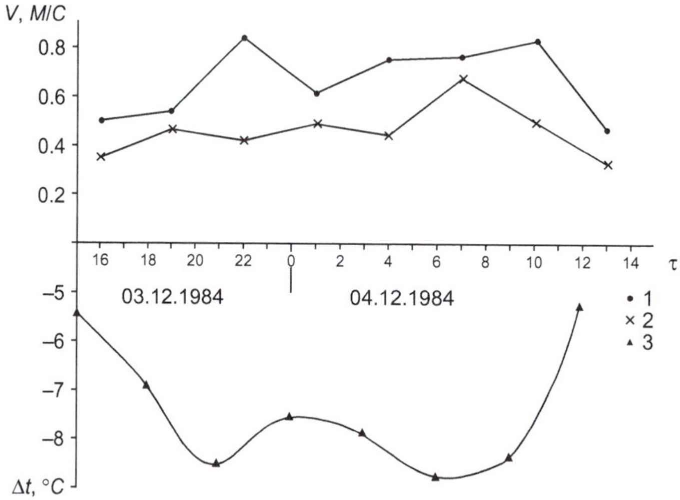

FIG. 4.1.6

Shows the daily change in wind velocity in Skhvava Cave and air temperatures outside of the cave (3). 1—approximately 42 m below the surface, 2—approximately 50 m below the surface.

If part of the inclined cave close to entrance has good circulation (for example, in Bol'shoj Buzluk Cave in Crimea it occurs at a depth of about 40 m below the surface), and sometimes is warmed by the sun, there will be large annual fluctuations in the annual temperature. In the summer air temperature in entrance parts of caves is above freezing, but it falls close to zero in the cave interior where it is kept at a stable level because cold air stagnates there and a lot of heat is expended to melt ice. In the summer, air temperature in a cave is lower than on a surface, in the winter it is higher than on a surface. Annual amplitudes of air temperature fluctuations decrease with cavity depth (Fig. 4.1.4).

Vertical cavities. Cavities of this type represent pits and mines with vertical trunks in which seasonal and perennial snow-ice formations occur. The mouth of these cavities should not be too narrow or it will be blocked by snow during winter. In vertical caves, as in descending inclined cavities, cold, heavy air flows in during the winter, and in the summer there is air stagnation. But the basic source of cold in vertical cavities is snow, which penetrates into caves in the winter through an entrance aperture. Thus the snow, which has accumulated in vertical caves during winter does not completely melt in the summer.

If the air temperature in the upper part of a pit is close to the temperature of a surface outside the cave, as depth increases air fluctuation amplitude will decrease (Fig. 4.1.4). As in the summer in pits with snow, air temperature is close to zero, and MAAT at the bottom parts of these caves is negative.

Formation of a cold reserve

Winter air cooling cavities leads to areas where NTA are formed, that is, cold reserves are formed. The cold reserve is equivalent to the amount of heat needed for cooled air to return to a condition where temperatures of the bedrock and the air are equal. For cave glaciation to occur, it is necessary that air and rock temperatures in cavities are below 0°C. Thus the cold reserve in caves can be short-term, seasonal, or perennial. That can serve as the reason for the formation of corresponding glaciation. For a perennial cold reserve in a cave, it is necessary that the annual arrival of heat to a cave be no more than the cold reserve available in it. The cold reserve is calculated as:

$$Q = Q_1 + Q_2 + Q_3 + Q_4, \tag{4.1.2}$$

where Q—cold reserve; Q_1—heat carried out of a cave; Q_2—heat carried into a cave with water and air; Q_3—heat arriving in a cave from the bedrock Q_4—heat of water phase transitions. The cold reserve in caves is formed due to frozen cavity walls, cold water inputs, and the formation of an ice reserve. Thus in polar areas cold comes more from the cave walls, while in nonpolar areas cold comes more from snow and ice than from the cave walls.

The cold reserve in caves changes during the year (before the beginning of winter it is minimum, and at the end of winter it is maximum), and along the cave length: the cold reserve of rocks decreases from the entrance to the inside of cavities, while the distribution of ice related cold reserves strongly depends on the dimension and position of water inflows. Direct calculation of cold reserves, both for all caves as a whole and for its parts, is complicated, because the cold reserve in caves also depends on the form and structure of the cavity and also on outside temperatures. It is possible to estimate the dimension of cold reserve at a specific spot in a cave based on the MAAT at this point. If the long-term MAAT is positive but close to zero in this part of a cave, so cold reserves are adequate for only a seasonal glaciation. If MAAT in a cave is equal to zero or negative, it will cause perennial cave glaciation. The lower the MAAT of a specific place in a cave, the more stable glaciation will be at this point.

Let's consider the distribution of cold reserves in caves of different morphologies.

"Horizontal" caves. It is common for them to carry out heat from an NTA zone during winter, and carry out more heat than comes in during the summer. In the Kungurskaya Ice Cave (Fig. 4.1.7) the sum of mean daily air temperatures at the entrance during the period of ascending air draught in 1984 was about −1904 degrees, and the sum of mean daily air temperatures flowing into the cooled cavity from

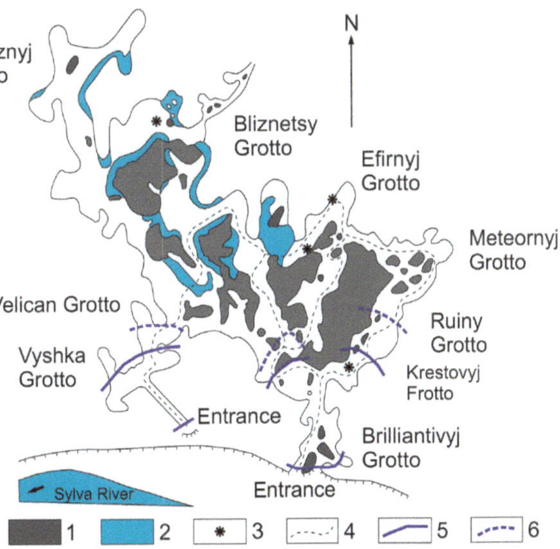

FIG. 4.1.7

The scheme of Kungurskaya Cave glaciation (Ural Mountains). 1—rocks; 2—lakes in cave and river outside cave; 3—organ tubes; 4—touristic roads; 5—boundary of permanent glaciation; 6—boundary of seasonal glaciation.

inside a cave due to a descending air draught in summer was +970 degrees, that is, approximately twice as much heat was carried out than was carried into the cooler part of the cave (not measuring the heat from the cave walls). In 1969 the sum of negative air temperatures at the entrance of the Kungurskaya Ice Cave was −2405 degrees, and the sum of air temperatures moving in the opposite direction was about +830 degrees, that is, cooling had exceeded warming by almost three times. We will consider the distribution of cold reserves from different caves.

In Dachstein Cave in Austria the sum of the mean monthly air temperatures in 1920–25 at cave entrances during the period of ascending air was −28.3° (Fig. 4.1.8), and the sum of mean monthly air temperatures from the cave interior where air descends was +16.7° that corresponds to an excess of cooling over warming by almost 1.5 times (Saar, 1956).

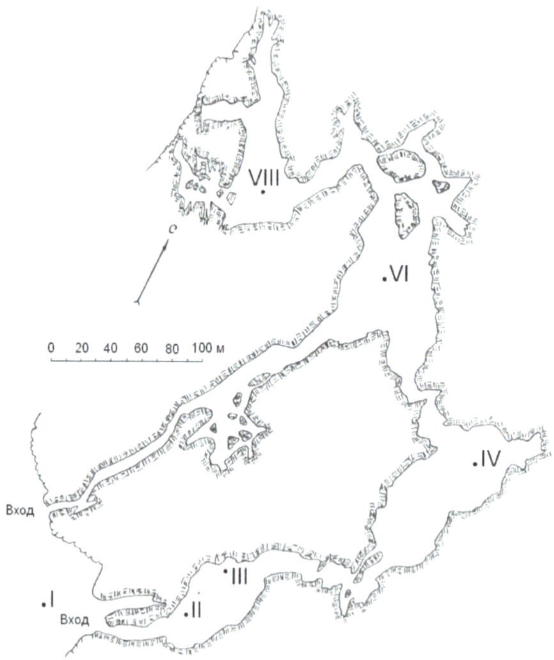

FIG. 4.1.8

Plan of horizontal Dachstein Cave (Austria). Numbers—points of weather stations (Saar, 1956), data is shown in Table 4.1.2.

Let's consider the distribution of MAAT versus the length of these caves (Tables 4.1.1 and 4.1.2).

As we can see in both caves, there are zones of steady negative MAAT, and in the Kungurskaya Cave the length of the NTA zone is approximately 200 m from entrance, and in Dachstein it is about 400 m. Thus the boundary of permanent glaciation situated at isotherm is +0.2°C, and seasonal glaciation at isotherm is about +1.0°C.

We can find similar situations in horizontal caves of Pinega (Shavrina and Guk, 2005).

Inclined caves. The almost total lack of air movement in inclined caves in summertime is the reason of that these cavities are more effective at storing cold than horizontal caves.

Table 4.1.1 Distribution of Air Temperatures in the Kungurskaya Cave (Ural Mountains) in 1984 (Data of the Kungursky Permanent Establishment by the Mine Institute of Ural Division of the Russian Academy of Sciences) [Mavlyudov, 2008]

Place of Investigations (Fig. 4.1.7)	Distance From Entrance, м	$\sum T_{day}$, °C	$\sum T_{month}$, °C	$\bar{T}_{Jan}$, °C	$\bar{T}_{year}$, °C	K^o
Outside cave	0	319	9.5	−11.2	0.78	0.69
Briliantovyj Grotto	50	−1073	−35.4	−4.8	−2.9	0.49
Dante Grotto	130	−512	−16.9	−2.0	−1.4	0.29
Krestovyj Grotto	210	53	2.1	−0.3	0.2	0.06
Ruiny Grotto	330	858	28.3	2.0	2.4	−0.66

Table 4.1.2 Distribution of Air Temperatures in the Dachstein Cave (Austria) in 1920–25 [Saar, 1956]

Point of Measurements (№ of Point) (Fig. 4.1.8)	Distance From Entrance, m	$\sum T_{month}$, °C	$\bar{T}_{Jan}$, °C	$\bar{T}_{year}$, °C	K^o
I	0	31.9	−14.7	2.6	0.71
II	65	−29.5	−10.6	−2.4	0.64
III	100	−20.1	−8.8	−1.7	0.59
IV	300	−11.5	−3.4	0.9	0.36
VI	445	19.2	0.9	1.6	−0.17

Therefore MAT in the area of glaciation is close to zero or negative, and inclined caves with ice occur considerably more to the south than horizontal ice caves. We will consider the distribution of MAT in an inclined descending cave in Demenovskaya Cave, Slovakia (Table 4.1.3).

Table 4.1.3 Distribution of Air Temperatures in 1978 in the Demenovskaya Cave (Slovakia) [Hallas, 1983]

Place of Measurements	$\sum T_{month}$, °C	$\bar{T}_{Jan}$, °C	$\bar{T}_{year}$, °C	K^o
Outside cave	75.0	−4.4	6.3	0.32
100 m	−2.4	−0.8	−0.2	0.22
300 m	9.6	0.0	0.8	0.0

As MAAT in the area of NTA in a cave is close to zero (−0.2°C) its cold reserve occurs both in the cold reserve of walls and in the ice (Hallas, 1983). A similar phenomenon is observed in other caves with ice in Slovakia: in the Dobshinskaya Cave MAAT in the zone of glaciation is equal to −0.3°C with an external MAAT of about +8.0°C (Droppa, 1964), in the Silitskaya Cave, those values are, 0°C and 6.8°C (Roda et al., 1974). Thus in the Dobshinskaya Cave the cold reserve mainly is connected with cold reserves in ice where its volume is equal to 145,000 m^3 instead of with the cold reserve of the walls.

Vertical cavities. Temperature conditions in these cavities in general are similar to conditions in inclined descending caves. The only difference is that water in pits is generally already in a solid phase and is cooled to negative temperatures (i.e., there is almost no water cooling and heat generation by ice crystallization). Regular measurements of air temperature in vertical cavities with ice is rare. But it can be done using the following approach. Table 4.1.4 shows the generalized results of separate observations in Snedznaya Cave System in the Caucasus from 1979 for 1982 (Dublyanskij and Ilukhin, 1982; Mavlyudov, 1980, 1981, 1988c; Mavlyudov and Morozov, 1984; Mavlyudov, 2016) are shown (Fig. 4.1.9).

Table 4.1.4 Distribution of Air Temperatures in Snedznaya Cave System (Caucasus)

Place of Measurements, Depth From Surface, m	$\sum T_{month}$, °C	$\overline{T}_{Jan}$, °C	$\overline{T}_{year}$, °C	$K°$
0	24.0	−5.0	2.0	0.55
40	14.0	−4.6	1.2	0.48
100	−6.1	−3.5	−0.5	0.4
190	−1.8	−2.0	−0.15	0.27
230	0.0	−0.5	0.0	0.08

In vertical cavities the cold reserve is mainly connected with the accumulation of snow-ice formations, therefore such cavities can only be in areas where a lot of snow penetrates into them.

However, using MAAT in caves for an understanding of situations with cold reserves in them is inconvenient, because very often temperatures in them are unknown. To measure temperatures requires a special regiment of meteorological observations in the cave. Estimating the cold reserve in cavities is also possible by knowing the mean minimum air temperatures in caves (i.e., mean air temperature of the coldest month, and the background temperature of the rock, which can be procured from the literature (Frolov, 1976, or something similar), and calculating the temperature index of glaciation of separate parts of the cave. Do not confuse this with the general temperature index of glaciation (cave glaciation index, CGI), we name it $K°$. As $K°$ is practically linearly connected with the MAAT of a corresponding cave place (Fig. 4.1.10), it can practically give the same information about the cold reserve in the cave.

The index $K°$ shows the degree of cooling of a specific place in a cavity. Thus the cooling maximum is reached at $K°$ value closely to one, and a minimum—when it approaches zero.

Permanent cave glaciation is developed within values of $K°$ from 1 to 0.05. In areas with cold winters, $K°$ gradually decreases from the entrance to the inside of cavities. In areas with warm winters, when frosts alternate with warmth, $K°$ inward from cave entrances first increases, then decreases. This is explained by selective air penetration with different temperatures occurring in caves during cooling. In Tables 4.1.1–4.1.4 values of $K°$ in caves of different morphology are presented.

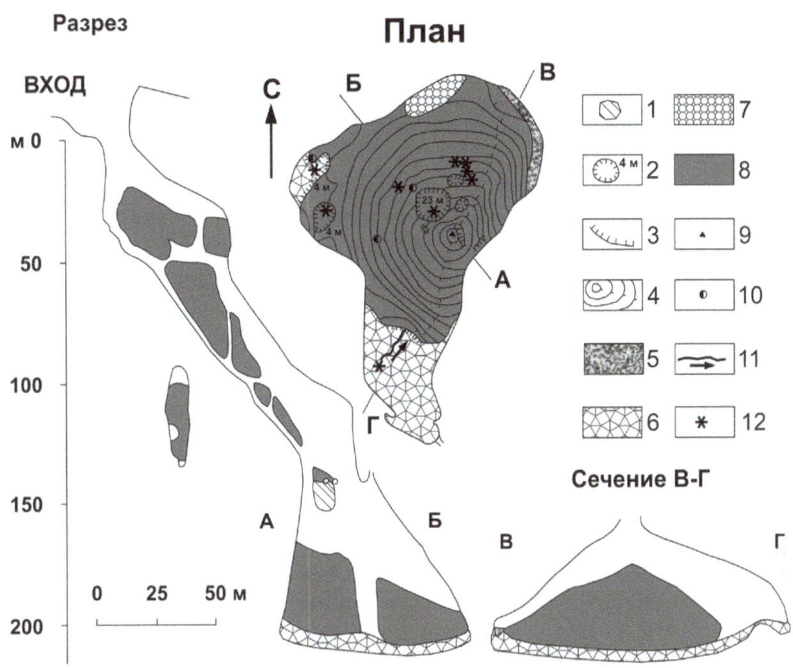

FIG. 4.1.9

Snedznaya Cave System (Caucasus) which is about 200 m deep. 1—cave walls and the rock blocks in them; 2—pits and their depth; 3—rock contours, 4—ice contours, 5—collapsed sediments, 6—alluvial sediments, 7—collapsed moraines, 8—snow and ice on cross sections, 9—summit of snow-ice cone, 10—ice stalagmites, 11—rivers and water courses, 12—drops (Mavlyudov, 1988c).

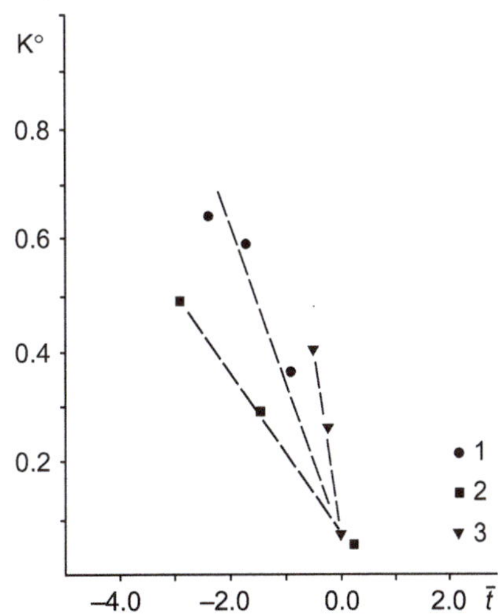

FIG. 4.1.10

Comparison of MAAT in caves (t) with the temperature index of glaciation of separate parts of a cave (K°). Caves: 1—Dachstein (Saar, 1956), 2—Kungurskaya, 3—Snezhnaya Cave System.

Water inflow into cavities

Cave glaciation can occur only when water penetrates into frozen cavities. This occurs when the heat brought in by the water is less than the cold reserve in the cave. It is necessary to distinguish the impact of the seasonal heat reserve introduced by water and the annual cold reserve in the cave. If both the annual and seasonal heat reserves are less than the cold reserve in the cave, then perennial glaciation occurs (at positive ice mass balance). If the heat reserve of seasonal water inflow is less, but the annual one is more, then the cold reserve in the cave in the cavity causes seasonal glaciation.

The amount of water entering a cave also defines the morphology, distribution, ice reserve and ice mass balance in cavities. The quantity of water in caves depends on: (1) the amount of solid and liquid precipitation penetrating directly into cavities or being transformed into groundwater; (2) conditions of condensation and sublimation of water vapor in the cavities defined mainly by their morphology and the climatic conditions outside the cave.

4.1.2.3.2 Secondary factors affecting cave glaciation

Atmospheric pressure. In some studies it is shown that a change of atmospheric pressure can cause air movement in caves (Ford and Gullingford, 1976). In this study it is observed that cave air always responds to changes of atmospheric pressure, but the velocity of rising air in a cavity is usually small. Wind velocity increases in caves with a big volume and/or for caves with a small entrance but rapid changes in atmospheric pressure. For caves with a volume about $3 \times 10^5 \text{м}^3$ rapidly falling external pressure, with a speed of 5 M, creates an air flow out of the entrance with a velocity less than $0.5\,\text{m}^3/\text{s}$. Dmitriev (1980b) considers baric winds in caves as one of the reasons for their glaciation. Sotskova (1981) noticed that wind velocities in caves caused by a change of atmospheric pressure can reach 0.5 m/s.

It is clear that baric winds can both strengthen and weaken processes of cave warming and cooling. Winter baric winds directed into caves will cool them if the air outside of the cave is colder, and warm them if the outside air is warmer than in the cave. Summer winds directed into cave entrances will always be a warming factor. However, if the volume of a cave is large enough, it is clear that baric winds cannot continue for a long time and will quickly stop. Although these cave winds are not thoroughly studied, it is possible to assume that they will not have a large influence the formation of cold reserves in caves.

Atmospheric air circulation. This factor influences the glaciation of different types of caves unequally. Winds directed to the lower cave entrance can influence the glaciation of horizontal caves. Thus in winter there is an additional cave cooling effect if the outside air is colder than the air in a cavity, and there is additional cave warming if outside air is warmer than the air in a cavity. During the summer air directed into the lower cave entrance blocks air flowing out of the cave, which reduces the warming of NTA zones. This is true in the case of the steady, long influence of constantly directed winds. The wind pressure in horizontal caves can be defined by an analogy to wind pressure in artificial tunnels (Kirish and Ushakov, 1983):

$$H = \gamma V^2 \cos^2 \alpha / 2g(3), \qquad (4.1.3)$$

where α—the angle between the tunnel entrance and a dominant wind direction, degrees; γ—relative air density, n/m^3; V—wind velocity, m/s; H—wind pressure, Pascals. The analysis of the formula (4.1.3) shows that wind direction essentially influences wind pressure in caves. If the wind pressure at $\alpha = 0$ for 1 at angles 30°, 45° and 60°, the wind pressure will be 0.75; 0.5; 0.25 accordingly. Influence of wind velocity on wind pressure is even more as it depends on the square of wind velocity.

In inclined descending caves, the influence of wind on cavity temperature fields will be smaller than in horizontal ones, because in inclined cavities, two additional factors will have an influence on wind pressure: the angle of the cavity floor with respect to horizontal, and the additional effect of the wind on the character of air circulation in the cave. The first factor will reduce the size of wind pressure with increasing floor inclination stated by the cosine law, and the second factor interferes with the wind during the winter. As in inclined caves, with a descending draught located at the cavity floor, and ascending at the ceiling, the wind blowing into the cave entrance will strengthen the bottom air current in the cave in winter and weaken upper air circulation. That will lead to bursts of air movement in the entrance part of the cave. The influence of wind will not reach the lowest levels of the caves. In the summer, in the absence of air movement in cavities, wind can easily blow inside a cave, but as its temperature will almost always be above cave air temperature, the warming influence will mainly be felt along the ceiling. Therefore ice formations will degrade faster on the ceilings of such caves.

Vertical cavities are the extreme example of inclined descending caves when their floor inclination is close to 90°. Therefore, in agreement with the cosine law, wind influence should be absent. Only the upper parts of these cavities will be effected by the wind. Wind influence on air circulation in vertical cavities will be important only where there is a hydraulic connection between several pits. Thus there is a natural air draught between lower and upper pits where air currents will change direction. Changing air currents are also observed in horizontal caves during the transition period from winter to summer when air movement is unstable.

Humidity is not a determining factor of cave glaciation. It controls moisture redistribution in cavities, its condensation, sublimation, and removal as a result of water and ice evaporation. In inclined descending caves and pits, in the winter moisture is directed out of the cave, and in the summer it is directed into the cave. In horizontal caves in winter there is moisture transport from lower to upper entrances, and afterward into the atmosphere, but partial condensation of moisture on snow and cold rocks leads to the return of a small amount of moisture back into the cavity (Malkov and Frants, 1981). In the summer it is possible to detect moisture transport from upper to lower entrances. In winter, at lower entrances, air is undersaturated with respect to moisture due to water and ice evaporation. In the summer, at the upper entrance, air is supersaturated with condensation moisture. In summer air transport from the internal part of caves in an NTA area leads to moisture condensation on ice, increasing its melting. There is also summer moisture sublimation forming deposits on the surface of cold ice.

Solar radiation. Its influence on cave glaciation is insignificant, being limited to mainly cave entrance parts. In horizontal caves the influence of solar radiation is completely excluded. In inclined descending caves, solar radiation can effect the entrance and sometimes internal parts of caves when the sun is at certain angles (for example, in Bolshoj Buzluk Cave in Crimea in May). Snow at the bottom of pits is influenced only by diffuse radiation, and with increasing pit depth, the quantity of diffuse radiation decreases exponentially (Mavlyudov and Vturin, 1988).

Structure of cavities. For glaciation to occur in caves it is important not only for the cave to have favorable shapes, but also structure: their extent, the difference of heights between entrances, or between the entrance and the remote part of the cave, and passage cross-sections. All these parameters influence the character of air circulation in cavities and the dimensions of cold reserves in them. Each cave type has unique structural features that promote glaciation.

Horizontal caves: The NTA area in such caves can exist at a considerable distance from the PTA zone, therefore, for glaciation to occur, cavity length should not be less than 200–300 m.

The difference in height between lower and upper entrances defines the velocity of air movement in a cave, so also does the extent of NTA and PTA zones. The less difference in elevation between entrances, the slower the air movement will be. Therefore in order to cool these caves, colder conditions are necessary. The greater the difference in elevation, the smaller the *K* at which caves glaciation will be possible and vise versa (Fig. 4.1.11).

Wind velocity in horizontal caves is calculated using the formula (Golod and Golod, 1974):

$$V = \sqrt{2gh \cdot \frac{T_{entr} - T_{exit}}{T_{entr}(1 + \sum \xi_i)}}, \quad (4.1.4)$$

where *V*—wind velocity; *g*—free fall acceleration; *h*—elevation difference between cave entrances; T_{entr}—air temperature entering the cave; T_{exit}—air temperature flowing from the cave; ξ_i—local resistance to air movement inside the cave.

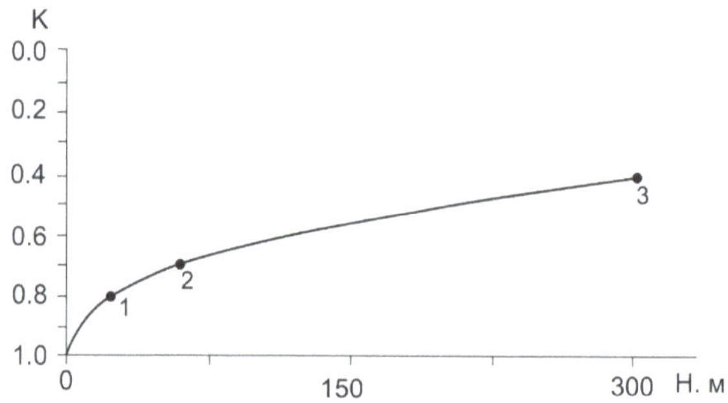

FIG. 4.1.11

Comparison of the difference in elevation between entrances of horizontal caves (*H*) and the temperature index of cave glaciation (*K*). Caves: 1—Kulogorskaya (the Pinego-Kulojskoe plateau), 2—Kungurskaya (Urals Mountains), 3—Dakhstein (Alpes).

All other conditions remaining the same, we will consider wind velocity in the cave at $h=20$ m for 1 (caves of the Pinego-Kulojskoe plateau in the Arkhangelsk Region), for $h=60$ m (the Kungurskaya Cave in Ural Mountains) according to the formula (4.1.4), and wind velocity will be 1.7 and 3.9 at $h=300$ m (Dakhstein Cave, Austria; Saar, 1956), closely approximating reality. However, the greater *h* is, the greater the difference between air temperatures at the entrance and exit of a cave, so wind velocity in the caves will be even higher. In reality, wind velocity in the Kungurskaya Cave is about 3 m/s (Lukin, 1965) and in Dakhstein Cave it is up to 10 m/s (Saar, 1956).

The diameter of a cave tunnel determines the resistance to air movement. If the diameter is small the resistance will be greater, wind velocity will be less, and vice versa. Local resistance in horizontal caves can be calculated from the formula (Atkinson et al., 1983):

$$\xi = \frac{2fL}{D}\left\{1 + \frac{1}{2}\frac{h}{L}\left(\frac{D}{d}\right)^5\right\}, \quad (4.1.5)$$

where $f=0.04$–0.1; *L*—cave length; *h*—elevation difference; *D*—diameter of a horizontal gallery; *d*—diameter of vertical channels. If $f=0.1$; $L=1000$ m; $h=20$ m; $D=2$ m; $d=1$ m, $\xi=132$. Reducing the diameter of the lower tunnel two times (at invariable *d*), increases resistance by 1.5 times, or reduces wind velocity 20%. If $D=2$ m, and d is decreased to 0.5 m, resistance will also increase 1.5 times. If D is increased to 4 m at $d=1$ m, resistance will fall 1.3 times compared to the first case, that is, wind velocity in the cave will increase 15%. Naturally, if the diameter of the tunnel is less, the more cold air will be necessary for cave glaciation, and vice versa. If climatic conditions remain the same outside the cave, and the cave shape and wind velocity stay constant but the diameter of the tunnel varies, the minimum length of NTA zone (l) will vary (Mavlyudov, 1985): at $D=1$ m, $l=1$; at $D=0.5$ m, $l=0.8$; At $D=2$ m, $l=1.3$; at $D=4$ m, $l=1.75$. Thus, for horizontal caves in different climatic conditions, conditions for glaciation will be favorable for different sized cavities. If the climate outside the cave is colder, the size of the cavities will have less influences on glaciation, and if the climate outside the cave is warmer, the size of cavities necessary for glaciation will be more selective.

Inclined descending caves: For short caves, outside climatic conditions have a great influence and only seasonal glaciation occurs. The maximum extent of NTA zone in inclined descending caves is not established yet, but that it exists was mentioned by Kruber (1915). Among caves with ice known to the author, the maximum length of NTA zones does not exceed 120 m (Kemple and Kets-Kemple, 1979); and its minimum size is not more than 5–10 m.

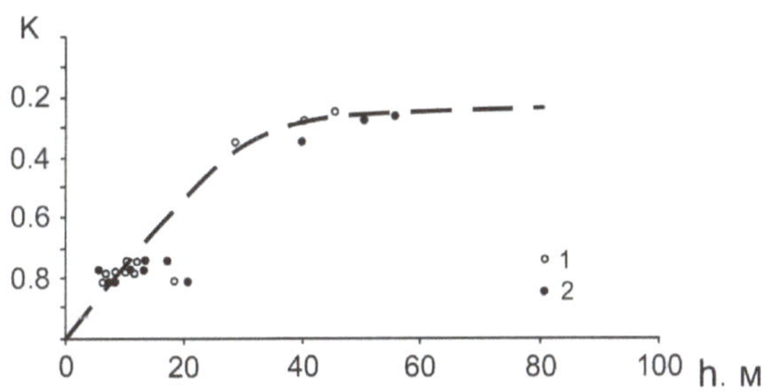

FIG. 4.1.12

Correlation of depth (h) of permanent ice (1) and an ice ledge (2) in inclined descending caves, versus a temperature index of cave glaciation (K).

The difference in elevations between the entrance and lower points of a cave creates conditions for thermal air circulation in the cavity. The difference in heights elevations necessary for glaciation is not the same for different climatic conditions, and it is less for areas with colder winters (Fig. 4.1.12).

Theoretically, in permafrost areas, glaciation of inclined descending caves should begin at the surface. But as summer air temperatures outside become positive, cave glaciation begins at once from the entrance (for example, in Putnikov Cave in Pamir or in caves of Spitsbergen). The maximum known difference in elevations (outside of permafrost areas) is found in Crimea—about 80 m (in Bolshoj Buzluk Cave) that probably is close to the size limit.

There is data (Lobanov, 1981), which state that the height of cave galleries influences air circulation in inclined descending caves. Kruber (1915) thought passage size is not the main thing, but rather it is the outside climate. The Influence of gallery height consists of the following: the greater the difference between heights in a cave, the greater the discharge of air into the cave, that is, wind will move faster in a taller gallery. If the cave entrance gallery height is small, it leads to increasing friction between two meeting air streams (the descending stream of cold air at floor level and the ascending warm air stream along the ceiling), that will lead to a reduction of air flowing into the cave, and that will reduce cavity cooling. Naturally, the smaller the height of the cave passage, the more friction there will be between air currents. The minimum height of a gallery was observed in the Kurgazaksky cave in Ural Mountains—about 0.25 m. The threshold sizes of cave galleries for glaciation development depends on the following geographical factors: the colder the climate, the smaller the passage cross-section that will be sufficient for cavity cooling. Apparently, for this reason, all inclined descending caves with permanent ice that are located close to the southern boundary of their distribution area have high ceilings (Table 4.1.5). It is necessary to add that the height of the entrance passage should exceed the annually accumulating, or drifting thickness of, snow in it. Otherwise, the internal cavity can be blocked by snow, and then it will stop cooling.

Table 4.1.5 Ceiling Heights in Entrance Areas of Some Inclined Descending Caves

Cave Name	MAAT of Area, °C	*K*	Minimum Height of Ceiling, m	Cross Section Area, m^2	Author
Skhvava (Caucasus)	7.0	0.25	7.0	15.0	Mavlyudov, 2008
Bolshoj Busluk (Crimea)	6.4	0.28	15.0	100.0	Dublyanskij and Lomaev, 1980
Cilitskaya (Slovakia)	6.8	0.35	10.0	250.0	Roda et al., 1974
Dobshina (Slovakia)	8.0	0.25	2.0	44.0	Hallas, 1983

Vertical cavities: The maximum depth of snow and ice in such cavities depends on regional precipitation and the quantity of solid precipitation. With an increase in precipitation and amount of uplift in an area, there will be less snow melt in the pits causing the average annual air temperature to fall. Therefore, the higher the pits are situated in mountains, the more probability there will be for more snow in them in the summer, and the smaller the depth of the pits will have to be.

Naturally, the climatic conditions in upland areas influence the snow and ice accumulation in pits. Therefore, in different highlands, the lower boundary of distribution, and the maximum depth of pits with ice, are very different. In the Dinaric Mountains, at the coast of Adriatic sea, there is a pit with snow about 80 m deep, and at an elevation of about 600 m (A. Mihevch, the oral message). On the Western Caucasus the minimum elevation for pits with snow and ice is about 1800 m a.s.l., and their depth reaches 100 m. The minimum depth of pits, within a few meters, would be marked by a snow line (about 2700 m a.s.l.) (Mavlyudov and Vturin, 1988; Mavlyudov, 1989a). The maximum depth of pits with permanent snow and ice usually does not exceed 200 m (Mavlyudov, 1989a).

The diameter of a pit controls the dimension of natural and snowstorm snow concentration in them. The smaller the diameter of the pit, the greater the natural concentration of snow in it (at similar snow thickness) (Mavlyudov and Vturin, 1988). Snow concentration by snowstorms depends more on the form of the entrance section of the pit: in a roundish pit, snowstorm concentration will be more than in a fissure if the fissure deflects the dominating winds in another direction by its smaller diameter. A bigger diameter pit leads to more free air exchange with outside air as a result of blowing winds, and a smaller diameter pit can cause snow blockage, which prevents snow accumulation in the pit. Dublyanskij (1977) considers that the minimum diameter of pits accumulating snow in Crimea is 0.5 m. We observed, on the Caucasus, pits without snow blockage with entrance diameters less than 0.3 m. Apparently the minimum diameter for pits with snow accumulation varies from place to place depending on snow conditions, its fluidity, the temperature, humidity, and the wind conditions of the district. The maximum diameter of pits is difficult to define, but apparently it should not be much more then the depth of the pit, otherwise the pit loses its advantages of snow accumulation over cavities with other forms.

The cave entrance influences the cold reserve in cavities through: (1) orientation of the cave entrance (entrances) in relation to exposure; (2) entrance orientation in relation to directions of dominating winds; (3) absolute height of the entrance.

As southern slopes in the northern hemisphere receive more solar heat than northern slopes, it is clear that caves situated on north slopes are more favorable for cave glaciation. Here the mediated influence of solar radiation on cave glaciation is observed. For example, snow melt on northern slopes occurs over longer periods of time than on southern ones, and also, slow water flow into cavities is more favorable for an accumulation of bigger ice volumes inside. Also entrances with a northern orientation are more favorable for cave glaciation. On northern slopes the forest boundary is situated lower than on southern ones. This means that on northern slopes snowstorm conditions and snow accumulation in cavities extends to lower elevations than on the southern slopes. So the entrance orientation is not the only reason for cave glaciation.

The elevation of cave entrances defines climatic conditions in their vicinity. As air temperature decreases with increasing altitude in mountains, it is clear that this will be the best condition for cave glaciation. Increasing solid precipitation with increasing height is favorable for snow accumulation in pits and in inclined cavities.

Vegetation. The influence of vegetation on cave glaciation is not well studied but the literature indicates that a decrease in vegetation near caves has led to a degradation of glaciation in the Askinskaya Cave in Urals Mountains (Kudryashov and Salikhov, 1968; Lobanov, 1981). The vegetation (first of all lignosa) regulates snow accumulation (by weakening snowstorm redistribution of snow) (Dublyanskij and Lomaev, 1980), makes spring snow melting more gradual, and stabilizes soil preventing the filling of the upper entrances of horizontal caves with friable material. In certain works (Vincent, 1974) the presence of lignosa in vicinities of caves is considered as one of principal reasons of cavity glaciation. The dense lignosa promotes preservation of lower air temperatures under the shade of trees in the summertime (Dublyanskij and Lomaev, 1980). It is possible to illustrate this by an example: in the Perm district in the Ural Mountains there are two caves with permanent ice located near each other (both inclined descending cavities): Andronovskaya and Kladbishchenskaya. The first is on an open southern slope on the bank of the Kamsky water basin, the second on northern slope in dense fir tree forest in a flood gully. In the Andronovskaya Cave permanent ice begins from a depth of about 12 m, and in Kladbishchenskaya Cave ice begins from a depth of about 4 m.

Thermal properties of rocks. In 1885 Yu. Listov considered that thermal properties of rocks are one of the conditions necessary for the existence of caves with ice (Listov, 1885). He wrote that rock favorable for such caves should be bad heat conductors, but needed to have a big thermal capacity.

The analysis of thermal properties of rocks in which caves form (Table 4.1.6) shows that the difference in thermal properties of sedimentary and igneous rocks is not large enough to be an advantage in cold accumulation.

Table 4.1.6 Thermal Characteristics of Some Rocks (Dorofeeva, 1986; Dzidziguri et al., 1966)

Rock	Specific Heat, kJ/kg degrees	Coefficient of Heat Conduction, Vt/m degrees	Elevation, kg/m^3
Limestone	0.9 (0.4–1.0)	2.5 (0.9–3.3)	2500
Marble	0.92	1.3–3.0	2700–2800
Gypsum	0.9–1.0	2.0 (0.7–3.7)	2300
Basalt	0.92	3.5	2800
Granite	0.92	2.2	2700
Ice	2.2	2.25	917

But heat properties of ice is another matter. Heat conductivity in ice is about the same as in the surrounding rocks, but its specific thermal capacity is almost 2.5 times higher. This means that at identical heat fluxes, for example, for ice and limestone, the first heats up in less than a second. But it also means that to cool ice it is necessary to take away almost 2.5 times more heat than from the limestone.

We have considered the factors regulating cave glaciation, but we must also identify the factors preventing the formation of cave glaciation. They are: (1) plentiful water flowing into caves; (2) air movement in inclined and vertical cavities which form the upper entrances of multientrance cave systems (many pits of the Alps, Crimea and of Caucasus; inclined descending Lednevaya Cave in Ural Mountains); (3) absence of any kind of water in caves (caves of southeast Pamir).

4.1.2.4 STABILITY OF CAVE GLACIATION

As we saw above, cave glaciation is influenced by many factors. Naturally, there is the question: can we consider all these factors to predict cavity glaciation and its interaction with changing external conditions, or is it impossible? We will analyze the conditions necessary for cave glaciation. Among them are conditions that are invariable (or are almost invariable) in time, and those that are strongly variable. Among "invariable" conditions are: all morphological features of cavities, thermophysical properties of rocks, and vegetation. Also, conventionally invariable conditions (that vary within certain known limits), include: (1) water inflow into caves, especially for horizontal and inclined cavities, as it is in large degree regulated by the thickness of the overlying rocks; (2) air humidity; (3) surface wind; (4) atmospheric pressure; (5) solar radiation. Thus, cave glaciation will be influenced mainly by: air temperature fluctuation and quantity of precipitation. From them will depend: air movement in caves (i.e., cavity cooling) and the quantity of water that penetrates into the cavity (i.e., quantity of possible snow-ice formations). These two factors are paramount when we speak about the possibility and stability of cave glaciation (regionally the cave shape will also be important). How they vary from each other is a principal reason for the stability or instability of cave glaciation.

Not all of the winter cold in caves exists in ice. It is clear that in the presence of water inflow into a cave, even given a general warming of the cave ice, the quantity of ice in a cave, can increase. This means that increased ice formation in a cave cannot be unambiguously connected with a change in air temperature outside the cave. We will see the same picture both with climate cooling and with the presence of water in the cave (Table 4.1.7).

Table 4.1.7 Cave Glaciation versus Changes in Outside Air Temperature and Water Inflow

Water Inflow	Temperature	Cold Reserve	Ice Quantity	Ice Melting	Ice Accumulation	Ice Evaporation	Ice Sublimation
C	—	+	+	—	C	+	C
+	—	+	+	—	+	+	+
—	—	+	—	—	—	—	—
+	C	C	+	+	+	+	+
—	C	C	—	—	—	—	—
C	+	—	C	+	C	—	C
+	+	—	+	+	+	C	C
—	+	—	—	+	—	C	C

Comments: **C**: *constant, +: increasing, —: decreasing.*
Sublimation = ice "condensation."

From Table 4.1.7 it is clear that glaciation of individual cavities depends mainly on air temperature, and ice quantity which in turn depend on the quantity of water inflow (at conditions such that heat flux from water is less than a cold reserve in the cavity).

Stability of glaciation is defined by the long-term cold reserve in caves, that is, the cold reserve in cavities with permanent glaciation is much higher than the annual heat flux (including interannual variations). If there is an excess of cold reserve in a cavity, there will be a heat flux in the thermal balance, so that glaciation will be more stable and it will be less likely to change due to changes in the outside climate or changes in water inflow.

The stability of cave glaciation is expressed in a dynamic balance of arrival and consumption of heat in caves (over many years). Therefore the dynamic balance in development of glaciation is shown in quasistability between all of its defining elements: (a) the extent of NTA zone in horizontal caves; (b) the position of an ice ledge in inclined descending caves; (c) the position of a snow line inside the pits.

Understanding the mechanism of dynamic balance in the development of cave glaciation means not only understanding the essence of this phenomenon, but also means revealing the conditions at which the given state can occur (Pozdnyakov, 1988).

It is necessary to distinguish the stability of glaciation of a single cave and caves of a specific region, and in the latter case it is defined by glaciation preservation in a group of caves during a change in climate.

The established dynamic equilibrium condition of cave glaciation can be disrupted due to certain changes in: external climate, water inflow to caves, and cavity structure. Change of external air temperature is reflected in cave glaciation through a change in their cold reserves (Table 4.1.7). The biggest influence is a change in the external air temperature on horizontal cave glaciation: climate warming leads to a reduction of the NTA zone length, climate cooling leads to its increase. The inertia of systems can last several years. Oscillation of external air temperatures from year to year leads to fluctuations in the dimension of the seasonal glaciation zone and, to a lesser degree, the permanent glaciation zone. The change of external air temperature during these long time periods leads to a transition in temperature conditions which displaces permanent and seasonal glaciation boundaries in a cavity. With climate cooling cold reserves in caves increase, and glaciation boundaries are displaced inside caves; during warming the cold reserve in caves decreases, and glaciation boundaries move to the cave exit.

The character of change in cave glaciation, and the scale of change of external air temperature, can be estimated by modeling calculations (Mavlyudov, 1985). Decreasing of the mean January, and mean annual, air temperatures by 1°C will lead to an increased length of the NTA zone in such caves as Kungurskaya in the Ural Mountains by 140 m. Decreasing both temperatures by 2°C will lead to an increase in the length of the NTA zone by 225 m (calculations are made for a cavity in limestone with the following fixed: morphology, and thermophysical properties of rock and air).

The influence of a change in water inflow into cavities is clearly visible in Table 4.1.7. We can illustrate it with the example of the Bidzhinskaya Cave glacier fluctuation in Kuznetski Ala-Tau (Dmitriev, 1979). Dating moraine complexes in the cave has shown that the growth of the cave glacier occurred during periods of maximal humidity outside of the cave.

Change of cavity form can result in an increase or in a decrease of the cold reserve in a cave without depending on external climatic conditions or on water inflow.

4.1.2.5 INTERACTION OF CAVE GLACIATION WITH THE EARTH'S SURFACE

The cryosphere of the Earth as cold and ice cover exists in close interrelationship with the atmosphere, hydrosphere, lithosphere and biosphere.

Interaction with lithosphere. After the cavity has undergone glaciation, the character of its evolution will change. And its development can, or will be, accelerated or slowed down, but more often its development is characterized by the presence of both processes at once. In those parts of caves where ice is permanent and cave permafrost is present, there is cavity conservation and a delay of cavity development. In other parts of caves where ice is seasonal, or the influence of seasonal cold occurs, or increased frost weathering is observed, aggressive melt water that can dissolve rock can occur more intensively, that is, acceleration of cavity development can occur.

As mentioned above, there are cavity types whose formation is connected with the interaction of rocks and water formed by snow and ice melting in cavities: these are nival-corrosion cavities. On occasion talus is frozen, creating a steep slope in galleries of caves (in Skhvava Cave in Caucasus floor inclination is up to 40–45°) that slow down the movement of clastic rock material to lower levels. Cave glaciers deepen the bed, shift friable material, form end moraines, and move fragments of rocks downwards in the direction of ice movement in the form of bulk moraines (Dmitriev, 1977a,b, 1979, 1980b). A similar situation is connected with icings slipping along slopes.

Frozen cavity walls cause a considerable temperature change in the rock surrounding the cave passage, creating permafrost. In the Brilliantovyj Grotto of Kungurskaya Cave mentioned above, the thickness of frozen walls reaches 7 m around the cave (Dorofeev, 1981a,b). The thickness of frozen rock decreases inside the cave, and a further 250 m from the entrance permanent permafrost is not present, but a reduced wall temperature (in relation to the rock temperature) is observed in almost all cave galleries (except for a reserved part). Thus in many places rock temperature at the gallery floor is almost always lower than at the ceiling.

Frost weathering causes blocks to collapse from gallery ceilings that lead to a gradual increase in the height of the entrance. In inclined descending cavities it increases the difference of elevations and leads to more cave cooling and glaciation. In horizontal caves, raising the entrance level, on the one hand, weakens the draught between two entrances as differences in elevations between them decrease. On the other hand, it improves ice conservation in a cave in the summer because it slows air flowing out from its lower entrance.

Ice accumulation in caves can play a regulating role: at constrictions, or at low ceilings. Ice growth reduces their cross sections, which inevitably causes a drop in wind velocity and less cavity cooling. Cave galleries can sometimes be closed by ice completely, separating cavities from external influences (caves Shartashinskaya and Ylasyn in the Ural Mountains, Kulogorskaya in Pinega).

Cave glaciation changes the character of cavities by adding secondary sediments: flowstone, alluvial and residual sediments, collapse, glacial, sell, and solifluction sediments. Flowstones either do not grow at all, are oppressed, or start to change their shape. For example, some specific forms of stalactites in caves show evidence of gelifraction processes that are frost weathering (Povara and Diaconu, 1974); with freezing processes, associated with the formation of cubic cave pearls (pizoliths) (Roborge and Garon, 1983), and cryomineral growth (Andrejchuk et al., 2013). Frost destruction of flowstones is also noted (caves: Mariinskaya and Askinskaya in the Ural Mountains). Detection of similar forms in friable cave sediments (as well as the accumulations of gypsum or dolomitic flour) testifies to periods when, there was glaciation in the cave (Kosygin, 1977; Pissart et al., 1988).

Cave glaciation can lead to cavity liquidation (i.e., filling with ice). That often occurs in areas of permafrost and on glaciers.

Interaction with hydrosphere. Cave glaciation can influence drainage from karst uplands. First, the snow and ice of karst cavities have a fixed water content, except for some periods of water circulation. These periods can sometimes last several thousand years. It was found that out that on periods when more water is excluded from the water cycle, the total volume of water in karst cavities in the area decreases. Secondly, snow melting in karst cavities (especially in mountains) promotes a decrease of the peak and an increase in the duration of spring floods caused by snow melt (the degree of influence depends on the number of cavities with snow and ice in the cathment area). This happens because not all snow in karst cavities is involved in melting at the same time, but occurs gradually during warm periods: snow on rock surfaces melts first, then melts on thicker snowfields on rock surfaces, and in karst dolines, but only after snow in karst pits and shafts begin to melt. In spite of the fact that the number of pits and shafts is less than the number of dolines, snow melting in pits and shafts provide summer drainage from karst massifs in addition to condensation moisture in cavities.

Cave glaciation changes the character of underground water circulation in the vicinity of cavities. The presence of frozen places in cavities changes the character of water flowing into the cavity, in particular it can be observed filling cracks in the ceiling and walls with ice. This, on the one hand, keeps walls from collapsing (in some parts of caves fragments along the ceiling can be cemented by ice—especially in caves with gypsum), but, on the other hand, glaciation can divert water inputs to less frozen parts of caves or remove it completely. This causes a volume reduction of ice formation in cavities. Frozen cavities can serve also as absorbers of underground water streams, transforming water into ice. It is possible for small water streams to flow in cavities of considerable dimensions which have a significant cold reserve (for example, at lower level of Sumgan-Kutuk Cave in Urals Mountains in the 20th century).

Crevasses in mountain glaciers in the upper part of warm firn zones in an ice formation can absorb all superficial drainage, transforming it into ice.

Snow and ice melting in caves, in connection with the melting of superficial snowfields in mountains, is the cause of temperatures decreasing in underground waters (including in caves), and that leads to cooling all of the aerated zone in the karst massif. For example, at the karst spring at Mchishta (Caucasus, elevation about 70 m a.s.l.) the discharge was fed by meltwater from superficial snowfields and cave ice from an elevation of about 1800–2600 m a.s.l. The water temperature is about 9.8°C at MAT, while the surrounding area is about 14°C.

Interaction with the air. Cave glaciation changes cavity climate, because it, in many respects, depends on the temperature and hydrological conditions of the rock. Today changes in climate due to geography in nonice caves is known. On the basis of statistical analysis of air temperatures in caves, equations were derived for the calculation of thermal conditions in cavities, which depend on latitude and the absolute height of districts where there are caves. For North America, the equation is (Moore and Sullivan, 1978):

$$T = 38 - 0.6L - 0.002h, \tag{4.1.6}$$

and for Europe (Choppy, 1977):

$$T = 54.3 - 0.9L - 0.006h, \tag{4.1.7}$$

where T—air temperature inside a cave, °C; L—geographical latitude, degrees; h—absolute height, m. Cave glaciation can change these variables, lowering air temperatures in entrance areas of horizontal caves and frequently along the whole length of inclined and vertical cavities. This is especially common in mountain areas. For example, we will compare air temperatures of two entrances in Snedznaya Cave System (Caucasus), which are located at almost the same elevation (at 2000 and at 2040 m), but one entrance is the subject of permanent glaciation to a depth of about 200 m below the surface, and in the second permanent ice is not present. In the first entrance, 100 m below the surface, MAAT was about −0.5°C and in the second it was +3.8°C. This means that, because of glaciation, air temperature in the first case is more than 4°C less than in the second one. If using the formula (4.1.7) we calculate the height at which air temperature will equal zero at the latitude of Caucasus, it will be about 2600 m. However, caves with permanent ice (and with negative MAAT) are known with heights of about 1800 m (vertical cavities), and from 1300 m for inclined descending ones.

In a region with a continental climate, glaciation of inclined descending caves is possible if the MAAT outside of caves is equal to 12°C, that is, at rock temperatures about 15°C. We see how big the imbalance can be of rock temperature fields caused by cave glaciation.

Cave glaciation often changes the climate of areas close to the cave entrances. Near the lower entrances of horizontal caves during summer air circulation, there are areas where MAAT is lower than in surrounding areas (Golod and Golod, 1974). The lowered MAAT is observed also in entrance depressions of inclined and vertical cavities.

Interaction with the biosphere. Cave glaciation locally forms a climate colder than in surrounding areas. It defines the biota in caves with ice. The influence of cave glaciation on biota is twofold: (1) suppression, pauperization of specific structure, occurrence of more cold adapted kinds of flora and fauna because of climate severity (similar to the climate of cold deserts); (2) attraction of biota to caves with ice (to their entrance parts) because of the moisture and food source, especially in summertime in drought prone areas.

The cold climate at lower entrances of horizontal caves leads to a reduction in plant vegetation, as well as a time lag in their flowering and fruits maturing. In a zone of cooling by air flowing from the entrance in the Kinderlinskaya Cave in the Ural Mountains, it was observed that primary growth of firs and oppressed grassy vegetation occurred, in spite of the fact that a mixed forest with rich underbrush and magnificent herbage grew around the area. The stagnation of cold air in dolines and entrance parts of caves also slows down vegetation, and the cold climate promotes preservation of endemic flora. At the end of September 1985, in the doline of Kutukskaya-1 Cave (Ural Mountains), we observed what looked like a seasonal progression of grassy vegetation: from the previous autumn conditions above, to summer in the middle, and early-spring at the bottom of the doline closest to ice. In the doline-like extension of the entrance pit of Snedznaya Cave System, mostly cold-moisture-shade trees were located, like rowan and birch, while the subalpine vegetation that is normally around the trees was absent. Naturally, following the geographical and elevation zonation, the contrast of vegetation in dolines, and in surrounding areas decreases from south to the north.

Cave glaciation also influences fauna. Thus in glaciation zones, insects and warm-blooded animals (cheiroptera) are not found; the absence of sunlight leads to oppression of alga, mushrooms and bacteria. At the same time, in the summer, snow-ice formations, especially in entrance parts of caves, harbor some kinds of insects (in particular, bugs). The quantity of species living in caves with ice decreases from south to north.

The examples considered here do not cover all the interactions cave glaciations have with the Earth's surface, but they show that this interaction is diverse in character (and that it is more often local testifies only to an insufficient level of study), and changes brought by glaciation are considerable (because of the large contrast between a zone of cave glaciation and an external climate).

4.1.3 ICE GENESIS IN ICE CAVES

4.1.3.1 ZONES OF ICE FORMATION IN CAVES

Conditions of ice formation in caves are still poorly studied. Only recently have zones of ice formation in caves been mentioned in the literature. Dmitriev (1977a,b) noticed that on cave glaciers in Kuznetski Ala-Tau it is possible to distinguish two zones of ice formation: moist firn (=warm firn zone) and an ice feeding zone (=congelation zone) in conditions of seasonal congelation zone outside the caves. Mavlyudov and Vturin (1988) show that on permanent cave snow masses in the Western Caucasus warm infiltration (=warm firn) and congelation zones of ice formation are present as well as in seasonal congelation zones outside caves.

Local areas of highly irregular relief can cause snow and ice to accumulate in karst features at elevations below the normal base of the cryogenic zone. One similar possibility for snowfields located in relief depressions Shimskij (1959) is specified. But other correlations of ice formation zones in caves and out of them are also possible.

Summarizing the direct and indirect data about ice formation conditions in caves is shown in Table 4.1.8.

Table 4.1.8 Correlation of Ice Formation Zones in Caves and Out of Them

Zones of Ice Formation Outside Caves	Zones of Ice Formation in Caves		
	Horizontal	Incline Descending	Vertical
Seasonal-congelation (SC)	SC, C, SF	SC, C	SC, C, WF
Congelation (C)	SC, C	SC, C	SC, C, CF
Warm firn (WF)	SC, C	—	SC[a], C, WF, CF
Cold firn (CF)	C	—	CF[a], SF
Snow-firn (SF)	—	—	SF[a], S
Snow (S)	—	—	S[a]

[a] *Only in cavities inside glaciers.*

The congelation zone is a widespread zone of ice formation in caves. It is caused by the absence of snow accumulation in the majority of caves (horizontal and inclined descending). Thus conditions of ice formation in caves can be equated to external conditions of ice formation and also corresponds to higher and lower zones of the cryosphere.

A special case of ice formation, specific only to caves is ice formation on the ceilings of some horizontal caves in a zone of negative MAAT and constantly negative cavity wall temperatures. Here sublimation ice crystals recrystallize on the roof of the cavity during the winter. Sublimation crystals become monolithic ice under the influence of heat allocated to water vapor that condenses on ice in the summer. Total melting of ice does not occur because of the big cold reserve in the surrounding rocks. We observed formation of this ice on part of a gallery in Kungurskaya Cave located opposite the entrance into Polarnyj Grotto (Fig. 4.1.7). Apparently, in this case, conditions of ice formation are similar to snow-firn formation outside of caves.

From Table 4.1.8 it is apparent that conditions of ice formation in vertical cavities have the most variations. Thus in caves and rocks ice formation occurs in seasonal-congelation to cold-firn zones, and as well as in glacial crevasses up to the snow zone. In some cases, ice formation in caves occurs contrary to general climatic laws. For example, the conditions of a maritime climate with cool summers, cool but not cold winters and big quantity of precipitation can occur in caves as typical conditions (similar external), if there is a lot of snow and there are no low air temperatures, that is atypical, there is no snow but there are low air temperatures. All this is caused by the ability of different caves not to receive a snow or an abundance of solid precipitation, or on the contrary, to concentrate snow in small quantities of solid precipitation. This can also be typical in a continental climate where there is not enough snow but there are low air temperatures; and atypical conditions occurring when there is a lot of snow and low air temperatures. Therefore, almost similar ice formation zones can originate in caves in areas with different climates. The caves of Caucasus and of Kuznetskyj Ala-Tau can serve as examples of warm firn zone conditions. Apparently, ice formation zones in caves are not limited by what is listed in the table, and further studying of cave glaciation will reveal new versions of ice formation zones in caves.

As we have seen, ice formation zones in caves do not completely correspond to conditions outside of the caves, but have their own features, because ice formation zones in caves are formed in conditions sharply distinct from what exist outside of the caves. Therefore studying ice formation zones in caves allows us to expand the representation about ice formation zones in the cryosphere.

4.1.3.2 SNOW-ICE FORMATIONS

Cave ice is a special class of natural ice having various geneses, but which exists in a cavity and is located below the earth's surface but is connected with the open air. Cave ice belongs to the category of subaerial—underground ice.

There is a question: why it is necessary to separate cave ice from underground ice? According to (Kotlyakov, 1984), underground ice is ice of any origin, which is a part of the lithosphere and situated under the earth's surface. However, the cave ice is not included directly in the lithosphere (an exception is permafrost in caves). Besides, the concept of underground ice means its formation under the earth's surface. It is the lack of light that is reflected in the composition and structure of cave ice, which provides the basis for their separation into an independent class of natural ice, intermediate between subaerial and underground.

4.1.3.3 CLASSIFICATION OF SNOW-ICE FORMATIONS IN CAVES

Classification of cave ice is still poorly developed. Classifications have been proposed by karstologists and glaciologists (Vturin, 1975; Dmitriev, 1980b; Maksimovich, 1947; Shimskij, 1955). Karstologist Maksimovich (1947) identified three types of cave ice: atmogenic (sublimation), hydrogenic, and heterogenic (mixed). Glaciologist Shimskij (1959) considered cave ice to be underground ice which he separated into groups of termokarstic cave ice, which are divided into: "flowstone", sublimation, infiltration, and congelation ice. Glaciologist Dmitriev (1980a,b) has come to the conclusion that a uniform classification of cave ice is absent because karstologists do not use the experience of glaciologists, and geocryologists have confidence in the competence of karstologists. V.E. Dmitriev gives the most comprehensive classification of cave ice, in which cave ice is divided into: congelation, sublimation, and sedimento-metamorfic. In his classification, V.E. Dmitriev did not always use consistent criteria.

In the present work, we make an attempt to generalize contemporary data about cave ice and to further develop its classification, and make our own additions, such as: sources of cave ice formation, separate types of cave ice, etc. (Mavlyudov, 1989b, 2001).

Cave ice can be subdivided in a number of ways.

Compared to the classifications of natural ice (Vturin and Vturina, 1984; Koreisha, 1984), we composed a matrix (Table 4.1.9), and hierarchical (Fig. 4.1.13) classification of snow-ice cave formations.

Table 4.1.9 Matrix Classification of Snow-Ice Formations (SIF) in Caves

I. By Place of Formation	II. By Aggregate State of Water	III. By Main Processes of Ice Formation	IV. By Composition of SIF	V. By Ice Salinity	VI. By age
1. On boundary of air and rock 2. On boundary of air and water 3. On boundary of water and rock 4. Into rock 5. Into water	1. Solid 2. Liquid 3. Gaseous	1. Congelation 2. Sublimation 3. Sedimentation 4. Metamorphism	1. Icy 2. Firn-icy 3. Snow-firny 4. Snow-icy 5. Snow-firn-icy 6. Firny 7. Snowy	1. Fresh 2. Saltish 3. Salty	1. Ephemeral 2. Seasonal 3. Existing more then one year 4. Long-term (per annual)

The advantage of matrix classification is that it is multidimensional, considering cave ice from different sides and from the different points of view, which can be, and cannot be, connected with each other.

Each form of cave ice in this classification has a digital index referring to the column and line where it is located.

Icings—I.1; II.2; III.1; IV.1; V.1-3; VI.1-3
Ice of lakes—I.2; II.2; III.1; IV.1; V.1-3; VI.1-2
Ice in rocks—I.4; II.1; III.1; IV.1; V.1-3; V1.1-3
Snowfields—I.1; II.1-2; III.2; IV.2-4; V.1; VI.1-3
Glaciers—I.1; II.2; III.2; IV.4; V.1; VI.3
Ice breccia—I.1; II.1; III.1; IV.1; V.1-3; VI.1-3
Hoarfrost—I.1; II.3; III.3; IV.1; V.1; VI.1-3.

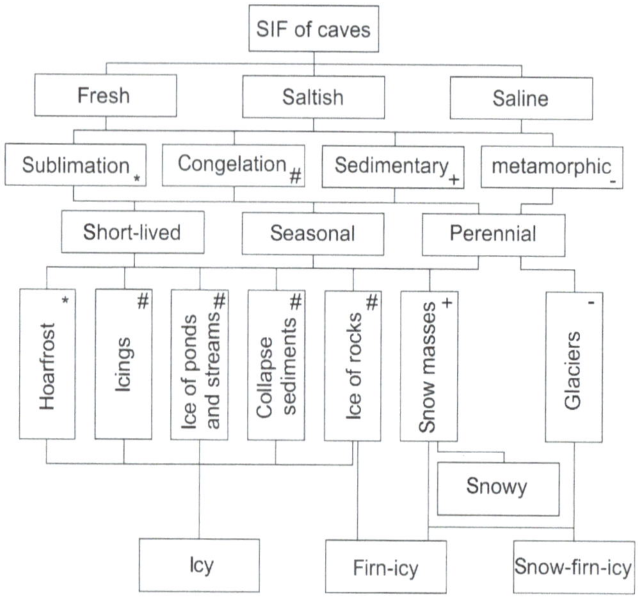

FIG. 4.1.13

Hierarchical classification of snow-ice cave formations (similar symbols in small squares show an interconnection). SIF—snow-ice formations.

Hierarchical classification is more compact and shows interrelations of groups of cave ice. Congelation ice is subdivided into "flowstone" ice (icings), ice of lakes and rivers, ice in rocks. It is possible to make more divisions in each group (Fig. 4.1.13) (Mavlyudov, 1989c).

Congelation ice can be formed from water representing atmospheric precipitation, infiltration, influation, condensation water, and water in rocks (Fig. 4.1.13).

Sedimentary ice is formed from atmospheric precipitation, snowstorm and avalanche snow, broken snow cornices, and also ice collapses, both in caves and in their entrances (collapses of icings and glacial ice), and glaciers flowing into cave entrances. All kinds of water, but in insignificant quantity, can also take part in the formation of sedimentary ice. The first four sources feed cave snowfields, which, in part, can give rise to cave glaciers (metamorphic ice). Collapse ice, which has not melted in caves during the summer, will be cemented into ice breccia. Sublimation ice in caves occurs as hoarfrost formed around sources of warm, humid air from the cave interior, or from lateral galleries in frozen cavities; sublimation ice is divided by its form of accumulation (Fig. 4.1.14).

4.1.3.4 DETAILS OF ICE STRUCTURE IN CAVES OF DIFFERENT MORPHOLOGIES

Horizontal caves. This type of cavity forms only congelation and sublimation ice. The ratio of quantities of this ice, and a specific set of ice forms in each cave, is defined by its structure and the character of the water inflow, depending on the fissures in the bedrock containing the cave. For example,

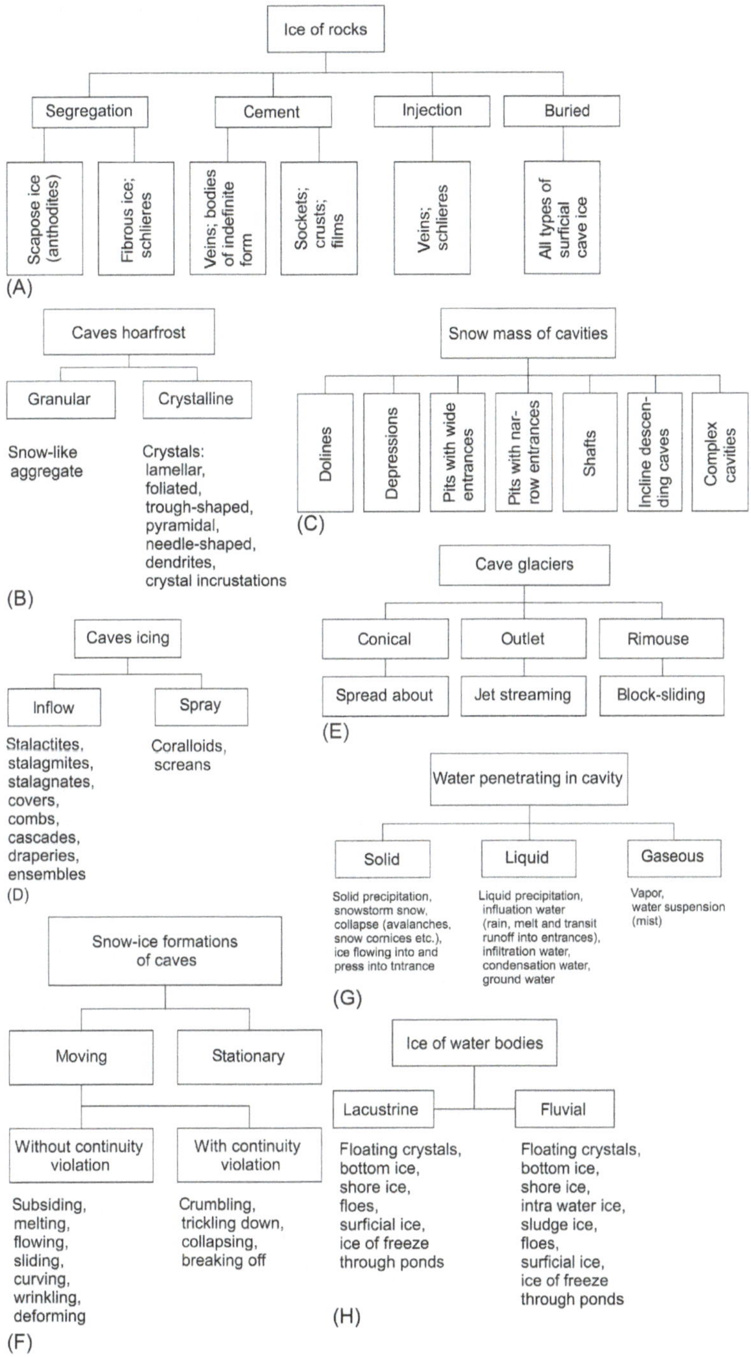

FIG. 4.1.14

Classifications of snow-ice formations of caves into categories.

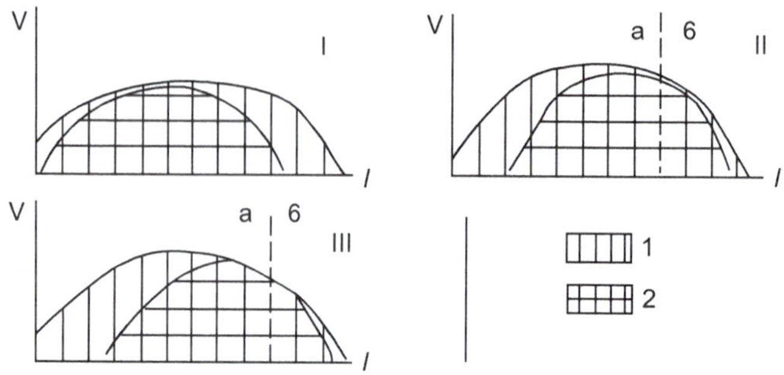

FIG. 4.1.15

Relative quantity of snow-ice formations (V) versus the cave length (*l*): I—horizontal; II—inclined descending (a—small extent, b—big extent); III—vertical (a—small depth, b—big depth). 1—seasonal ice, 2—perennial ice.

in the Kungurskaya Cave in the Ural Mountains, the quantity of congelation ice occurring as icings is estimated at about 98%, and the quantity of sublimation ice, which occurs as various forms of skeletal crystals is about 2%; in the Kinderlinskaya Cave in the Ural Mountains, the long-term ice is exclusively congelation (also icings) (Mavlyudov, 1988a).

Inclined caves. A cavity descending from the entrance promotes accumulation of snow and an absence of big volumes of fluid and stagnant water. These caves contain cogelation, sedimentary, and sublimation ice. Congelation ice occurs as icings, ice in rocks, and sometimes as lake ice. Sedimentary and sublimation ice is more often seasonal. This means that perennial ice in these caves is almost totally congelation.

Vertical cavities. Pits are fine collectors of solid precipitation, therefore the prevailing types of ice in them are sedimentary and metamorphic ice. Congelation ice is rare, and sublimation ice occurs only in single instances. In these caves all types of sedimentary-metamorfic ice, and congelation ice occur as icings and, more rarely, as ice in rocks.

The distribution of ice in caves of different morphologies is shown in Fig. 4.1.15.

4.1.3.5 CHEMICAL COMPOSITION OF SNOW-ICE FORMATIONS IN CAVES

The chemical composition of cave ice depends, first of all, on the chemistry of the water, which reflects the environment in which this water is formed. However, inside caves the ice chemistry changes in time and space. Characterized by salt content, cave ice is divided into fresh (to 1000 mg/L), saltish (1000–5000 mg/L), and salty (more than 5000 mg/L).

Every form considered above for cave ice types (congelation, sublimation, and sedimentary-metamorfic), has their own chemistry inside caves. Congelation ice is the most variable, since practically any water can be its source. Atmospheric or melt water leads to the formation of ice with low mineralization (to 50–100 mg/L). Data on the chemistry of cave ice is covered in many papers (Andrejchuk, 1994; Birzhevaya, 2001; Maksimovich and Shumkov, 1964; Maksimovich and Panarina, 1966; Yashenko, 1965; Dublyanskij et al., 1992).

REFERENCES

Andrejchuk, V.N., 1994. Mineralization of snow-ice formations of Kungurskaya Ice Cave. In: Abstracts of IV All-Ural Conference of Underground Water of Ural and Adjacent Areas, Perm, pp. 118 (in Russian).

Andrejchuk, V.N., Kadebskaya, O.I., Chajkovskij, I.I., 2013. Cryogenic mineral formations of Kungurskaya Cave. Silesian University-Mine institute UB RAS, Sosnovets-Perm. 128 pp.

Atkinson, T.C., Smart, P.L., Wigley, T.M.L., 1983. Climate and natural radon levels in Castleguard cave, Columbia Icefield, Alberta, Canada. Arct. Alp. Res. 15 (4), 487–502.

Birzhevaya, I.A., 2001. Chemical composition of ice from Kungurskaya Ice Cave. In: Scientific Progress of Students at the Turn of the Century. Proceedings of Conference, Perm, pp. 6–10 (in Russian).

Choppy, J., 1977. La tempérâture des cavités, en founction de la latitude et de l'altitude. Spelunca 3, 117–118.

Dmitriev, V.E., 1977a. Ice formation on cave glaciers. Nature and Economy of Siberia, pp. 19–20. Yakutsk (in Russian).

Dmitriev, V.E., 1977b. Snowfields of karst forms of relief. Nature and Economy of Siberia. pp. 36–37. Yakutsk (in Russian).

Dmitriev, V.E., 1979. Specifics of caves glaciers oscillations of Kuznetskij Alatau. Materialy glatsiologicheskikh issledovanij (Data Glaciol. Stud.) 36, 181–185 (in Russian).

Dmitriev, V.E., 1980a. To question of complex approach to caves ice study. Voprosy geographii Sibiri (Quest. Sib. Geogr.) 13, 99–102. Tomsk.

Dmitriev, V.E., 1980b. Caves glaciation as a part of Earths glaciosphere. Karst of Siberia and Far East, pp. 130–145. Vladivistok (in Russian).

Dorofeev, E.P., 1981a. Interrelation of air and rocks temperature in Kungurskaya Cave. Accumulation of Winter Cold in Mountain Rocks and Its Usage in National Economy, pp. 55–56. Perm (in Russian).

Dorofeev, E.P., 1981b. Permafrost of Kungurskaya Cave. Accumulation of Winter Cold in Mountain Rocks and Its Usage in National Economy, pp. 65–67. Perm (in Russian).

Dorofeev, E.P., Mavlyudov, B.R., 1993. Dynamics of glaciation of Kungurskaya Cave. Peshery (Caves) 23-24, 131–140 (in Russian).

Dorofeeva, R.P., 1986. Thermal conduction of Earth crust at South of Eastern Siberia. Geol. Geophys. 10, 85–94 (in Russian).

Droppa, A., 1964. Ice caves of Chekhoslovakiya. Peshery (Caves) 4 (5), 85–94 (in Russian).

Dublyanskij, V.N., 1963. About role of snow in karst formation and in feeding of karst water. In: Proceedings of the USSR Academy of Sciences, Series Geographical, vol. 2, pp. 69–75 (in Russian).

Dublyanskij, V.N., 1977. Karst Caves and Shafts of Mountain Crimia. Nauka, Leningrad. 182 pp. (in Russian).

Dublyanskij, V.N., Ilukhin, V.V., 1982. The Largest Karst Caves and Shafts of the USSR. Nauka, Moscow. 137 pp. (in Russian).

Dublyanskij, V.N., Lomaev, A.A., 1980. Karst caves of Ukraine. Naukova Dumka, Kiev. 180 pp. (in Russian).

Dublyanskij, V.N., Shutov, V.I., 1967. Corrosion-nival karst cavities of Mountain Crimia. Proc. Geogr. Soc. USSR 99 (6), 468–476 (in Russian).

Dublyanskij, V.N., Andrejchuk, V.N., Bersenev, Yu.I., et al., 1992. Chemical composition of ice of karst cavities of the USSR. Eng. Geol. 5, 119–123.

Dzidziguri, A.A., Duganov, G.V., Oniani, Sh.I., et al., 1966. Thermalphysic Characteristics of Mountain Rocks and Methods of Their Detection. Metsniereba, Tbilisi. 228 pp. (in Russian).

Ezhov, Yu.A., 1990. Snow caves. Problems of geometeorology and accumulation of winter cold, pp. 35–37. Sverdlovsk (in Russian).

Ford, T.D., Gullingford, C.H.D. (Eds.), 1976. The Science of Speleology. Academic Press, London, New York, San Francisco. 503 pp.

Ford, D.C., Williams, P.W., 1989. Karst Geomorphology and Hydrology. Unwin Hyman, London. 601 p.

Frolov, N.M., 1976. Hydrogeothermia. Nedra, Moscow. 280 pp. (in Russian).

Golod, V.M., Golod, M.P., 1974. Microclimate of gypsum caves of Pinezhye. Caves of Pinego-Severodvinskoj karst area, pp. 128–154. Leningrad (in Russian).
Gvozdetskij, N.A., 1954. Karst. Geografgiz, Moscow. 352 pp. (in Russian).
Gvozdetskij, N.A., 1968. Types of ice caves in the USSR. In: Actes du IV Congress Internationale de Speleologie en Yougoslavie, Ljubljana, vol. 3. (in Russian).
Gvozdetskij, N.A., 1972. Problems of Karst Study and Practice. Mysl', Moscow. 392 pp. (in Russian).
Hallas, J., 1983. Niectore posnatky s merania teploty horninoveho plasta v Dobsnskej l'adovej jaskyni a Demenovskej l'adovej jaskyni. Slovensky Kras 21, 79–91.
Kemple, S., Kets-Kemple, C., 1979. Fire and ice atop Hawaii. NSS News 8, 185–188.
Kirish, B.F., Ushakov, K.Z., 1983. Mine and Industrial Aerology. Nedra, Moscow. 256 pp. (in Russian).
Koreisha, M.M., 1984. Outline of matrix claccification of natural ice. Materialy glatsiologicheskikh issledovanij (Data Glaciol. Stud.) 51, 39–44 (in Russian).
Kosygin, Yu.A. (Ed.), 1977. Forms of Geological Bodies (Terminological Reference Book). Nedra, Moscow. 246 p. (in Russian).
Kotlyakov, V.M. (Ed.), 1984. Glaciological Glossary. GIMIZ, Leningrad. 528 pp. (in Russian).
Kruber, A.A., 1915. Karst area of Mountain Crimea, Moscow. 319 pp. (in Russian).
Kudryashov, I.K., Salikhov, R.A., 1968. Reasons of cold preservation and ice formation in Askinskaya Cave. Trans. Bashkir Branch Geogr. Soc. USSR 5, 68–79 (in Russian).
Listov, Yu, 1885. Caves - ice-houses. Data Russ. Geol. 12, 105–280. Sankt-Petersburg (in Russian).
Lobanov, Yu.E., 1981. Negative temperature anomalies and morphology of caves in carbonate massifs. Accumulation of Winter Cold in Mountain Rocks and Its Usage in National Economy, pp. 28–30. Perm (in Russian).
Lukin, V.S., 1965. Temperature anomalies in caves of Preduralie and critical analysis of theories of underground cold. Peshery (Caves) 5 (6), 164–172 (in Russian).
Lukin, V.S., 1969. Quantative expression of processes of water evaporation and condensation of water vapour in Gypsum-Angydrit massifs of Ufimskoe Plateau. Zemlevedenie 8 (48), 213–218. Moscow: MGU (in Russian).
Maksimovich, G.A., 1947. Caves ice. Trans. Geogr. Soc. USSR 79 (5), 237–250 (in Russian).
Maksimovich, G.A., 1963. Foundation of Karstology, vol. 1. Perm, 445 pp. (in Russian).
Maksimovich, G.A., Panarina, G.A., 1966. Chemical composition of caves ice in Perm region. Chem. Geogr. Hydrochem. 6 (7), 33–37. Perm (in Russian).
Maksimovich, G.A., Shumkov, V.M., 1964. Chemical composition of caves ice. Chemical Geography and Hydrochemistry 3 (4), 43–46. Perm (in Russian).
Malkov, V.N., Frants, N.A., 1981. To a question about winter condensation. Accumulation of winter cold in mountain rocks and its usage in national economy, pp. 97–99. Perm (in Russian).
Mavlyudov, B.R., 1980. Specifics of structure of snow ice formations in Snezhnaya Shaft in Western Caucasus. Materialy glatsiologicheskikh issledovanij (Data Glaciol. Stud.) 40, 189–193 (in Russian).
Mavlyudov, B.R., 1981. Conditions of snow and ice formation in karst shaft Snezhnaya. Accumulation of winter cold in mountain rocks and its usage in national economy, pp. 62–64. Perm (in Russian).
Mavlyudov, B.R., 1985. Geographical peculiarities of spreading of caves with ice. Materialy glatsiologicheskikh issledovanij (Data Glaciol. Stud.) 54, 193–200 (in Russian).
Mavlyudov, B.R., 1988a. Glaciation in caves of Ural. Materialy glatsiologicheskikh issledovanij (Data Glaciol. Stud.) 61, 123–129 (in Russian).
Mavlyudov, B.R., 1988b. Melting of snow and ice in karst cavities. Materialy glatsiologicheskikh issledovanij (Data Glaciol. Stud.) 64, 123–128 (in Russian).
Mavlyudov, B.R., 1988c. Structural peculiarities of the snow and ice accumulations in the Snezhnaya Propast' in the Western Caucasus. In: Avsyuk, G.A. (Ed.), Data of Glaciological Studies. vol. 40. Amerind Publishing Co. Pvt. Ltd., New Delhi, pp. 313–319 (translation from Russian) (in English).
Mavlyudov, B.R., 1989a. Snow and ice in karst cavities of Khipstinskij massif (Western Caucasus). In: Problems of Karst of Mountain Countries. Metsniereba, Tbilisi, pp. 205–208 (in Russian).

Mavlyudov, B.R., 1989b. Caves glaciation of the Soviet Union. Author's abstract of PhD dissertation. Moscow. 23 pp. (in Russian).
Mavlyudov, B.R., 1989c. Snow and ice formation in caves and their regime. In: International Congress of Speleology, 13–20 August 1989. Proceedings, vol. 1, pp. 295–297 (in English).
Mavlyudov, B.R., 1989d. Cave glaciation. In: International Congress of Speleology, 13-20 August 1989. Proceedings, vol. 1, pp. 298–300 (in English).
Mavlyudov, B.R., 1992. Cavities in natural ice. Problems of Pceudokarst (Abstracts), pp. 56–58. Perm (in Russian).
Mavlyudov, B.R., 1994. Caves climatic systems. In: Questions of Physical Speleology. MPTI, Moscow, pp. 6–24 (in Russian).
Mavlyudov, B.R., 1997. Dynamics of glaciation of Kungurskaya Cave for 25 years. Materialy glatsiologicheskikh issledovanij (Data Glaciol. Stud.) 82, 193–198 (in Russian).
Mavlyudov, B.R., 2001. Classification of snow-ice formations in caves. Peshery (Caves) 27-28, 97–108 (in Russian).
Mavlyudov, B.R., 2006. Internal Drainage Systems of Glaciers. Institute of Geography RAS, Moscow. 396 pp. (in Russian).
Mavlyudov, B.R., 2008. Caves Glaciation. Institute of Geography RAS, Moscow. 290 pp. (in Russian).
Mavlyudov, B.R., 2016. Cave system Snezhnaya-Mezhennogo-Illyuziya, the Western Caucasus. Bol. Geol. Min. 273 (1), 219–235.
Mavlyudov, B.R., Morozov, A.I., 1984. Snezhnaya shaft. Peshery (Caves) 19, 15–25. (in Russian).
Mavlyudov, B.R., Vturin, B.I., 1988. Snow-ice formations of karst cavities of Khipstinskij massif. Materialy glatsiologicheskikh issledovanij (Data Glaciol. Stud.) 62, 148–153 (in Russian).
Moore, G.W., Sullivan, F.S.C., 1978. Speleology. In: The Study of Caves. Zephyrus Press Inc., Teaneck. 150 pp..
Pissart, A., Vliet-Lanoe, B.V., Ek, C., Juvigue, E., 1988. Des traces de glace de segregation dans la Grotte de Romouchamps (Belgique): consequences en ce qui conserne la sedimentation et la paleoclimatologie. Ann. Soc. Geol. Belg. 111 (1), 125–133.
Povara, J., Diaconu, G., 1974. Deroulement du processue de gelifraction dane le milieu souterrain. Trav. Inst. Speol. "E. Racovitza" 13, 139–146.
Pozdnyakov, A.V., 1988. Dynamic Equilibrium in Relief Formation. Nauka, Moscow. 207 pp.
Roborge, I., Garon, D., 1983. The occurrences of an unusual type of pisolite: the cubic cave pearls of Castleguard cave, Columbia Icefield, Alberta, Canada. Arct. Alp. Res. 15 (4), 517–522.
Roda, S., Rajman, L., Erdos, M., 1974. Vyskum mikroklimy a dynamiky sal'adnonia v Silickej l'ednici. Slovensky Kras 12, 157–174.
Saar, R., 1956. Eishohlen, ein meteorologisch-geophysikalisches phanomen. Geogr. Ann. 38 (1), 1–63.
Shavrina, E.V., Guk, E.V., 2005. Modern dynamics of ice formations in Pinega caves. In: Mavlyudov, B.R. (Ed.), Glacier Caves and Glacial Karst in High Mountains and Polar Regions. Institute of Geography RAS, Moscow, pp. 113–117.
Shimskij, P.A., 1955. Foundation of Structural Ice Study. Publishing House of USSR Academy of Sciences, Moscow. 492 pp. (in Russian).
Shimskij, P.A., 1959. Foundation of geocryology, Vol. 1. Moscow, 321 pp. (in Russian).
Shutov, Yu.I., Sevast'yanov, E.M., 1983. Elevation zonation of nival corrosion in Mountain Crimea. Geol. J. 43 (6), 73–75. (in Russian). Kiev: Naukova Dumka.
Sokolov, D.S., 1962. Main Conditions of Karst Development. Gosgeoltekhizdat, Moscow. 322 pp. (in Russian).
Sotskova, L.M., 1981. Air conditions of karst cavities. Accumulation of winter cold in mountain rocks and its usage in national economy, pp. 11–12 Perm (in Russian).
Thornbury, W.D., 1954. Principles of Geomorphology. J. Wiley & Sons Inc/Chapman & Hall Ltd., New York/ London. 618 pp..
Vincent, W.V., 1974. Environmental influence of the glaciers of the Pryor Mountains. Bull. NSS 36 (3), 13–21. Montana.

Vturin, B.I., 1975. Underground Ice of the USSR. Nauka, Moscow. 214 pp. (in Russian).
Vturin, B.I., Vturina, E.A., 1984. Classification of natural ice of the Earth. Materialy glatsiologicheskikh issledovanij (Data Glaciol. Stud.) 50, 93–97 (in Russian).
Williams, G.P., Gold, L.W., 1976. Ground Temperatures. http://irc.nrc-cnrc.gc.ca/pubs/cbd/cbd180_e.html.
Yashenko, R.V., 1965. About ice composition of Kungurskaya Cave. Peshery (Caves) 5, 27–28 (in Russian).

FURTHER READING

Wigley, T.M.L., Brown, M.C., 1976. Cave meteorology. In: Ford, T.D., Gullingford, C.H.D. (Eds.), The Science of Speleology. Academic Press, London, New York and San Francisco, pp. 329–344.

CHAPTER

ICE SURFACE MORPHOLOGY

4.2

Pavel Bella

State Nature Conservancy of the Slovak Republic, Slovak Caves Administration, Liptovský Mikuláš, Slovakia
Catholic University in Ružomberok, Ružomberok, Slovakia

CHAPTER OUTLINE

4.2.1 INTRODUCTION

In caves, the volume and surface morphology of ice bodies results from the extent of glaciation and interannual or seasonal changes of perennial ice fill. Knowledge about large- and small-scale forms of ice morphology is important for a study of the glaciation and dynamics of ice caves in different natural settings. Problems concerning morphology of ice surfaces in caves has not yet been studied systematically or with any complexity. Only a few articles or studies related to the ice surface morphology of selected ice caves have been published (e.g., Marshall and Brown, 1974; Schroeder, 1977; Lauriol et al., 1988; Racoviţă and Onac, 2000; Bella, 2003, 2007; Citterio et al., 2004a,b; Shavrina and Guk, 2005; Onac et al., 2007; Gomez Lende et al., 2014). Studies on some ablation forms in ice caves are more or less sporadic (Curl, 1966; Cigna and Forti, 1986; Bella, 2005). Descriptions of ice decoration forms in caves are more frequent (Beer and Hassinger, 1902; Hauser and Oedl, 1923; Racoviţă, 1927; Kyrle, 1929; Kunský, 1939; Racoviţă and Viehmann, 1966; Şerban, 1970; Racoviţă and Onac, 2000; Mavljudov, 2008 and many others). More comprehensive genetic and morphological classifications, as well as basic descriptions of ice surfaces in caves, are given further in this chapter.

Ice Caves. https://doi.org/10.1016/B978-0-12-811739-2.00033-4

4.2.2 ICE SURFACES IN CAVES: FORMATION, SPATIAL POSITION, AND DURATION

Depending on climatic conditions, the ice fill forms in static vertical and inclined caves (cold air traps), as well as in dynamic caves with two or more openings to the surface at different altitudes (with a seasonally changed unidirectional ventilation). Such deposits occur both in the temperate zone, in middle- to high-mountains, as well as in the subpolar climatic zone. From the speleogenetic point of view, these caves could be different origin: corrosional, fluvially modeled, volcanic (lava tubes), crevice, talus, or termokarst caves (Halliday, 1954 and others). Ice surfaces in ice caves could be of depositional or ablational origin (or a combination of the two).

Cave ice can originate from snow and firn (recrystallized snow, metamorphic ice), freezing of stagnant, flowing, or dripping meteoric waters (congelation ice), freezing of moisture when warmer, humid air contacts cold surfaces (ice sublimation crystals), freezing of moist sediments (ice in clastic sediments), and from frozen supercooled water emerging from microfissures on cave bedrock (Pulina, 1971; Marshall and Brown, 1974; Schroeder, 1977; Dmitriev, 1980; Marshall, 1981; Lauriol et al., 1988; Ford and Williams, 1989, 2007; Hill and Forti, 1997; Mavlyudov, 1989; Mavljudov, 2001, 2008; Yonge and MacDonald, 1999; Yonge, 2004; Luetscher, 2005; Perşoiu and Onac, 2012; Luetscher, 2013 and others). In some caves, ice is a result of the intrusion of glacier ice into a cave at glacier/rock contact (Ford et al., 1976). Within the study of ice surface morphology (not considering to ice sublimation crystals and other small depositions), ice-deposited forms relate to ice accumulation from the diagenesis of snow accumulated during the winter season and from the freezing of infiltration water (massive congelation ice).

Ablation is defined as a decrease of ice mass caused by natural processes or human impacts. Physical ablation relates to the snow and ice melting on the surface of glacier, running water, snow and ice sublimation, evaporation of melting water, ice melting at the contact with water, melting of glacier bottom at the contact with rock basement and internal melting of glacier (Jania, 1997, and others). Ablation is sensitive to the amount of absorbed air temperature, air humidity and air flow (Bennett and Glasser, 2009). Several ablation forms in ice caves originate as a consequence of dripping and flowing water or air flow and convection. In show caves, the ablation of ice can be caused by electric lighting or air flow enhanced by artificial trenches or tunnels excavated into the ice floor; as well as by the movement of visitors (Bella, 2003, 2007). Anthropogenic ablation forms contribute to the variety of cave ice morphologies. According to their position with respect to the ice body, supraglacial, intraglacial (englacial), and subglacial ablation forms can be classified. By their persistence in time, ephemeral, seasonal, interannual, and perennial forms are distinguished (Mavlyudov, 1989; Mavljudov, 2001, 2008).

Several similar or identical ablation forms are known in glacier caves (i.e., caves formed inside glaciers). Moulins, intraglacial cascaded and meandering stream channels, subglacial scalloped tunnels, sublimation scallops, and other ablation forms were described from glaciers and glacier caves (Pulina, 1982; Mavlyudov, 1991, Mavljudov, 2006; Pulina and Řehák, 1991; Schroeder, 1991; Eraso, 1992; Pulina et al., 2003; Fountain, 2005; Vatne and Irvine-Fynn, 2016 and others). Some of these ablation forms are observed also in subglacial caves formed by subglacial river of thermal water (Favre, 1985; Pulina et al., 2003; Řehák et al., 2004; Smart, 2004). Sublimation cavities and large scallops occur in firn caves formed by sub-ice fumaroles and warm air currents beneath ice filled volcanic craters (Kiver and Steele, 1975; Kiver and Mumma, 1975; Giggenbach, 1976; Sabroux et al., 1999; Badino and Meneghel, 2001; Meneghel and Badino, 2002).

4.2.3 LARGE-SCALE MORPHOLOGY OF ICE BODIES IN CAVES

Large-scale ice morphology in ice-filled caves relates to ice bodies of different sizes, for example, from several hundred to more than 110,000 m^3 in Dobšiná Ice Cave, Slovakia (Tulis and Novotný, 1995), or from a few dozen or hundred to 27,890 m^2 in Eisriesenwelt Cave, Austria (Petters et al., 2011). Sizes and shapes of ice bodies are controlled and limited by cave morphology, as well as by local climatic conditions. Mavljudov (2008) outlines basic relationships between the cave morphology (horizontal, inclined, and vertical caves) and the origin of ice in caves. From the morphologic and genetic viewpoints, the following morphogenetic types of ice bodies (large-scale ice morphologies) can be distinguished (Fig. 4.2.1):

(1) *Snow/firn cones and similar bodies in the bottom of abysses (shafts) or vertical entrance parts of caves* formed by intrusion of snow during winter (static caves with firn in sense of Luetscher and Jeannin, 2004a, in which cold air is trapped owing to its higher density), e.g., Tri brezna v Jamcach shafts, Slapenski lednik Cave and Veliki trški lednik Cave on the Nanos Plateau, Slovenia (Habič, 1964; Habe, 1971), Schwarzmooskogelhöhle Cave (Luetscher, 2005, 2013) and Hundsalm Ice Cave, Northern Calcareous Alps, Austria (Spötl et al., 2014), Volcano Room, Q5, Vancouver Island, Canada (Yonge, 2004) or some karstic pits near the Baikal Lake, Primorsky Range, Russia (Trofimova, 2007).

The immense ice cone in Solunska Glava 5 Abyss, Jakupica Mountain Massif, Macedonia (with the entrance at ca 2,240 m a.s.l.) occurs at depths from ca 200 m (at the base of the entrance shaft) to 370 m (the bottom of the Great Hall). The surface area of this ice cone, formed by snow accumulation and the freezing of infiltration waters, reaches ca 15,000 m^2. However, the ice thickness is unknown; at the foot of the ice cone is at least 3-4 m (Temovski, 2016; see also Šmoll and Szunyog, 2005; Mlejnek, 2009).

In some downward sloping cave entrance parts, *asymmetrical snow cones* or *tongue-shaped snow lobes* are formed by snow avalanche or wind-drift (subvertical snowdrift), e.g., the entrance part of Velika ledena jama v Paradani Cave (Habe, 1971; Mihevc, 2008), Dole pri Predmeji Cave and some other ice caves on the Trnovski gozd Plateau, Ledene jama pri Kunču Cave on the Mt. Rog at Kočevje Town, Slovenia (Habe, 1971), as well as the Peña Castil Ice Cave, Picos de Europa, Spain (Gomez Lende et al., 2014) or many caves in the Alps (Colucci et al., 2016 and others).

(2) *Conic or cylindrical firn/ice plugs or similar ice bodies in the upper part of abysses or vertical caves* formed by snowdrift during winter and snow recrystallization, and/or by the freezing of rainfall, melting and/or infiltration waters. From climatological and glaciological points of view, these abysses are similar to static caves with firn and congelation ice, e.g., Ice Abyss in the Červené vrchy Mountains (Hochmuth, 1979) and Great Ice Abyss on the Ohnište, Slovakia (Droppa, 1958), Lo Lc 1650 "Abisso sul Margine dell´Alto Bregai" Ice Cave, Central Italian Alps, Italy (Citterio et al., 2004a,b), Ledena jama Cave in Lomska Duliba, Vukušić Cave (Jelinić et al., 2001; Kern et al., 2008; Buzjak et al., 2014), and Ledenica Cave in Mt. Bukovi Vrh, Velebit Mountains, Croatia (Garašić, 1980, 2014), as well as Solunska jama, Slovačka jama and Ledenika abysses, Jakupica Mountain Massif, Macedonia (Šmoll and Szunyog, 2005; Šmoll and Sluka, 2007; Temovski, 2016). The upper and adjacent lower parts of larger vertical or subvertical caves appertain to statodynamic caves with firn and congelation ice, that

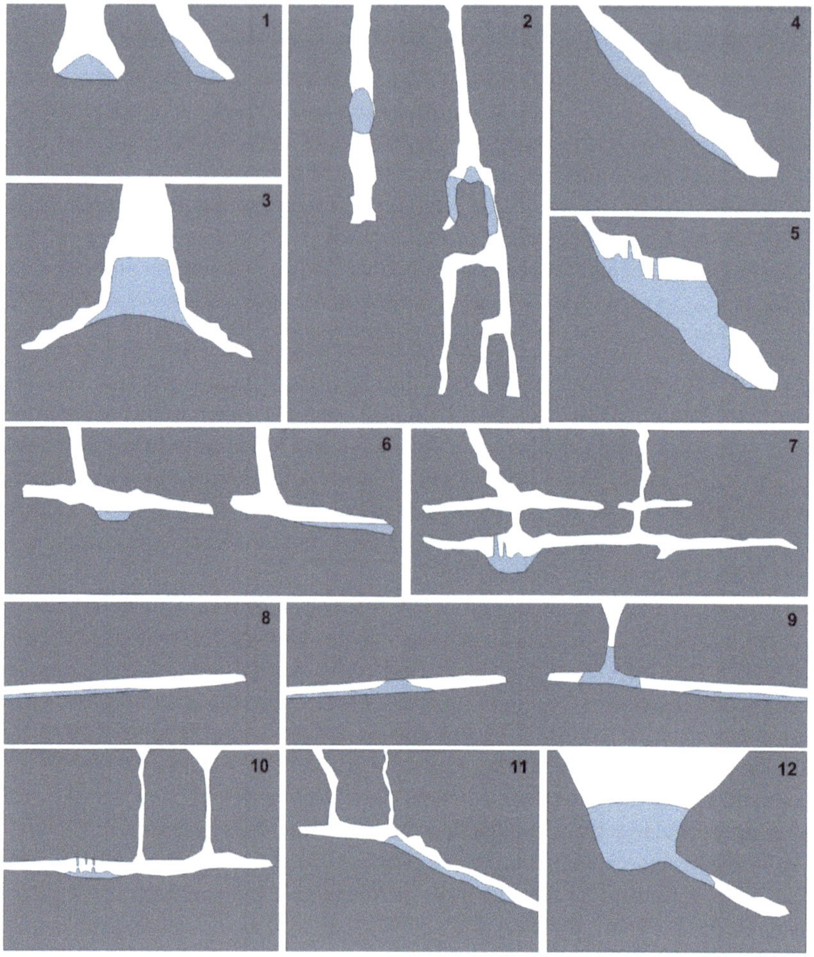

FIG. 4.2.1

Morphogenetic types of ice bodies (gross ice morphologies) in relation to cave bedrock morphology and cave thermodynamic regime: 1—snow cones in the bottom of abysses or vertical entrance parts of caves; 2—conic or cylindrical firn/ice plugs or similar ice bodies in the upper part of abysses or vertical caves; 3—glacier-like ice blocks with a flat top surface and steep walls (stump-like block or truncated cone) in the bottom of shafts; 4—downward sloping or cascading glacier-like ice block in the entrance or upper descending parts of caves; 5—downward sloping glacier-like ice block with large terraces, small ice-hillocks and mounds in the upper descending parts of caves; 6—flat ice floor bodies accumulated as an ice "pond" in the lowermost part of subhorizontal caves; 7—combined ice bodies consisting of vertical ice formations and floor ice in the lowermost part of subhorizontal or multilevelled caves; 8—slightly inclined floor ice sheets in horizontal and subhorizontal caves in the alpine and permafrost regions; 9—ice plugs in horizontal caves (without or with vertical shafts) in the alpine and permafrost regions; 10—combined ice bodies consisting of vertical ice formations and floor ice in the lowermost part of horizontal or subhorizontal caves with open vertical chimneys; 11—subhorizontal massive ice bodies or glacier-like ice block sloping to the lower entrance of vertical dissected caves with several entrances at different elevations; 12—glacier ice tongues/plugs intruced into cave passages (detailed explanation in the text).

occurs down to around 100-200 m below their entrances, e.g., Velika ledena jama v Paradani Cave, Trnovski gozd, Slovenia (Habe, 1971; Mihevc, 2008), Snežnaja Chasm, Western Kavkaz Mountains, Russia (Mavljudov, 1980; Mavljudov and Morozov, 1984), Schellenberger Ice Cave, Berchtesgadener, Northern Limestone Alps, Germany (Grebe et al., 2008 and others), Kremenčetskaja Cave, the eastern foothills of the Eastern Sayan Mountains, Russia (Fillipov, 2005), as well as many other caves in high-mountain areas.

(3) *Glacier-like ice blocks with a flat top surface and steep walls (stump-like block or truncated cone) at the bottom of shafts* (static caves with congelation ice, mostly from melting snow and firn). In the summer, the top surface of the ice block is melting and shallow supraglacial lake is formed, e.g., Scărişoara Ice Cave, Romania with the ice block exceeding 22 m in thickness. The cave ice is also melting at the ice/rock interface under the influence of geothermal heat. Following the development of large voids between the rock wall and the ice block, the vertical ice wall is further sculptured by dry air flow, as well as by the melted water running from the top of the ice block. The ice structures (folds) observed in the Little Reserve of this cave probably indicate ice flow toward the peripheral parts of the ice block (Racoviţă, 1927; Şerban et al., 1948; Racoviţă and Onac, 2000; Holmlund et al., 2005; Perşoiu, 2005; Onac et al., 2007 and others). Usually, lateral ice tongues are attached to the edges of these glacier-like ice blocks. In some cases, a snow cone is formed during winter on the almost-flat, top surface of the ice block, e.g., Monlési Ice Cave, Switzerland, (Luetscher and Jeannin, 2004b; Luetscher, 2005), Focul Viu Ice Cave, Bihor Mountains, Romania (Perşoiu et al., 2007) or Borţig Ice Cave, Apuseni Mountains, Romania (Kern et al., 2009).

(4) *Downward sloping or cascading glacier-like ice block in the entrance or upper descending parts of caves* (static caves with firn and/or congelation ice in sense of Luetscher and Jeannin, 2004a). In some caves, ground ice (or glacier-like block) is moving downward on the inclined rock basement, and, at the lower end of static caves, is melting by the action of warmer air circulating from lower non-glaciated cave parts, e.g., Silica Ice Cave, Slovakia (Böhm and Kunský, 1938; Roth, 1940; Rajman et al., 1985 and others). In other cases, shorter ice masses, formed from firn and/or seepage water, can be more or less stabilized on the inclined rough rock basement of cave entrance parts, e.g., Snežna jama na Raduni Cave, Slovenia (Mihevc, 2008), Booming Ice Chasm, Alberta, Canada (Yonge, 2014), Arnold Ice Cave, Oregon, USA (Halliday, 1954), or some ice caves in sulphate and carbonate rocks in Perm Region, Russia (Kadebskaya, 2008). Some descending caves of smaller cross-sections in Belomorsko-Kulojskoe Plateau, Russia are fully filled by firn (and congelation ice) or "ice siphons" (plugs) have been formed in narrower places of inclined passages (Shavrina and Guk, 2005).

(5) *Downward sloping glacier-like ice block with large terraces, small ice-hillocks and mounds in the upper descending parts of caves.* Flat or slightly inclined ice surfaces (large terraces) develop in cave sections where intense seepage of meteoric waters, stagnation of melted and seeping waters in lakes, and the subsequent freezing of lake water are seasonally or interannually repeated (statodynamic caves with congelation ice in sense of Luetscher and Jeannin, 2004a). In some cases, the formation of terraced flat ice surface is also conditioned by the barrier contact of the growing ice surface with the inclined cave ceiling, in front of which the floor ice has accumulated as horizontal layers, e.g., Dobšiná Ice Cave, Slovakia (Pelech, 1879; Balch, 1900; Bella, 2005). In this cave, the glacier-like ice block is moving downward to its lower part, at a maximum speed of 2–4 cm per year (Lalkovič, 1995; Tulis, 1997).

(6) *Flat ice floor bodies accumulated as an ice "pond" in the lowermost part of subhorizontal caves* (startic or statodynamic caves with congelation ice). In some caves, the pond ice is several meters thick. It originates in the floor depression dammed by rock walls, steep rock talus or ice cliffs in the vertical entrance part of static caves (lava tubes), e.g., Candelaria Ice Cave, New Mexico, USA (Dickfoss et al., 1997), or in their subhorizontal passages blocked by ice plugs, e.g., Crystal Falls Cave, Idaho, USA (Halliday, 1954), possibly also in the eroded or collapsed floor depressions below the steep inclined entrance part of statodynamic caves, e.g., Snežna jama na Raduni Cave, Slovenia (Mihevc, 2008). In Russia, flat ice floor bodies were described in Medeo Cave and Mariinskya Cave, Perm Region (Kadebskaya, 2008), and in some of Pinega caves (Shavrina and Guk, 2005). In several caves, flat ice floor is combined with seasonal ice stalagmites and stalactites, e.g., Kichmenskaya Cave, Perm Region, Russia (Kadebskaya, 2008). In areas with significant seasonal temperature variations, flat ice surfaces are formed seasonally by the freezing of ponded water, e.g., Grotte Valerie, Nahanni, Northwest Territory, Canada (Ford and Williams, 2007). Similarly, small shallow lakes in some cave near the Baikal Lake, Russia freeze in winter (Trofimova, 2007). In some of the Pinega caves, Belomorsko-Kulojskoe Plateau, Russia, "ice siphons" form in low passages due to the freezing of small streams during the pre-winter season. Winter floods are dammed by these siphons, and the long, flat ice floors originate on the long stream flood lakes (Shavrina and Guk, 2005). Long-term observations indicate that the "Ice Lake" in Geldloch Cave, Austria, can completely disappear and reform within a few years (Behm et al., 2009). The flat ice floor in Mauna Loa Ice Cave, Hawaii, USA, had been barricaded by the ice plug at the terminal end of the lava tubes; but this ice floor (called "Skating Rink") has disappeared during recent decades (Kempe and Ketz-Kempe, 1979; Pflitsch et al., 2016).

(7) *Combined ice bodies consisting of vertical ice formations and floor ice in the lowermost part of subhorizontal or multilevelled caves* (statodynamic or static caves with congelation ice). In caves with intensively fractured and karstified soluble bedrock, seeping meteoric waters are ponded only in places where floor depressions are covered by impermeable or poorly permeable layers. Therefore, at other places of these caves, mostly ice mounds, columns, stalagmites and stalactites have formed by the freezing of seeping meteoric waters, e.g., Ledenica Cave, Bulgaria (Popov, 1971). When frozen talus on the cave floor is filled by ice, floor ice can accumulate, partially as an ice "pond," e.g., in the lowermost part of the multilevelled Demänová Ice Cave, Slovakia (Strug et al., 2006; Strug and Zelinka, 2008).

(8) *Slightly inclined floor ice sheets in horizontal and subhorizontal caves in the alpine and permafrost regions.* Such floor ice, from several centimetres to over 1 m thick, in several places with elevated ice mounds or columns, is described from some Canadian caves. The caves Grande Caverne and Caverne Glacée 85, Yukon on the Arctic Circle at 900-1200 m a.s.l. have three internal climatic zones (>0°C, ≤0°C, <0°C). Floor ice sheets originates in the middle zone, where the cave floor dips toward the entrance (Lauriol et al., 1988). Floor ice sheets occur also in Canyon Creek Ice Cave, Canadian Rocky Mountains, Alberta (at 1768 m a.s.l.), in which floor ice becomes a shallow lake in August and September (Harris, 1979), as well as on the slightly sloping rock basement of the lower part of the Superman´s Gliterring Ice Palace, British

Columbia, Canada (Yonge, 2014). The slightly inclined perennial floor ice sheet covers the basalt floor of lava tubes in Fuji Ice Cave, Japan (Ohata et al., 1994). This later cave is a static one, with congelation ice, its upper entrance being formed by the collapse of the basalt roof. In a shallow, thermally-responsive cave named Caverne de l´Ours, Quebec, Canada, an inclined ice floor is seasonally formed in its entrance parts (in the freezing and sublimating zone) by the freezing of water slowly moving from the outside lake into the cave (Lacelle et al., 2009). In some lava tubes of the Lava Beds National Monument, along the California-Oregon border, USA, accelerated ice loss finally resulted in the disappearance of perennial ice, including the floor ice in Merrill Cave by 2006 (Fuhrmann, 2007; Kern and Thomas, 2014).

(9) *Ice plugs in horizontal caves (without or with vertical shafts) in the alpine and permafrost regions.* Massive ice blocking is known in some horizontal caves in the limestone mountains of Bear Cave and Tsi-it-toh-Choh in the basin of the upper Porcupine River on the Arctic Circle, Yukon, Canada, in which ice plugs were formed by gradual accumulation of floor ice from below, and accretion of hexagonal ice growing down from the cave roof (Lauriol et al., 1988, 2006; Lauriol and Clark, 1993). Also, L-shaped ice plugs are observed below shafts falling into these caves from the surface. The L-shape is given by the combination of the snow/firn fills in the vertical shafts, continued into horizontal passages (Lauriol et al., 2006). One of the passages in Coulthard Cave, located in the Crowsnest Pass area of southwestern Alberta, Canada (at an elevation of 2650 m a.s.l.), ends in massive ice blockages (Marshall and Brown, 1974; Brown and Marshall, 1975). Similarly, the lava tubes in the Aiyansh lava flow, the youngest volcanic features of British Columbia, Canada, are blocked by massive ice deposits (Marshall, 1975).

(10) *Combined ice bodies consisting of vertical ice formations and floor ice in the lowermost part of horizontal or subhorizontal caves with open vertical chimneys* (dynamic caves with congelation ice). Usually, on top of the ice floor, ice mounds, columns, stalagmites, and stalactites develop, elevating the general topography. In dynamic caves, congelation ice is formed by the freezing of infiltration waters in their cold zone during winter (due to the so-called chimney effect, in which several entrances at different elevations causes seasonally reversing air circulation between the entrances), e.g., the horizontal part of Kungur Ice Cave, Russia near its lower entrance (Mološmanova et al., 2005; Andrejčuk et al., 2013 and others).

(11) *Subhorizontal massive ice bodies or glacier-like ice block sloping to the lower entrance of vertical dissected caves with several entrances at different elevations* (dynamic caves with congelation ice as the result of the above described chimney effect). Perennial ice occurs in the cold zone adjacent to the lower entrance of caves with this morphology, e.g., Eisriesenwelt, Austria (Obleitner and Spötl, 2011; Schöner et al., 2011 and others). The total length of the Eisrieswenwelt is 42 km, its lower ice-covered part is slightly more than 1 km long and extends over a vertical span of 134 m (Millius and Petters, 2012). However, the occurrence of seasonal ice in the cold zone of dynamic caves is more frequent, e.g., Castleguard Cave, Alberta, Canada (Brown et al., 1971; Yonge, 2014).

(12) *Glacier ice tongues/plugs intruded into cave passages* at the glacier/rock contact were described in the vicinity of the Columbia Icefield, Alberta and British Columbia, Canada. The cave is sealed at one end by an extrusion of glacier ice 300 m below the icefield surface (Ford et al., 1976). The sealing of the cave conduit by glacier ice injection can also be observed in Milchbach Cave, Switzerland (Luetscher et al., 2011).

4.2.4 SMALL-SCALE ICE MORPHOLOGIES

For the systematic nomenclatures or classification of small-scale ice surface morphologies, it is sufficient to distinguish long-lasting and ephemeral forms. The long-lasting forms persist for several years (sometimes longer), while ephemeral forms are mostly visible during a shorter time interval (e.g., during and after heavy rains on the surface), but often are persistent during a whole year (Bella, 2003, 2007).

4.2.4.1 LONG-LASTING FORMS

4.2.4.1.1 Supraglacial Ice-Deposited Forms

Forms generated by the freezing of tiny water film or sheet wash water flow

Slightly inclined ice floors, in some places with ice micro-terraces or lobes (Fig. 4.2.2), and *evenly inclined or cascaded ice slopes* belong to the largest ice surfaces in ice caves. *Slightly inclined floor ice sheets* originate mostly in horizontal and subhorizontal ice caves, e.g., in Grande Caverne, Caverne Glacée 85, Canyon Creek Ice Cave, and some other Canadian caves in permafrost (Harris, 1979; Lauriol et al., 1988), or in Fuji Ice Cave (Ohata et al., 1994). *Evenly inclined or cascading ice tongues* form in strips of more concentrated and long-term sheet water flow. Larger and steeply downward-sloping massive ice covers are usually called *ice falls*, e.g., in Booming Ice Chasm (Yonge, 2014) and Peña Castil Ice Cave (Gomez Lende et al., 2014). *Small downward-sloping ice tongues* form on steep rock walls and are fed by water penetrating along faults and other structural discontinuities. *Ice curtains* are oblong ice formations hanging from ceilings or overhanging walls below karstified faults or inclined bedding-planes that control the seepage of meteoric waters. They are observed in Dobšiná Ice Cave (Small Hall, Collapsed Chamber), Peña Castil Ice Cave, and many other ice caves.

FIG. 4.2.2

Slightly inclined ice floors with micro-terraces, Dobšiná Ice Cave, Great Hall.

Photo: P. Bella.

Forms generated by the freezing of dripping water

Vertical ice formations originate in the periglacial cave environment, in places of the seepage of meteoric waters that are controlled by faults and other structural discontinuities within the cave roof. These ice speleothems consist of *ice stalactites, stalagmites, columns,* and *mounds* (Kyrle, 1929; Kunský, 1939; Racoviţă and Viehmann, 1966; Şerban, 1970; Mološmanova et al., 2005; Racoviţă and Onac, 2000; Mavljudov, 2008; Gomez Lende et al., 2014 and others).

The formation of ice stalagmites is controlled by air temperature, dripping rate, and temperature of seeping water. A bamboo-shaped stalagmite represents a particular type of ice stalagmite (Beer and Hassinger, 1902; Hauser and Oedl, 1923; Racoviţă, 1927; Kunský, 1939; Perşoiu and Onac, 2012 and others; Fig. 4.2.3). It is composed of two rhythmically alternating sections: (1) slightly milky, non-transparent (opaque) narrower portions formed at lower temperatures when dripping water freezes instantly, favoring growth in height (calcium carbonate granules, precipitated from the solution of calcium bicarbonate, remain enclosed in the stalagmite; Šumskij, 1955, Schoumskiy, 1957), (2) transparent ice knobs formed when higher temperatures at the apex of stalagmites allow dripping water to flow down the stalagmite before freezing, thus increasing their diameter (a precipitating white powder of fine granules of calcium carbonate are easily washed from the surface of the stalagmites by water film). Therefore these stalagmites are considered as ice thermoindicators, studied in detail in Scărişoara Ice Cave (Viehmann and Racoviţă, 1967, 1968).

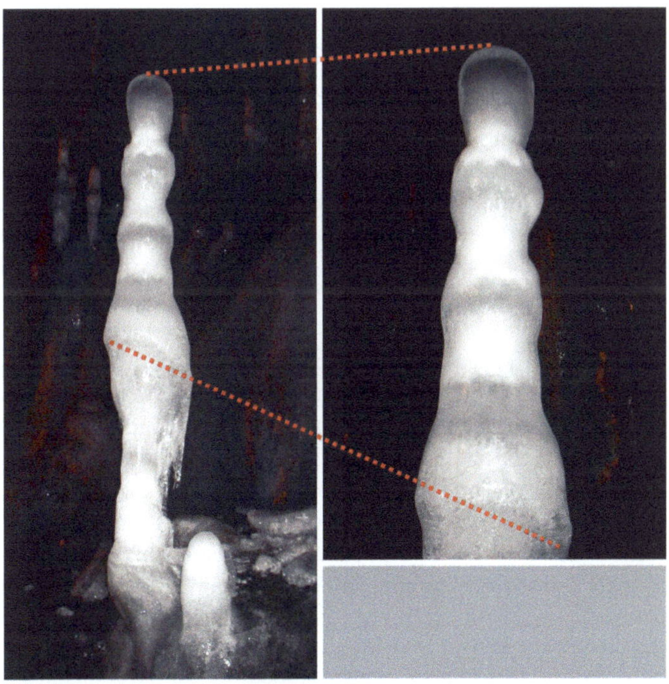

FIG. 4.2.3

Bamboo-shaped stalagmite, Scărişoara Cave, Great Reserve.

Photo: P. Bella.

Some ice stalagmites, over 5 m in height, may curve under their own weight during summer, when the ice plasticity increases due to the partial melting of the ice formations, e.g., in Scărișoara Ice Cave (Racoviță, 1927). When dripping water flow down and freezes mostly around the lower part of stalagmite-like formation, wider bell-shaped mounds are formed. Ice columns usually form below vertical karstified chimneys, e.g., in Dobšiná Ice Cave (in the cave's Great Hall and Small Hall).

4.2.4.1.2 Supraglacial Ablation Forms

Forms induced by air flow

Sublimation large scallops and flutes are asymmetrical hollows deepened into ice walls including walls of artificial trenches excavated into the ice floor for a tourist path, mostly in places of intensive air flow. These were formed through the sublimation of ice due to turbulent air flow in contact with ice (Curl, 1966; Cigna and Forti, 1986; Bella, 2003, 2007). In the longitudinal section, the steep sides of scallops are on the upstream part of the scallop. Analogically with scallops deepened into limestones or other soluble rocks, smaller scallops are formed by faster, turbulent flow. Scallops are longer (larger) in ice and snow than on limestone due to a much lower viscosity of air than water (Curl, 1966, 1974; Ford and Williams, 1989, 2007; Lauritzen and Lundberg, 2000, and others). Although present-day temperatures in several ice caves never exceed 0°C (thus being favorable for ice conservation), large scallop-like depressions in the ice suggest a continuous reduction of ice volume by sublimation. Sublimation scallops are observed in many ice caves, e.g., in Dobšiná Ice Cave (Bella, 2003, 2007; Fig. 4.2.4A) and Peña Castil Ice Cave (Gomez Lende et al., 2014). In Coulthard Cave, an experiment indicated a sublimation rate of 3 mm/yr (Marshall and Brown, 1974; Brown and Marshall, 1975). *Sublimation flutes* are asymmetrical hollows (same as scallops), but in cross-section. Their longer dimensions are oriented perpendicular to the direction of air flow. Sublimation flutes were described in Eisriesenwelt Cave (Curl, 1966).

Sublimation steep ice walls, from several meters to more than 15 m high, are one the most spectacular ice surfaces in ice caves. These smooth sublimation surfaces cut through the horizontal or subhorizontal layers of ice that record the formation of ice block (superposition principle). Excellent examples of sublimation steep ice walls (stratigrafied ice profiles) are in Dobšiná Ice Cave (Ruffíny's Corridor, Great Curtain in the Ground Floor; Fig. 4.2.4B), Scărișoara Ice Cave (Little Reserve) or Monlési Ice Cave. In some places, these ice walls can be slightly dissected by sublimation large scallops and similar hollows.

Ablation windows and smaller holes, as well as *ablation ice irregular protrusions,* are residues of larger ice formations (mostly ice stalagmites or columns), e.g., in the upper part of Great Hall in Dobšiná Ice Cave (Bella, 2003, 2007; Fig. 4.2.4C and D) or in the Great Hall of Scărișoara Ice Cave. Also, horizontal and subhorizontal *melt channels and bridges* produced by air fluxes are described from the Peña Castil Ice Cave (Gomez Lende et al., 2014).

Ablation oval mound-shaped elevations are visible as protrusions from floor ice as a result of ice sublimation due to air flow, e.g., in Dobšiná Ice Cave, in the northeastern part of Great Hall that is connected with the Collapsed Chamber (air flow circulation between glaciated and non-glaciated parts of the cave) or in Lo Lc 1650 "Abisso sul Margine dell´Alto Bregai" Ice Cave (Citterio et al., 2004a,b).

Forms generated by stagnant ponded water

Ablation bevels at the edge of flat ice floors present smooth ice surfaces inclined downwards the "ice ponds." They were formed by melting of ice caused by the water of shallow lakes that accumulated during phases of intensive seepage of meteoric waters. The first description was made from the Dobšiná Ice Cave, Collapsed Chamber (Bella, 2005; Fig. 4.2.5). The ablation bevels, generally exceeding 1 m in height,

FIG. 4.2.4

Supraglacial sublimation forms, Dobšiná Ice Cave: (A) sublimation large scallops, Small Hall; (B) sublimationally sculptured steep ice wall, Ground Floor; (C) sublimation window, Great Hall; (D) sublimation corner-shaped protrusion, Great Hall.

Foto: P. Bella.

display a steeply sloping face toward the lake that shaped them (supraglacial lakes are, on occasion, only several centimeters deep). The smooth bevels are not dissected by horizontal water-level notches, but sometimes display segments with varying degrees of inclination, probably linked to phases of unequal ablation intensity or the more intensive ice melting at the water-level, compared to its bottom (Bella, 2005, 2007). In Dobšiná Ice Cave, the ablation bevels between the Collapsed dóm Chamber and Great Hall have been reduced and dissected into several smaller oval mounds by ice sublimation due to air flow.

Anthropogenic forms

Small melting depressions near electric reflectors in show caves are negative human impacts resulted from their development for tourism, e.g., in the Great Hall and some other places near the tourist path in Dobšiná Ice Cave. New technologies (with less heat emission) of cave illumination eliminated these melting depressions.

Artificial trenches and notches of tourist path excavated into the ice floor and ice walls are the most visible features of human impacts in ice caves that were developed for tourism. These trenches and notches

FIG. 4.2.5

Ablation bevels at the edge of flat ice floor, Dobšiná Ice Cave, Collapsed Chamber.

Foto: P. Bella.

have been dug during the construction of tourist paths. In Dobšiná Ice Cave, one of these trenches, that led from the cave entrance to lower glaciated parts (excavated along the western edge of Small Hall in the beginning of 1970s), was markedly enlarged by ice sublimation due to changed air circulation. For this reason, it was repeatedly artificially glaciated in the second part of the 1990s (Bobro et al., 1995b; Zelinka, 1996).

4.2.4.1.3 Supraglacial Compounded Ice-Deposited/Ablation Forms

Forms generated by the repeated ice melting and freezing of stagnant ponded water

Flat ice floors develop in cave parts where the intensive seepage of meteoric waters, melting of floor ice, lake accumulation and stagnation of waters alternate with the freezing of lake water (Bella, 2005, 2007).

FIG. 4.2.6

Flat ice floor surfaces: (A) Scărişoara Cave, Great Hall; (B) Dobšiná Ice Cave, Collapsed Chamber.

Photo: P. Bella.

Flat floor ice surfaces are observed on the top position of ice stump-like ice block (in the Great Hall of Scărişoara Ice Cave; Fig. 4.2.6A) or the near-entrance part of downward sloping ice block (in the Small Hall of the Dobšinská Ice Cave), in the lower position of ice blocks (the ice surface barricaded by a cave rock roof; the former ice surface in the Great Hall of the Dobšinská Ice Cave, also in its Collapsed Chamber at the contact of glaciated and non-glaciated parts of the cave; Fig. 4.2.6B), on the floor of melting and lowering ice block that completely covers the cave ground (Shoshone Ice Cave, Idaho, USA), in the ponded floor erosion or dammed depressions (Snežna jama na Raduni Cave, Candelaria Ice Cave), in the sloping floor flooded due to ice plug at the end of the passage (Mauna Loa Ice Cave) or in small shafts completely filled by ice (Peña Castil Ice Cave).

In Scărişoara Ice Cave, the top surface of the ice block covers approximately 3000 m^2. In the summer, this ice surface is melting and 10–15 cm deep supraglacial lake is formed (Racoviţă and Onac,

2000). In Dobšiná Ice Cave, the flat ice floor in the lower position existed also in the Great Hall. Here, Pelech (1879) wrote about the flat ice surface of 1726 m^2 used for summer skating. Also, several old postcards show the ice flat surface with figure-skaters. However, in Dobšiná Ice Cave, this horizontal surface has changed, as meteoric dripwater freezed on top of the existing flat ice (below karstified chimneys), to form a slightly inclined surface, from the inflow point to the lower areas represented by the artificial channels excavated along the tourist paths (last time, this flat ice surface was used for skating in 1947–1952).

Seasonal flat ice floors originate also through the freezing of water in bedrock depressions (Trofimova, 2007; Kadebskaya, 2008) or stream flood lakes dammed by "ice siphons" (Shavrina and Guk, 2005).

In some caves, perennial (or subannual) flat ice floor surfaces have melted during last decades. In Merrill Cave, Northeastern California, USA, a ponded ice has been partially reduced, until completely disappeared (Fuhrmann, 2007; Kern and Thomas, 2014). Although in Mauna Loa Ice Cave, Hawaii, USA, a perennial ice barrier still blocks the lava tube at its terminal end, a previously present ice floor (estimated 260 m^2) has disappeared (Pflitsch et al., 2016).

4.2.4.1.4 Intraglacial Ablation Forms

Forms generated by air flow

Sublimation tunnels at the ice/rock contact were produced by air flow within the ice block of the Lo Lc 1650 "Abisso sul Margine dell´Alto Bregai" ice cave. Its walls are dissected by shallow large scallop-like depressions (Citterio et al., 2004a,b).

Sublimation large scallops and flutes on ice walls of artificial tunnels excavated through the ice block. In Dobšiná Ice Cave, two tunnels on the tourist trail (between Hell Abyss and Ground Floor, and between Ground Floor and Ruffíny's Corridor) have been excavated. Air flow is more intense in the narrower parts of these tunnels. The genesis and enlargement of large sublimation scallops and flutes on the walls and ceiling of artificial tunnels is due to air flow (Bella, 2003, 2007; Fig. 4.2.7) that is enhanced by the movement of visitors (Bobro et al., 1995a). The asymmetric shape of sublimation scallops corresponds with the direction of air flow through these artificial tunnels (Piasecki et al., 2005, Fig. 4.2.7).

Sublimation ceiling pockets in the artificial tunnels excavated through the ice block. In Dobšiná Ice Cave, shallow, oval, plate-shaped or bowl-shaped ceiling hollows have formed by small air turbulence in the tunnel descending from the Peklo Abyss to the Ground Floor (Bella, 2007).

Anthropogenic forms

Ablation ceiling cupola above the electric reflector in the artificial cavity excavated into the ice body. This widening ablation form generated by ice melting is visible in Dobšiná Ice Cave, in the cavity named Chapel, accessible from the lower edge of the Ruffíny's Corridor (Bella, 2003).

Artificial tunnels excavated through the ice block during the development of the Dobšiná Ice Cave for tourism. These tunnels are integrated into the current air circulation in the glaciated part of the cave (Piasecki et al., 2005). Their walls and ceiling were remodeled and enlarged by ice sublimation (Bella, 2003, 2007).

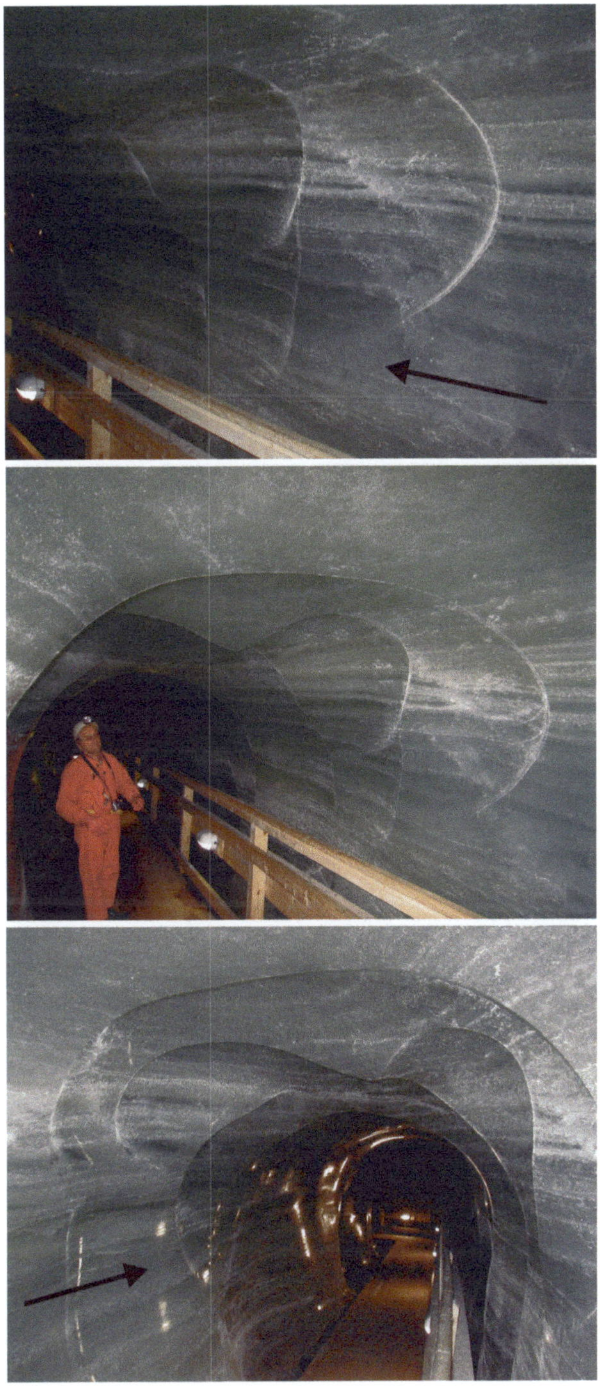

FIG. 4.2.7

Sublimation large scallops and flutes deepened into the walls and ceiling of artificial tunnel between Ground Floor and Ruffíny's Corridor, Dobšiná Ice Cave (the *arrows* show the direction of air flow).

Photo: P. Bella.

4.2.4.1.5 Subglacial Ablation Forms

Forms generated by the ice sublimation due to air flow

Sublimation cavities at the contact of ice block with bedrock. In Dobšiná Ice Cave, the subglacial cavity with sublimation ice crystals under the Great Curtain at the Ground Floor (Fig. 4.2.8), as well as the narrow slit-shaped cavity in the Ruffíny's Corridor (several centimeters to max. 1 m wide) was formed and enlarged by ice sublimation between the ice block and bedrock (Bella, 2003, 2007). The air flow from lower unknown cave parts (supposed to exist below the ice block and glaciated parts) or the adjacent Dry Chamber is indicated also by the coating of sublimation ice crystals formed on the rock wall and ice ceiling surface of these subglacial cavities (air flows through underlying rock blocks and debris). Periodically, the upper parts of the narrow slit-shaped cavity in the Ruffíny's Corridor are filled by new ice, mostly formed by the freezing of sheet wash water flow. The basal melting of the ice block in Dobšiná Ice Cave was not studied in detail, however, studies performed in Scărișoara Ice Cave have shown that the ablation of the base of ice block is a result of the ice melting generated by geothermal heat flux and of the ice sublimation due to the circulation of cold and dry air under the ice block. Subglacial tunnels are enlarged mostly by ice sublimation (Perșoiu, 2004, 2005).

4.2.4.2 EPHEMERAL ICE FORMS

These morphologies originate mostly in the upper ice-covered parts of caves with seasonal temperature variations, thin overlying rocks with karstified tectonic faults or other structural discontinuities that control the seepage of waters from melting snow or rainfalls. Ice-deposited forms are usually formed at the end of winter and persist during spring and sometimes, partially, until the beginning of summer. Physical ablation forms are formed mostly during summer and persist until the end of winter, after which they are usually filled and covered by new ice. Anomalies of ice stratigraphy (depressions filled

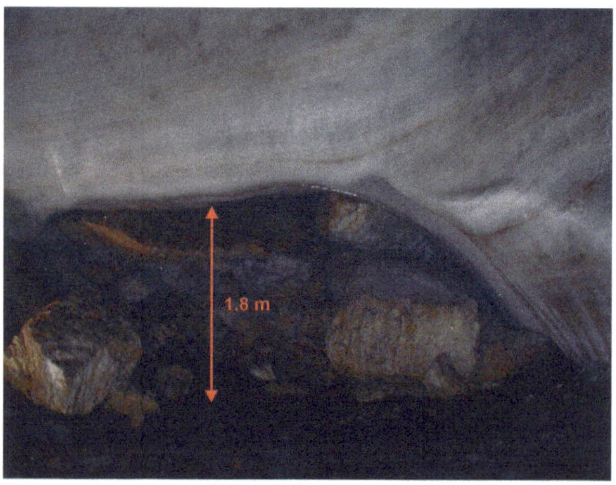

FIG. 4.2.8

Subglacial sublimation cavity, Dobšiná Ice Cave, Ground Floor.

Photo: P. Bella.

by newer ice) are related to former ablation forms. The ablation forms in Dobšiná Ice Cave described below were observed mainly during summer and autumn 2002 and winter 2002/2003 after very intense rainfalls in summer 2002.

4.2.4.2.1 Supraglacial Ice-Deposited Forms

Forms formed by the freezing of thin water film or sheet wash water flow

Seasonally *ice curtains and flags*, as well as *ice organ tubes* form in the entrance part of ice caves, e.g., in Peña Castil Ice Cave (Gomez Lende et al., 2014).

Forms formed by the freezing of dripping water

Ice bamboo-shaped stalagmites seasonally formed in the entrance part of ice caves. Seasonal *ice stalactites* hanging from fractured, karstified and frost weathered roof, occur mainly in the entrance breakdown parts of ice caves. Ice stalactites in the entrance parts are covered by sublimation ice crystals during the end of winter and at the beginning of spring, e.g, in the Small Hall of the Dobšiná Ice Cave.

4.2.4.2.2 Supraglacial Ablation Forms

Forms generated by dripping water

Ablation dripping pits occur as small hollows deepened into the cave ice below fractures and other structural discontinuities on the ceiling that allow the seepage of meteoric waters. According to the intensity, impact strength, and temperature of dripping waters, shallow plate pits, deeper bowl-shaped pits, and larger well-shaped pits (with vertical or overhanging walls) can be distinguished (Bella, 2003, 2007, Fig. 4.2.9). Small ablation outflow runnels lead downward from larger dripping pits.

Ablation dripping multipit depressions consist of several adjoining well-shaped pits that occur in rows or in star-clusters, e.g., in Dobšiná Ice Cave, Small Hall and Great Hall.

Ablation ice spires present small residues of ice stalagmite reduced and remodeled by dripping water, e.g., in Dobšiná Ice Cave, Small Hall and Great Hall.

Forms generated by dripping and stagnant water

Ablation shallow pans are shallow plate-shaped depressions (resembling kamenitzas) formed on horizontal to subhorizontal ice floor surfaces (Fig. 4.2.9A). The flat bottom and steep-to-overhanging walls of these shallow depressions result from the repeated thawing and freezing of waters (in small ice lakes). In some cases, small inflow and outflow runnels occur on opposing sides of these ablation pans. Ablation pans at the ice/rock contact have also been described, e.g., in Lo Lc 1650 "Abisso sul Margine dell´Alto Bregai" ice cave (Citterio et al., 2004a,b).

Ablation dripping kettle-shaped holes are deeper ice floor depressions that originated in places of intensive seepage of meteoric waters, e.g., in Dobšiná Ice Cave, in the upper part of the Great Hall (Fig. 4.2.9B). Ablation outflow runnels occur at the lower side of these holes.

Forms generated by falling/running or intensively dripping water

Ablation vertical half-tube grooves on ice columns originate below karstified vertical chimneys with occasional intensive seepage of meteoric waters, e.g., in the Great Hall of Dobšiná Ice Cave (Fig. 4.2.10A).

Ablation shaft-like depressions deepened by falling/running seepage water below intensively karstified vertical chimneys. In Dobšiná Ice Cave (Small Hall, Great Hall), these shapes were observed as

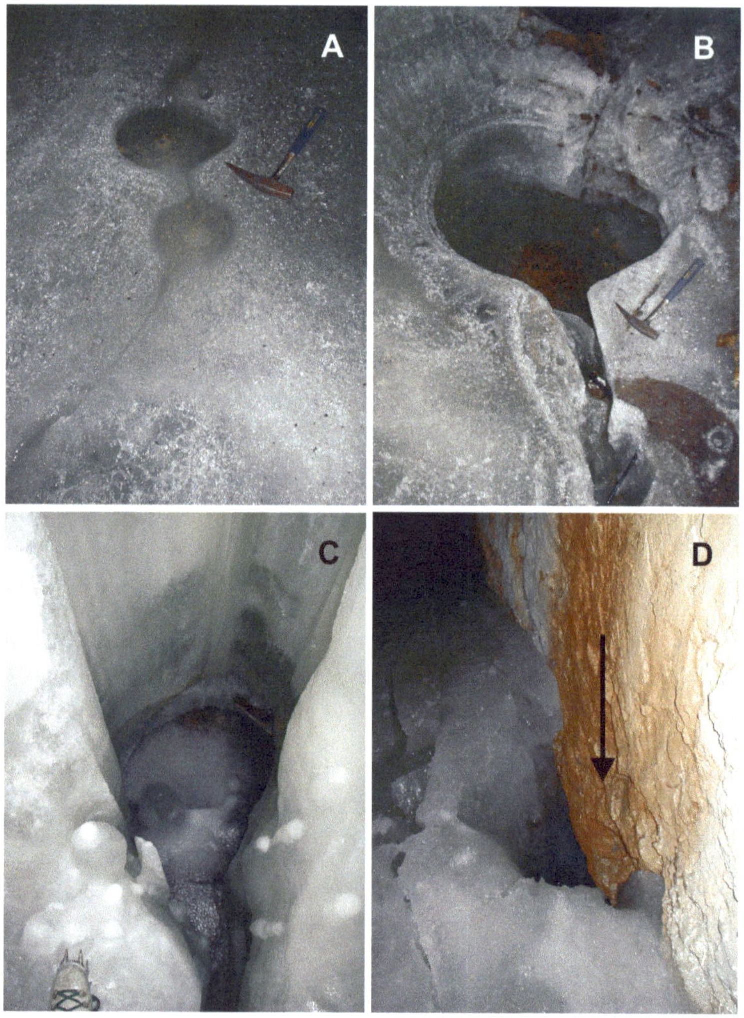

FIG. 4.2.9

Ablation pans and deeper pitholes, Dobšiná Ice Cave: (A) shallow pans, the sector between Small Hall and Great Hall; (B) dripping kettle-shaped hole, Great Hall; (C) shaft-like depression, Great Hall; (D) contact shaft-like depression, between Small Hall and Hell Abyss.

Foto: P. Bella.

c.1.3–1.8 m deep depressions with steep walls (Bella, 2003, 2007; Figs. 4.2.9C and 4.2.10B). An outflow runnel is gradually cut downward into the floor ice from the lower edge of these depressions (outflow runnels are sloping from the bottom of shaft-like depressions to lower ice-covered cave parts). Some ablation shaft-like depressions also formed at the contact of cave ice with a rock wall (*contact ablation shaft-like depressions*), e.g. in Dobšiná Ice Cave (Bella, 2003, 2007; Fig. 4.2.9D). They are deepened into the inclined ice slope and are narrowing downward along rock walls (outflow channels were not developed).

FIG. 4.2.10

Ablation runnels and grooves: (A) vertical half-tube groove on ice column, Dobšiná Ice Cave, Great Hall; (B) continuig meandering runnel deepened into the ice floor, Dobšiná Ice Cave, Great Hall; (C) outflow runnel leading from temporary lake on seasonally thawing flat ice floor, Scărișoara Ice Cave, Great Hall; (D) meandering runnel on the slightly inclined ice floor with debris, Fuji Ice Cave, Japan.

Photo: P. Bella.

Forms generated by sheet wash water flow

Ablation parallel rills (ripples) form on the steep walls of artificial trenches of tourist paths cut into the ice floor, e.g., in Dobšiná Ice Cave, between Small Hall and Great Hall (Bella, 2003, 2007).

Forms generated by channeled water flow

Shallow meandering runnels on slightly inclined (almost flat) ice floor were observed in the Collapsed Chamber of the Dobšiná Ice Cave, where the formation of mini-meanders, and in some places also braided channel patterns, was conditioned by mechanical obstacles, mainly the small remains of decaying wood (Bella, 2003, 2007). Deeper meandering *supraglacial water channel (bédière)* were documented on the top of ice block in the Lo Lc 1650 "Abisso sul Margine dell´Alto Bregai" ice cave (Citterio et al., 2004a,b).

Ablation outflow, straight or meandering, channels leading from shaft-like depressions deepened by falling/running seepage water or from larger kettle-shaped dripping holes have also been observed. Ablation runnels, several centimeters to about 1.5 m deep, were occasionally observed in the Great Hall of the Dobšiná Ice Cave (Bella, 2003, 2007; Fig. 4.2.10B). Seasonally, by the end of summer, small runnels form on the flat top surface of the glacier-like ice block in the Scărişoara Ice Cave which drain water from the Great Hall to the Little Reserve (Racoviţă and Onac, 2000; Fig. 4.2.10C).

Forms generated by the upward expansion of freezing stagnant ponded water

Small ice bulges with radial cracks occur as upward-expanding dome-shaped positive (against the ice floor) features, up to about 1-2 m in diameter and up to several centimeters in height. Their origin is in the expansion of freezing water in the deepest place of shallow supraglacial lake (or in the shallow large dripping pan) during winter, e.g., in the Small Hall of the Dobšiná Ice Cave (near the entrance zone, where large seasonal temperature changes occur, Bella, 2007; Fig. 4.2.11). Similarly shaped ice mounds, but larger in size, have been studied on the surface of glaciers. An explanation of the origin of ice mounds in the "ice lake" could be the refreezing of an accumulated pocket of melt water within the ice (Van Autenboer, 1962).

Anthropogenic forms

Artificial channels excavated into the floor ice along a tourist path for the outflow of seeping waters to lower cave parts, e.g., in the upper glaciated parts (Small Hall, Great Hall) of the Dobšiná Ice Cave.

4.2.5 CONCLUSIONS

Large-scale morphologies of ice fills in caves resulted from different natural settings of the cave ice. The spatial position, morphometry and morphology of caves and their entrances control microclimatic conditions in caves, as well as snowdrift or surface runoff (from rainfall and meltwater) into the underground. In some caves, large ice morphologies are influenced by the measure of structural and geological discontinuities of overlying bedrock and its spatial configuration (the control of waters seeping through cave roof). Morphological changes of large-scale ice morphologies are caused by weather and climate changes at the surface; and, in some cases, also by human impacts in both glaciated and nonglaciated sectors of caves.

Smaller ice-deposited and ablation forms originated mostly on the top surface and inclined to steep sides of ice blocks, and also, in some cases, below ice blocks. Compared to glacier caves, intraglacial and subglacial ablation conduits and tunnels formed by running water are not known in ice-filled cave,

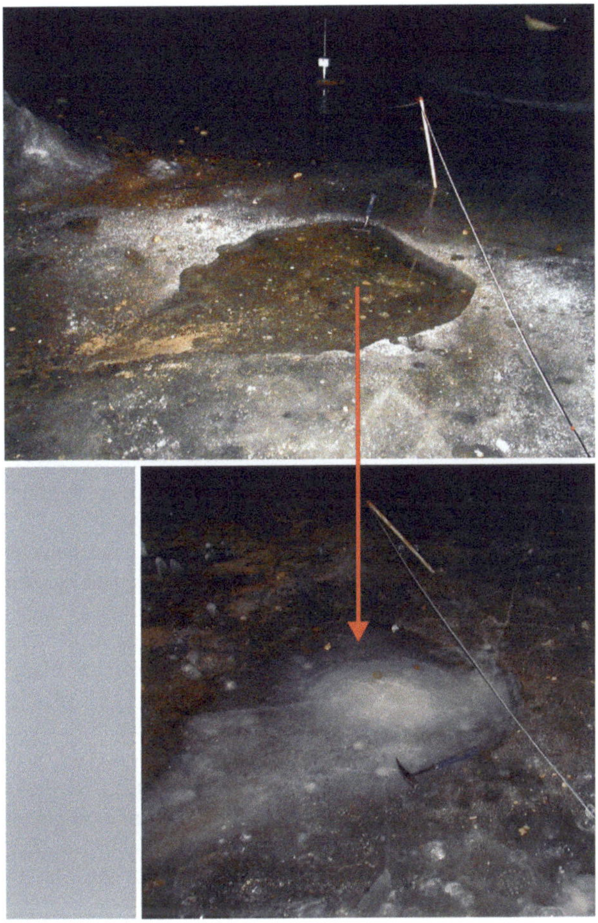

FIG. 4.2.11

Small ice bulge with radial cracks originated in the place of occasional (seasonal) lake, Dobšiná Ice Cave, Small Hall.

Foto: P. Bella.

nor are the ablation flat ceilings described by Mavlyudov (2005) in a subglacial channel. Occasional streamlets from seeping meteoric waters, active mostly during heavy rains, form only supraglacial channels on inclined ice surfaces. Although the thickness of ice floor in some caves reaches several meters (up to 25-30 m), ice bodies are not fractured enough for deep water penetration (nor are the available water volumes large enough to allow for genesis of such features by meting). Therefore, occasional streamlets flow only on inclined ice surfaces. The internal drainage of surface glaciers, including intraglacial (englacial) channels, is connected with ice structures, mainly fractures resulted from the stress and strain determines how glaciers flow downslope (Stenborg, 1968, 1969; Fountain et al., 2005a,b and others). Large flat ice floors were described only in the chambers and halls of ice-filled caves (Bella, 2005, 2007). In ice-filled caves, intraglacial ablation forms are sporadic (e.g., tunnels in the ice formed by air circulation), but large sublimation scallops are sculpted by air flowing through artificial tunnels in show caves. Human impacts influence the formation of ice-deposited forms (ice cones or tongues in places where are mouths of artificial grooves regulating outflow), as well as ablation forms (artificial tunnels, trenches or grooves).

ACKNOWLEDGMENTS

The work was supported by the Project of the Scientific Grant Agency VEGA No. 2/0193/15.

REFERENCES

Andrejčuk, V.N., Kadebskaja, O., Čajkovskij, I.I., 2013. Kriogennye mineral'nye obrazovanija Kungurskoj ledjanoj peščery [Cryogenic mineral formations of Kungur Ice Cave]. Silezskij universitet – Gornyj institut UrO RAN, Sosnovec – Perm, 128.

Badino, G., Meneghel, M., 2001. Caves in the Glaciers of Terra Nova Bay (Victoria Land, Antarctica). Proceedings of the 13th International Congress of Speleology, 1, Brazil, pp. 187–190.

Balch, E.S., 1900. Glacières of Freezing Caverns. Allen, Lane and Scott, Philadelphia, PA, 337 p.

Beer, E., Hassinger, H., 1902. Das Geldoch im Ötscher (die Seelucke – eine Höhle). Zeitschrift der deutschen und österreichischen Alpenvereins 23, 117–149.

Behm, M., Dittes, V., Greilinger, R., Hartmann, H., Plan, L., Sulzbacher, D., 2009. Decline of cave ice – a case study from the Austrian Alps (Europe) based on 146 years of observation. In: White, W.B. (Ed.), Proceedings of the 15th International Congress of Speleology, 3. Kerrville, TX, pp. 1413–1416.

Bella, P., 2003. Glaciálne ablačné formy v Dobšinskej ľadovej jaskyni [Glacial ablation forms in the Dobšiná Ice Cave]. Aragonit 8, 3–7 (in Slovak).

Bella, P., 2005. Rovné podlahové ľadové povrchy v zaľadnených jaskyniach (Dobšinská ľadová jaskyňa, jaskyňa Scărișoara) [Flat floor ice surfaces in ice caves (Dobšiná Ice Cave, Scărișoara Cave)]. Aragonit 10, 12–16 (in Slovak).

Bella, P., 2007. Morphology of ice surface in the Dobšiná Ice Cave. In: Zelinka, J. (Ed.), Proceedings of the 2nd International Workshop on Ice Caves, Demänovská Dolina, Slovak Republic, May 8–12, 2006. SSJ, Liptovský Mikuláš, pp. 15–23.

Bennett, M.R., Glasser, N.F., 2009. Glacial Geology: Ice Sheets and Landforms. Wiley-Blackwell, 385 p.

Bobro, M., Hančuľák, J., Zelinka, J., 1995a. Súčasné mikroklimatické pomery v Dobšinskej ladovej jaskyni [Current microclimatic conditions in the Dobšiná Ice Cave]. In: Bella, P. (Ed.), Ochrana ľadových jaskýň, Proccedings, Dobšinská Ľadová Jaskyňa, September 21–22, 1995. SSJ, Liptovský Mikuláš, pp. 29–34. (in Slovak).

Bobro, M., Krepelka, P., Zelinka, J., 1995b. Rekonštrukcia vstupného priestoru v Dobšinskej ladovej jaskyni vo vzťahu k úbytku ľadu [Reconstruction of the entrance space in the Dobšiná Ice Cave in relation to ice decrease]. In: Bella, P. (Ed.), Ochrana ľadových jaskýň, Proccedings, Dobšinská Ľadová Jaskyňa September 21–22, 1995. SSJ, Liptovský Mikuláš, pp. 35–36. (in Slovak).

Böhm, J., Kunský, J., 1938. Silické lednice [Silická ľadnica Cave]. Sborník Československé společnosti zeměpisné 44, 129–133 (in Czech).

Brown, M.C., Ford, D.C., Wigley, T.M.L., 1971. Ice caves of the Canadian Rockies. Caves Karst 13 (5), 38.

Brown, M.C., Marshall, P., 1975. Ice in Coulthard Cave. In: Panoš, V. (Ed.), Proceeding of the 6th International Congress of Speleology (Olomouc 1973). vol. 1. Academia, Praha, pp. 371–375.

Buzjak, N., Paar, D., Dubovečak, V., Bočić, N., 2014. The influence of karst topography to ice cave occurrence – example of Ledena Jama in Lomska Duliba (Croatia). In: Land, L., Kern, Z., Maggi, V., Turri, S. (Eds.), Proceedings of the Sixth International Workshop on Ice Caves, August 17–22, Idaho Falls, ID. NCKRI Symposium 4, National Cave and Karst Research Institute, Carlsbad (NM). pp. 17–23.

Cigna, A.A., Forti, P., 1986. The speleogenetic role of air flow caused by convection. Int. J. Speleol. 15 (1–4), 41–52.
Citterio, M., Turri, S., Bini, A., Maggi, V., 2004a. Some observations on the structure and morfology of an in the "Abisso sul Margine dell´Alto Bregai" cave (Grigna settentrionale, Italian Alps). In: Eraso, A. (Ed.), Proceedings of the 6th Symposium of the UIS-GLACKIPR «Glacial Caves and Karst in Polar Regions», September 3–8, 2003. Ny-Ålesund, Svalbard, Norway. Madrid, pp. 13–19.
Citterio, M., Turri, S., Bini, A., Maggi, V., Pelfini, M., Pini, R., Ravazzi, C., Santilli, M., Stenni, M., Udisti, R., 2004b. Multidisciplinary approach to the study of Lo Lc 1650 "Abisso sul Margine dell´Alto Bregai" ice cave (Lecco, Italy). Theor. Appl. Karstol. 17, 27–44.
Colucci, R.R., Fontana, D., Forte, E., Potleca, M., Guglielmin, M., 2016. Responce of ice caves to weather extremes in the southeastern Alps, Europe. Geomorphology 261, 1–16.
Curl, R.L., 1966. Scallops and flutes. Trans. Cave Res. Group Great Brit 7 (2), 121–160.
Curl, R.L., 1974. Deducing flow velocity in cave conduits from scallops. Bull. Natl. Speleol. Soc. 36 (2), 1–5.
Dickfoss, P.V., Betancourt, J.L., Thompson, L.G., Turner, R.M., Tharnstrom, S., 1997. History of ice at Candelaria Ice Cave, New Mexico. Bull. New Mex. Bur. Min. Mineral Resour. 156, 91–112.
Dmitriev, V.E., 1980. Oledenenie peščer kak časť gljaciosfery zemli [Cave Glaciation as a Part of the Earth´s Glaciosphere]. In: Krasnov, E.V. (Ed.), Karst Sibiri i Daľnego Vostoka, Vladivostok. pp. 130–145. (in Russian).
Droppa, A., 1958. Geomorfologický charakter priepastí na Ohništi [Geomorphological character of abysses on the Ohnište]. Slovenský kras 1, 14–23 (in Slovak).
Eraso, A., 1992. Internal glacier melting and naled ice generated by air circulation. Proposal of an enthalpy-entropy diagram for quantitative calculations.. In: Pulina, M., Eraso, A. (Eds.), Proceedings of the 2nd International Symposium of Glacier Caves and Karst in Polar Regions, Miedzygórze – Veľká Morava, February 10–16, 1992. Sosnowiec, pp. 29–42.
Favre, G., 1985. Kverkfjöll, rivières géothermiques sous la glace, Islande. Spelunca 17, 11–17.
Fillipov, A.G., 2005. The age of the relic firn plug in the Kremenshetskaya Cave, Eastern Siberia. In: Mavlyudov, B.R. (Ed.), Glacier Caves and Clacial Kast in High Mountains and Polar Regions (Proceedings of the 7th GLACKIPR Symposium, Azau – Kabardino-Balkarija (Caucasus), Russia, September 6–12, 2005), Moscow. pp. 98–100.
Ford, D.C., Harmon, R.S., Schwarz, H.P., Wigley, T.M.L., Thompson, P., 1976. Geo-hydraulic and thermometric observations in the vicinity of the Columbia Icefield, Alberta and British Columbia, Canada. J. Glaciol. 16 (74), 219–230.
Ford, D.C., Williams, P.W., 1989. Karst Geomorphology and Hydrology. Chapmam & Hall, London, 601 p.
Ford, D.C., Williams, P.W., 2007. Karst Hydrogeology and Geomorphology. Wiley, Chichester, 562 p.
Fountain, A.G., 2005. Glacier Caves. In: Culvier, D.C., White, W.B. (Eds.), Encyclopedia of Caves. Elsevier Academic Press, pp. 271–275.
Fountain, A.G., Jacobel, R.W., Schlichting, R., Jansson, P., 2005a. Fractures as the main pathways of water flow in temperate glaciers. Nature 433 (7026), 618–621.
Fountain, A.G., Schlichting, R., Jansson, P., Jacobel, R.W., 2005b. Observations of englacial water passages: a fracture-dominated system. Ann. Glaciol. 40, 25–30.
Fuhrmann, K., 2007. Monitoring the disappearance of a perennial ice deposit in Merill Cave. J. Cave Karst Stud. 69 (2), 256–265.
Garašić, M., 1980. Ledenica u Bukovom vrhu na Velebitu [Ledenica Cave in Mt. Bukovi Vrh on Velebit Mt.]. Naše planine 11–12, 278–280 (in Croatian).
Garašić, M., 2014. New research in cave Ledenica in Bukovi Vrh on Velebit Mt. in Croatian Dinaric karst. In: Land, L., Kern, Z., Maggi, V., Turri, S. (Eds.), Proceedings of the Sixth International Workshop on Ice Caves, August 17–22, Idaho Falls, ID. NCKRI Symposium 4, National Cave and Karst Research Institute, Carlsbad (NM). pp. 31–32.
Giggenbach, W.F., 1976. Geothermal ice caves on Mt. Erebus, Ross Island, Antartica. N. Z. J. Geol. Geophys. 19 (3), 365–372.

Gomez Lende, M., Berenguer, F., Serrano, E., 2014. Morphology, ice types and thermal regime in a high mountain ice cave. First studies applying terrestrial laser scanner in the Peña Castil Ice Cave (Picos de Europa, Northern Spain). Geogr. Fis. Din. Quat. 37 (2), 141–150.

Grebe, C., Ringeis, J., Pflitsch, A., 2008. Study of temperature and airflow in the Schellenberger Ice Cave (Berchtesgadener, Limestone Alps, Germany). In: Kadebskaya, O., Mavludov, B., Pyatunin, M. (Eds.), Proceedings of the 3rd International Workshop on Ice Caves, Kungur, Russia, May 12-17, 2008. Perm State University, Perm. pp. 26–29.

Habe, F., 1971. Die Eishöhlen im Slowenischen Karst. Slovenský kras 9, 135–145.

Habič, P., 1964. O podzemeljskich ledenikih na Nanosu [About underground glaciers on the Nanos Plateau]. Naše jame 5 (1963), 19–29 (in Slovenian).

Halliday, W.R., 1954. Ice caves of the United States. Bull. Natl. Speleol. Soc. 16, 3–28.

Harris, S.S., 1979. Ice caves and permafrost zones in southwest Alberta. Erkunde 33, 61–70.

Hauser, E., Oedl, R., 1923. Eisbildungen und meteorologische Beobachtungen. Speläologisches Jahrbuch 4, 17–47.

Hill, C., Forti, P., 1997. Cave Minerals of the World, second ed. National Speleological Society, Huntsville, AL, 463 p.

Hochmuth, Z., 1979. Ľadová priepasť v Červených vrchoch [Ice Abyss in the Červené vrchy Mountains]. Slovenský kras 17, 133–137 (in Slovak).

Holmlund, P., Onac, B.P., Hansson, M., Holmgren, K., Mörth, M., Nyman, M., Perşoiu, A., 2005. Assessing the palaeoclimate potential of cave glaciers: the example of the Scărişoara Ice Cave (Romania). Geogr. Ann. 87 A (1), 193–201.

Jania, J., 1997. Glacjologia [Glaciology]. Wydawnictwo Naukowe PWN, Warszawa, 359 p. (in Polish).

Jelinić, I., Horvatinčić, N., Božić, V., 2001. Ledena Jama u Lomskoj Dulibi [The Ledena jama Cave (The Ice Pit) in Lomska Duliba]. Senjski zbornik 28, 5–28 (in Croatian).

Kadebskaya, O., 2008. Ice caves of Perm Region. In: Kadebskaya, O., Mavludov, B., Pyatunin, M. (Eds.), Proceedings of the 3rd International Workshop on Ice Caves, Kungur, Russia, May 12–17, 2008. Perm State University, Perm, pp. 94–99.

Kempe, S., Ketz-Kempe, C., 1979. Fire and ice atop Hawaii. Bull. Natl. Speleol. Soc. 37 (8), 185–188.

Kern, Z., Bočić, N., Horvatinčić, N., Fórizs, I., Nagy, B., László, P., 2008. Palaeoenvironmental records from ice caves of Velebit Mt. – Ledena pit and Vukušić ice cave, Croatia. In: Kadebskaya, O., Mavludov, B., Pyatunin, M. (Eds.), Proceedings of the 3rd International Workshop on Ice Caves, Kungur, Russia, May 12–17, 2008. Perm State University, Perm. pp. 108–113.

Kern, Z., Molnár, M., Svingor, É., Perşoiu, A., Nagy, B., 2009. High resolution, well preserved tritium record in the ice of Borţig Ice Cave (Bihor Mts., Romania). The Holocene 19 (5), 729–736.

Kern, Z., Thomas, S., 2014. Ice level changes from seasonal to decadal time-scales observed in lava tubes, Lava Beds National Monument, NE California, USA. Geogr. Fis. Din. Quat. 37 (2), 151–162.

Kiver, E.P., Mumma, M.D., 1975. Mount baker firn caves, Washington. Explor. J. 53, 84–87.

Kiver, E.P., Steele, W.K., 1975. Firn caves in the volcanic craters of Mount Rainer, Washington. Bull. Natl. Speleol. Soc. 37 (3), 45–55.

Kunský, J., 1939. Některé formy ledových krápníků [Some forms of ice stalactites and stalagmites]. Rozpravy II. Třídy České akademie, 49 (22), 8 p. (in Czech).

Kyrle, G., 1929. Beiträge zur Kenntins der Eiskeulen in Höhlen. Mitteilungen der Geographischen Gesellschaft in Wien 72, 13.

Lacelle, D., Lauriol, B., Clark, I.D., 2009. Formation of seasonal ice bodies and associated cryogenic carbonates in Caverne de l'Ours, Quebec, Canada: Kinetic isotope effects and pseudo-biogenic crystal structures. J. Cave Karst. Stud. 71 (1), 48–62.

Lalkovič, M., 1995. On the problems of the ice filling in the Dobšiná Ice Cave. Acta Carsol. 24, 313–322.

Lauriol, B., Carrier, L., Thibaudeau, P., 1988. Topoclimatic zones and ice dynamics in the caves of the Northern Yukon, Canada. Arctic 41 (3), 215–220.

Lauriol, B., Clark, I.D., 1993. An approach to determine the origin and age of massive ice blockages in two arctic caves. Permafr. Periglac. Process. 4 (1), 77–85.

Lauriol, B., Prévost, C., Lacelle, D., 2006. The distribution of diatom flora in ice caves of the northern Yukon Territory, Canada: relationship to air circulation and freezing. Int. J. Speleol. 35 (2), 83–92.

Lauritzen, S.E., Lundberg, J., 2000. Solutional and erosional morphology. In: Klimchouk, A.B., Ford, D.C., Palmer, A.N., Dreybrodt, W. (Eds.), Speleogenesis. Evolution of Karst Aquifers. National Speleological Society, Huntsville, AL, pp. 408–426.

Luetscher, M., 2005. Processes in ice caves and their significance for paleoenvironmental reconstructions. SISKA, La Chaux-de-Fonds, 154 p.

Luetscher, M., 2013. Glacial processes in caves. In: Shroder, J., Frumkin, A. (Eds.), Treatise on Geomorphology, vol. 6, Karst Geomorphology. Academic Press, San Diego, CA, pp. 258–266.

Luetscher, M., Jeannin, P.-Y., 2004a. A process-based classification of alpine ice caves. Theor. Appl. Karstology 17, 5–10.

Luetscher, M., Jeannin, P.-Y., 2004b. The role of winter air circxulations for the presence of subsurface ice accumulations: an example from Montlési Ice Cave. Theor. Appl. Karstology 17, 19–25.

Luetscher, M., Hoffmann, D., Frisia, S., Spötl, C., 2011. Holocene glacier history from alpine speleothems, Milchbach cave, Switzerland. Earth Planet. Sci. Lett. 302, 95–106.

Marshall, P., 1975. Ice-bloked tubes in the Aiyarsh Flow, British Columbia. Arct. Alp. Res. 7 (4), 399–400.

Marshall, P., 1981. The formation and age of ice in caves. Biul. Peryglac. 28, 79–84.

Marshall, P., Brown, M.C., 1974. Ice in Coulthard Cave, Alberta. Can. J. Earth Sci. 11 (4), 510–518.

Mavljudov, B.R., 1980. Osobennosti stroenija snežno-ledjanych nakoplenij v propasti Snežnoj na Zapadnom Kavkaze [Peculiarities of the formation of snow-ice deposits in Snežnaja Chasm in the Western Kavkaz Mountains]. Materialy gljaciologičeskich issledovanij 40, 189–193 (in Russian).

Mavljudov, B.R., 2001. Klassifikacija snežo-ledovych obrazovanij v peščerach [Classification of cave snow-ice formations]. Peščery 27–28, 97–107 (in Russian).

Mavljudov, B.R., 2006. Vnutrennie drenažnye sistemy lednikov [Internal drainage systems of glaciers]. Rossijskaja akademija nauk. Moskva, Institut geografii, 396 p. (in Russian).

Mavljudov, B.R., 2008. Oledenenie peščer [Caves glaciation]. Rossijskaja akademija nauk. Moskva, Institut geografii, 290 p. (in Russian).

Mavljudov, B.R., Morozov, A.I., 1984. Propasť Snežnaja [The Snežnaja Chasm]. Peščery 19, 15–25 (in Russian).

Mavlyudov, B.R., 1989. Snow and ice formation in caves and their regime. In: Kósa, A. (Ed.), Proceeding of the 10th International Congress of Speleology (Budapest, August 13–20, 1989), vol. 1. Magyar Karszt- és Barlangkutató Társulat, Budapest. pp. 295–297.

Mavlyudov, B.R., 1991. The influence of air flows on glacier caves forming. In: Eraso, A. (Ed.), Proceedings of the 1st International Symposium of Glacier Caves and Karst in Polar Regions, Madrid, October 1–5, 1990. pp. 199–206.

Mavlyudov, B.R., 2005. About new type of subglacial channels, Spitsbergen. In: Mavlyudov, B.R. (Ed.), Glacier Caves and Glacial Karst in High Mountains and Polar Regions (Proceedings of the 7th GLACKIPR Symposium, Azau – Kabardino-Balkarija (Caucasus), Moscow, Russia, September 6–12, 2005), pp. 54–60.

Meneghel, M., Badino, G., 2002. Ice Caves of Terra Nova Bay (Victoria Land, Antarctica). Nimbus 23–24 (7), 130–133.

Mihevc, A., 2008. Distribution and characteristics of ice caves in Slovenia. In: Kadebskaya, O., Mavludov, B., Pyatunin, M. (Eds.), Proceedings of the 3rd International Workshop on Ice Caves, Kungur, Russia, May 12–17, 2008. Perm State University, Perm. pp. 61–64.

Millius, J., Petters, C., 2012. Eisriesenwelt – from laser scanning to photo-realistic 3D model of the biggest ice cave on Earth. In: Jekel, T., Car, A., Strobl, J., Griesebner, G. (Eds.), GI_Forum 2012: Geovizualisation, Society and Learning, Wichmann Verlag, Heidelberg, Salzburg. pp. 513–523.

Mlejnek, R., 2009. Propast Solunska glava 5 v pohoří Jakupica (Makedonie) [The Solunska glava 5 Abyss in the Jakupica Mountains (Macedonia)]. Speleofórum 28, 88–89 (in Czech).

Mološmanova, N.E., Maksimovič, N.G., Dubljanskij, V.N., Mavljudov, B.R., Nazarova, U.V., 2005. Otloženija. In: Dubljanskij, V.N. (Ed.), Kungurskaja ledjanaja peščera: opyt režimnych nabljudenij [Kungur Ice Cave: An Experience of Regime Observations]. UrO RAN, Ekaterinburg. pp. 179–237.

Obleitner, F., Spötl, C., 2011. The mass and energy balance of ice within the Eisriesenwelt Cave, Austria. Cryosphere 5 (1), 245–257.

Ohata, T., Firukawa, T., Higuchi, K., 1994. Glacioclimatological study of perennial ice in the Fuji Ice Cave, Japan. Part 2. Interannual Variation and Relation to Climate. Arct. Alp. Res. 26 (3), 238–244.

Onac, B.P., Perşoiu, A., Racoviţă, G., Tămaş, T., Viehmann, I., 2007. Scărişoara Ice Cave and Its Surroundings. Studia, Cluj Napoca, 88 p.

Pelech, J.E., 1879. Valley of Stracena and the Dobschau ice-cavern. Trubner & Co., Ludgate Hill, London, 31 p.

Perşoiu, A., 2004. Ice speleothemes in Scărişoara Cave: dynamics and controllers. Theor. Appl. Karstology 17, 71–76.

Perşoiu, A., 2005. Evidence of basal melting of the ice block from the Scărişoara Ice Cave. In: Mavlyudov, B.R. (Ed.), Glacier Caves and Glacial Karst in High Mountains and Polar Regions (Proceedings of the 7th GLACKIPR Symposium, Azau – Kabardino-Balkarija (Caucasus), Moscow, Russia, September 6–12, 2005). pp. 109–112.

Perşoiu, A., Feier, I., Citterio, M., Turri, S., Maggi, V., 2007. Preliminary data on air temperature in Focul Viu Ice Cave (Bihor Mts., Romania). In: Zelinka, J. (Ed.), Proceedings of the 2nd International Workshop on Ice Caves, Demänovská Dolina, Slovak Republic, May 8–12, 2006. SSJ, Liptovský Mikuláš, pp. 62–64.

Perşoiu, A., Onac, B.P., 2012. Ice in caves. In: White, W., Culver, D.C. (Eds.), Encyclopedia of Caves. Elsevier Academic Press, pp. 399–404.

Petters, C., Milius, J., Buchroithner, M.F., 2011. Eisriesenwelt: Terrestrial laser scanning and 3D visualisation of the largest Ice Cave on Earth. Proceedings European LiDAR Mapping Forum, Salzburg, Austria, 10 p.

Pflitsch, A., Schörghofer, N., Smith, S.M., Holmgren, D., 2016. Massive ice loss from the Mauna Loa Icecave, Hawaii. Arct. Antarct. Alp. Res. 48 (1), 33–43.

Piasecki, J., Sawiński, T., Zelinka, J., 2005. Spatial differentiation of the air temperature in the entrance collapse of Dobšinská Ice Cave as contribution to the recognition of the problem of air exchange between cave and the surface. Slovenský kras 43, 81–96.

Popov, V., 1971. Ledjanye obrazovanija peščery «Ledenika» (Vračanska planina) [Ice formations of Ledenika Cave (Vračanska Plateau)]. Slovenský kras 9, 151–158 (in Russian).

Pulina, M., 1971. Typy ľadu v tatranských jaskyniach [Types of ice in the Tatra caves]. Slovenský kras 9, 57–70.

Pulina, M., 1982. Karst-related phenomena at the Bertil Glacier, West Spitsbergen. Kras i speleologia 4 (XIII), 67–82.

Pulina, M., Řehák, J., 1991. Glacial caves in Spitsbergen. In: Eraso, A. (Ed.), Proceedings of the 1st International Symposium of Glacier Caves and Karst in Polar Regions, Madrid, October 1–5, 1990. pp. 93–117.

Pulina, M., Řehák, J., Schroeder, J., 2003. Les cavités glaciaires sous le regard spéléologique. Karstologia 42 (2), 23–36.

Racoviţă, E., 1927. Observations sur la glacière naturelle dite "Gheţarul de la Scărişoara". Bull. Soc. Hist. Natur. Sci. Cluj 3, 75–108.

Racoviţă, G., Onac, B.P., 2000. Scărişoara Glacier Cave. Editura Carpatica, Cluj-Napoca, 139 p.

Racoviţă, G., Viehmann, I., 1966. Observaţii asupra dinamicii stalagmitelor de gheaţă din Gheţarul de la Scărişoara – sala Biserica [Observations sur la dynamique des stalagmites de glace de la grotte Gheţarul de la Scărişoara – salle de d'Eglise]. Lucratile Institutului de Spéologie "Emile Racovita" 5, 43–65 (in Romanian).

Rajman, L., Roda, Š., Ščuka, J., 1985. Výskum dynamiky ľadu v Silickej ľadnici [Research of ice dynamics in the Silická ľadnica Cave]. Slovenský kras 23, 253–260 (in Slovak).

Řehák, J., Řehák, J.j., Řehák, S., 2004. Špicberky 2003 [Spitcbergen 2003]. Speleofórum 23, 57–59 (in Czech).

Roth, Z., 1940. Několik geomorfologických poznámek o Jihoslovenském Krasu a o Silické Lednici [Some geomorphological notes on the South-Slovak Karst and Silická ľadnica Cave]. Rozpravy II. tř. ČA 49 (8), 1–24 (in Czech).

Sabroux, J.C., Richon, P., Faivre-Pierret, R.X., 1999. Radon monitoring in a geothermal ice cave of Mt. Erebus, Antartica. In: Barone, N., Bonaccorso, R., Licitra, G. (Eds.), Proceedings of the IXth International Symposium on Vulcanospeleology, Catania, Italy. pp. 224–227.

Schöner, W., Weyss, G., Mursch-Radlgruber, E., 2011. Linkage of cave-ice changes to weather patterns inside and outside the cave Eisriesenwelt (Tennengebirge, Austria). Cryosphere 5 (3), 603–616.

Schoumskiy, P.A., 1957. Principes de glaciologie structurale. La pétrographique de la glace comme méthode d´étude glaciologique. Annales du Centre d´Études et de Documentation Paléontologique 22, 309 p.

Schroeder, J., 1977. Les formes de glaces des grottes de la Nahanni, Territoires du Nord-Ouest, Canada. Can. J. Earth Sci. 14 (5), 1179–1185.

Schroeder, J., 1991. Les cavités du Hansbreen creusées par les eaux de fonte. Svalbard, 77° Lat. N. In: Eraso, A. (Ed.), Proceedings of the 1st International Symposium of Glacier Caves and Karst in Polar Regions, Madrid, October 1–5, 1990. pp. 21–33.

Şerban, M., 1970. Morphologie comparée des stalagmites de glace de la grotte Scărişoara (Roumanie). Travaux de L´Institut de Spéologie "Emile Racovitza" 9, 35–60.

Şerban, M., Coman, D., Givulesco, R., 1948. Decouvertes récentes et observations sur la glacière naturelle dite Ghearul de la Scărişoara. Bull. Soc. Hist. Natur. Sci. Cluj X, 174–210.

Shavrina, E.V., Guk, E.V., 2005. Modern dynamics of ice formations in Pinega Caves. In: Mavlyudov, B.R. (Ed.), Glacier Caves and Clacial Kast in High Mountains and Polar Regions (Proceedings of the 7th GLACKIPR Symposium, Azau – Kabardino-Balkarija (Caucasus), Moscow, Russia, September 6–12, 2005). pp. 113–117.

Smart, C., 2004. Glacier caves and glacier pseudokarst. In: Gunn, J. (Ed.), Encyclopedia of Caves and Karst Sciences. Fitzroy Dearbon, New York, London, pp. 385–388.

Šmoll, J., Sluka, M., 2007. Slovenská jama (Slovačka jama, Ceki 1) na Karadžici [Slovenská jama Abyss (Slovačka jama, Ceki 1) on the Karadjica]. Spravodaj Slovenskej speleologickej spoločnosti 38 (4), 50–54 (in Slovak).

Šmoll, J., Szunyog, J., 2005. Priepasť Solúnska jama [Solunska jama Abyss]. Spravodaj Slovenskej speleologickej spoločnosti 36 (1), 51–56 (in Slovak).

Spötl, C., Reimer, P.L., Luetscher, M., 2014. Long-term mass balance of perennial firn and ice in an Alpine cave (Austria): constraints from radiocarbon-dated wood fragments. The Holocene 24 (2), 165–175.

Stenborg, T., 1968. Glacier drainage connected with ice structures. Geogr. Ann. 50A (1), 25–53.

Stenborg, T., 1969. Studies of the internal drainage of glaciers. Geogr. Ann. 51A (1–2), 13–41.

Strug, K., Piasecki, J., Szymanowski, M., Sawiński, T., Zelinka, J., 2006. Quantitative character of the bottom ice in the Demänovská Ice Cave (Slovakia). In: Bella, P. (Ed.), Výskum, využívanie a ochrana jaskýň, Proceedings of the 5th Scientific Conference, Demänovská Dolina, September 26–29, 2005. SSJ, Liptovský Mikuláš. pp. 167–174.

Strug, K., Zelinka, J., 2008. The Demänovská Ice Cave – the volume balance of the ice monolith in 2003 – 2007 (Slovakia). Slovenský kras 46 (2), 369–386.

Šumskij, P.A., 1955. Osnovy strukturnogo ledovedenija [Principles of Structural Glaciology]. Izdavateľstvo AN SSSR, Moskva, 492 p. (in Russian).

Temovski, M., 2016. Ice caves in Macedonia. Abstracts of the 7th International Workshop on Ice Caves, Postojna, Slovenia, May 16–22, 2016.

Trofimova, E.V., 2007. Genesis and morphological peculiarities of cave ice deposits of Lake Baikal. In: Zelinka, J. (Ed.), Proceedings of the 2nd International Workshop on Ice Caves, Demänovská Dolina, Slovak Republic, May 8–12, 2006. SSJ, Liptovský Mikuláš, pp. 77–81.

Tulis, J., 1997. Pohyb ľadu v Dobšinskej ľadovej jaskyni [Ice movement in the Dobšiná Ice Cave]. Aragonit 2, 6–7 (in Slovak).

Tulis, J., Novotný, L., 1995. Čiastková správa o morfometrických parametroch v zaľadnených častých Dobšinskej ľadovej jaskyne [Partial report on morphomertric parameters in the glaciated parts of Dobšiná Ice Cave]. In: Bella, P. (Ed.), Ochrana ľadových jaskýň, Procedings, Dobšinská Ľadová Jaskyňa, September 21–22, 1995. SSJ, Liptovský Mikuláš, pp. 25–28 (in Slovak).

Van Autenboer, T., 1962. Ice mounds and melt phenomena in the Sør-Rondane, Antarctica. J. Glaciol. 4 (33), 349–354.

Vatne, G., Irvine-Fynn, T.D.L., 2016. Morphological dynamics of an englacial channel. Hydrol. Earth Syst. Sci. 20 (7), 2947–2964.

Viehmann, I., Racoviță, G., 1967. Les stalagmites de glace thermoindicatrices. Actes du Colloque International Spéléologie CISNA´67, Hann-sur-Lesse, Belgique, pp. 92–96.

Viehmann, I., Racoviță, G., 1968. Stalagmitele de gheață termoindicatore [Les stalagmites de glace thermoindicatrices]. Dări de seamă ale Comitetului Geologic 44 (1), 353–363. (in Romanian).

Yonge, C.J., 2004. Ice in caves. In: Gunn, J. (Ed.), Encyclopedia of Cave and Karst Science. Fitzroy Dearborn, New York, London, pp. 435–437.

Yonge, C.J., 2014. The fate of ice caves in North-Western Canada as climate warms: what will we lose? Report, RCGS – Ice Cave, 21 p.

Yonge, C.J., MacDonald, W.D., 1999. The potential of perennial cave ice in isotope palaeoclimatology. Boreas 28 (3), 357–362.

Zelinka, J., 1996. Rekonštrucia vstupných častí Dobšinskej ľadovej jaskyne [Reconstruction of the entrance parts of the Dobšiná Ice Cave]. Aragonit 1, 15–16 (in Slovak).

FURTHER READING

Anderson, C.H., Halliday, W.R., 1969. The Paradise Ice Caves, Washington: an extensive glacier cave system. Bull. Natl. Speleol. Soc. 31 (3), 55–72.

CHAPTER

ICE DYNAMICS IN CAVES

4.3

Aurel Perşoiu

Emil Racoviță Institute of Speleology, Cluj-Napoca, Romania

CHAPTER OUTLINE

4.3.1 INTRODUCTION

Ice accumulations are highly dynamic occurrences in caves, changing their morphology in response to the cave morphology and climatic conditions on scales ranging from days to millennia. Ice in caves occurs mainly as a result of water freezing and to a lesser extent of snow densification and diagenesis. The latter processes occur in cave entrances, where snowpacks up to 100m in thickness occur, being subsequently compacted to form crystalline ice. Transformation of snow into ice does not occur as a "pure" compaction; freezing of water that percolates through the snow mass also contributes to the genesis of perennial ice masses. The results of these processes are large bodies of perennial ice that will subsequently deform and flow gravitationally towards the lower parts of the caves that host them. Contrary to these, ice formed by the freezing of water takes a large variety of shapes and sizes, from millimeter-scale, ephemeral ice crusts on the walls of caves, to large ice bodies, tens of thousands of cubic meters in size. Climatic factors strongly determine their dynamics on time scales ranging from hours to millennia, while the caves' morphology acts on scales on the order of centuries to millennia. A third, less-obvious process that leads to ice formation and subsequent dynamics, is that of ice sublimation, that is, the formation of ice crystals on the undercooled walls of caves, as warm and moist air is pushed out from the inner sections of these caverns.

As a result of the above summarized processes, two classes of ice forms develop in caves (Fig. 4.3.1): perennial ice bodies and seasonal to multiannual ice speleothems (stalagmites, stalactites, ice crusts, and crystals).

While most authors agree that the later are true speleothems (secondary mineral deposits formed in caves, *sensu* Hill and Forti, 1997) as they are minerals (crystalline H_2O), they are of secondary genesis (formed by the freezing of water) and they have formed in caves; no consensus exists on whether the large, perennial, ice bodies accumulated in caves are speleothems, glaciers, both, or neither.

Ice Caves. https://doi.org/10.1016/B978-0-12-811739-2.00034-6

FIG. 4.3.1

Various ice bodies in caves—large ice block, semiperennial stalagmites, ice domes and seasonal stalactites (Scărișoara Ice Cave, Romania).

Here I argue that the large, perennial ice bodies found in caves and that have formed through water freezing are both speleothems and glaciers, while the similar ones that have formed by snow diagenesis are glaciers only, not speleothems, regardless of the fact that they are in caves. Snow diagenesis, regardless where it occurs, is a pure glacial process: snow is being slowly compacted and as air is squeezed out, ice develops. Infiltration and freezing of water within the ice body speeds up the compaction of the snow pack, and the end result is a layered mass of ice, with variable density (higher for the congelation ice), that also incorporates organic and inorganic sediments. While the end result is a large ice body *inside a cave*, the processes that led to its genesis are restricted to the caves' entrances, or, at most, to those areas that are directly fed by snow (e.g., the bottom of long, dipping slopes below cave entrances), thus being external to the cave environment. Freezing of liquid water to form perennial, large ice bodies can only occur (with the notable exception of permafrost) in single entrance, descending caves that act as cold air traps, maintaining negative temperatures throughout the year. In these caves, ice can form either as thin sheets of water freeze on sloping surfaces, or as shallow lakes freeze in autumn and winter. Most commonly, both these processes occur in caves (Fig. 4.3.2). In summer, lakes up to 30 cm deep can accumulate on top of existing ice bodies and will start to freeze, top to bottom, during a cooling period in late autumn through early winter to form a layer of "lake ice," 1–30 cm thick. Warmer intervals during winter could lead to snow melting and water infiltration in caves to form thin sheets of "floor ice" (Perșoiu and Pazdur, 2011) on top of the existing lake ice. The end result, on a multiannual scale, is a layered body of secondary genesis crystalline H_2O, formed by a cave-specific processes—a speleothem.

Once formed, the large perennial ice bodies to be found in caves display a dual dynamic: they both accumulate and lose ice on top, as a result of both short and long term climatic variability and also, like any large surface glacier, flow under the influence of gravity. This later behavior occurs for both types (in terms of genesis) of large cave ice masses, resulting in flow-specific structures—ice thinning along the flow path, folds, push-moraines in front of the advancing (and melting) ice tongues, etc. (Fig. 4.3.3).

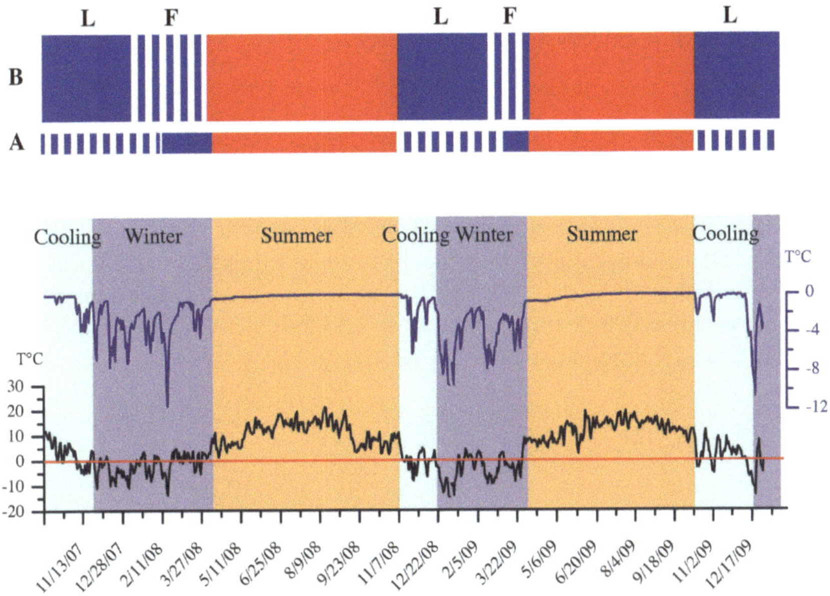

FIG. 4.3.2

Genesis of ice in caves through the freezing of water: (A) Formation of seasonal forms (stalagmites, stalactites, columns) occur intermittently throughout the winter, intensifying in early spring, when temperatures inside the caves are still negative and large amounts of water (released by melting snow) are available. With increasing outside temperature, warm water will start melting the ice. (B) Perennial ice bodies form by the freezing of lakes (accumulated in summer) in the early periods of cave cooling (lake ice-L) and, additionally, freezing of infiltrating water (floor ice-F) on top of existing lake ice, during winter.

FIG. 4.3.3

The underground glacier of Scărișoara Ice Cave. Layers deformed by flow are visible on both sides of the vertical wall, and an ice tongue extends on the lower-left side of the image. A push-moraine is visible in front of the advancing glacier.

Thus, while in terms of their genesis, only large, perennial ice blocks formed by the freezing of water can be termed speleothems, contrary to the ice accumulations formed by snow diagenesis, both these two types of cave ice bodies are glaciers, in terms of postgenesis behavior.

The dynamics of all ice formations in caves are the result of external climatic influences, cave morphology, or the combination of the two. Whereas the former acts on time scales ranging from days to millennia, the latter is only active on centennial scales—and the dynamics of cave ice will be analyzed further along these temporal differentiations (Fig. 4.3.4).

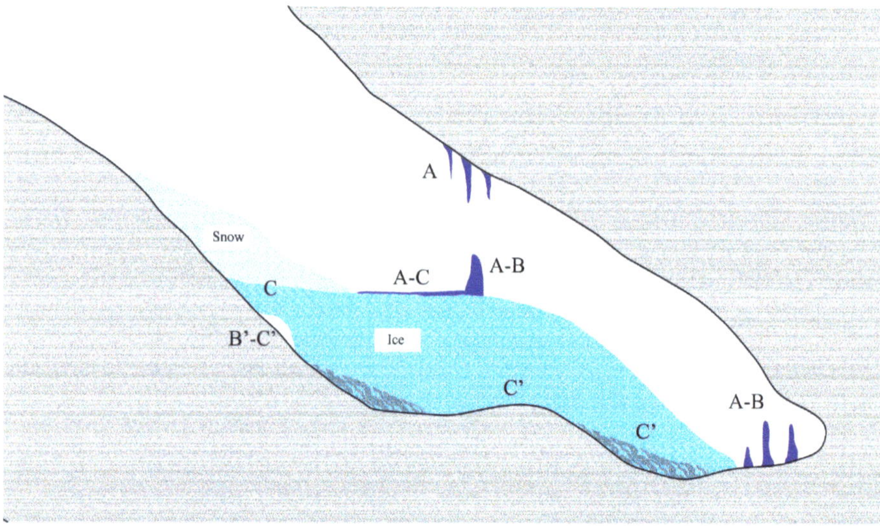

FIG. 4.3.4

Conceptual model of an ice cave showing the temporal range of dynamics for various ice formations: (A) subannual dynamics (accumulation and degradation); (B) multiannual dynamics (accumulation and ablation); (C) decadal to millennial dynamics (accumulation and ablation); (B′) multiannual dynamics—ablation only; (C′) decadal to millennial dynamics—flow and ablation only.

For clarity in the subsequent text, I will use the term "speleothem" for stalagmites, stalactites, and columns only, and "glacier" for perennial ice blocks.

4.3.2 SUBANNUAL DYNAMICS

Accumulation and ablation of ice in caves on daily to monthly time scales affects all ice speleothems—stalagmites, stalactites, and columns, as well as the seasonally occurring hoar frost and the large, perennial glaciers.

4.3.2.1 HOAR FROST

Hoar frost is the most dynamic form of ice in caves. Two processes are acting in conjunction to allow for the development of hoar frost: substratum (rock wall) undercooling and circulation carrying moisture towards the cold walls. The process starts in early winter, when external cold air, triggered by its higher density, sinks into the caves, leading to the slow undercooling of the walls. As this cold air reaches the caves, it pushes out the warmer and moister cave air, which will condensate on the wall as liquid water. Continuous inflow of cold air, combined with water evaporation from the rock will cool the walls to temperatures below 0°C, thus setting the scene for the initiation of hoar frost development (Pflitsch et al., 2007). Continuous inflow of cold air leads to the development of a bidirectional air circulation, with colder air swiping on the surface of ice and pushing out the warm and moist air along the ceilings, resulting in the deposition and growth of large ice crystals (Fig. 4.3.5). The moisture feeding the growth of crystals has a dual origin: the in situ moisture existing in the warmer sector of the caves, and moisture resulting from the sublimation of cave ice under the influence of cold and dry inflow of cave air, flowing on top of the ice. The hoar frost deposits develop and survive until early to mid-summer, when inflow of cold air ceases and both air and rock start to warm, leading to rapid degradation of the crystals. Warming of the walls results in crystals falling to the ground, where they either melt away, as in most cases, or, where temperatures are still below 0°C, they can accumulate on top of the existing ice to contribute to the growing of the ice block (Marshall and Brown, 1974). As a result of these processes, the dynamics of hoar frost in ice caves follows a simple, annual cycle, being present between mid- and late winter (when they reach a maximum) and absent throughout the rest of the year. The size of the crystals is dependent on the intensity of wall undercooling, the strength of air circulation, and amount of moisture available, reaching a maximum of 50–60 cm, and usual values of 5–15 cm (Fig. 4.3.5).

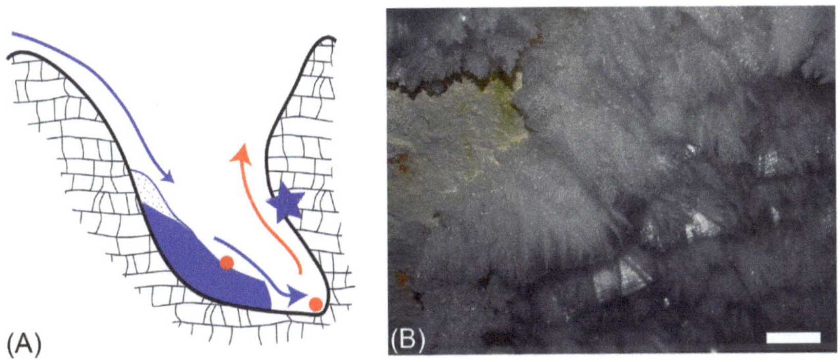

FIG. 4.3.5

Mechanism of hoar frost deposition in ice caves (A) and close-up view of the resulting ice crystals (B) in Scărişoara Ice Cave. The *blue* star marks the location of hoar frost deposits, and the two *red* dots indicate the sources of moisture. Length of scale bar is 3 cm.

4.3.2.2 ICE SPELEOTHEMS (STALAGMITES, STALACTITES, AND COLUMNS)

Ice stalagmites and stalactites are common occurrences in cave entrances during the cold season in mid-to-high-latitude and/or altitude caves. They form as drip water freezes and usually melt as the temperature of the water feeding them becomes positive. In caves with perennial ice, these stalagmites can be perennial, the negative temperatures preserved throughout the year in the vicinity of underground glaciers helping their survival. However, differences exist between stalagmites, on one hand, and stalactites and columns, on the other, as the latter melt earlier and usually completely, the warm water leading to their detachment from the ceiling and collapse.

The dynamics of these ice formations have been studied in great detail in Scărişoara Ice Cave, Romania (SIC), by Viehmann and Racoviţă (1968), Racoviţă et al. (1987), and Racoviţă (1994), and it is summarized below, as a "template" for other caves, as well.

In Scărișoara Ice Cave, the dynamics of ice speleothems were studied separately for stalagmites, ice massifs, and ice crusts on the cave floor (Fig. 4.3.6), on both annual and subannual time scales.

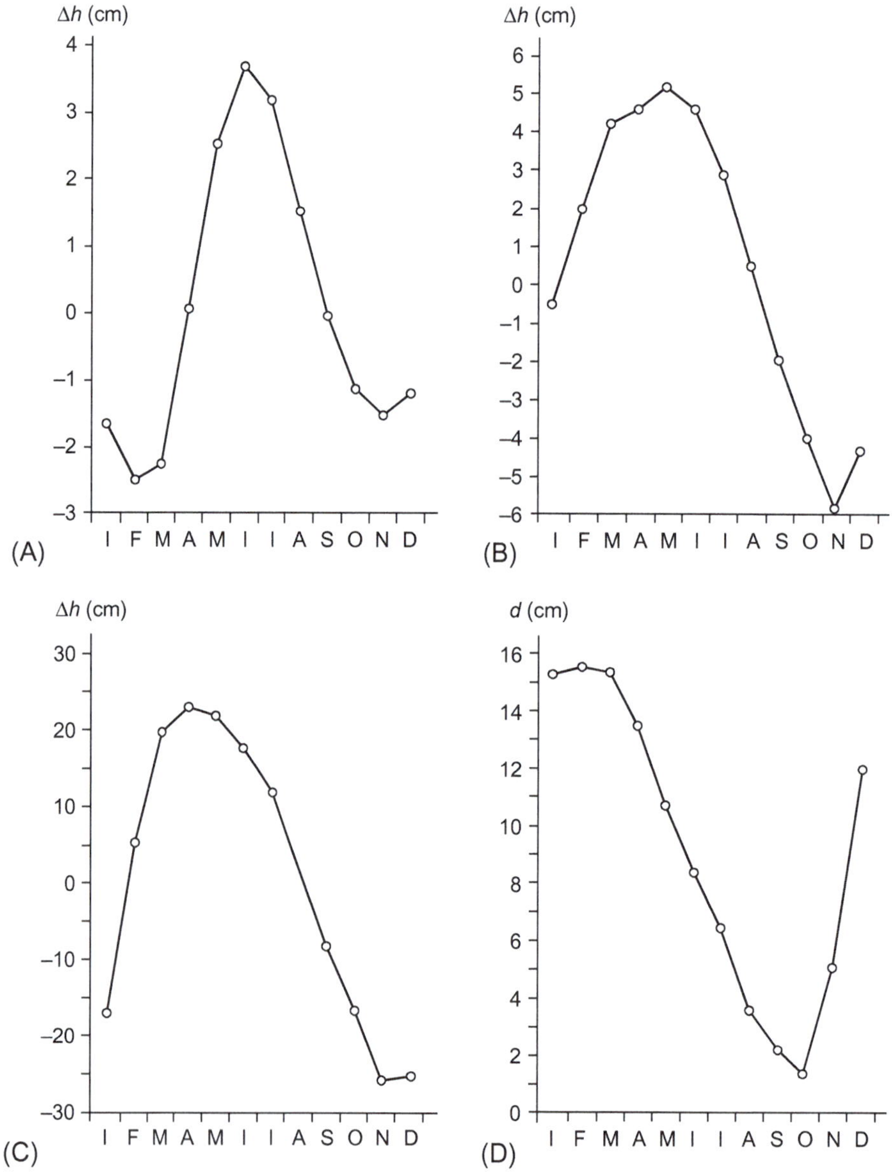

FIG. 4.3.6

Seasonal variations (10 years average) of the upper face of the ice block in SIC (A), ice massifs (B), ice stalagmites (C), and floor ice crust (D). The ice massifs are similar to the one visible on the left in Fig. 4.3.1.

Modified from Racoviță, G., 1994. Eléments fondamentaux dans la dynamique des spéléothèmes de glace de la grotte de Scărișoara, en relation avec la météorologie externe. Theor. Appl. Karstol. 7, 133–148.

For all these speleothems, a maximum was reached between March and June, with the larger ice bodies having a delayed onset of melting. Ice crusts of the floor (Fig. 4.3.6D) grow outwards from the edges of the ice block, thinning with increased distance, their dynamics being controlled by the inflow of cold air which sweeps the upper surface of the ice block and leads to water freezing. However, this genetic mechanism renders them extremely vulnerable to melting, a process which initiates once the inflow of cold air ceases in early spring (Figs. 4.3.2 and 4.3.6D). The melting of the larger ice forms, both stalagmites and ice massifs (Fig. 4.3.6B and C), is delayed by the larger thermal inertia of the ice (Perşoiu et al., 2011). However, regardless of their size and shape, all ice forms in caves (including the large ice blocks) reach a minimum at the end of the melting season, just before the onset of freezing, usually in November (Figs. 4.3.2 and 4.3.6). Both growing and melting of ice is a complex process, controlled by the variable interplay between air temperature and precipitation amount and distribution, thus there is no clear correlation with either of these two (Perşoiu et al., 2011). The effect of precipitation amounts reaching ice caves on ice dynamics is strongly dependent of air temperature during winter, with water input superimposed on below 0°C conditions leading to rapid ice build-up, while the same input occurring during periods with positive temperature anomalies resulting in ice loss. However, in summer, temperature doesn't play too important a role, as the latent heat of the ice prevents temperature inside the caves reaching above 0°C. Thus, water input is the main factor leading to ice ablation, the heat delivered to the ice by inflowing warm water being the main factor in the ablation of ice.

A peculiar type of subannual ice dynamics of ice is that of the "thermoindicator" speleothems (Viehmann and Racoviță, 1968, Fig. 4.3.7). These form in winter, during periods of alternating cold and mild weather, with the translucent, bulky sections developing when temperatures are between 0°C and −2°C, and drip water will freeze slowly, expelling gas and calcite impurities. When temperatures drop below −3°C, dripping water will tend to freeze quickly, incorporating both air bubbles and in situ precipitated cryogenic cave calcite (Žák et al., 2008), hence the semiopaque appearance of the ice.

4.3.3 MULTIANNUAL TO CENTENNIAL ICE DYNAMICS

Long-term dynamics of ice has been observed in a few caves, and monitoring programs exist in even fewer ones (Kern and Perşoiu, 2013). In Scărişoara Ice Cave, a monitoring program was in place for ice speleothems as well (summarized in Racoviță, 1994) between 1965 and 1992, the results indicating an inverse relationship between the lowering of the upper face of the ice black (mass loss) and increase in the height of stalagmites. Racoviță et al (1987) suggested that this contrasting behavior is the result of long-term periodicity in the dynamics of ice that could be possibly linked to climatic conditions outside the cave. However, owing to the short time series available (53 years between 1963 and 2017) and long-term suggested periodicity (50+ years) no such link has been found yet (Perşoiu and Pazdur, 2011), although photographic evidence from 1947 (Şerban et al., 1948) suggest that ice speleothems at that time were taller than in the 1960s.

The dynamics of the glaciers in caves has been investigated sporadically in the past century, by combining various sources of data: photographic evidence, markers in the ice and/or cave walls, absolute dating, regular monitoring programs (Kern and Perşoiu, 2013 and references therein). However, with a few exceptions, monitoring programs longer than a few years are very rare (Ohata et al., 1994; Rachlewicz and Szczucinski, 2004; Fuhrmann, 2007; Perşoiu and Pazdur, 2011; Kern and Thomas, 2014). With little exception, most monitoring programs indicate a continuous and accelerated melting of cave ice, especially in the past 20–30 years (Fig. 4.3.8).

FIG. 4.3.7

Thermoindicator speleothems in Scărișoara Ice Cave.

The causes of the decadal to centennial mass balance changes in caves can be grouped in three classes: (1) morphologic, (2) climatic and (3) endogenic.

The morphologic ones are the consequences of changes in the dynamics of airflows and associated heat transfer in caves related to *changes in the morphology of the open spaces* as ice grows and shrinks (while the morphology of the rock walls in ice caves could changes as a result of collapses, no such occurrences have been reported so far in the literature). These changes in turn affect the climate of the caves and the behavior of ice, resulting in long-term modifications of the ice morphology and dynamics. In Scărișoara Ice Cave (Romania) such a case was described by Șerban et al. (1948), with the ice block switching between several phases of growth and decay (Fig. 4.3.9).

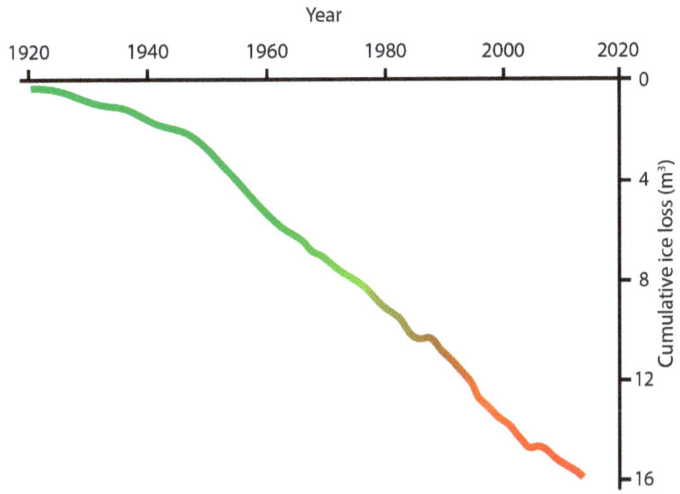

FIG. 4.3.8

Cumulative ice loss in ice caves in the Northern Hemisphere during the last 100years.

From Kern, Z., Perşoiu, A., 2013. Cave ice—the imminent loss of untapped mid-latitude cryospheric paleoenvironmental archives. Quat. Sci. Rev. 67, 1–7.

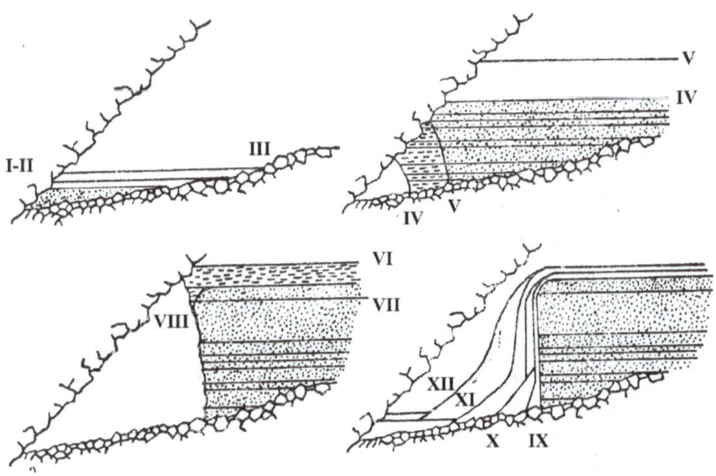

FIG. 4.3.9

Growth and retreat of the ice block in Scărişoara Ice Cave (Romania). Original sketches by Mihai Şerban.

In phases I–III, ice grew through water freezing in layers of "lake ice," however the growth of ice progressively isolated the lowermost sections of the incipient ice block, and the constant geothermal heat flux (70 mW/m^2, Demetrescu and Andreescu, 1994) led to the melting of the ice and progressive growth towards the surface of the resulting opening (phases IV–VIII in Fig. 4.3.8). Once this opening reached the surface, inflowing water started to freeze on the vertical, exposed ice wall, progressively closing the opening (phases IX–XII) reinitiating a new cycle. Presently, the ice block is a stage similar to phase IX (Fig. 4.3.3), with newly formed ice growing on top of the exposed, older, ice.

A secondary dynamic process related to melting near the base is that of ice flow, leading to progressive growth of folds in the ice (Fig. 4.3.10).

FIG. 4.3.10

Ice flow and folding in Svarthammarhola, Norway.

The climatic conditions affecting the dynamics of cave glaciers have are acting in a more predictable fashion, with simple, annual cycles (Fig. 4.3.6A) superimposed on long-term trends, generally pointing towards continuous melt (Fig. 4.3.8). While increasing air temperature (and possibly diminishing precipitation during the growing season) are likely the main causes leading to general ice loss, observations of a peculiar ice loss case in the Romanian Carpathians suggest that cave glaciers might respond nonlinearly to climatic conditions: in “Adevăratul Gheţar de la Vârtop” Ice Cave, a 6 m thick, ~2000 m^3 ice block vanished completely between 2008 and 2012, its demise being possibly triggered by summer inflow of warm water that lead to catastrophic disintegration once the glacier was dissected in smaller ice blocks.

Only a few ice caves have ages going back more than a thousand years, so that the dynamics over centennial to millennial scales is difficult to assess. Studies in Swiss (Luetscher, 2005), Austrian (Spötl et al., 2014) and Romanian (Perşoiu et al., 2017) caves have shown that ice in these caves responded sensitively to the main climatic swings during the past 2000 years, with ice accumulating in caves during cold and/or wet periods (the Dark Ages Cold Period of the 4–7th centuries AD and the Little Ice Age, between the 14th and 19th centuries AD), and melting during the medieval Warm period (between the 9th and 13th centuries AD). Stoffel et al. (2009) and Perşoiu et al (2017) have further shown that long-term changes in ice mass balance and dynamics could be controlled by large-scale circulation patterns affecting Europe and these could also be recorded by the isotopic composition of cave, thus being a valuable tool in reconstructing past climatic conditions.

Geothermal heat delivered to the base of cave glaciers (Fig. 4.3.4C) has been shown (Tulis and Novotný, 2003; Luetscher, 2005; Perşoiu, 2005) to lead to variable rates of melt: 8 cm/year in Monlesi Ice Cave (Switzerland), 1.53 cm/year in Scărişoara ice Cave (Romania) and 1 cm/year in Dobsinska Ice Cave (Slovakia). However, these basal melting rates were calculated based on limited observation, and the resulting mass turnover rates are not in agreement with the age of the cave glaciers (Kern, this volume), thus suggesting that either the estimates are wrong (by a factor of 10 in Scărişoara Ice Cave), or that basal melting was not happening at a constant rate during the existence of the ice blocks.

REFERENCES

Demetrescu, C., Andreescu, M., 1994. On the thermal regime of some tectonic units in a continental collision environment in Romania. Tectonophysics 230 (3–4), 265–276.

Fuhrmann, K., 2007. Monitoring the disappearance of a perennial ice deposit in Merrill Cave. J. Caves Karst Stud. 69, 256–265.

Hill, C.A., Forti, P., 1997. Cave Minerals of the World, second ed., National Speleological Society Inc, Huntsville, Alabama, pp. 217–225.

Kern, Z., Perşoiu, A., 2013. Cave ice—the imminent loss of untapped mid-latitude cryospheric paleoenvironmental archives. Quat. Sci. Rev. 67, 1–7.

Kern, Z., Thomas, S., 2014. Ice level changes from seasonal to decadal time-scales observed in lava tubes, lava beds national monument, NE California, USA. Geogr. Fis. Din. Quat. 37, 151–162.

Luetscher, M., 2005. Processes in Ice Caves and their Significance for Paleoenvironmental Reconstructions. Swiss Institute for Speleology and Karst Studies, La Chaux-de-Fonds, Switzerland, p. 154.

Marshall, P., Brown, M.C., 1974. Ice in Coultard Cave, Alberta. Can. J. Earth Sci. 11, 510–518.

Ohata, T., Furukawa, T., Osada, K., 1994. Glacioclimatological study of perennial ice in the Fuji ice cave, Japan. Part 2: interannual variation and relation to climate. Arct. Alp. Res. 26, 238–244.

Perşoiu, A., 2005. Evidence of basal melting of the ice block from Scărişoara Ice Cave. In: Mavlyudov, B.R. (Ed.), Glacier Caves and Glacial Karst in High Mountains and Polar Regions, Institute of Geography of the Russian Academy of Sciences, Moscow, pp. 109–112.

Perşoiu, A., Pazdur, A., 2011. Ice genesis and its long-term mass balance and dynamics in Scărişoara Ice Cave, Romania. Cryosphere 5 (1), 45–53.

Perşoiu, A., Onac, B.P., Perşoiu, I., 2011. The interplay between air temperature and ice dynamics in Scărişoara Ice Cave, Romania. Acta Cardiol. 40 (3), 445–456.

Perşoiu, A., Onac, B.P., Wynn, J.G., Blaauw, M., Ioniţă, M., Hansson, M., 2017. Holocene winter climate variability in Central and Eastern Europe. Sci. Rep. 7, 1196.

Pflitsch, A., Piasecki, J., Sawiński, T., Strug, K., Zelinka, J., 2007. Development and degradation of ice crystals sediment in Dobšinská Ice Cave (Slovakia). In: Zelinka, J. (Ed.), Proceedings of the 2nd International Workshop on Ice Caves, Demänovská dolina, Slovak Republic, pp. 38–49.

Rachlewicz, G., Szczucinski, W., 2004. Sesonal, annual and decadal ice mass balance changes in the ice cave Jaskinia Lodowa w Ciemniaku, the Tatra Mountains. Poland. Theor. Appl. Karstol. 17, 11–18.

Racoviţă, G., 1994. Eléments fondamentaux dans la dynamique des spéléothèmes de glace de la grotte de Scărişoara, en relation avec la météorologie externe. Theor. Appl. Karstol. 7, 133–148.

Racoviţă, G., Şerban, M., Viehmann, I., 1987. Tendances de long terme dans la dynamique des formations de glace de la Grotte de Scărişoara (Monts du Bihor). Theor. Appl. Karstol. 3, 143–164.

Şerban, M., Coman, D., Givulescu, R., 1948. Découvertes récentes et observations sur la glacière naturelle dite, Gheţarul de la Scărişoara. Bull. Soc. Sci. Cluj X, 174–210.

Spötl, C., Reimer, P.J., Luetscher, M., 2014. Long-term mass balance of perennial firn and ice in an Alpine cave (Austria): constraints from radiocarbon-dated wood fragments. The Holocene 24, 165–175.
Stoffel, M., Luetscher, M., Bollschweiler, M., et al., 2009. Evidence of NAO control on subsurface ice accumulation in a 1200 yr. old cave-ice sequence, St. Livres ice cave, Switzerland. Quat. Res. 72, 16–26.
Tulis, J., Novotný, L., 2003. Changes of glaciation in the Dobšinská Ice Cave. Aragonit 8, 7–9.
Viehmann, I., Racoviţă, G., 1968. Stalagmitele de gheaţă termoindicatoare. Dări de Seamă Comit. Geol. LIV, 353–363.
Žák, K., Onac, B.P., Perşoiu, A., 2008. Cryogenic carbonates in cave environments: a review. Quat. Int. 187 (1), 84–96.

CHAPTER

DATING CAVE ICE DEPOSITS

5

Zoltán Kern

Institute for Geological and Geochemical Research, Research Centre for Astronomy and Earth Sciences, MTA, Budapest, Hungary

CHAPTER OUTLINE

5.1 INTRODUCTION

Why do we need to date cave ice? This question might be put by a person who sees in cave ice deposits only spectacular, glittering "decoration" in a weird, chilly "cellar." However, the importance of dating to scientists (paleoclimatologist or paleoenvironmental researchers) is obvious, because they would like to decipher the information about past subsurface and surface conditions encoded in the various environmental archives (chemical composition, pollen grains, macrofossils, etc.) to be found in cave ice deposits. If we want to place the changes found in these environmental archives, as analyzed along particular cave ice profiles, into a time frame and link them to well (or less well) known events from history of the Earth, and of humans on it, a numerical estimate of their age is essential. This dating procedure consists of two main tasks. Firstly, ages have to be assigned to certain levels (reference horizons) along the cave ice profile. Secondly, a model should be developed to assign an estimated age to the ice levels in between the reference horizons, and even out of their range, if needed. The first task is the focus of this chapter. The latter task, so-called age-depth modeling, is a general discipline, has no any specialty for cave ice profiles. However, it is necessary if the temporal fluctuations in any biological species, or chemical compound measured along any sediment sequence, are to be interpreted. If numerical age data have been assigned to a number of reference horizons in a cave ice sequence then

Ice Caves. https://doi.org/10.1016/B978-0-12-811739-2.00005-X

appropriate on-line (e.g., OxCal, Bronk Ramsey, 2008) or desktop (e.g., MOD-AGE, Hercman and Pawlak, 2012) tools are available for age-depth modeling.

It should be noted that there is no special age determination technique designed especially for cave ice deposits. The methods to be applied can be adapted from the usual toolbox of geochronology developed for other terrestrial sedimentary records (Walker, 2005; Rink and Thompson, 2015). Since the methods mentioned in the next few pages might be well-know from other studies not connected to cave ice, any detailed technical description is beyond the scope of this contribution and the theoretical background will be introduced as briefly as possible. However, some space will be devoted to special problems which might be regarded as unique to the cave ice environment.

It must be also kept in mind that any age determination must be interpreted in the context of stratigraphy which can show remarkable difference among cave ice profiles (Fig. 5.1).

5.2 DATING METHODS

5.2.1 DIRECT DATING—LAYER COUNTING

Cave ice deposits are usually layered. The clearly visible layered structure observed in the ice wall of Dobšiná Ice Cave inspired the earliest hypothesis on layer counting in cave ice sequences (Hanzlik and Ulrich, 1938).

The speleoglaciological processes, which can potentially create seasonal banding in cave ice, derived from firn and congelation ice, which are somewhat different. Firn layers are usually winter deposits and are usually capped with a relatively thinner layer of organic rich detritus, representing deposition during the snow free season. These pairs of layers taken together can be interpreted as annual. The layers of congelation ice typically consist of large, columnar ice crystals. The elongation of these columnar ice crystals belonging to the same layer is parallel and their size is also similar. The usual season of congelation ice development is spring, when water enters into the freezing cave environment. If erosion takes place in the season following this spring ice accumulation, organic and clastic material, and the potentially released cryominerals as well, can form a distinct seasonal marker in the sequence.

However, assuming that annual bands can be designated difference in the age between two levels can be ascertained by simply counting the number of annual bands between the given horizons. In this respect, therefore, the concept of dating by counting the layers of cave ice is similar to any other annually/seasonally layered sediment.

There are rather few published records available on layer counting in cave ice deposits. Counting of the visible layers, thought to be annual, in Ice Cave in Cemniak resulted in a figure of ~400 layers in the mid-1990s. The ~400 years age for that sequence roughly agreed with the palynological information (Rygielski et al., 1995), and no large discrepancy was observed between this and the radiometric results recently obtained from the same deposit (Hercman et al., 2010).

Based on the 4133 counted ice layers, the age of the Dobšiná Ice Cave has been estimated to have an age of ~5000 years (Droppa, 1960). This estimate, however, has not been confirmed by the recent radiocarbon ages derived from the deposit (Gradziński et al., 2016).

Finally, at the Monlesi Ice Cave, the stratigraphical age model (i.e., layer counting) underestimated the age of the basal ice when compared with the result of all other applied methods (Luetscher et al., 2007).

FIG. 5.1

Cave ice deposits with different stratigraphical complexity and different abundance of impurity horizons. (A) Ice Cave in Cemniak (Photo: Michał Gradziński, detailed description: Hercman et al., 2010), (B) Bortig Ice Cave (Photo: Balázs Nagy, detailed description: Kern et al., 2009), (C) Veronica Ice Cave (Photo: Manuel Gómez Lende, detailed description: Gómez-Lende, 2015), (D) Walkin-Ice Cave System (Photo: Greg Horne, detailed description: Chapter 15), (E) Hs4 ice block (Photo: Bernard Hivert, detailed description: Gómez-Lende, 2016), (F) Strickler Cavern (Photo: Jeff Munroe).

Special points: Detection of annual banding can be more difficult in congelation ice because congelation ice layers often represent sub-seasonal periods of active ice accumulation, associated, for instance, with specific recharge events. Layer counting is also a challenge in the case of mixed profiles consisting of both metamorphosed and congelation ice layers.

5.2.2 INDIRECT DATING

5.2.2.1 Mass turnover

If the basal melting rate of an ice deposit seems to be constant (that is, seems to be forced predominantly by the geothermal heat flux) then the age of the deepest layer can be estimated as the ratio of the maximum ice thickness and the basal melting rate. The existence of a long-term balance between annual mean accumulation at the ice surface and basal melting is also a presumption.

The basal melting rate averaged from 5 years of measurements at three fixed points in the Monlesi Ice Cave resulted in a $8 \pm 2\,\mathrm{cm\,a^{-1}}$ rate of lowering. Given the 12 m thickness of the Monlesi glacier, complete mass turnover was estimated to take between 120 and 200 years (Luetscher et al., 2007).

The basal melting rate in the Dobšiná Ice Cave was estimated to *c*.1 $\mathrm{cm\,a^{-1}}$ from the lowering of the entrance opening to the artificial Kapluka Cavity from 1.8 to 0.87 m over 100 years (Tulis and Novotný, 2003). The maximum ice thickness of the deposit is 26.5 m (Novotný and Tulis, 1996) so the estimated mass turnover time of the deepest ice layer in the Dobšiná Ice Cave is ~2650 years. A recent study, however, based on radiocarbon dates from bat remains taken from multiple levels of the deposit, suggested only *c*.1280 cal. year BP (c.AD 670) for the oldest ice at present exposed below dated samples (Gradziński et al., 2016). Inferred ice accumulation rates, however, exceed the basal melting rate, the discrepancy of which plausibly explains the overestimation of the mass turnover time in this cave.

Finally, based on multiannual instrumental observations, the basal melting rate was determined as $1.54\,\mathrm{cm\,a^{-1}}$ in the Scărişoara Ice Cave (Romania) (Perşoiu, 2005). Given the ~22 m maximum thickness of the Scărişoara glacier (Holmlund et al., 2005), the complete mass turnover can be estimated to take ~1430 years. In contrast to the case of Dobšiná Ice Cave, it is far below the estimation of an age-depth model based on numerous radiocarbon data (Perşoiu et al., 2017). The reverse relation between the estimated mass turnover time and the statistically estimated age suggest that the long-term ice accumulation rate is below the basal melting rate in this cave.

5.2.2.2 Dating the last ~100 years by anthropogenic and/or short-lived radionuclides

Short-lived radionuclides are widely applied as dating tools and represent an objective control in age-accumulation models in various environments (Carroll and Lerche, 2003).

Tritium (^3H)

Among the aforementioned radiochemical tracers, tritium (^3H, $t_{1/2} = 12.32$ years) is probably the most frequently used one. Natural tritium is produced in the upper troposphere and lower stratosphere by the influence of cosmic radiation. However, a much greater amount was produced in the course of the atmospheric testing of thermonuclear weapons during the second half of the 20th century. Dispersion of the anthropogenic tritium into the atmosphere started in 1951. However, major deposition of anthropogenic tritium started only in 1953 in Europe. The maximum deposition peak in the Northern Hemisphere was in 1963, when the annual average of the tritium content in atmospheric precipitation was a thousand fold greater than natural occurrence. The maximum deposition in the Southern

Hemisphere was in 1965. Additional minor peaks worth also mentioning (e.g., Central Europe 1975) and can be considered as additional reference horizons in high resolution serial tritium measurement of cave ice sequence.

Tritium measurements have been applied quite frequently to constrain the chronology of cave ice sequences thought to preserve ice deposited over the past 50 years (Horvatinčić, 1996; Borsato et al., 2006; Luetscher et al., 2007; Kern et al, 2007a,b, 2011). Although tritium records were usually conforming to other age constraints, there are some exceptions (e.g., Pavuza and Mais, 1999; Fórizs et al., 2004).

The most complete tritium records, preserving detailed patterns of the characteristic changes of the atmospheric tritium concentration, have been presented from the upper 2.5 m section of the ice deposit of Grotta del Castelletto di Mezzo (Borsato et al., 2006) and from the Borţig Ice Cave (Kern et al., 2009). The excellent preservation of the high-resolution tritium record of the atmospheric precipitation have been confirmed by the replicated record measured in two parallel ice cores in the Borţig Ice Cave (Fig. 5.2).

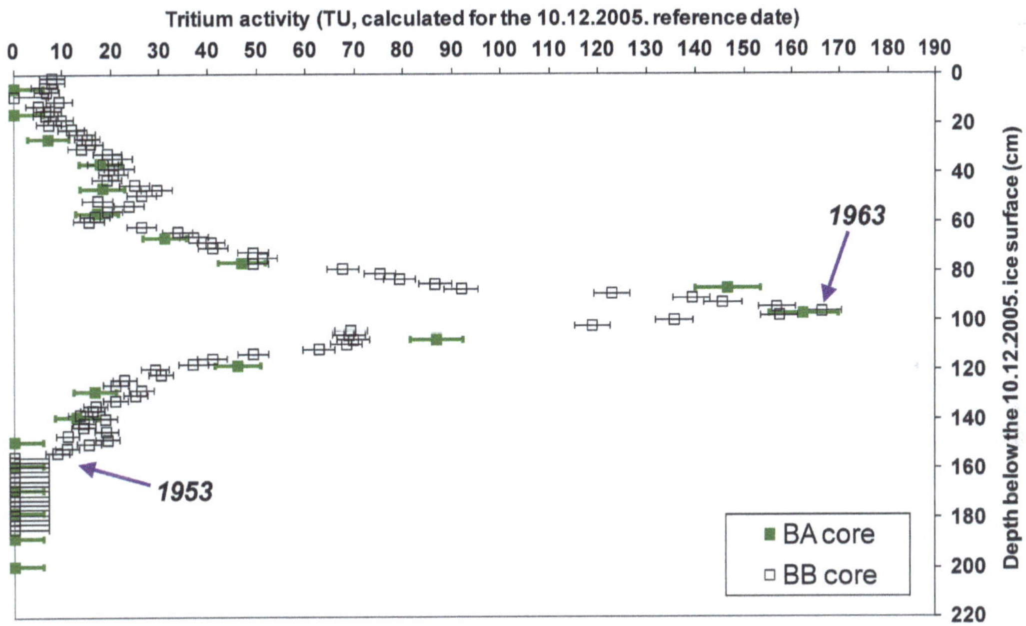

FIG. 5.2

Tritium depth profiles of parallel ice cores (*green squares*: BA, 10 cm sections: Kern et al., 2007a, *open squares*: BB, 2 cm sections: Kern et al., 2009) from Borţig Ice Cave. Assigned calendar dates based on the two most characteristic marker horizons, such as the start of atmospheric thermonuclear bomb tests (1953) when the anthropogenic tritium surplus appeared in the atmosphere and the sharp peak (1963) related to the sudden decrease of the anthropogenic emission following the Nuclear Test Ban Treaty, are indicated.

Special points: The method dates the water from which the ice layer was frozen. If the water supply was directly seasonal precipitation, then the age of the water equals the age of the deposition of the ice layer. However, if the origin of the water is the drip water from the karstic conduits, then a significant lag could exist between the date of precipitation event and the deposition of water originating from it as an ice layer in the ice cave due to the storage and travel time of the infiltrated water through the vadose zone (Kluge et al., 2010). This delay needs to be taken into account in a close analysis.

One must use the original tritium date if stable isotopes or water chemistry data are to be analyzed from the ice, as these parameters are very likely to coeval with the tritium age of the water.

The time scale of tritium dating theoretically can be extended by a few decades beyond the era of anthropogenic tritium emission. If the ^{3}H-^{3}He ingrowth method (detection limit ~0.1 TU; Palcsu et al., 2010) is applied, low levels of tritium content remained from the natural tritium activity (~5 TU in Central Europe) after decay, through 6–7 half-life tritium levels can still be detected.

Radiocaesium (^{137}Cs) and Americium-241 (^{241}Am)

These technogenic radioisotopes can provide reliable radiometric marker horizons over the past half-century in undisturbed sedimentary sequences. Neither the radioactive caesium isotope ($t_{1/2}$ = 30 years), nor Americium-241 isotope ($t_{1/2}$ = 432.2 years), occur naturally in the environment. They have been produced and released by anthropogenic processes. Their most important global environmental source was the fallout from atmospheric thermonuclear weapon tests (from 1954 to 1963), which peaked in the early-1960s and declined rapidly in terms of intensity after the Nuclear Test Ban Treaty in 1963. Afterwards, the majority of Eurasia was affected by a subsequent deposition of ^{137}Cs due to the accident at the Chernobyl Nuclear Power Plant (26 April 1986) (IAEA, 1991; De Cort et al., 1998). The magnitude of the Chernobyl Peak in the Eastern European sediments frequently exceeds the mid-1960s level. In case of areas affected by both fallout events, the presence of ^{241}Am can help distinguish between the time markers, as this was exclusively related to the fallout from nuclear weapon tests (cf. Cambray et al., 1989).

To my knowledge, there has been only one attempt to search for the mentioned ^{137}Cs reference horizons in a cave ice deposit. Ten samples were analyzed from a 2 m long cave ice core from Ledena Pecina (Montenegro) and no meaningful results were obtained, despite the fact that parallel tritium measurements argued for modern water above ~0.9 m in the core (Kern et al., 2007b). It is highly probable that the ^{137}Cs was absorbed by organic materials during the infiltration process (e.g., in the topsoil).

Special points: More promising conditions for the application of these anthropogenic radionuclides might occur when the cave ice originates from direct deposition of atmospheric precipitation. If the main water supply is seepage water then the ^{137}Cs contained in the fallen precipitation is prone be absorption by organic materials during the infiltration process (e.g., in the topsoil).

Radiolead (^{210}Pb)

Fallout-based ^{210}Pb is one of the most important means of dating recent (<150 years) sediments (Appleby, 2001, 2008). This method is based on the escape of gaseous ^{222}Rn from the lithosphere into the atmosphere. This radioactive radon isotope decays to ^{210}Pb. This radioactive lead ($t_{1/2}$ = 22.3 years) is removed by precipitation from the atmosphere and deposited into sediments where it subsequently decays to the stable ^{206}Pb isotope. A general decrease of ^{210}Pb was observed with depth, but anomalously elevated ^{210}Pb activities were recorded from some ice samples which contained debris in Monlesi Ice Cave (Luetscher et al., 2007). It is a warning sign that soil-derived debris probably corrupts the application of the ^{210}Pb method for dating ice samples with a high content of clastic sediments. However, estimates based on ^{210}Pb analyses from clear ice samples in Monlesi Ice Cave gave results comparable to those from other methods (Luetscher et al., 2007), thus warranting further application.

5.2.2.3 Radiocarbon (^{14}C)

Radiocarbon analysis is a well-known Quaternary dating method, and this is the technique most frequently applied to the dating cave ice sequences, too. The principle can be summarized as follows: living organisms absorb traces of ^{14}C during their uptake through the food chain and via metabolic processes. This provides a supply of ^{14}C that compensates for the decay of the existing ^{14}C in the organism, establishing an equilibrium between the ^{14}C concentration in living organisms and that of the atmosphere.

When the organism dies, this supply is cut off, and the ^{14}C concentration of the organism starts to decrease due to radioactive decay.

There are two principal measurement methods. Decay counting involves measuring ^{14}C using either gas proportional or liquid scintillation counters, and accelerator mass spectrometry (AMS). The reader is referred to the literature dedicated to this theme to get insight into the analytical details or to sample specific pretreatments (Taylor et al., 1992; Hua, 2009).

A widely known difference between these methods which is of great practical importance is in the quantity of material required for dating. While 0.1–2 mg of carbon is sufficient for AMS, 0.5–2 g or more of carbon is required for the radiometric method (Jull and Burr, 2006; Molnár et al., 2013).

Probably ^{14}C dating of organic macroremains is the most often applied approach in cave ice deposits. Usually vegetal macroremains are analyzed (e.g., Achleitner, 1995; Horvatinčić, 1996; Fórizs et al., 2004; Holmlund et al., 2005; Maggi et al., 2008; Sancho et al., 2012; Spötl et al., 2014). However, occasionally insect (Citterio et al., 2005; Hercman et al., 2010) or other animal remains (Dickfoss et al., 1997; Yonge and MacDonald, 1999; Gradziński et al., 2016) have been also analyzed.

An interesting approach has been published on ice core derived samples from Eisreisenwelt (May et al., 2011). Although radiocarbon dating on small particulate organic matter separated from the cave ice samples rendered inconclusive, probably due to a background contamination introduced by the applied antifreeze drilling liquid, a crude estimate of a basal ice age in the order of several thousand years could be provided. Despite the modest success, an updated version of this technique probably has more potential (Kern et al., 2016), since the analogue method for surface ice cores has been considerably developed recently (Uglietti et al., 2016).

Finally, when large sets of organic remains are available from a certain ice cave, the temporal distribution of the remains, e.g., the aggregated probability distribution of the calibrated ages (Spötl et al., 2014), or, combined with the dendrochronologically constrained felling date of the larger trunks (see the next section and Stoffel et al., 2009), might indicate major halts in the ice accumulation.

Special points: A crucial prerequisite is the presence of a sufficient amount of organic carbon in the ice. This is an important prerequisite. The otherwise generally easily conducted radiocarbon dating is not feasible in some endogenic Alpine cave ice deposits due to the lack of a sufficient amount of organic carbon in the ice.

A potential source of error or uncertainty can be the spurious radiocarbon activity of the detritus feeding speleofauna, due to the consumption of aged organic litter. They can thus be ^{14}C depleted relative to atmosphere, and this will result in older than expected ^{14}C ages (Hatté and Jull, 2015). This, in turn, can introduce a shift in the radiocarbon age of the organism similar to that of the reservoir effect of aquatic organisms. To avoid this potential source of error, the superficial terrestrial organic remnants (the foliage and branches of trees and shrubs) is recommended for use in cave ice dating.

An additional factor that needs to be taken into account when a radiocarbon date is to be interpreted, is the potential residence time of the matter before it was deposited in the ice. For example, the decomposition rate of wood can be fairly different depending on species and environmental (e.g., climatic) conditions. It can range from a couple of years to a century. Hence, a piece of wood can be older by a few decades than the deposition date of the stratum where it was found. The decomposition rate of leaves or insects' corpus are much faster, hence, the best temporal agreement between the deposition date of a particular cave ice layer and captured organic matter can be expected from these kinds of remains (see also Section 5.3.1)

A related bias can be introduced when pieces of disintegrating larger trunks are deposited onto the ice surface. A radiocarbon date for these reworked wood fragments could again predate the actual age of the hosting ice layer.

Thick horizons rich in organic material are frequently observed, especially in firn deposits. These horizons usually originate from a longer erosion period and are an aggregate of mixtures of both current organics accumulated during the negative mass balance period and older material released during the erosion of the ice strata deposited before the erosion period. The radiocarbon result of a bulk sample from such a horizon will very likely give mixed results of these components, which can give an older age than the actual age of the negative mass balance period. Separating, again, only leaves, especially those in better condition, might help to avoid the reworked older material. Or, if multiple measurements are affordable, the distribution of single-sample ^{14}C results from such a thick organic horizon might reveal the age clusters of both the current and the reworked material.

5.2.2.4 Dendrochronology

Some of the cave ice deposits enclose large quantities of woody macrofossils. Şerban and Racoviţa (1987) suggested first applying dendrochronology in the dating of ice layers in the Scărişoara's ice block.

In line with the general methodological rules of dendrochronology, the wide-narrow pattern of the sequence of tree-ring widths gained from a log trapped in cave ice is to be synchronized with the local/regional reference chronology, ideally with that of the same tree species as the analyzed sample (Speer, 2010). The procedure of this synchronization is called crossdating (Stokes and Smiley, 1996). The accuracy of the crossdating procedure is evaluated by different statistics. It is a fact well-known among dendrochronologists, though still worth mentioning, that samples consisting of only a few rings are useless for dendrochronological dating. The short sequence of tree rings is more prone to provide seemingly better statistics for a false position along the reference chronology than for the real date.

The first successful tree-ring based datings of cave ice-bound logs were presented from the St. Livres Ice Cave (Schlatter et al., 2003) and from the Focul Viu Ice Cave (Kern et al., 2004). However, these studies still neglected the use of startigraphical information. The ringwidth record of a larch log which emerged at the retreating edge of the ice deposits in the Ledena Jama pri Planini Viševnik (Slovenia) has been successfully synchronized to the local larch master chronology (Staut et al., 2016).

A much advanced application of dendrochronology in cave ice study was presented for the St. Livres Ice Cave, Swiss Jura Mts (Stoffel et al., 2009). Combining radiocarbon data with dendrochronological analyses on 45 samples harvested from subfossil logs trapped in the cave ice, four major deposition gaps could be identified and dated to the fourteenth, fifteenth, mid-nineteenth, and late nineteenth centuries.

Special points: A potentially non-trivial antagonism is worth discussing briefly here in relation to tree-ring dating of cave ice captured trunks. The more rings the sample has, the greater the potential for dendrochronological dating. However, the more rings a sample has, in general, the greater its size, i.e., diameter of the trunk, tends to be larger, as well. This results in a strange paradox, as for larger sample, consisting of more rings, we could expect a more accurate chronological match from dendrochronological dating; however, the larger a sample, the less accurate will be the linking of the provided date to a fixed horizon into the sequence. In other words the accuracy of the more precise dendrochronological date can be biased by stratigraphical uncertainty. Two factors should be considered to minimize this uncertainty. Firstly, the date of the outermost ring of the dendrochronologically dated sample needs to be corrected for the potentially missing sapwood rings and by the potential residence time of the sample out of the cave. Secondly, the corrected date should refer to the level of the lowermost point of the trunk in the ice stratigraphy and not to the level of the mid-point of the sample.

5.3 SOME PRACTICAL ASPECT

5.3.1 SAMPLE SELECTION FOR RADIOCARBON DATING: THE BIGGER, THE BETTER?

When the ice accumulation is located relatively close to the cave entrance, organic material can settle onto the cave ice surface in various sizes, ranging from microscopic pollen grains to large trunks. Aggrading cave ice deposits will cover these remnants, though the required burial time will clearly depend on the height of the object measured perpendicularly to the ice surface, assuming uniform net accumulation over the cave ice surface. This means that thin/small objects can be buried quickly, for instance by the end of the following season, while a decade might be needed for a ~1 m diameter tree trunk to be completely covered by cave ice (Fig 5.3).

Smaller objects (e.g., leaf, cone) can be more suitable for radiocarbon dating because their smaller size provides a better confined reference level in the sequence, and their presumably shorter surface residence time (see Section 5.2.2.3) secures a closer coincidence between their radiocarbon date and their deposition onto the past ice surface. Both factors are valuable in reducing the uncertainty of a developing age-depth calibration in any cave ice profile.

5.3.2 A POTENTIAL METHOD—CRYOGENIC CAVE CARBONATE (CCC) LAYERS

Precipitation of cryogenic cave carbonate (hereafter CCC) proceeds at the freezing point and involves rapid or slow freezing of water, rapid or slow CO_2 outgassing from a solution, and locally, partial water evaporation. Depending on the freezing rate, and the thickness of the freezing water layer, either fine-grained (CCCfine), or coarsely crystalline (CCCcoarse) precipitation can be formed (for details see Chapter 6).

Lauriol and Clark (1993) approximated the age of two Arctic cave ice deposits based on the radiocarbon activity of their enclosed cryogenic powder. Derived age estimates were supported by independent ^{14}C dates on associated animal and plant remains. However, the validity of their assumptions, and the performance of the proposed calculation scheme, still remained untested for other cave ice deposits.

Larger aggregates of CCCcoarse have been successfully dated by the U-series technique, although these were rarely observed embedded in cave ice sequences. The occurrence of CCCfine is a usual phenomenon in sequences of congelation ice. CCCfine frequently forms quite thick interbedded layers in cave ice sequences, offering another potential for cave ice dating.

Hitherto, only one study has tested the suitability of fine-grain CCC for U-series dating. Analyzed samples from Eisriesenwelt (using TIMS in the Heidelberg lab) were too rich in ^{232}Th and, given its young age, could not be dated (Spötl, 2008).

FIG. 5.3

Different burial conditions for small size (leaf, insect) samples and a large branch or trunk. Top: Deposition of different sized remains at the same time to the same ice surface. Middle: Subsequent season with a positive mass balance (i.e., net accumulation) is usually sufficient to cover the small size remains. Bottom: Much thicker deposition, laid down by a series of positive mass balance seasons (six in this example), should cover larger size remains. In addition, the larger size remains usually represent a less accurate position in the stratigraphy further complicating the utilization of their derived age estimates in age-depth modeling.

Despite this negative experience, however, careful pre-cleaning to remove potential clay contamination can help to get better results. Another potential problem could be the co-occurrence of microscopic host-rock particles in the CCC accumulation. The separation of CCCfine from host-rock particles is recommended since the old host-rock carbonate could bias the results towards a greater age.

5.4 CONCLUSION OR WHAT IS THE RECOMMENDED DATING STRATEGY IN CAVE ICE PROFILES?

Following the above overview, the conclusion cannot be anything other than that a uniform protocol is hardly recommended for the dating of every cave ice deposit. The peculiarities of the studied cave ice section have to be taken into account, and the dating strategy should be designed accordingly. Abundant organic material, for instance, provides a good opportunity for radiocarbon analysis, although a careful selection, focusing on small size and ephemeral samples (such as foliage, small branches, or insects) is recommended on account of their better defined stratigraphical position and shorter expected decomposition time. The conduct of a taxonomical analysis is also recommended before the radiometric dating because the species' ecological demand might contribute to the paleoenvironmental interpretation of the section enclosing the dated specimen, and ^{14}C results from a member of the potentially detritus feeding speleofauna could be less suitable for accurately dating the studied cave ice deposit. Individual samples collected from undisturbed stratigraphical locations near to the base of the cave ice deposit can provide useful information for the antiquity, i.e., the expected maximum age, of the deposit. If the temporal changes of the cave ice mass balance history should be detected (potentially betrayed by characteristic stratigraphical patterns), then a series of age determination is needed, obviously.

Tritium activity measurements can have the most potential to detect ice deposited from atmospheric precipitation after the early-1950s. The most characteristic changes of the atmospheric tritium concentration, such as the start of atmospheric thermonuclear bomb tests (1953) and the sharp peak (1963/1964) related to the sudden decrease of the anthropogenic emission following the Nuclear Test Ban Treaty, could provide distinct time markers in cave ice record. Radiocarbon analysis of surface derived ephemeral plant (e.g., foliage and small branch) and animal (e.g., insects) remains is currently the most potentially accurate dating approach for the older cave ice deposits.

Ideally, it is recommended that various methods be applied, and a series of dates be established, for any cave ice profile. The various dating methods could help to detect a systematic bias present in any particular technique. In addition, serial dating could help to detect outliers.

ACKNOWLEDGEMENTS

Michał Gradziński, Chas Yonge, Greg Horne, Manuel Gómez Lende, and Jeff Munroe are acknowledged for their photo contribution in Fig. 5.1. Thanks to Christoph Spötl for sharing his experience on U-series dating of CCCfine. Thanks for support from the "Lendület" program of the Hungarian Academy of Sciences (LP2012-27/2012). This is contribution No.44 of *2 ka Palæoclimatology* Research Group.

REFERENCES

Achleitner, A., 1995. Zum Alter des Höleneises in der Eisgruben-Eishöhle im Sarstein (Oberösterreich). Die Höhle 46 (1), 1–5.

Appleby, P.G., 2001. Chronostratigraphic techniques in recent sediments. In: Last, W.M., Smol, J.P. (Eds.), Tracking Environmental Change Using Lake Sediments, vol. 1. Dordrecht, Netherlands, pp. 171–203. https://doi.org/10.1007/0-306-47669-X_9.

Appleby, P.G., 2008. Three decades of dating recent sediments by fallout radionuclides, a review. The Holocene 18 (1), 83–93. https://doi.org/10.1177/0959683607085598.

Borsato, A., Miorandi, R., Flora, O., 2006. I depositi di ghiaccio ipogei della Grotta dello Specchio e del Castelletto di Mezzo (Dolomiti di Brenta, Trentino): morfologia, età ed evoluzione recente. Studi Trent. Sci. Nat. Acta Geol. 81 (2004), 53–74.

Bronk Ramsey, C., 2008. Deposition models for chronological records. Quat. Sci. Rev. 27, 42–60.

Cambray, R.S., Playford, K., Carpenter, R.C., 1989. Radioactive fallout in air and rain: results to the end of 1988. Her Majesty's Stn. Off., London, Rep. AERE-R 13575.

Carroll, J., Lerche, I., 2003. Sedimentary Processes: Quantification Using Radionuclides. In: Radioactivity in the Environment. Elsevier Science & Technology, Oxford, UK, pp. 282.

Citterio, M., Turri, S., Perşoiu, A., Bini, A., Maggi, V., 2005. Radiocarbon ages from two ice caves in the Italian Alps and the Romanian Carpathians and their significance. In: Glacier Caves and Glacial Karst in High Mountains and Polar Regions. Institute of Geography of the Russian Academy of Sciences, Moscow, pp. 87–92.

De Cort, M., Dubois, G., Fridman, S.D., Gemenchuk, M.G., Izrael, Y.A., Jansses, A., Jones, A.R., Kelly, G.N., Kvasnikova, E.V., Matveenko, I.I., Nazarov, I.M., Pokumeiko, Y.M., Sitak, V.A., Stukin, E.D., Tabachny, L.Y., Tsaturov, Y.S., Avdyushin, S.I., 1998. Atlas of Caesium Deposition on Europe after the Chernobyl Accident. EUR report nr. 16733 EC, Office for Official Publications of the European Communities, Luxembourg. p. 63.

Dickfoss, P.V., Betancourt, J.L., Thompson, L.G., Turner, R.M., Tharnstrom, S., 1997. History of ice at Candelaria Ice Cave, New Mexico. Bull. New Mex. Bur. Min. Mineral Resour. 156, 91–112.

Droppa, A., 1960. Dobšiná Ice Cave. Šport, Bratislava. 115 pp. (in Slovak, German and French summary).

Fórizs, I., Kern, Z., Nagy, B., Szántó, Zs, Palcsu, L., Molnár, M., 2004. Environmental isotope study on perennial ice in the Focul Viu Ice Cave, Bihor Mts., Romania. Theor. Appl. Karstol. 17, 61–69.

Gómez-Lende, M., 2015. Las cuevas heladas en Picos de Europa:clima, morfologías y dinámicas. (Ph.D. thesis) Universidad de Valladolid. 663 pp.

Gómez-Lende, M., 2016. Cuevas heladas en el Parque Nacional de Picos de Europa, fronteras subterráneas del hielo en el Macizo Central. In: Organismo Autónomo de Parques Nacionales (O.A.P.N)-Ministerio de Medio Ambiente y Medio Rural y Marino. ISBN: 978-84-8014-891-7. 244 pp.

Gradziński, M., Hercman, H., Peresviet-Soltan, A., Zelinka, J., Jelonek, M., 2016. Radiocarbon dating of fossil bats from Dobšina Ice Cave (Slovakia) and potential palaeoclimatic implications. Ann. Soc. Geol. Pol. 86, 341–350. https://doi.org/10.14241/asgp.2016.016.

Hanzlik, S., Ulrich, F., 1938. The ice cave of Dobsina in Czechoslovakia. IAHS Publ. 23, 695–698.

Hatté, C., Jull, A.J.T., 2015. ^{14}C in plant macrofossils. In: Rink, J.W., Thompson, J.W. (Eds.), Encyclopedia of Scientific Dating Methods. Springer Science + Business Media, Dordrecht, pp. 127–132. https://doi.org/10.1007/978-94-007-6304-3.

Hercman, H., Gąsiorowski, M., Gradziński, M., Kicińska, D., 2010. The first dating of cave ice from the Tatra Mountains, Poland and its implication to palaeoclimate reconstructions. Geochronometria 36, 31–38.

Hercman, H., Pawlak, J., 2012. MOD-AGE: an age-depth model construction algorithm. Quat. Geochronol. 12, 1–10.

Holmlund, P., Onac, B.P., Hansson, M., Holmgren, K., Mörth, M., Nyman, M., Perşoiu, A., 2005. Assessing the palaeoclimate potential of cave glaciers: the example of the Sčarişoara Ice Cave (Romania). Geogr. Ann. 87A, 193–201.

Horvatinčić, N., 1996. Izotopina mjerenja u ledu, Jama Ledenica, Velebit. In: Kubelka, D., Kovač, J. (Eds.), Proceedings of the Third Symposium of the Croatian Radiation Protection Association, Zagreb, pp. 297–302.

Hua, Q., 2009. Radiocarbon: a chronological tool for the recent past. Quat. Geochronol. 4 (5), 378–390. https://doi.org/10.1016/j.quageo.2009.03.006.

IAEA, 1991. The International Chernobyl Project: an overview, assessment of radiological consequences and evaluation of protective measures. Report by an International Advisory Committee. IAEA, Vienna, Austria. Soil Use Manage. 7, 169–176.

Jull, A.J.T., Burr, G.S., 2006. Accelerator mass spectrometry: is the future bigger or smaller? Earth Planet. Sci. Lett. 243 (3), 305–325. https://doi.org/10.1016/j.epsl.2005.12.018.

Kern, Z., Fórizs, I., Kázmér, M., Nagy, B., Szántó, Z.s., Gál, A., Palcsu, L., Molnár, M., 2004. Late Holocene environmental changes recorded at Gheţarul de la Focul Viu, Bihor Mts, Romania. Theor. Appl. Karstol. 17, 51–60.

Kern, Z., Molnár, M., Perşiou, A., Nagy, B., 2007a. Radiochemical and stratigraphic analysis of two ice cores from Borţig Ice Cave, Apuseni Mts, Romania. In: Zelinka, J. (Ed.), Proceedings of the 2nd International Workshop on Ice Caves, Liptovský Mikuláš, Slovakia. pp. 65–69.

Kern, Z., Surányi, G., Molnár, M., Nagy, B., Balogh, D., 2007b. Investigation of natural perennial ice deposits of Durmitor Mts, Montenegro. In: Zelinka, J. (Ed.), Proceedings of the 2nd International Workshop on Ice Caves, Liptovský Mikuláš, Slovakia, pp. 70–73.

Kern, Z., Molnár, M., Svingor, É., Perşoiu, A., Nagy, B., 2009. High resolution, well preserved tritium record in the ice of Borţig Ice Cave, Bihor Mountains, Romania. The Holocene 19, 729–736. https://doi.org/10.1177/0959683609105296.

Kern, Z., Széles, E., Horvatinčić, N., Fórizs, I., Bočić, N., Nagy, B., 2011. Glaciochemical investigations of the ice deposit of Vukušić Ice Cave, Velebit Mountain, Croatia. Cryosphere 5, 485–494. https://doi.org/10.5194/tc-5-485-2011.

Kern, Z., Molnár, M., Pavuza, R., 2016. Radiocarbon analysis of carbonaceous particulate matter from the Saarhalle ice block, Dachstein-Mammoth cave (Mammuthöhle), Austria. In: Mihevc, A., Hajna Zupan, N., Gostincar, P. (Eds.), 7th International Workshop on Ice Caves: Program Guide and Abstracts. Karst Research Institute ZRC SAZU, Postojna, pp. 48–49.

Kluge, T., Riechelmann, D.F.C., Wieser, M., Spötl, C., Sültenfuß, J., Schröder-Ritzrau, A., Niggemann, S., Aeschbach-Hertig, W., 2010. Dating cave drip water by tritium. J. Hydrol. 394 (3), 396–406.

Lauriol, B., Clark, I.D., 1993. An approach to determine the origin and age of massive ice blockage in two arctic caves. Permafr. Periglac. Process. 4, 77–85.

Luetscher, M., Bolius, D., Schwikowski, M., Schotterer, U., Smart, P.L., 2007. Comparison of techniques for dating of subsurface ice from Monlesi ice cave, Switzerland. J. Glaciol. 53 (182), 374–384.

Maggi, V., Turri, S., Bini, A., Udisti, R., 2008. 2500 year of history in Focul Viu Ice Cave. In: Kadebskaya, O., Mavlyudov, B.R., Pyatunin, M. (Eds.), Proceedings of the 3rd International Workshop on Ice Caves, Kungur, pp. 11–15.

May, B., Spötl, C., Wagenbach, D., Dublyansky, Y., Liebl, J., 2011. First investigations of an ice core from Eisriesenwelt cave (Austria). Cryosphere 5, 81–93. https://doi.org/10.5194/tc-5-81-2011.

Molnár, M., Janovics, R., Major, I., Orsovszki, J., Gönczi, R., Veres, M., Leonard, A.G., Castle, S.M., Lange, T.E., Wacker, L., Hajdas, I., Jull, A.J.T., 2013. Status report of the new AMS C-14 sample preparation lab of the Hertelendi Laboratory of Environmental Studies (Debrecen, Hungary). Radiocarbon 55, 665–676.

Novotný, L., Tulis, J., 1996. Výsladky najnovších výskumov v Dobšinskej ľadovej jaskyni. Slovensky kras 34, 139–147.

Palcsu, L., Major, Z., Köllő, Z., Papp, L., 2010. Using an ultrapure ^{4}He spike in tritium measurements of environmental water samples by the ^{3}He-ingrowth method. Rapid Commun. Mass Spectrom. 24, 698–704. https://doi.org/10.1002/rcm.4431.

Pavuza, R., Mais, K., 1999. Aktuelle höhlenklimatische Aspekte der Dachstein-Rieseneishöhle. Die Höhle 50 (3), 126–140.

Perşoiu, A., 2005. Evidence of basal melting of the ice block from Scărişoara Ice Cave. In: Mavlyudov, B.R. (Ed.), Glacier Caves and Glacial Karst in High Mountains and Polar Regions. Institute of Geography of the Russian Academy of Sciences, Moscow, pp. 109–112.

Perşoiu, A., Onac, B.P., Wynn, J.G., Blaauw, M., Ionita, M., Hansson, M., 2017. Holocene winter climate variability in Central and Eastern Europe. Sci. Rep. 7 (1), 1196.

Rink, W.J., Thompson, J.W., 2015. Encyclopedia of Scientific Dating Methods. Springer Science + Business Media, Dordrecht.

Rygielski, W., Siarzewski, W., Wieliczko, P., 1995. Variability of the ice deposit in ice cave on Mount Ciemniak in the West Tatra Mountains. Quaestiones Geographicae 17 (18), 55–64.

Sancho, C., Belmonte, A., López-Martínez, J., Moreno, A., Bartolomé, M., Calle, M., Santolaria, P., 2012. Potencial paleoclimático de la cueva helada A294 (Macizo de Cotiella, Pirineos, Huesca). Geogaceta 52, 101–104.

Şerban, M., Racoviţa, Gh., 1987. Sur l'information paléoclimatique comprise dans le dépôt souterrains de glace de la Grotte de Scărişoara (Monts de Bihor) et sa valorisation pour une prognose de long terme. Theor. Appl. Karstol. 3, 133–143.

Schlatter, F., Stoffel, M., Monbaron, M., Luetscher, M., 2003. A new approach to dating firn accumulation in an ice cave in the Swiss Jura mountains. In: Brandova, Haeberli (Eds.), 8th International Conference on Permafrost, Extended Abstracts, pp. 143–144.

Speer, H.J., 2010. Fundamentals of Tree-Ring Research. The University of Arizona Press, Tucson.

Spötl, C., 2008. Kryogene Karbonate im Höhleneis der Eisriesenwelt. Die Höhle 59, 26–36.

Spötl, C., Reimer, P.J., Luetscher, M., 2014. Long-term mass balance of perennial firn and ice in an Alpine cave (Austria): constraints from radiocarbon-dated wood fragments. The Holocene 24, 165–175. https://doi.org/10.1177/0959683613515729.

Staut, M., Vreča, P., Merela, M., Brenčič, M., 2016. Recent fluctuations of ice deposits in the cave Ledena Jama pri Planini Viševnik, NW Slovenia. In: Mihevc, A., Hajna Zupan, N., Gostincar, P. (Eds.), 7th International Workshop on Ice Caves: Program Guide and Abstracts. Karst Research Institute ZRC SAZU, Postojna, pp. 73–74.

Stoffel, M., Luetscher, M., Bollschweiler, M., Schlatter, F., 2009. Evidence of NAO control on subsurface ice accumulation in a 1200 yr old cave-ice sequence, St. Livres ice cave, Switzerland. Quat. Res. 72, 16–26. https://doi.org/10.1016/j.yqres.2009.03.002.

Stokes, M.A., Smiley, T.L., 1996. An Introduction to Tree-Ring Dating. The University of Arizona Press, Tucson.

Uglietti, C., Zapf, A., Jenk, T.M., Sigl, M., Szidat, S., Salazar, G., Schwikowski, M., 2016. Radiocarbon dating of glacier ice: overview, optimisation, validation and potential. Cryosphere 10, 3091–3105. https://doi.org/10.5194/tc-10-3091-2016.

Taylor, R.E., Long, A., Kra, R.S., 1992. Radiocarbon after Four Decades: An Interdisciplinary Perspective. Springer-Verlag, New York.

Tulis, J., Novotný, L., 2003. Changes of glaciation in the Dobšinská Ice Cave (in Slovak). Aragonit 8, 7–9.

Walker, M., 2005. Quaternary Dating Methods. John Wiley and Sons, Chichester, England.

Yonge, C.J., MacDonald, W.D., 1999. The potential of perennial cave ice in isotope palaeoclimatology. Boreas 28, 357–362.

CHAPTER

CRYOGENIC MINERAL FORMATION IN CAVES

6

Karel Žák*, Bogdan P. Onac[†,‡], Olga I. Kadebskaya[§], Michal Filippi*, Yuri Dublyansky[¶], Marc Luetscher[¶,||]

Institute of Geology of the Czech Academy of Sciences, Prague, Czech Republic[*] University of South Florida, Tampa, FL, United States[†] Emil Racoviță Institute of Speleology, Cluj-Napoca, Romania[‡] Mining Institute of the Ural Branch of the Russian Academy of Sciences, Perm, Russia[§] University of Innsbruck, Innsbruck, Austria[¶] Austrian Academy of Sciences, Innsbruck, Austria[||]

CHAPTER OUTLINE

6.1 INTRODUCTION

Ice caves, and caves that were iced in the past, contain a specific genetic group of secondary mineral formations that are not observed in caves that never experienced freezing temperatures. These mineral accumulations were formed by the segregation of solutes during the freezing of salt-containing solutions. They belong to a large family of speleothems (Moore, 1952), but differ in many morphological, mineralogical, and geochemical aspects from the common speleothem types occurring in non-iced caves. All minerals precipitated in caves because of freezing temperatures belong to the *cryogenic cave minerals* (CCMs) group. The most frequent CCMs are *cryogenic cave carbonate* (CCC) and *cryogenic cave gypsum* (CCG), whereas the others are more rare. The natural accumulations of CCMs are rarely composed of only one mineral; nevertheless, either carbonate or gypsum typically represents a dominant component of them.

One of the most common morphological features of CCMs is their loose character. They form free particles dispersed within the ice, accumulations of primary or secondary deposits on the ice surface, or layers on the cave floor, where ice melted or disappeared because of sublimation. Generally, CCMs are not firmly attached to a solid ground or to each other. The CCMs usually have various morphologies and sizes of individual crystals and aggregates. The sizes range from several microns

Ice Caves. https://doi.org/10.1016/B978-0-12-811739-2.00035-8

to several centimeters. In the case of the CCC, two subtypes, fine-grained powdery precipitates (CCC_{fine}) and larger crystals and aggregates (CCC_{coarse}), were defined, although the exact size boundary separating the two groups is difficult to define (abbreviations were introduced by Luetscher et al., 2013). The size of individual CCC_{fine} crystals and aggregates is typically below 1 mm, which can be adopted as an approximate boundary between CCC_{fine} and CCC_{coarse}. Nevertheless, the largest aggregates of CCC_{fine} can reach above this limit, and the smallest CCC_{coarse} grains can be below it—the size distributions of both types partly overlap. More importantly, the CCC_{fine} and the CCC_{coarse} form under (and serve as markers of) different environmental conditions. Apart from calcite, the cryogenic carbonates also comprise other mineral phases, including ikaite ($CaCO_3 \cdot 6H_2O$) and monohydrocalcite ($CaCO_3 \cdot H_2O$).

The existence of CCMs has been known for more than 280 years. Johann Georg Gmelin, a German natural scientist, was probably the first person to mention the "gypsum powder" on ice surfaces, which he observed in 1733 in the Kungur Ice Cave, located in the Ural Mountains in Russia (Kozlova and Naumkin, 2014) and which he assumed formed during the water-freezing process. The number of papers describing CCMs and explaining their formation processes began to grow, especially after it was noted that CCC displayed an unusual carbon and oxygen isotope compositions (Clark and Lauriol, 1992). Altogether, several hundred papers, conference abstracts, and unpublished reports on CCMs had been published or presented by the end of 2016. However, the study of CCMs is still at an empirical level. While the data on carbon and oxygen stable isotopes can help identify various types of CCC, distinguishing between CCG and other types of gypsum is more difficult and requires a detailed study of the crystal morphology. Additional experimental, mineralogical and theoretical work needs to be done to better understand the minerogenetic processes occurring in ice caves.

6.2 FREEZING OF MINERALIZED AQUEOUS SOLUTIONS IN CAVES—THEORETICAL PRINCIPLES AND SUBDIVISION OF THE ENVIRONMENTS

Salts are generally only slightly soluble in ice, with a solubility in the range of micromoles per liter or less (e.g., Petrenko and Whitworth, 1999). Freezing of aqueous solutions results in salt rejection, a process in which the majority of dissolved ions are rebuffed by the advancing ice-water front during ice crystal growth. This results in a significant increase of salt concentration in the unfrozen part of the solution. As the ice formation proceeds, some dissolved compounds reach their saturation limits and precipitate as cryogenic minerals. Gases present in solutions are also rejected by the ice, and, as a result of diminishing volume of unfrozen liquid, either escape into the cave atmosphere or are entrapped as bubbles (gaseous inclusions) in the ice. Some highly soluble compounds (e.g., NaCl, organic acids, etc.) form small liquid inclusions of brine, trapped at the ice crystal boundaries. The natural ice produced by the freezing of salt-enriched water is therefore a complex substance, containing, in addition to ice crystals, solid (CCMs), liquid (brine), and gaseous inclusions (Mulvaney et al., 1988; Petrich and Eicken, 2010; Souchez and Lorrain, 1991). Chemical analyses of filtrated meltwater (e.g., Mavlyudov, 2008) thus represent a bulk composition: ice crystals + brine inclusions + those CCMs that are redissolved + soluble part of allogenic particles.

The presence of dissolved salts has another well-known consequence, which is lowering the freezing point of the solution. For ideal solutions, the freezing-point depression can be estimated using the simple linear relationship

$$\Delta T_F = K_F \cdot m \cdot i$$

where ΔT_F is the freezing point depression (temperature difference in degrees between the freezing point of pure water and that of the solution), K_F is the cryoscopic constant ($1.853\,K \cdot kg \cdot mol^{-1}$ for water), m is the molality (moles of solute per kg of solvent), and i is the van't Hoff factor (for ionic compounds, the number of solute particles after dissociation, e.g., 2 for NaCl). This equation gives acceptable results for diluted solutions (for calculation of the freezing-point depression of concentrated solutions; see, e.g., Ge and Wang, 2009). For usual karst groundwater with total dissolved solids (TDS) below $1\,g \cdot L^{-1}$, the freezing-point depression is low, at the onset of the freezing generally less than 0.1°C. It can be higher during the late phase of freezing, when the residual water fraction is more concentrated, or for waters in the gypsum karst, especially if the host lithology also includes the salt rock. Seawater with an average salinity of $34\,g \cdot L^{-1}$ begins to freeze at temperatures of approximately −1.86°C.
The freezing process is complicated by the fact that temperature gradients (because of the release of latent heat) and compositional gradients (because of salt rejection) occur just at the freezing front. As a result, the originally planar freezing front becomes uneven (Worster and Wettlaufer, 1997). Roughness of the growing ice surface favors trapping bubbles of released gasses and the inclusion of unfrozen residual brine, unless the brine gravitationally flows away from the ice through open pores, as is common during seawater freezing (Petrich and Eicken, 2010). Freezing of seawater and NaCl solutions is by far the most studied case of salt rejection, both experimentally (Loose et al., 2009; Papadimitriou et al., 2004) and theoretically, using various computational approaches (Marion, 2001; Marion et al., 2009, 2010; Vrbka and Jungwirth, 2005, 2007). Field observations and theory were reviewed by Petrich and Eicken (2010). The distribution of soluble impurities in ice was also studied in Antarctic and Greenland polar ice (e.g., Baker et al., 2005; Barnes and Wolff, 2004; Faria et al., 2010).

The chemistry of natural groundwater in limestone karst (without significant gypsum/rock salt presence) is dominated by Ca^{2+} and HCO_3^-, with TDS usually not exceeding $1\,g \cdot L^{-1}$ (Ford and Williams, 2007). Several theoretical and/or experimental studies of freezing of low-ionic-strength calcium bicarbonate water were performed (Clark and Lauriol, 1992; Ek and Pissart, 1965; Fairchild et al., 1996; Hallet, 1976; Killawee et al., 1998; Pulina, 1990; Socki et al., 2001, 2010). Solid cryogenic carbonate mineral phases resulted from all these experiments.

The most detailed freezing experiments with diluted calcium bicarbonate solutions were performed by Fairchild et al. (1996) and Killawee et al. (1998). During these tests, with a constant linear ice growth rate of 3 to $8\,mm \cdot h^{-1}$, the chemical composition of ice produced in each individual freezing step was determined along with the evolution of the chemical composition of the residual solution. The produced carbonate precipitates were studied with respect to their morphology, mineralogy, chemistry, and C and O isotopic composition. Cryogenic precipitates were mostly represented by calcite, with traces of vaterite and/or aragonite. With respect to their morphology, the precipitates included rhombohedral (some of them skeletal) crystals, their aggregates, and hollow spheres. The theoretical presumption that incorporation of ions into the ice lattice depends on the ion size was also supported during these experiments. The effective segregation coefficient (k_{eff}) during the experiments of Killawee et al. (1998) was defined as

$$k_{eff} = C_{ice} / C_{iw}$$

where C_{ice} is the concentration of a species in the ice, and C_{iw} is the concentration of a species in the initial water. Segregation coefficients were different for individual ions (k_{eff} of $Mg^{2+} > Ca^{2+} > Sr^{2+}$); they changed also depending on the proportion of ice and residual solution and freezing rate. Mg^{2+}, which has the smallest ion size, is slightly more easily incorporated into the ice than larger ions of Ca^{2+} and Sr^{2+}. CCC should therefore have a lower Mg/Ca ratio than speleothems produced in the same cave but in a nonfreezing environment. Under real field circumstances, crystals or aggregates with significantly higher Mg content, or even Mg minerals, are sometimes observed in the CCC_{fine} accumulations (Bazarova et al., 2016), which may reflect either Mg enrichment at ice surfaces during repeated cycles of new ice formations and their sublimation or the preferential consumption of Ca during carbonate crystallization. The Mg^{2+} partitioning between calcite/ikaite and residual water may thus be more significant than between ice and water.

Whereas freezing of karst water in limestone caves produces CCC (mostly calcite and/or ikaite), freezing of sulfate-rich solutions in gypsum caves is responsible for CCG along with small quantities of other minerals (e.g., calcite, ikaite, celestine, barite, and silica-unidentified phase). In gypsum karst, mineralization of groundwater can be higher, reaching several $g \cdot L^{-1}$, especially if halite is present (Bock, 1961; Calaforra, 1998; Ford and Williams, 2007).

In addition to water freezing, other processes also support mineral precipitation in caves with high airflow. Caves with dynamic airflow during winter are either vertically steep, down-sloping caverns (into which cold dense air sinks and remains trapped) or caves with two entrances located at different elevations (Perşoiu and Onac, 2012; more details in Chapters 3 and 4). In such caves, the heat transfer between the walls and the cold air causes a drop in the relative humidity of the caves' air. The potential for the evaporation of water thus increases even at low temperatures inside ice caves. Freezing and evaporation have similar effects on the chemical composition of residual solutions and are thus difficult to distinguish.

As already mentioned, not all dissolved salts precipitate as solid phases during freezing. Although some ions are more readily incorporated into the ice (F^-, NH_4^+), others (e.g., Na^+, Cl^-, humic/fluvic acids) remain in the residual liquid. When ice stalactites form in caves within forested karst terrains, the drops of unfrozen solution on their tips are sometimes brown in color as a result of cryogenically concentrated organic compounds, mostly humic/fulvic acids (Žák et al., 2010). If freezing is complete, small liquid inclusions containing these substances form along ice crystal boundaries.

In the above-mentioned caves with high airflow, the frozen water (ice) is commonly converted directly into water vapor (see Chapter 4). This process, known as ice sublimation, creates concentrations of CCMs formerly suspended within the ice on the retreating ice surface. Small carbonate and/or sulfate particles inside the ice, which are almost invisible to the naked eye, form conspicuous accumulations of a whitish powder on the ice surface. Such deposits can accumulate gravitationally around the bases of ice stalagmites and columns, at the bases of cave ice blocks, and under the ablated ice block faces. Cave ice sublimation rates can be very high, with values reaching c. 20 mm of ice per year, as reported from caves in the Pamir Mountains, and up to 35 mm per year, as reported in the northern part of the Russian Platform (Mavlyudov, 2008). Ice sublimation may lead to the deposition of cryogenic minerals through the evaporation of brines that were trapped in the ice as liquid inclusions.

The behavior of individual CCMs differs during ice melting. The CCC typically survives as a solid phase, because part of the CO_2 released during mineral crystallization escapes into the cave atmosphere and is ultimately removed from the cave by natural ventilation. Only a small fraction of this CO_2 remains trapped as gas bubbles in the cave ice. During ice melting, part of the CO_2 released from bubbles in ice also degasses; the aggressiveness of meltwater is therefore too low to redissolve all the CCC. In contrast, CCG can be dissolved again during the ice melting process, because the precipitation and dissolution of gypsum is not controlled by any gaseous phase release/input. Nevertheless, in the presence of ice sublimation, various amounts of CCG may survive ice retreat. These powdery CCG deposits may become wetted during the warm season and undergo later changes, including recrystallization.

To describe these complex processes, Andrejchuk and Galuskin (2008) subdivided the occurrences of CCMs into (1) primary *cryomineral inclusions* dispersed in the cave ice; (2) *cryogenic aggregates*, formed during repeated ice growth/melt cycles and ice diagenesis; and (3) *cryogenic deposits*, represented by secondary accumulations of CCMs.

The processes of the CCMs' formation differ in the cave parts with perennial ice (called *glacial zone* throughout this chapter) and in sections where the ice occurs only seasonally (*periglacial zone*). Microclimatic conditions similar to the periglacial zone are also met seasonally in the entrance sections of caves, which are not iced permanently. Therefore seasonal formation of CCMs and icicles can occur there. The ice cave climatology characterizing the glacial and periglacial zones is discussed in detail in Chapter 3.

It follows from the preceding discussion that the mineralogy, morphology, and particle size of CCMs are largely controlled by two sets of factors: physical conditions and chemical conditions. The first one includes temperature, air humidity, airflow changes during the year, and the thickness of the water layer subjected to freezing. A combination of these physical factors governs the phase changes of water: freezing, evaporation, and ice sublimation. The second one is represented by the chemical composition of the karst waters subjected to freezing and by the composition of the cave atmosphere (especially its CO_2 content). Combining these physical and chemical factors determines three major groups of cave environments, in which CCMs form most frequently. The calcium bicarbonate-dominated waters are present in environments (i) and (ii) and calcium sulfate-dominated waters are present in environment (iii), as shown here:

(i) Ice caves in carbonate rocks (both their glacial and periglacial zones, seasonally also cave entrances of non-iced caves) characterized by intense winter air circulation and by rapid freezing of thin layers of low-TDS calcium bicarbonate waters. These environments produce CCC_{fine} (mainly) and, more rarely, also other fine-grained CCMs. The majority of ice caves described in the second part of this book belong to this type of environment. The periglacial zones of ice caves located in limestone karst, where the freezing temperature occurs only seasonally, in addition to CCC_{fine}, also host another specific speleothem type, cryogenic cave pearls, whose precipitation is caused by both cryogenic and non-cryogenic mechanisms. Details on these environments are given in Section 6.3.

(ii) Caves in carbonate rocks where slow freezing of low-TDS calcium bicarbonate karst waters in larger water pools usually deeper inside the caves (or in restricted environments, e.g., ice-covered water pools) leads to precipitation of CCC_{coarse}. Other coarse-grained CCMs are rarely present. The formation of CCC_{coarse} typically occurs in permafrost conditions, near the 0°C isotherm. The vast majority of documented CCC_{coarse} sites comes from caves that are not being iced presently, thus indicating freezing conditions during past glaciations. Details on these environments are given in Section 6.4.

(iii) Ice caves in gypsum-rich sedimentary sequences (both their glacial and periglacial zones, seasonally also cave entrances of non-iced caves in gypsum) characterized by intense winter air circulation and by rapid freezing of high TDS sulfate-dominated karst waters. CCG is the most common mineral in these environments. The periglacial zones of these gypsum ice caves are characterized by the greatest mineralogical and morphological variability of CCMs, including ephemeral sulfate and borate minerals. Details on these environments are given in Section 6.5.

6.3 CRYOGENIC MINERALS RELATED TO RAPID FREEZING OF LOW-TDS WATER IN LIMESTONE CAVES

CCC_{fine} dispersed in ice or formed directly on ice surfaces commonly transforms into secondary accumulations. If there is a period of seasonal ice melting (frequently related to higher influx of drip water), the flowing melt-water usually transports the CCC_{fine} along ice surfaces and deposits it either on flat sections of the ice bodies or on the cave floor outside the perennial ice accumulations. The thickness of these layers is usually several millimeters, but locally they can reach a few centimeters or more (Scărișoara Ice Cave, Romania, Racoviță and Onac, 2000; Eisriesenwelt Ice Cave, Austria, Spötl, 2008; general descriptions of these caves are in Chapters 25 and 13). Buildups of CCC_{fine} are common on the ablated surfaces of ice speleothems (Fig. 6.1B–C).

Accumulations of cryogenic powder are rarely pure. Besides CCMs, these accumulations contain a mixture of particles of different origins, commonly fragments of limestone loosened by frost shattering (see Fig. 6.1A), pollen, and macroplant remains (Feurdean et al., 2011; Pop and Ciobanu, 1950), diatoms (Lauriol et al., 2006), fly ash particles, fragments of insects hibernating in the cave (and though rarely, of bat bones), and other organic matter transported into the cave by air flow and flowing and percolating water, fallen gravitationally or transported by visitors.

The presence of white carbonate powder on the surface of cave ice was certainly noticed during the early visits of ice caves in the 19th century, but to our knowledge, the first explanation of CCC_{fine} as a cryogenic product in limestone caves was published by Kunský (1939) from caves in the Tatra Mountains (Belianské Tatras) near the border between Slovakia and Poland. From that time, powdery cryogenic precipitates were noticed and/or studied by many researchers (e.g., Bazarova et al., 2014a,b; Clark and Lauriol, 1992; Ek and Pissart, 1965; Lacelle, 2007; Lacelle et al., 2006, 2009; Lauriol et al., 1988, 2006; Luetscher et al., 2007, 2013; May et al., 2011; Viehmann, 1958, 1959; Pulina, 1990; Pulinowa and Pulina, 1972; Savchenko, 1976; Spötl, 2008; Teehera et al., 2017; Žák et al., 2004, 2008, 2010, 2013).

Optical and scanning electron microscopy (SEM) revealed that the CCC_{fine} typically comprise a varied mixture of individual crystals and a wide range of their more or less complex aggregates. The typical size of crystal aggregates ranges from 30 μm to 1 mm. Crystals usually appear in various deformed rhombohedrals and other not easily recognizable morphologies with tabular, lathy, and

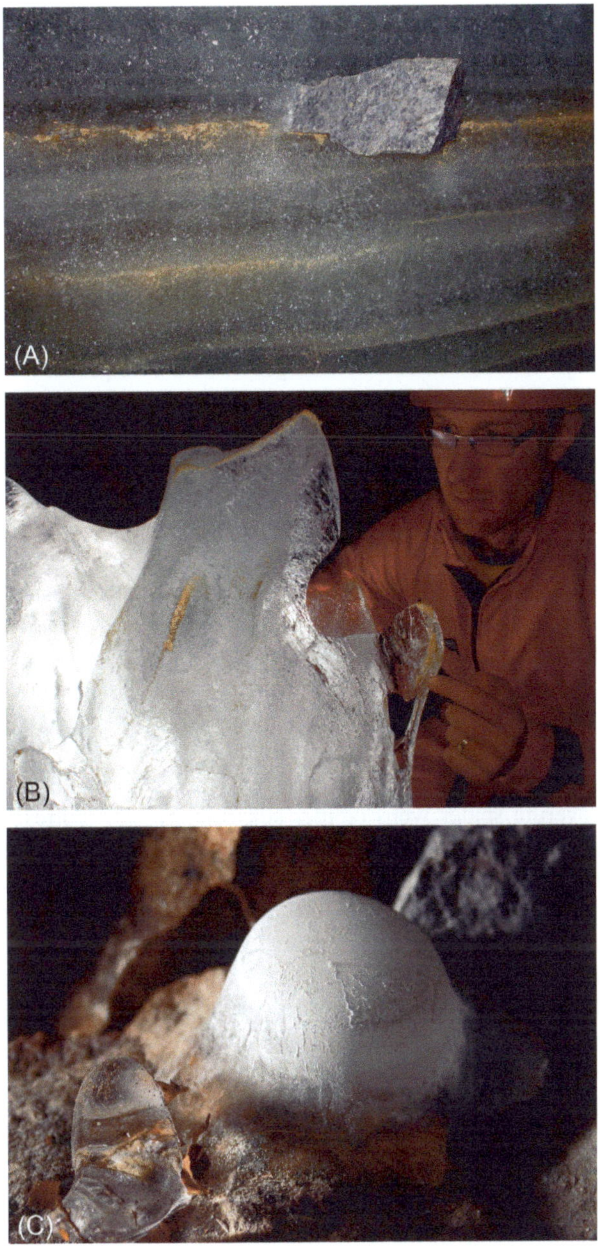

FIG. 6.1

Typical occurrences of CCC_{fine} in the field: (A) layers of CCC_{fine} enclosed in the ice together with a limestone clast (length 20cm), which fell from the cave roof because of frost shattering, Eisriesenwelt Cave, Austria (photo by K. Žák); (B) secondary accumulation of CCC_{fine} on congelation ice in the entrance section of Suchá Cave, Demänovská Valley, Slovakia (photo by M. Filippi); and (C) secondary accumulation of CCC_{fine} dominated by ikaite on the surface of an ice stalagmite (height of the stalagmite 7cm) in Koda Cave, Bohemian Karst, Czech Republic (photo by M. Filippi).

acicular habits, commonly exhibiting parallel or subparallel growth. Sometimes, crystals develop curved edges and skeletal or dendritic morphologies (Spötl, 2008). The latter morphologies suggest fast growth, with edges expanding faster than at crystal faces. Aggregates representing pseudomorphs (e.g., calcite after ikaite) are typically porous (Andreychouk et al., 2013). It is assumed that aggregates containing spherical cavities formed around bubbles of CO_2 gas (Andreychouk et al., 2013; Spötl, 2008).

Some indicia and remarks in the published studies suggest that pseudomorphism after unstable (or metastable) minerals (e.g., ikaite or vaterite) could play a more important role in CCC morphology than previously recognized (cf., Grasby, 2003; Lacelle et al., 2009). Sometimes, calcite crystal faces can show pitted surfaces and etched edges, probably as a result of calcite redissolution during periods of ice melting (Lacelle et al., 2009). Given the aforementioned, longer warmer periods with a slow temperature increase could be responsible for structural changes in CCC precipitates and thus recrystallization, overgrowths, and morphological distortions of some minerals.

The crystals' aggregates usually consist of concentrically grown individuals grouped in (hemi) spherical nodules, rosettes, and hedgehog-like formations. In addition, the spheres are frequently connected into chains. Besides spherulitic aggregates, sheaf-like ones and micro-flakes are also common in the CCC_{fine}. At some localities, micro-flakes dominate the cryogenic powder. They usually form distinct flat aggregates, very similar to cave rafts known from non-iced caves (cf. Hill and Forti, 1997); thus they are called here raft-like formations. Sometimes, the raft-like aggregates display a smooth side and a coarser crystalline opposite side. The traditional interpretation that the coarse crystalline side is the lower one (Spötl, 2008), derived from observations done on cave rafts of non-iced caves, is probably not valid in all cases in the ice caves. Andreychouk et al. (2013) considered the smooth surface as being the lower one, sitting on the ice. Typical morphologies of crystals and aggregates of CCC_{fine} are shown in Fig. 6.2.

In addition, Lacelle et al. (2009) identified a very specific morphological variety of cryogenic carbonate aggregates covering spider webs (spider silk calcite). This corresponds with information presented by Murase et al. (2001) stating that spider webs may act as sites of ice and calcite crystal nucleation, especially near the cave entrances.

The mineralogical composition of cryogenic powder is relatively simple, dominated in a majority of cases by one or more carbonate minerals. Low temperature of formation and increased mineralization of the solution are the conditions under which, in addition to calcite, other carbonate minerals that are metastable at higher temperature (ikaite, vaterite, monohydrocalcite) can form. Their detection requires sampling, transportation, and XRD mineral identification at a low temperature or traditional thermoanalytical methods. The analytical difficulties are probably the reason why these minerals have been detected so far in only a limited number of caves. Over the past 10 years, ikaite has been documented from the Scărișoara Ice Cave in the Bihor Mountains, Romania (Onac, 2008; Onac et al., 2011), on the surface of seasonal ice decorations in Koda Cave, Czech Republic, (Žák et al., 2010) and in caves of the Baikal Lake surroundings, Russia (Bazarova et al., 2014a,b). In these environments, ikaite occurs as white or yellowish powdery accumulation, formed by morphologically varied crystals and crystal aggregates typically smaller than 1 mm. The ikaite appears to be much more common in ice caves than was previously thought, as indicated by recent morphological studies of CCC_{fine} in caves of the Ural Mountains (Chaykovskiy and Kadebskaya, 2014; Kadebskaya and Chaykovskiy, 2012, 2013).

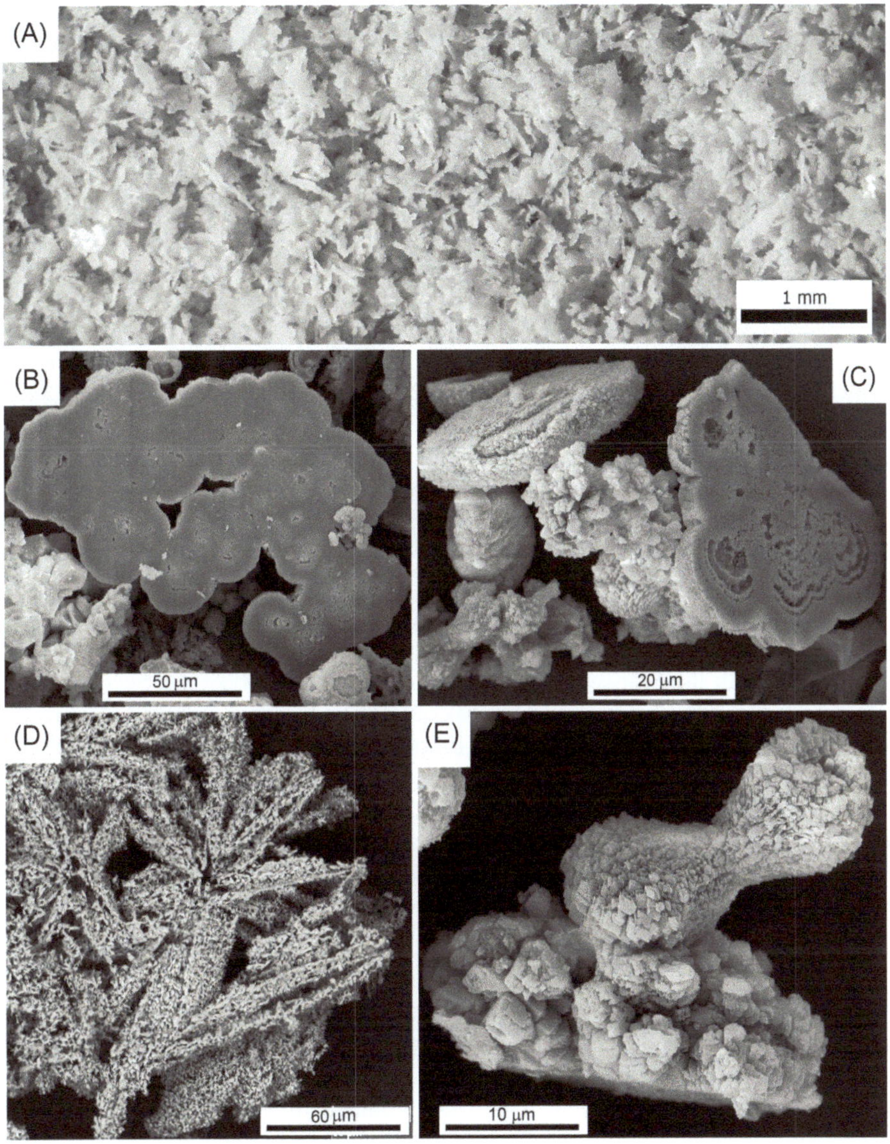

FIG. 6.2

Typical occurrences of CCC_{fine} and morphologies of crystals and crystal aggregates: (A) general photo of CCC_{fine} dominated by raft-like aggregates, collected from the ice surface in Demänovská Ice Cave, Slovakia; (photo by M. Filippi); (B, C) SEM image of the two examples of raft-like aggregates with globular upper and flat lower parts and other forms from ice surface in Scărişoara Ice Cave, Romania (SEM images by M. Filippi); (D) skeletal textures (with hollow internal parts; pseudomorphs after ikaite?) typical of Medeo and Eranka caves, Northern Ural Mountains (SEM image by O. Kadebskaya); and (E) SEM image of the sheaf-like aggregate consisting of very small rombohedral crystals, Scărişoara Ice Cave, Romania (SEM image by M. Filippi).

Ikaite ($CaCO_3 \cdot 6H_2O$) is a monoclinic carbonate that is metastable at temperatures approximately above 4°C (for stability conditions at high pressure, see Marland, 1975a, b; for stability conditions at low pressure, see Bischoff et al., 1993a). Ikaite was first detected in nature as a main component of underwater carbonate towers growing on the floor of the Ika Fjord (also spelled Ikka Fjord; Southwest Greenland) where cold groundwater discharges into near-freezing seawater (Buchardt et al., 1997; Pauly, 1963). Ikaite can form from cold solutions where pH and total mineralization are sufficient to support stability of carbonate and where Ca, bicarbonate, and/or carbonate ions are present. Some authors considered the presence of phosphate an important factor in ikaite precipitation and stabilization (Bischoff et al., 1993a; Selleck et al., 2007). When formed near the Earth's surface, ikaite is always metastable. If an ikaite sample is left at laboratory temperature, it decomposes within hours to other carbonate phases, mostly to monohydrocalcite ($CaCO_3 \cdot H_2O$) and anhydrous carbonate phases (calcite ± vaterite or aragonite). Suess et al. (1982), Shearman and Smith (1985), Jansen et al. (1987), Greinert and Derkachev (2004), and Selleck et al. (2007) observed ikaite formation in deep-sea sediments, usually in the form of idiomorphic crystals and aggregates, probably related to seepage of methane-rich diagenetic fluids. The glendonites (rhomb-shaped cross-section crystals of calcite and/or stellate clusters of well-developed pyramid faces), frequently found in marine sediments, were earlier interpreted as calcite pseudomorphs after the thenardite (Na_2SO_4; Kemper and Schmitz, 1975, 1981). Since the discovery of ikaite in deep-sea sediments, glendonites have been considered to be pseudomorphs of calcite after the ikaite (Selleck et al., 2007; Swainson and Hammond, 2001; similar morphologies were observed in the Scărişoara Ice Cave, Romania, Onac, 2008; see Section 6.4). Ikaite was also detected in salty springs near Shiowakka (Hokkaido, Japan) where monohydrocalcite is precipitated in the summer, whereas ikaite forms in the winter along with ice (Ito, 1996). It also forms (and formed in the past) during the winter season around groundwater discharging points on the bottom of saline Mono and Pyramid lakes in the United States (Bischoff et al., 1993a,b; Council and Bennettt, 1993; Whiticar and Suess, 1998). Omelon et al. (2001) proved ikaite's presence as a winter precipitate of mineralized springs on the Axel Heiberg Island in the Canadian Arctic. Occurrences of ikaite related to seawater freezing have been documented in the Southern Ocean around Antarctica, the Arctic Ocean (e.g., Dieckmann et al., 2008), and two fresh water lakes in southern Patagonia, Argentina (Oehlerich et al., 2010).

Vaterite, an unstable hexagonal $CaCO_3$, can also form in cold, highly mineralized environments (Grasby, 2003) under atmospheric pressure (Albright, 1971). It usually takes a spherical shape with crystals ranging from 0.05 to 2 mm (Kralj et al., 1990; Vecht and Ireland, 2000). Vaterite was observed as a cryogenic precipitate in freezing experiments conducted by Fairchild et al. (1996). This mineral has not been reported in natural ice cave environments, but its formation in them is probable. In a Canadian cave, Lacelle et al. (2009) observed spherical structures in cryogenic carbonate powder that morphologically resembled vaterite. While studying the seasonal ice in caves of the Belianske Tatras, Slovakia, Kunský (1939) suggested that the white mineral powder on the surface of ice could be vaterite (ikaite was not known at that time). Another mineral in cold ice cave environments is monohydrocalcite. It was first described by Onac (2001), as found in the Scărişoara Ice Cave, and was later listed among other CCMs by Andrejchuk and Galuskin (2008). Whether it forms primarily by cryogenic precipitation or because of ikaite breakdown is not presently clear.

Specific cryogenic mineralogy was observed in ice caves in the Baikal Lake area in Russia. Because of local climatic conditions and a presence of discontinuous permafrost, numerous caves in the western surroundings of the Baikal Lake host seasonal or perennial ice. The caves are developed in an Upper Proterozoic sedimentary sequence that, in addition to limestone and dolomite, includes sporadic gypsum layers. Besides CCC (ikaite, Bazarova et al., 2014a,b) and rare gypsum and barite (for barite occurrence in gypsum ice caves, see Section 6.5), lansfordite ($MgCO_3 \cdot 5H_2O$) was found in the cryogenic precipitates of the Khrustalnaya and Bolshaya Onotskaya caves (Bazarova and Gutareva, 2011; Bazarova et al., 2016). The cryogenic mineral formations dominated by lansfordite are widespread in the first cave, forming the largest accumulations in its lower part near the bases of ice stalagmites. It forms relatively large (up to 200 μm) tabular crystals that combine a rhombic prism and two pinacoids. In the Bolshaya Onotskaya Cave, the cryogenic powder covers the surface of ice stalagmites and falls down, generating small accumulations at the foot of ice stalagmites. An XRD investigation showed that the cryogenic powder in this cave consists predominantly of lansfordite, sometimes with an admixture of ikaite and calcite and minor amounts of hydromagnesite, $Mg_5(CO_3)_4(OH)_2 \cdot 4H_2O$. Bazarova et al. (2016) supposed that ikaite crystallization reduced the Ca^{2+} content of the solution and thus provided favorable conditions for the precipitation of Mg minerals. The Mg^{2+} can also be concentrated in the surface layer of ice undergoing repeated cycles of growth and melt, because the segregation coefficient between water and ice is different from Ca^{2+} (see Section 6.2).

The XRD study of cryogenic precipitates formed during the winter season on the surface of icicles in the entrance section of the Koda Cave (Bohemian Karst, Czech Republic) indicated that apart from ikaite, rapidcreekite ($Ca_2(SO_4)(CO_3) \cdot 4H_2O$) is probably present (Žák et al., 2010), but more analytical work is needed for its definitive confirmation as a cryogenic mineral.

The periglacial zones of limestone-hosted ice caves can contain, in addition to the previously mentioned fine-grained CCMs, another conspicuous speleothem type: cryogenic cave pearls (CCPs). The existence of this specific type of rounded secondary carbonate speleothems was recorded very early by Brückmann (1728) during his visit to the Demänovská Ice Cave in Slovakia. Based on an occurrence in the Scărișoara Ice Cave in Romania, CCPs were described and studied in detail by Viehmann (1958, 1959, 1962, 1963, 1967), Viehmann and Moțiu (1974) and Viehmann (1993). Current petrographic and geochemical methods were applied on samples from Romanian and Slovakia ice caves by Žák et al. (2013), who supported their classification as a specific and genetically distinct type of cave pearls.

The majority of published occurrences of CCPs are in caves in the Carpathian Mountains. These mountains, extending across Central and Eastern Europe, are located in a temperate climatic zone and do not accommodate surface glaciers. Under the present-day climate, very restricted patches of sporadic and shallow permafrost exist on the northern slopes of the highest peaks (especially in the High Tatra Mountains in Slovakia). The Carpathian caves hosting CCPs are located at elevations between 700 and 1200 m a.s.l. in the territory of Romania and Slovakia. Their periglacial zones are wide and are characterized by seasonal temperature oscillations around the freezing point. Ice speleothems form during winter and early spring there and usually disappear almost completely during late spring and summer.

The best-known occurrences of CCPs are found in the Scărișoara Ice Cave (see also Chapter 25) and in the Demänovská, Suchá, Stanišovská, and Malá Stanišovská ice caves (or seasonally ice caves) in Slovakia (see also Chapter 29). The seasonal freezing represents a necessary condition for the formation of CCPs. Without this cold period, only ordinary cave floor speleothems (stalagmites, continuous flowstone layers) will form. In addition to the CCPs in the Carpathian caves, CCPs were observed in the periglacial environment of some other caves hosting perennial or seasonal ice deposits. One such example is the Kinderlinskaya (also known as Pobeda) Cave in the Southern Urals (Kadebskaya and Chaikovsky, 2014). The position of CCPs in the field and their isotope data are similar to those in the caves of Carpathians.

CCPs occur in spatially restricted accumulations or extensive pearl fields (usually a few centimeters up to 0.5 meters in thickness) within the scree covering the cave floor in the periglacial zone of these caves (Fig. 6.3). The CCPs sites are located in the cave floor areas, where the ice stalagmites and columns form during the winter, and where drip water is available. Any liquid water is, however, drained away through talus, hindering the formation of water pools. Locally, this periglacial zone of ice caves exhibits a patterned ground. A similar size-sorting effect has also been observed within some of the CCPs' accumulations. Freezing/thawing processes are responsible for moving the pearls, preventing their cementation on the floor or between individual pearls. It was directly observed and documented that the newly forming ice crystals and floor ice are capable of lifting and rotating the CCPs.

FIG. 6.3

Occurrences of cryogenic cave pearls (CCPs) in the periglacial zone of ice caves: (A) typical position of accumulations of CCPs in the area with cryogenic limestone freed, photographed after melting of bottom ice and ice stalagmites, Demänovská Ice Cave, Slovakia, width of photo c. 2 m; (B) size-sorted CCPs with fragments of limestone, pearls size up to 6 mm, field photo Stanišovská Cave, Low Tatra Mountains, Slovakia; (C) small CCPs with rough crystalline surface, pearls' size up to 1 mm, Demänovská Ice Cave, Slovakia; and (D) small CCPs with smoother surface, pearls' size up to 2 mm, Demänovská Ice Cave, Slovakia (photos by M. Filippi).

CCPs show high porosity, ranging from 7.6% to 22.6% (Žák et al., 2013). In the center, they frequently contain aggregates of radial carbonate crystals (cryogenic micro-pearl) or several such aggregates cemented together acting as a nucleus. The concentric layering of CCPs is less obvious as compared to normal cave pearls. Morphologically, CCPs are irregular or spherical, ranging from less than 1 mm to ~30 mm. Three basic groups can be distinguished (Žák et al., 2013): (1) CCPs without cortex formed by many coarse-crystalline nuclei, (2) CCPs formed by coarse-crystalline nuclei encased by a concentric cortex, and (3) CCPs without a clearly defined nucleus, consisting especially of a concentrically zoned cortex. The surface of the larger CCPs is macroscopically smooth, whereas the smallest CCPs (<1 mm) are typically more coarsely crystalline (Fig. 6.3). In addition, polyhedral pearls and local pearl aggregates are common. In the Scărişoara Ice Cave, cementation of thousands of individual pearls has led to the formation of an oolite-like layer, for which Viehmann (1963) used the term "pearl conglomerate."

Results of U-series and radiocarbon dating indicate that the pearls are primarily Holocene in age, with their growth continuing into the present (Žák et al., 2013). Stable carbon and oxygen isotope studies indicate the formation of an initial crystal aggregate by cryogenic precipitation. After that, pearls grow by addition of new carbonate layers, either at water freezing conditions or at temperatures slightly above 0°C. Carbon and oxygen isotope data of larger CCPs fall along a line, which links the field of C and O isotope composition of common speleothems with that of highly kinetically fractionated CCC_{fine}, formed during rapid freezing of water on the ice surface (details on the isotope chemistry of CCPs are given in Section 6.6). Micro-pearls show the highest $\delta^{13}C$ values and are formed cryogenically.

In both the periglacial and glacial zones of ice caves, it is common to find soft accumulations of fine-grained, dominantly fibrous mineral matter, known as moonmilk. The genetic processes of moonmilk are very complex and widely discussed (Hill and Forti, 1997). Whether the moonmilk subtype occurring in cold cave environments is genetically related to the ice presence or not is not clear and is not discussed in this chapter.

6.4 CRYOGENIC CARBONATES (CCC_{COARSE}) FORMED BY SLOW FREEZING OF LOW-TDS WATER IN LIMESTONE CAVES

This setting is characterized by a low-TDS calcium bicarbonate water similar to that producing CCC_{fine}, but by different freezing conditions. Whereas CCC_{fine} forms during the rapid freezing of a thin water film on the ice surface, the CCC_{coarse} is supposed to crystallize more slowly during freezing of larger water bodies. The CCC_{coarse} represents the most frequently studied cryogenic cave mineral because of its paleoclimatic significance. CCC_{coarse} is mainly documented in caves presently free of ice and is included in this chapter as an important indicator of former cave glaciations.

The cryogenic origin of CCC_{coarse} was documented by four types of observation: field position and character of CCC_{coarse} sites in the caves, morphology of the aggregates, U-series age determinations of CCC_{coarse} fitting to glacial conditions, and C and O isotope data that provide evidence of freezing conditions (Section 6.6). Since the majority of the cavities in which the CCC_{coarse} was found are largely isolated from the seasonal temperature oscillations at the surface, the only process that could have cooled them down to freezing temperatures is the presence of permafrost during cold climatic phases.

Deciphering the mechanisms of CCC_{coarse} formation and understanding the geographic distribution of caves hosting these mineral accumulations took several decades. During the second half of the 20th century, the intensity of cave exploration increased in Central Europe (Germany, Poland, Czech Republic, and Slovakia). New caves were discovered not only in mountain regions but also in lowlands and uplands (hilly regions with elevations generally below 600 m a.s.l.). Many of the cave passages were completely filled with sediments, which had to be excavated during the exploration work. Other caves that were not connected with the surface were discovered during operations in limestone quarries. The majority of these caves are associated with very old karstification phases taking place during periods when the groundwater level was much higher than it is today and are now hydrologically inactive. Accumulations of CCC_{coarse}, freely deposited on the bottom of cavities, have been discovered in some of these caves. All of these caves presently have cave temperatures above the freezing point and are free of ice. Skřivánek (1954), probably the first one to notice this specific speleothem type, found these carbonates in Srbsko Caves in the Bohemian Karst close to Prague, Czech Republic. Unfortunately, the site was not studied in detail and was subsequently destroyed by the cave's visitors. Similar types of cave carbonates were discovered later in Slovakia (Tulis and Novotný, 1989), Poland (Urban and Złonkiewicz, 1989), and Germany (Erlenmeyer and Schudelski, 1992; Schmidt, 1992). Using radiocarbon dating and stable isotope analyses, Durakiewicz et al. (1995) first recognized that these speleothems must have formed during glacial conditions, in the related freezing temperatures.

After the publication of a new genetic model (Žák et al., 2004), numerous other CCC_{coarse} sites were described, all located in Eurasia (Bartolomé et al., 2015; Bazarova et al., 2014a,b; Chaykovskiy et al., 2014; Colucci et al., 2017; Dublyansky et al., 2014, 2015a, 2017; Luetscher et al., 2013; Meissner et al., 2010; Onac et al., 2011; Orvošová, 2015; Orvošová et al., 2014; Richter and Niggemann, 2005; Richter and Riechelmann, 2008; Richter et al., 2008, 2009a,b, 2010a,b, 2011, 2013, 2015; Spötl and Cheng, 2014; Žák et al., 2006, 2008, 2011, 2012). The U-series ages of CCC_{coarse} and C and O stable isotope data are contained in majority of these papers.

With respect to the position and general characteristics of CCC_{coarse} sites, it can be summarized that their individual components are neither connected nor attached to the substrate on which they are

FIG. 6.4

Field photo of a typical CCC_{coarse} occurrence. Crystals and aggregates with sizes up to 35 mm are deposited on the surface of limestone block, Jaskyňa Studeného Vetra (Cold Wind Cave), Low Tatra Mountains, Slovakia; the scale bar is 10 cm (photo by M. Majer).

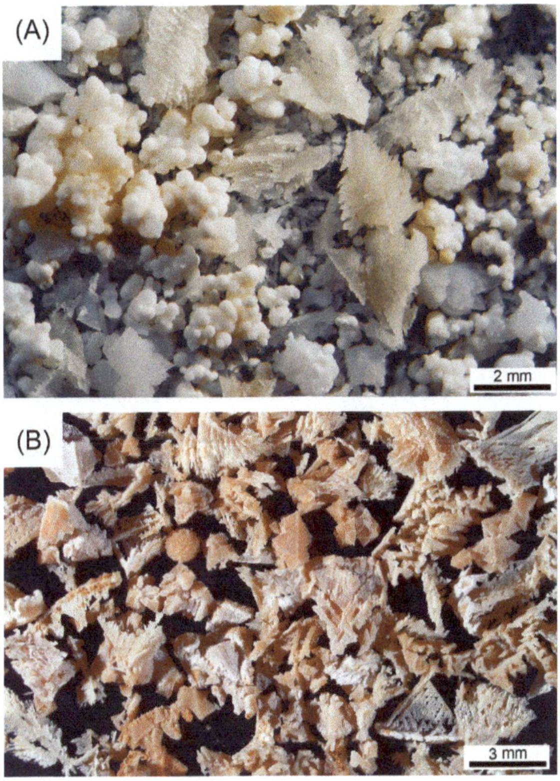

FIG. 6.5

Typical morphologies of CCC_{coarse} mixtures of various morphological forms: (A) Jaskyňa Studeného Vetra (Cold Wind Cave), Low Tatra Mountains, Slovakia; and (B) Portálová Cave, Bohemian Karst, Czech Republic (photos by M. Filippi).

deposited, a feature typical of all CCMs (Fig. 6.4). The material typically forms loose accumulations (Fig. 6.5), which do not cover the whole floor of the cave chambers, but occur in restricted areas only, typically covering a few square meters or less. The thickness of these deposits usually ranges between 1 and 50 mm. The CCC_{coarse} is usually found in large chambers or corridors, typically in their central parts, that is, not close to the cavity walls. The underlying material comprises either limestone blocks that fell from the cave roof or clastic cave sediments of various types.

The habits of CCC_{coarse} crystals and the morphology of the aggregates are similar to those in CCC_{fine} powders. The CCC_{coarse} crystals' larger dimensions, however, enable easy observation on a large number of specimens. As a result, numerous local terms describing these forms were developed in German, Polish, Russian, Czech, and Slovak languages. Generally, the CCC_{coarse} can be classified as several end-member types, each differing in characteristic morphology, its position in sequence of crystallization, and (probably) by slightly different (unfortunately, still not fully understood) environments of precipitation. A number of "transitional" types can be found in CCC_{coarse} accumulations. As of now, the following types (groups) of CCC_{coarse} can be roughly distinguished: (1) single crystals and their random or organized intergrowth/aggregates (Fig. 6.6), (2) fine- to coarse-crystalline, more or less (hemi-) spherical forms (Fig. 6.7), and (3) planar raft-like aggregates (Fig. 6.8).

FIG. 6.6

Typical morphologies of CCC_{coarse} aggregates of coarse crystals: (A) Hačova Cave, Slovakia, size 10 mm; (B) Novoroční (New Year) Cave, Bohemian Karst, Czech Republic, size 13 mm (photos by M. Filippi).

Single crystals (several tenths of millimeter up to several millimeters in size) usually exhibit rhombohedral or scalenohedral (or combinations thereof) morphology and have tabular, lathy, or acicular habits. Individuals are rarely perfectly/regularly developed; they commonly form twins and aggregates. Crystals also often exhibit skeletal growth, resulting in hopper and serrated habits that may be characteristic of the entire crystals or only certain faces or edges. In addition, various aggregates with complex morphologies such as rosettes, sheaves, chains, dendrites, or interpenetrating formations are typical (Durakiewicz et al., 1995; Richter et al., 2013).

Spherulites, represented by variously distorted spherical or hemispherical formations, are the second large group of CCC_{coarse}. In the literature, they are known by various names: "Zopfsinter" by Erlenmeyer and Schudelski (1992), "hemispheres" by Tulis and Novotný (1989) and Forti (1990),

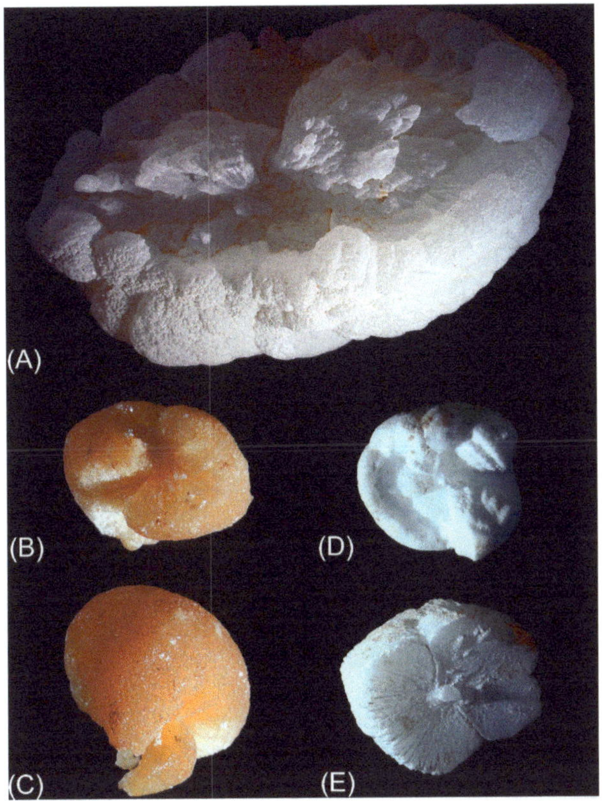

FIG. 6.7

Typical morphologies of CCC_{coarse} spherulitic forms: (A) basal part of a 27 mm large aggregate, Jaskyňa Studeného Vetra (Cold Wind Cave), Low Tatra Mountains, Slovakia; (B, C) examples from Chelosiowa Jama Cave, Poland, size up to 9 mm; and (D, E) Stratenská Cave, Slovakia, size up to 7 mm (photos by M. Filippi).

"cave caps" by Hill and Forti (1997), and "cupula-spherolites" by Richter and Riechelmann (2008). Terms such as stellar-, rosette-, hedgehog-, dumbbell-, dish-, and fungiform-shaped can also be found in published papers (Richter et al., 2008, 2013; Urban and Złonkiewicz, 1989).

The spherulites may have either smooth or fine- to coarse-crystalline surfaces. They occur individually, but can also be arranged in planar or chain-like aggregates, or the spherulites overgrow other types of CCC_{coarse}. Spherulitic CCC_{coarse} has polycrystalline fibrous internal structure, built up of subradially oriented crystallites. Crystalline fibers branch at small angles, which do not appear to be crystallographically controlled. According to Meakin and Jamtveit (2010) spherulites form by two mechanisms: they either grow outward from a nucleus, splitting and branching to fill the available space, or they first form a fiber (crystal individual) that evolves through crystal splitting into a sheaf that continues to grow and spread until it becomes spherical. Examples of both growth mechanisms can be observed for CCC_{coarse}.

Flat raft-like formations represent the third significant group of CCC_{coarse} morphologies. They are characterized by a planar appearance. Orientations of crystals in such aggregates vary. In most cases,

FIG. 6.8

Typical morphologies of CCC_{coarse} planar and raft-like aggregates: (A) Na Javorce Cave, Bohemian Karst, Czech Republic, size 11 mm; (B) Javoříčko Caves, Moravia, Czech Republic, size 7 mm; and (C) raft-like form with two types of crystals (the white are later), Na Javorce Cave, Bohemian Karst, Czech Republic, size 10 mm (photos by M. Filippi).

crystals are arranged at steep angles to the raft's plane (Fig. 6.8). In some cases, however, rafts are formed by stellar or rosette-like aggregates composed of crystals oriented parallel to the raft's plane (Spötl and Cheng, 2014; Žák et al., 2011). The calcite crystals on raft-like plates are often elongated in the direction of their *c* axes. The sizes of crystals vary from several tenths of micrometers to several millimeters. In contrast to raft-like aggregates of the CCC_{fine}, the coarse-grained ones do not usually have a smooth side (Žák et al., 2011; CCC_{coarse} raft-like aggregates with flattened lower side reported by Richter et al., 2013).

From a mineralogical point of view, these aggregates are represented by low-Mg calcite. Based on the Žák et al. (2004) model and stable isotope studies (Section 6.6), it follows that the sequence of formation of these aggregates begins with the single crystals and their simple aggregates, whereas more complex, often spheroidal aggregates form toward the end of the depositional sequence. The isotopically most evolved type of CCC_{coarse}, formed at the final stages of crystallization, is represented by regular compact hemispheres with tight radial arrangement of crystals, rarely by spheres of sizes up to several centimeters (see Fig. 6.7).

The published models (Richter et al., 2010a; Žák et al., 2004, 2012) envisage freezing of water in a cave environment at least partly isolated from the outside atmosphere (or in a local micro-environment isolated from the cave atmosphere). This isolation can be caused by a layer of congelation ice covering the entire surface of a water pool. The analysis of carbonate C and O isotopes (cf. Section 6.6) indicate that, during the freezing process, the water pond is not subject to significant evaporation and that gases were at least to some degree confined either by the ice cover or by the cave walls. The current understanding is that CCC_{coarse} typically forms near the 0°C isotherm during the influx of water into a cavity located within permafrost. This occurs especially during permafrost degradation (thawing) during periods of surface warming (Richter et al., 2010a; Žák et al., 2004, 2012).

The aggradation and degradation of permafrost have a significant inertia with respect to more rapid climatic changes at the surface. During permafrost growth, the potential for CCC_{coarse} formation is reduced, because the frozen ground prevents infiltration of water into the caves. In contrast, during a warmer period, the permafrost thaws, leaving relic permafrost at depth. Soil respiration during these warm phases enhances the total dissolved solids (TDS) of the infiltrating water. Penetration of this mineralized water into caves located within the relic permafrost represents the most suitable conditions for the formation of CCC_{coarse}. Based on thermal modeling, Pielsticker (2000) showed that an ice fill of a cave in permafrost typically has a short lifespan, occurring only during a permafrost destruction phase. Ice fills the voids in a cave when thawing of the permafrost is locally accelerated by water penetrating along water-conducting zones, which can be a rather rapid process that forms a small pool inside the still-frozen cave.

The preceding model also explains the position of the CCC_{coarse} findings within the cavities. The ice morphology in a cave located in a permafrost zone is different from the ice in ice caves in a nonpermafrost setting. Within the permafrost, the cave walls have subzero temperatures, and the ice does not melt on or near the walls, but may form on them (cf. Lauriol et al., 1988). Frequently, ice completely plugs cave passages, but open karst conduits are common even in permafrost environments (Ford and Williams, 2007, p. 421–427). If a solution enters an empty cavity situated in a permafrost zone, cave ice may form. Depending on the discharge, water ponds form on the surface of this cave ice. Freezing occurs in all directions: from the cavity floor, the cavity walls, and on the pool's surface. The CCC_{coarse} forms in pools of slowly freezing water and can eventually be trapped within the ice. After the ice melts, the cryogenic carbonates are deposited on the cavity floor. Despite increasing evidence for Holocene CCC_{coarse} deposits (see group (ii)), actively forming sites have not yet been observed.

Based on their geographic position, cave altitude, and age, the known CCC_{coarse} sites can be subdivided into several groups, each with a specific paleoclimatic inference:

(i) The first large group is represented by caves located in lowlands, uplands, and lower mountains of Central Europe (Germany, Czech Republic, Slovakia, and Southern Poland), an area that today is free of permafrost. All these sites are arranged in an ice-free, E–W extended zone, located during the last glacial period between the southern extension of the Scandinavian Ice Sheet and the northern limit of the alpine glaciers. This periglacial zone of the last glacial period had practically no surface glaciers (except for small glaciers in local mountains), but had sporadic or widespread permafrost (French, 2007; Vandenberghe et al., 2014). The ages and depths of CCC_{coarse} occurrences in Germany, Czech Republic, Slovakia, and Poland were summarized by Žák et al. (2012), who concluded that the Last Glacial Maximum (LGM) permafrost locally reached depths of at least 65 m in these areas, and the presence of relic permafrost has been documented at depths of more than 400 m below the surface in Northeast Poland, about 500 km to the north of the studied caves (Honczaruk and Śliwiński, 2011; Szewczyk and Nawrocki, 2011). The CCC_{coarse} ages of this Central European area plot typically at the transitions from cold to warmer climates (e.g., from stadial to interstadial during the Weichselian glaciation period or at the transition from the Weichselian glaciation to the Holocene geological epoch; Orvošová et al., 2014; Žák et al., 2012).

(ii) The second group of localities represents sites in the caves located at high mountain elevations in the Pyrenees, Alps, and Carpathians, characterized by much cooler climates and the presence of Holocene mountain permafrost of various extents and thicknesses (Krainer and Ribis, 2012; Schöner et al., 2012). The ages of these CCC_{coarse} sites correspond both to pre-Holocene glacial periods and Holocene climatic oscillations. Žák et al. (2009) described sites in the Cold Wind Cave (Low Tatra Mountains, Slovakia; cave entrance at 1678 m a.s.l.), where CCC_{coarse} was formed repeatedly during the Saalian and Weichselian glacials. Žák et al. (2011, 2012) studied an occurrence at a depth of 285 m below the nearest surface in the Mesačný Tieň Cave (cave entrance 1767 m a.s.l.), High Tatra Mountains, Slovakia, with a U-series age of 17.13 ± 0.12 ka, which represents the deepest known occurrence of Pleistocene CCC_{coarse}. Richter et al. (2009b) described an occurrence at a depth of 229 m below surface in the Glaseishöhle Cave in Steinernes Meer, Northern Calcareous Alps, Germany, which is not dated yet and may date from the Pleistocene geological epoch. The known Holocene sites occur at even higher altitudes. Luetscher et al. (2013) described an occurrence in the Leclanché Cave located in the Western Swiss Alps at an elevation of 2620 m a.s.l., formed during several episodes of alpine permafrost degradation associated with warm climate intervals between 129 BC and AD 1249. Similarly, Spötl and Cheng (2014) described a site in a cave with an entrance at 2558 m a.s.l. in the western part of the Austrian Alps, which has morphological similarities to the cold-trap ice caves. The CCC_{coarse} U-series age indicated a formation around 2600 BP. This age corresponds to a cooling following the second peak of the Holocene climatic optimum (cf. McMichael, 2012), suggesting that the relationship between CCC_{coarse} formation and permafrost may be more complex than initially thought. Although most of these samples were found after the cave ice melted, CCC_{coarse} findings were also reported in situ from two ice outcrops in the Pyrenees and the Alps (Bartolomé et al., 2015; Colucci et al., 2017). The age of the site in the Pyrenees occurs at an elevation of 2780 m a.s.l. and corresponds to the Medieval Climate Anomaly (Bartolomé et al., 2015).

(iii) A specific distribution of CCC_{coarse} localities was observed in the Ural Mountains, Russia, almost a 2000 km-long N–S extended mountain range passing several climatic zones. CCC_{coarse} from several caves of Northern and Central Ural were studied and dated by Chaykovskiy et al. (2014). At these locations, CCC_{coarse} formed during several periods between 125.3 ± 1.2 and 13.4 ± 0.27 ka BP, that is, not only during the Weichselian but also the Eemian interglacial. The periods of CCC_{coarse} formation in caves of the Ural Mountains were shown to stretch as far back as 700 ka BP according to preliminary reports of Dublyansky et al. (2014, 2015a). More recently, Dublyansky et al. (2017) reported CCC_{coarse} of the Marine Isotope Stage 3 age in two caves in the Southern Urals (at 53 degrees North).

(iv) Until now, the geographic distribution and ages of CCC_{coarse} sites distributed in the Eastern Sayan Mountains, the Kuznetsk Alatau Mountains, the Lake Baikal area, and the Sikhote-Alin Mountains in the Russian Far East have been poorly understood. Bazarova et al. (2014a, 2014b) described CCC_{coarse} occurrences in the Okhotnichya Cave in the area west of Lake Baikal. The area around this cave has a present mean air temperature of −2.7°C (Bayanday village) and a slowly degrading discontinuous permafrost with a thickness of 40 to 60 m. The site is located near the entrance, in the lower part of a down-sloping cave section.

All CCC_{coarse} findings, which formed in relation to the existence of permafrost, have been documented in Europe and Asia; no occurrences of CCC_{coarse} have been reported in North America, which is probably related to sharply different distribution of permafrost during the late Pleistocene epoch (Vandenberghe et al., 2014). Although areas located south of the Eurasian continental glaciation display wide periglacial permafrost zones, permafrost zone was much narrower in North America.

A very specific CCC_{coarse} locality briefly mentioned by Onac et al. (2011) was from the Scărişoara Ice Cave in Romania. Large CCC_{coarse} aggregates (rosettes up to 4.7 cm) with shapes similar to the "glendonites" in marine environments have been found protruding out of an ice tongue extending from the main ice block in the periglacial zone of the cave. Their rather unusual low $\delta^{13}C$ values (average −13.3‰) suggest formation of these crystal aggregates in combination with carbon derived from organic matter, probably from decomposition of organic debris commonly present in the ice in the Scărişoara Ice Cave. The age of the ice block is Holocene (Perşoiu, 2011; Perşoiu et al., 2017).

The majority of the CCC_{coarse} sites are dominated by crystals and aggregates of carbonate minerals and other minerals are reported quite rarely. One such case mentioned by Dublyansky et al. (2017) was in the Shulgan-Tash Cave of South Ural, Russia, where abundant cryogenic gypsum and accessory barite were observed along with CCC_{coarse}.

6.5 CRYOGENIC CAVE MINERALS RELATED TO RAPID FREEZING OF HIGH-TDS WATER IN GYPSUM CAVES

Compared to limestone ice caves, in gypsum ice caves, the formation of CCMs typically involves solutions with higher mineralization (usually more than $0.8\,g \cdot L^{-1}$ TDS), dominated by dissolved sulfate ($SO_4^{2-} > HCO_3^-$). CCMs can also form during the winter in the cave entrance sections of non-iced gypsum caves. In these environments with high winter ventilation, the solutions freeze rather rapidly as thin water films on the ice surface or in very shallow water pools.

By far, the most common cryogenic precipitate is represented by fine-grained gypsum powder, which regularly contains minor admixtures of other minerals. The typical sequence of crystallization during progressive freezing of a single batch of solution in gypsum ice caves is gypsum → carbonate → celestine. In Russia and Ukraine, where the most important and extensively studied gypsum-hosted ice caves are located, the cryogenic gypsum powder is commonly called *gypsum flour*, a synonym for the term *gypsum powder* preferred in this book.

A major difference between the behavior of CCC and CCG is that during ice melting, the gypsum powder can be partly (or in some cases totally) redissolved, whereas the carbonates typically survive ice melting without a substantial redissolution. Therefore CCG forms the most conspicuous secondary accumulations associated with cave ice sublimation. Partial dissolution and recrystallization of the CCG powder lead to the formation of a wide variety of crystal and aggregate morphologies and sizes.

CCG recrystallization occurs in the glacial zone of a cave when liquid water influx follows a period of ice sublimation and the related accumulation of gypsum powder on the ice surface. This is because the temperature of the glacial zone of the most frequently studied ice cave in gypsum (Kungur Ice Cave, Russia; Andreychouk et al., 2013; Dublyansky, 2005; Kadebskaya and Tchaikovsky, 2013) remains below 0°C year-round (Fig. 6.9).

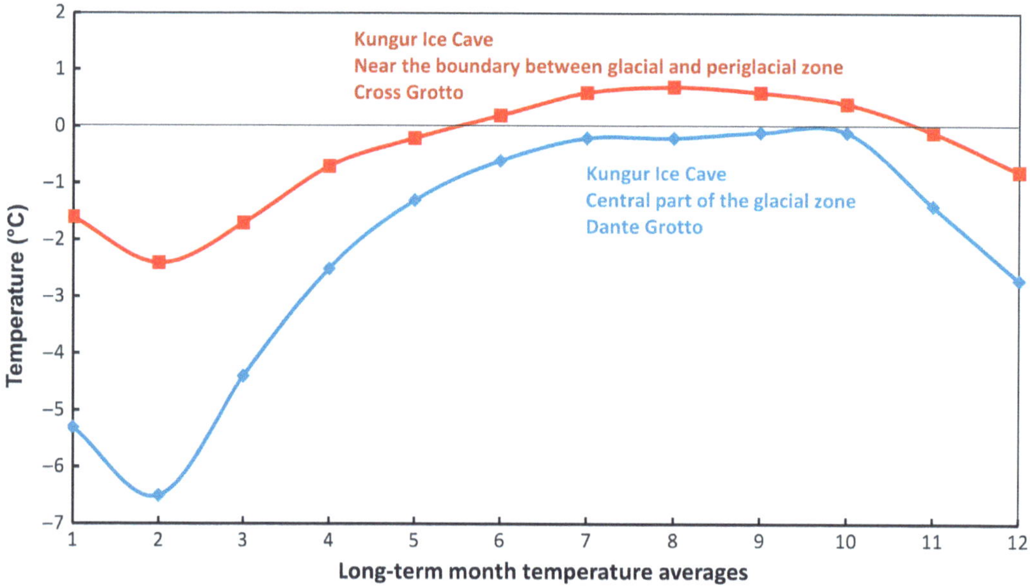

FIG. 6.9

Seasonal changes of temperature (monthly means, 1951–94) in the glacial and periglacial zone of the Kungur Ice Cave, Russia.

Data after Dublyansky (2005).

In contrast, seasonal changes of air temperature, relative humidity, and airflow rate and direction produce highly diversified microenvironments in the periglacial zone of the Kungur Ice Cave. Ice grows, melts, and undergoes sublimation there, in addition to which water evaporation and condensation also occur. This transitional environment is characterized by wide changes in the TDS of the solutions, which are strongly oversaturated at some sites. The partial segregation of the dissolved ions (either cryochemically, or by ion consumption during sequential crystallization of minerals, or by sorption and desorption by clayey sediments) produces the largest mineralogical and morphological variability of secondary minerals in the periglacial zone of gypsum caves, including mineral phases such as mirabilite ($Na_2SO_4 \cdot 10H_2O$) and blödite $Na_2Mg(SO_4)_2 \cdot 4H_2O$, or the boron minerals ulexite $NaCaB_5O_6(OH)_6 \cdot 5H_2O$ and inyoite $CaB_3O_3(OH)_5 \cdot 4H_2O$ (Andreychouk et al., 2013). The morphology and mineralogy of cryogenic precipitates of glacial and periglacial cave zones are given mainly based on the studies conducted in the Kungur Ice Cave.

The Kungur Ice Cave is located in the western foothills of the Central Urals (Andreychouk et al., 2013; see also Chapter 26). The cave developed within a 90 m thick sequence of sulfate-rich rocks dated to the Lower Permian age. The presence of powdery gypsum on the surface of ice was noticed as early as the 18th century by J. G. Gmelin, and it was studied again in the late 19th century (in 1880, by archaeologist I. S. Polyakov, who introduced the term "gypsum flour"; in 1884 by the mineralogist E. S. Fedorov, who suggested the idea of gypsum powder formation during ice sublimation). Although the formation of gypsum powder in relation to ice growth/melting/sublimation was postulated in these historical studies (see also Maksimovich, 1947), its cryogenic origin was definitely confirmed by Andrejchuk and Galuskin (2001). The studies undertaken in the Kungur Ice Cave and other gypsum caves with the presence of CCMs in Russia and Ukraine (e.g., Andrejchuk et al., 2004; Andreychouk et al., 2009; Potapov et al., 2008) were summarized by Andreychouk et al. (2013).

The new ice and a fresh portion of fine-grained CCMs form in the glacial zone of the Kungur Ice Cave depending on the quantity of percolating drip water and temperature. The highest rate of new congelation ice formation is observed during the spring surface snowmelt and the first part of summer in the periglacial and also partly in the glacial cave zone, depending also on the delay of water percolating into the cave. Some parts of the glacial zone have a permanent influx of percolating water and the largest new ice formation during the winter (e.g., the Polar Grotto; the term "grotto" is traditionally used in the Kungur Ice Cave for individual cave chambers), whereas other parts with several-meters-thick permafrost zones around the cavities are without percolation. Highly mineralized waters (0.985 to 1.407 $g \cdot L^{-1}$, Andreychouk et al., 2013), which dissolved gypsum in the sedimentary layers above the cave, percolate into the glacial and periglacial parts of the cave and freeze.

The largest ice sublimation occurs during the winter, which is promoted by a strong circulation of very cold and dry air. Therefore the coldest part of the winter is when most of the secondary accumulation of cryogenic gypsum powder happens on ice surfaces and at the base of steep or vertical surfaces of the perennial ice blocks. This CCMs buildup is due to ice sublimation. The cryogenic gypsum powder forms a thin layer covering the surface of practically all congelation ice deposits in the Kungur Ice Cave. The thickness of this secondary gypsum powder can locally exceed 10 cm in ice depressions and below steep ablation ice walls, where it accumulates gravitationally. From a morphological point of view, the CCG is usually dominated by tabular to lathy gypsum crystals and spherulitic aggregates (Fig. 6.10). The crystals commonly display hollows, which frequently have regular geometric shapes. They probably represent perimorphs after melted ice crystals.

Scanning electron microscopy was used extensively by Andrejchuk and Galuskin (2001), who showed that the cryogenic powder sampled between the Brilliant and Polar grottos of the Kungur Ice Cave consists of two morphologically distinct types of gypsum. The first is represented by small isolated crystals (usually 100 to 200 μm) and aggregates of parallel-grown crystals ranging from 300 to 400 μm in length. The second aggregate type (size 50 to 100 μm) is represented by spherulites (Fig. 6.10).

The studies on cryogenic gypsum powders from ice caves located in gypsum-rich sediments of the Bukovina Region (Andrejchuk et al., 2004) and the Pinezhsky District (Potapov et al., 2008) confirmed all previous morphological observations made on the CCG, including not only the existence of the parallel-grown crystals but also crystal twins (both swallow-tail and fish-tail types) and aggregates of small crystals forming together the shape of a larger crystal, resembling the poikilitic texture found in igneous rocks. The major part of the gypsum powder was represented by subhedral crystals ranging from only 1 to 30 μm. Twins, parallel crystals, aggregates with random orientation of crystals, and spherulitic aggregates appeared to be about 5–20 times larger than the single crystals. It was also confirmed that the spherulites, which usually are related to a rather irregular growth, actually formed at a later stage, thus covering older individual crystals/aggregates. The typical morphological evolution (microcrystals → twins → spherulites) documents the cryogenic segregation of the dissolved mineral substance from the water during its gradual oversaturation (Andreychouk et al., 2013).

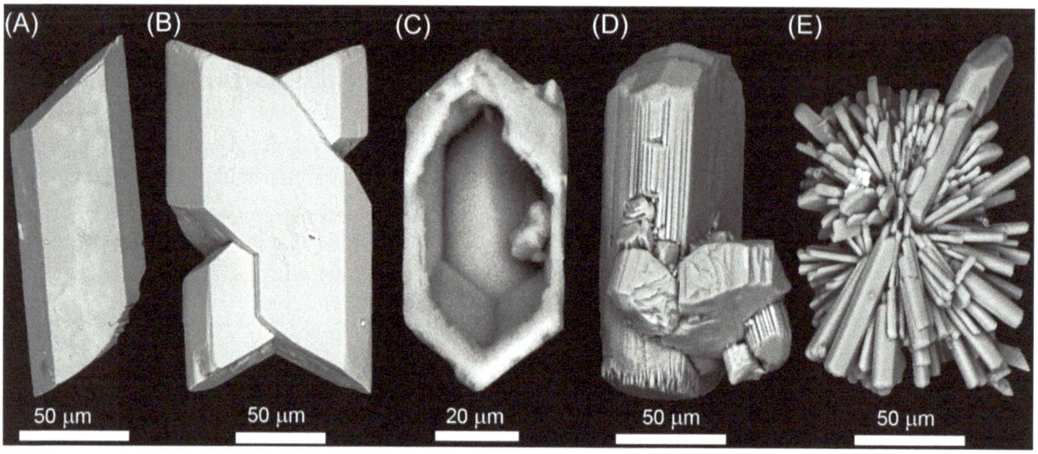

FIG. 6.10

The morphology of crystals and aggregates of cryogenic gypsum, glacial zone of the Kungur Ice Cave: (A) tabular gypsum crystal; (B) gypsum crystal twins; (C) gypsum crystal with a large hollow; (D) an aggregate of flat tabular crystals; (E) spherulitic aggregate; (A, C, D, E) from a layer of perennial ice in the corridor between Brilliant and Polar grottos; and (B) from cryogenic powder on the ice surface, Cross Grotto (SEM images by O. Kadebskaya).

In addition to the prevailing gypsum crystals and aggregates (95 to 98 vol. %), the cryogenic powder from the glacial zone of the Kungur Ice Cave also contains carbonates (calcite and more rarely dolomite, together 1 to 3 vol. %) and celestine (1 to 5 vol. %; Andreychouk et al., 2013). In addition, barite crystals and cotton-wool-like aggregates of silica (unidentified phase, possibly low tridymite; Andreychouk et al., 2013) are observed as accessory minerals. The calcite crystals from the cryogenic powder are represented by rhombohedrons (often in parallel and/or sheaf-like growths) and spherulites (Fig. 6.11). When dolomite is present, its crystals are combinations of rhombohedrons and pinacoids. Some of these morphologies are similar to those described for the CCC_{fine} in limestone caves.

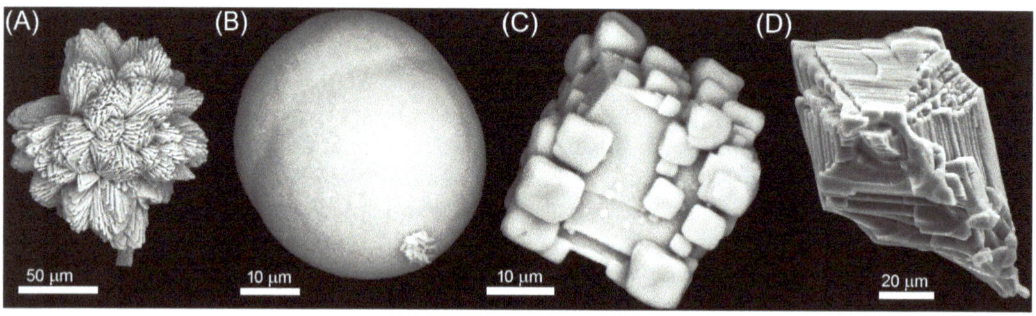

FIG. 6.11

The morphology of cryogenic calcite from the glacial zone of the Kungur Ice Cave: (A) aggregate of split crystals; (B) calcite spherulite; (C) skeletal crystal; (D) crystal of laminar external structure; (A, D) from a layer of perennial ice in the corridor between the Brilliant and Polyar grottos; and (B, C) from cryogenic powder from ice surface, Cross Grotto (SEM images by O. Kadebskaya).

The celestine occurs as a late precipitate from an oversaturated solution, when a large proportion of Ca^{2+} was already removed by earlier gypsum and carbonate precipitation. The separation of Ca^{2+} and Sr^{2+} is probably also supported by their different ion sizes and thus different segregation coefficients during water freezing (cf. Killawee et al., 1998). Celestine crystals, like the calcite and gypsum observed in cryogenic powders of the Kungur Ice Cave, show variable morphologies: from initial microscopic crystals to spherulitic aggregates of individual tabular crystals reaching up to 80 μm (Fig. 6.12); the crystal habit also depends on the supersaturation level of solution.

FIG. 6.12

Morphology of celestine crystals from the glacial zone of the Kungur Ice Cave, from a layer of perennial ice in the corridor between the Brilliant and Polyar grottos: (A) aggregate of single crystals; and (B) aggregate of split crystals (SEM images by O. Kadebskaya).

Besides the crystal and aggregate morphologies previously described, larger gypsum crystal aggregates (up to 4 mm) have been found at the surface of perennial cave ice deposits in the Kungur Ice Cave, the Ledyanaya Volna Cave (Arkhangelsk Region), and the Ledyanoy Paporotnik Cave (Northern Kazakhstan). These white and yellow crystal aggregates occur in up to several-centimeter-thick deposits. Microscopic investigations revealed that they consist of the usual tabular crystals and acicular sub-individuals building up three-dimensional or flat fan-shaped accumulations. Usual tabular crystals (0.2 to 0.5 mm) were observed either as separate individuals or as individuals overgrown by splitting crystals. The crystals could both overgrow the aggregates and serve as a nucleus for them. Such interrelations support the alternating formation of regular crystals with straight edges and with splitting crystals. This finding is in agreement with ideas about the repeated growth of gypsum aggregates. Zonal aggregates consisting of three generations (formed during three years) were observed (Fig. 6.13).

The study of these acicular aggregates using electron microscopy showed that the majority of crystals composing the aggregates are in fact represented by twins having a complex internal skeletal structure with numerous hollows (Fig. 6.14). The skeletal growth of gypsum crystals results from the presence of internal partitions, which are oriented parallel to the edges of two or three pinacoids ({010}, {100},{001}) and two prisms ({120} {111}), and also to the outer edges of the crystal. Such complex internal structures of gypsum crystals was not found earlier within the mineral objects.

FIG. 6.13

Larger gypsum forms: single and twinned crystals and intergrowths; regular overgrowths on the acicular individuals, acicular, skeletal, dendritic, and zonal aggregates; Kungur Ice Cave, First Grotto (photos by O. Kadebskaya).

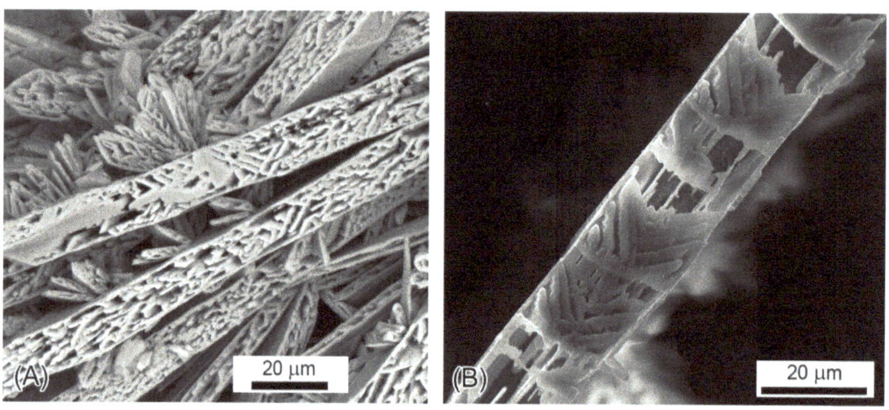

FIG. 6.14

Skeletal structure of gypsum crystals, with presence of numerous internal partitions, oriented parallel to major crystal edges; Kungur Ice Cave, Polyar Grotto: (A) general view of the aggregates; and (B) detail of internal structure (SEM images by O. Kadebskaya).

After Kadebskaya, O., Tchaikovsky, I., 2013. Skeleton crystals of cryogenic gypsum from Kungur Ice Cave, Ural Mountains, Russia. In: Filippi, M., Bosák, P. (Eds.), Proceedings of the 16th International Congress of Speleology, vol. 3. Czech Speleological Society, Prague, pp. 454–458.

Flat aggregates of gypsum microcrystals kept on the water surface because of surface tension were observed during the autumn of 2010 in the Krestoviy (Cross) Grotto of the Kungur Ice Cave (Tchaikovskiy et al., 2015). These gypsum rafts covered the water in a 1 cm deep pool and were an equivalent of gypsum rafts of non-ice caves.

The environment of periglacial zones of ice caves in gypsum differs from the environment of glacial zones by seasonal temperature oscillations above and below the freezing point (see Fig. 6.9), by only seasonal presence of ice, and by repeated partial dissolution of accumulations of CCG. This results in much wider morphological and mineralogical variability of CCMs in the periglacial zone. Larger gypsum aggregates were collected in the periglacial zone of the Kungur Ice Cave in the Scandinavian Grotto. Their formation is mainly associated with the summer period. The gypsum powder released from the ice by sublimation (during winter) or by melting caused by influx of warmer water is wetted by the melt and infiltration waters and may recrystallize, which generally leads to enlargement of the aggregates. The latter ones consist of spherulites with a narrow "root" that penetrates into the cave floor's sediment or ice (cf. Fig. 6.13). Alternation of normal crystal growth with phases of enhanced crystal splitting is characteristic of the periglacial zone (Andreychouk et al., 2013). Another type of gypsum comprises unconsolidated fragments of crystalline crusts (1.5 to 6.0 mm thick) made up of skeletal, subparallel, or interwoven individuals. They evidently formed within a water layer located on a clay surface under significant water evaporation that kept the solution oversaturated. Genetically, these gypsum occurrences of the periglacial zone were formed by a combination of cryogenic and non-cryogenic processes. More details on their origin are available in Andreychouk et al. (2013).

The periglacial zone of the Kungur Ice Cave also hosts accumulations of fibrous sulfate minerals such as mirabilite and blödite. These deposits are seasonal (ephemeral) and were observed during the winter at sites where the temperature is close to the freezing point. It is generally known that for the precipitation of fibrous minerals, an important factor is the presence of loose sediments or porous bedrock (Hill and Forti, 1997). The mineralized solutions move *per ascensum* toward the surface through the pores of clayey material covering the sulfate rock. Hairy crystals form during this process, because of capillary pressure from below, and thus have an extrusive character. This process is responsible for the occurrence of fibrous gypsum or mirabilite crystals anywhere in the cave (floor, wall, ceiling) as long as an underlying porous layer exists. Because their formation mainly relates to water evaporation with a possible pre-concentration of the solutions by cryogenic processes or ion sorption in clays, these minerals cannot be considered as primarily cryogenic. Potapov et al. (2008) confirmed that these hairy accumulations change their mineralogy seasonally in response to changes in solution concentrations because of water condensation and evaporation. Typically, blödite occurs in the summer and mirabilite in the winter.

The studies on the morphology and chemical composition of mineral deposits found in the Kungur Ice Cave's Smielych Grotto in the periglacial zone enabled distinguishing between two environments in which sulfate and borate minerals form (Andreychouk et al., 2013; Kadebskaya and Kalinina, 2015). These minerals are non-cryogenic, but previous cryochemical processes followed by sorption and desorption from clays along with water evaporation and condensation are probably important in promoting solutions of suitable composition. The first environment is represented by relatively poorly ventilated cave sections where mainly spherulites of ulexite and rare acicular mirabilite crystals form on polymineral porous horizontal and inclined surfaces. When acicular mirabilite is removed from a cave and transferred into a warm room, these crystals start to bend and decompose into a white powder of thenardite.

Ulexite crystals form as radially arranged spherulites 0.5 to 1.0 mm wide, and as larger (up to 2 mm) spherical aggregates with a zonal structure (acicular crystals forming more porous surface layer and a denser partly fibrous core). The acicular crystals attach to each other forming bouquets resembling wet moss (Fig. 6.15). The chemical microanalyses indicate wide variability in the ulexite composition, from the theoretical sodium end-member in the outer hairy fringe to a K-rich phase in the central part. Rare minor domains corresponding to a calcium borate known as inyoite also have been confirmed (Chaykovskiy and Kadebskaya, 2012). The separation of boron from the other elements is probably due to sorption and desorption on clay minerals, which is a well-known process (e.g., Hingston, 1964; Keren and Gast, 1981).

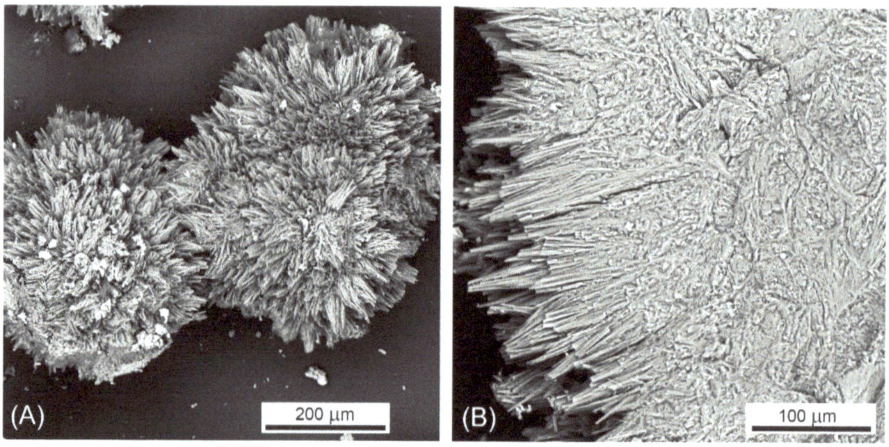

FIG. 6.15

Ulexite aggregates from the periglacial zone of the Kungur Ice Cave: (A) general view of the aggregates; and (B) internal structure of the aggregate (SEM images by O. Kadebskaya).

6.6 STABLE ISOTOPE CHARACTERISTICS OF CRYOGENIC CAVE MINERALS

Cryogenic formation of carbonate minerals is accompanied by large carbon and oxygen stable isotope fractionation (all C and O isotope data are given relative to V-PDB). Over the past 20 years, more than 40 papers and conference abstracts were published on carbon and oxygen isotope fractionations of CCC (e.g., Chaykovskiy et al., 2014; Clark and Lauriol, 1992; Durakiewicz et al., 1995; Lacelle, 2007; Lacelle et al., 2006, 2009; Lauriol et al., 2006; Luetscher et al., 2013; May et al., 2011; Meissner et al., 2010; Onac, 2008; Onac et al., 2011; Orvošová, 2015; Richter and Niggemann, 2005; Richter and Riechelmann, 2008; Richter et al., 2008, 2009a,b, 2010a,b, 2011, 2013, 2015; Spötl, 2008; Spötl and Cheng, 2014; Žák et al., 2004, 2006, 2008, 2009, 2010, 2011, 2012, 2013).

These studies reported hundreds of C and O isotope determinations on CCC; their isotope systematics is relatively well understood. In the usual $\delta^{13}C$ versus $\delta^{18}O$ plot, the CCC data covered extensive fields. The CCC's isotopic variability was much greater than that of carbonate speleothems from non-iced caves (Fig. 6.16). Furthermore, the C and O stable isotope signatures of CCC_{fine} and CCC_{coarse}

differed significantly, each being controlled by a distinct isotope fractionation mechanism and genetic processes, displaying different isotopic trends on the plot as a result. Some of the CCC_{coarse} sample sets showed uniquely low $\delta^{18}O$ values (as low as −25‰) and simultaneously high $\delta^{13}C$ values (up to +6‰). No other natural carbonate samples possessing similar isotopic composition are known. Some of the CCC_{fine} samples are unique as well, reaching extremely high $\delta^{13}C$ values (up to +17‰; Clark and Lauriol, 1992), which are probably the most positive values recorded in the literature for any type of carbonates.

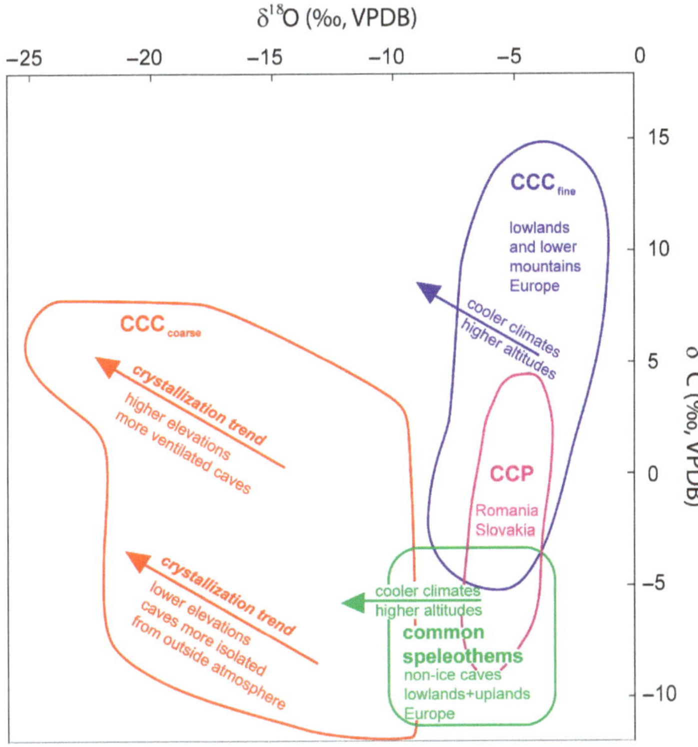

FIG. 6.16

Stable isotope data of cryogenic cave carbonates. The term "common speleothems" refers to secondary carbonate formations of non-iced caves that are not influenced by strong kinetic isotope fractionations effects.

The $\delta^{13}C$ and $\delta^{18}O$ values of the CCC are generally controlled by a complex set of parameters that include

- the $\delta^{18}O$ value of initial water and $\delta^{13}C$ of initial dissolved HCO_3^- before onset of freezing;
- oxygen isotope fractionations between water, forming ice, and, for CCC_{fine}, also between liquid water and escaping water vapor; and
- complex carbon isotope fractionation between carbonates solute species (HCO_3^-, CO_3^{2-}), escaping CO_2, and formed solid carbonates.

Isotope fractionation can occur under equilibrium or kinetic regimes, and the depositional environment (solution + solid phases + cave atmosphere) can correspond to semi-closed or open systems (with CO_2 and water vapor freely escaping). Under the latter setting, the C isotope systematic is controlled by Rayleigh-type fractionation.

During the *formation of CCC_{fine}*, carbonate C and O isotope composition is controlled mainly by kinetic isotope effects under open-system conditions. The oxygen isotope systematics of the water-ice system is controlled by water freezing and water evaporation, under these two processes the isotope fractionation has the opposite effects on the residual solution $\delta^{18}O$. If the freezing proceeds slowly and under isotopic equilibrium between the water and ice, the ice preferentially incorporates the heavier isotope ^{18}O (e.g., Jouzel and Souchez, 1982; Souchez and Jouzel, 1984; Suzuoki and Kimura, 1973; field data, e.g., Gibson and Prowse, 1999). This isotopic fractionation results in a progressive depletion of the ^{18}O isotope (thus decreasing $\delta^{18}O$ values) in the residual solution. If the freezing is rapid, the difference between the $\delta^{18}O$ values of the liquid water and ice becomes smaller than under equilibrium conditions (Souchez et al., 2000). Evaporation, which usually accompanies water freezing during CCC_{fine} formation, has an opposite fractionation effect; that is, the lighter ^{16}O isotope preferentially escapes into the vapor phase, and the residual water becomes ^{18}O-enriched. If acting simultaneously, the processes of water freezing and water evaporation can therefore cancel out (partly or entirely) the isotopic effects of each other.

Nevertheless, during the relatively rapid process of CCC_{fine} formation, the dehydration of HCO_3^- and degassing of CO_2 follow kinetic isotope fractionation. The new oxygen isotope equilibrium between the water and HCO_3^- and CO_3^{2-} ions cannot be achieved because of the high rate of the process. The empirical $\delta^{18}O$ data from CCC_{fine} show that the combined effects of water evaporation and kinetic fractionation during the HCO_3^- dehydration strongly outweigh the equilibrium isotope fractionations.

Isotopic fractionation of carbon is also kinetic during the CCC_{fine} formation. The isotopically "lighter" CO_2 escaping from the solution shifts the $\delta^{13}C$ of the residual solution (and cryogenic carbonates precipitated from it) to progressively higher values. Because the cave atmosphere of ice caves typically shows dynamic circulation during the winter, the system is fully open with respect to escaping CO_2. Thus the residual fraction of HCO_3^- in the water before final freezing can become extremely "heavy" because of the Rayleigh-type fractionation. The $\delta^{13}C$ values of cryogenic carbonate powder sampled on the ice surface can reach up to +17‰.

Empirical data from CCC_{fine} usually exhibit steep positive trends in the $\delta^{13}C$-$\delta^{18}O$ space, with the $\delta^{13}C$ showing much greater variability than the $\delta^{18}O$ (Fig. 6.16). These trends are similar to those described for speleothems formed by rapid precipitation (frequently accompanied by evaporation, e.g., in cave entrance sections; Hendy, 1971; Mickler et al., 2006). Lacelle et al. (2006) demonstrated that the degree of kinetic C and O isotope fractionation depends on the freezing rate but also on the attainment of calcite saturation. More recently, Sade and Halevy (2017) exploited the C and O isotope data from CCC_{fine} collected from all available published papers to test the magnitudes of oxygen isotope kinetic fractionation at 0°C during CO_2 (de)hydration and (de)hydroxylation, which are major reactions in the CO_2-H_2O-$CaCO_3$ system. It was found that the largest observed CCC_{fine} kinetic isotope fractionations are close to the theoretically possible kinetic limit.

The much slower *formation of CCC_{coarse}* in conditions limiting water evaporation and partly restricting escape of CO_2 is associated with different mechanisms of isotope fractionation. For oxygen isotopes, the equilibrium (or close-to-equilibrium) fractionation between the liquid water and ice governed by Rayleigh-type conditions shifts the $\delta^{18}O$ data of the residual liquid phase to progressively lower values. This effect was confirmed by measuring strongly depleted δD values in residual water trapped in fluid inclusions of the CCC_{coarse} (Dublyansky et al., 2015b). Isotopic equilibration between the water and HCO_3^- captures this trend, which is therefore reflected in a progressively decreasing $\delta^{18}O$ of CCC_{coarse} along the crystallization process (Luetscher et al., 2013; Žák et al., 2004). The released CO_2 remains (at least temporarily) in contact with liquid phase; thus the shift of $\delta^{13}C$ during the carbonate crystallization to heavier values is not so pronounced. Nevertheless, partial removal of CO_2 from the system is necessary to maintain conditions of calcite precipitation (this can be realized by the regular entrapment of CO_2 rich bubbles into the advancing ice or by sudden release along ice cracks).

Various levels of closeness/openness of the system with respect to CO_2 escape result in a relatively wide range of $\delta^{13}C$ values measured for CCC_{coarse} (cf. Fig. 6.16). Factors restricting the degassing of CO_2 are given either by the newly formed ice on the surface of cave pools or directly by the cave walls in caves largely isolated from the outside atmosphere (see Section 6.4). In the $\delta^{13}C$–$\delta^{18}O$ space, the CCC_{coarse} isotope data have shown negatively sloped trends, with consecutively precipitating CCC_{coarse} aggregate types characterized by progressively lower $\delta^{18}O$ values and concurrently slightly higher $\delta^{13}C$ values. The temporal evolution of the data in these trends was determined by analyzing core-to-rim C and O isotope profiles within larger crystals and aggregates (Luetscher et al., 2013; Žák et al., 2004), and was also mathematically modeled by Žák et al. (2004).

Kluge et al. (2014a) examined the C and O isotope systematics of CCC_{coarse} in more detail using the clumped isotope technique. Carbonate clumped isotope analysis (reported as Δ^{47}) can be used to determine the formation temperature of carbonate minerals (e.g., Eiler, 2007, 2011). This approach is based on the temperature-dependent overabundance of ^{13}C–^{18}O bonds in the crystal lattice compared to a stochastic distribution. Carbonate clumped isotope measurements yielded apparent temperatures between 3°C and 18°C (while the real temperature of CCC_{coarse} crystallization is close to 0°C), thus confirming isotopic disequilibrium (Kluge et al., 2014a). The incomplete isotope equilibrium probably corresponds to HCO_3^- decomposition and CO_2 degassing steps.

Some specific geochemical features of CCC_{coarse} have also been studied. Kluge et al. (2014b) examined its formation process using water and noble gases extracted from inclusions in these crystals. Noble gas concentrations derived from fluid inclusions deviated significantly from commonly observed concentrations in atmospheric air, surface water, groundwater, and common types of stalagmites. The authors concluded that the gradual freezing process led to a partitioning of the noble gases between the cave ice and the remaining water, resulting in a pronounced overabundance of heavy noble gases in the liquid phase, which was preserved in the fluid inclusions of the cryogenic calcite crystals.

Only limited sulfur and oxygen isotope fractionations occur between the dissolved and solid sulfate during precipitation of sulfate minerals (e.g., gypsum), regardless of whether precipitation is due to oversaturation triggered by water evaporation (Bottrell and Newton, 2006; Claypool et al., 1980) or by water freezing. If the entire dissolved sulfate content is precipitated during the water freezing process as cryogenic gypsum, the $\delta^{34}S$ and $\delta^{18}O$ of the dissolved sulfate and of the sulfate in CCG should be identical. The exchange of oxygen isotopes between the water and dissolved sulfate at low temperature and neutral or higher pH is so slow that no changes in the sulfate $\delta^{18}O$ values can be expected during the process of water freezing (cf. Brunner et al., 2005). The $\delta^{34}S$ and $\delta^{18}O$ values of sulfate of cryogenic gypsum should therefore be similar to those of dissolved sedimentary sulfate, except the rare case when bacterial sulfate reduction occurs. No detailed sulfur and oxygen isotope study of cryogenic cave gypsum has been published. Analyses of

$\delta^{18}O$ or δD of the hydration water of cryogenic gypsum are more promising, because they can provide information on the isotopic composition of water during gypsum crystallization and on the fractionations between the mother solution and the hydration water (e.g., Evans et al., 2015; Herwatz et al., 2016) during gypsum crystallization at low temperature.

ACKNOWLEDGMENTS

The preparation of this chapter was partly financed through the institutional support RVO67985831 of the Institute of Geology of the Czech Academy of Sciences. B.P. Onac research on ikaite was supported by the NSF grant AGS-1103108.

REFERENCES[1]

Albright, J.N., 1971. Vaterite stability. Am. Mineral. 56, 620–624.

Andrejchuk, V., Galuskin, E., 2001. Cryogenic mineral formations of the Kungur Ice Cave. Peshchery (Caves) 27–28, 108–116. Perm State University, Russia (in Russian with English abstract).

Andrejchuk, V., Galuskin, E., 2008. Cryomineral formations of caves: introduction into problem. Speleol. Karstol. 1, 67–80. Simferopol (in Russian with English abstract).

Andrejchuk, V., Galuskin, E., Ridush, B., 2004. Cryomineral formations from North Bukovinian Caves. In: Citterio, M., Turri, S. (Eds.), 1st International Workshop in Ice Caves, February 28–March 3, 2004. Department of Earth Sciences "Ardito Desio", University of Milano, Milano, pp. 37–38.

Andreychouk, V., Ridush, B., Galuskin, E., 2009. Pionerka Cave: conditions and specificity of cryogenic mineral formation. Speleol. Karstol. 2, 54–69 (in Russian with English abstract).

Andreychouk, V.N., Kadebskaya, O.I., Chaykovski, I.I., 2013. Cryogenic mineral formations of Kungur Ice Cave. University of Silesia and Mining Institute of Ural Branch of Russian Academy of Sciences, Sosnowiec and Perm (in Russian with English summary).

Baker, I., Illiescu, D., Obbard, R., Chang, H., Bostick, B., Daghlian, C.P., 2005. Microstructural characterization of ice cores. Ann. Glaciol. 42, 441–444.

Barnes, P.R.F., Wolff, E.W., 2004. Distribution of soluble impurities in cold glacial ice. J. Glaciol. 50 (170), 311–324.

Bartolomé, M., Sancho, C., Osácar, M.C., Moreno, A., Leunda, M., Spötl, C., Luetscher, M., López-Martínez, J., Belmonte, A., 2015. Characteristics of cryogenic carbonates in a Pyrenean ice cave (northern Spain). Geogaceta 58, 107–110.

Bazarova, E.P., Gutareva, O.S., 2011. Novye dannye o mineralogii peshcher Ikrutskoy oblasti (New data on mineralogy of caves in the Irkutsk Region). In: Stroenie litosfery i geodinamika, Materialy XXIV Vsyerossiyskoy molodiezhnoy konferencii. Irkutsk, 19–24 April 2011. Institut Zemnoy Kory SO RAN, Irkutsk, pp. 62–63 (in Russian).

Bazarova, E., Kononov, A., Gutareva, O., 2014a. Cryogenic Mineral Formation in the Okhotnichya Cave in the Primorsky Mountin Ridge (Western Baikal Region, Russia). Speleo Int. Mag. 2014 (1), 116–130.

Bazarova, E.P., Kononov, A.M., Gutareva, O.S., Nartova, N.V., 2014b. Specific characteristics of cryogenic mineral formations of Okhotnichya Cave in Pre-Baikal area (Irkutsk Region). Kriosfera Zemli (Earth's Cryosphere) XVIII (3), 62–71.

[1]Papers in Russian which contain English version of the author's names and the title are given in the published transliteration. The BGN/PCGN transliteration is used for papers published in Russian only. Language is indicated in the parentheses for publications in languages other than English, French or German.

Bazarova, E.P., Kadebskaya, O.I., Kononov, A.M., Ushchapovskaya, Z.F., 2016. Conditions and characteristics of cryogenic mineral formation in the caves of southern part of Siberian Platform Folding Margins (Western Pre-Baikal Region and Eastern Sayan). Vestnik Permskogo Universiteta 31 (2), 22–34. Perm (in Russian with English summary).

Bischoff, J.L., Fitzpatrick, J.A., Rosenbauer, R.J., 1993a. The solubility and stabilization of ikaite ($CaCO_3 \cdot 6H_2O$) from 0°C to 25°C – environmental and paleoclimatic implications for thinolite tufa. J. Geol. 101 (1), 21–33.

Bischoff, J.L., Stine, S., Rosenbauer, R.J., Fitzpatrick, J.A., Stafford Jr., T.W., 1993b. Ikaite precipitation by mixing of shoreline springs and lake water, Mono Lake, California, USA. Geochim. Cosmochim. Acta 57 (16), 3855–3865.

Bock, E., 1961. On the solubility of anhydrous calcium sulphate and gypsum in concentrated solutions of sodium chloride at 25° C, 30° C, 40° C, and 50° C. Can. J. Chem. 39, 1746–1751.

Bottrell, S.H., Newton, R.J., 2006. Reconstruction of changes in global sulfur cycling from marine sulfate isotopes. Earth Sci. Rev. 75, 59–83.

Brückmann, E.F., 1728. Francisci Ernesti Brückmanni, med. doct., Epistola itineraria IIItia de Bellaris lapideis liptoviensibus hungaricis. Wolffenbuttelae.

Brunner, B., Bernasconi, S.M., Kleikemper, J., Schroth, M.H., 2005. A model for oxygen and sulfur isotope fractionation in sulfate during sulfate reduction processes. Geochim. Cosmochim. Acta 69 (20), 4773–4785.

Buchardt, B., Seaman, P., Stockmann, G., Vous, M., Wilken, M.V.U., Düwel, L., Kristiansen, A., Jenner, C., Whiticar, M.J., Kristensen, R.M., Petersen, G.H., Thorbjørn, L., 1997. Submarine columns of ikaite tufa. Nature 390, 129–130.

Calaforra, J.M., 1998. Gypsum karstology. Ciencia y Tecnologia Monografia 3. Instituto de Estudios Amerienses, Univ, Almería (in Spanish with English summaries and captions).

Chaykovskiy, I.I., Kadebskaya, O.I., 2012. Mineralogiya perechodnoy klimaticheskoy zony Kungurskoy ledyanoy peshchery (Mineralogy of transitional climatic zone of the Kungur Ice Cave). Nauchnye chteniya pamyati P.N. Chirivinskovo, Perm, pp. 100–105 (in Russian).

Chaykovskiy, I.I., Kadebskaya, O.I., 2014. Mineralnye obrazovaniya peshchery Eranka na Severnom Urale (Mineral formations of the Eranka Cave on the Northern Ural). In: Problemy mineralogii, petrografii i metallogenii. vol. 17. Nauchnye chteniya pamyati P.N. Chirvinskogo, Perm, pp. 92–107 (in Russian).

Chaykovskiy, I.I., Kadebskaya, O.I., Žák, K., 2014. Morphology, composition, age and origin of carbonate spherulites from caves of Western Urals. Geochem. Int. 52 (4), 336–346.

Clark, I.D., Lauriol, B., 1992. Kinetic enrichment of stable isotopes in cryogenic calcites. Chem. Geol. 102, 217–228.

Claypool, G.E., Holser, W.T., Kaplan, I.R., Sakai, H., Zak, I., 1980. The age curves of sulfur and oxygen isotopes in marine sulfate and their mutual interpretation. Chem. Geol. 28, 199–260.

Colucci, R.R., Luetscher, M., Forte, E., Guglielmin, M., Lenaz, D., Princivalle, F., Vita, F., 2017. First alpine evidence of in situ coarse cryogenic cave carbonates (CCC_{coarse}). Geogr. Fis. Din. Quat. 40.

Council, T.C., Bennettt, P.C., 1993. Geochemistry of ikaite formation at Mono Lake, California: Implications for the origin of tufa mounds. Geology 21, 971–974.

Dieckmann, G.S., Nehrke, G., Papadimitriou, S., Gottlicher, J., Steininger, R., Kennedy, H., Wolf-Gladrow, D., Thomas, D.N., 2008. Calcium carbonate as ikaite crystals in Antarctic sea ice. Geophys. Res. Lett. 35 (8), L08501.

Dublyansky, V.N. (Ed.), 2005. Kungurskaya Ledyanaya Peshchera: opit rezhimnych nablyudeniy (Kungur Ice Cave: experience from regime observations). UrO RAN, Ekaterinburg (in Russian).

Dublyansky, Y., Kadebskaya, O., Luetscher, M., Cheng, H., Chaykovskiy, I., Spötl, Ch., 2014. Preliminary data on the Pleistocene history of the permafrost in Central Urals (Russia) derived from cryogenic cave carbonates. In: Proceedings of the Conference The Quaternary of the Urals. Global Trends and Pan-European Quaternary records, 10–16 September 2014. INQUA-SEQS, Ekaterinburg, pp. 43–44.

Dublyansky, Y., Kadebskaya, O., Cheng, H., Luetscher, M., Spötl, C., 2015a. Cryogenic cave carbonates as an archive of Late Pleistocene permafrost in the Ural Mountains: preliminary results. Geophys. Res. Abstr. 17. EGU2015-4095.

Dublyansky, Y., Luetscher, M., Spötl, C., Töchterle, P., Kadebskaya, O., 2015b. First results on stable isotopes in fluid inclusions in cryogenic carbonates from Ural Mountains (Russia). Geophys. Res. Abstr. 17. EGU2015-4061.

Dublyansky, Y., Kadebskaya, O., Menshikova, E., Demeny, A., Spötl, C., 2017. Polymineral cryogenic deposits in caves of the Southern Ural, Russia, as tracers of past permafrost. Abstracts of Climate Change: the Karst Record 8, Austin (in press).

Durakiewicz, T., Hałas, S., Migaszewski, M., Urban, J., 1995. Origin of "calcite groats" in the Chelosiowa Cave near Kielce (Holy Cross Mts.) inferred from petrographic and isotopic investigations. Geol. Q. 39, 75–94.

Eiler, J.M., 2007. "Clumped-isotope" geochemistry – the study of naturally-occurring, multiply-substituted isotopologues. Earth Planet. Sci. Lett. 262, 309–327.

Eiler, J.M., 2011. Paleoclimate reconstruction using carbonate clumped isotope thermometry. Quat. Sci. Rev. 30, 3575–3588.

Ek, C., Pissart, A., 1965. Depôt de carbonate de calcium par congélation et teneur en bicarbonate des eaux résiduelles. C. R. Acad. Sci. Paris 260 (9), 929–932.

Erlenmeyer, M., Schudelski, A., 1992. Der Malachitdom und die anderen Höhlen im Steinbruch Düstertal. In: Geologisches Landesamt Nordrhein-Westfalen (Eds.), Der Malachitdom. Ein Beispiel interdisziplinärer Höhlenforschung im Sauerland. Geologisches Landesamt Nordrhein-Westfalen, Krefeld, pp. 39–68.

Evans, N.P., Turchyn, A.V., Gázquez, F., Bontognali, T.R.R., Chapman, H.J., Hodell, D.A., 2015. Coupled measurements of d18O and dD of hydration water and salinity of fluid inclusions in gypsum from the Messinian Yesares Member, Sorbas Basin (SE Spain). Earth Planet. Sci. Lett. 430, 499–510.

Fairchild, I.J., Killawee, J.A., Spiro, B., Tison, J.-L., 1996. Calcite precipitates formed by freezing processes: kinetic controls on morphology and geochemistry. In: Proceedings of Fourth International Symposium on the Geochemistry of Earth's Surface, Ilkley, England. pp. 178–183.

Faria, S.H., Freitag, J., Kipfstuhl, S., 2010. Polar ice structure and the integrity of ice-core paleoclimate records. Quat. Sci. Rev. 29 (1–2), 338–351.

Feurdean, A., Perşoiu, A., Pazdur, A., Onac, B.P., 2011. Evaluating the palaeoecological potential of pollen recovered from ice in caves: a case study from Scărişoara, Romania. Rev. Palaeobot. Palynol. 165, 1–10.

Ford, D., Williams, P., 2007. Karst hydrogeology and geomorphology. John Wiley & Sons, Chichester.

Forti, F., 1990. La grotta Malachitdom: un ambiente mineralogicoeccezionale che rischia di andare distrutto. Riv. Ital. Mineral. Paleontol. 67, 3–7.

French, H.M., 2007. The periglacial environment, third ed. J. Wiley & Sons, Chichester.

Ge, X., Wang, X., 2009. Calculations of freezing point depressions, boiling point elevation and vaporization enthalpies of electrolyte solutions by a modified three-characteristic parameter correlation model. J. Solut. Chem. 38, 1097–1117.

Gibson, J.J., Prowse, T.D., 1999. Isotopic characteristics of ice cover in a large northern river basin. Hydrol. Process. 13, 2537–2548.

Grasby, S.E., 2003. Naturally precipitating vaterite (μ-$CaCO_3$) spheres: Unusual carbonates formed in an extreme environment. Geochim. Cosmochim. Acta 67, 1659–1666.

Greinert, J., Derkachev, A., 2004. Glendonites and methane-derived Mg-calcites in the Sea of Okhotsk, Eastern Siberia: implications of a venting-related ikaite/glendonite formation. Mar. Geol. 204, 129–144.

Hallet, B., 1976. Deposits formed by subglacial precipitation of $CaCO_3$. Geol. Soc. Am. Bull. 87 (7), 1003–1015.

Hendy, C.H., 1971. The isotopic geochemistry of speleothems I. The calculation of the effects of different modes of formation and the isotope composition of speleothems and their applicability as palaeoclimatic indicators. Geochim. Cosmochim. Acta 35, 801–824.

Herwatz, D., Surma, J., Staubwasser, M., 2016. Triple oxygen isotopes of gypsum hydration water. In: Goldschmidt Conf. Abstr. 1110.

Hill, C.A., Forti, P., 1997. Cave Minerals of the World, second ed. National Speleological Society, Huntsville, AL.

Hingston, F.J., 1964. Reactions between boron and clays. Soil Res. 2, 83–95.

Honczaruk, M., Śliwiński, Ł., 2011. Results of hydrogeological research of the deep permafrost zone in the Udryń PIG 1 borehole. Biul. Państw. Inst. Geol. 445, 203–216.

Ito, T., 1996. Ikaite from cold spring water at Shiowakka, Hokkaido, Japan. J. Mineral. Petrol. Econ. Geol. (Japan) 91, 209–219.

Jansen, J.H.F., Woensdregt, C.F., Kooistra, M.J., van der Gaast, S.J., 1987. Ikaite pseudomorphs in the Zaire deep-sea fan: an intermediate between calcite and porous calcite. Geology 15, 245–248.

Jouzel, J., Souchez, R.A., 1982. Melting-refreezing at the glacier sole and the isotopic composition of the ice. J. Glaciol. 28 (98), 35–42.

Kadebskaya, O.I., Chaikovsky, I.I., 2014. Mineral formations of Pobeda cave (Bashkortostan) associated with the formation and thawing of multi-year ice. Izvestiya RAN, Seriya geograficheskaya 2014 (3), 66–72 (in Russian with English abstract).

Kadebskaya, O.I., Chaykovskiy, I.I., 2012. Specifika krio- i mineralogeneza peshchery Medeo (Severnyy Ural) (Specifics of the cryo- and mineralforming processes of the Medeo Cave (Northern Ural)). Peshchery 35, 41–47. Perm (in Russian).

Kadebskaya, O.I., Chaykovskiy, I.I., 2013. Mineralogicheskaya i izotopnaya typizatsiya karbonatnych obrazovaniy peshcher Zapadnogo Urala (Mineralogical and isotopic classification of carbonate formations of the caves of Western Ural). In: Problemy mineralogii, petrografii i metallogenii. vol 16. Nauchnye chteniya pamyati P.N. Chirvinskogo, Perm, pp. 303–311 (in Russian).

Kadebskaya, O.I., Kalinina, T.A., 2015. Usloviya formirovaniya bornykh mineralov v karstovykh i iskustvennykh podzemnych polostyakh (Conditions of the formatio of borate minerals in karst and artificial underground cavities). Peshchery 38, 68–75 Perm (In Russian).

Kadebskaya, O., Tchaikovsky, I., 2013. Skeleton crystals of cryogenic gypsum from Kungur Ice Cave, Ural Mountains, Russia. In: Filippi, M., Bosák, P. (Eds.), Proceedings of the 16th International Congress of Speleology. vol. 3. Czech Speleological Society, Prague, pp. 454–458.

Kemper, E., Schmitz, H.H., 1975. Stellate nodules from the Upper Deer Bay Formation (Valanginian) of the Arctic Canada. Geological Survey of Canada, Ottawa. 109–119. Paper 75–1C.

Kemper, E., Schmitz, H.H., 1981. Glendonite – Indikatoren des polarmarinen Ablagerungsmilieus. Geol. Rundsch. 70 (2), 759–773.

Keren, R., Gast, R.G., 1981. Effects of wetting and drying, and of exchangeable cations, on boron adsorption and release by montmorillonite. Soil Sci. Soc. Am. J. 45, 478–482.

Killawee, J.A., Fairchild, I.J., Tison, J.-L., Janssen, L., Lorrain, R., 1998. Segregation of solutes and gases in experimental freezing of dilute solutions: Implications for natural glacial systems. Geochim. Cosmochim. Acta 62, 3637–3655.

Kluge, T., Affek, H.P., Zhang, Y.G., Dublyansky, Y., Spötl, C., Immenhauser, A., Richter, D.K., 2014a. Clumped isotope thermometry of cryogenic cave carbonates. Geochim. Cosmochim. Acta 126, 541–554.

Kluge, T., Marx, T., Aeschbach-Hertig, W., Spötl, C., Richter, D.K., 2014b. Noble gas concentrations in fluid inclusions as tracer for the origin of coarse-crystalline cryogenic cave carbonates. Chem. Geol. 368, 54–62.

Kozlova, N.N., Naumkin, D.V., 2014. The names of the scientists in the space of Kungur Ice cave. Geopolitika i Ekogeodinamika Regionov 10 (1), 623–632 (in Russian with English abstract).

Krainer, K., Ribis, M., 2012. A rock glacier inventory of the Tyrolean Alps (Austria). Aust. J. Earth Sci. 105, 35–47.

Kralj, D., Brecevic, L., Nielson, A.E., 1990. Vaterite growth and dissolution in aqueous solution: 1. Kinetics of crystal growth. J. Cryst. Growth 104, 793–800.

Kunský, J., 1939. Některé formy ledových krápníků (Some forms of ice speleothems). Rozpravy II. Třídy České akademie XLIX (22), 1–8 (in Czech).

Lacelle, D., 2007. Environmental setting (micro)morphologies and stable C-O-isotope composition of cold climate carbonate precipitation. A review and evaluation of their potential as paleoclimatic proxies. Quat. Sci. Rev. 26 (11–12), 1670–1689.

Lacelle, D., Lauriol, B., Clark, I.D., 2006. Effect of chemical composition of water on the oxygen-18 and carbon-13 signature preserved in cryogenic carbonates, Arctic Canada: implications in paleoclimatic studies. Chem. Geol. 234 (1–2), 1–16.

Lacelle, D., Lauriol, B., Clark, I.D., 2009. Formation of seasonal ice bodies and associated cryogenic carbonates in Caverne de l'Ours, Québec, Canada: kinetic isotope effects and pseudo-biogenic crystal structures. J. Cave Karst Stud. 71 (1), 48–62.

Lauriol, B., Carrier, L., Thibaudeau, P., 1988. Topoclimatic zones and ice dynamics in the caves of the Northern Yukon, Canada. Arctic 41 (3), 215–220.

Lauriol, B., Prévost, C., Lacelle, D., 2006. The distribution of diatom flora in ice caves of the northern Yukon Territory, Canada: relationship to air circulation and freezing. Int. J. Speleol. 35, 83–92.

Loose, B., McGillis, W.R., Schlosser, P., Perovich, D., Takanashi, T., 2009. Effects of freezing, growth, and ice cover on gas transport processes in laboratory experiments. Geophys. Res. Lett. 36, L05603.

Luetscher, M., Bolius, D., Schwikowski, M., Schotterer, U., Smart, P.L., 2007. Comparison of techniques for dating of subsurface ice from Monlesi ice cave, Switzerland. J. Glaciol. 53, 374–384.

Luetscher, M., Borreguero, M., Moseley, G.E., Spötl, C., Edwards, R.L., 2013. Alpine permafrost thawing during the Medieval Warm Period identified from cryogenic cave carbonates. Cryosphere 7 (4), 1073–1081.

Maksimovich, G.A., 1947. Peshchernye ldy (Cave ice). Izv. Vsesoyuz. geogr. obshchestva 1947 (5), 537–549 (in Russian).

Marion, G.M., 2001. Carbonate mineral solubility at low temperatures in the Na-K-Mg-Ca-H-Cl-SO_4-OH-HCO_3-CO_3-H_2O system. Geochim. Cosmochim. Acta 65 (12), 1883–1896.

Marion, G.M., Millero, F.J., Feistel, R., 2009. Precipitation of solid phase calcium carbonate and their effect on application of seawater S-A-T-P models. Ocean Sci. 5 (3), 285–291.

Marion, G.M., Mironenko, M.W., Roberts, M.W., 2010. FREZCHEM: a geochemical model for cold aqueous solutions. Comput. Geosci. 36 (1), 10–15.

Marland, G., 1975a. Phase-equilibria in system calcium carbonate-water. Geochim. Cosmochim. Acta 39 (8), 1193–1197.

Marland, G., 1975b. The stability of $CaCO_3.6H_2O$ (ikaite). Geochim. Cosmochim. Acta 39, 83–91.

Mavlyudov, B.R., 2008. Caves glaciation. Russian Academy of Sciences, Institute of Geography, Moscow (in Russian with abstract in English).

May, B., Spötl, C., Wagenbach, D., Dublyansky, Y., Liebl, J., 2011. First investigations of an ice core from Eisriesenwelt cave (Austria). Cryosphere 5 (1), 81–93.

McMichael, A.J., 2012. Insights from past millennia into climatic impacts on human health and survival. Proc. Natl. Acad. Sci. U. S. A. 109 (13), 4730–4737.

Meakin, P., Jamtveit, B., 2010. Geological pattern formation by growth and dissolution in aqueous systems. Proc. R. Soc. A. 466, 659–694.

Meissner, P., Richter, D.K., Immenhauser, A., 2010. Cryogenic calcites in caves of the Rhenish Slate Mountains – a new proxy for palaeoclimate reconstructions? Schriftenreihe der Deutschen Gesellschaft für Geowissenschaften 68, 385. Darmstadt.

Mickler, P.J., Stern, L.A., Banner, J.L., 2006. Large kinetic isotope effects in modern speleothems. Geol. Soc. Amer. Bull. 118, 65–81.

Moore, G.W., 1952. Speleothem – a new cave term. Natl. Speleol. Soc. News 10 (6), 2.

Mulvaney, R., Wolff, E.W., Oates, K., 1988. Sulphuric acid at grain boundaries in Antarctic ice. Nature 331, 247–249.

Murase, N., Ruike, M., Matsunaga, N., Hayakawa, M., Kaneko, Y., Ono, Y., 2001. Spider silk has an ice nucelation activity. Naturwissenschaften 88, 117–118.

Oehlerich, M., Mayr, C., Griesshaber, E., Ohlendorf, C., Zolitschka, B., Sánchez-Pastor, N., Kremer, B., Lücke, A., Oeckler, O., Schmahl, W., 2010. Occurrences of ikaite and pseudomorphs after ikaite in Patagonian lakes – crystal morphologies and stable isotope composition. Geophys. Res. Abstr. 12. EGU2010-3375.

Omelon, C.R., Pollard, W.H., Marion, G.M., 2001. Seasonal formation of ikaite ($CaCO_3.6H_2O$) in saline spring discharge at Expedition Fiord, Canadian High Arctic: Assessing conditional constraints for natural crystal growth. Geochim. Cosmochim. Acta 65 (9), 1429–1437.

Onac, B.P., 2001. Mineralogical studies and Uranium-series dating of speleothems from Scărişoara Glacier Cave (Bihor Mountains, Romania). Theor. Appl. Karstol. 13–14, 33–38.

Onac, B.P., 2008. Ikaite in the Scărişoara ice deposit: precipitation and significance. In: Turri, S. (Ed.), 3rd International Workshop in Ice Caves, IWIC-III, Volume of Abstracts. Perm State University, Perm, p. 28.

Onac, B.P., Wynn, J.G., Citterio, M., 2011. Ikaite in the Scărişoara Ice Cave (Romania): origin and significance. Geophys. Res. Abstr. 13. EGU2011-5188.

Orvošová, M., 2015. The importance of cryogenic cave carbonates in the palaeoclimatic studies and in the study of other cryogenic phenomena in the caves of Slovakia. Comenius University, Bratislava. Unpublished PhD work. (in Slovak).

Orvošová, M., Deininger, M., Milovský, R., 2014. Permafrost occurrence during the Last Permafrost Maximum in the Western Carpathian Mountains of Slovakia as inferred from cryogenic cave carbonate. Boreas 43 (3), 750–758.

Papadimitriou, S., Kennedy, H., Kattner, G., Dieckmann, G.S., Thomas, D.N., 2004. Experimental evidence for carbonate precipitation and CO_2 degassing during sea ice formation. Geochim. Cosmochim. Acta 68 (8), 1749–1761.

Pauly, H., 1963. "Ikaite", a new mineral from Greenland. Arctic 16, 263–264.

Perşoiu, A., 2011. Paleoclimatic significance of perennial ice accumulations in caves: an example from Scărişoara Ice Cave, Romania. Unpublished PhD Thesis, University of South Florida, Tampa, FL.

Perşoiu, A., Onac, B.P., 2012. Ice in caves. In: White, W.B., Culver, D.C. (Eds.), Encyclopedia of caves. Academic Press, Chennai, pp. 399–404.

Persoiu, A., Onac, B.P., Wynn, J.G., Blaauw, M., Ionita, M., Hansson, M., 2017. Holocene winter climate variability in Central and Eastern Europe. Sci. Rep. 7, 1196. https://doi.org/10.1038/s41598-017-01397-w.

Petrenko, V.F., Whitworth, R.W., 1999. Physics of Ice. Oxford University Press, Oxford.

Petrich, C., Eicken, H., 2010. Growth, structure and properties of sea ice. In: Thomas, D.N., Dieckmann, G.S. (Eds.), Sea Ice. Wiley-Blackwell, Oxford, pp. 23–77.

Pielsticker, K.-H., 2000. Höhlen und Permafrost Thermophysikalische Prozesse von Höhlenvereisungen während des Quartärs. Bochumer geologische und geotechnische Arbeiten 55, 187–196.

Pop, E., Ciobanu, I., 1950. Pollen analysis in the ice from Scărişoara Cave. Analele Academiei Republicii Populare Romane, Seria Geologie, Geografie, Biologie, Ştiinţe Tehnice şi Agricole 3, 23–50 (in Romanian).

Potapov, S.S., Parschina, H.V., Titov, A.T., Rakin, V.I., Nizovskii, A.I., Shavrina, E.V., Kadebskaya, O.I., 2008. Cryomineral formations of the caves in Pinezhya. Mineralogiya tekhnogeneza, Nauchnoe izdanye, IMin UrO RAN 9, 18–43.

Pulina, M., 1990. Geomorphological effects of the cryochemical process. Quaestiones Geographicae 13/14, 99–112.

Pulinowa, M.Z., Pulina, M., 1972. Phenomenes cryogenes dans les grottes et gouffres des Tatras. Biul. Peryglac. 21, 201–235.

Racoviţă, G., Onac, B.P., 2000. Scărişoara Glacier Cave. Monographic study. Editura Carpatica, Cluj-Napoca.

Richter, D.K., Niggemann, S., 2005. Kryogene Calcite in Höhlen des Rheinischen Schiefergebirges. Mitteilungen des Verbandes der deutschen Höhlen- und Karstforscher 51 (4), 129–132. München.

Richter, D.K., Riechelmann, D.F.C., 2008. Late Pleistocene cryogenic calcite spherolites from the Malachitdom Cave (NE Rhenish Slate Mountains, Germany): origin, unusual internal structure and stable C-O isotope composition. Int. J. Speleol. 37 (2), 119–129.

Richter, D.K., Neuser, R.D., Voigt, S., 2008. Kryogene Calciumpartikel aus der Heilenbecker Höhle in Ennepetal (NE Bergisches Land/Nordrhein-Westfalen) (Cryogenic calcite particles from the Heilenbecker Cave in Ennepetal, NE Bergisches Land/North-Rhine Westphalia). Höhle 59 (1–4), 37–47.

Richter, D.K., Dreyer, R., Niggemann, S., Pielsticker, K.-H., 2009a. Kryokalcite in der Großen Sunderner Höhle (Sauerland) – ein weiterer Beleg für die vormalige Eishöhle. Mitteilungen des Verbandes der deutschen Höhlen- und Karstforscher 55 (3), 80–85.

Richter, D.K., Voigt, S., Neuser, R.D., 2009b. Kryogene Calcite unterschiedlicher Kristallform und Kathodolumineszenz aus der Glaseishöhle am Schneiber (Steinernes Meer/Nationalpark Berchtesgaden, Deutschland). Höhle 60 (1–4), 3–9.

Richter, D.K., Meissner, P., Immenhauser, A., Schulte, U., Dorsten, I., 2010a. Cryogenic and non-cryogenic pool calcites indicating permafrost and non-permafrost periods: a case study from the Herbstlabyrinth-Advent Cave system (Germany). Cryosphere 4, 501–509.

Richter, D.K., Schulte, U., Mangini, A., Erlenmeyer, A., Erlenmeyer, M., 2010b. Mittel- und oberpleistozäne Calcitpartikel kryogener Entstehung aus der Apostelhöhle südöstlich Brilon (Sauerland, NRW). Geologie und Paläontologie in Westfalen 78, 61–71.

Richter, D.K., Mischel, S., Dorsten, I., Mangini, A., Neuser, R.D., Immenhauser, A., 2011. Zerbrochene Hohlensinter und Kryocalcite als Indikatoren fur eiszeitlichen Permafrost im Herbstlabyrinth-Adventhohle-System bei Breitscheid-Erdbach (N-Hessen). Höhle 62, 31–45.

Richter, D.K., Meyer, S., Scholz, D., Immenhauser, A., 2013. Multiphase formation of Weichselian cryogenic calcites, Riesenberg Cave (Süntel/NW Germany). Z. Dt. Ges. Geowiss. (German J. Geosci.) 164 (2), 353–367.

Richter, D.K., Goll, K., Grebe, W., Niedermayr, A., Platte, A., Scholz, D., 2015. Weichselzeitliche Kryocalcite als Hinweise für Eisseen in der Hüttenblässerschachthöhle (Iserlohn, NRW). E&G Quat. Sci. J. 64 (2), 67–81.

Sade, Z., Halevy, I., 2017. New constraints on kinetic isotope effects during $CO_{2(aq)}$ hydration and hydroxylation: Revisiting theoretical and experimental data. Geochim. Cosmochim. Acta 214, 246–265.

Savchenko, E.B., 1976. Mineral deposits on ice samples in the cave "Friendship". Peshchery 16, 21–24. Perm (in Russian).

Schmidt, F.-X., 1992. Mineralogische Besonderheiten aus dem Höhlensystem Kreiselhalle-Malachitdom. In: Geologisches Landesamt Nordrhein-Westfalen (Ed.), Der Malachitdom. Ein Beispiel interdisziplinärer Höhlenforschung im Sauerland. Geologisches Landesamt Nordrhein-Westfalen, Krefeld, pp. 91–104.

Schöner, W., Boeckli, L., Hausmann, H., Otto, J.-C., Reisenhofer, S., Riedl, C., Seren, S., 2012. Spatial patterns of permafrost at Hoher Sonnblick (Austrian Alps) – Extensive field-measurements and modelling approaches. Aust. J. Earth Sci. 105, 154–168.

Selleck, B.W., Carr, P.F., Jones, B.G., 2007. A review and synthesis of glendonites (pseudomorphs after ikaite) with new data: assessing applicability as recorders of ancient coldwater conditions. J. Sediment. Res. 77, 980–991.

Shearman, D.J., Smith, A.J., 1985. Ikaite, the parent mineral of jarrowite-type pseudomorphs. Proc. Geologist. Assoc. 96, 305–314.

Skřivánek, F., 1954. Jeskyně na Chlumu v Českém krasu (Caves at Chlum in the Bohemian Karst). Československý kras 7, 25–34 (in Czech).

Socki, R.A., Romanek, S., Gibson Jr., E.K., Golden, D.C., 2001. Terrestrial aufeis formation as a Martian analog: clues from laboratory produced C-13 enriched cryogenic carbonate. Lunar Planet. Sci. 32, 2032.

Socki, R.A., Niles, P.B., Fu, Q., Gibson Jr., E.K., 2010. Cryogenic carbonate formation on Mars: clues from stable isotope variations seen in experimental studies. In: 41st Lunar and Planet. Sci. Conf. 2526.pdf.

Souchez, R.A., Jouzel, J., 1984. On the isotopic composition in δD and $\delta^{18}O$ of water and ice during freezing. J. Glaciol. 30 (106), 369–372.

Souchez, R.A., Lorrain, R.D., 1991. Ice composition and glacier dynamics. Springer Verlag, Berlin, Heidelberg.

Souchez, R., Jouzel, J., Lorrain, R., Sleewaegen, S., Stiéverand, M., Verbeke, V., 2000. A kinetic isotope effect during ice formation by water freezing. Geophys. Res. Lett. 27 (13), 1923–1926.

Spötl, C., 2008. Kryogene Karbonate im Höhleneis der Eisriesenwelt. Höhle 59 (1–4), 26–36.

Spötl, C., Cheng, H., 2014. Holocene climate change, permafrost and cryogenic carbonate formation: insights from a recently deglaciated, high-elevation cave in the Austrian Alps. Clim. Past 10, 1349–1362.

Suess, E., Balzer, W., Hesse, K.-F., Muller, P.J., Ungerer, C.A., Wefer, G., 1982. Calcium carbonate hexahydrate from organic rich sediments of the Antarctic shelf: precursor for glendonites. Science 216, 1128–1131.

Suzuoki, T., Kimura, T., 1973. D/H and $^{18}O/^{16}O$ fractionation in ice-water systems. J. Mass Spectroscop. Soc. Japan 21, 229–233.

Swainson, I.P., Hammond, R.P., 2001. Ikaite, $CaCO_3.6H_2O$: Cold comfort for glendonites as paleothermometers. Am. Mineral. 86, 1530–1533.

Szewczyk, J., Nawrocki, J., 2011. Deep-seated relic permafrost in northeastern Poland. Boreas 40, 385–388.

Tchaikovskiy, I., Kadebskaya, O., Kalinina, T., 2015. Mineral pellicles on the lakes surface of warm and cold zone in Kungur Ice Cave. Acta Carsol. 44 (1), 101–106.

Teehera K.B., Jungbluth, S.P., Onac, B.P., Acosta-Maeda, T.E., Hellebrand, E., Misra, A.K., Pflitsch, A., Rappé, M.S., Smith, S.M., Telus, M., Schorghofer, N., under review. Cryogenic minerals in Hawaiian lava tubes: A geochemical and microbiological exploration. Geomicrobiol. J. https://doi.org/10.1080/01490451.2017.1362079.

Tulis, J., Novotný, L., 1989. System of Stratenská Cave. Osveta Publishers, Martin (in Slovak with English summary).

Urban, J., Złonkiewicz, Z., 1989. Wstępne wyniki badań geologicznych jaskini Chelosiowa Jama w Jaworzni koło Kielc (First results of geological research in the Chelosiowa Jama Cave in Jaworznia near Kielce). Geol. Q. 33 (2), 367–369 (in Polish).

Vandenberghe, J., French, H.M., Gorbunov, A., Marchenko, S., Velichko, A.A., Jin, H., Cui, Z., Zhang, T., Wan, X., 2014. The Last Permafrost Maximum (LPM) map of the Northern Hemisphere: permafrost extent and mean annual air temperatures, 25–17 ka BP. Boreas 43, 652–666.

Vecht, A., Ireland, T.G., 2000. The role of vaterite and aragonite in the formation of pseudo-biogenic carbonate structures: implications for martian exobiology. Geochim. Cosmochim. Acta 64, 2719–2725.

Viehmann, I., 1958. Perlele de cavernă din Gheţarul de la Scărişoara. Com. Geol., Dări de Seamă ale Şedinţelor XLV, 283–295.

Viehmann, I., 1959. Příspěvky k vývoji jeskynních perel (Un nouveau processus de génèse des perles de caverne). Československý kras 12, 177–185 (in Czech with French summary; the volume 12/1959 was in fact published in 1960).

Viehmann, I., 1962. Perlele de cavernă din Gheţarul de la Scărişoara. DS Şed. Com. Geol. 45, 283 (in Romanian).

Viehmann, I., 1963. Un nou proces de geneza a perlelor de caverna. Lucr. Inst. Speol. "E. Racoviţă" I–II (1962–1963), 295–303 (in Romanian with Russian and French abstracts).

Viehmann, I., 1967. Nouvelle contribution à la genèse des perles de caverne dans les grottes à glace. Lucr. Inst. Speol. "E. Racoviţă" VI, 141–147 (in Romanian with French abstract).

Viehmann, I., 1993. Note sur des perles de caverne particulières. Bull. Soc. géogr. Liège 29, 105–107.

Viehmann, I., Moţiu, A., 1974. Modifications morphologiques obtenues expérimentalement sur des perles des cavernes. Trav. Inst. Spéol. "Emile Racovitza" XIII, 201–203.

Vrbka, L., Jungwirth, P., 2005. Brine rejection from freezing salt solutions: a molecular dynamics study. Phys. Rev. Lett. 95, 148501-1–148501-4.

Vrbka, L., Jungwirth, P., 2007. Molecular simulations of water freezing: brine rejection and homogeneous nucleation. In: Kuhs, W.F. (Ed.), Physics and Chemistry of Ice. Royal Society of Chemistry, Cambridge, pp. 627–634.

Whiticar, M.J., Suess, E., 1998. The cold carbonate connection between Mono Lake, California, and the Bransfield Strait, Antarctica. Aq. Geochem. 4, 429–454.

Worster, M.G., Wettlaufer, J.S., 1997. Natural convection, solute trapping, and channel formation during solidification of saltwater. J. Phys. Chem. B 101 (32), 6132–6136.

Žák, K., Urban, J., Cílek, V., Hercman, H., 2004. Cryogenic cave calcite from several Central European caves: age, carbon and oxygen isotopes and a genetic model. Chem. Geol. 206, 119–136.

Žák, K., Onac, B.P., Perşoiu, A., 2006. Cryogenic formation of carbonate in caves – a review. In: Onac, B.P., Tămaş, T., Constantin, S., Perşoiu, A. (Eds.), Archives of Climate Change in Karst. vol. 10. Karst Waters Inst. Spec. Publ. Leesburg, Virginia, pp. 238–241.

Žák, K., Onac, B.P., Perşoiu, A., 2008. Cryogenic carbonates in cave environments: a review. Quat. Int. 187, 84–96.

Žák, K., Hercman, H., Orvošová, M., Jačková, I., 2009. Cryogenic cave carbonates from the Cold Wind Cave, Nízke Tatry Mountains, Slovakia: Extending the age range of cryogenic cave carbonate formation to the Saalian. Int. J. Speleol. 38 (2), 139–152.

Žák, K., Skála, R., Filippi, M., Plášil, J., 2010. Ikaite – little known mineral of iced caves: occurrence in seasonal cave ice formations of the Koda Cave (Bohemian Karst). Bulletin mineralogicko-petrologického oddělení Národního muzea 18 (1), 109–115 (in Czech with English abstract).

Žák, K., Šmída, B., Filippi, M., Živor, R., Komaško, A., Vybíral, S., 2011. New localities of cryogenic cave carbonates in the Czech Republic and Slovakia. In: Bosák, P., Geršl, M., Novotná, J. (Eds.), Speleofórum. vol. 30. Czech Spelological Society, Prague, pp. 103–110 (in Czech with English abstract).

Žák, K., Richter, D.K., Filippi, M., Živor, R., Deininger, M., Mangini, A., Scholz, D., 2012. Coarsely crystalline cryogenic cave carbonate – a new archive to estimate the Last Glacial minimum permafrost depth in Central Europe. Clim. Past 8, 1821–1837.

Žák, K., Orvošová, M., Filippi, M., Vlček, L., Onac, B.P., Perşoiu, A., Rohovec, J., Světlík, I., 2013. Cryogenic cave pearls in the periglacial zone of ice caves. J. Sedment. Res. 83 (2), 207–220.

CHAPTER

ICE CAVE FAUNA

7

Sanda Iepure
IMDEA Water Institute, Alcala de Henares, Spain

Over the past decade, studies of cave life have largely focused on cave ecosystems free of ice. However, caves enclosing ice, either perennial or seasonal, have attracted the interest of biologists as a means of providing an analogy to extraterrestrial subsurface life on planets such as Mars (Boston, 2011; Williams et al., 2009).

Ice caves present a particularly challenging environment for the existence of lifeforms. They are found in various climatic regions on Earth, from high latitude subpolar regions in Canada (Yukon territory), Norway and Siberia to Ross Island in the Antarctic region, high altitudes in the Altai Mountains in Russia (often under permafrost conditions), and the Spanish Pyrenees and Picos de Europa in the Mediterranean region (Yonge, 2004; Tebo et al., 2015; Bartolomé et al., 2015; Gómez-Lende, 2015). There also are reports of a few ice caves in hot arid climates in Southwestern United States and Central Turkey (Perşoiu and Onac, 2012).

Ice caves with perennial and seasonal ice present a mosaic of microhabitats commonly found in caves, i.e., fissures, cracks, microfissures, pools (temporary or permanent), and subterranean rivers and lakes. Nevertheless, ice caves also have habitats suitable for fauna formed in the periglacial sectors, such as various ice formations (ice walls, ice stalagmites, and floor ice) and cryogenic calcite powders. The environmental conditions in ice caves are particularly characterized by the variability in temperature, ranging from 0°C down to −14°C (Perşoiu et al., 2011). The presence of perennial ice inside galleries causes a drop in the air temperature extending from the cave entrance to a few hundreds of meters, resulting in the freezing of drip water and the formation of speleothems with short or long lifespans (Perşoiu and Onac, 2012). The particular physical and environmental attributes of ice caves significantly influence the colonization of terrestrial and aquatic habitats by subterranean fauna. The low temperature and other limiting factors in the subterranean realm (i.e., absence of light, high humidity, and reduced availability of food resources) influence the number of organisms that may actively colonize ice caves (Racoviță, 2000).

Studies on cave fauna have a long history, but biospeleological surveys on caves with perennial or seasonal ice are still exceptional. The small number of studies reflects the perception that the subterranean habitats and environmental conditions of ice caves are too harsh to sustain terrestrial and aquatic populations. Yet biological communities in ice caves consist of a variety of organisms, from bacteria to cold-adapted terrestrial and aquatic invertebrates and vertebrates (Table 1).

Terrestrial invertebrates in ice caves are mainly represented by arthropods (Table 1). Several groups of coleopteran insects (beetles), araneids (spiders), and gryllobattids (crickets) include cold-adapted species that are reported in ice caves in Europe, the United States, and Canada (Lauriol et al., 2001, 2006).

Ice Caves. https://doi.org/10.1016/B978-0-12-811739-2.00007-3

Table. 7.1 Ice cave fauna (1 - Scarisoara Glacier Cave, Romania; 2 - Jabukovcem Cave, Croatia; 3 - Dobšinská Ice Cave, Slovakia; 4 - Tana delle Sponde Cave, Italy; 5 - Koppenbrüler-Höhle, Austria; 6 - Snezna Ice Cave, Slovenia; 7 - Kungur Ice Cave, Russia; 8 - Ledjanaka Ice Cave, Russia; 9 - Sam's Point Ice Cave, USA; 10 – Wilson Ice Cave, USA; 11 - Sawyer Ice Cave, USA; 12 - Three Level Ice Cave, USA; 13 - Woodville Ice Cave, Canada; 14 - Minasville Ice Cave, Canada; 15 - Castleguard Cave; 16 - Collingwood Cave, Canada; 17 - Khabarovsk Ice Cave, Russia).

Taxonomical group	Species	Cave
TERRESTRIAL		
Pseudoscorpions	*Neobisium maderi* Beier, 1938	2
	Oligolophus sp.	13, 14
Arachnida	*Nesticus racovitzai* Dumitrescu, 1979	1
	Troglohyphantes racovitzai Dumitrescu & Georgescu, 1970	1
	Oligolophus sp.	13,14
Acarina	*Traegardhia dalmatina gigantea* Willmann, 1941	1
	Ixodes vespertilionis Koch, 1844	2
	Veigaia cerva (Kramer, 1876)	3
Copepoda	*Speocyclops infernus* (Kiefer 1930)	6
	Bryocamptus sp.	
	Arcticocamptus cuspidatus var. *ekman* (Schmeil, 1893); *Elaphoidella* sp.	3
	Diacyclops crassicaudis brachycercus Kiefer, 1927	13
Amphipoda	*Niphargus; Gammarus pulex fossarum* Margalef, 1951	3
	Crangonyx chlebinkovi ssp. *maximovitchi* Pankov & Pankova, 2004	7, 8
	Stygobromus allegheniensis (Holsinger, 1967)	9
	Stygobromus canadensis Holsinger, 1980	15
Isopoda	*Salmasellus steganothrix* Bowman, 1975	15
Syncarida	*Bathynella glacialis* Birstein & Ljovuschkin, 1967	17
	Bathynella natans Vejdovský, 1882	3
Ostracoda	*Cypridopsis subterranea* Wolf, 1919	3
Diplopoda	*Biokoviella mauriesi* Mrsic 1992	2
	Allorhiscosoma sphinx (Verhoeff, 1907)	3
Collembola	*Onychiurus multiperforatus* Gruia, 1973	1
	Onychiurus granulosus multisetis Gruia, 1971	
	Onychiurus armatus (Tullberg, 1869)	
	Onychiurus sibiricus (Tullberg, 1876)	
	Onychiurus rectopapillatus Stach, 1933	
	Onychiurus variotuberculatus Stach, 1934	
	Oncopodura crassicornis Shoebotham, 1911)	
	Tomocerus minor (Lubbock, 1862)	
	Hypogastrura crassaegranulata dobsinensis (Stach, 1949)	3
	Megalothorax sp. nov.	
	Isotomidae	6
	Protaphorura pseudarmatus (Folsom, 1917) (Onchiuridae)	13, 14
	Folsomia stella Christiansen and Tucker, 1977	

Table. 7.1 Ice cave fauna (1 - Scarisoara Glacier Cave, Romania; 2 - Jabukovcem Cave, Croatia; 3 - Dobšinská Ice Cave, Slovakia; 4 - Tana delle Sponde Cave, Italy; 5 - Koppenbrüler-Höhle, Austria; 6 - Snezna Ice Cave, Slovenia; 7 - Kungur Ice Cave, Russia; 8 - Ledjanaka Ice Cave, Russia; 9 - Sam's Point Ice Cave, USA; 10 – Wilson Ice Cave, USA; 11 - Sawyer Ice Cave, USA; 12 - Three Level Ice Cave, USA; 13 - Woodville Ice Cave, Canada; 14 - Minasville Ice Cave, Canada; 15 - Castleguard Cave; 16 - Collingwood Cave, Canada; 17 - Khabarovsk Ice Cave, Russia). ***Continued***

Taxonomical group	Species	Cave
Coleoptera	*Pholeuon* (s. str.) *proserpinae glaciale* Jeannel, 1923	1
	Bathysciomorphus globosus Miller, 1855	
	Astagobius (*Astagobius angustatus deelemani* Pretner, 1970	
	Nebria sp.	
	Haplotropidius taxi (Muller, 1903) *Spelaites grabowskii* Apfelbeck, 1907	2
	Choleva nivalis (Kraatz, 1856)	3
	Neobathyscia mancinii Jeannel 1924	4
	Neobathyscia pasai Ruffo, 1950	
	Arctaphaenops styriacus Winkler, 1933	5
	Glacicavicola bathyscioides Westcott, 1968	-
	Gennadota canadensis Casey, 1906	13, 14
	Corticaria serrata (Paykull, 1798)	
Insecta	*Triphosa dubitata* (Linnaeus, 1758)	3
	Imagos of caddisflies; fragments of Nematocera (diptera)	
	Entognata, Lathriidae and Hydracarina	6
	Diptera	7
	Ceuthophilus maculatus (Harris, 1835)	14
	Larvae of Tipuliadae, *Trichocera*, and Sciaridae	
	Odonata and Plecoptera	16
Grilobatida	*Grylloblatta gurneyi* Kamp, 1963	10, 11, 12

In Europe, ice caves Coleoptera are represented by species of the genera *Pholeuon* and *Bathysciomorphus*. Jeannel (1923) described one Leptodirinae species of *Pholeuon*, *P.* (s. str.) *proserpinae glaciale*, from the Scărişoara Glacier Cave in Northwest Romania (Jeannel, 1923). Extensive population and ecological studies on this species performed over more than 40 years show that this species develops large populations (more than 33,000 individuals) and lives in the cave sectors where the air temperature is above 0°C, whereas its presence in frozen areas is assumed to be accidental (Racovita, 1969; Racoviţă, 1980; Racoviţă and Serban, 1975; Fejer and Moldovan, 2013). Scărişoara Cave has a cold microclimate restricting the life conditions for fauna; therefore the structure of the whole terrestrial biocoenosis is founded on a very limited selection of biotopes. Besides *Pholeuon*, other groups of arthropods populate the cave, such as araneids (*Nesticus racovitzai* Dumitrescu, 1979 and *Troglohyphantes racovitzai* Dumitrescu & Georgescu, 1970), springtails (*Oncopodura crassicornis* Shoebotham, 1911, *Onychiurus* spp. and *Tomocerus minor* (Lubbock, 1862)), Acarina (*Ixodes vespertilionis* Koch, 1844), and a large diversity of Collembola (*Onychiurus multiperforatus* Gruia, 1973, *O. armatus* (Tullberg, 1869), *O. sibiricus* (Tullberg, 1876), *O. granulosus multisetis* Gruia, 1971 (troglobite), *O. rectopapillatus* Stach, 1933, *O. variotuberculatus* Stach, 1934 and *Oncopodura crassicornis* Shoebotham, 1911)) (Table. 7.1) (Racoviţă and Onac, 2000).

The ice caves scattered throughout the Dinaric Mountains in Slovenia and Croatia host a few leptodirine troglobite species, such as *Bathysciomorphus globosus* (Miller, 1855), which are reported as developing populations at temperatures below 6°C (Juberthie and Decu, 2012; Lobl and Lobl, 2015). The ice caves in Slovenia with permanent ice deposits and that are located at higher elevations are populated by troglobiotic and troglophilic species of Coleoptera Carabidae from the genus *Nebria* and troglomorphic Coleoptera Leptodirinae from the genus *Astagobius* (*Astagobius angustatus deelemani* Pretner, 1970). The ice caves of Croatia (called ledenice) host two Coleoptera species reported from the Ledenica pod Jabukovcem Cave: *Haplotropidius taxi* (Muller, 1903) and *Spelaites grabowskii* (Apfelbeck, 1907) (Pretner, 1970). The Jabukovcem Cave is located in a deep dolina, favoring snow and ice retention until the summer months when the temperature inside the cave maintains an average of 3°C. Based on the findings in this cave, a new family of Diplopoda, Biokoviellidae, and a new species, *Biokoviella mauriesi* (Mrsic 1992), have been described (Gottstein-Matočec et al., 2002; Antić et al., 2016). The Jabukovcem Cave is also known as the *locus typicus* for three other new arthropod species: *Neobisium maderi* (Beier, 1938) (Pseudoscorpiones), *Traegardhia dalmatina gigantea* (Willmann, 1941) (Acarina), and *Speoplanes giganteus* (J. Müller, 1911) (Coleoptera).

The Dobšinská Ice Cave is one of the most important caves in Slovakia because of the large amount of ice felling deposits (Kováč et al., 2014). The terrestrial fauna of the Dobšinská Ice Cave was found to be particularly rich and diverse with these reported species: *Triphosa dubitata* (Linnaeus, 1758); imagos of caddisflies; *Hypogastrura crassaegranulata dobsinensis* (Stach, 1949) (a springtail glacial relict); *Choleva nivalis* (Coleoptera); and *Allorhiscosoma sphinx* (Diplopoda), with the later (*A. sphinx*) found in the deepest parts of the cave (Kováč et al., 2014; Kriesch, 1875). In Italy, the Tana delle Sponde Cave shows occasional ice formation during winter. The cave hosts two cold-resistant Coleoptera Cholevidae: *Neobathyscia mancinii* (Jeannel 1924) and *N. pasai* (Ruffo, 1950) (Lencioni et al., 2010). These two cold-tolerant species of *Neobathyscia* are assumed to be descended from a cold stenothermal ancestor inhabiting cold temperate forests during glacial and periglacial periods. Ice caves in the northeastern parts of Dachstein in the Austrian Alps host one of the troglomorphic coleopteran species of Trechini, *Arctaphaenops styriacus* (Winkler, 1933), found in ice-free parts of the Koppenbrüler-Höhle (Daffner, 1993; Winkler, 1993).

The terrestrial ice cave fauna from North America is characterized by a particular troglobite, the Leptodirinae from the genus *Glacicavicola* (Westcott, 1968). The species of this genus, *G. bathyscioides* (Westcott, 1968) from lava tube caves with perennial ice in the eastern Snake River Plain in Idaho, does not have close epigean relatives and is considered as a glacial relict (Westcott, 1968). The ice caves investigated in Nova Scotia and New Brunswick, Canada (Woodville Ice Cave and Minasville Ice Cave) are populated by the rove beetles (Staphylinidae) *Gennadota canadensis* (Casey, 1906) and *Corticaria serrata* (Paykull, 1798), the araneid *Oligolophus*, and the gryllobattids (Moseley, 2007). Gryllobattids, also known as ice bugs, ice crawlers, and rock crawlers, are a poorly known group of terrestrial insects restricted to cold and extreme habitats. *Grylloblatta gurneyi* (Kamp, 1963) has been found in three North American caves with permanent ice and at low elevations ranging from 300 to 1000 m (Wilson Ice Cave, Sawyer Ice Cave, and Three Level Ice Cave) (Kamp, 1979; Jarvis and Whiting, 2006).

Although bats have been reported to develop relatively large populations in ice caves, they are not strictly adapted to low temperatures. A large number of species have been reported in the Scărişoara Glacier Cave in Romania, that is, *Myotis myotis* (Borkhausen, 1797), *M. oxygnathus* (Monticelli, 1885), *M. mystacinus* (Kuhl, 1817), *M. brandtii* (Eversmann, 1845), *M. dasycneme* (Boie, 1825), *M. daubentonii* (Kuhl, 1817), *Nyctalus noctula* (Schreber, 1774), *Plecotus austriacus* (Fischer, 1829), and *Vespertilio murinus* (Linnaeus, 1758) (Onac et al., 2010). From findings in the Dobšinská Ice Cave in Slovakia, 10 species of bats have been reported, among them *Myotis mystacinus*, *M. brandtii*, *Barbastella barbastellus* (Schreber, 1774), and *Rhinolophus hipposideros* (Bechstein, 1800), the latter being generally frequent in caves in the temperate region of Europe (Kováč et al., 2014; Bobakova, 2002; Obuch, 2012). The present knowledge on bats in ice caves in Canada is restricted to the Woodwille Ice Cave in Nova Scotia, where three species have been recorded: *Myotis lucifugus* (Le Conte, 1831), *M. septentrionalis* (Trovessart, 1897), and *Pipistrellus subflavus* (Menu, 1984) (Moseley, 2007).

Groundwater invertebrates from ice caves are rarely mentioned. Currently, only a few species of crustaceans (copepods, amphipods, syncarids, and isopods), insects, and Acarina have been reported (Table 1). In Europe, published records on groundwater species are available for ice caves in northern Slovenia, Slovakia, and the Russian Far East (Pankov and Pankova, 2004; Papi and Pipan, 2011; Sidorov, 2008; Sidorov et al., 2012; Juberthie et al., 2016). The Snezna Cave is located in the Julian Alps in Slovenia at 1500 m a.s.l. The maximum temperature registered in the cave is 4°C (Zupan-Hajna et al., 2008). An extensive study on the epikarst fauna in the Snezna Ice Cave indicates the presence of two copepod species, *Speocyclops infernus* (Kiefer 1930) (cyclopoid) and *Bryocamptus* sp. (harpacticoid); one amphipod from the genus *Niphargus*; insects such as Entognata, Lathriidae, and Isotomidae (Collembola); and Hydracarina (Papi and Pipan, 2011). These studies also show that there is no difference between the abundance of epikarst copepod populations in the Snezna Ice Cave and the epikarst copepod populations in caves located at lower altitudes in the Dinaric Mountains (Papi and Pipan, 2011). In the Dobšinská Ice Cave (Slovakia), several groups of arthropods have been reported in the nonglaciated part of the cave: *Cypridopsis subterranea* (Wolf, 1919) (ostracod); *Arcticocamptus cuspidatus* var. *ekman* (Schmeil, 1893) and *Elaphoidella* sp. (harpacticoid); *Gammarus pulex fossarum* (Margalef, 1951) and *Niphargus* sp. (amphipods); *Bathynella natans* (Vejdovský, 1882) (Syncarida); *Veigaia cerva* (Kramer, 1876) (Acarina); *Arrhopalites aggtelekiensis* (Stach, 1929); *Protaphorura janosik* (Weiner, 1990); *Megalothorax* sp. nov. (collembolan); and fragments of Nematocera (diptera) (Kováč et al., 2014).

The groundwater fauna in ice caves in the Russian Far East are known only from two caverns: Kungur Ice Cave located in the Ural Mountains and Ledjanaka Ice Cave located in the Samara basin (Khabarovsk district) (Sidorov, 2008; Juberthie et al., 2016). The Kungur Ice Cave is one of the most investigated caves in Russia, being known as the *locus typicus* for the endemic stygobite amphipod *Crangonyx chlebinkovi* ssp. *maximovitchi* (Pankov and Pankova, 2004). This species was recently found in other groundwater habitats in the Priuralye karst area (caves, wells, small springs, seeps) and appears to live under constant low water temperatures (0–5.5°C) (Sidorov et al., 2012). The Kungur Ice Cave is also inhabited by springtails (with unknown taxonomic affiliation) and Diptera (Sidorov, 2008). The only Syncarida (the species of this crustacean group all being stygobites) reported so far from Russian ice caves is *Bathynella glacialis* (Birstein & Ljovuschkin, 1967), reported from the Khabarovsk Ice Cave (Juberthie et al., 2016; Birstein and Ljovuschkin, 1967; Schminke, 1986; Camacho, 2006).

In Canada and North America, the ice caves host several species of copepods, amphipods, insects, and water mites (Hydracarina) (Moseley, 2007). The ponds and streams in the Woodville Ice Cave in Nova Scotia (Canada) are populated by the cold-stenotherm cyclopoid *Diacyclops crassicaudis brachycercus* (Kiefer, 1927), a species most likely to have evolved from a glacial relict ancestor (Moseley, 2007; Tash, 1971; Reid, 1992). The same cave hosts several collembolan species, including *Protaphorura pseudarmatus* (Folsom, 1917) (Onchiuridae) and *Folsomia stella* (Christiansen and

Tucker, 1977) (Isotomidae); insects such as *Ceuthophilus maculatus* (Harris, 1835) (Perlidae) and *Gennadota canadensis* (Staphylinidae); larvae of Tipuliadae, *Trichocera*, and Sciaridae; and Acarina Uropodidae and Histiostmatidae. *Folsimia candida* (Isotomidae) has also been reported from ponds in the Minasville Ice Cave, whereas insects such as Odonata and Plecoptera have been found in the Collingwood Cave, both caves are situated in Nova Scotia (Moseley, 2007).

The ice caves in North America are well known to host one of the most diverse genera of amphipod crustaceans, *Stygobromus* (Cope, 1872), with only a few such species reported so far in Eurasia (Holsinger, 1967, 2010). Most *Stygobromus* species are currently reported in caves located in nonglaciated areas of North America, and only a few species appear to have survived in areas covered by ice during the last glaciation (Holsinger, 1967). *Stygobromus allegheniensis* (Holsinger, 1967), a large species fully depigmented and eyeless, has been reported in Sam's Point Ice Cave, in the United States (Cahill et al., 2015) and in ice caves located in the Shawangunk Ridge in New York (Espinasa et al., 2015). In Sam's Point Ice Cave, *S. allegheniensis* inhabits pools that freeze only at the surface; the individual organisms likely remain at the bottom of the pool to avoid being frozen. The metabolism of *S. allegheniensis* seems to be adjusted to survive at low temperatures, similarly to the European species *Niphargus rhenorhodanensis* (Schellenberg, 1937), a cold-adapted species of amphipod that accumulates cryoprotective molecules such as glycerol and free amino acids, enabling it to endure extreme thermal conditions (Issartel et al., 2006; Colson-Proch et al., 2009). *S. allegheniensis* is not restricted to Sam's Point Ice Cave and is also found in caves in Maryland, Pennsylvania, and New York, being rather common in caves developed in the glaciated Appalachian Plateau region (Holsinger, 1967). *Stygobromus* is also present in ice caves in Canada, with *S. canadensis* (Holsinger, 1980) found in a series of pools located at about 2 km in the Castleguard Cave (Holsinger et al., 1983). The cave also hosts an asellid isopod crustacean, *Salmasellus steganothrix* (Bowman, 1975), and are abundant in pools rich in sediment (Holsinger et al., 1983).

Fossils preserved in cave sediments are usually excellent archival proxies that enable the impact of past environmental changes on faunal biodiversity to be detected (Sasowsky and Mylroie, 2004). It was only recently discovered that ice caves preserve important vestiges of past living fauna; however, these remains are currently found only in sediments and not in ice blocks (Obuch, 2012). One of the ice caves with a significant accumulation of skeletal remains (named thanatocoenoses) is the Dobšinská Ice Cave in Slovakia, where remnants of bats (hibernating within the cave); birds (*Turdus philomelos* (Brehm, 1831), *Parus major* (Linnaeus, 1758), *Loxia curvirostra* (Linnaeus, 1758), and other *Passeriformes*); amphibians (*Bufo bufo* (Linnaeus, 1758) and *Rana temporaria* (Linnaeus, 1758)); and the lizard *Lacerta vivipara* (Jacquin, 1787) have been retrieved from cave sediments (Obuch, 2012). In the Scarişoara Glacier Cave (Romania), the only skeletal remains found are bones of *Rupicapra* found in the "Great Reservation," close to the cave entrance (Racoviță and Onac, 2000).

REFERENCES

Antić, Ž.D., Tvrtko, D., Tonći, R., Luka-Lučić, R., Slobodan-Makarov, E., 2016. Taxonomic status of the family Biokoviellidae Mršić, 1992 (Diplopoda, Chordeumatida): reconsideration, with a description of one new species. Eur. J. Taxon. 205, 1–23.

Bartolomé, M., Sancho, C., Osácar, C., Moreno, A., Leunda, M., Oliva, B., Spötl, C., Luetscher, M., López-Martínez, J., Belmonte, A., 2015. Characteristics of cryogenic calcites in Pyrenean ice caves (Northern Spain). Geogaceta 58, 115–118.

Birstein, J.A., Ljovuschkin, S.I., 1967. Biospeleologica sovietica 33. The order Bathynellacea (Crustacea, Malacostraca) in the U.R.S.S. I. Family Bathynellidae. Bull. Mosk. Ova. Ispyt. Prir. Biol (Bull. Moscow Soc.) 72 (4), 51–66.

Bobakova, L., 2002. Quantitative and qualitative characteristics of bat fauna of the Dobsinska Ice Cave (E Slovakia). Lynx (Prague) 33, 47–58.

Boston, P.J., 2011. Extraterrestrial caves. In: Gunn, J. (Ed.), Encyclopedia of Cave and Karst Science. Fitzroy-Dearborn Publishers, Ltd., London, pp. 355–358.

Cahill, A., Kavanagh, A., McCahill, A., Scott, A., Espinasa, L., 2015. Phylogenetic analysis of several new populations of Stygobromus allegheniensis (Allegheny Cave Amphipod) in the Ice Caves of the Shawangunk Ridge, NY. In: The Northeast Natural History Conference, Springfield, MA.

Camacho, A.I., 2006. An annotated checklist of Syncarida (Crustacea, Malacostraca) in the world. Zootaxa 1374, 1–54.

Colson-Proch, C., Renault, D., Gravot, A., Douady, C.J., Hervant, F., 2009. Do current environmental conditions explain physiological and metabolic responses of subterranean crustaceans to cold? J. Exp. Biol. 212 (12), 1859–1868.

Daffner, H., 1993. Die Arten der Gattung Arctaphaenops Meixner, 1925 (Coleoptera: Carabidae). Koleopt. Rundschau 63, 1–18.

Espinasa, L., McCahill, A., Kavanagh, A., Espinasa, J., Scott, A., Cahill, A., 2015. A troglobitic amphipod in the Ice Caves of the Shawangunk Ridge: behavior and resistance to freezing. Subterr. Biol. 15, 95–104.

Fejer, A., Moldovan, O.T., 2013. Population size and dispersal patterns for a *Drimeotus* (Coleoptera, Leiodidae, Leptodirini) cave population. Subterr. Biol. 11, 31–44.

Gómez-Lende, M., 2015. Las Cuevas Heladas en Picos de Europa. Tesis Doctoral, Universidad de Valladolid. 663 pp.

Gottstein-Matočec, S., Bakran-Petricioli, T., Bedek, J., Bukovec, D., Buzjak, S., Franičević, M., Jalžić, B., Kerovec, M., Kletečki, E., Kralj, J., Kružić, P., Kučinić, M., Kuhta, M., Matočec, N., Ozimec, R., Rađa, T., Štamol, V., Ternjej, I., Tvrtković, N., 2002. An overview of the cave and interstitial biota of Croatia. Nat. Croat. 11 (Suppl 1), 1–112.

Holsinger, J.R., 1967. Systematics, speciation, and distribution of the subterranean amphipod genus *Stygonectes* (Gammaridae). U.S. Natl. Mus. Bull. 259, 1–176.

Holsinger, J.R., 2010. *Stygobromus* – Distribution Map. Old Dominion University (Retrieved June 24).

Holsinger, J.R., Mort, J.S., Recklies, A.D., 1983. The subterranean crustacean fauna of Castleguard Cave, Columbia Icefields, Alberta, Canada and its zoogeographic significance. Arct. Alp. Res. 15, 543–549.

Issartel, J., Voituron, Y., Odagescu, V., Baudot, A., Guillot, G., Ruaud, J.P., Renault, D., Vernon, P., Hervant, F., 2006. Freezing or supercooling: how does an aquatic subterranean crustacean survive exposures at subzero temperatures? J. Exp. Biol. 209 (17), 3469–3475.

Jarvis, K.J., Whiting, M.F., 2006. Phylogeny and biogeography of ice crawlers (Insecta: Grylloblattodea) based on six molecular loci: designating conservation status for Grylloblattodea species. Mol. Phylogenet. Evol. 41, 222–237.

Jeannel, R., 1923. Révision du genre Choleva. L'Abeille. J. Entomol. 32, 1–160.

In: Juberthie, C., Decu, V. (Eds.), 2012. Encyclopaedia Biospeologica. Vol. II. Société de Biospéologie, Moulis, France, pp. 913–920.

Juberthie, C., Sidorov, D., Decu, V., Mikhaljova, E., Semenchenko, K., 2016. Subterranean fauna from Siberia and Russian Far East. Encyclopedia Biospeologica (Siberia-Far East Special Issue). Ecol. Montenegrina 7, 507–529.

Kamp, J.W., 1979. Taxonomy, distribution, and zoographic evolution of *Grylloblatta* in Canada (Insecta: Notoptera). Can. Entomol. 111, 27–38.

Kováč, L., Elhottová, D., Mock, A., Nováková, A., Krištůfek, V., Chroňáková, A., Lukešová, A., Mulec, J., Košel, V., Papáč, V., Luptáčik, P., Uhrin, M., Višňovská, Z., Hudec, I., Gaál, L., Bella, P., 2014. The Cave Biota of Slovakia, Liptovský Mikuláš: State Nature Conservancy SR. Slovak Caves Administration.

Kriesch, J., 1875. Allataniutazasi jelentések az 1870 és 1872-ij évről. Matematikai és természettudományi kőzlemények 10, 201–220.
Lauriol, B., Prevost, C., Deschamps, E., Cinq-Mars, J., Labrecque, S., 2001. Faunal and archaeological remains as evidence of climatic change in freezing caverns. Yukon Territory, Canada. Arctic 54 (2), 135–141.
Lauriol, B., Prevost, C., Lacelle, D., 2006. The distribution of diatom flora in ice caves of the northern Yukon Territory, Canada: relationship to air circulation and freezing. Int. J. Speleol. 35 (2), 83–92.
Lencioni, V., Bernabò, P., Latella, L., 2010. Cold resistance in two species of cave-dwelling beetles (Coleoptera: Cholevidae). J. Therm. Biol. 35 (7), 354–359.
Lobl, I., Lobl, D. (Eds.), 2015. The Catalogue of Palaearctic Coleoptera Provides Information About All Beetles Occurring in Europe, North Africa and Asia North of the Tropics. Brill Publisher, Leiden/Boston.
Moseley, M., 2007. Acadian biospeleology: composition and ecology of cave fauna of Nova Scotia and southern New Brunswick, Canada. Int. J. Speleol. 36 (1), 1–21.
Obuch, J., 2012. Assemblages of bats in deposits of the Dobšinská Ice Cave, Slovenský raj National Park, Slovakia. Vespertilio 16, 205–210.
Onac, B.P., Racovita, G., Brad, T., Ciubotarescu, C., Damm, P.E., Tamas, T., Iepure, S., Borda, D. (Eds.), 2010. Atlasul pesterilor din Munţii Apuseni. Munţii Bihor - 1. Bazinul Arieşului Mare. University of South Florida Karst Studies Series, 290 pp.
Pankov, N.N., Pankova, N.V., 2004. The biology of the cave amphipod Crangonyx chlebinkovi Borutzky, 1928 (Gammaridae) with the description of a new species from Kungur Ice cave. Peshchery: mezhvuzovskiy sbornik nauchnykh trudov, Perm: Perm University, vol. 29–30, pp. 141–149 [in Russian].
Papi, F., Pipan, T., 2011. Ecological studies of an epikarst community in Snežna jama na planini Arto—an ice cave in north central Slovenia. Acta Carsologica 40, 505–513.
Perşoiu, A., Onac, B.P., 2012. Ice in caves. In: White, W.B., Culler, D.C. (Eds.), Encyclopedia of Caves, second ed. Elsevier, Burlington, MA, pp. 399–404.
Perşoiu, A., Onac, B.P., Wynn, J., Bojar, A.V., Holmgren, K., 2011. Stable isotopes behavior during cave ice formation by water freezing in Scarisoara Ice Cave, Romania. J. Geophys. Res. 116, D02111. https://doi.org/10.1029/2010JD014477.
Pretner, E., 1970. Remarks to the catalogue of Laneyrie's new classification of subfamily Bathysciinae (Coleoptera) and explanation to the catalogue of subfamily Bathysciinae.
Racovita, G., 1969. Recherches sur l'écologie du Bathysciinae troglobionte *Pholeuon proserpinae glaciale* Jeann. 5 Int. Congr. Speleol., Stuttgart 4 (B 14), 1–6.
Racoviţă, G., 1980. Étude écologique sur les coléoptères Bathysciinae cavernicoles. Mémoire de Biospéologie 6, 1–199.
Racoviţă, G., 2000. Ice cave in temperate regions. In: Wilkens, H., Culvier, D.C., Humphreys, W. (Eds.), Subterranean Ecosystems. Elsevier, Amsterdam, pp. 561–568.
Racoviţă, G., Onac, B.P. (Eds.), 2000. Scarisoara Glacier Cave. Monographic Study. Editura Carpatica, Cluj-Napoca.
Racoviţă, G., Serban, M., 1975. Recherches sur la biologie de la population de coléoptères cavernicoles de la Peştera cu apă din Valea Leşului. Ann. Spéliol. 30, 351–363.
Reid, J.W., 1992. Redescription of *Diacyclops nearcticus* (Kiefer, 1934) and description of four similar new congeners from North America, with comments on *D. crassicaudis* (G.O. Sars, 1863) and *D. crassicaudis* var. *brachycercus* (Kiefer, 1927) (Crustacea: Copepoda). Can. J. Zool. 70, 1445–1572.
Sasowsky, I.D., Mylroie, J., 2004. Studies of Cave Sediments. Physical and Chemical Records of Paleoclimate. Kluwer Academic, Plenum Publishers, New York.
Schminke, H.K., 1986. Syncarida. In: Botosaneanu, L. (Ed.), Stygofauna Mundi. A Faunistic, Distributional and Ecological Synthesis of the World Fauna Inhabiting Subterranean Waters. E.J. Brill/Dr. W. Backhuys, Leiden, pp. 389–404.
Sidorov, L.P., 2008. Cave Collection of Scientific Papers. vol. 31. Perm State University, Perm, Russia.

Sidorov, L.P., Pankov, N.N., Holsinger, J.R., 2012. Distribution and ecology of the subterranean amphipod Crangonyx chlebnikovi Borutzky, 1928 (Crustacea: Crangonyctidae), with lectotype designation and comments on morphology of the lateral cephalic lobe. Arthropoda Sel. 21 (2), 149–160.

Tash, J.C., 1971. The zooplankton of fresh and brackish waters of the Cape Thompson area, northern Alaska. Hydrobiologia 38, 93–121.

Tebo, B.M., Bradley, M., Davis, R.E., Anitori, R.P., Connell, L.B., Schiffman, P., Staudigel, H., 2015. Microbial communities in dark oligotrophic volcanic ice cave ecosystems of Mt. Erebus, Antarctica. Front. Microbiol. 11 (6), 179.

Westcott, R., 1968. A new subfamily of blind beetle from Idaho ice caves with notes on its bionomics and evolution (Coleoptera: Leiodidae). Cotntr. Sci., Los Angeles County Mus. Nat. Hist. 141, 1–14.

Williams, K.E., Toon, O., Heldmann, J., Mellon, M., 2009. Ancient melting of midlatitude snowpacks on Mars as a water source for gullies. Icarus 200, 418–425.

Winkler, A., 1993. Eine zweite höhlenlaufkäferart aus den nordalpen: arctaphaenops styriacus sp. n. Koleopt. Rundsch 19, 237–240.

Yonge, C., 2004. Ice in caves. In: Gunn, J. (Ed.), Encyclopedia of Cave and Karst Science. Fitzroy-Dearborn Publishers, Ltd., London, pp. 435–437.

Zupan-Hajna, N., Mihevc, A., Pruner, P., Bosák, P., 2008. Paleomagnetism and Magnetostratigraphy of Karst Sediments in Slovenia. Založba ZRC/ZRC Publishing, Ljubljana. pp. 266.

CHAPTER

MICROBIAL LIFE IN ICE CAVES

8

Cristina Purcarea

Institute of Biology, Bucharest, Romania

CHAPTER OUTLINE

8.1 INTRODUCTION

In contrast with other parts of the Earth's cryosphere, ice caves represent unique icy habitats constituting secluded and relatively stable light-deprived ecosystems of low nutrient content and constant low temperature (Sattler et al., 2002; Margesin et al., 2004). Although the extensive investigations of microcosms from various glacial environments (Robinson, 2001; Sattler et al., 2002; Hodson et al., 2007; Margesin and Miteva, 2010) covering Arctic (Miteva, 2008; Miteva et al., 2009; Anesio et al., 2009; Anesio and Laybourn-Parry, 2012) and Antarctic (Abyzov, 1993; Yao et al., 2006, 2008) glaciers, ice caps (Christner et al., 2003; Yang et al., 2016), mountain glaciers (Jungblut et al., 2010), snow (Miteva et al., 2009; Maccario et al., 2015), permafrost (Rivkina et al., 2004), sea ice (Deming, 2009), Antarctic lake ice (Priscu et al., 1998; Abyzov et al., 2001; Murray et al., 2012), and mountain lakes (Felip et al., 1995), very little is known so far about the diversity, activity, and environmental role of total and culturable microorganisms colonizing the perennial ice accumulated in ice caves. The interest in understanding the ice caves' microbial world relates to the fairly inaccessible settings of this type of habitat with a more restricted anthropogenic and environmental impact, which has led to an anticipated

Ice Caves. https://doi.org/10.1016/B978-0-12-811739-2.00008-5

relatively conserved microbial diversity. Corroborating data on the structural and functional diversity of bacteria, archaea, and fungi in ice caves' sediments and perennial ice deposits could contribute to understanding their role in this glacial environment and provide a source of cold-adapted microbial strains with enhanced applicative properties. Moreover, ice caves were recently recognized as a proxy for palaeoclimate reconstitution (e.g., Perşoiu and Onac, 2012); therefore investigation of the cave ice microbiome could lead to the identification of microbial biomarkers for climate variations. Unraveling the microcosm of this type of frozen habitat could also be a model in astrobiology studies.

A first report on the presence of bacteria in ice caves dates back to the last century (Pop, 1949), revealing the presence of nitrifying bacterial communities in limestone sediments from the Scărişoara Ice Cave, Romania (Racovita and Onac, 2000). This limestone cave, hosting one of the largest and oldest underground perennial ice deposits identified so far (Holmlund et al., 2005; Feurdean et al., 2011), is one of the most investigated ice caves in the world (Racovita, 1927; Racovita, 1994; Onac et al., 2007; Perşoiu et al., 2017; Perşoiu and Pazdur, 2011; Perşoiu et al., 2011). Apart from the palaeoclimatic significance of the isotopic composition of ice (Perşoiu et al., 2011) and the pollen trapped inside the ice block (Feurdean et al., 2011), the geochemistry and microbiota of the ice could also be added as putative indicators of past environmental and anthropogenic activities.

The ice deposits formed during the last millennium in the Scărişoara Ice Cave and the ice stalagmites from the Little Reservation area were recently investigated (Hillebrand-Voiculescu et al., 2013, 2014; Itcus et al., 2016) in order to characterize the microbial diversity and chronosequence and the impact of geochemical parameters on spatial and temporal variations of ice microcosm in this habitat.

Earlier studies on Austrian alpine ice caves showed the existence of bacterial and fungal communities and isolated the first ice cave bacterial strain (Margesin et al., 2003, 2004). Diatom communities were also investigated from ice deposits in Canadian ice caves in the Yukon Territory (Lauriol et al., 2006). Also, several cold-active autotrophic bacteria were isolated from ice-rock interfaces in the South Ice Cave (Oregon, United States) and characterized (Popa et al., 2012).

Sediments and ice deposits from other types of ice-containing caves were screened to unravel the presence and diversity of microbial life in such underground habitats and to isolate psychrophilic and psychrotolerant strains of particular metabolic activities. Volcanic Ice Cave systems in Mount Erebus, Antarctica, were recently investigated in order to characterize the bacterial and archaeal community structure in the sediments from three caves. In addition, Arctic subglacial cave formations were analyzed for the presence of fungi in glacier ice caves (Butinar et al., 2011).

The recent knowledge on the presence, abundance, diversity, and functional activity of microbial communities and isolated strains from karst ice caves and ice-containing caves represents a small glimpse into understanding the environmental role and adaptation mechanisms of this particular glacial microcosm.

8.2 SAMPLE COLLECTION

Sediment samples from ice caves were initially investigated to identify the presence of bacteria in the limestone area of the Scărişoara Ice Cave, Romania, (Pop, 1949). Fifty years later, carbonate-rich deposits from the Austrian alpine Eisriesenwelt Werfen ice cave were collected and used for isolation of a novel bacterial species (Margesin et al., 2004). In another study of Alpine ice cave microbiota, moonmilk samples were collected aseptically from the Hundalm cave (Austria) to investigate the archaeal community in the 80-cm depth profile of the cave floor (Reitschuler et al., 2014). Sediments were also sampled from three volcanic ice caves located in the proximity of Mount Erebus (Antarctica) to analyze the microbial diversity from these icy habitats (Connell and Staudigel, 2013; Tebo et al., 2015). Thermally active sediments were collected from the floor of the Harry's Dream, Warren, and Hubert's Nightmare ice caves to investigate their bacterial diversity (Tebo et al., 2015) and from the Warren Cave (Connell and Staudigel, 2013) to unravel the role of a fungal community in this habitat.

Ice samples of different ages, covering 1 (AD 2012), 400 (399 cal. yrs. BP), and 900 (887 cal. yrs. BP) years old ice (Perşoiu and Pazdur, 2011), light exposure regimes (dark, direct sunlight (S) and indirect sunlight (L) exposures), and organic contents (clear ice (I) and organic rich (O) sediment) were collected from the perennial ice block of the Scărişoara Ice Cave for investigating the presence and diversity as well as the spatial and temporal variability of microorganisms from this habitat (Hillebrand-Voiculescu et al., 2014; Itcus et al., 2016). Ice coring (Fig. 8.1) was carried out by using both a manual

FIG. 8.1

Cave ice sampling in Scărişoara Ice Cave (Romania): (A) floor of the Great Hall area representing the ice block surface formed by recent ice; (B) vertical ice coring using an electrical drill, Great Hall area; (C) horizontal ice coring with a manual ice drill, Little Reservation ice wall; and (D) ice core (photos by C. Purcarea).

auger (Hillebrand-Voiculescu et al., 2014; Itcus et al., 2016) and an electric ice drilling system (Itcus et al., 2015). The coring was carried out vertically from the top of the glacier, in the Great Hall area (Fig. 8.1A), for collecting the recent ice samples 1-S and 1-L (Fig. 8.1B), or horizontally, from the previously ^{14}C-dated exposed ice layers of the ice block sidewall in the Little Reserve area (Fig. 8.1C), for the 400-O, 900-O and 900-I samples (Perşoiu and Pazdur, 2011). The manual auger (Fig. 8.1C) of 5-cm diameter and 50-cm length was initially used for collecting ice samples from different cave locations in the Great Hall and Little Reserve cave areas. The surface (about 20 cm) of the ice block was removed and briefly flamed, and the drilling auger was sterilized by a 96% ethanol treatment and surface flaming to ensure aseptic sampling conditions (Itcus et al., 2016). The cave ice samples were transported frozen in autoclaved flasks and stored at −20°C.

Water samples from the supraglacial pond formed during warm seasons in the Great Hall area of the Scărişoara Ice Cave on the ice block surface were collected in a 1-L sterile flask, from a sun-exposed location, to identify the presence and resilience of phototrophic microorganisms during ice block formation (Hillebrand-Voiculescu et al., 2014).

Permanent ice samples were also collected from the rock-ice interface of the South Ice Cave, Oregon Cascades, lava tube (Popa et al., 2012). These surface ice deposits, together with rock fragments, were aseptically preserved in sterile plastic bags and kept frozen for further isolation of cryophilic iron-oxidizing bacterial strains.

Arctic glacier ice from the Kongsfjorden coast, Western Spitsbergen (Svalbard), was also collected aseptically and used for isolation and characterization of fungal strains from a series of frozen habitats, including subsurface ice from cryokarst formations (Butinar et al., 2011).

Recently formed ice stalagmites (Perşoiu and Pazdur, 2011) from the Little Reserve area of the Scărişoara Ice Cave (Hillebrand-Voiculescu et al., 2013) were screened for the presence of cultured and uncultured bacteria. Ice samples were collected aseptically by flaming for 5–10 s both the ice stalagmite surface and the coring drill after a 96% ethanol treatment, as previously described (Itcus et al., 2016).

8.3 MICROBIAL ABUNDANCE

The microbial cell density of ice *cave sediments* from three volcanic caves located near the summit of Mount Erebus was measured by qPCR of SSU ribosomal RNA gene using both bacterial and archaeal SSU-specific primers (Tebo et al., 2015). Under these conditions, no archaeal amplicons were obtained for any of the locations. The bacterial cell density was $1.6–2.8 \times 10^6$ cells g^{-1} sediment in Hubert's Nightmare and Harry's Dream caves, while the Warren Cave, characterized by a higher air temperature, had a 20-fold higher (40×10^6 cells/g sediment) bacterial biomass.

An earlier quantitative study of archaeal communities from calcite deposits of the alpine Hundalm ice cave, Austria (Reitschuler et al., 2014), was carried out to study the aerobic and anaerobic enrichments in different media at 10°C and 25°C by qPCR using 787F/1059R specific primers (Yu et al., 2005). Under these conditions, the archaeal abundance (10^7 copies mL^{-1}) was 10-fold higher in the deepest (60–80 cm) sediments and near-surface (0–20 cm) moonmilk deposits than that from the intermediary sediments (20–40 and 40–60 cm) (Reitschuler et al., 2014).

A lower microbial biomass was identified in perennial *cave ice* deposits. Investigation of five ice samples in the Scărişoara Ice Cave of different ages and light exposures indicated the presence of 2.4×10^4–2.9×10^5 cells mL^{-1} melted ice, when SYBR Green I-labeled uncultured samples were measured by epifluorescence microscopy (Itcus et al., 2016). The cell content of the ice block layers appeared to be relatively conserved (10^5 cells mL^{-1} range) in 1-year-old ice exposed to sunlight in a direct and indirect manner, as well as the 400-year-old ice, and it decreased by 10-fold in 900-year-old ice deposits (2.4–2.9×10^4 cells mL^{-1}) of high and low organic content (Itcus et al., 2016). Microbial flow cytometry measurements of these ice samples using SYBR Green I labeling showed a similar biomass values range (10^4–10^5 cells mL^{-1} melted ice) in the Scărişoara ice block up to 900-year-old deposits, with a two- to threefold lower cell content in clear ice as compared with organic-rich ice samples of the same age (Itcus et al., in prep). The live/dead cell content of these ice samples, estimated by ethidium homodimer-1 staining (del Giorgio and Gasol, 2008), indicated a 68%–78% cell viability decrease with age and reduction of organic sediment content (Itcus et al., in prep).

Cell density of cultured heterotrophic bacteria from the ice block of Scărişoara ice cave ranged within 50–7.8×10^4 CFU mL^{-1} when cultivated on R2A medium (Reasoner and Geldreich, 1985) at 4°C and 15°C (Itcus et al., 2016). In this habitat, an exponential decrease of the cultured bacterial cell content with the age of ice was observed. Also, the microbial community cultivated at low temperature (4°C) appeared to be more abundant than the one at 15°C in all analyzed ice samples, indicating a higher viability of cold-active bacteria independent of ice age throughout the cave ice block. Meanwhile, the high organic content of the ice sediments appeared to increase the culturable cell viability, particularly for the older ice layers (Itcus et al., 2016).

Meanwhile, a cultured heterotrophic microbial community from ice stalagmites formed in a light-deprived area (Little Reservation) of the cave at 4°C and 15°C (using a Luria-Bertani (LB) medium) indicated the presence of 0.5×10^2–0.31×10^3 CFU ml^{-1} melted ice (Hillebrand-Voiculescu et al., 2013).

8.4 BACTERIAL COMMUNITIES

Culture-dependent and independent studies were carried out for identifying, isolating, and characterizing autotrophic (chemolitotrophic and phototrophic) and heterotrophic bacteria from ice caves. Recent and old ice deposits, moonmilk, and sediments from ice caves were used for genomic DNA extraction, gene amplification and sequencing, or cultivation at different temperatures in various media, leading to identification of uncultured (total) bacterial communities, cultured microbiota, and individual bacterial strains.

8.4.1 UNCULTURED BACTERIA

The presence of microbial communities in ice stalagmites from the Little Reservation dark area of the Scărişoara Ice Cave, Romania, was revealed by epifluorescence microscopy (Hillebrand-Voiculescu et al., 2013). SYBR Green I and live/dead ethidium homodimer staining (del Giorgio and Gasol, 2008) of melted ice indicated the dominance of live cells. Moreover, transparent exopolysaccharide particles were evidenced in the melted cave ice sample (Hillebrand-Voiculescu et al., 2013), as expected for frozen habitats where these particles play a stabilizing role (Wingender et al., 1999).

A comparative study of the microbial community structure from an Austrian Alpine ice cave and Antarctic glacier caves, based on clone library construction and sequencing, showed a lower bacterial diversity with a high representation of cyanobacteria in the Alpine ice cave relative to the glacier cave (Standhartinger et al., 2010).

A pioneering report on the spatial and temporal variability of microbial communities in perennial cave ice revealed the presence of bacteria and microbial eukaryotes of Scărişoara cave (Romania) ice sediments of up to 900 years old from the ice block, based on 16S rRNA gene PCR amplification using B8F/B1525R primers (Röling et al., 2001; Hillebrand-Voiculescu et al., 2014).

Using genomic DNA from the Scărişoara cave, 454 pyrosequencing of 16S rRNA amplicons of the five ice samples, 1-S, 1-L, 400-O, 900-O, and 900-I, led to unraveling for the first time the bacterial diversity in recent and old (up to 900 years old) deposits in a perennial cave ice block (Itcus et al., 2015; Itcus et al., in prep). More than 16,000 OTUs corresponding to 4,557 bacterial and archaeal species were determined in ice layers of different ages, of different exposures to light and dark conditions, and of high and low organic sediment contents, showing a heterogeneous phyla distribution across the ice block. *Proteobacteria* representatives were identified in all samples, but presented a high relative content in both 1-S and 900-O ice, *Firmicutes* OTUs were dominant in the 400-O ice sample, while *Actinobacteria* phylum were found mainly in 900-year-old ice, particularly in the clear ice layers (900-I). The relative content of phototrophs also varied throughout the cave ice layers, with *Cyanobacteria* phylum dominating the sunlight-exposed recent ice (sample 1-S), *Chlorobi* being mainly in 1-year old ice located in the center of the Great Hall area, under indirect sunlight exposure (sample 1-L), and *Chloroflexi* being found in organic-rich ice strata (400-O and 900-O). An equilibrated distribution of *Cyanobacteria* and *Chloroflexi* phyla was observed in the 900-year-old clear ice sample (900-I) (Itcus et al., in prep).

Bacterial diversity from sediments of Antarctic volcanic caves (DOVE) located in the proximity of Mount Erebus was investigated by SSU rRNA gene cloning and sequencing (Tebo et al., 2015). The 16S rRNA sequences obtained for Harry's Dream (102), Warren (82), and Hubert's Nightmare (78) caves led to the identification of 34, 11, and 18 Operational Taxonomic Units (OTUs), respectively, most of them being homologous to phylotypes from cold environments, caves, and volcanic habitats. The presence of phototrophic *Cyanobacteria* and *Chloroflexi* phyla was noticed in the light-exposed sediments of Harry's Dream cave, with a major representation of the *Ktedonobacteria* class. A high relative content of OTUs belonging to *Betaproteobacteria* and *Acidobacteria* was characteristic for this cave, in addition to the presence of unclassified bacteria. Warren Cave sediments were characterized by the dominance of *Chloroflexi* and *Acidobacteria*, with a small content of *Alphaproteobacteria* (*Hyphomicrobium* spp.) and *Planctinomycetes* taxa. Interestingly, the colder and dark cave, Hubert's Nightmare, contained a great abundance of phyla representatives belonging to *Proteobacteria* (both *Alpha-* and *Betaproteobacteria*), *Bacteroidetes*, *Verrucomicrobia*, and *Actinobacteria* (Tebo et al., 2015). The high representation of *Chloroflexi* in these cave habitats appears to be unusual, considering the previously reported Antarctic (Cary et al., 2010; Pearce et al., 2013) and other environmental (Huber et al., 2006) metagenomes. Chemoautotrophy based on carbon fixation RubisCO *cbbl* gene screening marked the presence of chemoautotrophic microorganisms in this habitat. Representatives of both Green Type (green algae, cyanobacteria, purple bacteria) and Red Type (non-green algae, including *Alpha-* and *Betaproteobacteria*) Form I species were found in all three caves, where the latter one was dominant. The ATP citrate lyase *aclB* gene, important for CO_2 fixation in aerobes and anaerobes, was absent in all investigated cave sediments (Tebo et al., 2015). The corroborated structural and functional diversity data from these Antarctic volcanic ice caves suggested that Warren Cave represents a low-carbon environment as a particular case for studying the microbial metabolism and adaptation mechanisms in extreme oligotrophic conditions (Tebo et al., 2015).

8.4.2 CULTURED BACTERIA

The first mentioning of bacteria in calcareous sediments from the *Cathedral* area of the Scărişoara Ice Cave (Pop, 1949) referred to enrichments of rock surface deposits in media used for nitrifying bacteria in the presence and absence of ammonium salts. The microbial growth monitored between 11 and 180 days under laboratory conditions indicated the development of coccoidal populations associated with nitrite-to-nitrate conversion. The author suggested that the high nitrification bacterial activity led to the formation of a dark detritus layer on virgin limestone rocks. Moreover, the environmental low temperatures are expected to enhance the nitrification activity of the cave microbial community (Pop, 1949).

Seasonally formed ice stalagmites from the Little Reservation area of the Scărişoara Ice Cave (Racovita and Onac, 2000) were investigated in order to identify the cultured bacterial community embedded in these recent ice formations (Hillebrand-Voiculescu et al., 2013). Melted ice samples were cultivated at 4°C and 15°C in LB medium for 7 and 26 days, respectively. The microbial community obtained at 4°C was used for constructing a 16S rRNA gene library using B8F/B1525R primers (Röling et al., 2001). After *HinfI/HaeIII* and *RsaI* ARDRA analysis, 20 of the 101 clones were sequenced, corresponding to 14 bacterial OTUs. Among these, the dominant taxa in the light-deprived recent ice stalagmites belonged to *Pseudomonas* (*Proteobacteria*) and *Paenibacillus* (*Firmicutes*) genera (Hillebrand-Voiculescu et al., 2013).

A recent report on cultured bacterial chronosequence from an ice cave investigated the structural and functional diversity of bacterial communities from ice deposits up to 900 years old in the Scărişoara Ice Cave, Romania (Itcus et al., 2016). Five samples of 1-, 400-, and 900-year-old ^{14}C-dated ice layers (Perşoiu and Pazdur, 2011) were collected from the ice block surface characterized by direct (1-S) and indirect (1-L) sunlight exposure and from organic rich layers (400-O and 900-O) or clear ice sediments (900-I). Geochemical analyses showed a neutral to slightly basic pH (7.48–8.03) that increased in older ice strata, electrical conductivity (EC) decreasing exponentially with the ice's age, and a high content of total organic carbon (TOC) and total nitrogen (TN) in 1-S and 400-O organic rich ice sediments. Colonies obtained from all five ice samples after inoculation of LB medium at 4°C for 39 days (Fig. 8.2) are currently under investigation by SSU rRNA gene sequencing (unpublished data). Microbial growth at 4°C and 15°C of various liquid media (Bidle et al., 2007; Itcus et al., 2016) of melted ice samples

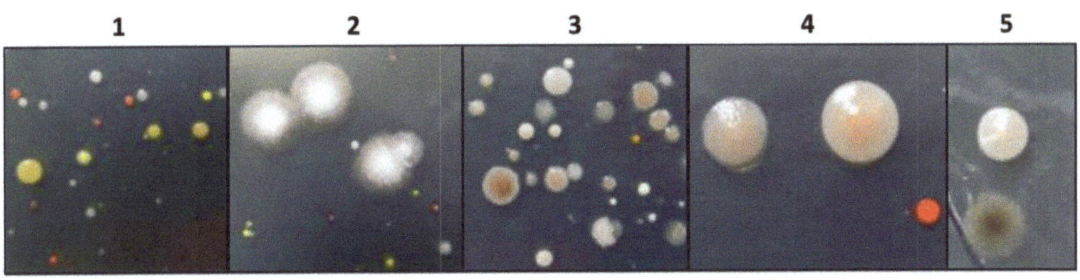

FIG. 8.2

Microbial colonies from Scărişoara Ice Cave (Romania): Melted ice samples up to 900 years old from the perennial ice block layers (1) 1-year old, direct sunlight exposure; (2) 1-year old, indirect sunlight exposure; (3) 400 years old, organic rich ice sediment; (4) 900 years old, organic rich ice sediment; and (5) 900 years old, clear ice (see details in Itcus et al., 2016); cultured at 4°C in LB medium.

revealed different microbial community structures across the ice block, based on growth parameters of ice microbiota (Itcus et al., 2016). The functional diversity analysis measured by using the BIOLOG Eco Plates system (Garland and Mills, 1991; Lehman et al., 1995) showed a higher Shannon-Weaver diversity index and Richness calculated parameters in the cases of the 1-S and 400-O ice layers, suggesting a higher functional diversity of microbiota cultivated at both high and low temperatures for these ice communities. Moreover, PCA plot analyses of the distribution of ice-contained microbiota based on carbon source utilization showed the clustering of microbial communities from the cave ice block depending on the ice age, independently of the sediment content of the ice substrate, and the correlation of ice geochemical parameters EC, TOC, and TN with the functional diversity of 1-S and 400-O ice samples microbiota. The pH appeared to mainly affect the microbial communities' structure of 900-year-old ice of both high and low organic content (Itcus et al., 2016). PCR-DGGE profile of ice bacterial 16S rRNA gene fragments from the five ice samples cultivated at 4°C and 15°C in different growth media (e.g., LB in the presence and absence of 10% glucose, T1 (Bidle et al., 2007) and T2 (Bidle et al., 2007)) showed that they sustained their structural diversity, and the amplicon sequencing led to the identification of 68 distinct OTUs. Among these, 18 taxons appeared to be homologous to bacteria from cold environments, and 8 taxons with strains from cave habitats. Eight OTUs belonging to *Pseudomonas*, *Serratia*, and *Rahnella* and uncultured clone homologues were also found in different cave ice layers, confirming the partial microbial conservation across the cave ice block. The relative abundance of the identified cultured bacteria showed the dominance of *Proteobacteria* in recent ice exposed to direct sunlight (1-S) in the Great Hall area of the cave (Itcus et al., 2016), in accordance with that in the dark-exposed ice stalagmite (Hillebrand-Voiculescu et al., 2013). A high representation of *Gammaproteobacteria* was characteristic for both 900-O and 900-I old ice strata, while members of *Bacteroidetes* were mainly identified in 400-year-old ice. *Actinobacteria* was found in the central part of the ice block surface (sample 1-L), such as the cold-active *Arthrobacter* strain [Accesion number KF853212] related to the previously reported *A. psychrophenolicus* species from another alpine ice cave (Margesin et al., 2004).

Cultured cyanobacteria in the sunlight exposed 1-year old ice sample (1-S) in the Scărişoara cave were obtained after cultivation in a BG_{11} medium (Rippka et al., 1979) at 7°C for 2 months and were visualized by bright-field microscopy based on chlorophyll autofluorescence (Hillebrand-Voiculescu et al., 2013).

The occurrence of a karst-water *Rahnella* strain in 400- and 900-year-old ice deposits in the Scărişoara Ice Cave could constitute a putative climatic bacterial biomarker, considering the representation of the spruce (*Picea abies*) endophytic species in the cave's surrounding forest during warmer (Little Ice Age, sample 400-O) and colder (Medieval Warm Period, samples 900-O and 900-I) periods (Feurdean et al., 2011; Itcus et al., 2016).

8.4.3 ISOLATED BACTERIAL STRAINS

Psychrotolerant phenol-degrading bacterial strains belonging to *Arthrobacter* genus (AG30 and AG31) were the first reported isolates from alpine ice caves (Margesin et al., 2003, 2004). A novel bacterial species, *Arthrobacter psychrophenolicus*, was isolated in the Austrian alpine ice cave, Eisriesenwelt Werfen, and was characterized (Margesin et al., 2004). This aerobic, facultatively cryophilic strain characterized by a 1°C–25°C growth temperature interval is capable of phenol degradation and mannitol and phenylacetate assimilation, but not glucose. The 16S rRNA gene sequence of this strain indicated the closest phylogenetic species, *A. sulfureus* DSM 20167 (X83409) (Margesin et al., 2004).

Among the 29 aerobic bacteria isolated from a mixture of ice and rock fragments from the South Ice Cave (Oregon, United States), a cold-active *Pseudomonas* sp. HerB strain capable of autotrophic growth based on neutrophilic iron oxidation under microaerophilic conditions was characterized (Popa et al., 2012). Enrichment studies led to the identification of bacterial strains able to use olivine as an energy source, corresponding to two cryophilic *Brevundimonas* and two *Acidovorax* species, and seven mesophilic *Pseudomonas* species. Among these, the novel *Pseudomonas* sp. HerB strain showed a neutrophilic growth in the presence of bicarbonate, Fe(II), and low oxygen (1.6%) content, with an optimal temperature of 12°C–14°C. Considering the particular environment where this microbe thrives (Jepson et al., 2007) and the presence of olivine-contained rocks on Mars (Hoefen et al., 2003; Edwards et al., 2008), this bacterial strain living at the cave ice-rock interface, in a low-temperature (0°C) mineral habitat deprived of light and of neutral pH, could constitute a putative terrestrial microbial candidate for evolutionary and exobiology studies (Popa et al., 2012).

8.5 ARCHAEAL COMMUNITIES

The occurrence and diversity of archaea in moonmilk deposits in the Hundalm ice cave (Austria) and of both aerobically and anaerobically cultured archaeal communities obtained from the cave ice sediments were investigated by PCR-DGGE analysis and sequencing, using enrichments at 10°C and 25°C in oligotrophic and methanotrophic media (Reitschuler et al., 2014). The data showed that the archaeal population in the cave surface sediments was dominated by *Nitrosopumilus maritimus* and *N. gargensis* belonging to *Thaumarchaeota*, while the deeper sites contained mainly anaerobic (methanotrophic) *Euryarchaeota* representatives. The physiological activity of bacterial and archaeal mixed populations cultivated under aerobic and anaerobic conditions indicated glucose utilization as the main carbon source under anaerobic growth, and methane and short-chain organic acid production by fermentative microorganisms (Reitschuler et al., 2014). Interestingly, the presence of ammonia-oxidizing archaea in moonmilk deposits of this alpine ice cave is significantly lower than that from several other Austrian alpine caves with no perennial ice (Reitschuler et al., 2016).

No archaea were detected in the volcanic ice caves sediments from Mount Erebus, Antarctica (Tebo et al., 2015).

The first report of archaea in permanent ice deposits from caves identified members of this domain in old ice strata of the perennial ice block harbored in the Scărişoara Cave (Itcus et al., 2015). PCR-DGGE analysis of archaeal 16S rRNA amplicons using Arch340F(GC) and Arch516R primers (Øvreås et al., 1997) showed the presence of archaeal OTUs in 900-year-old ice (Purcarea, unpublished data). 16S rRNA gene 454 pyrosequencing revealed the existence of 0.1%–1% archaeal communities in the old ice strata, with the Miscellaneous Crenarchaeota Group (MCG) dominant in 400-year-old ice, and the presence of *Candidatul Nitrosphaera* (*Thaumarachaeota*) in organic-rich contained ice samples 400-O and 900-O, while the clear ice sample 900-I mainly contained *Euryarchaeaota*, represented by *Methanomicrobia* and *Thermoplasmata* phyla (Itcus et al., 2015; Brad et al., in prep).

8.6 FUNGAL COMMUNITIES

Among the reported data on microbial diversity from ice caves, few studies focused on eukaryotic microorganisms. Hillebrand-Voiculescu and collaborators (2014) revealed the existence of these microorganisms in the ice block of Scărişoara Ice Cave, based on 18S rRNA gene amplification using Euk1A/Euk516R primers (Diez et al., 2001). Microbial eukaryotes were found in recent ice samples collected from the Great Hall areas of the cave exposed to sunlight in a direct (sample 1-S) and indirect (sample 1-L) manner, and in 900-year-old ice samples containing organic sediments (sample 900-O).

Cultured phototrophic and heterotrophic eukaryotes were also obtained from the recent ice sample 1-S in the Scărişoara Ice Cave and from water samples collected from the supraglacial pond (SGP) that accumulated on top of the cave's ice block due to precipitation and infiltration during the warm period (Hillebrand-Voiculescu et al., 2014). Filamentous eukaryotes from these melted ice and water samples when cultivated at 7°C in BG_{11} medium (Rippka et al., 1979) were observed by epifluorescence and light microscopy (Hillebrand-Voiculescu et al., 2014). While both 1-S and SGP were collected from the same area prior to and after freezing, the presence of eukaryotic microorganisms in these samples represented a preliminary step in investigating the resilience of microbiota during the ice block's formation (Perşoiu and Pazdur, 2011).

The first study of alpine ice cave fungi consisted of the isolation and characterization of two hydrocarbon-degrading strains in the Eisriesenwelt Werfen Ice Cave, Austria, in order to isolate cold-active enzymes of fungal origin (Margesin et al., 2003). The isolated *Cryptococcus* sp. and *Heterobasidiomycete* psychrophilic strains showed a maximum growth temperature of 20°C and 15°C in R2A medium, respectively, with a higher phenol degradative activity than that of the homologous fungal strains isolated in an alpine glacier's foot or cryoconites (Margesin et al., 2003).

The fungal community structure of ice samples collected from the Scărişoara Ice Cave (Romania) was recently investigated by PCR-DGGE analysis and sequencing of 18S rRNA amplicons (Brad et al., in prep), using FF390 and FR1-GC fungal-specific primers (Vainio and Hantula, 2000). The amplicon distribution of ice samples up to 900 years old, as well as that of the corresponding enrichments at 4°C and 15°C in various heterotrophic media T1, T2, LB, and LBG media (Bidle et al., 2007; Itcus et al., 2016), indicated a spatial and temporal variability of fungal communities throughout the cave's ice block, a higher diversity of cold active strains, and the conserved presence of the psychrophilic yeast *Mrakia stokesii* specific for glacial habitats (Buzzini et al., 2012; Brad et al., in prep).

A previous study on glacier ice microbiota (Butinar et al., 2011) revealed the presence of the psychrotolerant Archiascomycete species *Protomyces inouyei*, a rare species of this ubiquitous fungal genus, in ice deposits in an Arctic glacier cave located in Kongsfjorden fjord, Spitsbergen, in the Svalbard archipelago in Norway.

Investigation of the fungal community in soil sediments of the volcanic Warren Ice Cave (Mount Erebus, Antarctica) led to the identification of 61 fungal species, based on ITS clone library sequences (Connell and Staudigel, 2013). These taxons belong to *Ascomycota* and *Basidiomycota* phyla, unlike the exposed Antarctic soils in the McMurdo Dry Valleys, which are dominated by *Basidiomycota* (Connell et al., 2006, 2008; Fell et al., 2006). Among these, *Dothideomycetes* and *Eurotiomycetes* were the most representative classes of *Ascomycota*, while *Agaricomycetes* and *Exobasidiomycetes* were the dominant classes among *Basidiomycota* clones. Among the identified taxons, the presence of several *Malassezia* species representing the anthropogenic impact fungal biomarker indicated human contamination of this cave (Connell and Staudigel, 2013).

8.7 DIATOMS

The first study of ice cave diatoms, unicellular algae usually found in humid environments, reported their presence and distribution in ice formations and cryogenic calcite from three ice caves (Balch, 1900) and from a slope cave (Mitter, 1983) in the Yukon Territory, Canada (Lauriol et al., 2006). The content and diversity of diatom flora from ice stalagmites and floor ice and plugs, cryogenic calcite deposits, and cave grus from Grande Caverne, Caverne'85, and Caverne des Méandres ice caves were analyzed by chemical treatment, cell enumeration, and taxonomical characterization. The diatom community found in these ice caves included 96 different strains that exhibited a large heterogeneity dependent on the water vapor's condensation regime along the cave's walls, with the dominance of *Achnanthes linearis*, *A. minutissima*, *Cymbella lateens*, *C. microcephala*, and *C. minuta* in the ice stalagmites, of *Eunotia praerupta*, *Hantzschia amphioxys*, *Orthoseira roeseana*, and *Stauroneis obtusain* in the cryogenic calcite deposits, and of *Orthoseira* spp. in the grus from the slope cave (Lauriol et al., 2006). The analysis of the origin of diatoms in these caves could contribute to identifying putative anthropogenic pollution biomarkers.

8.8 CONCLUSIONS

For the last decade, microbial ecology investigations of icy ecosystems showed an increased interest in determining the diversity, conservation, and role of ice cave microcosms. The limited number of studies so far, as discussed herein, focused on characterizing the microbial community structure and isolating cold-adapted bacterial strains taken from ice cave sediments or permanent ice deposits of alpine ice caves, lava tube caves, volcanic Antarctic caves and Arctic glacier subsurface formations. In addition, metabolic diversity inquiries of these unique microbial ecosystems began to unravel the identities and functions of the key players in this underground glacial habitat.

Further studies on reconstituting the structural and functional cave ice microcosms of alpine and glacier ice caves are required to explore the impact of climate and environmental pollution on these types of habitat, to identify putative palaeoclimate microbial biomarkers, and to isolate novel cold-active bacterial strains of high bionanotechnological applicative potential.

ACKNOWLEDGMENTS

Data reported by the author on the Scărişoara Ice Cave were supported by the UEFISCDI PN-II-ID-PCE-2011-3-074 grant and H2020 ELAC2014_DCC-0178 Joint Program.

REFERENCES

Abyzov, S.S., 1993. Microorganisms in the Antarctic ice. In: Friedman, E.I. (Ed.), Antarctic Microbiology. John Willey & Sons, Inc., New York, pp. 265–295.

Abyzov, S.S., Mitskevitch, I.N., Poglazova, M.N., Barkov, N.I., Lipenkov, V.Y., Bobin, N.E., Kudryashov, B.B., Pashkevich, V.M., Ivanov, M.V., 2001. Microflora in the basal strata at Antarctic ice core above the Vostok lake. Adv. Space Res. 28, 701–706.

Anesio, A.M., Laybourn-Parry, J., 2012. Glaciers and ice sheets as a biome. Trends Ecol. Evol. 27 (4), 219–225.

Anesio, A.M., Hodson, A.J., Fritz, A., Psenner, R., Sattler, B., 2009. High microbial activity on glaciers: importance to the global carbon cycle. Glob. Chang. Biol. 15, 955e960.

Balch, E.S., 1900. Glacières or Freezing Caverns. Allen, Lane and Scott, Philadelphia, PA. p. 337.

Bidle, K.D., Lee, S., Marchant, D.R., Falkowski, P.G., 2007. Fossil genes and microbes in the oldest ice on Earth. Proc. Natl. Acad. Sci. U. S. A. 104, 13455–13460.

Brad, T., Itcus, C., Pascu, M.D., Persoiu, A., Purcarea, C., Living fungi in 900 yrs old cave ice. The Cryosphere (Unpublished manuscript).

Butinar, L., Strmole, T., Gunde-Cimerman, N., 2011. Relative incidence of ascomycetous yeasts in arctic coastal environments. Microb. Ecol. 61, 832–843.

Buzzini, P., Branda, E., Goretti, M., Turchetti, B., 2012. Psychrophilic yeasts from worldwide glacial habitats: diversity, adaptation strategies and biotechnological potential. FEMS Microbiol. Ecol. 82 (2), 217–241.

Cary, S.C., McDonald, I.R., Barrett, J.E., Cowan, D.A., 2010. Ontherocks: the microbiology of Antarctic Dry Valley soils. Nat. Rev. Microbiol. 8, 129–138.

Christner, B.C., Mosley-Thompson, E., Thompson, L.G., Reeve, J.N., 2003. Bacterial recovery from ancient glacial ice. Environ. Microbiol. 5 (5), 433–436.

Connell, L.B., Staudigel, H., 2013. Fungal diversity in a dark oligotrophic volcanic ecosystem (DOVE) on Mount Erebus, Antarctica. Biology 2, 789–809.

Connell, L.B., Redman, R.S., Craig, S.D., Rodriguez, R.J., 2006. Distribution and abundance of fungi in the soils of Taylor Valley, Antarctica. Soil Biol. Biochem. 38, 3083–3094.

Connell, L.B., Redman, R.S., Craig, S.D., Scorzetti, G., Iszard, M., Rodriguez, R.J., 2008. Diversity of soil yeasts isolated from south Victoria land, Antarctica. Microb. Ecol. 56, 448–459.

del Giorgio, P.A., Gasol, J.M., 2008. Physiological structure and single-cell activity in marine bacterioplankton. In: Kirchman, D.L. (Ed.), Microbial Ecology of the Oceans. J Wiley & Sons, Inc., Hoboken, NJ, pp. 243–298.

Deming, J.W., 2009. Extremophiles: cold environments. In: Schaechter, M. (Ed.), Encyclopedia of Microbiology. Elsevier, Oxford, pp. 147–158.

Diez, B., Pedrós-Alió, C., Marsh, T.L., Massana, R., 2001. Application of denaturing gradient gel electrophoresis (DGGE) to study the diversity of marine picoeukaryotic assemblages and comparison of DGGE with other molecular techniques. Appl. Environ. Microbiol. 67 (7), 2942–2951.

Edwards, C.S., Christensen, P.R., Hamilton, V.E., 2008. Evidence for extensive olivine-rich basalt bedrock outcrops in Ganges and Eos chasmas, Mars. J. Geophys. Res. 113, E11.

Felip, M., Sattler, B., Psenner, R., Catalan, J., 1995. Highly active microbial communities in the ice and snow cover of high mountain lakes. Appl. Environ. Microbiol. 61 (6), 2394–2401.

Fell, J.W., Scorzetti, G., Connell, L.B., Craig, S.D., 2006. Biodiversity of micro-eukaryotes in Antarctic dry valley soil with <5% soil moisture. Soil Biol. Biochem. 38, 3107–3119.

Feurdean, A., Persoiu, A., Pazdur, A., Onac, B.P., 2011. Evaluating the palaeoecological potential of pollen recovered from ice in caves: a case study from Scarisoara Ice Cave, Romania. Rev. Palaeobot. Palynol. 165, 1–10.

Hillebrand-Voiculescu, A., Rusu, A., Itcus, C., Persoiu, A., Brad, T., Pascu, D., Ardelean, I., Onac, B.P., Purcarea, C., 2013. Bacterial 16S-rRNA gene clone library from recent ice stalagmites of Scărișoara cave. Rom. J. Biochem. 50 (2), 109–118.

Garland, J.L., Mills, A.L., 1991. Classification and characterization of heterotrophic microbial communities on the basis of patterns of community level sole-carbon-source utilization. Appl. Environ. Microbiol. 57, 2351–2359.

Hillebrand-Voiculescu, A., Itcus, C., Ardelean, I., Rusu, A., Persoiu, A., Brad, T., Popa, E., Onac, B.P., Purcarea, C., 2014. Searching for cold-adapted microorganisms in the underground glacier of Scarisoara Ice Cave, Romania. Acta Carsol. 43 (2–3), 319–329.

Hodson, A., Anesio, A.M., Tranter, M., Fountain, A., Osborn, M., Priscu, J.C., Sattler, J., Laybourn-Parry, B., 2007. Glacial ecosystems. Ecol. Monogr. 78, 41–67.

Hoefen, T.M., Clark, R.N., Bandfield, J.L., Smith, M.D., Pearl, J.C., Christensen, P.R., 2003. Discovery of olivine in the Nili Fossae region of Mars. Science 302, 627–630.

Huber, J.A., Johnson, H.P., Butterfield, D.A., Baross, J.A., 2006. Microbial life in ridge flank crustal fluids. Environ. Microbiol. 8, 88–99.

Holmlund, P., Onac, B.P., Hansson, M., Holmgren, K., Morth, M., Nyman, M., Perşoiu, A., 2005. Assessing the paleoclimate potential of cave glaciers: the example of the Scarisoara Ice Cave (Romania). Geogr. Ann. A. 87, 193–201.

Itcus, C., Pascu, M.D., Hillebrand-Voiculescu, A., Brad, T., Persoiu, A., Onac, B.P., Purcarea, C., 2015. Prokaryotic community structure across the ice block of Scarisoara cave determined by 454 pyrosequencing. In: 6th International Conference on Polar and Alpine Microbiology. B-32. http://polaralpinemicrobiology2015.prf.jcu.cz/data/uploads/pam_2015_boa_small.pdf.

Itcus, C., Pascu, M.D., Brad, T., Persoiu, A., Purcarea, C., 2016. Diversity of cultured bacteria from the perennial ice block of Scarisoara Ice Cave, Romania. Int. J. Speleol. 45, 89–100.

Itcus, C., Pascu, M.D., Brad, T., Hillebrand-Voiculescu, A., Persoiu, A., Purcarea, C., Prokaryotic community structure of Scarisoara Cave ice block by 454 pyrosequencing of 16S RNA gene. (Unpublished manuscript).

Jepson, S.M., Priscu, J.C., Grimm, R.E., Bullock, M.A., 2007. The potential for lithoautotrophic life on Mars: application to shallow interfacial water environments. Astrobiology 7, 342–353.

Jungblut, A.D., Lovejoy, C., Vincent, W.F., 2010. Global distribution of cyanobacterial ecotypes in the cold biosphere. ISME J. 4, 191–202.

Lauriol, B., Prevost, C., Lacelle, D., 2006. The distribution of diatom flora in ice caves of the northern Yukon Territory, Canada: relationship to air circulation and freezing. Int. J. Speleol. 35, 83–92.

Maccario, L., Sanguino, L., Vogel, T.M., Larose, C., 2015. Snow and ice ecosystems: not so extreme. Res. Microbiol. 166 (10), 782–795.

Lehman, R.M., Colwell, F.S., Ringelbergb, D.B., Whiteb, D.C., 1995. Combined microbial community level analyses for quality assurance of terrestrial subsurface cores. J. Microbiol. Methods. 22, 263–281.

Margesin, R., Miteva, V., 2010. Diversity and ecology of psychrophilic microorganisms. Res. Microbiol. 162, 346–361.

Margesin, R., Gander, S., Zacke, G., Gounot, A.M., Schinner, F., 2003. Hydrocarbon degradation and enzyme activities of cold-adapted bacteria and yeasts. Extremophiles 7, 451–458.

Margesin, R., Schumann, P., Sproer, C., Gounot, A.-M., 2004. Arthrobacter psychrophenolicus sp. nov., isolated from an Alpine ice cave. Int. J. Syst. Evol. Microbiol. 54, 2067–2072.

Miteva, V., 2008. Bacteria in snow and glacier ice. In: Margesin, R., Schinner, F., Marx, J.-C., Gerday, C. (Eds.), Psychrophiles: From Biodiversity to Biotechnology. Springer-Verlag Berlin Heidelberg, pp. 31–50.

Miteva, V., Teacher, C., Sowers, T., Brenchley, J., 2009. Comparison of the microbial diversity at different depths of the GISP2 Greenland ice core in relationship to deposition climates. Environ. Microbiol. 11 (3), 640–656.

Mitter, P., 1983. Frost features in the karst regions of the West Carpathian Mountains. In: 4th International Conference of Permafrost (Fairbanks, AL), Washington, DC, July 17–22, pp. 861–865.

Murray, A.E., Kenig, F., Fritsen, C.H., McKay, C.P., Cawley, K.M., Edwards, R., Kuhn, E., McKnight, D.M., Ostrom, N.E., Peng, V., Ponce, A., Priscu, J.C., Samarkin, V., Townsend, A.T., Wagh, P., Young, S.A., Yung, P.T., Doran, P.T., 2012. Microbial life at −13 °C in the brine of an ice-sealed Antarctic lake. Proc. Natl. Acad. Sci. U. S. A. 109 (50), 20626–20631.

Onac, B.P., Perşoiu, A., Racovita, G., Tamas, T., Viehmann, I., 2007. Scărişoara. Studia, Cluj. 84.

Øvreås, L., Forney, L., Daae, F.L., Torsvik, V., 1997. Distribution of bacterioplankton in meromictic lake Sælenvannet, as determined by denaturing gradient gel electrophoresis of PCR-amplified gene fragments coding for 16S rRNA. Appl. Environ. Microbiol. 63, 3367–3373.

Pearce, D., Hodgson, D., Thorne, M., Burns, G., Cockell, C., 2013. Preliminary analysis of life within a former subglacial lake sediment in Antarctica. Diversity 5, 680–702.

Perşoiu, A., Onac, B.P., 2012. Ice in caves. In: Culver, D.C., White, W. (Eds.), Encyclopedia of Caves. Elsevier Science Publishing Co Inc, United States, pp. 399–404.

Perşoiu, A., Pazdur, A., 2011. Ice genesis and its long-term mass balance and dynamics in Scarisoara Ice Cave, Romania. Cryosphere 5, 45–53.

Perşoiu, A., Onac, B.P., Wynn, J.G., Bojar, A.V., Holmgren, K., 2011. Stable isotope behavior during cave ice formation by water freezing in Scarisoara Ice Cave, Romania. J. Geophys. Res. Atmos. 116, D02111.

Perşoiu, A., Onac, B.P., Wynn, J.G., Blaauw, M., Ionita, M., Hansson, M., 2017. Holocene winter climate variability in Central and Eastern Europe. Sci. Rep. 7 (1), 1196–1203.

Pop, E., 1949. Bacterii nitrificante in pestera de la Scarisoara. Bul. Stiint. 1, 901–907.

Popa, R., Smith, A.R., Popa, R., Boone, J., Fisk, M., 2012. Olivine-respiring bacteria isolated from the rock-ice interface in a lava-tube cave, a Mars analog environment. Astrobiology 12 (1), 9–18.

Priscu, J.C., Fritsen, C.H., Adams, E.E., Giovannoni, S.J., Paerl, H.W., McKay, C.P., Doran, P.T., Gordon, D.A., Lanoil, B.D., Pinckney, J.L., 1998. Perennial Antarctic lake ice: an oasis for life in a Polar desert. Science 280, 2095–2098.

Racovita, E.G., 1927. Observations sur la glaciere naturelle dite "Gheţarul de la Scărişoara". Bull. Soc. Sci. Cluj 3, 75–108.

Racovita, G., 1994. Bilan climatique de la grotte glacière de Scărişoara (Monts du Bihor, Roumanie), dressé sur dix années d'observations. Travaux de l'Institut de Spéologie "E. Racovitza" 33, 107–158.

Racovita, G., Onac, B.P., 2000. Scarisoara Glacier Cave. Monographic studt. Ed. Carpatica, pp. 1–140.

Reasoner, D.J., Geldreich, E.E., 1985. A new medium for the enumeration and subculture of bacteria from potable water. Appl. Environ. Microbiol. 49, 1–7.

Reitschuler, C., Lins, P., Wagner, A.P., Illmer, P., 2014. Cultivation of moonmilk-born non-extremophilic Thaum and Euryarchaeota in mixed culture. Anaerobe 29, 73–79.

Reitschuler, C., Spötl, C., Hofmann, K., Wagner, A.O., Illmer, P., 2016. Archaeal distribution in moonmilk deposits from Alpine Caves and their ecophysiological potential. Microb. Ecol. 71, 686–699.

Rippka, R., Deruelles, J., Waterbury, J.B., Herdman, M., Stanier, R.Y., 1979. Generic assignments, strain histories and properties of pure cultures of cyanobacteria. Microbiology 111, 1–61.

Rivkina, E., Laurinavichius, K., McGrath, J., Tiedje, J., Shcherbakova, V., Gilichinsky, D., 2004. Microbial life in permafrost. Adv. Space Res. 33, 1215–1221.

Robinson, C.H., 2001. Cold adaptation in Arctic and Antarctic fungi. New Phytol. 151 (2), 341–353.

Röling, W.F.M., van Breukelen, B.M., Braster, M., Lin, B., van Verseveld, H.W., 2001. Relationships between microbial community structure and hydrochemistry in a landfill leachate-polluted aquifer. Appl. Environ. Microbiol. 67, 4619–4629.

Sattler, B., Wille, A., Waldhuber, S., Sipiera, P., Psenner, R., 2002. Various ice ecosystems in alpine and polar regions—an overview. In: Lacoste, H. (Ed.), Proceedings of the First European Workshop on Exo-Astrobiology, 16–19 September 2002, Graz, Austria, ESA SP-518. ESA Publications Division, Noordwijk, ISBN: 92-9092-828-X, pp. 223–226.

Standhartinger, B., Sattler, B., Psenner, R., 2010. Bacterial biodiversity in ice and glacier caves in the Alps and Antarctica [PS2-A.58]. https://elsevier.conference-services.net/resources/247/1976/pdf/Oslo2010_2096.pdf.

Tebo, B.M., Davis, R.E., Anitori, R.P., Connell, L.B., Schiffman, P., Staudigel, H., 2015. Microbial communities in dark oligotrophic volcanic ice cave ecosystems of Mt. Erebus, Antarctica. Front. Microbiol. 6, 179–192.

Vainio, E.J., Hantula, J., 2000. Direct analysis of wood-inhabiting fungi using denaturing gradient gel electrophoresis of amplified ribosomal DNA. Mycol. Res. 104, 927–936.

Wingender, J., Neu, T.R., Flemming, H.-C., 1999. What are bacterial extracellular polymer substances? In: Wingender, J., Neu, T.R., Flemming, H.-C. (Eds.), Microbial Extracellular Polymer Substance. Springer, Berlin, pp. 1–19.

Yang, G.L., Hou, S.G., Le Baoge, R., Li, Z.G., Xu, H., Liu, Y.P., Du, W.T., Liu, Y.Q., 2016. Differences in bacterial diversity and communities between glacial snow and glacial soil on the chongce ice cap, West Kunlun Mountains. Sci. Rep. 6, 36548–36555.

Yao, T., Xiang, S., Zhang, X., Wang, N., 2006. Microorganisms in the Malan ice core and their relation to climatic and environmental changes. Glob. Biogeochem. Cycles 20 (1), GB1004.

Yao, T.D., Liu, Y.Q., Kang, S.C., Jiao, N.Z., Zeng, Y.H., Liu, X.B., Zhang, Y.J., 2008. Bacterial variabilities in a Tibetan ice core and their relations with climate change. Glob. Biogeochem. Cycles 22 (4), GB4017.

Yu, Y., Lee, C., Hwang, S., 2005. Group-specific primer and probe sets to detect methanogenic communities using quantitative real-time polymerase chain reaction. Biotechnol. Bioeng. 89 (6), 670–679.

FURTHER READING

Anesio, A.M., Mindl, B., Laybourn-Parry, J., Hodson, A.J., Sattler, B., 2007. Viral dynamics in cryoconite holes on a high Arctic glacier (Svalbard). J. Geophys. Res. 112, G04S31.

Gremberghe, I.V., Van Wichelen, J., Van der Gucht, K., Vanormelingen, P., D'hondt, S., Boutte, C., Wilmotte, A., Vyverman, W., 2008. Covariation between zooplankton community composition and cyanobacterial community dynamics in Lake Blaarmeersen (Belgium). FEMS Microbiol. Ecol. 63, 222–237.

Margesin, R., Neuner, G., Storey, K.B., 2007. Cold-loving microbes, plants, and animals—fundamental and applied aspects. Naturwissenschaften 94, 77–99.

CHAPTER

9

PALEOCLIMATIC SIGNIFICANCE OF CAVE ICE

Aurel Perşoiu

Emil Racoviţă Institute of Speleology, Cluj-Napoca, Romania

CHAPTER OUTLINE

9.1 INTRODUCTION

If we are to understand the ongoing climatic changes, adapt to them, and/or mitigate their effects, a better understanding of the climate system and of the mechanisms behind its variability is required, and given the shortness of the instrumental climate record, proxy-based reconstructions are needed. In recent years, reconstructing past climate variability during the past millennium advanced in major leaps, however, the majority of the reconstructions are based on biological proxies, thus being biased towards the warm season, with only a limited set capturing winter climate (e.g., Fohlmeister et al., 2013). One of the best candidates for reconstructing past winter climatic conditions is the isotopic composition of snow accumulated as ice caps and mountain glaciers. In Europe, glaciers archiving past winter climate variability exist in the Alps and the Scandinavian Mountains, only. These glaciers are difficult to date and cover only a short time span due to fast mass turnover, thus being of limited use for paleoclimatic studies. However, mountains in Europe harbor numerous *underground glaciers in caves* (Luetscher, 2013; Kern and Perşoiu, 2013), that have accumulated by either snow digenesis or freezing of water over the past millennia (Luetscher et al., 2007; Perşoiu and Pazdur, 2011; May et al., 2011) and *that harbor a wealth of information on past winter temperature variability* (Perşoiu et al., 2017), *NAO dynamics* (Stoffel et al, 2009), *winter precipitation* (Spötl et al., 2014), *atmospheric circulation* (Kern et al., 2011), and past vegetation dynamics (Feurdean et al., 2011). In the following sections, I will assess the suitability of the various proxies harbored by cave ice for paleoclimatic research and outline the main study direction that could lead to the best results in terms of paleoclimatic reconstructions. One of the main issues when trying to use ice deposits of caves in paleoclimatic studies is that of the age of ice and its determination. This topic is extensively investigated by Kern (this volume) and thus will not be discussed here.

Ice Caves. https://doi.org/10.1016/B978-0-12-811739-2.00009-7

9.2 CAVE GLACIERS AND THEIR AGE

The presence of ice itself in a cave is sometimes regarded as a tentative indicator of climatic conditions. However, the occurrence of ice caves is controlled by (1) the presence of karstified rocks (i.e., caves), (2) specific cave morphologies, and (3) peculiar combination of climatic conditions (e.g., Eraso, 1991; Perşoiu, 2004). Ice caves have only been found so far in the Northern Hemisphere (Fig. 9.1), at latitudes ranging between north of the Arctic Circle (in Canada and Norway), and south of the Tropic of Cancer (in Hawaii), and altitudes between ~250 m asl (Walkin-Ice Cave System, Canada) and 3590 m asl (Dark Star Cave, Uzbekistan). While their distribution follows a general trend of decreasing altitude with increasing latitude (see Mavlyudov, this volume), a distinction can be made between the two different types of ice caves: vertical pits, with snow accumulating at their bottom, follow more closely the previously mentioned rule (and thus the temperature gradient), while for caves with congelation ice, the passages' morphology and the timing of negative temperatures and precipitation allow for a more erratic distribution. In Europe, such caves are found generally at mid-altitudes, where both temperatures are negative for a long-enough period so that (1) dripping water would freeze to ice and (2) summer warming will be insufficient to completely melt the accumulated ice. Recent observations (Kern and Perşoiu, 2013) have shown that, with very few exceptions, all ice caves are losing their perennial ice accumulations at an accelerated rate, and thus glaciation could be out of equilibrium with the present climatic conditions. Additional support for this hypothesis comes from the age of the cave glaciers, with most of the studies (Stoffel et al., 2009; Perşoiu and Pazdur, 2011; Spötl et al., 2014) indicating that ice in caves (1) existed, started to accumulate and/or had an increase in accumulation rate during the colder Little Ice Age (roughly between AD 1300 and 1800), and (2) was missing and/or partly melted during the precedent (~ AD 800–1300) Medieval Warm Period.

9.3 OXYGEN AND HYDROGEN ISOTOPE COMPOSITION OF ICE

The pioneering study of the stable isotope composition on cave ice was published from Scărişoara Ice Cave (Şerban et al., 1967). Recent observational records of the stable isotopic composition of cave ice have been published recently from several different locations in Europe: Focul Viu Ice Cave in Romania (Fórizs et al., 2004), LoLc 1650 Ice Cave in Italy (Citterio et al., 2004), Dobšiná Ice Cave in Slovakia (Clausen et al., 2007), Scărişoara Ice Cave in Romania (Perşoiu et al., 2011), Monlesi and St. Livres Ice Caves in Switzerland (Stoffel et al., 2009; Luetscher et al., 2007), and Vukusic Ice Caves in Croatia (Kern et al., 2011). Each of these studies linked the isotopic composition of cave ice to winter temperature variability outside the cave, and, additionally, the d-excess parameter in cave ice has been unambiguously linked to changes in the moisture sources feeding the precipitations in the area.

However, the intricacies of interpreting measured isotopic variation will necessarily be different from cave to cave, because the processes involved will vary with the local climate, cave morphology, timing of ice accumulation, etc. For example, Yonge and MacDonald (1999) have shown that unlike surface ice bodies, there is an inverse relationship between surface air temperature and stable oxygen isotopic ratios of cave ice. Citterio et al. (2004) attributed the intra-layer trend of isotopic depletion to the freezing process of a shallow lake. And finally, annual cycles and seasonal fluctuation of $\delta^{18}O$ values within the annual layers were linked to surface temperature in Monlesi Ice Cave (Luetscher, 2005; Luetscher et al., 2007).

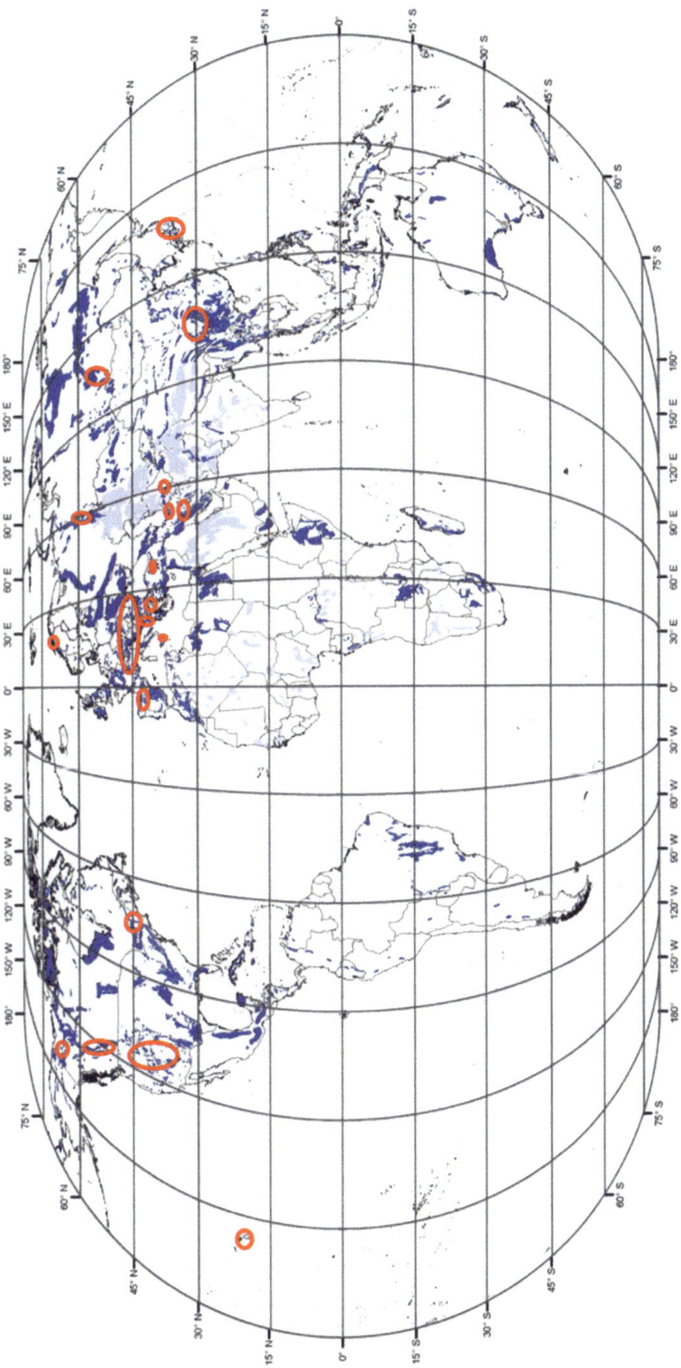

FIG. 9.1

Areas with perennial ice accumulations in caves. The background map shows the distribution of the main karst areas.

Modified from Ford, D.C., Williams, P., 2007. Karst Hydrogeology and Geomorphology, Wiley, New York, 576 p. ISBN: 978-0-470-84996-5.

Following the study of Şerban et al. (1967) that reported the first isotope data from ice caves (without calibration, hence impossible to compare with modern data), the next paper to discuss the isotope composition of cave ice as a possible climatic proxy was published by Yonge and Macdonald (1999). These authors have suggested the high values of $\delta^{18}O$ and $\delta^{2}H$ in cave ice [enriched compared to the precipitation values and plotting on a Local Meteoric Water Line (LMWL) below the Global Meteoric Water Line (GMWL)] were due to a complex genetic process, involving deposition of isotopically-enriched hoar frost and subsequent incorporation of melt-water derived from the hoar frost deposits in the newly formed cave ice.

This early study was followed in 2004 by a series of three papers emerging from the 1st International Workshop on Ice Caves (IWIC). Citterio et al. (2004) were first to report a decreasing trend of $\delta^{18}O$ values with depth in an ice section that formed by the slow, top-to-bottom freezing of a shallow lake. The authors attributed this decreasing trend to the preferential incorporation of heavy isotopes in the ice and the continuous depletion of the remaining water reservoir in ^{18}O and ^{2}H isotopes. This hypothesis was further tested by Perşoiu et al. (2011) in Scărişoara Ice Cave, where both observational and modeling data were used to show how ice that formed under kinetic conditions (with the subsequent loss of the initial climatic signal carried by the isotopic composition of ice) could be still used in paleoclimatic studies. In this later study, both the lake water (before the onset of freezing) and the resulting ice layer (three separate cores) were analyzed for their stable isotope composition. The ice had a clear depletive trend in both $\delta^{18}O$ and $\delta^{2}H$ from top to bottom, due to the preferential incorporation of heavy isotopes in the ice during freezing (e.g., Jouzel and Souchez, 1982), a consequence of the relatively high diffusion coefficient of H^1H^2O and $H_2^{18}O$ molecules in water ($10^{-5}\,cm^2/s$), compared to ice ($10^{-11}\,cm^2/s$). However, the isotopic composition of this first layer of ice was much lower than the theoretical one, calculated using the equilibrium fractionation factors of 1.003 for oxygen (O'Neil, 1968), and 1.0208 for hydrogen (Arnason, 1969) and the lake water as initial input. These less than maximally enriched isotopic values for the first ice layer could be explained (Perşoiu et al., 2011) by either (1) nonequilibrium ice formation during fast freezing of water at the onset of the process, which leads to a smaller degree of separation of the isotope species between the solid and liquid phases (Souchez et al., 1987; Jouzel et al., 1999), and (2) trapping and subsequent freezing of water during initial crystal growth (O'Neil, 1968; Jouzel et al., 1999). These suggestions are consistent with observations of ice genesis in the cave, as follows.

The freezing of water in the cave proceeds in two main stages. In the first stage, ice needles and crystals develop on the surface of water, under the influence of very fast cold air inflow, in the form of turbulent air avalanches, triggered by the higher density of the external, colder air. Ice needles coalesce to form a thin layer of ice that isolates the remaining water and slows down the freezing processes. In the second stage, slow freezing of the remaining water occurs, with a freezing front that moves downwards, parallel to the ice-water interface, until the entire column of water solidifies. Within the water column, two isotopically distinct parts can be separated (Souchez and Jouzel, 1984; Souchez et al., 1987): a boundary layer at the ice-water interface, and the remainder of the water, which is isotopically homogenous. The isotopic composition of the boundary layer is continuously modified by two processes: freezing of water and thus preferential incorporation of heavy isotopes in the ice, and the diffusion of heavy isotopes from the bulk of the liquid. A steady state can develop in which the rate of heavy isotope incorporation in ice equals the flux from the bulk of remaining water. This state lasts until the final stage of freezing, when the water from the bottom of the column, strongly depleted of heavy isotopes, freezes.

Similar to the findings of Yonge and Macdonald (1999), the ice samples align on a LWML with a slope of 6.55, lower than the slope of the local or global precipitation (both ~8, Craig, 1961; Bojar et al., 2009). This low value of the slope for the $\delta^{18}O$-$\delta^{2}H$ relationship in ice is characteristic for ice formed by freezing of water (Jouzel and Souchez, 1982; Souchez and Jouzel, 1984), with the slope itself being termed a "freezing slope" by these authors. The two papers have also shown that the initial isotopic composition of water before freezing can be reconstructed by intersecting the freezing slope with the LMWL, the intersection point giving the values for $\delta^{18}O$ and $\delta^{2}H$ of the initial water (on the assumption that the water that subsequently froze to form the ice is derived from precipitation that plots on the LMWL on a $\delta^{18}O$-$\delta^{2}H$ diagram). This later finding suggests that, although kinetic processes are at play during the formation of cave ice from a standing pool of water, the initial isotopic composition of water (and hence of the climatic signal) can be safely reconstructed and further used to build time series of past climate variables (Perşoiu et al., 2011, 2017). Similar observations were obtained from Eisriesenwelt cave (Austria) by May et al. (2011), who show that ice in this cave has formed by the slow freezing of standing water, with the isotopic composition of the resulting ice reflecting this process: a highly enriched first layer, depletion of heavy isotopes with increasing depth and samples from single layers aligned along a "freezing slope," lower than the slope of precipitation water (Fig. 9.2).

The first authors using the isotope composition of cave ice to make paleoclimatic reconstructions were Kern et al. (2004) and Fórizs et al. (2004), studying the Focul Viu Ice Cave (Romania). These authors used the "lake ice" model as well, but due to the limited number of samples they have analyzed for both $\delta^{18}O$ and $\delta^{2}H$ and the lack of observational data on the timing of ice formation, they

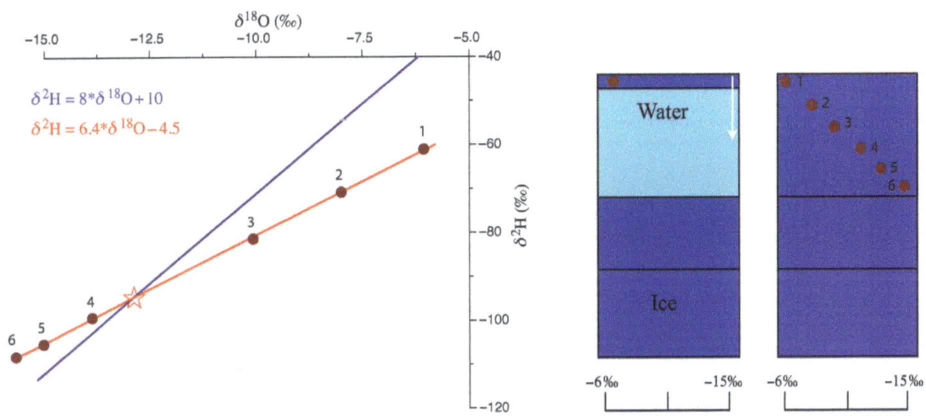

FIG. 9.2

Theoretical plot of the stable isotopes behavior during water freezing in the "lake ice" model. The left panel shows a plot of $\delta^{18}O$ against $\delta^{2}H$, with the *blue line* representing the Local Meteoric Water Line of precipitation outside the cave, and the *red* one, the "freezing slope" of samples from ice that formed by the downward freezing of a stagnant pool of water (right). The *red star* shows the isotopic values of this pool of water, before freezing. The first layer of ice to form is enriched in heavy isotopes compared to the original water (number 1), and the last one is depleted (number 6). The right panel shows a conceptual model of this process, with dark *blue* layers representing ice, and light *blue* ones, water. In the left column, the ice below the water layer was formed in the previous year. Numbers in the right panel correspond to those in the left one. The scale at the bottom of the two columns in the right panel are for $\delta^{18}O$ ($\delta^{2}H$ has a similar pattern).

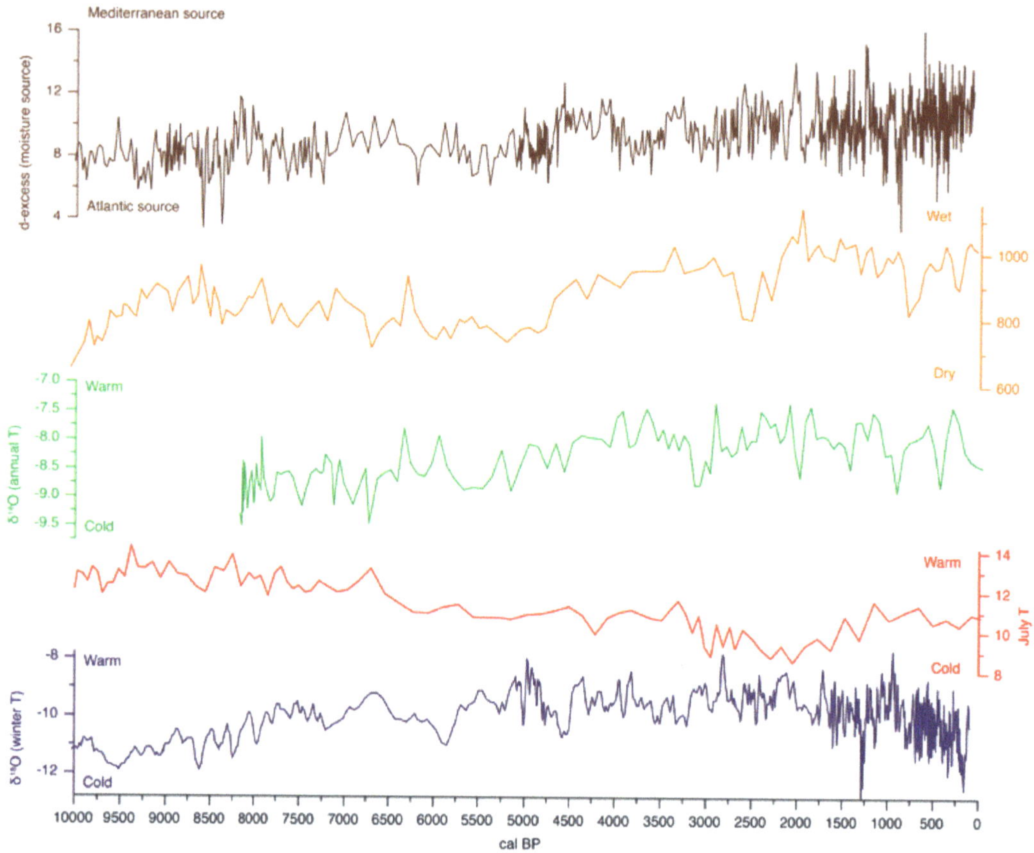

FIG. 9.3

Multiproxy climatic reconstruction in East-central Europe. Winter temperature *(blue)* and moisture source changes *(brown)* from Perşoiu et al. (2017), July temperature *(red)* from Tóth et al. (2015), annual temperature *(orange)* from Draguşin et al. (2014), and precipitation amount *(green)* from Feurdean et al. (2008).

could not reliably asses the paleoclimatic information in the ice. A newer study was recently published by Perşoiu et al. (2017), the authors reconstructing winter temperatures and sources of moisture for the past 10,500 years (Fig. 9.3) with multidecadal resolution. By analyzing radiocarbon dated profiles from ice deposits in caves in Northern Spain, Bartolomé et al. (2017) have shown that the $\delta^{18}O$ values of water could be used to reconstruct past solar variability, with low isotopic values corresponding to periods of reduced solar activity.

Another possible source of past climatic information hosted by the perennial ice accumulations in caves is the isotopic composition of cryogenic cave calcite (CCC, Žák et al., 2008). Water freezing in caves is accompanied by degassing of CO_2, and precipitation of CCC under strong kinetic conditions. These kinetic processes could lead to the alteration of the original putative climatic signal carried by the isotopic composition of CCC. To test whether the climatic signal is still preserved in the isotope composition of CCC, we have analyzed oxygen and hydrogen isotopes in an ice core extracted from Scărişoara Ice Cave (Romania, Bădăluţă et al., 2017), as well as oxygen and carbon isotopes in CCC from the same core. The water isotopic values in the ice core show low values until AD 900, higher values between AD 900 and ~1300 (Medieval Warm Period, MWP), and again lower values after AD 1300 (Little Ice Age, LIA), reaching their minimum after AD 1800. However, $\delta^{13}C$ and $\delta^{18}O$ values in CCC have higher values for samples from the MWP than those from the LIA. CCC results from the deposition of $CaCO_3$ from $Ca(CO_3)_2$, the latter being a product of rock dissolution by carbonic acid. The main source of CO_2 to form carbonic acid is soil CO_2, produced by root respiration. Previous studies have shown that $\delta^{18}O$ of this CO_2 is in equilibrium with the $\delta^{18}O$ of water, so that the higher (lower) $\delta^{18}O$ values of CCC could reflect warmer (colder) conditions during the MPW (LIA). The interpretation of $\delta_{13}C$ values of CCC is less straightforward. Higher $\delta^{13}C$ values in soil CO_2 are determined by moisture limitation on plants, either due to low moisture or higher evaporative conditions. While the MWP was warmer in the study area, conflicting data exists on precipitation, with studies suggesting both drier and wetter conditions, so that it is difficult to interpret our carbon isotope data. Apart from the direct climatic influence, the depth of soil could have also played a part, as deeper soils (as expected under birch forests that dominated during the MWP) would have had more enrichment in the heavy isotopes with depth, than the thinner soils of the LIA (formed under mostly spruce forests). Further, drought/higher temperatures could also influence the kinetics of the reaction, which can be large enough to overprint any soil signal in $\delta^{13}C$.

9.4 ORGANIC REMAINS TRAPPED IN ICE

Caves act as natural traps for pollen and surface-derived organic matter and these can be locked in the ice, providing evidence of past vegetation changes (Feurdean et al, 2011). In 1950, Pop and Ciobanu (1950) attempted to reconstruct the vegetation around the Scărişoara Ice Cave (Romania) by counting the pollen grains found in the subterranean ice block. While they correctly identified pollen taxa, lack of age control (radiocarbon dating had yet to be developed) led to an erroneous recreation of the landscape around the cave. This pioneering study was followed half a century later by a more precise one (Feurdean et al., 2011) that used both pollen and macrofossils recovered from the ice to reconstruct the dynamics of vegetation and its interaction with the local human population during the last 1000 years. In the Pyrenees, pollen entrapped in the ice of A294 cave provided (Leunda et al., 2017) a compelling story of past environmental changes between 5680 and 2230 cal BP, the good preservation of both pollen and macrofossils allowing for the reconstruction of the dynamics of the tree line in the area and of human-environmental interactions.

9.5 CONCLUSIONS

Cave glaciers preserve a large variety of candidate proxies for both past climate and environmental changes, however, understanding how the climatic signal is preserved is generally more difficult than in the case of surface ice accumulations (Perşoiu et al., 2011). The best candidates for past climate reconstructions are the stable isotope composition of cave ice and the pollen and macrofossils archived in the ice deposits. Findings in various ice caves call attention to the fact that it is hardly possible to create a uniformly valid model for the isotope composition of cave ice, due to its complex origin. The general opinion is that detailed monitoring in each individual cave is recommended for reasonable calibration of the geochemical information archived in cave ice (Fórizs et al., 2004; Holmlund et al., 2005). While pollen is routinely found in ice, it is usually scarce in terms of grain counts, and thus is difficult to use as a reliable proxy of past vegetation dynamics. In the light of recent advances in DNA analysis, the excellent preservation of macrofossils in ice could be the next hot topic in ice cave research, as it could shed light on past vegetation assemblages, plant use, diet, agriculture, thus being a valuable proxy of past ecology and human-environment interactions (e.g., Schlumbaum et al., 2008).

ACKNOWLEDGMENTS

The author acknowledges support from projects PNII-RU-TE-2014-4-1993 and EraNet-LAC ELAC2014/DCC0178.

REFERENCES

Arnason, B., 1969. The exchange of hydrogen isotopes between ice and water in temperate glaciers. Earth Planet. Sci. Lett. 6, 423–430.

Bădăluţă, C., Ersek, V., Piotrowska, N., Perşoiu, A., 2017. Oxygen and carbon stable isotopes in cryogenic cave calcite (CCC)—possible proxy for past climate changes. In: Geophysical Research Abstracts Vol. 19, EGU2017-1554-1, Vienna, Austria, April 23–28.

Bartolomé, M., Sancho, C., Moreno, A., Belmonte, Á., Leunda, M., Delgado-Huertas, A., Oliva-Urcía, B., Cacho, I., Stoll, H., Edwards, R.L., Cheng, H., 2017. Is the climate signal adequately recorded in the $\delta^{18}O$ isotope composition from ice cave deposits? Climate variations during the little ice age and the industrial era inferred from Pyrenean ice deposits and stalagmites. In: Abstract Book, 5th PAGES Open Science Meeting, Zaragoza. pp. 179–180.

Bojar, A.-V., Ottner, F., Bojar, H.-P., Grigorescu, D., Perşoiu, A., 2009. Stable isotope and mineralogical investigations on clays from the Late Cretaceous sequences, Hateg Basin, Romania. Appl. Clay Sci. 45, 155–163.

Citterio, M., Turri, S., Bini, A., Maggi, V., 2004. Observed trends in the chemical composition, d18O and crystal sizes vs. depth in the first ice core from the LoLc 1650 "Abisso sul Margine dell'Alto Bregai" ice cave (Lecco, Italy). Theor. Appl. Karstol. 17, 45–50.

Clausen, H.B., Vrana, K., Hansen, S.B., Larsen, L.B., Baker, J., Siggaard-Andersen, M.-L., Sjolte, J., Lundholm, S.C., 2007. Continental ice body in Dobšiná Ice Cave (Slovakia)—Part II. Results of chemical and isotopic study. In: Zelinka, J. (Ed.), Proceedings of the 2nd International Workshop on Ice Caves, Liptovsky Mikulás. pp. 29–37.

Craig, H., 1961. Isotopic variations in meteoric waters. Science 133, 1702–1703.

Dragusin, V., Staubwasser, M., Hoffmann, D.L., Ersek, V., Onac, B.P., Veres, D., 2014. Constraining Holocene hydrological changes in the Carpathian-Balkan region using speleothem $\delta^{18}O$ and pollen-based temperature reconstructions. Clim. Past 10, 1363–1380. https://doi.org/10.5194/cp-10-1363-2014.

Eraso, A., 1991. Diagrama para el calculatiòn de la ablaciòn glaciar y formaciòn de hielo en las cavernas glaciares. In: Atti e Mem. Comm. Grotte "E. Boegan". 30. pp. 37–52.

Feurdean, A., Klotz, S., Mosbrugger, V., et al., 2008. Pollen-based quantitative reconstructions of Holocene climate variability in NW Romania. Palaeogeogr. Palaeoclimatol. Palaeoecol. 260, 494–504.

Feurdean, A., Perşoiu, A., Pazdur, A., Onac, B.P., 2011. Evaluating the palaeoecological potential of pollen recovered from ice in caves: a case study from Scărişoara Ice Cave, Romania. Rev. Palaeobot. Palynol. 165, 1–10. https://doi.org/10.1016/j.revpalbo.2011.01.007.

Fohlmeister, J., Vollweiler, N., Spötl, C., Mangini, A., 2013. COMNISPA II: update of a mid-European isotope climate record, 11 ka to present. The Holocene 23, 749–754. https://doi.org/10.1177/0959683612465446.

Fórizs, I., Kern, Z., Szántó, Z., Nagy, B., Palcsu, L., Molnár, M., 2004. Environmental isotopes study on perennial ice in the Focul Viu Ice Cave, Bihor Mountains, Romania. Theor. Appl. Karstol. 17, 61–69.

Holmlund, P.E.R., Onac, B.P., Hansson, M., Holmgren, K., Morth, M., Nyman, M., Perşoiu, A., 2005. Assessing the palaeoclimate potential of cave glaciers: the example of the Scărişoara Ice Cave (Romania). Geogr. Ann. Ser. A 87, 193–201.

Jouzel, J., Souchez, R.A., 1982. Melting refreezing at the glacier sole and the isotopic composition of the ice. J. Glaciol. 28, 35–42.

Jouzel, J., Petit, J.R., Souchez, R., Barkov, N.I., Lipenkov, V.Y., Raynaud, D., Stievenard, M., Vassiliev, N.I., Verbeke, V., Vimeux, F., 1999. More than 200 meters of lake ice above subglacial lake Vostok, Antarctica. Science 286, 2138–2141.

Kern, Z., Perşoiu, A., 2013. Cave ice—the imminent loss of untapped mid-latitude cryospheric palaeoenvironmental archives. Quat. Sci. Rev. 67, 1–7.

Kern, Z., Fórizs, I., Nagy, B., Kázmér, M., Gál, A., Szánto, Z., Palcsu, L., Molnár, M., 2004. Late Holocene environmental changes recorded at Ghetarul de la Focul Viu, Bihor Mountains, Romania. Theor. Appl. Karstol. 17, 51–60.

Kern, Z., Forizs, I., Pavuza, R., et al., 2011. Isotope hydrological studies of the perennial ice deposit of Saarhalle, Mammuthöhle, Dachstein Mts, Austria. Cryosphere 5, 291–298.

Leunda, M., Sancho, C., Bartolomé, M., Belmonte-Ribas, Á., Gómez, D., Gil-Romera, G., Moreno, A., Oliva-Urcia, B., González-Sampériz, P., 2017. The last ice caves of western Mediterranean mountains and its potential for palaeoenvironmental reconstructions: an announced disappearance in the Pyrenees. In: Abstract Book, 5th PAGES Open Science Meeting, Zaragoza, pp. 297–298.

Luetscher, M., 2005. Processes in Ice Caves and Their Significance for Paleoenvironmental Reconstructions. Swiss Institute for Speleology and Karst Studies, La Chaux-de-Fonds, Switzerland, p.154.

Luetscher, M., 2013. Glacial processes in caves. In: Shroder, J., Frumkin, A. (Eds.), Treatise on Geomorphology, vol. 6. Academic Press, San Diego, CA, pp. 258–266.

Luetscher, M., Bolius, D., Schwikowski, M., Schotterer, U., Smart, P.L., 2007. Comparison of techniques for dating of subsurface ice from Monlesi ice cave, Switzerland. J. Glaciol. 53 (182), 374–384.

May, B., Spötl, C., Wagenbach, D., Dublyansky, Y., Liebl, J., 2011. First investigations of an ice core from Eisriesenwelt cave (Austria). Cryosphere 5, 81–93. https://doi.org/10.5194/tc-5-81-2011.

O'Neil, J.R., 1968. Hydrogen and oxygen isotope fractionation between ice and water. J. Phys. Chem. 72, 3683–3684.

Perşoiu, A., 2004. Ice speleothemes in Scărişoara Cave: dynamics and controllers. Theor. Appl. Karstol. 17, 71–76.

Perşoiu, A., Pazdur, A., 2011. Ice genesis and its long-term mass balance and dynamics in Scărişoara Ice Cave, Romania. Cryosphere 5 (1), 45–53.

Perşoiu, A., Onac, B.P., Wynn, J.G., Bojar, A.-V., Holmgren, K., 2011. Stable isotope behavior during cave ice formation by water freezing in Scărişoara Ice Cave, Romania. J. Geophys. Res. Atmos. 116, D014477. https://doi.org/10.1029/2010JD014477.

Perşoiu, A., Onac, B.P., Wynn, J.G., Blaauw, M., Ioniţă, M., Hansson, M., 2017. Holocene winter climate variability in Central and Eastern Europe. Sci. Rep. 7, 1196. https://doi.org/10.1038/s41598-017-01397-w.

Pop, E., Ciobanu, I., 1950. Analize de polen în ghiaţa dela Scărişoara. In: Analele Academiei Republicii Populare Române, seria Geologie, Geografie, Biologie, Ştiinte Tehnice şi Agricole, III, pp. 23–50.

Schlumbaum, A., Tensen, M., Jaenicke-Després, V., 2008. Ancient plant DNA in archaeobotany. Veg. Hist. Archaeobotany 17, 233–244. https://doi.org/10.1007/s00334-007-0125-7.

Şerban, M., Blaga, L., Blaga, L., Chifu, A., Ciobotaru, T., 1967. Contribuţii la stratigrafia depozitelor de gheaţă din Gheţarul de la Scărişoara. In: Lucr. Inst. Speol. "E. Racoviţă". VI, pp. 107–140.

Souchez, R.A., Jouzel, J., 1984. On the isotopic composition in delta-d and delta-o-18 of water and ice during freezing. J. Glaciol. 30, 369–372.

Souchez, R., Tison, J.L., Jouzel, J., 1987. Freezing rate determination by the isotopic composition of the ice. Geophys. Res. Lett. 14, 599–602.
Spötl, C., Reimer, P.J., Luetscher, M., 2014. Long-term mass balance of perennial firn and ice in an Alpine cave (Austria): constraints from radiocarbon-dated wood fragments. The Holocene 24, 165–175. https://doi.org/10.1177/0959683613515729.
Stoffel, M., Luetscher, M., Bollschweiler, M., et al., 2009. Evidence of NAO control on subsurface ice accumulation in a 1200 yr old cave-ice sequence, St. Livres ice cave, Switzerland. Quat. Res. 72, 16–26.
Tóth, M., et al., 2015. Chironomid-inferred Holocene temperature changes in the South Carpathians (Romania). The Holocene 25, 569–582. https://doi.org/10.1177/0959683614565953.
Yonge, C.J., Macdonald, W.D., 1999. The potential of perennial cave ice in isotope palaeoclimatology. Boreas 28, 357–362.
Žák, K., Onac, B.P., Perşoiu, A., 2008. Cryogenic carbonates in cave environments: a review. Quat. Int. 187 (1), 84–96.

CHAPTER

THE MANAGEMENT OF ICE SHOW CAVES

10

Friedrich Oedl[*,†,‡]

Eisriesenwelt Cave, Salzburg, Austria[*] *International Show Caves Association (ISCA), Ancona, Italy*[†] *Austrian Caving Association VÖH, Vienna, Austria*[‡]

CHAPTER OUTLINE

10.1 INTRODUCTION

Fascination with subterranean cave systems and their rich variety of formations led to them being opened up to the public and developed commercially more than 100 years ago. Paths were laid and equipped with lighting systems, and general infrastructure was created to facilitate access to the cave systems. Show caves number well over 1000 worldwide (see http://www.showcaves.com, which quotes a total of 1255 show caves so far). In contrast, very few ice caves on Earth provide access to the public. There are only an estimated 20 ice caves worldwide (Fig. 10.1) that function as show caves. It should be mentioned here that this number does not include glacier caves, irrespective of whether they are naturally or artificially created.

Ice Caves. https://doi.org/10.1016/B978-0-12-811739-2.00011-5

Austria	4	Eisriesenwelt
		Dachstein Rieseneishöhle
		Eiskogelhöhle
		Hundsalm Eishöhle
China	3	Ningwu ice cave
		Summer Ice Cave Chongqing
		Ice Cave in Songluo County / Shennongjia
Germany	1	Schellenberger Eishöhle
Japan	1	Narusawa Ice Cave
Romania	1	Scărişoara Cave
Russia	2	Kungur Ice Cave
		Caves of Pinega
Slovakia	2	Demänovská Ice Cave
		Dobšinská Ice Cave
Slovenia	1	Snežna Jama
USA	5	Lava Beds
		Boy Scout Cave / Craters of the moon
		Niter Ice cave
		Bandera Ice cave
		Big Ice Cave—Montana
	20	

FIG. 10.1

List of ice show caves in the world.

10.2 SHOW CAVE MANAGEMENT WITH SPECIAL REFERENCE TO THE EISRIESENWELT IN AUSTRIA

When considering the fact that very few ice caves function as show caves, we must define which system of ice caves is being referred to. It is especially important to differentiate between dynamic and static ice caves. In the majority of cases, we are talking about static ice caves. This article, however, is predominantly about the management of a dynamic ice cave using the example of the Eisriesenwelt in Austria, the largest and most famous ice cave in the world with public access. The key facts and figures relating to this cave are as follows:

The entrance to the Eisriesenwelt is in the Northern Limestone Alps (Kalkalpen) near Salzburg at an altitude of 1641 m, and is located approximately 1000 m above the valley floor. The first entry into the cave is documented as being in 1879 with guided tours beginning in 1920. As a result of the exposure and the hazards in the winter, visitors can gain access only during the six summer months. In the period up to 2016, more than seven million guests had been recorded, with a current annual total of 180,000 visitors. The daily maximum is around 3000 visitors, accommodated on 70 guided tours per day.

10.3 MANAGEMENT FROM A HISTORICAL PERSPECTIVE

The biggest challenge related to enabling general public access to the Eisriesenwelt revolved around creating a path that didn't require visitors to have mountain climbing skills in order to ascend 1000 m over difficult terrain and precipitous rocks to the cave's entrance. In addition, it was necessary to create a series of paths within the cave. Due to the area open for visitation being completely covered by ground ice, in the early years of access, a vast number of steps were cut into the bare ice, and the cave could be visited only by those wearing crampons. The steepest sections, which were vertical in places, were secured only by a safety rope. Flights of wooden steps were soon constructed for which a method with the flexibility to adjust to the permanent changes along the path had to be developed. Every year the height of the ground ice changes according to the amount of water flowing in during the springtime. One example of the change in elevation is Hymir's Hall where the height of the ice has increased by 10 m! In order to keep the paths completely free of ice, channels were dug by hand alongside the walkways in the spring as soon as water started flowing in. This diverted the water flow, but created the additional problem of having to remove the ice and rubble and deposit it outside the areas visited.

The lighting in the Eisriesenwelt has always been provided by Davy lamps, which are distributed to the visitors. This is also the case for the Hundalm Ice Cave in Austria and the Schellenberger Ice Cave in Germany. This initial form of lighting has been retained, as has the use of magnesium strips used by the guide to illuminate the ice figures.

On warm summer days especially, there is strong air movement within the cave, which is typical for dynamic ice caves. The heavy drafts can reach wind speeds of up to 100 km/h in the entrance area prompting the need and construction of a weather door. As a result, the drafts during summer have substantially decreased, meaning that the Davy lamps are no longer extinguished so regularly. The reduced draft also increased the comfort level of the visitors, because a permanent draft accentuates the feeling of being cold.

In the 1960s a tunnel was built through the ground ice to enable visitors to walk in a circuit through the largest ground ice formation (Great Wall). This tunnel is 150 m long and was dug from the bottom upward with the use of a petrol-driven compressor and a pneumatic hammer to break ice. Cutting through the ice with power saws was not introduced until later.

10.4 THE KEY ELEMENTS OF ICE SHOW CAVE MANAGEMENT

The development of places of natural beauty for the general public always involves a conflict between economic efficiency and protection of the natural environment. The more sensitive the area, the greater the challenge to find not only the best solution but also one that can be retained in changing situations where dynamic development is concerned. Caves, and especially ice caves, are without a doubt one of the most sensitive areas. However, show caves are in direct competition with other leisure attractions such as fun parks and cultural or sporting activities. Therefore it is also necessary to position the show cave business within the general tourist market. Finally, show caves provide a unique insight for a wide audience into the basic scientific knowledge surrounding mountain and karst formations.

10.4.1 PATHWAYS

Because the ice conditions in the caves can change from year to year or on a seasonal basis, the construction and maintenance of pathways in ice caves pose a particular challenge. Whereas it is usual to have gravel tracks and concrete paths in most caves, especially those consisting of limestone features such as stalactites and stalagmites, there are very few places to lay out such paths in ice caves resulting in only two viable methods for walkway construction. The first method is to develop pathways and steps on a flexible apparatus so that they can be adjusted according to the ice levels. The second alternative is, within reason, to change the ground ice formations so that water flowing can be diverted away from walkways. Theoretically, pathways could also be fixed to the cave ceiling, but this would likely result in an unacceptable depreciation of the cave's visual appearance. There is no example to date of an ice cave resorting to this solution.

In the case of the Eisriesenwelt, with almost 100% ice coverage on the ground, there is a further challenge with the first half of the trail consisting of sections with steep incline. In order to maintain a steady flow while touring the caves, the gradient must be as constant as possible. Due to the large difference in gradient, groups innately elongate over the course of the tour causing a detrimental loss in fluency and efficiency. In addition, it has become evident that the path must be wide enough for children to be led by the hand at the side (required width approximately 1 m).

With regard to the material used for pathways, experience and various trials show that wooden steps are still the most suitable solution. Compared to metal or synthetic materials, wood has distinct advantages when it comes to ice removal and slip resistance.

The static supports of the walkway consist mostly of corrosion-resistant steel. To avoid icing up in spring when water flows into the cave, the pathways in potentially dangerous areas have been set at a specific height so that the ice can expand without impedance.

10.4.2 LIGHTING

Only a few show caves throughout the world are without electric lighting and instead use hand-held Davy lamps. The latter form of lighting was retained in the Eisriesenwelt to ensure that the presentation of the cave's environment is as original and natural as possible. Using this method, the visitors carry the lamps through the cave, resulting in a wide variety of experiences and impressions. Electric lighting in caves can be a significant problem when the units emit heat in close proximity to the ice formations. New technology, specifically LED lamps, is the actually standard system by installing electric lighting in show caves, which offers spectacular effects on one side and needs only a little part of energy as it was necessary by using halogen lamps or incandescent bulbs in former days on the other side.

10.4.3 MAINTENANCE/CLEANING

Because the fascination with ice caves depends to a large extent on the cleanliness of the ice, dirty deposits must be removed annually before the ice begins to build up in Nov. or Dec. Pollution is caused directly by the visitors and the abrasive wear of the wooden steps. In addition, the ice surface is polluted by the natural build up of dust, which stems from water entering the cave or being deposited by drafts. For this reason, the ice sections of the Eisriesenwelt cave are completely cleaned on an annual basis. This cleaning is done both by hosepipe washing and, in the dry periods, by sweeping. The storage of large water quantities is necessary for the washing and cleaning that must exclusively take place at the end of the tourist season and before the start of the freezing period.

Maintenance work, especially the upkeep of the pathways, can be carried out only when the guided tours have concluded and should be completed before the frost period to avoid any pollution of the cave ice. Consequently, each year this cleaning must take place within a few weeks.

General maintenance and repair also includes the removal of past construction and store rooms which requires special care when involving the extraction of old posts from the ice.

10.4.4 VISITOR MANAGEMENT

To ensure that visitors receive the best possible experience, many different aspects must be considered. With regard to the infrastructure, it is important that guests be properly informed onsite so that they have a complete understanding of location and timing (length of stay, physical requirements, necessary equipment, etc.). It has become standard practice for all important tourist attractions to offer guided tours in foreign languages. On a logistical level, consideration must be given to the division of groups by language preference so that waiting times can be minimized. At the Eisriesenwelt, for example, the visitor's language is electronically registered at the main desk so that the parties can be divided into language groups at the cave entrance.

Filming and photography are forbidden in the Eisriesenwelt. This is partly due to the security, but is predominantly for organizational reasons. Without this ban, there would be consistent delays making an orderly tour unfeasible.

A few attempts have been made (not in Eisriesenwelt) to incorporate multimedia elements and an acoustic experience in order to make show caves more attractive. However, these have not proved successful. The question of whether the installation of Wi-Fi in show caves is useful will have to be the subject of further analysis and examination.

10.5 MANAGEMENT OF NATURE CONSERVATION

As mentioned earlier, development automatically disrupts the natural state of any cave. In order to keep this disruption to a minimum, installations such as pathways and lighting should be constructed as to change the natural appearance as little as possible. On the other hand, the introduction of visitors can influence the climate within the cave. This influence can result in a change to the wind circulation through the creation of artificial entrances and the alteration of the cave profile by installations. Similarly, the introduction of visitors can also result in a change of temperature in the cave. A further point to be considered is whether the introduction of organic materials can influence the biological balance of the cave.

10.5.1 MECHANICAL INTERFERENCE

Apart from areas in the ice caves where ice build up occurs annually, great care must be taken when interfering with the ice. Making artificial incisions and openings realizes the threat of creating new drafts which can result in an undesired loss of ice (ice diffusion). In the Eisriesenwelt, the placement of a weather door at the entrance has had a positive influence on the ice development. On the other hand, there has been a noticeable decline in the ice level in the artificially created ice corridor where the draft causes an extension of the tunnel.

10.5.2 INCREASE IN CAVE TEMPERATURE CAUSED BY VISITORS

The extent of visitors' influence on the temperature conditions inside a cave has been the subject of many studies. Experiments have proved that the air temperature directly adjacent to the visitor can increase by up to half a degree centigrade for short periods of time. However, the amount of heat energy must be considered in relation to the total air volume in the specific cave along with other influencing factors. In dynamic caves, there is a pronounced natural exchange of air which means that even large numbers of visitors influence the temperature at a minimum. Tests have shown that even in static ice caves with a much lower level of air exchange, such as the Schellenberger Ice Cave in Germany and the Scărişoara Ice Cave in Romania, the influence of visitors on the temperature is minimal. An experiment in the Scărişoara Ice Cave showed that air temperature increased by ca. 2°C in the vicinity of a large (20+ persons) group of visitors. Despite the temperature increase, the influence was seen only at a distance less than 2 m, and it disappeared within minutes of the group's departure. In days with 1000+ visitors, no increase in air temperature was noticed. This is, perhaps, because the heat realized by the human bodies dissipated toward the ceiling. The temperature levels inside the cave are influenced to a significantly greater extent by natural causes such as warm rain water.

It should also be noted that worldwide global warming influences cave climates in that, similar to surface ice in glaciers, subterranean ice levels have consistently decreased with enhanced rates of ice melting over the past decade.

10.5.3 INFLUENCE OF DEVELOPMENTAL MEASURES ON THE BIOLOGICAL SYSTEM OF CAVES

Ice caves do not have a particularly sensitive biological system because their natural temperature level is around or just under freezing point. Consequently, the general recommendation to avoid the installation of biological materials (specifically wooden pathways, etc.) was not extended to ice caves.

10.6 FURTHER ASPECTS OF CAVE MANAGEMENT

10.6.1 SAFETY

Show cave operators, with an exceptionally high level of liability, regard visitor safety as extremely important at all stages. Arrangements for preventing accidents include the safety of pathways with regard to damage that might be incurred through slipping, stumbling, or falling. Special attention should also be paid to lighting, to alternatives for turning around, to alarms, and to equipment for aid and assistance. In ice caves, special consideration must also be given to ice formations on the ceiling above the area where the guided tours are operating. Dangerous ice formations must be removed in a timely manner as to prevent potential hazards from falling ice features.

10.6.2 TRAINING

Cave guides face a number of challenges with the most significant obstacles being as follows: physical fitness, language abilities (pronunciation/foreign languages), subject knowledge relating specifically to the cave in question, and a general knowledge of geology and climate. In order to meet these challenges, guides must be constantly trained and monitored.

10.6.3 MARKETING

An integral part of the operating business is ensuring good presentation to the public and organizing a wide variety of marketing measures; thus being able to position the product successfully alongside the many other offerings for leisure activities.

10.7 SCIENTIFIC RESEARCH

In order to ascertain the development of the climate within show caves, it is imperative to collect sufficient quantities of measurements and data. Observations made over longer periods of time provide us with a sufficient basis for the prediction of future developments in ice caves. Scientific studies are used both to determine the age of the ice deposits and to analyze the substances found therein so that we can draw conclusions about climatic developments in the past.

Show caves tend to provide an excellent opportunity for permanent data gathering. Guides knowledgeable about the studied phenomena regularly visit the caves, and thus can provide their own managers with data to better nurture "their" caves.

The three largest and most visited show caves hosting underground ice bodies (Eisriesenwelt in Austria, Dobšiná Ice Cave in Slovakia and Scărişoara Ice Cave in Romania) have been subjected to scientifc studies since the early 1920s. These studies, summarized in the relevant chapters in this book, deal with cave climate, ice morphology and dynamics, and the palaleoclimatic significance of the ice.

Recently, close cooperation between scientists and managers has led to studies tailored toward the needs of both groups, with the results discussed at dedicated meetings (e.g., Scientific Research on Show Caves, organized by Park Skocjanske jame, Grotta Gigante S.A.G., Karst Research Institute ZRC SAZU, and Universita degli Studi di Trieste in Postojna, in 2012) and in publications (e.g., Schauhöhlen und Wissenschaft in Österreich, "Speldok-15," by the Department of Karst and Cave Research at the Natural History Museum in Vienna).

10.8 POSTSCRIPT

The unique nature of caves provides full justification for making them accessible to both explorers and interested members of the public. In order to make these natural phenomena accessible, developmental measures are necessary. However, these procedures can be managed in such a way as to minimize their effect on nature. The existence of show caves gives us a unique opportunity to provide a "window" for the public to enjoy wonderful natural phenomena from both a visual and scientific point of view all the while increasing their awareness and understanding of cave and karst exploration. The tasks listed here, being part of successful and responsible show cave management, are intended to define the key aspects as it pertains to the authors view and are not intended to be exhaustive.

FURTHER READING

Beckert, R. Results About Eisriesenwelt Scan Comparison, University of Dresden, Germany.

Results of IWIC – International Workshop on Icecaves, Held Every Two Years Since 2004, as a Special Committee of UIS. http://geomatic.disat.unimib.it/home/valter/icecave/index.html.

Perşoiu, A. Pictures and Graphic Concerning Scarisoara Cave, Emil Racovita Institute of Speleology, Romanian Academy Clinicilor, Cluj Napoca, Romania.

Spötl, C., Wagenbach, D., Obleitner, F., May, B., Behm, M., Schöner, W., Hausmann, H., Pavuza, R., Thaler, K., Schöner, M., 2008. AUSTRIA*ICE*CAVES*2100.

PART

II

ICE CAVES OF THE WORLD

CHAPTER

GEOGRAPHY OF CAVE GLACIATION

11

Bulat R. Mavlyudov

Institute of Geography of the Russian Academy of Sciences, Moscow, Russia

Cave glaciation (perennial or temporary) is a wide-spread phenomenon on Earth. For example, approximately 10% of all karst cavities known in the former USSR have been subjected to cave glaciation. Outside permafrost areas, an even higher proportion of glacier caves have developed in selected regions: on Bzyb Ridge in the Caucasus, that proportion reached 50% (Dmitriev and Chujkov, 1982) and in northern Albania, more than 40% (Stoev et al., 1997). Similar situations are possible in other mountain areas. Short-term and seasonal cave glaciation occurs in practically all areas where there are negative air temperatures. Despite numerous descriptions of snow and ice in cavities, the geographical "rules" of cave glaciation are not yet fully investigated, because cave glaciation depends on many factors with complex interactions. Nevertheless, some attempts to reveal the systematics of cave glaciation have been undertaken, and we will examine them in more detail.

Cave glaciation systematics are revealed in different ways. It has been noticed that the air temperature outside caves is a major factor influencing cave glaciation. In fact, the colder the climate, the greater the probability of having permanent ice in caves. Therefore, as we move in the direction of the Earth's poles, the probability of finding cave glaciation increases.

Cave glacial boundaries can be determined in different ways. The simplest way is to delineate regions that host caves with permanent ice. Such analysis has shown that in European Russia, the southern boundary of cave glaciation is located at approximately 50°N (Golod and Golod, 1975). Another way to determine boundaries of cave glaciation is to use the analysis of positive and negative temperatures distribution (Harris, 1981), as done for different regions in Canada (Fig. 11.1).

Caves with ice in Canada lie between a southern boundary with sporadic permafrost and a northern boundary with continuous permafrost; thus caves with permanent ice have a MAT isotherm ranging from 0°C to 6°C at their southern boundary. Most of these caves are in sporadic permafrost zones. However, this method does not firmly define cave glaciation boundaries inside this zone. In areas where caves with permanent ice are located, attempts were made to find connections between the air temperature and the region's altitude and latitude. It has been shown that the boundary of caves with ice distribution in European mountains occurs at the 10°C isotherm (Karakostanoglou, 1989; Fig. 11.2).

Fig. 11.2 shows that caves with permanent ice are located mainly in fields in which the boundaries are parallel with air MAT isotherms of 0°C and 10°C, but they also can be found in areas with MAT isotherms up to −3°C. Nevertheless, this method does not enable us to make more detailed divisions of zones with ice caves.

Ice Caves. https://doi.org/10.1016/B978-0-12-811739-2.00012-7

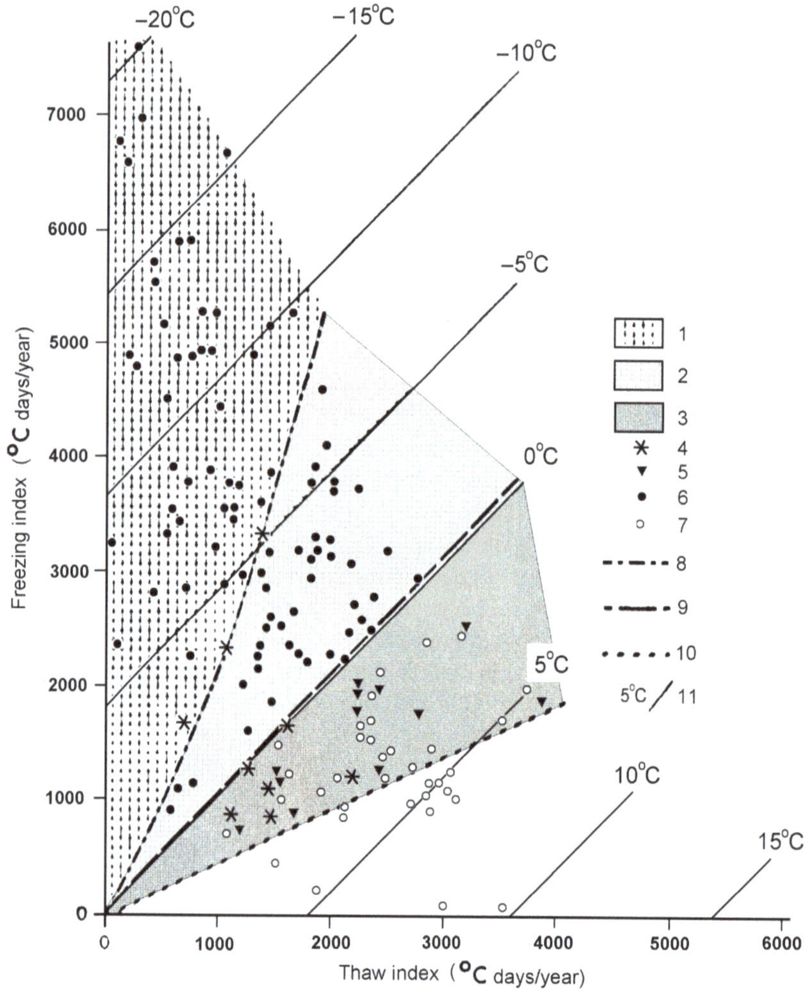

FIG. 11.1

Ratio of caves with permanent ice and zones of permafrost distribution, based on one example in Southern Alberta, Canada (Harris, 1981). Zones of permafrost distribution: (1) continuous, (2) discontinuous, and (3) sporadic; findings of ice: (4) permanent ice in caves, (5) ice under peat, (6) permafrost, and (7) permafrost absence; boundaries of permafrost zones distribution: (8) continuous, (9) discontinuous, and (10) sporadic; and (11) isotherms of mean annual temperature (MAT).

Previous attempts to determine zones of cave glaciation were not successful because the temperature of the rock massifs containing the caves was not considered. However, it is necessary to take this temperature into account for large areas, especially within continents.

Glaciation can occur in all caves where rock temperatures are constantly negative, whereas when the temperatures are positive, cave glaciation is possible only under certain conditions. These conditions occur in the winter when external air temperatures are low enough to produce negative temperatures

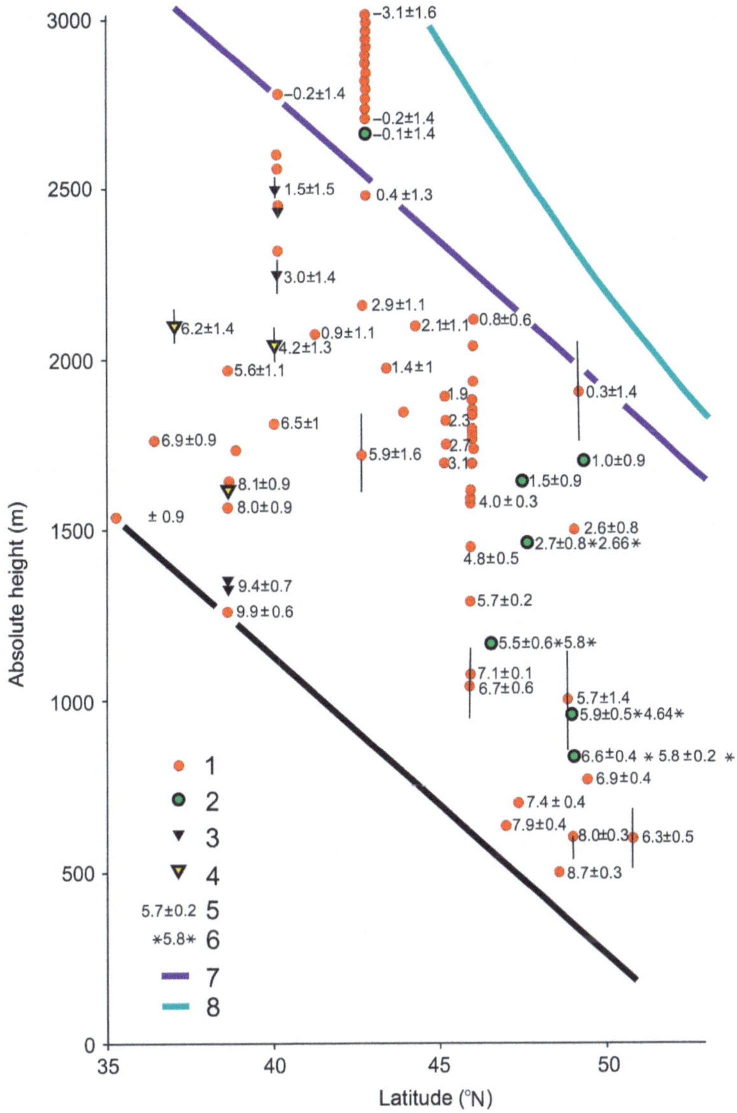

FIG. 11.2

Caves with permanent ice, depending on their altitudinal and latitudinal positions, based on one example of European caves with permanent ice (Karakostanoglou, 1989): (1) static caves with ice, (2) dynamic caves with ice, (3) dolines with permanent ice at bottom, (4) dolines with snow existing more than year, (5) air MAT outside of caves in °C (according to meteorological stations from 1950 to 1960), (6) air MAT measured directly in caves (on place), (7) snow line, and (8) isotherm of MAT=0°C.

in the rocks around the caves' passages through which the air moves. It is possible to use mathematical models of the thermal regime in caves to find an external winter air-temperature to rock-temperature ratio that is favorable for cave glaciation (Golod and Golod, 1974; Mavlyudov, 1985). These models overly simplify the situation in (horizontal) caves and require many assumptions; nevertheless, they give a reasonable picture of the interactions between external and cave temperature fields. The success of modelling is connected with the fact that ice in caves of this type forms in much smaller quantities than the caves' cold reserves allow.

The characteristics of interactions between external and cave temperature can be determined by analyzing the mean (January for the northern hemisphere and July for the southern hemisphere) minimal external winter air-temperature to rock-temperature ratio. This ratio can be calculated by using the cave glaciation index (CGI or *K*) (Mavlyudov, 1989a,b), which is based on the analysis of data retrieved from modeling calculations and observations in caves in different regions (see Chapter 4.1). CGI shows the possible degree to which a rock massif will cool in winter. The higher the CGI value, the greater the possibility that a cavity will cool sufficiently for glaciation to occur. Cave glaciation can develop in areas where the CGI values are from 0 to 1 or higher. The CGI value is 1 at continuous permafrost boundary where the temperature of the rock massif is equal zero ($T_m = 0$) and CGI is zero where the mean January air temperature is equal to zero. An analysis of the existing data on caves with ice shows that (1) in areas where CGI values are from 0 to 0.25, seasonal cave glaciation exists; (2) in areas where CGI values are from 0.25 to 1.0, permanent glaciation of caves also exists; and (3) in areas where CGI values are greater than 1, permanent glaciation of the majority of caves develops. It is also possible to find ephemeral ice at the entrance of some caves in areas where negative temperatures are possible at any time.

Note that CGI boundary parameters are determined on the basis of generalized data of individual measurements. Unfortunately, meteorological measurements are done only in some caves with ice, therefore it is impossible to statistically check and compare cave glaciation boundaries.

Modeling thermodynamic calculations (Lykov, 1967) have been used to compute the quantitative values of the length of the negative temperature anomaly (NTA) zone in horizontal caves with a given morphology (Mavlyudov, 1985). The climatic conditions related to the length of the NTA zone in horizontal caves (Lukin, 1965) shows the main characteristics of cave glaciation, according to latitudinal zones and altitudinal levels, as outlined below.

With the use of thermodynamic calculations, a map of cave glaciation in an area in the former USSR has been constructed (see Fig. 11.3). Because the southern boundary of seasonal cave glaciation occurs at the zero isotherm of the mean January air temperature ($K=0$), it appears that cave glaciation is probably possible in almost all areas of the former USSR and that the northern boundary of seasonal glaciation in caves can be located at the zero isotherm of the mean July temperature. This boundary is connected with ice melting, which is possible only when the external air has positive temperatures, as is the case for most areas in the former USSR. The boundary of seasonal cave glaciation in mountains can be similarly attained, but boundaries can be located at different levels on slopes with different orientations. For example, in the Caucasus Mountains, the boundary for seasonal cave glaciation can be at an absolute height of about 500 m on a southern slopes, while on the northern slopes at 100 m. The upper limit of seasonal cave glaciation in the Caucasus can be located at an altitude of about 4000 m a.s.l.

The position of the southern boundary of permanent cave glaciation varies for cavities of different types. On the Russian Plain, the boundary for horizontal caves is $K=0.7$; for inclined, descending caves, is $K=0.25$; and for vertical caves, is $K=0.3$. These boundaries reflect climatic limitations, only; as the topography also plays a role: in mountains, the boundary of permanent glaciation of inclined, descending caves is $K=0.25$, while on the Russian Plain, it would be necessary for the boundary to

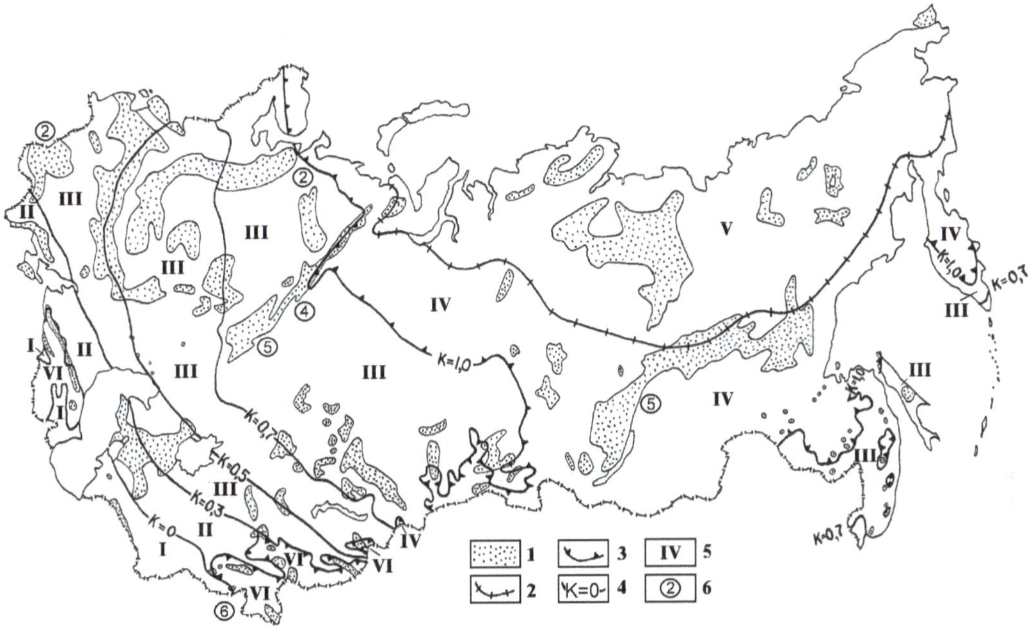

FIG. 11.3

Map of cave glaciation on area of the former USSR (Mavlyudov, 1989a). 1 – karst rocks; permafrost boundaries: 2 – continuous; 3 – discontinuous and sporadic; 4 – isolines of CGI and their values; 5 – number of cave glaciation zone (V – of all caves, IV – of majority caves, III – of separate caves, II–III – seasonal cave glaciation, I – cave glaciation is absent); 6 – difference (a) between MAT and temperature of "neutral zone" of caves.

be located at $K=0.5$ (there are no orographic opportunities for cave glaciation on the plain, as already at $K = 0.5$ the minimal depth of permanent ice in caves should be no less than 25 m). With the reduction of the CGI value, this depth will grow, and cave glaciation in the plain can't develop (for natural cavities). Determination of permanent glaciation boundaries for vertical cavities is important only for mountain areas where these cavities can exist.

According to the CGI, it is possible to estimate temperature intervals favorable for cave glaciation. Table 11.1 shows the maximal values of the external mean January air temperature at the given air MAT at which cave glaciation is still possible.

Table 11.1 Extremely High Mean January Air Temperatures (°C) at Which Cave Glaciation is Possible

	MAT										
Cave Types	**−2.0**	**−1.0**	**0.0**	**1.0**	**2.0**	**3.0**	**4.0**	**5.0**	**6.0**	**7.0**	**8.0**
Horizontal	−2.3	−4.7	−7.0	−9.7	−11.7	−14.0	−16.3	−18.7			
Inclined			−1.0	−1.3	−1.7	−2.2	−2.7	−3.0	3.3	−3.7	−4.0
Vertical			−1.3	−1.7	−2.1	−2.6	−3.0	−3.4	−3.9	−4.3	−4.7

The cave glaciation limit for horizontal caves in the Russian Plain is located at an isotherm of external MAT of about +4.0°C. For inclined, descending caves, the limit is located at an isotherm of about +10.0°C, which is comparable to the boundary of cave glaciation for inclined caves in Europe with a similar isotherm of +10.0°C (Karakostanoglou, 1989). In continental climates, such as in East Kazakhstan, a limit of distribution of inclined caves with ice will be at isotherm +12.0°C. For vertical cavities in which snow accumulates in winter, the temperature is not significant, but according to the available data, it probably does not exceed +6.0°C (though local exceptions are possible).

On the basis of the preceding information and also using data of air temperatures taken from the climate directories of the USSR (Spravochnik, 1966–1972) a map (cave glaciation of the former USSR) (Mavlyudov, 1997) was made with the following fields: glaciation of all caves, constant glaciation of caves, and seasonal glaciation of caves (Fig. 11.3, Table 11.2).

Table 11.2 Zones of a Cave Glaciation

Name of Zone	CGI
1. Glaciation of all caves	>1.0
2. Permanent glaciation of some caves:	
(a) Inclined and vertical	0.25–1.0
(b) Horizontal	0.7–1.0
3. Seasonal cave glaciation:	
(a) Inclined and vertical	0,0–0,25
(b) Horizontal	0.5–0.7

It is important to note that seasonal cave glaciation is developed in a cave's zone of constant glaciation; that is, these zones overlap (Fig. 11.4).

An analysis of boundaries of cave glaciation zones drawn on the map shows that, as a whole, they are positioned close in latitude but are dependent on their position of continentality (difference between continental and marine climates—for example, the Baltic's (northern former USSR) seaside climate has mild, warm winters, and Kazakhstan's (southern former USSR) continental climate has long, cold winters). Therefore, within the limits of the West Siberian Plain and the Relief of Kazakh with small knolls, constant glaciation of horizontal caves is possible in significant areas to the south with a permafrost boundary up to 48°N, and in the western part of the European Plain possible glaciation of horizontal caves is limited south at 62°N (Mavlyudov, 1989a,b). In mountain areas, boundaries of cave glaciation zones are situated similarly to boundaries in the plain areas, but the boundaries vary for different mountain ridges (Table 11.3, Fig. 11.5).

A zone of permanent glaciation of some caves (in particular, horizontal caves) can be detailed independently of the length of cave passages with constant negative temperatures (zone of constant glaciation, Mavlyudov, 1985).

For areas in Western Europe, the position of the southern and lower boundaries of caves with constant ice distribution has been shown to correspond to the +10°C isotherm (see Fig. 11.2) (Karakostanoglou, 1989) according to the following equation:

$$T°C = 37.43 - 0.52L(N) - 0.06A, \tag{11.1}$$

where $L(N)$ is northern latitude (degrees) and A is absolute altitude (m).

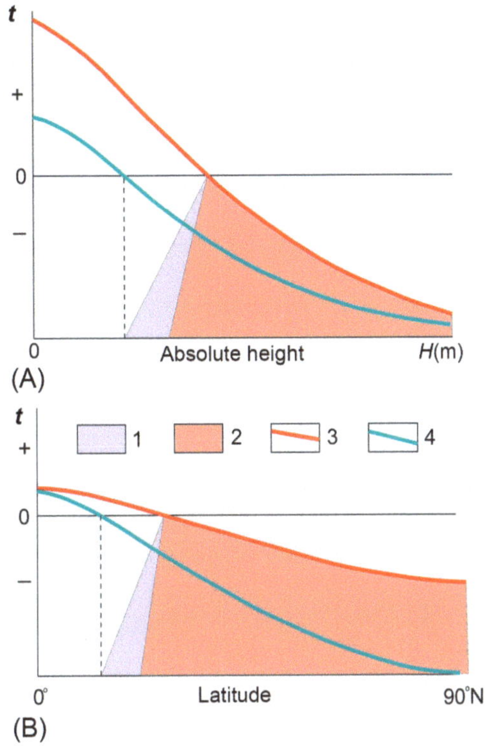

FIG. 11.4

Overlapping of zones of seasonal cave glaciation (1) and constant cave glaciation (2) depending on absolute height (A) and latitude (B) of districts (scheme); (3) mean July temperature and (4) mean January temperature.

Table 11.3 Boundaries of Cave Glaciation Zones in Mountains in m a.s.l.

Zone Number[a]	Caucasus		Tien-Shan			
	Western	Eastern	Northern	Eastern	Central	Pamir
1	3000	3600	3250	3100	3600	3750
2a	1150–3000	1350–3600	250–3250	1000–3100	1850–3600	1850–3750
2b	2250–3000	2700–3600	2250–3250	2500–3100	2850–3600	3000–3750
3a	550–1150	600–1350	0–250	0–1000	900–1850	1300–1850
3b	1750–2250	2100–2700	1350–2250	1900–2500	2350–2850	2500–3000

[a] *See Table 11.2.*

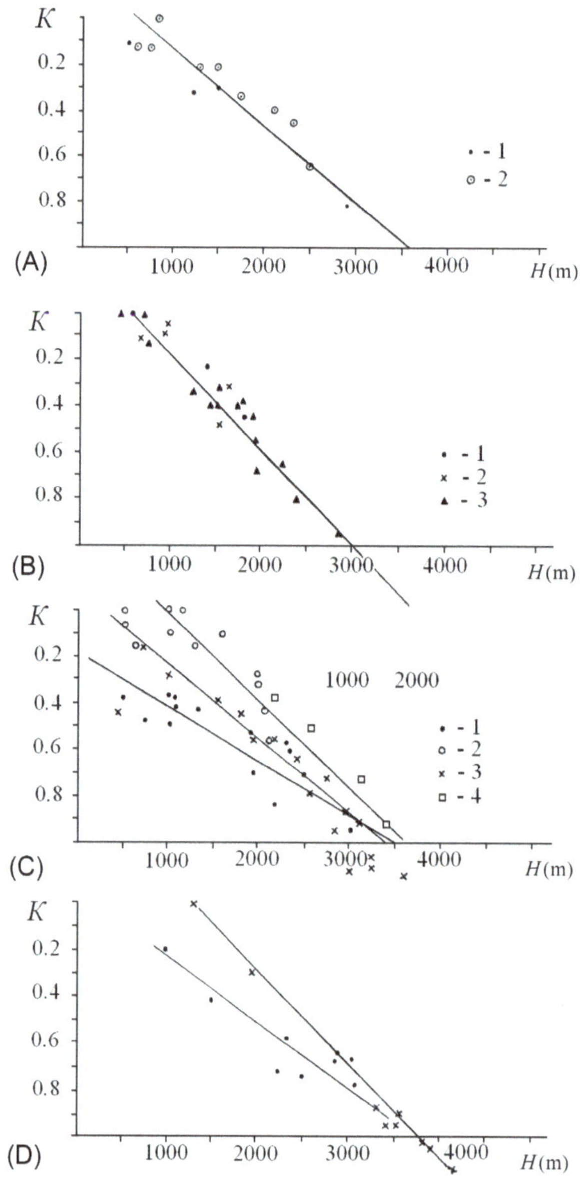

FIG. 11.5

Connection of CGI (*K*) and district heights (*H*): (A) East Caucasus—(1) Dagestan and (2) Azerbaidzhan; (B) Western Caucasus—(1) Krasnodar region, (2) Abkhazia, and (3) Georgia; (C) Tien-Shan—(1) Kazakhstan, (2) Uzbekistan, (3) Kirghizia, (4) Tajikistan; and (D) Pamiro-Alaj—(1) Kirghizia and (2) Tajikistan.

A comparison of the boundaries of cave glaciation zones with boundaries of possible zones for the construction of artificial ice storehouses of the Krylov system shows their general similarities (Glushnev, 1973; Kanaev and Chekotillo, 1952; Krylov, 1951; Shelokov, 1967), which is an indirect confirmation of the appropriateness of our construction.

Using the same method, we constructed a map of cave glaciation across the globe (Fig. 11.6). As the figure shows, we took the following differences between the temperature of a neutral layer (rock massif temperature) and the MAT: for an area of Asia, $a=3°C$; for an area of Europe, $a=0°C$; and for area of North America, $a=3°C$. It is clear that averaging of such large areas doesn't provide a real picture of cave glaciation distribution, but on the other hand, such averaging removes unnecessary details and smooths contours of the fields. However, to get a more exact estimation of a particular cave's glaciation conditions, it is necessary to use local MAT, T_{Jan}, and a.

Considering the map of cave glaciation (Fig. 11.6), we can say that it appears that the main areas where cave glaciation can exist are in North America and Eurasia.

In North America, the southern boundary of cave glaciation is not uniform. In the eastern part of the country, the boundary is located approximately 40°N; in the western part, in the area of the Rocky Mountains, the boundary reaches down to 35°N and then sharply turns north up to 50°N. Therefore the strip of cave glaciation distribution in North America outside of the permafrost zone ranges from 1000 to 2000 km.

In Eurasia the southern boundary of cave glaciation distribution is 25°N, but it is associated with the location of the highest mountain complex in the world (Pamir, Tien-Shan, Hindu Kush, and Himalayas). The air in this region typically has very low humidity, so not all caves contain ice, but rocks in cave channels are in a frozen condition. For example, despite negative air and rock temperature and an entrance at a height of about 4600 m, only small ice quantities can be found in the Rangkulskaya Cave in Pamir. Caves with ice in the Himalayas are also known. For example, there is the famous Amarnath Cave in the Kashmir, which is located at a height of about 3888 m. Large ice stalagmites (up to 3 m in height) in this cave melt only at the end of August. In Turkey and Iran, conditions for ice caves exist within the limits of high-mountainous areas. Therefore the boundary of possible cave glaciation on our map is located at approximately 30°N. Despite the fact that there are favorable climatic conditions for cave glaciation in all northern Japan, we find them only at the foot of the Fuji Mountain (volcanic caves) and in the north on Honshu Island. But because of unfavorable conditions (high thermal flux from the ground and absence of large descending caves) at the north of Japan only caves with seasonal glaciation are found (E.V. Isenko, personal communication).

In the greater part of Eurasia, the southern boundary of seasonal cave glaciation crosses from Japan to the Mediterranean Sea between 25°N and 35°N, turning sharply north, after rounding the Alps (Thus all Scandinavia is situated in a zone with climatic parameters favorable for cave glaciation).

Because of its climate, Australia is almost completely free of cave glaciation, with the possibility for only seasonal cave glaciation in the high mountains. Cave glaciation could occur in the mountains on the South Island in New Zealand, where glaciers exist. The majority of the areas in Africa are not favorable for cave glaciation, with the only exceptions being Mount Kilimanjaro and the Atlas Mountains. There are climatic conditions for cave glaciation on Mount Kilimanjaro, but only for the Atlas are these conditions realized as pits with long-term snow accumulation.

Cave glaciation in South America has been poorly studied. However, in agreement with our map, it is possible to find cave glaciation at places in the upper parts of the Andes Mountains, and some years ago, we actually received confirmation of caves with ice at high altitude in South American.

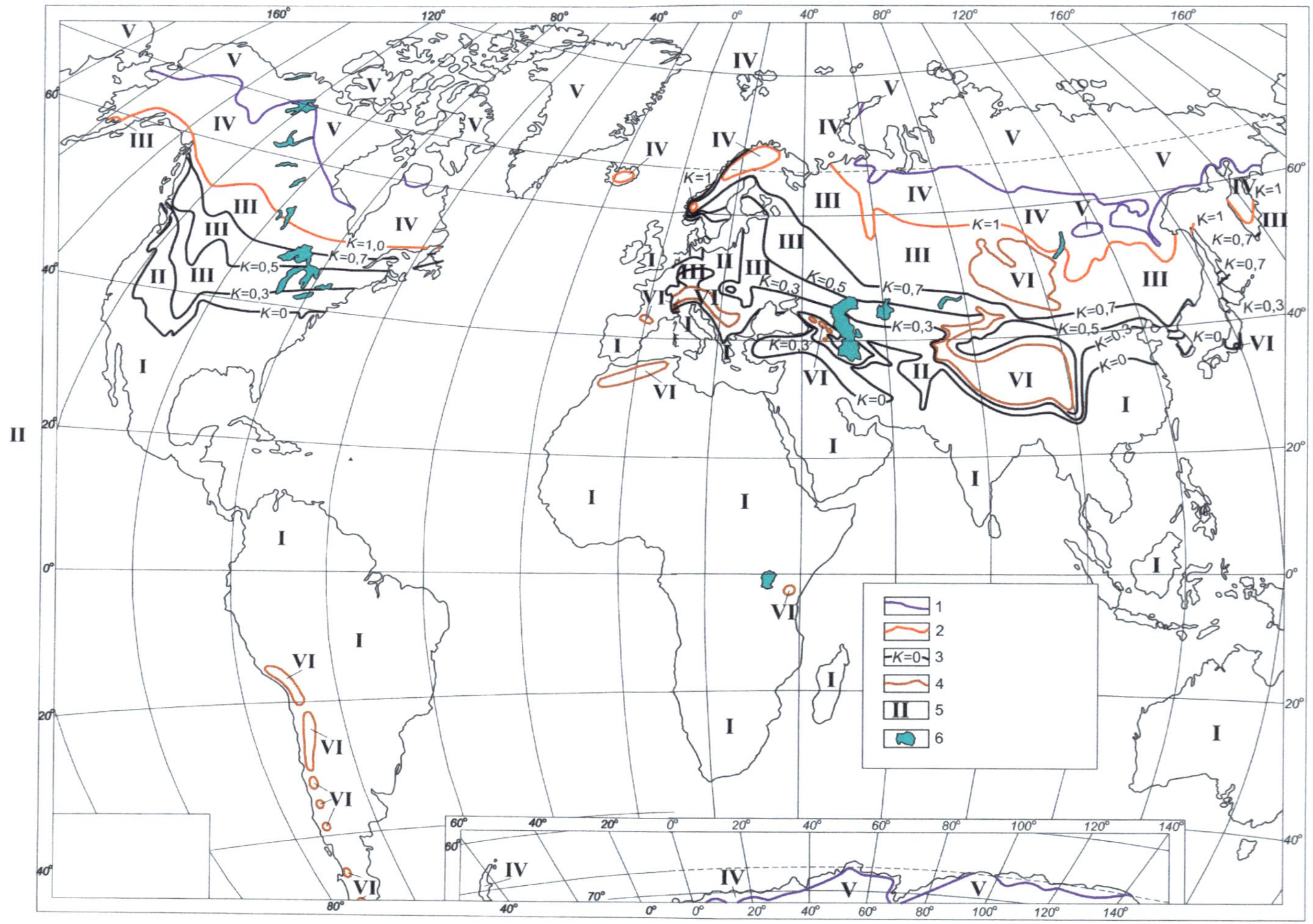

FIG. 11.6

Map of cave glaciation around the world. Boundaries of permafrost: (1) continuous, (2) discontinuous and sporadic, (3) contour lines of CGI and their values, (4) contour lines of CGI in mountains, (5) number of cave glaciation zone (V, of all caves; IV, of majority of caves; III, of separate caves; II–III, of seasonal cave glaciation; I, cave glaciation is absent); and (6) lakes (Mavlyudov, 2008).

For example, Qaqa Mach'ay Cave was researched in central Peru at a height of about 4930 m (https://en.wikipedia.org/wiki/Qaqa_Mach'ay). The cave originated as a glacial water absorber and was formed at the base of bedding crack in limestone. Caves with a significant quantity of permanent ice were investigated up to a depth of 125 m.

In Antarctica, conditions favoring cave glaciation exist practically everywhere. Only in the northern part of the Antarctic Peninsula does seasonal cave glaciation occur. Our research on King George Island in Antarctica from 2007 to 2017 showed that cave ice can occur only in glacial cavities, because other cavities on the island are almost absent. Cave ice can be seasonal and exist less than 1 year. It is presented usually as icings, frost crystals, and frozen rocks. There is no doubt that if any caves in Antarctica are found, they will be filled partly or completely by ice.

The comparison of boundaries drawn on a map of caves with ice and real caves distribution shows quite good similarity for all regions. It is possible to explain that local discrepancies are due to local features of cave glaciation. So, for horizontal caves in the Alps (Eisrisenweld, Dahstein), the glaciation origin at MAT at the lower entrance is about 5°C (Saar, 1956). It is connected with large elevation differences between cave entrances. As modeling calculations were done on the differences between entrances of about 90 m, it became clear that in the Alps at a difference of more than 300 m, conditions for the cavities' ventilation will be essentially best and that the glaciation can occur at high winter air temperatures outside of the caves.

Figs. 11.3 and 11.6 include the same areas of the former USSR but the position of the same boundaries on both maps does not always coincide, which is connected with different data sources that lay at the base of the two maps. For the map shown in Fig. 11.3, we used data of all meteorological stations available for areas of the former USSR from 1960 to 1980 whereas for the map in Fig. 11.6, we used maps of the MAT and average January air temperatures presented in published works (Fisiko, 1964; Gavrilova, 1998). For a geographical analysis of cave glaciation in area of the former USSR, it is preferable to use boundaries shown on Fig. 11.3.

Our experience of constructing a cave glaciation map showed that the CGI has enough generality for the construction of regional and global maps of cave glaciation, in both plains and mountains. In addition, it appears that CGI can be used to construct paleogeographical maps of cave glaciation in the past. Such maps can be constructed if the values of the MAT, T_{Jan}, a, deviations of temperature of the coldest month and the MAT between the modern values and the corresponding past date values are known (e.g., Mavlyudov, 2008).

REFERENCES

Dmitriev, V.E., Chujkov, V.D., 1982. Geograficheskie zakonomernosti oledeneniya pesher (Geographical regularities of caves glaciation). In: Sostoyanie, zadachi i metodyizucheniya glubinnogo karsta of the USSR. (Condition, Tasks and Methods of Deep Karst Studying in the USSR), pp. 136 Moscow (In Russian).

Fisiko-geograficheskij atlas mira (Physico-Geographical Atlas of the World). 1964. Academy of Sciences of the USSR, Moscow. 298 p. (In Russian).

Gavrilova, M.K., 1998. Klimat kholodnykh regionov Zemli (Climate of Cold Regions of the Earth). Publishing House of the Siberian Branch of the Russian Academy of Science, Yakutsk. 206 p. (In Russian).

Glushnev, M.P., 1973. Bezmashinnye kholodilniki, [Refrigerators without machines]. Research Institute of Agriculture of Northern Zauralye, Moscow, 65 p. (In Russian).

Golod, V.M., Golod, M.P., 1974. Microclimate of gypsum caves of Pinedzie. In: Peshery Pinego-Severodvinskoj karstovoj odlasti. [Caves of Pinego-Severodvinsk Karst Area], Leningrad, pp. 128–154. (In Russian).

Golod, V.M., Golod, M.P., 1975. Analiz formirivaniya ledyanykh pesher, [Analysis of ice caves formation]. In: Usloviya i sadachikarstovo-speleologicheskikh issledovanij. Conditions and Problems of Karst-Speleological Researches, Moscow, pp. 37–38. (In Russian).
Harris, S.A., 1981. Climatic relationships of permafrost zones in areas of low winter snow-cover. Arctic 34 (1), 64–70.
Kanaev, A.F., Chekotillo, A.M., 1952. Ledyanye sklady i ikh ispolsovanie (Ice Storehouses and Their Use). Publishing house of the Academy of Sciences of the USSR, Moscow. 112 p. (In Russian).
Karakostanoglou, I., 1989. Some remarks on the genesis and geography of static ice shafts. In: International Congress of Speleology, 13–20 August 1989. Proceedings. vol. III, pp. 713–715.
Krylov, M.M., 1951. Ledyanye izotermicheskie sklady, [Ice Isothermal Storehouses]. Publishing House of USSR Academy of Sciences, Moscow, 88 p. (In Russian).
Lukin, V.S., 1965. Temperatyrnye anomalii v pesherakh Predural'ya i kriticheskij analis teorij peshernogo kholoda (Temperatures anomalies in caves of Predural and critical analysis of underground cold theories). Peshery (Caves) 5 (6), 164–172 (In Russian).
Lykov, A.V., 1967. Teoriya teploprovodnosti (Theory of Heat Conductivity). Vyshaya Shkola, Moscow. 599 p. (In Russian).
Mavlyudov, B.R., 1985. Geograficheskie zakonomernosti rasprostraneniya pesher so l'dom (Geographical regularities of caves with ice distribution). Materialy Glaciologicheskikh Issledovanij (Data of Glaciological Studies) 54, 193–200 (In Russian).
Mavlyudov B.R., 1989a. Oledenenie pesher Sovetskogo Souza [Caves glaciation of the Soviet Union]. Abstract of PhD thesis. Moscow. 23 p. (In Russian).
Mavlyudov, B.R., 1989b. Caves glaciation. In: International Congress of Speleology, 13–20 August 1989. Proceedings. vol. 1, pp. 298–300 (In English).
Mavlyudov, B.R., 1997. Map «Caves glaciation of the USSR» (Scale 1:40000000). In: Kotlyakov, V.M. (Ed.), Atlas of Snow-Ice Resources of the World. Institute of Geography of the Russian Academy of Science or Institute of Geography RAS, Moscow, pp. 356. (in Russian).
Mavlyudov, B.R., 2008. Geography of caves glaciation. In: Proceedings of 3 International Workshop on Ice Caves, Kungur, pp. 38–44.
Mavlyudov, B.R., 2008. Oledenenie pesher (Caves Glaciation). Institute of Geography of the Russian Academy of Science, Moscow, 290 p. (in Russian).
Saar, R., 1956. Eishohlen, ein meteorologisch-geophysikalisches phanomen. Geogr. Ann. 38 (1), 1–63.
Shelokov, V.K., 1967. Ledyanye sklady (Ice Storehouses). Nauka, Moscow. 119 p.
Spravochnik po klimatu USSR (Climate Directory of the USSR). vols. 1–34. 1966–1972. Leningrad, Moscow part. 2 (in Russian).
Stoev, A., Muglova, P., Stoev, D., 1997. Temperature anomalies formation and secular instability research of ice of atmospheric origin in karst caves of North Albanian Alps. In: Proceedings of 12th International Congress of Speleology, La-Shau-de-Font. vol. 1, pp. 207–210.

CHAPTER

ICE CAVES IN SWITZERLAND

12

Marc Luetscher*,†, Pierre-Yves Jeannin†

Austrian Academy of Sciences, Innsbruck, Austria Swiss Institute of Speleology and Karst Studies (SISKA), La Chaux-de-Fonds, Switzerland†*

CHAPTER OUTLINE

12.1 INTRODUCTION

Switzerland comprises about 19% of karst covering a total surface area of more than 7900 km^2 (Wildberger and Preiswerk, 1997). Karst is primarily present in Mesozoic carbonate formations of the Jura Mountains and Helvetic nappes, although occasional features are reported from gypsum outcrops in the canton of Valais (Fig. 12.1). To date, close to 9000 caves have been added to the speleological inventory, totaling over 1300 km of surveyed cave passages, with the largest speleological network developing more than 200 km. Caves are distributed between 500 m and 3500 m above sea level (a.s.l.), covering a large array of climate zones. Overall, the climate is heavily influenced by the Atlantic Ocean, with westerlies delivering a mild and humid sea breeze to Switzerland (Meteoswiss, 2017). The average amount of precipitation ranges between 1500 and 2000 mm/year along the northern foothills. The mean annual air temperature at 1000 m a.s.l. typically averages 6.3°C (1981–2010) with monthly temperatures ranging between −2 and 15°C. During the winter season, snowfall is common above approximately 1200 m a.s.l., leading to a solid layer of snow between November and May.

The morphology of Alpine caves typically reflects a combination of vadose systems characterized by vertical shafts and meanders and (paleo-)phreatic cave passages. The vertical extension of these complex speleological networks commonly reaches several hundreds of meters, favoring intense

Ice Caves. https://doi.org/10.1016/B978-0-12-811739-2.00010-3

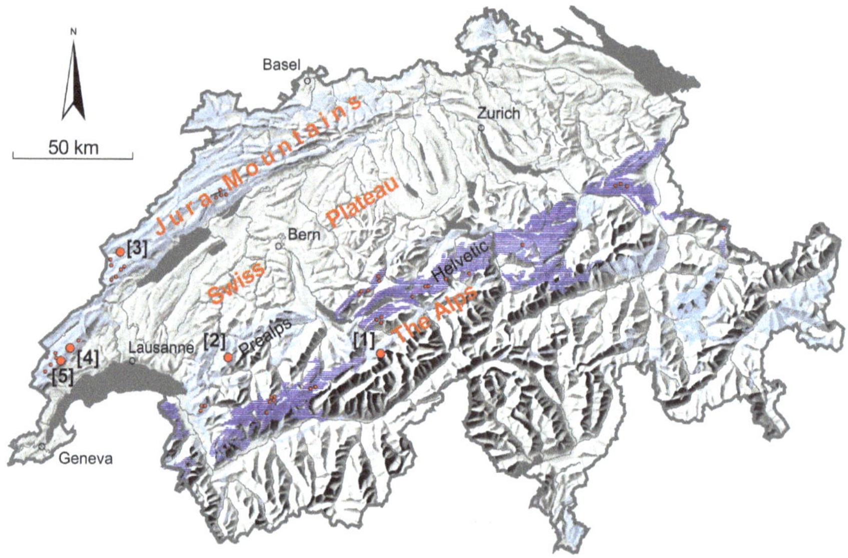

FIG. 12.1

Karst areas of Switzerland *(blue)* with some of the most significant ice caves *(red circles)* and some selected study sites mentioned in the text [1] Jochloch; [2] Diablotins ice cave; [3] Glacière de Monlési; [4] Glacière de St-Livres; [5] Glacière de Saint George. *Shaded areas* delineate high alpine karst with frequent snow pits (elevation model: Swiss Federal Office of Topography).

ventilation processes through forced convection. Beyond the cave entrance zone, still subject to seasonal temperature fluctuations ("heterothermic zone," sensu Luetscher and Jeannin, 2004b), the cave temperature reflects the surface mean annual air temperature at similar elevation. Cave monitoring including air and rock temperatures reveals the presence of permafrost above approximately 2300 m a.s.l. Although the number of high-elevation caves is limited, these sites are typically characterized by the absence of seepage water. There, cave ice is primarily present in form of hoarfrost which may reach up to 20 cm in thickness deposited on the cave walls (e.g., Jochloch, 3470 m a.s.l.; Häuselmann, 2004).

As soon as water finds its way through unfrozen conduits, congelation ice may form in the cold cave environment. Paradoxically, this process leads to the formation of cave ice, despite the progressive thawing of the host-rock, and has been reported from elevations as low as 2300 m a.s.l. (Borreguero et al., 2009). Below, the cave air temperature in the deep karst is generally positive, hindering the preservation of cave ice. The presence of ice caves s.l., therefore, denotes a local temperature anomaly associated with specific ventilation patterns (Luetscher and Jeannin, 2004b).

In a dynamically ventilated system with multiple cave entrances, cold air is drawn in at the lower entrance during the winter season, allowing for the congelation of seasonal water inlets. This ventilation process is reversed during the summer season, but maintains a thermal anomaly due to the difference in specific enthalpy between in- and outflowing air. In a preliminary model estimates, Luetscher (2005) suggested that at least 145 freezing days would be necessary for maintaining congelation ice at the lower entrance of a dynamic cave. In Switzerland, this would correspond to an elevation of approximately 1600 m a.s.l. under average climate conditions.

While perennial snow shafts and congelation ice formations may be a common feature at high elevation, they are rather unusual at lower elevation, where the mean annual air temperature is typically well above 0°C. There, ice caves are commonly associated with cold air trapping in single-entrance caves, leading to a thermal stratification during the summer season (Luetscher and Jeannin, 2004a). Such perennial cave ice formations are particularly spectacular in the Jura Mountains where they have been mentioned in numerous reports over the last few centuries.

12.2 HISTORICAL CONSIDERATIONS ON ICE CAVES

Subsurface ice features in the Jura Mountains were first mentioned in manuscripts of the 16th century (e.g., AEN, 1554; Poissenot, 1586), but archaeological remains suggest that at least some of these ice caves were known already during the Roman epoch (Girardot and Trouillet, 1885). Given the proximity of nearby settlements, cave ice was frequently exploited for personal use or regional trade by shepherds or local inhabitants. Browne (1865) reports that during the year 1863, 29 days of cave work allowed the extraction of approximately 40 m^3 of ice from St-Livres ice cave (Fig. 12.2) and probably twice as much from St. George ice cave. While some of this cave ice was likely sold in Lausanne and Geneva (Pictet, 1822), such volumes certainly remained negligible with respect to the ice extracted from the nearby Lake Brenet during winter. These caves nonetheless offered an original source of ice during the hot summer months. The cave ice extraction progressively stopped at the beginning of the 20th century with the ascent of refrigerators, but historical remains can still be observed (Dutruit, 1991; Hapka, 2002).

FIG. 12.2

Historical engraving of the Glacière de St-Livres (Browne, 1865). The cave has been exploited for its cave ice during the 19th century.

Easily accessible and well known by native people, ice caves of the Jura Mountains have been for hundreds of years the object of detailed descriptions (e.g., Boisot, 1686; Billerez, 1712). Recognizing these ice caves as a particular phenomenon, the interest of naturalists grew progressively throughout the 18th century, and temperature measurements were soon carried out in different caves (De Cossigny, 1750; Prévost, 1789). Hypotheses on ice formation were sometimes the object of vigorous discussions (Deluc, 1822), and the first consistent scientific explanations for the presence of cave ice based on field observations are attributed to Thury (1861), who underlined the major role of cold air circulation for the presence and the conservation of subsurface ice deposits. This abundant literature soon raised international interest (e.g., Browne, 1865; Balch, 1897) and encouraged detailed investigations in neighboring countries (e.g., Fugger, 1891; Bock, 1913; Saar, 1956). In comparison, systematic investigations of Swiss ice caves remained rare (Stettler and Monard, 1960), and it is only at the start of the 21st century that this topic gained renewed interest (Luetscher, 2005).

12.3 STATE OF KNOWLEDGE

Reviewing the main characteristics of ice caves across Switzerland, Luetscher and Jeannin (2004a) considered two main processes to classify ice caves according to (1) the ventilation dynamics, and (2) the cave ice petrography. Consequently, detailed field studies were initiated to understand and quantify mechanisms allowing for the formation and preservation of cave ice. Comparing historical surveys, photographic documents and field observations, Luetscher et al. (2005) concluded that most low-elevation ice caves in the Jura Mountains face a negative mass balance that increased markedly at the end of the 1980s. Detailed monitoring conducted at Monlési ice cave underlined the dominant role of winter air circulation in the energy balance of ice caves (Luetscher et al., 2008), suggesting that the decline of cave ice at low elevation primarily reflects increasing winter temperatures, together with reduced snow precipitation. Similar conclusions were drawn by Morard et al. (2010), who showed the predominant role of winter atmospheric conditions in controlling the seasonal cave-ice mass balance at Diablotins ice cave, in the Prealps. These results are consistent with major ablation phases identified from organic rich ice layers at St-Livres (Stoffel et al., 2009) which are interpreted to reflect contrasting regimes of winter moisture advection from the North Atlantic.

While a negative cave-ice mass balance can be largely attributed to the absence of annual accumulation, the progressive increase of the host-rock temperature due to climate change inexorably raises the conductive heat exchange at the ice-rock interface. Luetscher et al. (2008) anticipated, therefore, a rapid vanishing of the smallest cave ice deposits at low elevation, although successive cold-dry and wet-mild episodes may temporarily favor the formation of congelation ice. However, little information is available with respect to the range of actual mass turnover rates. A multi-parameter dating approach applied to Monlesi ice cave (Luetscher et al., 2007) demonstrated that at this site the basal cave ice was only between 120 and 158 years old. In contrast, dendrochronological studies supported by radiocarbon dates have revealed a cave ice sequence of approximately 1200 years at St-Livres ice cave (Stoffel et al., 2009), opening thrilling perspectives for studying ice caves as novel paleoenvironmental archives. But the evidence of former cave-ice deposits also has the potential to provide information about past climates (Luetscher, 2013). In particular, cryogenic cave carbonates found in presently deglaciated caves may indicate the presence of past cave ice deposits formed under different climate conditions.

12.4 SELECTED ICE CAVES

Among all the ice caves known in Switzerland, only a few sites were the subject of detailed investigations. Here we present the most significant study sites, giving an overview on the contrasting environments and thermodynamic context where cave ice is to be found.

12.4.1 JOCHLOCH—A HIGH-ELEVATION CAVE WITHIN THE PERMAFROST ZONE

At 3470 m a.s.l., Jochloch is the highest cave in Western Europe. A mean annual air temperature of −7.2°C favors the presence of continuous mountain permafrost which has been independently confirmed by borehole measurements (Permos, 2016). The cave was discovered in 1983 during construction of the Jungfraujoch train-station and opens in a narrow band of metamorphic limestone of metamorphic carbonates at the contact with the Aare crystalline rock (Häuselmann, 2004). The cave is an approximately 100 m long passage, the original morphology of which has been largely overprinted by frost shattering.

Monitoring between 2015 and 2017 revealed a nearly constant cave air temperature of −2.0°C. This is slightly warmer than the surrounding host-rock and consistent with a weak but distinct air flow transporting sufficient moisture to form hoarfrost on the cave walls (Fig. 12.3). The absence of congelation ice, however, suggests the cave was hydrologically inactive for a long time. U/Th dating of a small flowstone coating the cave wall suggests an open system behavior with respect to uranium (Moseley, personal communication) hindering any further chronological constraint. Although very much deteriorated, pollen from the cave sediment revealed the presence of *Carya*, possibly hinting towards an Early Pleistocene age (Groner, 2004).

FIG. 12.3

Hoarfrost at Jochloch, a cave located at 3470 m a.s.l. in the Alpine permafrost.

Photo courtesy of R. Shone.

12.4.2 DIABLOTINS ICE CAVE—A DYNAMICALLY VENTILATED ICE CAVE

The Gouffre des Diablotins (N 46°32′09″, E 07°09′43″, 2092 m a.s.l.) is located in the Vanil Noir massif of the Fribourg Prealps (Bovey, 1995). The mean annual air temperature recorded at the nearby Meteoswiss station Le Moléson (1974 m a.s.l.; 1981–2010) is 3.0°C with negative temperatures being recorded more than 170 days a year. February is typically the coldest month with the average daily temperature minima reaching −6.2°C.

The shaft system has a vertical extension of more than 650 m (Bovey, 1995). Two main entrances give access to the cave: while the upper one opens on top of the Rochers des Tours at an elevation of 2092 m a.s.l., the lower one opens at 2007 m a.s.l in a north-oriented rock face (Fig. 12.4). Both entrances are linked together by a 105-m vertical shaft and a 50-m long subhorizontal cave passage reaching the lower entrance. The L-shaped geometry of the cave favors a strong chimney effect (forced convection) with an ascending air flow during the winter season that reverses when the external air temperature crosses a thermal threshold of about +2°C (Morard et al., 2010).

The cave hosts the most important ice infill in the Fribourg Prealps, with a volume estimated at approximately 100 m^3 (Morard et al., 2010). Congelation ice extends discontinuously along the lower cave passage and the base of the vertical shaft. However, speleological observations revealed rapid changes in the cave ice geometry (Jutzet, 1991; Bovey, 1995) which may be partly associated with a re-routing of the water drainage (Morard et al., 2012) impeding the air flow. A monitoring campaign in 2009–10 confirmed the essential role of the winter air circulation on the cave ice mass balance with sublimation representing the dominant mechanism for cave ice ablation. In contrast, melting during the summer season plays only a secondary role on the overall cave ice mass balance (Morard et al., 2010).

12.4.3 MONLESI ICE CAVE—A STATODYNAMIC ICE CAVE WITH CONGELATION ICE

The Glacière de Monlesi (6°35′4″/46°56′18″, 1135 m a.s.l.) is the largest ice cave in the Swiss Jura Mountains and is by far the most studied. The cave opens in a closed depression on the edge of a wooded area. The outside mean annual air temperature measured at the nearby Meteoswiss station of La Brevine (1050 m a.s.l.; 1981–2010) is 4.9°C and annual precipitations reach ca 1600 mm. Negative temperatures are recorded >180 days a year with daily temperature minima averaging −10.4°C in February.

Three entrance shafts lead at −20 m to a chamber 40 × 20 × 15 m wide, partly filled with a massive congelation ice body, estimated, in 2002, at approximately 6000 m^3 (Luetscher and Wenger, 2002; Fig. 12.5). Air flux measurements have shown dynamic ventilation during the winter season with cold air being drawn in through the larger shaft and directed towards the smaller entrances (Fig. 12.6). This seasonal cooling favors frost shattering and the accumulation of numerous cryoclasts on the ice surface (Pancza, 1992). During the summer season, the cave chiefly acts as a cold air trap, leading to a thermal stratification (Luetscher and Jeannin, 2002). Still, measurements confirmed historical observations of a small oscillating air flow which, however, does not impact the thermal state of the cave. Based on a detailed monitoring program, Luetscher et al. (2008) determined the energy balance of the cave showing that forced convection largely controls the heat exchange between the cave and the surrounding environment. Stack measurements performed between 2001 and 2006 suggest a seasonal deposition rate ranging between 10 and 30 cm a^{-1}, though a major part of it melts during the course of the year (Luetscher, 2004). Overall, Monlési ice cave revealed a negative cave ice mass balance since 2001, consistent with increasingly warmer winters.

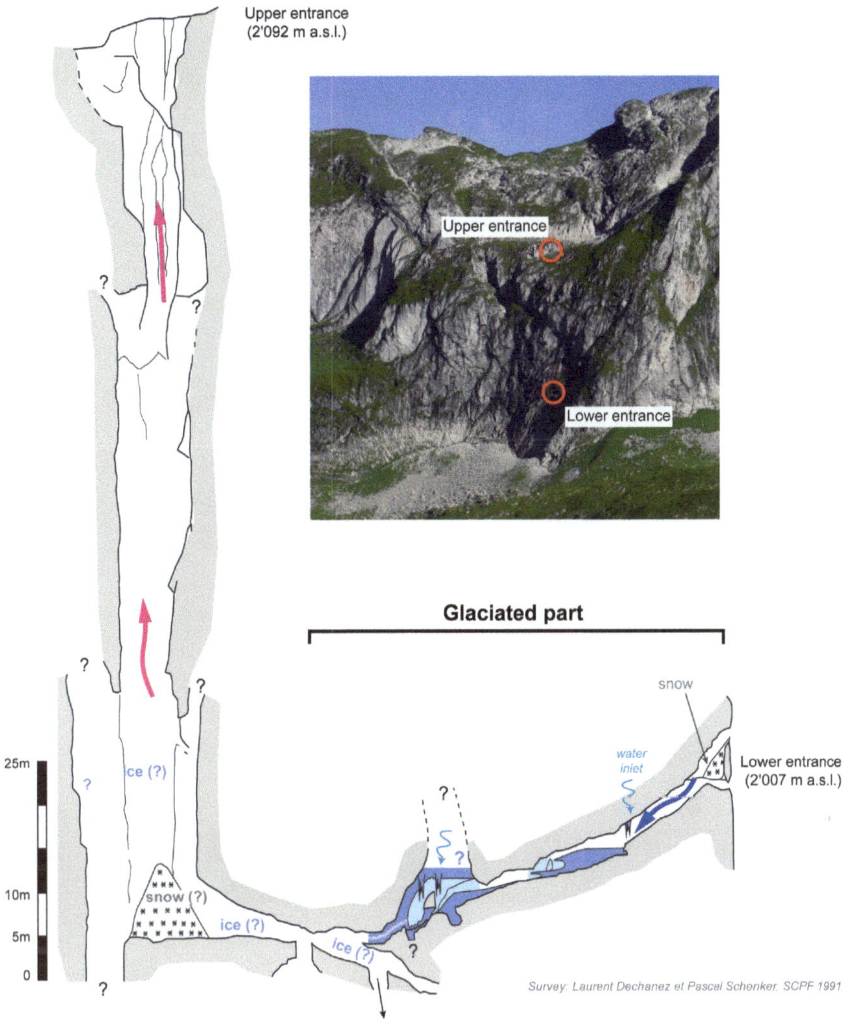

FIG. 12.4

Vertical cross-section of the Diablotins ice cave. Forced convection through the L-shaped cavity is at the origin of a thermal anomaly at the lower entrance allowing for the formation of congelation ice.

After Morard, S., Bochud, M., Delaloye, R., 2010. Rapid changes of the ice mass configuration in the dynamic Diablotins ice cave—Fribourg Prealps, Switzerland. The Cryosphere 4, 489–500.

Steam drilling conducted in 2002 revealed a minimum ice thickness of 8.5 m in agreement with the stratigraphy observed in a vertical outcrop between the ice body and the cave wall. No further study of the ice geometry was conducted yet, and substantial uncertainties remain on the actual ice volume. A multi-parameter dating attempt including measurements of mass turnover rates, radiocarbon ages, Tritium, and ^{210}Pb analyses suggested that the cave ice was a minimum of 120 and a maximum of

FIG. 12.5

Entrance shaft of the Glacière de Monlési.

Photo courtesy of R. Wenger.

158 years old (Luetscher et al., 2007). For the first time, it was thus possible to demonstrate active melting also at the base of the cave ice body, enhancing the mass turnover.

12.4.4 ST-LIVRES ICE CAVE—A TYPICAL COLD AIR TRAP (STATIC CAVE) WITH FIRN DEPOSIT

St-Livres ice cave (6°17′50″/46°33′47″, 1359 m a.s.l.) opens in a grove on the south-facing slope of a closed depression. A large collapsed doline (diameter = ~20 m) forms the unique entrance of this cave and leads to the deepest part of the cavity at −45 m (Fig. 12.7). The downsloping conduit acts as a thermal trap where convective air circulation is restricted to the winter season. When the external temperature falls below the cave air temperature, a strong bidirectional density driven air flow cools the

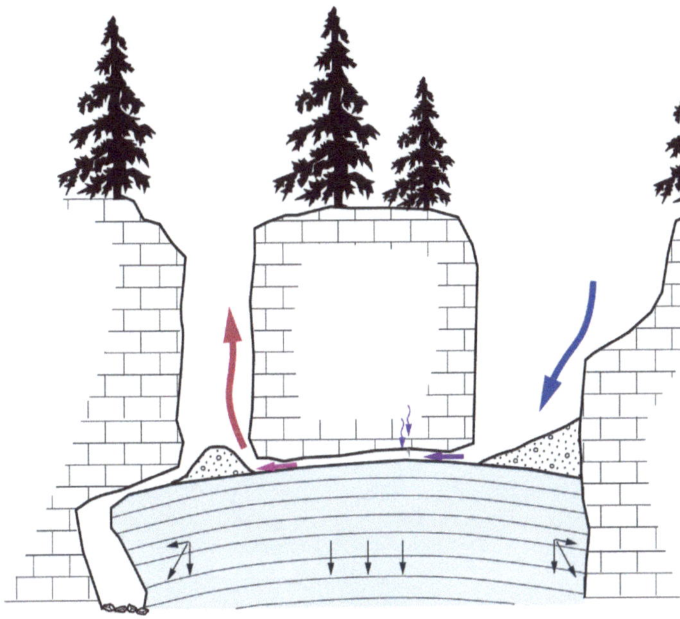

FIG. 12.6

Schematic ventilation during the winter season at Monlési ice cave. Air is drawn in through the larger shaft and directed towards the other entrances; the heat exchanged within the cave allows for freezing of water inlets. During the summer, when the outside air is warmer than the cave air, the cave acts as a cold air trap prompting a thermal stratification.

Adapted from Luetscher, M., 2004. Variations spatio-temporelles du volume de glace à la Glacière de Monlési (Boveresse/NE). Cavernes 2, 3–7.

system allowing for the preservation of cave ice. In contrast, the summer season is characterized by the absence of any significant subsurface air circulation and direct solar radiation barely reaches the cave interior. The higher density of the cold cave air therefore induces a thermal stratification resulting in a negative annual temperature anomaly as compared to the external atmosphere.

A volume of ice, estimated at about 1200 m^3 (Luetscher, 2005), occupies the base of the entrance shaft. It results essentially from the diagenesis of snow (i.e., firn) accumulated during winter, but local refreezing processes of infiltration water also contribute to the actual ice mass. The ice volume reaches its annual maximum during spring, from which point heat exchanges with the surrounding environment induce net ablation. The maximal extent of the ice volume is controlled by a temporary water inlet located at the extremity of the cave. Enhanced ablation at the cave ice front triggers a gravitational ice flow from the accumulation zone, at the base of the entrance-doline, to the lower part of the cave. This flow entrains detrital material comprising organic deposits and cryoclastic rock fragments (Stoffel et al., 2009) and delineates a complex stratigraphy along a vertical ice outcrop (Fig. 12.8). Dendrochronological analysis of 45 subfossil wood samples supported with radiocarbon dating suggests a maximum age of the cave ice of 1200 ± 50 ^{14}C yr BP (Stoffel et al., 2009). Four distinct deposition gaps dated to the 14th, 15th, mid-19th, and late 19th century were related to positive North Atlantic Oscillation anomalies (NAO+) and/or anthropogenic cave ice abstraction. Similarly, periods of rapid cave ice accumulation could be related to enhanced winter snowfall.

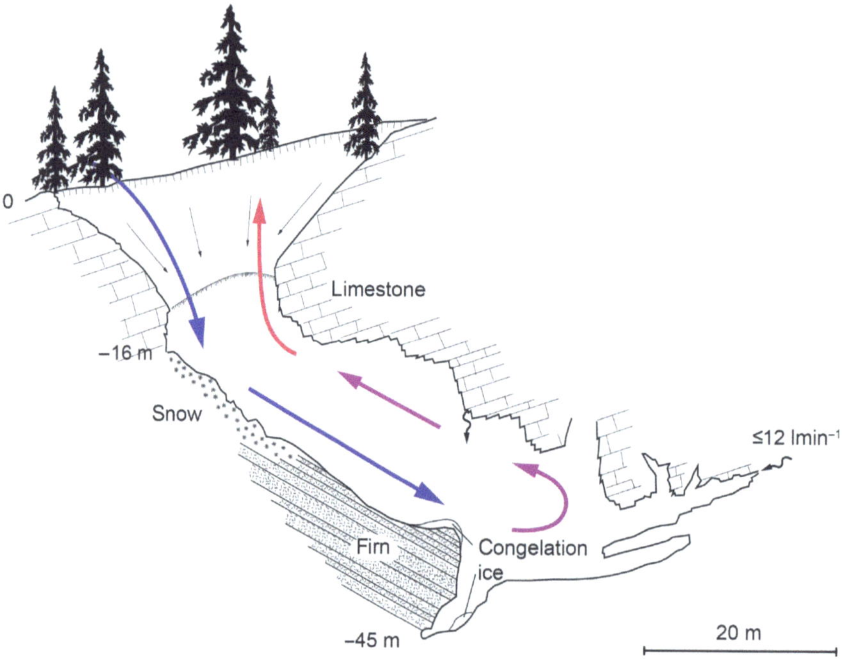

FIG. 12.7

The St Livres ice cave is a static cave hosting a perennial firn deposit. The bi-directional ventilation observed during winter leaves place to a thermal stratification during the summer season allowing for preservation of the cave ice.

After Stoffel, M., Luetscher, M., Bollschweiler, M., Schlatter, F., 2009. Evidence of NAO control on subsurface ice accumulation in a 1200 yr old cave-ice sequence, St Livres ice cave, Switzerland. Quat. Res. 72, 16–26.

FIG. 12.8

The ice front at St Livres ice cave reveals an inclusion-rich stratigraphy which has been radiocarbon dated to 1200 ± 50 yr BP.

Photo courtesy of R. Wenger.

12.4.5 FURTHER INVESTIGATION SITES

Many of the famous ice caves in the Jura Mountains which have been described during the 19th century (e.g., Browne, 1865; Balch, 1900) no longer host cave ice. Luetscher et al. (2005) attributed this general negative ice mass balance to reduced winter snowfall at low elevation, particularly marked since the 1990s. Still, some significant ice caves subsist and are currently part of a long-term monitoring program (Luetscher, 2007). Besides Monlesi and St-Livres, the Glacière de Saint-George (Brulhart, 2001) and Creux-de-Glace are probably the most famous ice caves in the Swiss Jura Mountains documented on a regular basis (Fig. 12.9). In the absence of similar initiatives in the Alps, only sporadic observations are accessible. Those include, among others, detailed observations in the Sanetsch area (Borreguero et al., 2009), Innerbergli (Bitterli and Häuselmann, 2010), Schrattenfluh (Hapka and Rotzer, 2003), Alpstein (Läubli and Graf, 1988), and in the Churfirsten (Rüegg, 2012).

In contrast to low elevation sites, the cave ice mass balance at higher elevation is less sensitive to fluctuations of the winter snowline than to a progressive increase of the cave host-rock temperature. Critically, the formation and preservation of cave ice depends on the annual freezing index and the regime of water infiltration (Luetscher, 2007). In presence of permafrost, the latter may significantly change with the progressive thawing of the fracture network (Fig. 12.10). While empirical observations from the Diablotins cave (Morard et al., 2010) and the Sanetsch area (Borreguero et al., 2009) concluded that the formation of cave ice may increase over short time scales, it will eventually disappear

FIG. 12.9

Photograph of St-Livres ice cave in 1978 and 2002 illustrating the rapid vanishing of many low-elevation cave ice deposits in the Jura Mountains.

Photo courtesy of R. Wenger.

FIG. 12.10

At high elevation, when mountain permafrost thaws, water infiltrates along the open fractures allowing for ice to form in a still frozen cave system.

Photo courtesy of R. Shone.

once the cave reaches a thermal equilibrium with the outside atmosphere. Evidence of this relict cave ice may nonetheless be traced by the study of cryogenic carbonates which have been, for instance, successfully dated in LeClanché cave (Luetscher et al., 2013).

12.5 CONCLUSION

The presence of perennial firn and ice is frequent in many Alpine caves in Switzerland (Fig. 12.11) and has never been the object of a systematic and exhaustive inventory. However, detailed investigations including energy and mass balance measurements were able to outline the main mechanisms at the origin and conservation of cave ice. Providing there is no unusual drought during the winter season, the distribution of ice caves largely reflects the number of annual freezing days. In a global warming context, most of the low-elevation ice caves are therefore threatened by a progressive vanishing.

While preliminary results underline the possible presence of multi-millennial deposits, the full potential for paleoenvironmental studies remains to be investigated. Obtaining precise chronological constraints based on radiocarbon dating of organic rich layers or, alternatively, on undissolved particles represents, therefore, a fundamental step forward. Many of these ice caves, however, face rapid morphological changes calling for action to recover this valuable archive. Although challenging, this endeavor will give access to a promising source of information for the winter half-year, which may be complementary to more classical environmental archives.

FIG. 12.11

The presence of perennial firn and ice is frequent in many Alpine caves and has not yet been the object of a systematic inventory.

Photo courtesy of R. Wenger.

REFERENCES

AEN, 1554. Reconnaissance du Val-de-Travers. vol. 6. Archives de l'Etat, Neuchâtel. p. 162.

Balch, E.S., 1897. Ice caves and the causes of subterranean ice. J. Franklin Inst. 3, 161–178.

Balch, E.S., 1900. Glacières or Freezing Cavernes. Allen, Lane & Scott, Philadelphia, PA. reprinted 1970 by Johnson Reprint Corp., New York: 337 p.

Billerez, 1712. Description de la Glacière naturelle du Comté de Bourgogne. Mém. Acad. pour 1712, 21–24.

Bitterli, T., Häuselmann, P., 2010. Die Höhlen des Innerberglis. Höhlenforschergemeinschaft Region Hohgant, 8, 440 p.

Bock, H., 1913. Matematische-physikalische Untersuchung der Eishöhlen und Windröhren. In: Bock, J., Lahner, G., Gaunersdorfer, G. (Eds.), Die Höhlen im Dachstein. Verein für Höhlenkunde in Österreich, Graz, 102–144.

Boisot, 1686. Lettre à Monsieur Dodard. Mém. Acad. pour 1686, 4–5.

Borreguero, M., Pahud, A., Favre, G., Heiss, G., Savoy, L., Blant, M., 2009. Lapi di Bou, recherches et explorations spéléologiques 1974–2009. Cavernes 53, 213 p.
Bovey, M., 1995. Gouffre des Diablotins. Actes du 10è Congrès national de Spéléologie. Breitenbach 1995, 170–172.
Browne, G.F., 1865. Ice Caves of France and Switzerland. Longmans, Green and co., London, 315 p.
Brulhart, D., 2001. La glacière naturelle de Saint-Georges (VD, Suisse): histoire et fluctuations d'un glacier souterrain. Bull. Soc. Vaud. Sci. Nat. 87, 261–276.
De Cossigny, 1750. Extrait d'une lettre écrite de Besançon à M. de Reaumur le 29 Novembre 1743 sur la Grotte qui se trouve à quelque distance de Besançon et qu'on nomme la Glacière. Mémoire de mathématique et de physique. Acad. Roy. Sci. 1, 195–211.
Deluc, J.-A., 1822. Des Glacières naturelles, et de la cause qui forme la glace dans ces cavités. Ann. Chim. Phys. 21, 113–127.
Dutruit, J., 1991. La Glacière de Correntanaz. Le Trou 52, 36–41.
Fugger, E., 1891–93. Eishöhlen und Windröhren. Jahresbericht K. uK. Oberrealschule Salzburg, 24–26.
Girardot, A., Trouillet, L., 1885. La Glacière de Chaux-les-Passavant. Mém. Soc. Emul. Doubs 5, 449–524.
Groner, U., 2004. Palynology and sediment data from a high alpine cave at Jungfraujoch. Eclogae Geol. Helv. 97, 237–243.
Hapka, R., 2002. Observations sur les piliers calcaires entourant la glacière de Monlési. Cavernes 46 (1), 5–8.
Hapka, R., Rotzer, S., 2003. Contributions à l'inventaire spéléologique du massif de la Schrattenfluh: travaux 2002–2003. Cavernes 47 (2), 18–29.
Häuselmann, P., 2004. Das Jochloch, die höchstgelegene Höhle Europas. Stalactite 54, 49–54.
Jutzet, J.M., 1991. Gouffre B15 aux Morteys (Préalpes fribourgeoises). Stalactite 41, 99.
Läubli, R., Graf, R., 1988. Schneehöhle D1. Höhlenpost 77, 18–19.
Luetscher, M., 2004. Variations spatio-temporelles du volume de glace à la Glacière de Monlési (Boveresse/NE). Cavernes 2, 3–7.
Luetscher, M., 2005. Processes in Ice Caves and Their Significance for Paleoenvironmental Reconstructions. SISKA, La Chaux-de-Fonds ISBN: 3-908495-19-9. 154 p.
Luetscher, M., 2007. Evolution des glacières dans un contexte climatique changeant. In: Actes du 12è Congrès national de Spéléologie, Vallée de Joux, pp. 75–80.
Luetscher, M., 2013. Glacial processes in caves. In: Shroder, J. (Editor in Chief), Frumkin, A. (Ed.), Treatise on Geomorphology. Karst Geomorphology, vol. 6. Academic Press, San Diego, CA, pp. 258–266.
Luetscher, M., Jeannin, P.-Y., 2002. Une année d'enregistrements de températures à la glacière de Monlési / Ein Jahr Temperaturmessungen in der Glacière de Monlési (NE). Stalactite 52, 27–29.
Luetscher, M., Jeannin, P.-Y., 2004a. A process-based classification of alpine ice caves. Theor. Appl. Karstol. 17, 5–10.
Luetscher, M., Jeannin, P.-Y., 2004b. Temperature distribution in karst systems: the role of air and water fluxes. Terra Nova 16, 344–350.
Luetscher, M., Wenger, R., 2002. Nouveau levé topographique de la glacière de Monlési. Cavernes 46 (1), 9–15.
Luetscher, M., Jeannin, P.-Y., Haeberli, W., 2005. Ice caves as an indicator of winter climate evolution—a case study from the Jura Mountains. The Holocene 15, 982–993.
Luetscher, M., Bolius, D., Schwikowski, M., Schotterer, U., Smart, P.L., 2007. Comparison of techniques for dating of subsurface ice from Monlesi ice cave, Switzerland. J. Glaciol. 53, 374–384.
Luetscher, M., Lismonde, B., Jeannin, P.-Y., 2008. Heat exchanges in the heterothermic zone of a karst system: Monlesi cave, Swiss Jura Mountains. J. Geophys. Res. 113, F02025.
Luetscher, M., Borreguero, M., Moseley, G., Spötl, C., Edwards, L.E., 2013. Alpine permafrost thawing during the Medieval Warm Period identified from cryogenic cave carbonates. Cryosphere 7, 1073–1081.
Meteoswiss, 2017. The climate of Switzerland. www.admin.ch (accessed 31.05.17.).
Morard, S., Bochud, M., Delaloye, R., 2010. Rapid changes of the ice mass configuration in the dynamic Diablotins ice cave—Fribourg Prealps, Switzerland. Cryosphere 4, 489–500.

Morard, S., Bochud, M., Delaloye, R., 2012. Evolution de la masse de glace et formation de poches d'eau dans la glacière dynamique des Diablotins (Préalpes fribourgeoises). In: Actes du 13è Congrès national de Spéléologie, Muotathal 2012. pp. 149–154.

Pancza, A., 1992. La gélivation des parois rocheuses dans une glacière du Jura neuchâtelois. Permafr. Periglac. Process. 3, 49–54.

Permos, 2016. PERMOS Database. Swiss Permafrost Monitoring Network, Fribourg, Switzerland. https://doi.org/10.13093/permos-2016-01.

Pictet, M.A., 1822. Mémoire sur les glacières naturelles qu'on trouve dans quelques grottes du Jura et des Alpes. Bibliothèque Universelle des Sciences. Belles-Lettres et Arts 20, 261–284.

Poissenot, B., 1586. Nouvelles histoires tragiques. G. Bichon, Paris.

Prévost, 1789. Une glacière naturelle près de Besancon. J. Genève, mars 1789, 43–44.

Rüegg, G., 2012. Les puits à neige du massif des Churfirsten; un état des lieux dans le Brisital et Gluristal. In: Actes du 13è Congrès national de Spéléologie, Muotathal 2012. pp. 155–158.

Saar, R., 1956. Eishöhlen, ein meteorologisch-geophysikalisches Phänomen. Untersuchungen an der Rieseneishöhle (R.E.H.) im Dachstein, Oberösterreich. Geogr. Ann. 38, 1–63.

Stettler, R., Monard, M., 1960. La Glacière de Monlési. Cavernes 1, 1–10.

Stoffel, M., Luetscher, M., Bollschweiler, M., Schlatter, F., 2009. Evidence of NAO control on subsurface ice accumulation in a 1200 yr old cave-ice sequence, St Livres ice cave, Switzerland. Quat. Res. 72, 16–26.

Thury, M., 1861. Etude des Glacières naturelles. Archives des Sciences de la Bibliothèque Universelle, Genève. pp. 1–59.

Wildberger, A., Preiswerk, C., 1997. Karst and Caves of Switzerland. Speleo Projects, Basel. 208 p.

CHAPTER

ICE CAVES IN AUSTRIA

13

Christoph Spötl*, Maximilian Wimmer†, Rudolf Pavuza‡, Lukas Plan‡

University of Innsbruck, Innsbruck, Austria Landesverein für Höhlenkunde in Oberösterreich, Linz, Austria† Natural History Museum Vienna, Vienna, Austria‡*

CHAPTER OUTLINE

13.1 INTRODUCTION

Austria is located in central Europe and comprises 29% of the Alps, also known as Eastern Alps. The high proportion of mountainous terrain reaching up to 3798 m a.s.l.—65% of the country's size—and the widespread distribution of karstifiable carbonate rocks (and in particular the presence of large uplifted karst plateaus) are the main reasons for the high abundance of ice-bearing caves in this country. The number of ice caves relative to the size of the country probably ranks among the highest worldwide.

About one fifth of Austria is made up of carbonate rocks, ranging from non-metamorphic limestones and dolomites to low- and medium-grade metamorphic marbles and calcareous schists. Although karst caves constitute by far the longest and deepest caves, caves unrelated to karst processes are also quite common in some regions, but few host cave ice.

16,300 caves are known and registered in the national cave cadaster (as of December 2016). They are unevenly distributed across the country, showing the highest abundance and density in the central and eastern segments of the Northern Calcareous Alps (NCA), as well as in the area north and northwest of Graz (Fig. 13.1).

Ice Caves. https://doi.org/10.1016/B978-0-12-811739-2.00013-9

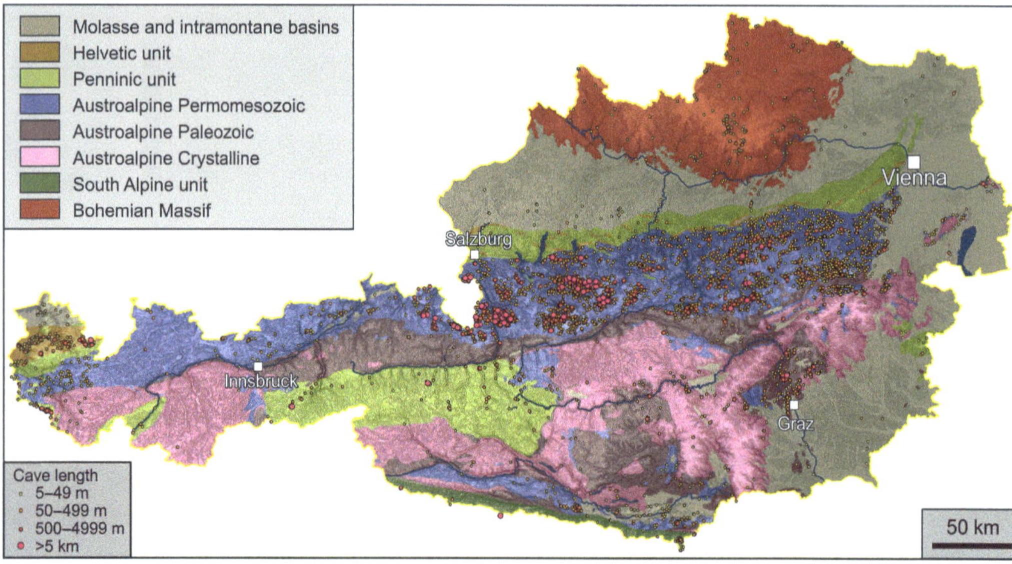

FIG. 13.1

Simplified geological map of Austria showing the distribution of caves as of 2017.

Schönberghöhlensystem, located in the western part of Totes Gebirge, is the longest cave in the European Union and comprises 146.7 km of passages (see below). Two additional caves are longer than 100 km, and 14 more caves exceed 20 km. Sixteen caves in Austria are deeper than 1 km, which reflects the thick vadose zone of uplifted karst massifs. Lamprechtsofen ranks as the deepest through-trip cave worldwide, with a vertical extent of 1632 m between the lower and the highest-lying entrance.

Thanks to the centralized cave cadaster operated by a web-based database system (Spelix), in which speleologists from 26 local caving clubs feed their exploration data, Austria has a comprehensive and well-maintained database of caves, including ice-bearing ones. 1200 caves classify as ice caves that host perennial ice, firn, or snow. Less well known is the temporal evolution of these caves, but qualitative observations clearly indicate that a number of ice-bearing caves has partly or completely lost their ice deposits in the past few decades.

13.2 ICE CAVE RESEARCH

Research on ice in subsurface cavities has a long tradition in Austria, as has speleology in general. An official expedition into Geldloch (part of the 29 km-long Ötscherhöhlensystem), a former ice-bearing cave in the Ötscher massif (Lower Austria), back in 1592 marks the first effort to explore these underground glaciers. Commanded by Reichart Freiherr von Strein on behalf of Emperor Rudolf II, the team managed to explore 860 m of cave passages and an official report was delivered to the authorities (Mais and Trimmel, 1992). Emperor Maria Theresa ordered another expedition into this cave in 1747. This time, a frozen lake stopped the explorers, but attempts were made to measure its temperature (Hartmann and Hartmann, 1984). Although these measurements were apparently flawed, the expedition leader,

Joseph Anton Nagel (later appointed mathematician at the imperial court) correctly concluded that ice formation occurs as a result of cooling of the cave during winter (Saar and Pirker, 1979). During the last few years Geldloch has become ice-free in autumn.

Eberhard Fugger, a high-school teacher who later became director of the Museum Carolino Augusteum in Salzburg (Jäger, 1919; Pillwein, 1919), pioneered instrumental research on ice caves in Austria and conducted multi-annual studies of cave air temperature and ice volume in Kolowrathöhle, now part of the 41 km-long Gamslöcher-Kolowrat-Salzburgerschacht-System, located in the Untersberg massif southwest of Salzburg (Fugger, 1888, 1891, 1892, 1893; Fig. 13.2). He visited this cave 35 times between 1876 and 1887 and also made systematic observations and measurements in two other ice caves in Untersberg, Großer Eiskeller, and Schellenberger Eishöhle (Klappacher and Mais, 1999), the latter now located in Germany. The two most prominent ice caves in Austria, Dachstein-Rieseneishöhle and Eisriesenwelt, were discovered in 1910 and 1912 (the latter actually 33 years earlier, but this went almost unnoticed), respectively, and immediately became focal points of research.

FIG. 13.2

Kolowrathöhle near Salzburg was one of the first targets of ice-cave research in the second half of the 19th century. Meanwhile this near-entrance part of the cave is largely ice-free. Copper engraving by Sebastian Stief from the middle of the 19th century.

Already in 1913, a comprehensive treatise on the physical principles of air flow and ice dynamics based on studies of caves in the Dachstein massif was published (Bock et al., 1913). Hauser and Oedl (1926) pioneered ice-cave research in Eisriesenwelt, which was the target of a dedicated expedition sponsored by the Austrian Academy of Sciences in 1921. Scientific studies in Dachstein-Rieseneishöhle were led by Kyrle and later by Saar, who conducted long-term changes in air and rock temperature, relative humidity and air flow speed and direction (Saar, 1955, 1956). Kral was the first to conduct systematic studies of pollen in ice from alpine caves and drew conclusions about the approximate age of these ice bodies (Kral, 1968; Schmeidl and Kral, 1969).

Research since the 1980s has focused on monitoring ice-volume changes in selected ice caves (e.g., Pavuza and Mais, 1999), assessing the 3D geometry of ice bodies using ground-penetrating radar (e.g., Hausmann and Behm, 2011), and attempts to constrain the age of ice by radiocarbon-dating of enclosed organic remains (e.g., Achleitner, 1995; Mais and Pavuza, 2000; Herrmann et al., 2010). Recent years have seen a concerted effort to study the meteorology and ice dynamics of Eisriesenwelt (see below), which included the first ice core drilling performed in an alpine cave (2007), and the study of cryogenic cave carbonates (CCC), both in previously glaciated parts (Spötl and Cheng, 2014; Pavuza and Spötl, 2017) and locally also within still-existing ice bodies (Spötl, 2008).

13.3 TYPES OF ICE CAVES IN AUSTRIA

Based on their geometry and the resulting seasonal air-flow pattern, ice caves in Austria range from sag-type caves lacking a lower entrance to caves with at least two entrances at different elevations that are large enough to allow significant air flow. While these geometries give rise to distinct modes of air exchange between the cave and the outside atmosphere—traditionally (albeit not fully correctly) referred to as static vs. dynamically ventilated ice caves—combinations of the two end members are common as well, in particular in large and complexly structured cave systems, such as Schönberg-Höhlensystem (see below). It is also well known that some ice caves undergo a change from one to the other type due to opening and closing of ice plugs (e.g., Großer Eiskeller in Untersberg or Eiskogelhöhle in Tennengebirge). Such cycles have not been studied in detail, but available observations suggest that they occur on timescales of 5–10 or more years in large Austrian ice caves (Klappacher, 1996; see below).

The majority of ice caves in Austria belong to the sag type and constitute pits of less than a few tenths of meters in depth, widespread on karst plateaus of the NCA. These caves are in general smaller than those showing dynamic ventilation and a significant proportion of their ice bodies formed by transformation of winter snow that fell into these subsurface cavities. The large underground glaciers, on the other hand, mostly developed in subhorizontally oriented caves characterized by multiple entrances and a resulting intensive and seasonally changing air-flow pattern.

13.4 DISTRIBUTION OF ICE CAVES IN AUSTRIA

Ice caves are unevenly distributed across the country and occur in the following karst regions (of decreasing abundance—Fig. 13.3):

- NCA: This tectonic unit stretches from Vorarlberg, the westernmost part of Austria, to the margin of the Vienna Basin in the east, shows the highest abundance and density of karst features and host 79% of the caves in Austria. All large ice caves of this country (and the largest of the entire Alps and possibly globally) can be found in the NCA, including Eisriesenwelt, Dachstein-Rieseneishöhle, Eiskogelhöhle, Schwarzmooskogel-Eishöhle and ice-bearing parts within the giant Schönberg-Höhlensystem (see below). Also, the largest ice bodies of sag-type caves can be found here, such as Kraterschacht in Sengsengebirge, which hosts a snow-firn-ice body of some 70 m in thickness (Weißmair, 1995, 2011), and Hochschneid-Eishöhle in Höllengebirge, which was recently explored and contains a firn and ice plug of about 70 m thickness. The NCA also host the lowest-elevation ice cave in this country, Schödlkogeleishöhle near Bad Mitterndorf (Styria), located at 939 m a.s.l., which is currently at the verge of transforming into a seasonal ice cave. Fig. 13.3 shows that within the NCA ice caves are particularly abundant between Salzburg and Styria, which reflects the presence of large karst plateaus in this region reaching up above the timberline (located between about 1600 and 1800 m a.s.l. in the NCA), with elevation ranging from about 1500 to almost 3000 m a.s.l.

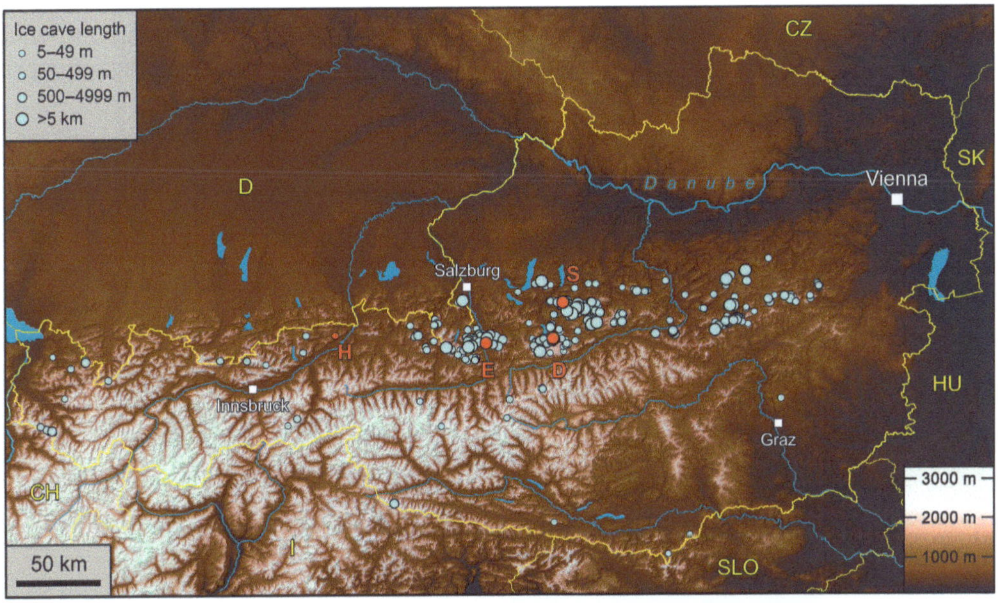

FIG. 13.3

Elevation map of Austria showing the distribution of ice caves differentiated according to their length. Note that the latter figure refers to the total length of the cave, not the length of the ice-bearing part. H, Hundsalm Eis- und Tropfsteinhöhle; E, Eisriesenwelt; D, Dachstein-Mammuthöhle; S, Schönberg-Höhlensystem.

- Central Alps: They form the main ridge of the Eastern Alps, include the highest summits, are partly glaciated and consist of metamorphic rocks. Although caves are known there up to about 2900 m a.s.l., most of these high-alpine caves are less than a few hundred (and often less than a few tens of) meters in length and no large ice bodies are known there. The scarcity of ice caves in the highest mountain ranges of the Eastern Alps may appear surprising, but is primarily due to the predominance of non-karstic silicate rocks.
- Southern Calcareous Alps: This narrow band of both non-metamorphic and low-grade metamorphic carbonate rocks straddles along the southern border of Austria. Karst abounds only in some regions, such as in the eastern Karawanken Mountains, and only few ice-bearing caves are known, most notably Obstanser Eishöhle. This cave opens at 2192 m a.s.l. near the western end of the Karnischer Hauptkamm, comprises 3.4 km of passages and contains some perennial ice behind its main entrance (Herrmann, 2017). Back in 1934, the then-known cave parts were

precisely surveyed using professional geodetic instruments (Killian, 1935). This dataset provides clear evidence that the ice volume has decreased since then, with an apparent acceleration since the early 1980s. Instrumental monitoring since 2008 has revealed a near-linear lowering of the ice height at two locations (Spötl et al., 2017). Based on this trend, Obstanser Eishöhle will likely completely loose its perennial ice in the near future.

The nationwide database on Austrian ice caves allows the exploration of the factors controlling their occurrence. The fact that ice caves are known to be present across an altitudinal range of 2 km clearly shows that elevation (and hence atmospheric temperature) is not the main controlling factor. Fig. 13.4 reveals that the highest abundance of ice caves occurs between about 1800 m and 2100 m a.s.l., which is well below the elevation of the 0°C annual isotherm of the atmosphere in the Alps (based on long-term measurements at meteorological stations).

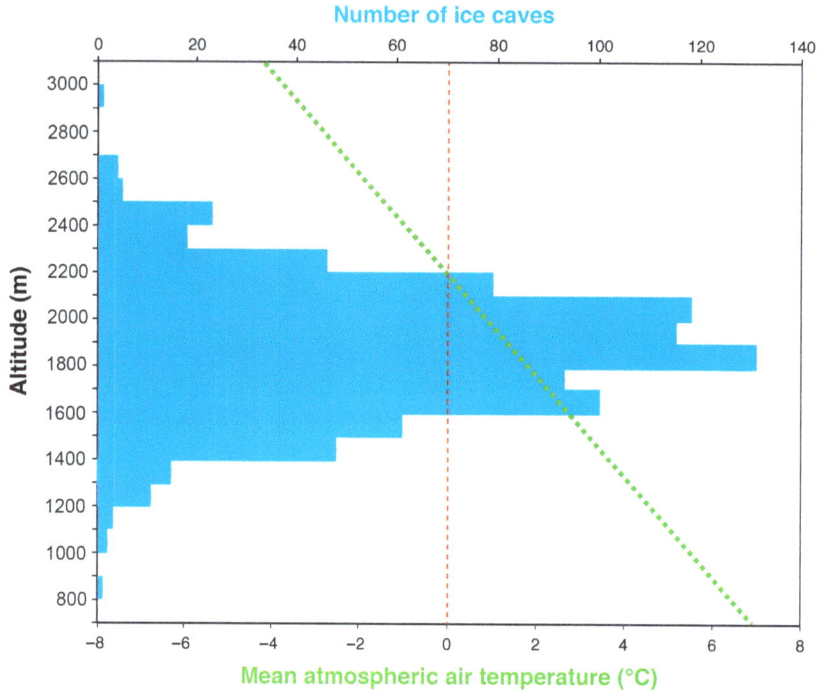

FIG. 13.4

Altitudinal distribution of ice-bearing caves in Austria. The *green dotted line* is the mean annual air temperature in the Eastern Alps and the *red dotted line* marks 0°C.

Ice caves are thus unrelated to Alpine permafrost, whose lower boundary is presently located at about 2500 m in the Central Alps. The altitudinal distribution of ice caves instead largely reflects a combination of the abundance of (uplifted) cave systems between about 1500 m and 2000 m a.s.l. and the decreasing likelihood of preserving cave ice at elevations below about 1400 m a.s.l.

The average lapse rate in the Eastern Alps is about 0.5°C/100 m, and the interior air temperature of most caves follows this trend. Ice-bearing cave parts are notable exceptions showing deviations of up to 4°C towards lower temperatures (Fig. 13.5).

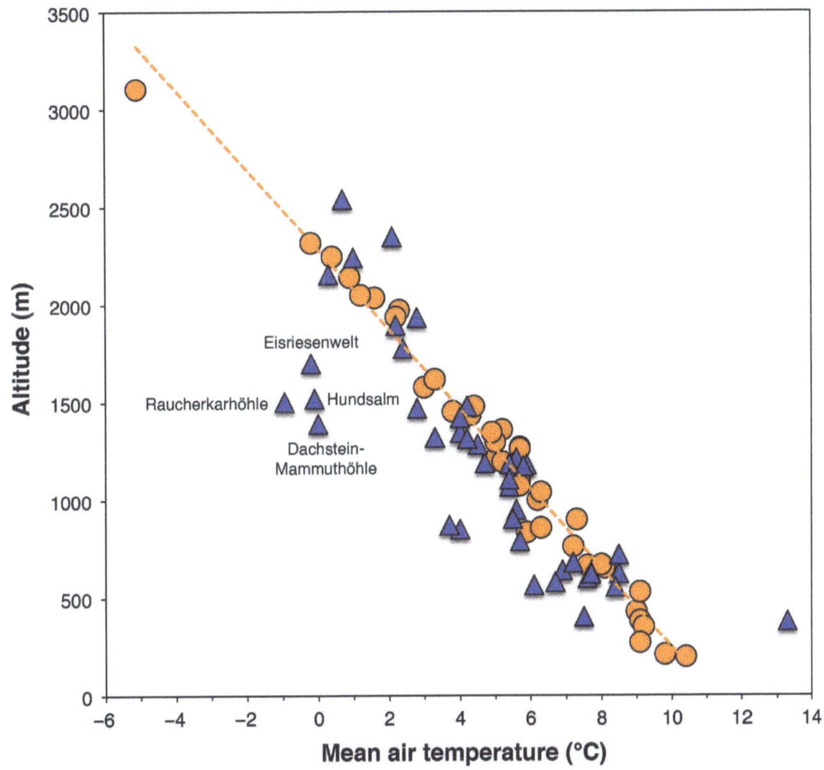

FIG. 13.5

The interior temperature of Austrian caves *(blue triangles)* corresponds approximately to the long-term average of the outside atmosphere *(orange dots)* at a given altitude. The overall cave air gradient follows that of the atmosphere (*orange line* based on data from selected climate stations in Austria; period 1981–2010). Ice-bearing cave (parts) deviate significantly from this altitude trend and four are labeled: the ice-bearing frontal part of Eisriesenwelt, the upper ice-bearing level of Hundsalm Eis- und Tropfsteinhöhle, Raucherkarhöhle (*Eisstadion*) as part of Schönberg-Höhlensystem, and the ice-bearing part of Dachstein-Mammuthöhle close to its western entrance.

Seven of Austria's 30 show caves contain perennial ice. These include the three large ice caves Eisriesenwelt, Dachstein-Rieseneishöhle and Eiskogelhöhle, as well as the small ice caves Prax-Eishöhle and Hundsalm Eis- und Tropfsteinhöhle as well as a small part of Frauenmauer-Langstein-Höhlensystem. Parts of Dachstein-Mammuthöhle aside the touristic route also contains some cave ice (see below).

Touristic use of ice caves in Austria started in 1845 in Kolowrathöhle near Salzburg. Guided tours into the two best known caves, Dachstein-Rieseneishöhle and Eisriesenwelt, commenced in 1912 and 1920, respectively. The former was already illuminated by electrical lights in 1928. In recent years, the

Austrian ice caves have been visited by a few hundred thousand guests per year, providing important local touristic and economic incentives.

In addition to ice caves, Austria hosts a few touristic, partly artificial, glacier caves, which—given the strong retreat of alpine glaciers—commonly exist only for a few years. An exception is Natur Eis Palast, an artificially modified glacier cave at 3200 m a.s.l., which provides insights into the geometry and morphology of open fractures in the uppermost part of Hintertux Glacier (Zillertal Alps).

13.5 EXAMPLES OF ICE CAVES IN AUSTRIA

In this chapter four ice caves are briefly presented, which are among the best studied in Austria.

13.5.1 EISRIESENWELT

As the name suggests this cave south of Salzburg hosts a tremendous ice body, which ranks among the largest cave ice accumulations worldwide. Eisriesenwelt opens on the western flank of Tennengebirge at 1657 m a.s.l., about 1150 m above the level of the adjacent Salzach Valley (47.511°N, 13.193°E). The cave comprises at least 42 km of predominantly horizontal passages (Pointner and Klappacher, 2016), which belong to a Miocene-age paleophreatic cave level.

The first approximately 1 km of this cave system contains perennial ice, reaching a thickness of up to several meters (Fig. 13.6) and comprising a surface area of 28,000 m^2. The ice starts right behind the entrance and consists of floor ice, a few several meter-tall stalagmites, stalactites and seasonal hoar frost.

FIG. 13.6

Layered ice containing fine crystalline cryogenic calcite particles exposed in the interior of the glaciated part of Eisriesenwelt.

Photo courtesy of Robbie Shone.

Eisriesenwelt shows a bidirectional air-flow pattern typical of dynamically ventilated caves. During the warm season, the draft at the otherwise-closed entrance gate reaches 13 m/s, while during winter cold outside air is drawn into the cave, warms upon contact with the rock walls and ascends through unexplored chimneys toward the plateau 400–500 m higher. During winter, the 0°C isotherm progressively moves further into the cave, and seasonal ice formations are locally present several hundred meters behind the main ice-bearing part. Detailed measurements show, however, that this simple air-flow pattern is often perturbed (Thaler, 2008; Obleitner and Spötl, 2011; Schöner et al., 2011). Temperatures at different depths in a 7-m-deep hole at *Eispalast* mirror these seasonal and shorter-term air temperatures changes with a slight delay, as do radon measurements, being particularly sensitive to ventilation changes of cave air (Gruber et al., 2014). The temperature at the base of the ice body at *Eispalast* varies between −0.1 and −0.5°C (Fig. 13.7), i.e., the ice is permanently frozen to bedrock (or gravel or loam covering bedrock).

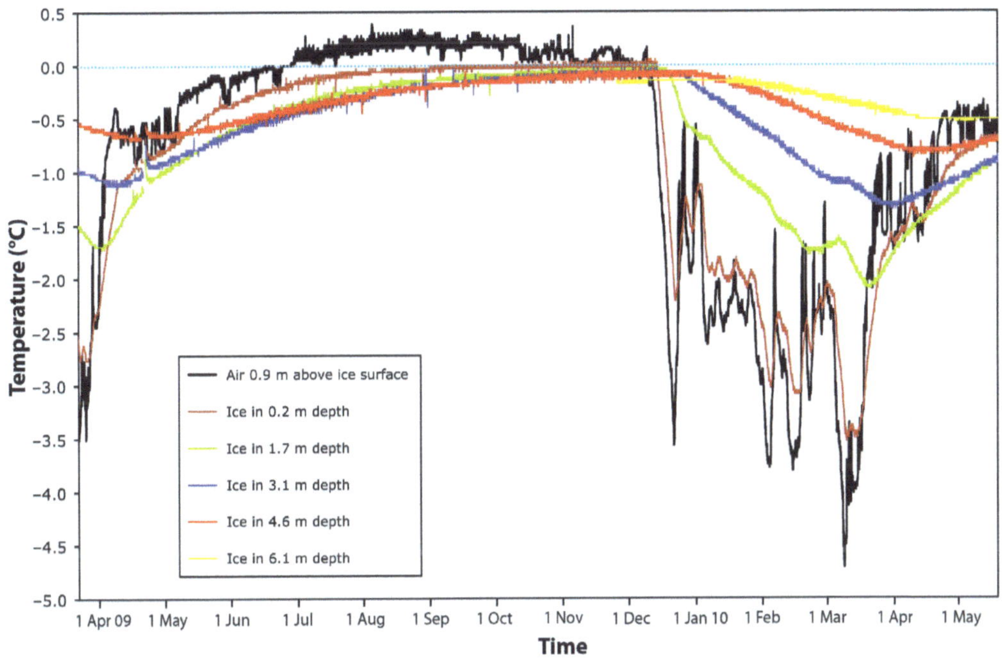

FIG. 13.7

Temperature distribution at different depths in the ice body at *Eispalast* near the inner limit of the glaciated part of Eisriesenwelt (about 500 m behind the entrance) over the course of about 1 year. Note delayed response of the ice temperature to seasonal cooling by cold winter air being drawn into the cave.

Since 1920, when this cave was developed as a touristic cave, the summer air flow out of the cave has been restricted by a door. This door remains open during the cold season allowing the cold winter air to freely enter the cave. Due to these artificial measures, the climate of this cave and hence its ice mass balance have been modified, in particular during summer and autumn. Only scattered reports exist about decadal-scale ice-volume changes (e.g., Abel, 1965; Gressel, 1965) and these in conjunction with photographic documents indicate that, e.g., a side passage called *Wimur*, leading to a second lower entrance at 1838 m a.s.l. also acts as a lower opening of the cave ventilation system. It was glaciated during the first half of the 20th century and became ice-free since then. A narrow restriction leading to *Eispalast* deeper into the cave (*Eistor*) has been progressively widened by ice ablation since the 1920s.

More recently, the ice cave was surveyed using terrestrial laser-scan techniques, providing an accurate and precise figure of the glaciated area (28,000 m^2) and a reference for future long-term mass balance monitoring (Petters et al., 2011; Milius and Petters, 2012).

In a large chamber called *Eispalast* at the inner end of the ice-bearing part, measurements using ground-penetrating radar (and verified using steam drilling) revealed an ice thickness between 1.5 and 7.5 m (Behm and Hausmann, 2007; Hausmann and Behm, 2011). There, an ice core was drilled to bedrock in June 2007, the first ice-core drilling ever performed in an alpine ice cave. Tritium analyses showed that ablation has led to an almost complete loss of bomb-derived tritium removing any ice accumulated since at least the early 1950s. Attempts to constrain the age of the 7.1 m-long ice profile were largely unsuccessful given the high purity of this ice. A crude estimate based on radiocarbon dating of particulate organic matter suggests a basal ice age in the order of several thousand years (May et al., 2011).

An earlier study by Fritz (1977) examined four samples of Eisriesenwelt (from *Großer Eisdom* near the entrance and from *Mörkgletscher* near the inner end of the ice-bearing cave part) and found low pollen concentrations and pollen spectra that included pollen of cultivated plants. The author concluded that these ice samples are likely less than a few hundred years old.

The ice of Eisriesenwelt commonly contains cryogenic cave carbonates (CCC), whose crystal size typically ranges from about 0.1 to 0.4 mm. At the ice cliff of *Mörkgletscher* several distinct layers of these carbonate particles are exposed (Spötl, 2008; Fig. 13.6), but attempts to date them using U-Th were unsuccessful. Cryogenic carbonates were also frequently encountered in the drill core in *Eispalast* (May et al., 2011). No coarse crystalline varieties of CCC were found in the ice-bearing part of this cave, reflecting its dynamic meteorology. Coarse crystalline CCC, however, was found in a remote and currently ice-free part of the Eisriesenwelt and yielded a late glacial age (C. Spötl and M. Luetscher, unpublished data).

13.5.2 SCHÖNBERG-HÖHLENSYSTEM

Located in the western part of Totes Gebirge, Austria's most extensive karst plateau, this giant cave system has 34 entrances between 1459 m and 1933 m a.s.l. and encompasses 146.7 km of passages and 1061 m of vertical extent (47.718°N, 13.787°E; Geyer et al., 2016). It is the result of a successful merger between Raucherkarhöhle (Zeitlhofer and Knobloch, 2008) and Feuertal-Höhlensystem (Jansky et al., 2008) in 2007. The complex geometry in combination with the multiple entrances gives rise to a dynamic ventilation pattern. While the general air flow direction in summer is from the high-lying Feuertal-Höhlensystem towards the lower lying Raucherkarhöhle (Fig. 13.8) and vice versa in winter, there are a variety of local complexities including areas of stagnant air.

The majority of Schönberg-Höhlensystem is ice-free. Perennial ice accumulations are only found near the entrances of Feuertal-Eishöhle (part of the Feuertal-Höhlensystem; *Gustave-Abel-Halle*), Raucherkarhöhle (*Eingangslabyrinth*) and Altarkögerlhöhle. The first two are well documented and are briefly presented here.

Feuertal-Eishöhle opens at 1718 m a.s.l. near the base of a large doline occupied by a steep snowfield, which does not completely disappear in late summer. This snowfield shows the transition into firn

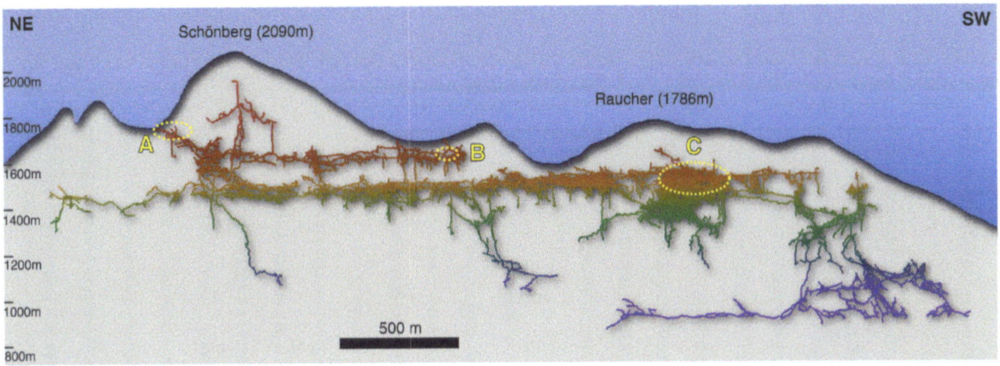

FIG. 13.8

Vertical section of Schönberg-Höhlensystem showing the location of the three ice-bearing parts. (A) Ice part of Feuertal-Eishöhle, (B) ice part of Altarkögerlhöhle, (C) ice part of Raucherkarhöhle. The depth distribution of this giant cave system is color-coded. Based on survey map by Harald Zeitlhofer.

in its deeper part and extends into the ice-bearing entrance hall of the cave. This 60×40 m wide hall also receives snow via two shafts reaching to the surface forming prominent snow cones. Photographs document that the ice level in this near-entrance hall did not change significantly over several decades (Fig. 13.9). This is confirmed by regular surveys of the ice height along traverses which show no trend since 1999 (Fig. 13.10).

FIG. 13.9

View from the ice floor in Feuertal-Eishöhle towards the snow cone of the doline in the background. The ice cover in this hall was roughly at the same level in 2016 (right) as in 1931 (left).

Right photo courtesy of Harald Zeitlhofer.

Meltwater that does not refreeze in the entrance hall finds its way via a narrow and strongly ventilated connection (*Bläser*, also referred to as *Eisbläser*) and a steep glaciated ramp into *Gustave-Abel-Halle*. In the past, this connection was repeatedly clogged by ice for several years disconnecting Feuertal-Eishöhle from the rest of Schönberg-Höhlensystem. During these episodes, the entrance hall was commonly occupied by a shallow ice lake. The ramp terminates with a prominent ca. 10 m-high ice wall towards *Gustave-Abel-Halle*. Photographs and survey data show that the volume and dimensions

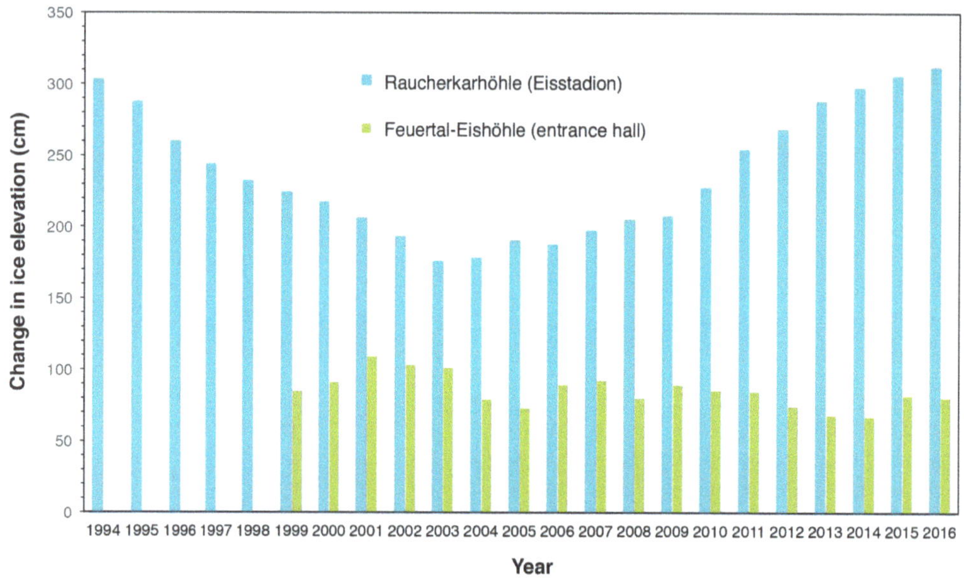

FIG. 13.10

Long-term evolution of ice height (approximately corresponding to ice thickness) in Feuertal-Eishöhle and Raucherkarhöhle. Each data point represents the mean of 10–11 measurements points along a traverse across the ice surface.

of this ice wall in 1928 was comparable to the situation in 2016. At the beginning of the 21st century, the cliff face retreated somewhat, exposing layered ice (Fig. 13.11).

In recent years, new ice has covered the cliff (Fig. 13.11) and the passage leading deeper into the cave was again buried by ice in 2016. Hence the ice in *Gustave-Abel-Halle* shows a cyclic behavior

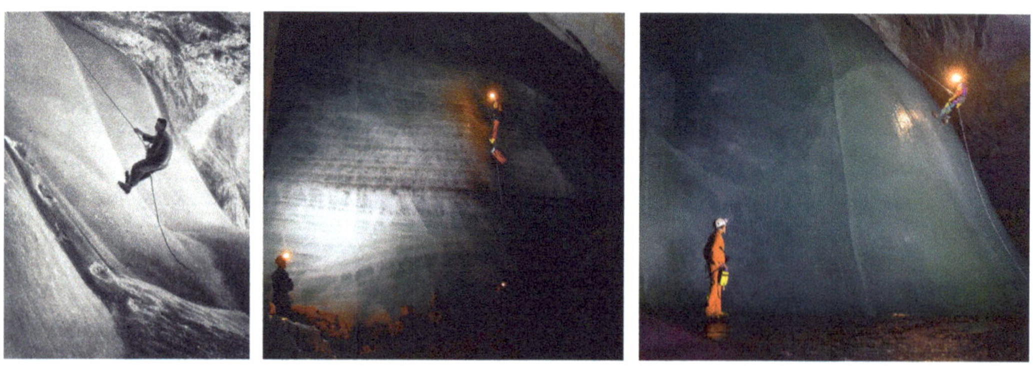

FIG. 13.11

The ice wall leading into *Gustave-Abel-Halle* of Feuertal-Eishöhle c.1931 (left; Abel, 1932), 2003 (middle) and 2016 (right).

Middle and right photos courtesy of Clemens Tenreiter and Harald Zeitlhofer, respectively.

with periods lasting 1–3 decades. Historical reports show that, e.g., the *Bläser* connection was closed in 1958, open in 1980, and closed again in the 1990s. Surveys of the ice body in the entrance hall show no systematic trend since the onset of these measurements in 1999 (Fig. 13.10).

In Raucherkarhöhle, the ice-bearing parts are also found close to the entrance and initially comprised several galleries including *Planer-Eishöhle*, which later was connected to the main cave system. Already at the time of discovery (1961), most ice formations showed clear signs of degradation. During the subsequent decades, the ice disappeared from all of these cave parts and small remnants only remained in *Pfeilerhalle*, *Pilzlinghalle*, *Rauhreifgang* and *Planer-Eishöhle*. The complete deglaciation of *Kleiner Rundgang* led to the discovery of new passages including *Eisstadion*, a 25×35 m wide hall. At the time of discovery (1993) about two thirds of this hall were occupied by a thick, layered ice body showing a near-horizontal surface. Currently, *Eisstadion* is the largest remaining ice accumulation in Raucherkarhöhle. Regular surveys started in 1994 and reveal a gradual decline ice height over the subsequent decade (Fig. 13.10). This trend, however, reversed in 2005, and since then, this cave part has shown a net ice increase. The reason for this decadal change in ice mass balance was a change in air circulation: Prior to 1998 air temperatures at *Eisstadion* were always slightly above freezing, even during winter (Fig. 13.12). In 1998, a new connection opened in the ice between *Eisstadion* and *Pilzlinghalle* due to wind ablation and dripping water.

This narrow hole widened in the following years and allowed the inflow of cold winter air into *Eisstadion* and *Großer Eissaal* (the latter was completely ice-free at that time). Already in 1998, new ice formation was

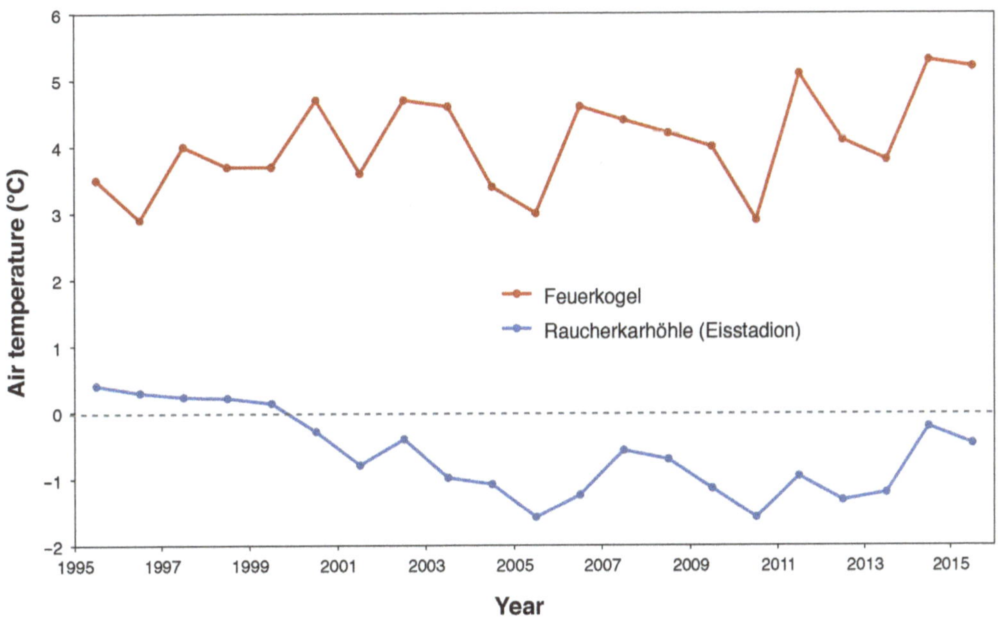

FIG. 13.12

Comparison between the decadal-scale cooling trend in the near-entrance parts of Raucherkarhöhle and the opposite trend of the outside atmosphere as recorded by the meteorological station at Feuerkogel (1618 m a.s.l.) located 13 km NNW of the cave (annual means).

observed locally. This new cycle of winter cooling terminated the decadal trend of ice decline at *Eisstadion* and initiated a new cycle of ice accumulation, albeit with a delay of several years (Fig. 13.10).

A model summarizing the complex interplay between cave air ventilation and ice mass balance was presented by Wimmer (2008). Its predictions regarding ice volume changes have been validated by observations made in the years since then. The positive ice volume trend continues. In *Großer Eissaal*, which had been ice-free for two decades, large ice columns as well as floor ice formed. A comparison of photographs demonstrates that there is already more ice in this chamber than in 1965 when it was discovered (Fig. 13.13).

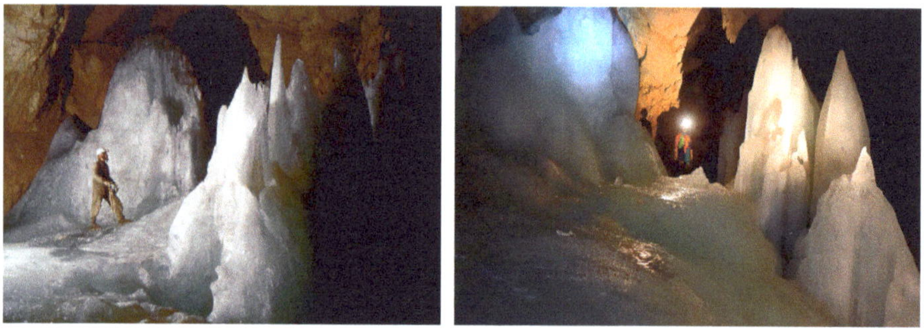

FIG. 13.13

Ice formations in *Großer Eissaal* at the time of discovery (1965, left) and in 2016 (right), soon after a new cycle of ice accumulation had commenced. In between, this cave chamber had been continuously deglaciated for 20 years.

Left photo courtesy of Helmuth Planer.

It is expected that this positive ice mass balance trend in *Eisstadion* and *Großer Eissaal* will continue for many years, possibly decades. This cyclic behavior appears to be only marginally influenced by the general warming trend in the Alps. On the other hand, other parts of Raucherkarhöhle are more vulnerable to climate warming, e.g., *Planer-Eishöhle*. This cave part shows an uninterrupted ice retreat since many years, reflecting its near-surface setting and insufficient winter cooling.

13.5.3 DACHSTEIN-MAMMUTHÖHLE

The Dachstein is situated in the central part of the NCA and comprises a major uplifted karst region 580 km^2 in size. It southern side reaches up to almost 3000 m a.s.l. and is partly glaciated. On the northern rim, the plateau shows a prominent escarpment towards Hallstätter See, a deep lake of glacial origin at 508 m a.s.l. Along this escarpment several large caves are known, which include from west to east Hirlatzhöhle (104.8 km long), Dachstein-Mammuthöhle (67.4 km), Mörkhöhle (4.2 km), Dachstein-Rieseneishöhle (2.7 km) and Schönberghöhle (9.3 km). High-discharge karst springs are located below the levels of these paleophreatic cave systems.

Dachstein-Mammuthöhle is currently the fourth longest cave in Austria with a vertical extension of 1.2 km (47.534°N, 13.708°E). The system is labyrinthic and comprises 21 entrances between 927 and 1828 m a.s.l. (Behm et al., 2016). Most paleo(epi)phreatic passages are located between 1250 m and 1550 m a.s.l. and show a long and complex speleogenetic history (e.g., Plan and Xaver, 2010).

Perennial cave ice is present near the western entrance (*Westeingang*), which opens at 1391 m a.s.l. Since the 1990s, this ice as well as that of the nearby Dachstein-Rieseneishöhle have been targets of annual ice level measurements in combination with cave air monitoring (Mais and Pavuza, 1999, 2000). In contrast to Dachstein-Rieseneishöhle, whose cave meteorology and hence ice volume and distribution are significantly modified by the show cave management, the ice in Dachstein-Mammuthöhle is fairly undisturbed. Ice is present in two places not far from the western entrance, *Feenpalast* and *Saarhalle*, located some 100–150 m apart from each other but well separated from the touristic parts of the cave with respect to the ventilation regime. In the 1970s and 1980s the thickness of this horizontally layered ice still reached a thickness of up to 15 m. Georadar measurements conducted in 2007 revealed a thickness of the ice bodies at *Feenpalast* and *Saarhalle* of only up to 9 and 6 m, respectively (Behm and Hausmann, 2007; Hausmann and Behm, 2011). Prominent reflectors in the radar data appear to be related to layers rich in fine-crystalline cryogenic carbonate particles which, however, have not yet been sampled and analyzed. A constant decline of 30–40 cm/year was observed in *Feenpalast* since the onset of systematic level measurements in the 1990s (Spötl and Pavuza, 2016, Fig. 13.14).

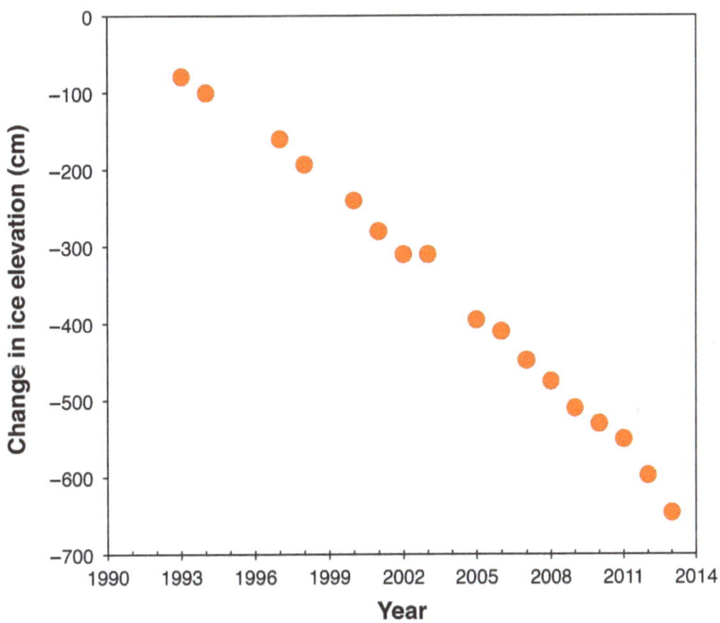

FIG. 13.14

Gradual decline of the ice height as monitored in *Feenpalast* of Dachstein-Mammuthöhle.

As there is also a significant lateral retreat of the ice body, the degradation of the ice is actually accelerating. If this trend persists the upper part of *Feenpalast* will be ice-free within the next 5 years.

A significantly slower but also almost linear decline of the ice of about 7–8 cm/year was observed in *Saarhalle* since the 1990s.

Organic material recovered close to the base of the ice at the retreating cliff in the highest part of *Feenpalast* yielded a radiocarbon age of 695 ± 35 BP (Mais and Pavuza, 2000; 1260–1389 cal AD, 2 sigma range) indicating that at least this part of *Feenpalast* was free of ice prior to the Little Ice Age. Two wood samples from deeper layers are slightly older (Table 13.1).

Fig. 13.15 shows a series of images taken at the ice cliff in *Feenpalast*, documenting the lowering of ice surface over the last two decades and the appearance of a prominent hole.

FIG. 13.15

Retreat of the ice body in *Feenpalast* over the course of 18 years. A small gallery with air flow led to the formation of the hole in the thin ice rim.

Photos from 1999, 2007, and 2010 courtesy of Heiner Thaler.

In September, 2009, a 5.3 m-long ice core was drilled at the site in *Saarhalle* showing the greatest ice thickness. During drilling, small amounts of liquid water were extracted at depths between 3.3 and 3.6 m. No melting of the ice, however, was observed at the base of the ice body. Tritium analyses of core samples indicate more than one source of water of different origin and age. The upper 1.2 m of ice most likely originated from precipitation fallen before the 1960s (based on Tritium data; Kern et al., 2011), i.e., younger ice layers have been lost by ablation similar to the situation at *Eispalast* of Eisriesenwelt (see above).

In nearby Dachstein-Rieseneishöhle there is some evidence that the ice level at the turn of the millennium was comparable to that during the 1930s, as documented by postcards (Mais and Pavuza, 1999). Unfortunately, such data are not available for the ice-bearing galleries of Dachstein-Mammuthöhle.

13.5.4 HUNDSALM EIS- UND TROPFSTEINHÖHLE

Located north of Wörgl in the lower Inn Valley of Tyrol (47.545°N, 12.027°E), this cave is an example of the many small sag-type caves in the NCA which contain perennial ice accumulations due to trapping of cold winter air. Hundsalm Eis- und Tropfsteinhöhle, however, is the only touristic ice cave in western Austria (which opened in 1967) and also the westernmost ice show cave in the western part of the Alps. The cave developed along a steeply dipping fault and has two adjacent shaft-type entrances, one higher than the other by a few meters. The lower of these two pits opens at 1520 m a.s.l. and is used to enter the cave via a staircase. The two shafts are crucial for the cooling of the cave during winter. Monitoring has shown that due to its slightly greater height (25 m) and larger diameter, the upper entrance acts as a "chimney" in winter, allowing cave air to ascend and exit the cave because of its lower density than the outside winter air. As a consequence, cold and dense outside air is drawn into the cave via the lower entrance (whose gated entrance is kept open during winter). This cold air cools the cave walls (Fig. 13.16) and the ice body and also results in a drying of the cave. While Hundsalm Eis- und Tropfsteinhöhle shows a dynamic air exchange at times when the outside air temperature drops below about −2°C, it acts as a "cold air trap" during the rest of the year with air cave air temperatures within a few tenths of a degree of the freezing point in the ice-bearing part and temperatures between 1.5 and 2.0°C in the southern, ice-free part.

Ice forms in two ways in this cave, as congelation ice and via transformation of snow and firn. The former results in the formation of seasonal ice stalagmites, stalactites and ice covering walls, but contributes only locally to the formation of floor ice. Snow enters the cave via both shafts and firn and grainy ice resulting from firn recrystallization constitutes a major part of the several meter-thick deposits in the northern part of the cave. Since 1967, when this cave was developed as a show cave, the snow and ice mass balance has been manipulated, because snow has been artificially introduced into the cave during mid-winter (in an attempt to preserve the ice body). In addition, the entrance shaft was slightly widened (and gated). Despite these measures, recent years have shown (a) an uninterrupted decline in the elevation of the main ice body (about 0.9 m since 2007), (b) a widening of the gap between the ice body and the cave wall (reaching 0.5–1 m), (c) generally less seasonal ice, and (d) a shorter "survival time" of these seasonal ice formations (disappearing in late summer). These observations in conjunction with long-term temperature data suggest that the cave will likely lose its perennial firn and ice deposits in the not too distant future, largely due to warmer and drier winters, which provide insufficient cooling of the cave to "survive" the warm season. A similar trend has been observed in other small ice

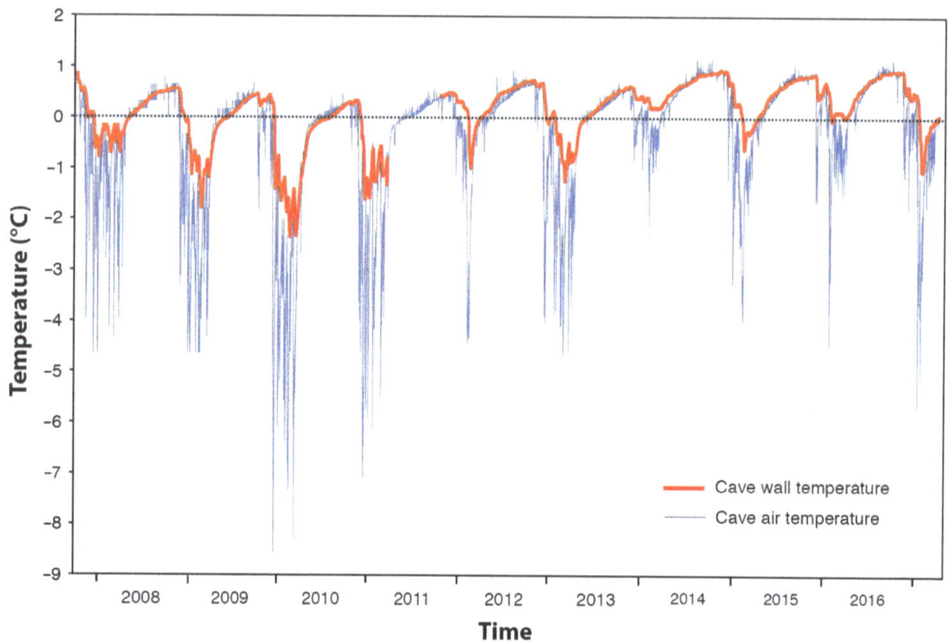

FIG. 13.16

During the last decade both the number of days with negative temperatures and the magnitude of winter cooling of the wall rock at Hundsalm Eis- und Tropfsteinhöhle as measured in a borehole at 1.3m depth have decreased. The extent of seasonal winter cooling of the rock surrounding the cave is a function of the cave air temperature (measured in the same chamber).

caves in Austria (Spötl and Pavuza, 2016), including Prax Eishöhle in Loferer Steinberge, which has transformed into a seasonal ice cave in the past few years.

Historical documents show that Hundsalm Eis- und Tropfsteinhöhle contained much more ice when discovered in 1921 (Spötl, 2013), a time when alpine glaciers were also advancing. Radiocarbon dating of 19 wood fragments embedded in the deeper part of the up to 7m thick firn and ice body provide clues about the long-term history (Spötl et al., 2014). This is the largest radiocarbon dataset for an Alpine ice cave and a summary of all dated samples currently available from ice caves in Austria is provided in Table 13.1. Although some of these fragments from Hundsalm Eis- und Tropfsteinhöhle may have been reworked during previous ablation episodes, the frequency distribution shows a clear maximum between the 15th and the 17th century (with samples dating back to the 13th century AD; Fig. 13.17), which coincides with the coldest centuries of the Little Ice Age, when East Alpine glaciers reached their Holocene maxima (Holzhauser et al., 2005; Nicolussi and Patzelt, 2000). An earlier period of ice build-up is dated to the 6th and 7th centuries AD, also coinciding with a cool climate (Büntgen et al., 2011) and positive glacier mass balances in the Alps. No samples date back to the first half of the Holocene. This is also confirmed by studies from other ice caves in Austria and the oldest dated wood samples are from the early part of the 5th millennium AD (Eisgruben-Eishöhle, Sarstein and Schneeloch, Schneealpe—Achleitner, 1995; Herrmann et al., 2010). This lack of wood fragments may be regarded as a hint that at least some Alpine ice caves were probably ice-free during most of the first

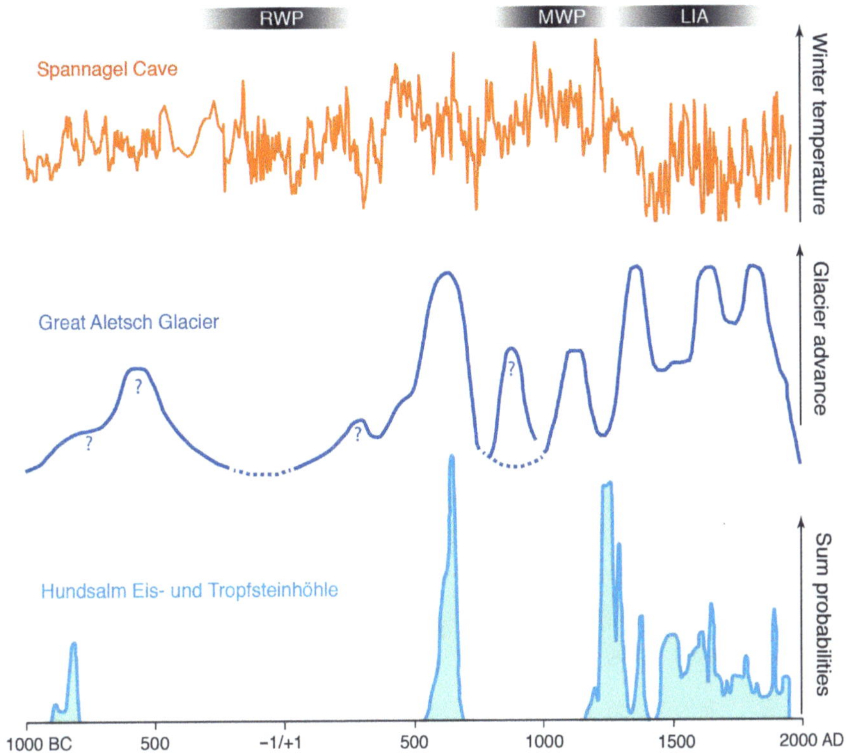

FIG. 13.17

Comparison between the abundance of radiocarbon-dated wood remains retrieved from the ice of Hundsalm Eis- und Tropfsteinhöhle (Spötl et al., 2014), length changes of Great Aletsch Glacier in Switzerland (Holzhauser et al., 2005) and winter temperatures reconstructed from oxygen isotopes in stalagmites from Spannagel Cave in western Austria (Mangini et al., 2005; Fohlmeister et al., 2013) over the past three millennia. RWP, Roman Warm Period; MWP, Medieval Warm Period; LIA, Little Ice Age.

half of the Holocene—a time interval characterized by generally negative mass balances also of Alpine glaciers (e.g., Joerin et al., 2008; Ivy-Ochs et al., 2009).

A microbiological study examined ice samples from the deep part of this ice body and found abundant bacteria (actinobacteria; Sattler et al., 2013). Cyanobacteria were also identified and were likely introduced from outside. Overall, the microbial community resembled communities known from accumulations of organic and inorganic material on Alpine glaciers (known as cryoconite).

As the name suggests, Hundsalm Eis- und Tropfsteinhöhle hosts both ice and speleothems, a combination which is rather uncommon. Studies have shown, however, that the speleothems in the ice-bearing part are ancient. The cave also contains a lower level, and the narrow connection between the two cave parts, which was artificially opened to allow exploration in 1984, was later closed by a door to prevent air exchange. This lower cave level has a temperature of 4.2±0.1°C, no ice, but active speleothem deposition, including abundant moonmilk (Reitschuler et al., 2012, 2015).

Table 13.1 Summary of Radiocarbon Dates of Organic Remains in East Alpine Ice Caves

Cave (Mountain Range)	Elevation of Main Entrance (m a.s.l.)	Lab Code	Conventional ^{14}C Age BP	Calibrated ^{14}C Age (2 Sigma)[a]	Reference
Eisgruben-Eishöhle (Sarstein)	1720				
		74,817	2210 ± 70	400–91 BC (0.99)	Achleitner (1995)
				67–65 BC (0.01)	
		74,818	4520 ± 50	3366–3088 BC (0.97)	Achleitner (1995)
				3058–3030 BC (0.03)	
Schneeloch (Schneealpe)	1450				
		LTL3900A[b]	2228 ± 50	395–184 BC (1.00)	Herrmann et al. (2010)
		LTL4709A[b]	242 ± 45	1495–1507 AD (0.01)	Herrmann et al. (2010)
				1511–1601 AD (0.21)	
				1616–1689 AD (0.39)	
				1729–1810 AD (0.31)	
				1925–1950[c] (0.08)	
		GrN-32288	4360 ± 30	3085–3064 BC (0.06)	Herrmann et al. (2010)
				3028–2904 BC (0.94)	
Beilstein Eishöhle (Hochschwab)	1308				
		GX-33495	700 ± 65	1216–1404 AD	Pavuza (unpubl.)
		GX-33634	200 ± 100	1495–1601 AD (0.15)	Pavuza (unpubl.)
				1616–1950[c] (0.85)	
		GrA-29972	135 ± 30	1672–1778 AD (0.43)	Pavuza (unpubl.)
				1799–1892 AD (0.41)	
				1907–1942 AD (0.16)	
Dachstein-Mammuthöhle (Dachstein)	1391				
			695 ± 35	1260–1316 AD (0.73)	Mais and Pavuza (1999), Kern et al. (2011)
				1354–1389 AD (0.27)	
			751 ± 45	1188–1300 AD (0.98)	Plan and Pavuza (unpubl.)
				1368–1381 AD (0.02)	
			1133 ± 40	776–794 AD (0.06)	Plan and Pavuza (unpubl.)
				799–988 AD (0.94)	
Kraterschacht (Sengsengebirge)	1531				
		LTL4510A	886 ± 45	1032–1224 AD (0.98)	Weißmair (2011)
				1234–1242 AD (0.02)	

Hundsalm Eis- und Tropfsteinhöhle (Brandenberg Alps)	1495				
HUN-H1		UBA-15860	326 ± 25	1485–1642 AD (1.00)	Spötl et al. (2014)
HUN-H2		UBA-15861	334 ± 25	1480–1640 AD (1.00)	Spötl et al. (2014)
HUN-H3		UBA-15862	38 ± 29	1695–1726 AD (0.19)	Spötl et al. (2014)
				1813–1838 AD (0.14)	
				1842–1853 AD (0.03)	
				1868–1918 AD (0.56)	
				1952–1955 AD (0.09)[c]	
HUN-H12		UBA-18582	250 ± 24	1528–1552 AD (0.06)	Spötl et al. (2014)
				1633–1669 AD (0.72)	
				1780–1798 AD (0.20)	
				1945–1951 AD (0.02)[c]	
HUN-H13		UBA-19071	800 ± 25	1189–1197 AD (0.02)	Spötl et al. (2014)
				1207–1274 AD (0.98)	
HUN-H14		UBA-19072	2664 ± 32	895–867 BC (0.14)	Spötl et al. (2014)
				863–795 BC (0.86)	
HUN-H16		UBA-20559	786 ± 31	1190–1196 AD (0.01)	Spötl et al. (2014)
				1207–1280 AD (0.99)	
HUN-H17		UBA-20560	790 ± 27	1210–1277 AD (1.00)	Spötl et al. (2014)
HUN-H18		UBA-21456	402 ± 29	1436–1521 AD (0.83)	Spötl et al. (2014)
				1575–1583 AD (0.01)	
				1590–1623 AD (0.15)	
HUN-H20		UBA-20561	1419 ± 30	582–661 AD (1.00)	Spötl et al. (2014)
HUN-H21		UBA-20710	1452 ± 30	560–650 AD (1.00)	Spötl et al. (2014)
HUN-H22		UBA-20888	361 ± 19	1455–1524 AD (0.56)	Spötl et al. (2014)
				1558–1564 AD (0.02)	
				1569–1631 AD (0.42)	
HUN-H25		UBA-21457	152 ± 29	1667–1708 AD (0.17)	Spötl et al. (2014)
				1718–1783 AD (0.34)	
				1796–1827 AD (0.12)	
				1832–1887 AD (0.18)	
				1911–1953 AD (0.19)[c]	

Continued

Table 13.1 Summary of Radiocarbon Dates of Organic Remains in East Alpine Ice Caves—cont'd

Cave (Mountain Range)	Elevation of Main Entrance (m a.s.l.)	Lab Code	Conventional ^{14}C Age BP	Calibrated ^{14}C Age (2 Sigma)[a]	Reference
HUN-H26		UBA-20889	688 ± 22	1272–1305 AD (0.78)	Spötl et al. (2014)
				1364–1384 AD (0.22)	
HUN-H27		UBA-21458	647 ± 32	1281–1328 AD (0.45)	Spötl et al. (2014)
				1341–1395 AD (0.55)	
HUN-H28		UBA-21459	172 ± 29	1660–1697 AD (0.18)	Spötl et al. (2014)
				1725–1815 AD (0.55)	
				1835–1877 AD (0.07)	
				1917–1952 AD (0.20)[c]	
HUN-H29		UBA-28124	333 ± 21	1485–1640 AD (1.00)	Spötl (unpubl.)
H-1		GrN-23952	1380 ± 30	607–680 AD (1.00)	Spötl et al. (2014)

[a] *Calibrated using CALIB 7.1 using the INTCAL13 database. Values in parentheses denote relative area under probability distribution.*
[b] Bone samples.
[c] Denotes influence of radiocarbon derived from nuclear testing.

13.6 OUTLOOK

Kern and Perşoiu (2013) made an attempt to compile historical data on ice volume change from caves in different parts of the world. For European ice caves, their data showed a decline of ice levels since about the middle of the 20th century and an acceleration of this trend in the 1980s. Observations from Austrian ice caves were not included in this compilation, but are generally consistent with these findings. A significant number of the 1200 caves hosting perennial ice, firn or snow in this mountainous country have either shown a strong decline in ice volume or a transition into a seasonal ice cave in recent decades. The larger the ice cave, the less prone they appear to be to the atmospheric warming trend of the 20th and 21st centuries, whose amplitude in the Alps is about twice the mean of the Northern Hemisphere (Auer et al., 2014). Some of the largest ice caves in Austria are also managed as show caves, and their meteorology and ice mass balance have been in part manipulated, e.g., by restricting the outflow of cold cave air during summer. Only scattered and discontinuous observations exist about Austrian ice caves prior to the 20th century (e.g., Fugger, 1888; Klappacher and Mais, 1999; Behm et al., 2009), but they consistently report markedly larger ice accumulations, reflecting the "glacier-friendly" climate of the Little Ice Age. Significant ice build-up during this period is also recorded by radiocarbon-dated organic remains in ice bodies from several East Alpine caves. These data also suggest generally small ice volumes during the preceding Medieval Warm Period and again positive mass balances during the Bronze Age. The emerging picture shows a broadly synchronous evolution of surface glaciers in the Alps (e.g., Holzhauser et al., 2005) and their counterparts in the subsurface of karstified mountains during the Holocene. Despite careful studies, none of the organic remains dates back to the first half of the Holocene, a time period when alpine glaciers were on average smaller than during the second half. Under current climate boundary conditions an increasing number of small- to medium-sized ice and firn bodies will likely disappear from East Alpine caves within the coming years. Only large caves at higher elevation and characterized by vigorous ventilation have a good chance to survive the next decades in spite of increasing winter temperatures.

ACKNOWLEDGMENTS

We are indebted to the following colleagues for providing photographs and data: Robbie Shone, Harald Zeitlhofer, Heiner Thaler, Helmuth Planer, Walter Klappacher, and Clemens Tenreiter.

REFERENCES

Abel, G., 1932. Die Feuertal-Eishöhle im Toten Gebirge und ihre Begehung. Der Bergfreund 5/6, 87–92.

Abel, G., 1965. Veränderungen am Höhleneis der Eisriesenwelt (Salzburg). Die Höhle 16, 105–106.

Achleitner, A., 1995. Zum Alter des Höhleneises in der Eisgruben-Eishöhle im Sarstein (Oberösterreich). Die Höhle 46, 1–5.

Auer, I., Foelsche, U., Böhm, R., Chimani, B., Haimberger, L., Kerschner, H., Koinig, K.A., Nicolussi, K., Spötl, C., 2014. Vergangene Klimaänderung in Österreich. In: Österreichischer Sachstandsbericht Klimawandel 2014 (AAR14). Austrian Panel on Climate Change (APCC). Austrian Academy of Sciences, Vienna, pp. 227–300.

Behm, M., Hausmann, H., 2007. Eisdickenmessungen in alpinen Höhlen mit Georadar. Die Höhle 58, 3–11.

Behm, M., Dittes, V., Greilinger, R., Hartmann, H., Plan, L., Sulzbacher, D., 2009. Decline of cave ice—a case study from the Austrian Alps (Europe) based on 416 years of observation. In: Proceedings 15th Int. Congr. Speleol., Kerrville, TX, July 19–26, 2009, vol. 3, pp. 1413–1416.

Behm, M., Plan, L., Seebacher, R., Buchegger, G., 2016. Dachstein. In: Spötl, C., Plan, L., Christian, E. (Eds.), Höhlen und Karst in Österreich. Oberösterreichisches Landesmuseum, Linz, pp. 569–588.

Bock, H., Lahner, G., Gaunersdorfer, G., 1913. Höhlen im Dachstein und ihre Bedeutung für die Geologie, Karsthydrographie und die Theorien über die Entstehung des Höhleneises. In: Verein für Höhlenkunde in Österreich, Graz. Reprint 2010 (Verein für Höhlenkunde Hallstatt-Obertraun).

Büntgen, U., Tegel, W., Nicolussi, W., McCormick, M., Frank, D., Trouet, V., Kaplan, J.O., Herzig, F., Heussner, K.-U., Wanner, H., Luterbacher, J., Esper, J., 2011. 2500 years of European climate variability and human susceptibility. Science 331, 578–582.

Fohlmeister, J., Vollweiler, N., Spötl, C., Mangini, A., 2013. COMNISPA II: update of a mid-European isotope climate record, 11 ka to present. The Holocene 23, 749–754.

Fritz, A., 1977. Pollenanalytische Untersuchungen von Alteisproben. Carnithia II 167/87, 217–226.

Fugger, E., 1888. Beobachtungen in den Eishöhlen des Untersberges bei Salzburg. Mitt. Ges. Salzburger Landeskunde 23, 1–99.

Fugger, E., 1891. Eishöhlen und Windröhren. Jahresbericht der k.k. Ober-Realschule in Salzburg 24, 1–70.

Fugger, E., 1892. Eishöhlen und Windröhren. Zweiter Theil. Jahresbericht der k.k. Ober-Realschule in Salzburg 25, 71–134.

Fugger, E., 1893. Eishöhlen und Windröhren. Dritter Theil (Schlus). Jahresbericht der k.k. Ober-Realschule in Salzburg 25, 135–223.

Geyer, E., Seebacher, R., Tenreiter, C., Knobloch, G., 2016. Totes Gebirge. In: Spötl, C., Plan, L., Christian, E. (Eds.), Höhlen und Karst in Österreich. Oberösterreichisches Landesmuseum, Linz, pp. 599–622.

Gressel, W., 1965. Veränderungen am Höhleneis der Eisriesenwelt (Salzburg). Die Höhle 16, 105–106.

Gruber, V., Ringer, W., Gräser, J., Aspek, W., Gschnaller, J., 2014. Comprehensive investigation of Radon exposure in Austrian tourist mines and caves. Radiat. Prot. Dosim. 162, 78–82.

Hartmann, H., Hartmann, W., 1984. Das Geldloch am Ötscher in Niederösterreich: Die Erforschungsgeschichte einer Höhle im Spiegel von vier Jahrhunderten. Die Höhle 35, 155–166.

Hauser, E., Oedl, R., 1926. Eisbildungen und meteorologische Beobachtungen. In: Angermayer, E., Asal, A., Czörnig-Czernhausen, W., Hauser, E., Lehmann, O., Oedl, R., Pia, J., Wettstein-Westersheim, O. (Eds.), Die Eisriesenwelt im Tennengebirge (Salzburg).—Speläologische Monographien. vol. 6. Speläol. Inst., Vienna, pp. 77–105.

Hausmann, H., Behm, M., 2011. Imaging the structure of cave ice by ground-penetrating radar. Cryosphere 5, 329–340.

Herrmann, E., 2017. Die Obstanser Eishöhle in den Karnischen Alpen, Osttirol. Die Höhle 68, 40–58.

Herrmann, E., Pucher, E., Nicolussi, K., 2010. Das Schneeloch auf der Hinteralm (Schneealpe, Steiermark): Speläomorphologie, Eisveränderung, Paläozoologie und Dendrochronologie. Die Höhle 61, 57–72.

Holzhauser, H., Magny, M., Zumbühl, H.J., 2005. Glacier and lake-level variations in west-central Europe over the last 3500 years. The Holocene 15, 789–801.

Ivy-Ochs, S., Kerschner, H., Maisch, M., Christl, M., Kubik, P.W., Schlüchter, C., 2009. Latest Pleistocene and Holocene glacier variations in the European Alps. Quat. Sci. Rev. 28, 2137–2149.

Jäger, V., 1919. Dr. Eberhard Fugger. Sein Werk. Mitt. Ges. Salzburger Landeskunde 59, 72–80.

Jansky, W., Tenreiter, C., Pürmayr, L., 2008. Das Feuertal-Höhlensystem als Teil des Schönberg-Höhlensystems. Die Höhle 59, 83–95.

Joerin, U.E., Nicolussi, K., Fischer, A., Stocker, T.F., Schlüchter, C., 2008. Holocene optimum events inferred from subglacial sediments at Tschierva glacier, eastern Swiss Alps. Quat. Sci. Rev. 27, 337–350.

Kern, Z., Perşoiu, A., 2013. Cave ice—the imminent loss of untapped mid-latitude cryospheric palaeoenvironmental archives. Quat. Sci. Rev. 67, 1–7.

Kern, Z., Fórizs, I., Pavuza, R., Molnar, M., Nagy, B., 2011. Isotope hydrological studies of the perennial ice deposit of Saarhalle, Mammuthöhle, Dachstein Mts, Austria. Cryosphere 5, 291–298.

Killian, K., 1935. Die Obstanser Eishöhle. Mitt. Höhlen- u. Karstforsch., 123–125.

Klappacher, W., 1996. Salzburger Höhlenbuch. vol. 6 Landesverein für Höhlenkunde, Salzburg.

Klappacher, W., Mais, K., 1999. Ice cave studies in Salzburg and the work of Eberhard Fugger 1842–1919. Slovesnsky Kras (Acta Carsologica Slovaca) 37, 115–130.

Kral, F., 1968. Pollenanalytische Untersuchungen zur Frage des Alters der Eisbildungen in der Dachstein-Rieseneishöhle. Die Höhle 19, 41–51.

Mais, K., Pavuza, R., 1999. Aktuelle höhlenklimatische Aspekte der Dachstein-Rieseneishöhle. Die Höhle 50, 126–140.

Mais, K., Pavuza, R., 2000. Hinweise zu Höhlenklima und Höhleneis in der Dachstein-Mammuthöhle (Oberösterreich). Die Höhle 51, 121–125.

Mais, K., Trimmel, H., 1992. 400 Jahre Forschung in den Höhlen des Ötschers (Niederösterreich). Die Höhle 43, 74–79.

Mangini, A., Spötl, C., Verdes, P., 2005. Reconstruction of temperature in the Central Alps during the past 2000 years from a $\delta^{18}O$ stalagmite record. Earth Planet. Sci. Lett. 235, 741–751.

May, B., Spötl, C., Wagenbach, D., Dublyansky, Y., Liebl, J., 2011. First investigations of an ice core from Eisriesenwelt cave (Austria). Cryosphere 5, 81–93.

Milius, J., Petters, C., 2012. Eisriesenwelt—from laser scanning to photo-realistic 3D model of the biggest ice cave on earth. In: Jekel, T., Car, A., Strobl, J., Griesebner, G. (Eds.), GI-Forum 2012: Geovisualisation, Society and Learning. Wichmann, Berlin, pp. 513–523.

Nicolussi, K., Patzelt, G., 2000. Untersuchungen zur holozänen Gletscherentwicklung von Pasterze und Gepatschferner (Ostalpen). Z. Gletscherk. Glazialgeol. 36, 1–87.

Obleitner, F., Spötl, C., 2011. The mass and energy balance of ice within the Eisriesenwelt cave Austria. Cryosphere 5, 245–257.

Pavuza, R., Mais, K., 1999. Aktuelle höhlenklimatische Aspekte der Dachstein-Rieseneishöhle. Die Höhle 50, 126–140.

Pavuza, R., Spötl, C., 2017. Neue Daten zu Vorkommen und Entstehung kryogener Calcite in ostalpinen Höhlen. Die Höhle 68, 100–106.

Petters, C., Milius, J., Buchroithner, M.F., 2011. Eisriesenwelt: terrestrial laser scanning and 3D visualisation of the largest ice cave on earth. In: Proceedings European LiDAR Mapping Forum, Salzburg, Austria. http://visual-geodata3d.com/img/Eisriesenwelt_Milius-Petters-Buchroithner_ELMF-2011.pdf.

Pillwein, E., 1919. Dr. Eberhard Fugger. Sein Leben.. Mit. Ges. Salzburger Landeskunde 59, 65–72.

Plan, L., Xaver, A., 2010. Geomorphologische Untersuchung und genetische Interpretation der Dachstein-Mammuthöhle (Österreich). Die Höhle 61, 18–38.

Pointner, P., Klappacher, W., 2016. Tennengebirge. In: Spötl, C., Plan, L., Christian, E. (Eds.), Höhlen und Karst in Österreich. Oberösterreichisches Landesmuseum, Linz, pp. 553–568.

Reitschuler, C., Schwarzenauer, T., Lins, P., Wagner, A.O., Spötl, C., Illmer, P., 2012. Zur Mikrobiologie von Bergmilch. Die Höhle 63, 3–17.

Reitschuler, C., Lins, P., Schwarzenauer, T., Spötl, C., Wagner, A.O., Illmer, P., 2015. New undescribed lineages of non-extremophilic archaea form a homogeneous and dominant element within alpine moonmilk microbiomes. Geomicrobiol. J. 32, 890–902.

Saar, R., 1955. Die Dachstein-Rieseneishöhle nächst Obertraun und ihre Funktion als dynamische Wetterhöhle. Jahrb. Oberösterr. Musealver. 100, 263–319.

Saar, R., 1956. Eishöhlen, ein meteorologisch-geophysikalisches Phänomen. Geogr. Ann. 38, 1–63.

Saar, R., Pirker, R., 1979. Geschichte der Höhlenforschung in Österreich. Die Höhle Beiheft 13, 1–96.

Sattler, B., Larch, P., Rambacher, J., Spötl, C., 2013. Das Eis der Hundsalm Eis- und Tropfsteinhöhle als Lebensraum für mikrobielle Gemeinschaften. Die Höhle 64, 15–24.

Schmeidl, H., Kral, F., 1969. Zur pollenanalytischen Altersbestimmung der Eisbildungen in der Schellenberger Eishöhle und in der Dachstein-Rieseneishöhle. Jahrb. Vereins zum Schutze der Alpenpflanzen und -tiere 34, 67–84.

Schöner, W., Weyss, G., Mursch-Radlgruber, E., 2011. Linkage of cave-ice changes to weather patterns inside and outside the cave Eisriesenwelt (Tennengebirge, Austria). Cryosphere 5, 603–616.

Spötl, C., 2008. Kryogene Karbonate im Höhleneis der Eisriesenwelt. Die Höhle 59, 26–36.

Spötl, C., 2013. Die Entdeckungsgeschichte und der ursprüngliche Zustand der Hundsalm Eis- und Tropfsteinhöhle. Höhlenkundliche Mitteilungen des Landesvereins für Höhlenkunde in Tirol 51, 22–33.

Spötl, C., Cheng, H., 2014. Holocene climate change, permafrost and cryogenic carbonate formation: insights from a recently deglaciated, high-elevation cave in the Austrian Alps. Clim. Past 10, 1349–1362.

Spötl, C., Reimer, P.J., Luetscher, M., 2014. Long-term mass balance of perennial firn and ice in an Alpine cave (Austria): constraints from radiocarbon-dated wood fragments. The Holocene 24, 165–175.

Spötl, C., Pavuza, R., 2016. Eishöhlen und Höhleneis. In: Spötl, C., Plan, L., Christian, E. (Eds.), Höhlen und Karst in Österreich. Oberösterreichisches Landesmuseum, Linz, pp. 139–154.

Spötl, C., Egger, M., Mangini, A., Pavuza, R., 2017. Das Karstsystem am Obstanser See (Kartitsch, Osttirol)—geowissenschaftliche Ergebnisse. Die Höhle 68, 59–78.

Thaler, K., 2008. Analyse der Temperaturverhältnisse in der Eisriesenwelt-Höhle im Tennengebirge anhand einer 12 jährigen Messreihe. Unpubl. MS thesis, Univ. Innsbruck.

Weißmair, R., 1995. Höhleneisbildung aus Schnee und Eisdynamik im Kraterschacht (Sengsengebirge, Oberösterreich). Die Höhle 46, 32–37.

Weißmair, R., 2011. Eisdatierung und Eisveränderungen im Kraterschacht (1651/24, Sengsengebirge, Oberösterreich) zwischen 1992 und 2009. Die Höhle 62, 27–30.

Wimmer, M., 2008. Eis- und Lufttemperaturmessungen im Schönberg-Höhlensystem (1626/300) und Modellvorstellungen über den Eiszyklus. Die Höhle 59, 13–25.

Zeitlhofer, H., Knobloch, G., 2008. Die Raucherkarhöhle (1626/55) als Teil des Schönberg-Höhlensystems. Die Höhle 59, 73–82.

CHAPTER

ICE CAVES IN MONTENEGRO AND BOSNIA AND HERZEGOVINA 14

Goran Barović*, Ditta Kicińska[†], Mihajlo Mandić[‡], Jasminko Mulaomerović[§]

University of Montenegro, Podgorica, Montenegro[*] Adam Mickiewicz University, Poznań, Poland[†] Geological Survey of Serbia, Belgrade, Serbia[‡] Center for Karst and Speleology, Sarajevo, Bosnia and Herzegovina[§]

CHAPTER OUTLINE

14.1 INTRODUCTION

Despite the fact that the area of Montenegro has been the target of much worldwide research, it has been only partially explored. The biggest drawback of previous research was a lack of systematically recording and publishing of the findings. The first official cave explorations were conducted by foreign researchers: Rovinski, Martel, Lahner, Gessman, Absolon, Komarek, Knirich, Remmy, Pretner, Hadzi, Brodar, and others. Systematic explorations in Montenegro started in the second half of the last century and were organized by the Institute for Geological Exploration, which hired exploration clubs from the former Yugoslav republics. With the help of those explorers, a domestic speleological staff formed and later carried out explorations independently. Today speleological explorations in Montenegro are partially arranged by the Agency for Environmental Protection, including the maintenance of a systematic cadastre of speleological objects. This cadastre database is created from data that exploration clubs send to the agency. Although a few thousands objects are

Ice Caves. https://doi.org/10.1016/B978-0-12-811739-2.00014-0

registered, many objects are not included in this database. In addition to our Montenegrin explorers, the area of Montenegro has been explored by speleologists from almost all the former Yugoslav republics and by explorers from France, England, Poland, and so on, individually or as participants in speleological expeditions. The results of those exploration are today in their clubs' archives, and a few of those reports were given to the appropriate institutions in Montenegro. The results of many explorations' studies are in different institutions that are interested in those activities, such as the Institute for Geological Research, Montenegrin Electric Enterprise, Bauxite Mines Niksic, former defense headquarters, and so on. These institutions were the financiers of individual projects within which the research projects were conducted. Because of the poor recordkeeping of the investigated features and their markings, it often happens that an object is "shot" more than one time.

The results of some explorations in which the studied topic is partially explained have been published in various scientific and other journals and at professional conferences, but all this is just a drop in the bucket compared to the actual state of researched features. The main task of the Speleological Organization and the Environmental Protection Agency is to collect information on all researched speleological features in the future. By determining the real state of exploration, further and better planning will be easier to conduct, and also studying some features repetitiously will be avoided, which is important because we still have many unexplored ones.

14.2 BASIC PHYSICAL AND GEOGRAPHICAL CHARACTERISTICS OF MONTENEGRO

14.2.1 RELIEF

"To the relief of Montenegro more than any other, the term karst is associated. In our language there are many synonyms for the karst (rocky place). By this is meant a set of forms of relief and characteristic appearances and flow of water in carbonate rocks" (Radojičić, 1996). Carbonate rocks form more than two-thirds of the area in Montenegro, but basically it consists of two types of rocks: limestone comprising more than 50% $CaCO_3$ and dolomite with more than 50% of $CaMg(CO_3)$, with some different other rocks that have impurities of different elements.

The territory of Montenegro is in the southeastern part of the Dinarides and has a very complex geological basis, on which erosion processes left a deep mark and caused the formation of a very dynamic relief. The relief of Montenegro can be divided into external and internal parts (according to the division of the Dinarides). The outer part is in the geospace of the deep karst, and the internal part is in the geospace of the fluvio-karst and fluvio-glacial relief. The main characteristic of the relief of Montenegro is that, on a very small geographic space, there are very large height differences, as noted in the following: "From the total area of Montenegro (13,812 km^2) to 200 meters above sea level is only 10% of the land, and between 200 and 100 m above sea level is 35%, and between 1000 and 1500 m above sea level is 40% and over 1500 m is 15% of the land" (Radojičić, 1996).

14.2.2 GEOTECTONIC RELATIONS WITH LITHOLOGICAL BASIS

The territory of Montenegro is divided into seven geotectonic zones: the Adriatic zone, the paraautochthonous, coastal fleece zone, the Budva cover zone, the deep karst zone, the Kuc overthrust zone, the Durmitor overthrust zone, and the Pljevlja overthrust zone. We will briefly describe the zones that, based on their characteristics, had a predisposition for the creation of speleological features addressed in this paper (Fig. 14.1).

FIG. 14.1

Map of basic tectonic zones of Montenegro.

14.2.2.1 The zone of deep karst

The zone of deep karst consists of Mesozoic limestone and spreads between the Budva zone in the south and the Kuc overthrust in the north. It is bordered by the Orijen, Lovcen, Sutorman, Rumija, Garac, Budos, Lisac, Njegos, and Somina mountain massifs. On the geospace of this zone, we can follow the syncline that spreads in the direction of the Duga-Nikisic field, the Bjelopavlic plain, the Podgorica-Skadar pit, and the Cetinje-Tresnjevo-Vilusi anticline. This geospace is seismically unstable, especially along a few rift lines.

14.2.2.2 Kuc overthrust

The geotectonic zone of the *Kuc overthrust* constructed the Golija-Vojnik-Maganik-Prekornica anticline, which is composed of carbonate rocks and shales. Shales and eruptives in the geospace of this zone are found in Niksic Zupa and are from the period between the Permian and Cretaceous geologic periods.

14.2.2.3 Durmitor zone

The *Durmitor zone* consists of Durmitor fleece, Mesozoic limestone, shale, and eruptives. The Durmitor-Bjelasica-Komovi-Prokletije anticline is in this zone. Drilling to great depths has shown that a large part of this area goes over the Kuc zone.

14.2.3 MAIN RELIEF UNITIES

Many explorers have studied the relief zones of Montenegro, but the study that especially showed the characteristics of the karst was done by B. Radojicic, and that work is presented here. The study showed that, in terms of relief zones in the territory of Montenegro, there are more areas of zoning: the Montenegrin coast, the area of deep karst, valleys of the middle part of Montenegro, the area of high mountains and plateaus, and Northeast Montenegro. Here we will also give some information about zones in which a greater number of speleological features are found.

14.2.3.1 The area of deep karst

The area of deep karst, also known as the area of Old Montenegro, is well separated from other geospaces. Toward the coast, Montenegro is bordered by the previously mentioned mountain range (Orjen-Lovcen-Sutorman-Rumija), and toward the valley in its central area, Montenegro, is bordered by the Garac-Budos-Zla gora-Njegos i Somina mountain range. The western border of this region is not natural, but is a generally accepted border between Montenegro and Hercegovina. This zone represents the most characteristic geospace of karst in the world, where all karst forms are developed in full measure. The thickness of limestone layers in this geographic space ranges over 400 m. An important feature of this place is Crkvice, which annually receives a maximum rainfall of 8000 mm, which was at one time the maximum rainfall in Europe (Radojičić, 1996). With all this rainfall, it is interesting that, in this entire geographic space, there is no surface water flow; all the water goes through the karst landscape, with only a minor portion evaporating. Special forms of relief in this geospace are karst fields, and in this part of Montenegro, there are several of them. The most important ones are the Cetinje field, Njegusi field, Dragaljsko field, and Grahovo field in which many speleological features have been examined. From the perspective of the topic being discussed, this geographical space is very distinctive because the entire area of Montenegro receives a large amount of rainfall, but has a negligible number of surface flows (Fig. 14.2).

14.2.3.2 The area of high mountains and plateaus

The area of high mountains and plateaus is dominant in the territory of Montenegro. The whole area is directed toward Dinara (which is on the border of Bosnia and Herzegovina and Croatia), though on the southeastern side, where two mountain ranges clearly branch out, the area turns in a

FIG. 14.2

Main relief units of Montenegro.

southeast/northwest direction. In the first branching mountain range, a geological structure of limestone prevails, which has resulted in a wide range of various karst landscapes, similar to the area of deep karst. The first range comprises Golija-Vojnik-Maganik-Prekornica. The other mountain range comprises Volujak with Maglic and Bioc-Ljubisa-Durmitor-Sinjajevina-Bjelasica-Komovi and Visitor-Prokletije. Two surfaces stand out in this area: the Piva mountain surface and the lake surface. This geospace is the hydrographic source of the most rivers in Montenegro, and is the geographical space with the most developed river network.

Previous cave exploration in the Montenegro area determined that most of the features that are completely or partly filled with snow and ice, particularly cave entrances, are located in the Durmitor and Prokletije mountain massifs. Based on the available data, in subsequent chapters, we will review the features of these two zones.

14.3 ICE CAVES IN THE DURMITOR MOUNTAINS

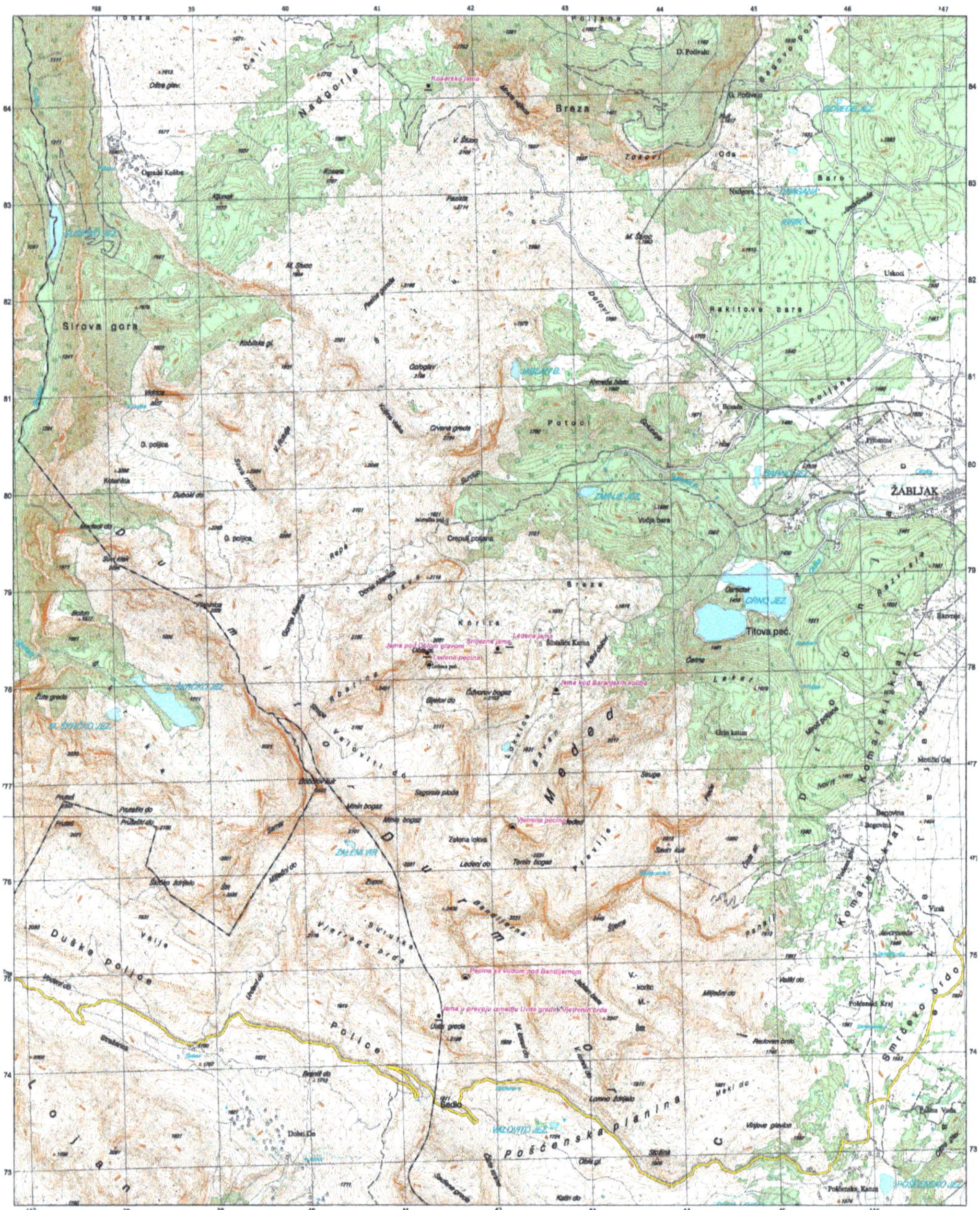

FIG. 14.3

Map of objects with snow and ice in Durmitor massif.

14.3.1 GROUP OF FACILITIES OF OBLA GLAVA VICINITY

14.3.1.1 Cave at Baranjske cottages

The Baranjske cave is located on the right side of the marked trail that leads from Žabljak toward the top of the Obla Glava, between Baranjske kolibe and Old Pasha in Lokvice. It is about 150m from Baranjske kolibe, but the entrance cannot be seen from the path. The entrance to the cave is an ellipse, a long axis oriented north, 7m in length and 3.5m in width. At the cave's entrance and to a depth of 4m, the channel is vertical, and it narrows to 1.5–1.2m, after which the dimensions of the channel gradually increase to the bottom of the cave. The bottom is initially covered with blocks and debris that, in some places, can be seen through layers of snow. The snow does not cover the entire surface of the bottom, but only the part just below the entrance. The cave falls to the north at an angle of about 30 degrees. The total depth of the cave is 18.5m. There are no accumulative forms, nor are there traces of water-forming ponds (Fig. 14.3).

14.3.1.2 The Ice Cave

The Ice Cave is the most famous speleological sight in the Durmitor National Park and one of its biggest tourist attractions. It is situated on the northeast slope of the Obla Glava, and the path to the Ice Cave is marked. The entrance to the cave is at the top of a moderate talus that is mostly covered with vegetation. The width of the entrance is 22m and the height is 20m. The real dimensions of the entrance are far greater, but it is largely filled with rock material of different sizes, which were created by the talus. The talus, by which you go down into the cave, was formed by the accumulation of broken rock and debris. The width of the talus is 35m, and the average incline is −45 degrees. From the entrance at the bottom of the talus, the channel narrows, and 7m from the entrance, the width is 13.5m and the height is 5m. In this part of the cave, the talus's material is bonded by clay and humus materials and can be seen only sporadically. During the winter and spring, this part is covered with layers of snow. At the bottom of talus, the channel retains the same dimensions, but after passing a group of columns and stalagmites located along the left wall, you enter the hall. The floor of the hall is covered with a permanent coating of ice that forms the stalagmites, whose height, depending on the time of year, reach 6m. In the recesses of ice on the bottom, a small reservoir of water that periodically freezes or melts forms lakes. As you go from the entrance, the height of the hall increases so that the central part is about 30m high. The entrance channel, about 15m long with a floor inclination of +30 degrees, continues from the hall. In this channel, the floor is completely covered with debris that is covered only in places with ice crust, created by the dripping water freezing. The total difference in height is −9m.

For the most part, there are no accumulative forms, and the hydrological function is reduced to that of dripping and leaking water. Signs of a significant accumulation of water are not noted in the hall.

14.3.1.3 Snowy Pothole

The Snowy Pothole cave is located northeast of Obla Glava, at the edge of the slope leading to the Korita plateau in the direction of the Obla Glava-Sibalica pasture. The diameter of the entrance is about 5m and thereafter continues as a vertical channel 90m deep. It cuts vertically at four places. The first vertical cut is 11m long, about 5m in diameter, and ends with a 4-meter-high snow cone. Through the cone, to the extension of the cave, you pass through an opening with a diameter of 1m that was created by melting and collapsing of snow near the wall. Then comes the new cascade, about 11m high, where the channel diameter reduces to 1.5–2m. At the bottom of this section is a thick ice plate with an inclination of 45 degrees to the north. At this point, the channel expands because of increased corrosion and forms a small room. The ice is compact, and dripping water forms ice stalagmites with diameters up to 0.5 m and heights of more than 1 m. Farther on, there is a new vertical with walls mainly covered with ice. The width of the channel is only 0.75 m, and it has an elongated shape in the direction of the initial crack. Dimensions of this part of the channel gradually increase with the depth, and the ice crust angles to the north, after which you come to the fourth section, which is 16 m long. This is also the most spacious part of the channel; it has a trapezoid shape, conditioned by two coupled systems of cracks, whose width and length are about 12 m. Sediments of ice are on the bottom. Next you come to the last vertical, which is 25 m long. This steep fractured part of the channel is completely covered in ice. At first, it narrows to 1 m in diameter, and at the tenth meter, it expands to 3 m and keeps these dimensions until the end of the passable part of the channel, where it becomes a crack closed with ice. The total depth of the cave is 90 m.

Except for snow in the cave's entrance and ice in its deeper parts, there are no phenomena related to hydrological activity or accumulative chemical formations (Fig. 14.4).

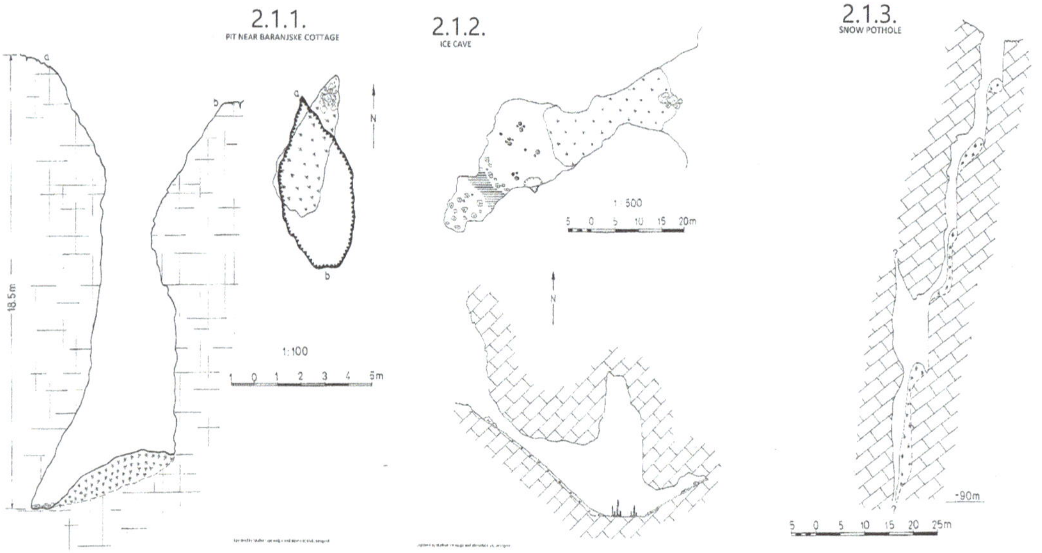

FIG. 14.4

Speleological profiles of cave at Baranjske cottages (2.1.1), the Ice Cave (2.1.2), and the Snowy Pothole (2.1.3).

14.3.1.4 The Ice Cave

The Ice Cave is close to the Snow Cave, at about 75 m toward the northeast in the direction of the Šibalić pasture. The facility has two entrances, with one below the other. The upper entrance is in the shape of a prolate, or horizontal crack, about 1.5 m long that continues in a vertical channel with a height of 12 m, through which you come to the lower entrance. This entrance also has small dimensions and has a triangular shape with a width of 2.5 m and a height that does not go above 1.5 m. Then there is 5-meter-long horizontal channel through which you arrive at the joint with the upper-entrance vertical. From the joint, the channel continues at an angle of −65 degrees for about 30 m. The channel's cross-section is elliptical, and its longer axis does not exceed 1.5 m. Through this channel, you enter the "Ice Hall," through the hole located in the ceiling. The length of the hall is 27 m, and the width is from 12 to 15 m. In the lowest part of the hall is an ice plate whose thickness reaches more than 7 m. This plate completely covers the floor of the hall, so we can assume that its real depth may be much greater. The total depth of the facility is 50 m.

In hydrological terms, dripping water is registered and is most intense in the "Ice Hall," which in combination with negative temperatures causes the formation of stalagmites on the ice panel and stalactites on the hall's ceiling.

14.3.1.5 The cave under Obla Glava

This cave is at the northeast foot of the Obla Glava, at the place where its foot passes into the Jamski plateau. The entrance is located on a small plateau at a relative height of 30 m above the Jamsko plateau. The plateau in which the cave resides is completely reshaped into a scarp with well-developed, deep scarps. You enter the cave through two of the scarps, which are 17 m apart. One of the entrances is very narrow and is physically impassible. The other entrance, the north one, is a scarp that extends into a steep channel that is 2 m wide and 5 m long. This curved channel leads to the first hall with a few steps made of collapsed blocks. At the bottom of the hall, whose diameter is 7.5 m, is an ice plate with a hole that leads to a manhole in the east part of the hall. The depth of this manhole is about 15 m, and the width gradually decreases and ends as a crack. The ice plate is about 2.5 m thick; it is tilted to the south and continues as the floor of the channel leading to deeper parts of the cave. This channel leads directly to another hall with a length of 22 m and a slope of −22 degrees. The width of the channel ranges from 5 to 6 m and the height from 2 to 3 m. An ice coating appears at 19 m, and farther on is the stem rock. In the second hall, the channel reaches a height of 2 m from the floor. This hall is the most spacious part of the cave. It is roughly round-shaped, with a diameter of 15 m and a height of more than 20 m. Like the first hall, it has larger dimensions at the ceiling level than at the floor level. On the floor, you can see two holes. The south one is a manhole with a diameter of 1.5 m and a depth of 12 m that ends in fallen rock material, mainly debris. The north manhole is the entrance into the graded part of the cave, which is at the bottom of the other hall. This part of the channel consists of several gradually arranged compartments, some of which are sloped and covered with ice. It is actually the biggest part of channel system, with a total length 80 m and a height difference of −62 m. The thickness of the ice ranges from 1.5 to 2 m. At the end of the tongue, which is built from granular snow, the channel narrows into a circular diameter, with a cross-section of length ranging from 1 to 10 m. The section ends with a ice threshold that is 4 m wide. It is an ice cap that prevents further progress, but it is likely that the channel continues. Through the narrow crack, on the left of the cap, it is possible to pass into a new vertical, 5 m high, that leads to a new cave channel. The length of this part is 12 m, the width of the channel is 3 m, and the height ranges from 2 to 3 m. At the end of this section, you come to a new vertical, which is passable for 10 m and then continues as a narrow, impassable crack. The total depth of the cave is 100 m.

In hydrological terms, except for the occurrence of ice and granular snow, there is no dripping water or significant traces of water flow down the walls.

14.3.1.6 Vjetrena cave

This cave is located in the south perimeter of the Lokvice cirque, at the foot of the northern slope of Terzin Bogaz. Its tunnel characteristic results in specific climatic conditions, the first being a very strong flow of air through the cave.

The cave has two entrances. The upper one is oriented in east direction, and it is near the top of the section, about 15 m high. From this entrance, going in a northeasterly direction, is a channel with a width of about 2 m and a height of 2.5 m. The channel continues in the same direction for about 18 m and then turns toward the east. It continues in this direction for the next 61 m. At about 17 m from the entrance, you come to a downward channel 20 m deep that ends in debris and accumulated water in the form of pond. Farther into the main channel, the dimensions decrease, and this part has the highest airflow velocity, which reaches 80 km/h (researcher's estimate). Narrowing of the channel is due to the collapse of blocks from the ceiling. The channel has a general decline of −40 degrees. After the next 15 m, you come to a hall with a ceiling that is more than 15 m high and that leads to two channels. The north channel is 8 m long, 3 m wide, and more than 3 m high. The floor of this channel is almost horizontal and ends in a narrowed crack. The east channel is about 18 m long, 2.5 m wide, and 3 m high. This channel steeply descends to the east, with a decline of −30 degrees. On the floor are blocks and debris, and the stream, formed during the research because of ice melting in the entrance section of the channel, runs between them. This channel ends with a siphon filled with mud. The last one, the north channel, has an entrance character and decline of +20 degrees. In the initial part, it is oriented to the south, but after a few meters, it turns toward the southeast. The total length of this channel is 25 m, and the width ranges from 3 m in the first part to 0.7 m at the seventh meter. Farther on, the width increases, and at the fifteenth meter, it is 5 m. The height of the channel varies from more than 5 m at the beginning to only 0.4 m at the end, which is in the zone of the higher entrance. The length of the channel is 105 m, and the height difference is 27 m.

The ice occurs at about 2 m after the entrance into the south channel after narrowing and covers the next 15 m of the channel. The thickness is more than 2 m. Other manifestations related to the phenomenon of water do not occur (Fig. 14.5).

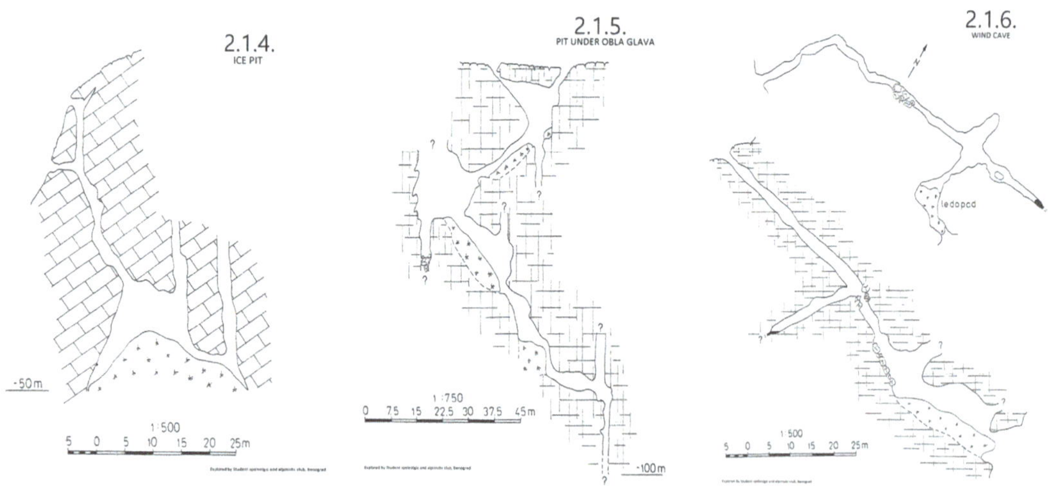

FIG. 14.5

Speleological profiles of Ice Pit (2.1.4), Pit under Obla Glava (2.1.5), and Wind Cave (2.1.6).

14.3.2 THE AREA OF VELIKI STUOC

14.3.2.1 Kosarska cave

This cave is located in the area of the Little Montenegro village in the lower part of the northern slope of Veliki Stuoc. You can approach the cave from the road from Zabljak to Little Montenegro.

The entrance into the cave is shaped like a modified rectangle with strong initial cracks. The dimensions of the entrance are 12 and 5 m. There is a vertical channel from the entrance to a depth of 50 m, where you come to a widening and a terrace 8 m wide covered with snow mixed with debris and blocks. After the terrace, the vertical continues taking on larger dimensions and extends east and west into two short channels that end after about 5–7 m. The depth at the bottom of the cave is −90 degrees, after which the dimensions gradually increase to a diameter of 10 m. The bottom is completely covered with an ice cap that is more than 10 m thick, along which you can go farther beside the walls, but there is no open connection to deeper parts of the cave. The total passable depth is −90 m. There are no hydrological activities, except for rarely dripping water.

14.3.3 FACILITIES IN THE ZONE OF VJETRENA HILLS

14.3.3.1 The cave with the ice under Bandijerna

This cave is located at the foot of the southern section of the Bandijerna in the area closest to Obrucine. The entrance is at the top of the rock creep that covers the entire foot of Bandjerine. It is partially on the right hidden by skerry, but it is easily seen from the path that goes from Sedla to Bobotov Kuk.

The entrance to the first part of the cave is 7 m high and 4.5 m wide. Just after the entrance, the height decreases to 2 m and the width to 4 m with the rung on the ceiling. The width of the channel gradually decreases, and after 9 m, it is 3.5 m and the height is 1 m. Debris and very big blocks are on the floor in this part of the cave. The walls are eroded, especially in the entrance where well-developed erosion grooves and scarps can be seen. The channel is oriented northeast for the next 8.5 m, and farther on, it has almost the same dimensions and orientation. The basic channel widens more at 16 m, and the average width there is about 5 m and the height 7 m. On the left, layers of ice and ice stalagmites partially cover the floor of the channel. At the end of this part of the channel, on the left, you see a small channel, which at the time of the first exploration was partially passable at 5 m. Further exploration has not been completed. In the forehead of the main channel, there is a basin, along which flows water that drips from the ceiling. This water forms a smaller accumulation, which the cave is named after.

The total depth of the entrance is +15 m, but the depth of the facility is much greater.

14.3.3.2 The cave in the saddle between Uvita Greda and Vjetrena Hills

This cave is located near the marked hiking trail, which goes from Sedla to Zelen Vir, about 20 m away from the exit of the trail to the Obrucine plateau.

The entrance has a southern exposition, viewed in a plan, a rectangular shape, and is 1.5 m high and 0.5 m wide. The channel continues vertically at 7.5 m, and its walls are covered with travertine malt. The width of the channel is 5 m, and the height is 7 m. This part of the channel ends at a terrace that steeply slopes toward the continuation of the channel, at an angle of about −70 degrees. The whole terrace is covered with a layer of debris and some large blocks. Farther on, there is a new vertical, 10.5 m long, and along it, the stem rock is decorated. The width of this part of the channel increases from 5 to 7 m, and the height from 3 to 5 m. At the bottom of the vertical, there is sediment of granular snow, about 1 m thick, which covers about 10 square meters of the floor. The channel then continues in the cave, and at 8.5 m has an inclination of −35 degrees. The width decreases, and the height falls at 3 m. On the seventh meter, there is a rung of accumulated rock material. It completely closes the profile of the channel at 1.5 m. After this rung, there are series of smaller cascades, and on the right is a niche, 5 m deep, which for about 1.5 m is subdued in relation to the floor of the channel. The floor of the channel is steeper in this part, and at 10 m, the fall is about −40 degrees. The passable part of the channel ends with piled blocks and debris that disguise a probable continuation of the cave. The total depth is −34 m.

In hydrological terms, in addition to the appearance of snow, sporadic trickling also appears, but there is no trace of any significant hydrological activity (Fig. 14.6).

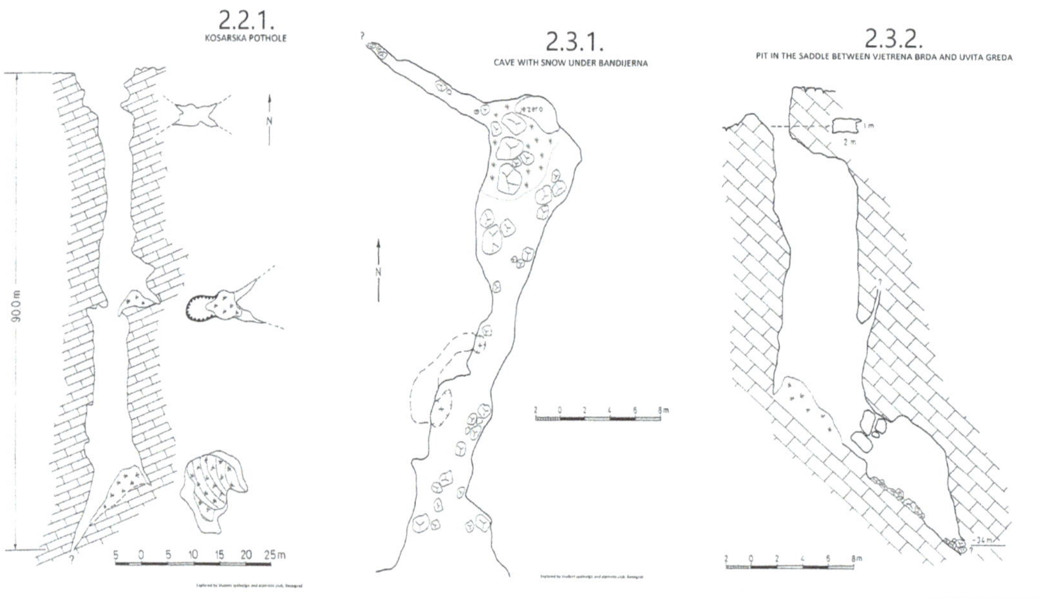

FIG. 14.6

Speleological profiles of Kosarska pothole (2.2.1), Cave with snow under Bandijerna (2.3.1) and Pit in the saddle between Uvita greda and Vjetreno brdo (2.3.2).

14.4 ICE CAVES IN THE PROKLETIJE MOUNTAINS

14.4.1 GENERAL INFORMATION ABOUT CAVES

The cave expeditions have discovered about 60 caves (Kicińska and Najdek, 2013a). Among these, the longest and deepest caves of the Belič and Kolata massifs are

- the Mining Cave/Górnicza Cave, 03 013 (depth of −585 m, length of 2083 m);
- the Ice Cave/LedenaPečina/JaskiniaLodowa, 03 110 (depth of −451 m, length of 2057 m);
- the Giant Cave/Gigant Cave, 03 113 (vertical extent of 306 m, length of 1871 m);
- the 03-061–T5 Cave System (depth of 242 m, length of 1131 m); and
- the Enthusiastic Cave System (depth of 225 m, length of 3098 m).

During the expeditions, several hundred entrances were checked. Most of them were terminated by blocks, snowy plugs, or narrow places. The caves of the Prokletije Mountains were shown to comprise Mesozoic limestones and dolomites that belong to the High Karst Unit (Djokič et al., 1976). The caves of the Belič and Kolata massifs developed along tectonic discontinuities (NW-SE and NE-SW) or bedding planes (cf. Shanov and Kostov, 2015). Most of the caves explored were simple systems of vertical shafts with short meanders leading to the next shaft. Some of them had short subhorizontal conduits. The caves studied were relatively poor in speleothems. The first speleothem U-Th dated in the Prokletije Mountains indicated that crystallization took place in the Eemian Interglacial period (Kicińska and Hercman, 2014).

14.4.2 ICE CAVES IN THE PROKLETIJE MOUNTAINS

In each expedition, many caves were reported to have snow, firn, and ice, which had not been investigated. The observations were conducted in the summer season (July and August). Three trips were organized in March and April. Snow, firn, or ice deposits were found in eight caves (Table 14.1). The entrances to these caves were located between about 1880 and 2080 m a.s.l. (Fig. 14.7). The mean

Table 14.1 Basic Morphometric Data of the Prokletije Mountains' Ice Caves

No.	Cave	Entrance Altitude	Length	Vertical Extent	Dominating Ice Type	Survey of Caves
1	Ice Cave (03 110)	1945	2057	−451	Snow, ice	I. Cvetkovič, F. Filar, A. Kasza, U. Kotewa, M. Kwiatkowski, P. Lulek, K. Maciąg, A. Maciejewska, M. Macioszczyk, Ł. Marciniak, P. Niziołek, I. Njunjić, K. Najdek, M. Parczewski, P. Szukała, Z. Tabaczyński
2	03 061–T5 Cave (03 067)	2034/2069	572/560	207/176	Snow, firn, ice	M. Borowiecka, P. Graczyk, B. Haremza, E. Hristova, D. Jankowiak, D. Kicińska, P. Kluza, S. Kozłowski, A. Łada, Ł. Marciniak, M. Milewska-Moult, A. Moult, K. Najdek, T. Pawłowski, M. Petkovič, P. Piechowiak, K. Przybyszewski, K. Piotrowski, P. Vucetič, Z. Tabaczyński
3	Mining Cave (03 313)	2019	2083	−585	Snow, firn	M. Filipiak, M. Gabryelewicz, R. Gałązkiewicz, D. Kicińska, P. Kluza, P. Lulek, K. Maciąg, Ł. Marciniak K. Najdek, S. Kozłowski, P. Szukała, Z. Tabaczyński, G. Žužić/Ł. Marciniak, K. Najdek

Continued

Table 14.1 Basic Morphometric Data of the Prokletije Mountains' Ice Caves—cont'd

No.	Cave	Entrance Altitude	Length	Vertical Extent	Dominating Ice Type	Survey of Caves
4	Ice Giant Cave (03 114)	2099	90	−23	Ice	A. Kasza, Z. Tabaczyński
5	Aladdin's Cave (03 149)	1960	156	16	Ice	S. Blagojevič, I. Budinski, I. Miskovič
6	Bolt from the Blue Cave (03 146)	1930	307	45	Ice	S. Blagojevič, I. Čvetkovič, B. Jankovič, I. Miskovič, N. Zbiljič
7	Ice Dragon Cave (03 003)	1881	396	−183	Snow, firn, ice	A. Kasza, P. Niziołek, M. Staniec, Z. Tabaczyński, E. Jachimkowska
8	The Hundred (03 001)–03 004	2003/1983	101/49	96/15	Snow	P. Niziołek, Z. Tabaczyński, A. Kasza, M. Staniec

annual air temperature (MAAT) was calculated on the basis of the measurements in Vermosh in Albania (6.7°C at 1152 m a.s.l.) and Kolašin in Montenegro (7.2°C at 945 m a.s.l.). Palmentola et al. (1995) and Milivojevič et al. (2008) stated that the temperature at about 2000 was, respectively, +1.4°C and +0.6°C. Most entrances faced the north. The explored caves were located in the Belič massif, according to the geographical distribution proposed by Mulić (2009) and the *topografska karta* 1:25,000 (TK 25).

14.4.2.1 Ice Cave (03 110)

This cave is located on the western side of the Scripa Valley. The cave has two entrances. The main (lower) entrance is located near the tourist path to the highest peak of Montenegro—Maja Kolata (2534 m a.s.l.). The upper entrance is located 30 m above lower entrance. The entrances face the east. The Ice Cave is the second deepest cave in the Prokletije Mountains (Fig. 14.8). From the lower entrance, strong, cold air blows during the summer, which is probably why the cave was named "Ledena Pečina" by the local people. During the winter, this entrance is covered with snow. Smaller or larger amounts of snow can be observed at the lower entrance at the end of August or later (though how much later was not known by the expedition). Ten meters from this entrance, the Ice Cave continues into a vertical shaft where less inclined walls are covered by large amounts of snow. At a depth of −40 m, there is a horizontal passage with a branch (the Lewe Gangi Passage) where the expedition found an icefall and stalagmites and a frozen pond (Photo 14.1). From 2010 through 2011, the icefall had a height of about 10 m and a width ranging from about 7 to 8 m. The stalagmites reached a height of 2.5 m. In the first 35 m of the Lewe Gangi Passage (with a width ranging from about 4 to 5 m), there was a frozen pond, which was formed by infiltrating water (Photo 14.2). The deepest part of the frozen pond had a depth ranging from about 40 to 50 cm (Mariusz Woźniak, personal information). Observations in 2015 showed that about 10% of the ice mass had melted and that the frozen pond took up less space, but that there was a large amount of snow in the vertical shaft near the entrance (Zbigniew Tabaczyński, personal information).

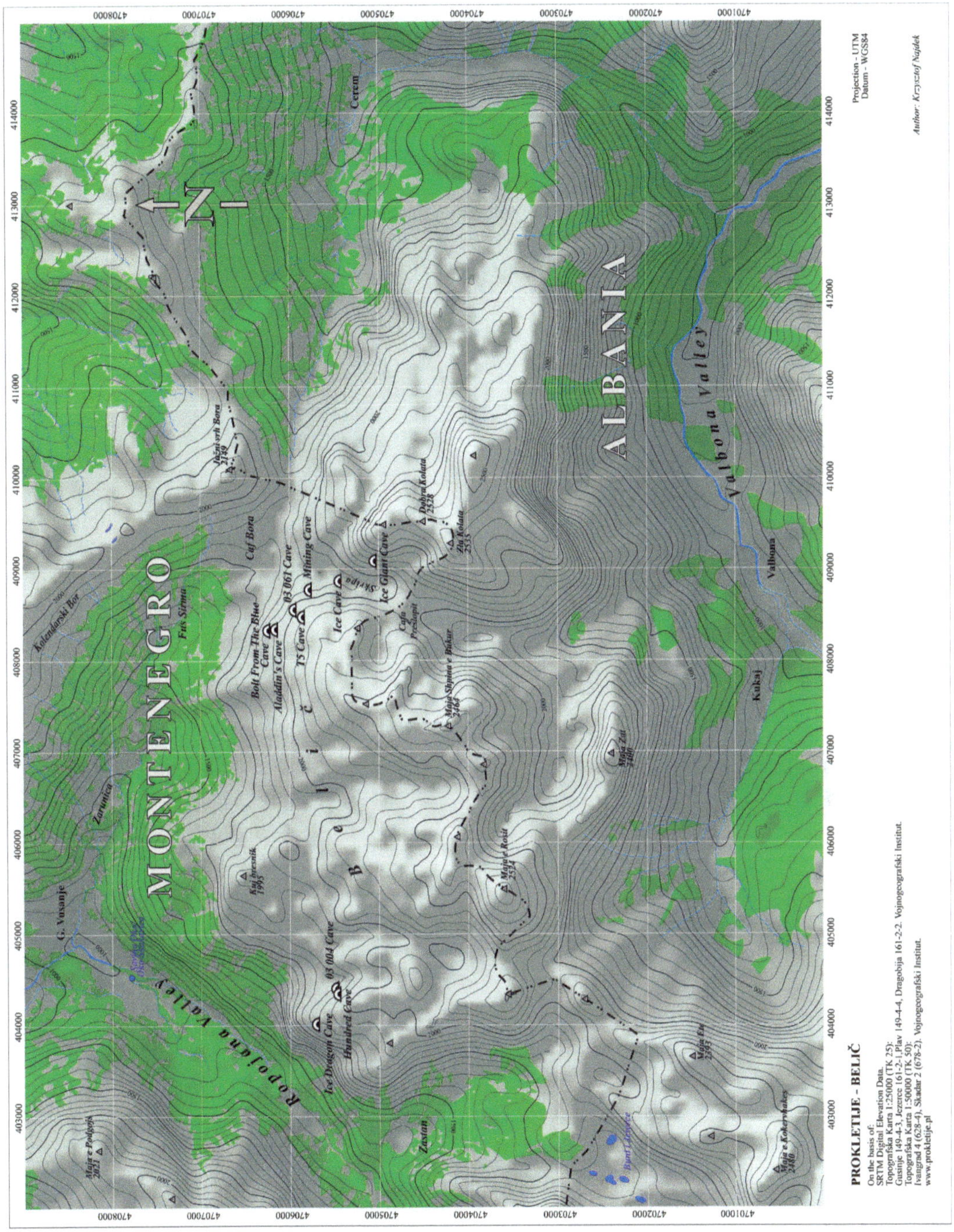

FIG. 14.7 Map of objects with snow and ice in the Prokletije Mountains (author: Krzysztof Najdek).

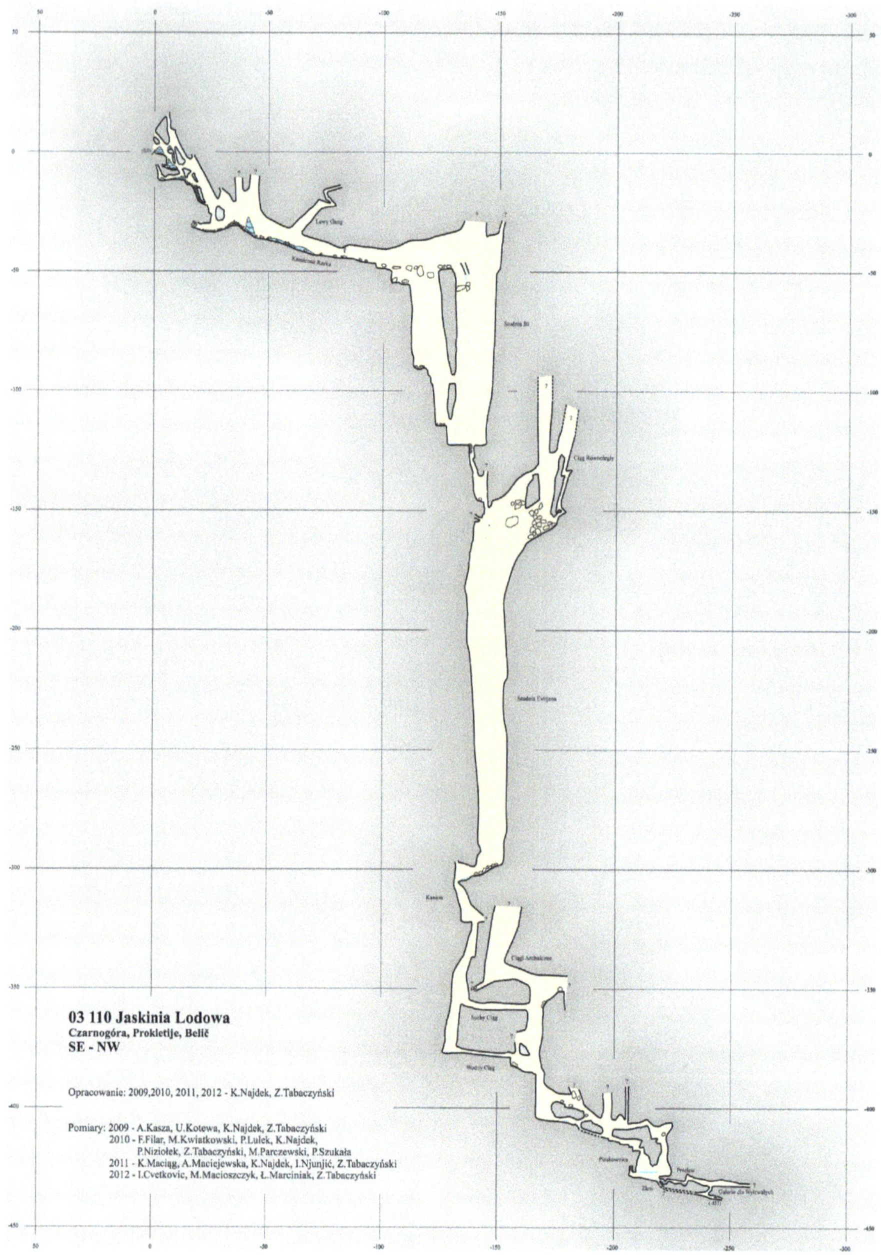

FIG. 14.8

The cross-sections of Ice Cave in the Prokletije Mountains.

PHOTO 14.1

The icefall and stalagmites in Ice Cave in the Prokletije Mountains (author: Mariusz Woźniak).

PHOTO 14.2

The frozen pond in Ice Cave in the Prokletije Mountains (author: Mariusz Woźniak).

14.4.2.2 The 03061–T5 (03067) cave system

The 03061 and the T5 caves were found in 2014. The connection to these caves was found in 2015 (Najdek et al., 2016). The system has two entrances. The entrances are located in the upper part of the eastern part of the Belič massif (west of the Scripa Valley). The entrance to the 03061 cave, which faces the northwest, was found to be partly filled with snow during July and the beginning of August

PHOTO 14.3

The ice cover in the 03061 Cave in the Prokletije Mountains (author: Marietta Milewska-Moult).

(2014–16). In the near-entrance part, a narrow meander leads into a vertical shaft. At a depth of −70 m are horizontal passages with ice cover, stalactites, and snow deposits (Photo 14.3). In 2014 some sites on the walls and floors of the passages were covered with ice (with lengths ranging from 5 to 6 m and a dozen meters wide) and stalactites (with a length of about 2 m). In 2016 the cavers observed lesser amounts of both snow and ice deposits, but the snow cover had a length of about 70 m. In some places where there was less snow, more ice stalagmites were found. The ice cover on the wall had a length ranging from 5 to 6 m, a width of 1 m, and a thickness ranging from 30 to 40 cm (Adam Łada, personal information).

The T5 Cave has a one shaft entrance. The cave is filled with snow, which in the lower parts leads into a firn from the beginning to a depth of about −60 m. In 2014 the walls were covered in ice to a depth of about 8 m (Karol Przybyszewski, personal information).

14.4.2.3 Mining Cave (03313)

Mining Cave is located on the upper-western side of the Scripa Valley (Kicińska and Najdek, 2013b). This is the deepest cave in the Prokletije Mountains in Montenegro. The shaft entrance faces northeast. The entrance shaft has a depth of 15 m, a width of 6 m, and a length of 9 m. These parameters cause the accumulation of large amounts of snow. In 2010 all the shaft was filled with snow. The cave was accessible only through a narrow passage that connected the upper part of the shaft with the bottom part. Since 2011 each summer it has been possible to descend by rope into this shaft to the deeper part of the cave. Every summer at least half of the shaft has been filled with snow. In the lower parts of the cave ice, firn, and snow deposits are not registered.

14.4.2.4 Ice Giant Cave (03114)

This cave is located in the western part of Maja Kolata. The cave has an entrance that faces the northwest. Ice Giant was found in 2009. The cave is developed horizontally as a single passage that at the end leads into a chamber measuring 40 m by 20 m (http://www.prokletije.pl/Mapa/Pokaz/03/114). All the chamber is filled with ice which is the state in the Prokletije Mountains in Montenegro. Observation of the ice in the marginal fissure indicated that the ice's thickness was at least 3 m in 2009 (Andrzej Kasza, personal information).

14.4.2.5 Aladdin's Cave (03 149)

Aladdin's Cave is located in the northeast part of the Belič massif. It is a horizontal cave with one entrance. The cave faces the southeast. Snow and ice deposits are in the cave. The ice cover is 10 m in length (http://www.prokletije.pl/Mapa/Pokaz/03/149).

14.4.2.6 Bolt from the Blue Cave

Bolt from the Blue Cave is located in the eastern part of the Belič massif. The cave has two entrances facing the northeast. Some sites in this cave are covered with ice (http://www.prokletije.pl/Mapa/Pokaz/03/146). The lower part of the cave is filled with ice at a length of least 15 m and a depth of 0.5 at 1 m (Aleksandar Pegan, personal information).

14.4.2.7 Ice Dragon Cave (03 003)

The Ice Dragon Cave is located in the western part of the Belič massif. The entrance faces southwest. The preliminary part of the cave is filled with snow deposits that lead into a firn. The lower part is situated in an icefall that in 2007 had a length of 50 m (at a depth of −30 to −80 m). Its width was 5 m, and the thickness was at least 5 m (http://www.prokletije.pl/Mapa/Pokaz/03/003).

14.4.2.8 The Hundred (03 001)–03 004 Cave system

The Hundred Cave system is located in the western part of the Belič massif. The caves were found and connected in 2007. The system has two shaft entrances. In the 03 004 Cave, snow was observed at a depth of −30 m and continued to a depth of about 96 m where the caves are connected (http://www.prokletije.pl/Mapa/Pokaz/03/001).

In the Prokletije Mountains, a lot of caves are filled with snow, firn, and ice deposits during the summer. The preceding caves were not observed in the autumn, so the ice has not been documented for an entire year. However, based on the size and distribution of the ice, firn, and snow deposits in these caves, it can be assumed that icing is a perennial phenomenon. Most caves have two entrances and are of a dynamic type. The Ice Giant and Ice Dragon caves are static caves with congelation ice and firn (cf. Luetscher and Jeannin, 2004). In many caves in the Prokletije Mountains, seasonal ice cover that develops during the spring has been observed.

14.5 ICE CAVES OF BOSNIA AND HERZEGOVINA

Karsts in Bosnia and Herzegovina are a part of the Dinaric Karst. In addition to the Dinaric Karst are the External Dinarides, which are in the areas southwest of the Gacko, Ključ, Bihać line. In Bosnia and Herzegovina, another karst, the Inner Dinarides, abuts the one in the northeast. The Inner Dinarides are shallower and mostly in isolated parts of the karst. According to the authors who wrote about karst in Bosnia and Herzegovina (Mulaomerović, 2014), there isn't a unique perspective on how many territories of Bosnia and Herzegovina are under karst, but they are estimated to be from 28% to even 65%.

Even though the first research of caves and karsts in Bosnia and Herzegovina were documented in the Middle Ages by scientists from Dubrovnik (16th century), serious and systematic research started after the Congress of Berlin, when the Austro-Hungarian Empire was mandated to occupy Bosnia and Herzegovina and establish its administration. After the administration was established, the Military Geographical Institute of Vienna started recording a cartography of the entire territory of Bosnia and Herzegovina (Malez and Lenardić-Fabić, 1988). During this recording, many kart phenomena were noticed. The mail goal of this research was to find water for economic development. Later, after the establishment of the National Museum in Sarajevo, the research spread into the fields of archeology, palaeontology, and cave fauna. The second period of more intensive speleological research in Bosnia and Herzegovina happened in the 1950s, when substantial research of karst territories were done in Yugoslavia and great hydrotechnical interventions were done in eastern Herzegovina, which included the research of caves and pits (Milanović, 1979).

However, despite all the activities on karsts, the ice caves were never the focus of the preceding research, mainly because of their geographical position—in the high mountains, which were not included in the hydrotechnical procedures. On the other hand, speleological research focused on the caves because of the easier exploration, a lack of adequate speleological equipment, and the complete lack of experience in the exploration of such speleological objects. Speleological research simply focused on the search for very long and beautiful (richly decorated) caves. No professional or scientific research of caves and pits has been done.

That is why there is no professional or scientific literature in Bosnia and Herzegovina referencing this area. There is only one paper, but it is more ethnographic and ethnological in character, and it deals with the exploitation of ice from caves as an economic activity (Mulaomerović, 1982). It is mostly about ice caves around Sarajevo, because the ice was needed for local hospitals, butcher shops, patisseries, and hotels. The extraction of ice from caves was especially developed in the Austro-Hungarian period, but it completely stopped about 10 years after World War II.

According to the *Cadastre of Speleological Objects of Bosnia and Herzegovina* (Mulaomerović et al., 2006), in Bosnia and Herzegovina there are over 4000 recorded speleological objects. Among these objects, more than 80 are referred to as Ledenica or Snježnjača (or some derivative of these names). It is difficult, for example, to talk about the real number of icicles because of insufficient research.

Nevertheless, conclusions can be made based on several ice caves on the outskirts of Sarajevo. In fact, in mid-July 1982, we explored the Ledenjača cave located on Romanija mountain (east of Sarajevo). Inside the cave were two large cones of ice below the vertical chimneys, and the rock at the end of the cave was also frozen. The temperature in that part of the cave was 0°C. At the same time, the temperature outside was a little bit over 20°C (Mulaomerović, 1982). A few years before that, Italian speleologists explored that cave, almost at the same time of the year (Brozzi, 2011), and there was no trace of ice. For several other pits and the Ledenica cave (on Bjelašnica mountain), we can confirm that in recent years, there has been no ice.

It is obvious that, at least when it comes to the aforementioned Ledenica cave on Romanija and caves with ice on Bjelašnica, uncontrolled deforestation has caused the change in the microclimate features that led to the ice melting, even during the spring.

The only two caves that we can safely say are filled with ice are the Ledenica located on the Treskavica mountain range and the Ledana pit (actually the cave with pit entrance) on Bobija mountain (Dujaković, 2004), both are in the central part of Bosnia and Herzegovina. But they are also explored only as an ethnographic site (Ledenica cave) or as a training ground for alpinists (Ledana pit).

Taking into account the state of speleology and karst research at institutes and faculties in Bosnia and Herzegovina, unfortunately, the ice caves will not be of the scientific interest for a long time.

Ditta Kicińska thanks colleagues from caving clubs (Poland, Serbia, Bosnia, and Hercegovina) and Krzysztof Najdek for preparing the map of the ice cave in the Prokletije Mountains.

REFERENCES

Brozzi, G., 2011. Grotta: Ledenjaca. Sottoterra 131, 55–56.

Djokič, V., Živaljevič, M., Petrovič, Z., 1976. Osnovna geološka karta i tumač za list Gusinje. Savezni geološki zavod, Beograd.

Dujaković, G., 2004. Pećine i jame Republike Srpske/Caves in the Republic of Srpska. Zavod za udžbenike i nastavna sredstva, Istočno Sarajevo.

Kicińska, D., Hercman, H., 2014. The first of a speleothems U-Th dating from the Prokletije Mountains (Montenegro). In: Materiały 48. Sympozjum Speleologicznego, Kletno, pp. 75.

Kicińska, D., Najdek, K., 2013a. Cave exploration of the Belič massif in the Prokletije Mountains (Montenegro). In: Proceedings of 16th International Congress of Speleology, Brno, vol. 2, pp. 165–167. http://digital.lib.usf.edu/SFS0050564/00001/search?search=speleology.

Kicińska, D., Najdek, K., 2013b. Prokletije 2009–2013. In: Kicińska, D. (Ed.), Polish Caving 2009–2013. Komisja Taternictwa Jaskiniowego Polskiego Związku Alpinizmu (Caving Commision of Polish Mountaineering Association). Wyd. Pracownia Kreatywna Bezliku, Kraków, pp. 32–34.

Luetscher, M., Jeannin, P.Y., 2004. A process-based classification of alpine ice caves. Theor. Appl. Karstol. 17, 5–10.

Malez, M., Lenardić-Fabić, J., 1988. Prilog poznavanju povijesti speleoloških istraživanja u Bosni i Hercegovini. Naš krš XIV (24–25), 169–176.

Milanović, P., 1979. Značaj speleologije u hidrotehničkim zahvatima na karstu Istočne Hercegovine – istorijski pregled. Naš krš V (7), 9–14.

Milivojevič, M., Menkovič, L., Čalič, J., 2008. Pleistocene glacial relief of the central part of Mt. Prokletije (Albanian Alps). Quat. Int. 190, 112–122.

Mulaomerović, J., 1982. Eksploatacija leda iz pećina Bosne i Hercegovine/Ice exploitation in the caves of Bosnia and Herzegovina. Naš krš VI (12–13), 107–114.

Mulaomerović, J., 2014. Speleoturizam u Bosni i Hercegovini. Centar za krš i speleologiju, Sarajevo.

Mulaomerović, J., Zahirović, D., Handžić, E., 2006. Katastar speleoloških objekata Bosne i Hercegovine. S.D. Speleo dodo, Sarajevo.

Mulić, R., 2009. Plavsko-Gusinjske Prokletije, četrdeset planinarskih staza. Planinarsko društvo "Karanfili", Gusinje. 213 pp.

Najdek, K., Kicińska, D., Kasza, A., 2016. Eksploracja jaskiniowa w górach Prokletije w latach 2015–2016 (Cave exploration of the Prokletije Mountains in 2015–2016). In: Mat. 50. Sympozjum Speleologicznego, Chęciny, pp. 132–133.

Palmentola, G., Baboci, K., Gruda, G., Zito, G., 1995. A note on rock glaciers in the Albanian Alps. Permafr. Periglac. Process. 6, 251–257.

Radojičić, B., 1996. Geography of Montenegro. UNIREKS, Nikšić. pp. 1–228.

Shanov, S., Kostov, K., 2015. Dynamic Tectonics and Karst. Springer, Heidelberg, New York, Dordrecht, London. pp. 27–35.

FURTHER READING

Radulović, M., 2000. Karst Hydrogeology of Montenegro. Republican Institute for Geological Research, Podgorica. pp. 1–271.

CHAPTER

ICE CAVES IN CANADA

15

Charles J. Yonge*, Derek Ford†, Greg Horne‡, Bernard Lauriol§, Jacques Schroeder¶

Yonge Cave and Karst Consulting Inc., Canmore, AB, Canada McMaster University, Hamilton, ON, Canada† Jasper National Park of Canada, Jasper, AB, Canada‡ Université d'Ottawa, Ottawa, ON Canada§ Université du Québec à Montréal, Montréal, QC, Canada¶*

CHAPTER OUTLINE

15.1 INTRODUCTION AND HISTORY

Canada is a cold country, the northern half mapped as having discontinuous to continuous permafrost conditions in the ground, and the southern half experiencing weeks to months of ice and snow cover at all elevations during the winter. The western provinces and northern territories (Alberta, British Columbia, the Northwest, Nunavut, and Yukon territories) have major mountain ranges with modern glaciers and icecaps. There are also extensive mountainous terrains (relief ≥1000 m a.s.l.) in Quebec, New Brunswick, Nova Scotia, and Newfoundland-Labrador. Almost all of the population lives in the southern quarter of the country, close to the border with the United States. There are few cavers or organized caving clubs, because distances are large and most of the mountain and northern regions have few roads or other trails to provide access. As a result, our knowledge of the numbers and characteristics of the natural caves of all kinds that exist in Canada must be considered provisional and incomplete. Some 700 caves are reported in the *Canadian Caver* index at the present time, but the true number is probably in the thousands. Given our climate, the large majority of these will display some ice dripstone and flowstone forms in the entrances' cold zones during the winter. Most fully melt the following summer, but some ablated masses survive for longer periods; they are unlikely to be documented. Defining an

Ice Caves. https://doi.org/10.1016/B978-0-12-811739-2.00015-2

"ice cave" as one that contains accumulated ice or firn known to be perennial, we cannot say how many there are in Canada but guess that there must be many hundreds at least. In this chapter we discuss 34 significant examples that have been documented and studied to some degree. Seven have received only mapping and/or photography (Fig. 15.1 and Table 15.1). All of them are found in the west, in the cordilleran alpine or in the northern permafrost regions. The basis for their selection is that they are more than a few meters long, have aesthetic merit, and have received some scientific study (Fig. 15.2).

FIG. 15.1

Significant Canadian Ice Caves.

Table 15.1 Significant Canadian Ice Caves

Name/Group	Area	Prov./Terr.	Lat.	Long.	El. (m)	Length (m)	Cave Type[a]	Rock	Study/Comments
Ni'iinlii Njik Caves	Old Crow	Yukon						Devonian limestone	
1. Tsi tse han Cave			66.3	−139.3	750	50	P		^{14}C dating
2. Grande Caverne Glacee			66.3	−139.3	750–950	90	P		Stable isotopes/diatoms/dating
3. Caverne des Meandres			66.7	−139.3	750–950	100	P		Diatoms
4. Bear Cave			66	−139.3	750–950	55+	P		Stable isotopes/diatoms
5. Caverne '85			66.8	−139.3	750–950	40	P		Stable isotopes/diatoms/dating
Nahanni Karst		NT			~700			Devonian limestone	
Grotte Valerie Ice Cave	Nahanni Canyon		61.3	−124.1		2002	P/CZ		
Mickey	Nahanni Canyon					2270	P/CT		Four areas of ice now sealed
Andree	Lafferty Creek					1205	CT		
Igloo	North Karst		61.5	−124.2		500–700	CT		Spectacular ice
Arnica	North Karst					242	CT		
Black Ice Cave	North Karst					?	?		
Kananaskis Caves	Kananaskis	AB						Mississippian, Livingstone limestone	
1. Canyon Creek			50.9	−114.8	1769	727	CZ		Stable isotopes
2. Plateau Mountain			50.3	−114.5	2226	410	P		Stable isotopes
Crowsnest Pass Caves:	Crowsnest Pass	AB/BC	49.5	−114.5				Mississippian, Livingstone limestone	
1. Booming Ice Chasm		AB			2189	961	CT		Stable isotopes
2. Glittering Ice Palace		AB			2226	286	P		Hoarfrost
3. Coultard		AB			2440	370	P		Ice petrography
4. Ice Chest		AB			2250	381	P		Extrusion ice, stable isotopes
5. Serendipity		AB			2165	473	P		Extrusion ice, stable isotopes

Continued

Table 15.1 Significant Canadian Ice Caves—cont'd

Name/Group	Area	Prov./Terr.	Lat.	Long.	El. (m)	Length (m)	Cave Type[a]	Rock	Study/Comments
6. Flop Pot		AB			2165	200	CT		Flood ice
6. Ice Hall		AB			2180	130	CT		Stable isotopes/dating
8. Rats Hole		BC			2455	130	P/CT		Stable isotopes
White Ridges Prov. Park	Vancouver Island	BC	49.8	−126.0				Upper Triassic	
1. Q5					1200	1867	CT		Stable isotopes
2. Projects Cave					1050	130+	CT		Stable isotopes
Individual Caves									
Castleguard Cave	Columbia Icefield, JNP	AB	52.1	−117.2	2016	21068	Active/relict CZ?	Middle Cambrian limestone, Cathedral/Eldon	Ice plug petrography
Disaster Point Ice Cave	Jasper National Park	AB	53.0	−118.1	1082	673	CT	Devonian Palliser Formation	Stable isotopes/paleontology
Ice Trap	Jasper National Park	AB	53.1	−118.3	2400	3002	P/CZ	Devonian Palliser Formation	Stable isotopes/dating
Mount Alberta Cave	Jasper National Park	AB	52.3	−117.5	3350	~150+	?	Middle Cambrian limestone, Cathedral/Eldon	
Icy Hope	Mt Sinwa, Taku, Stikine	BC	56.9	−133.3	1499	200	CT	Upper Triassic limestones of the Sinwa Fm.	Expedition reports 2014
Inverted Fridge	Inverted Ridge	BC	49.1	−114.8	1940	281	CT/CZ?	Mississippian, Livingstone limestone	Hoarfrost and ponded ice
Global Cooling	Inverted Ridge	BC	49.1	−114.8	1940	133	CT/CZ?	Mississippian, Livingstone limestone	Hoarfrost and ponded ice
Supermans Glittering Ice Palace (SGIP)	Mount Bisaro		49.6	−115.6	2250	21	CT	Mississippian, Livingstone limestone	Stable isotopes
Nakimu Cave	Glacier National Park	BC	51.3	−117.6	1576	6000	CT	Badshot marble Formation, Cambrian	Large icefall
Walkin-Ice Cave	Wood Buffalo Nat. Park	AB	59.8	−112.2	252	1624	Relict CT	Middle-Devonian gypsum	Stable isotopes/dating
Wedge Cave	Top of the World Prov. Pk.	BC	49.8	−155.4	2200	60	P/CT	Upper Cambrian limestone, Cathedral/Eldon?	Massive ice

[a] *CT, cold trap; CZ, cold zone; P, permafrost; ?, not known.*

FIG. 15.2

Booming Ice Cavern, the largest ice cave in Canada: view down the main 140 m shaft with climbers Adam Walker, followed by Nick Vieira and Christian Stenner. The much publicized ice shaft became the focus of summer ascents by ice climbers, which has led to management concerns. Inset: large side passage.

Photographs: (top) Francois-Xavier de Ruydts, (inset) Nick Vieira.

Regarding ice cave distribution, the highest known ice cave (~150+ m long) is at 3350 m a.s.l. on Mount Alberta, Jasper National Park and—as reported by J.-P. Kors of Parks Canada—the lowest is at 75 m a.s.l. in Hornaday Canyon in the North West Territories. The latter is also the most northerly one known at 69.15°, although further north in the Arctic islands, Banks Island has limestone with relict karst and may contain ice caves (possible because carbonate caves containing ice are known above 80° in Greenland). The most easterly ice cave known is the Walkin-Ice Cave System, Wood Buffalo National Park and the most westerly ice caves are Projects Cave and Q5 on Vancouver Island some 50 km from the Pacific Ocean. The most southerly ice cave (Inverted Fridge) lies close to the Unites States border, but this lacks significance as there are many ice caves further south in the Unites States. For details of the above see text, Figs. 15.3 and 15.21 and Table 15.1.

The discovery dates of the caves, especially those known for a long time, are incomplete. We do not have any direct evidence of ice caves being visited in prehistoric times, except perhaps the Ni'iinlii Njik Ice Caves of the Yukon Territory by the Dagoo Gwichen people. It is also likely that some of the Nahanni caves (Northwest Territories) would have been used by First Nation peoples, but an archeologist accompanying cave explorers in 2007 found no evidence, although Ford noted a wood and steel trap, presumably 20th century, abandoned in a cave entrance in the north karst there. The Walkin-Ice Cave in Wood Buffalo National Park may well have had early visitation, but the rapid erosion of its gypsum walls would have removed any pictographs and no artifacts have been found there. Further south in the Alberta Rockies' Plateau Mountain, Ice Cave and Canyon Creek Ice Cave may have been visited in these early years. Artifacts and pictographs are found on overhangs and, for example, in Rats Nest Cave, the latter containing artifacts dating back 3500 years to the local Pelican Lake culture.

Generally, and especially for ice caves in the alpine regions, exploration is recent (1900s onward). As documented examples, Nakimu Caves were discovered by C.H. Deutschmann of Revelstoke in 1904. In 1905 A.O. Wheeler surveyed them, finding ~1200 m of passages at that time, and part of these were developed as a show cave within the next few years (closed in 1933). The main cave contains one

FIG. 15.3

Location of the highest Ice Cave in Canada on Mount Alberta (main) with Jason Kruk inside the cave (inset). The cave was discovered on the first ascent of the North Face Route by Joshua Lavigne and Jason Kruk.

Photographs: John Scurlock (main) and Joshua Lavigne (inset).

FIG. 15.4

Ice cascade in Nakimu Caves, known to have existed since 1907 and probably much older. The photograph was taken under summer conditions (Jul. 1965) when local melting and re-freezing was occurring on the ice. Mike Goodchild of the KRG is on rappel.

Photograph: Derek Ford.

large ice mass now known to have persisted for at least 122 years (Fig. 15.4) but that is almost certainly much older (Little Ice Age?). Canyon Creek Ice Cave was discovered by Stan Fullerton in 1905 and Castleguard Cave by Cecil Smith in 1926 (the perennial ice plug by Mike Boon in 1970). From 1965 onward, Derek Ford's Karst Research Group (KRG) from McMaster University undertook cave and karst research in the Selkirk Mountains and Canadian Rockies and, after 1970, in the Nahanni region of the Mackenzie Mountains. French Canadian cavers and the Société Québécois de Spéléologie have

FIG. 15.5

Katie Graham, Current president of the Alberta Speleological Society, at the end of the Ice Crawls, a cold zone in Castleguard Cave. Katie is heading up a study of the cold zone, where ice has filled the passage in recent years preventing access to the cave.

Photograph: Charles Yonge.

worked in caves widely and during a similar period in the Canadian Rockies, Nahanni and the Yukon. The Alberta Speleological Society (ASS) from 1969 has done much work in the Canadian Rockies, in past times together with the KRG in the Crowsnest Pass and Castleguard (Fig. 15.5). The Society has recently made significant new ice cave discoveries: Ice Trap (1983), the Booming Ice Chasm (2008)—the largest ice cave in Canada—Inverted Fridge (2009), and Icy Hope (2013). Historical references are found in Thompson (1976) and Rollins (2004), and accounts of the recent discoveries are in various ASS newsletters and *The Canadian Caver* magazine.

15.2 CAVE ICE SCIENCE IN CANADA

Studies in this field have been sporadic during the past 40 years or so; Table 15.1 gives a brief summary of what has been done and where. In addition to the work of cavers, who have mapped the ice caves since 1969 and thus provided important baseline information, Ford (1974), Schroeder (1979), Ford and Williams (2007), and Yonge (2004) have described ice in Canadian caves, citing the many forms it takes and suggesting the processes leading to its formation.

The earliest study of ice in caves was the observation of ice in Grotte Valérie in a discontinuous permafrost zone by Ford in 1972. Further work on the various forms of ice there was undertaken by Schroeder (1977). Marshall and Brown (1974) at Coulthard Cave (Crowsnest Pass) used ice crystallography and stable isotopes to establish that massive ice in the cave was not glacially intruded. This view was supported with work in the 1980s by McMaster University on the ice plug in Castleguard and by Marshall (1975) on lava tube ice within the Aiyansh lava flow, British Columbia. Harris (1979) recognized a (mountainous) permafrost zone within Plateau Mountain Ice Cave and Yonge and MacDonald (1999) in Ice Chest and Serendipity, the latter modeling the process using stable isotopes.

A landmark paper by Wigley and Brown (1976) established the physics of cold zones in caves. Ford et al. (1976) and Atkinson et al. (1983) recognized an extensive cold zone in Castleguard Cave. Yonge (2014, 2015a) presented stable isotope evidence for cold zones in Grotte Valérie, the Ice Trap, and Canyon Creek Ice Cave.

Prior to the 1990s, few workers in Canada had used stable isotope (δ^2H, δ^{18}O, and δ^{13}C of HCO_3^-) analyses to look at the systematics of cave ice. Marshall and Brown (1974) were the first to use δ^2H and δ^{18}O, but reached no firm conclusions.

Lauriol and Clark (1993) proposed a mode of formation of cave ice in the Yukon (see below), using stable isotopes, ice fabric, and radiocarbon dating of cryogenic powder found on the ice. They found that although some of the ice dated back to the early Holocene Epoch, it nonetheless did not derive its source from glaciers, but more from ponded ice through melting and seepage and condensation in the permafrost zone. Therefore, from the preceding research, especially the crystallographic studies, it seems that massive perennial ice in ice caves is not from glacial intrusion in the general case for all ice caves. The exception is glaciers ("glacières"—Balch, 1900) within caves (e.g., Q5 described below). The interpretation of an atmospheric vapor for the origin of ice in Northern Yukon caves is supported by the studies of the diatom flora (Lauriol et al., 2006) and by the noble gas ratio from the air bubbles included in the ice (Utting et al., 2016). Clark and Lauriol (1992) were also able to radiocarbon date the bicarbonate in the ice (using extruded cryogenic calcite), by modeling the δ^{18}O and δ^{13}C. Lacelle et al. (2009) applied the work of Clark and Lauriol (1992) to seasonal ice in Caverne de l'Ours, Quebec (Lauriol and Bertrand, 2016). Yonge and MacDonald (1999) specifically modeled this process using δ^2H and δ^{18}O for two caves in the Crowsnest Pass region.

MacDonald (1994) studied 19 ice caves in the Western United States and Canada, generating δ^2H and δ^{18}O for all sites. Yonge (2015a,b) extended this data set to include five more sites in Canada plus a repeat visit to one (Canyon Creek Ice Cave). A further site, the Walkin-Ice Cave, was added later (Yonge, 2016). During this period, and in consideration of Luetscher and Jeannin's (2004) process-based classification of (alpine) ice caves, Yonge and MacDonald (2014) proposed a classification based on stable isotope systematics. Three classes of ice caves were suggested: *Cold Trap* (from direct snow or percolation), *Cold Zone* (humid air entering or exiting the cave), and *Permafrost* (intake and equilibrium freezing of moist summer air), recognizing that individual caves can display combinations of these types. This classification came about partly for assessing the suitability of cave ice types for proxy paleoclimatology.

With regard to dating for paleoclimatology, most of the perennial cave ice in Canada has been regarded as changeable and therefore of short duration. There are a few radiocarbon dates (trapped feces) from stratigraphic ice masses: 1000+ and 2500 year BP (Yonge and McDonald, 1999; Yonge, 2015b), and other masses may date back to the early Holocene based on stable isotope matching (Yonge, 2016), and using the ^{14}C of cryogenic calcite (Lauriol and Clark, 1993), back to >6000 year BP.

Finally, δ^2H and δ^{18}O data for 19 North American ice caves were compiled (Fig. 15.6), showing a difference in the Regional Meteoric Water Lines across the Continental Divide:

$$\delta^2\text{H} = 8.0\delta^{18}\text{O} + 4.1\%\text{VSMOW (East)} \quad \text{and} \quad \delta^2\text{H} = 8.0\delta^{18}\text{O} + 12.8\% \text{ (West)} \tag{15.1}$$

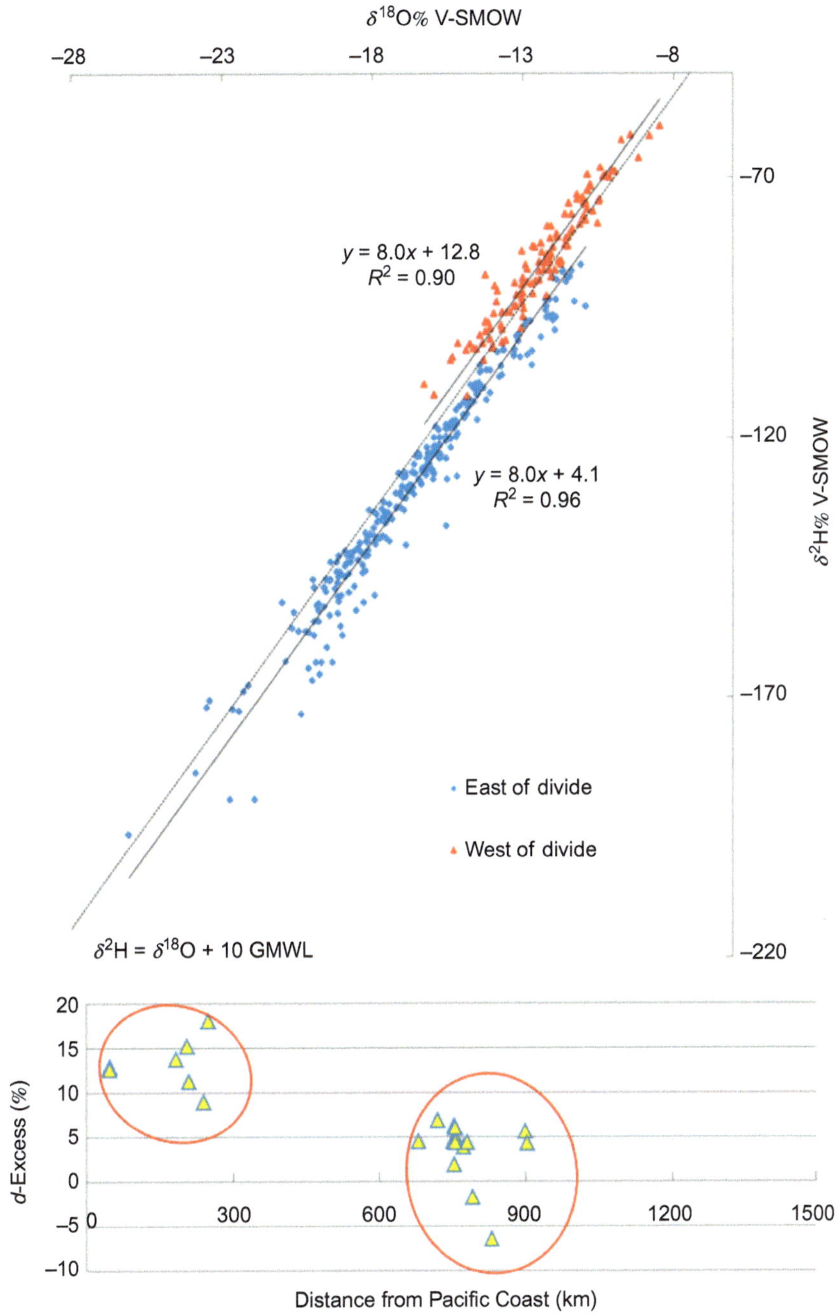

FIG. 15.6

$\delta^{18}O$-$\delta^{2}H$ of ice from 19 North American ice caves from East (<250 km from coast) and West (>750 km from coast) of the Continental Divide. Also shown below are the groupings *(red circles)* of cave-averaged *d*-excesses versus distance from the Pacific Coast.

Data from Yonge, C.J., MacDonald, W.D., 2006. The contrast in isotopic composition of cave ice across the Divide in Western North America. In: Archives of Climate Change in Karst. Proc. Symp. Climate Change: The Karst Record IV, Baile Herculane, Romania, vol. 10, pp. 26–28 (Special publication); Yonge, C.J., 2015a. The Systematics of Perennial Ice Found in Western North American Ice Caves. Thinking Mountains, Jasper, AB, p. 13 (Abstract and Paper).

By forcing these lines to a slope of 8, distinct deuterium excess (*d*-excess) values occur across the divide (+4.1‰ East and +12.8‰ West). That the ice caves east of the Great Divide on average yield a lower *d*-excess (4.1‰) appears to be a combination of the drier, more evaporative climate there and a greater tendency to precipitate snow and hoarfrost in the caves. The caves close to the coast are in high humidity regions where rain is more dominant, and the resulting ice (mainly from snow in cold traps) tends to plot closer to the GMWL (intercept of +12.8‰). Note that the easterly caves tend to plot with more negative δ-values on the Regional Meteoric Water Lines, which is indicative of the cooler climate on that side of the divide (Yonge and MacDonald, 2006).

Although our discussion here will be restricted to "perennial" ice, mention is made of significant studies of seasonal ice in caves where they have bearing on the understanding of the processes that lead to long-term accumulations of ice.

15.3 SELECTED ICE CAVES

15.3.1 YUKON TERRITORY

15.3.1.1 Ni'iinlii Njik Ice Caves, Old Crow Region

Ni'iinlii Njik ice caves are found at 66°30′N, in the northern Ogilvie Mountains on the Arctic Circle, where the Fishing Branch River nears its junction with the (Ch'oodèenjik) upper Porcupine River (Lauriol, 2016; Figs. 15.7 and 15.8). These mountains are within Eastern Beringia, a region of the northwestern part of America that remained unglaciated during the Quaternary Ice Age (Lauriol et al., 1997). It is a Devonian limestone area of caves, karst sinkholes, and springs (Utting et al., 2013). The cold, dry interiors of the caves there have preserved traces of Pleistocene plant and animal life from that last ice age (Lauriol et al., 2001).

The ice in the Fishing Branch River caves is a consequence of permafrost, and the ice's distribution is related to a "topoclimatic zonation" inside the caves (Lauriol et al., 1988). In general, near the entrance (Zone 1), both air and ground temperature are above 0°C under summer conditions and relative humidity is around 80%. Further inside the cave (Zone 2), relative humidity is near 100%; temperature near the floor is below 0°C but above 0°C at the ceiling. In this zone, the ice is abundant. At the deepest points in the cave (Zone 3), the air temperature is around −2°C (the same as the permafrost temperature, and 10°C below the mean air temperature of the region), humidity is 65% or less, and the walls are dry. The low water vapor in Zone 3 is explained by the fact that the atmosphere loses an important part of its moisture in Zones 1 and 2 as the air flow travels from outside and condenses on the walls. This dynamic air circulation occurs only in summer because snow tends to block the entrances in winter.

Massive and complete ice blockages are a common feature in Fishing Branch caves and more generally within cold caves in Canada. The presence of warmer air near the ceiling in entrance areas results in the melting of seasonal hoarfrost. The meltwater droplets form ice curtains on the walls and stalagmites on the floor that can coalesce to form the complete barrier. This phenomenon tends to occur a few meters from the entrance where the cave has a diameter less than 1 m, at 30–35 m from the entrance where the diameter is around 2–3 m, and as much as 100 m from the entrance where the diameter is around 4–5 m.

Bear Cave (Chii Ch'à'an) is 220 m long and has three main chambers (Fig. 15.9). It is within the second chamber (20 m long, 8 m wide, and ±10 m high) where ice is found. Clear pillars of ice rise from the floor of the chamber, and small hoar crystals coat the ceilings and walls, in places so thick as to form drapes. The third chamber is dry with a few very large hoar crystals, but contains copious and very old speleothems (Lauriol et al., 1997).

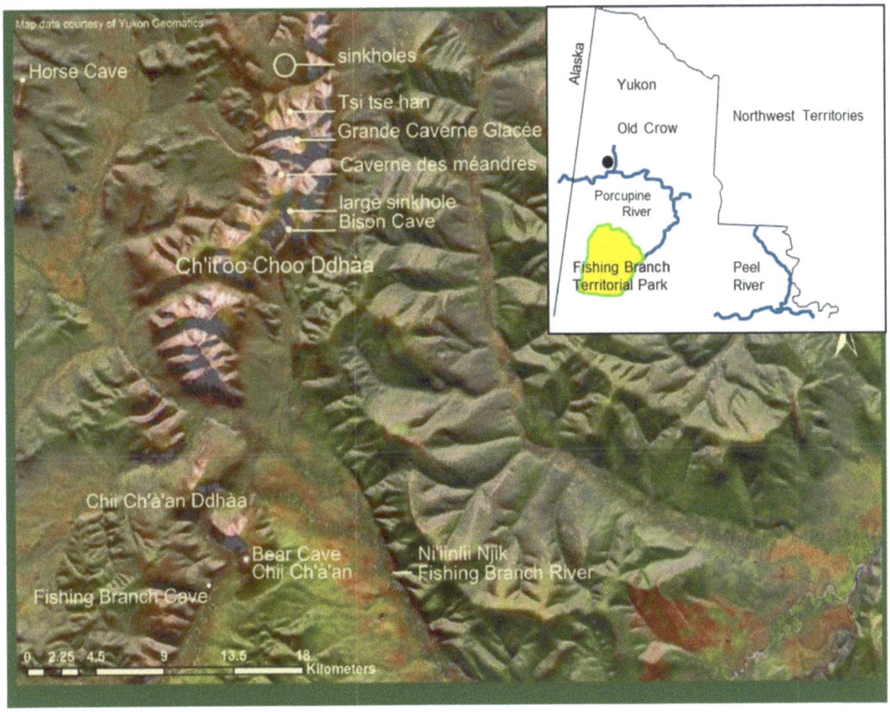

FIG. 15.7

Location of the Yukon Ice caves within Fishing Branch Territorial Park.

From Lauriol, B., 2016. The Ni'iinlii Njik Caves, Northern Yukon. In: Leduc, H. (Ed.), Notes on Geology and History. Government of Yukon, Yukon, ISBN: 978-1-55362-752-4.

FIG. 15.8

View of the bluffs above the Porcupine River of the Yukon Ice caves.

Photograph: Bernard Lauriol.

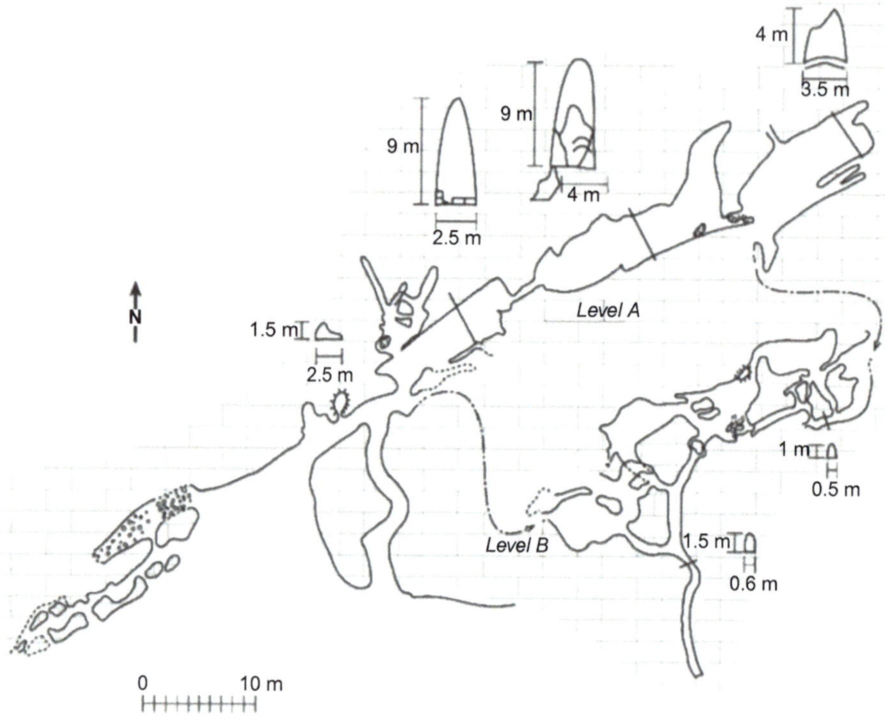

FIG. 15.9

Bear Cave map.

After From Lauriol, B., 2016. The Ni'iinlii Njik Caves, Northern Yukon. In: Leduc, H. (Ed.), Notes on Geology and History. Government of Yukon, Yukon, ISBN: 978-1-55362-752-4.

Grande Caverne Glacée (GCG) and Caverne '85: GCG is 110 m in length (Fig. 15.10). It contains well-preserved organic remains with a spruce wood fragment embedded in massive ice dating back to 7350 year BP. Caverne '85 is very similar in morphology, has massive ice, but is only 40 m long. Lauriol et al. (1988) and Lauriol and Clark (1993) surmised that ice formation, being sensitive to seasonal air circulation patterns, resulted in the condensation of warm, humid air flowing inward along the cave roof in summer, and sublimation by cold, dry air penetrating along the cave floor in winter. Massive ice blocking the two caves was shown to form by gradual accumulation of floor ice from below, plus accretion and regelation of hexagonal ice crystal masses growing down from the cave roof. Of great interest is that these workers used the radiocarbon activity of cryogenic powder found in the massive ice for dating, supporting their results by ^{14}C dates on associated animal and plant remains. Although the expulsion of calcite "flour" on the surface of the massive ice is a kinetic process leading to some of the highest ^{13}C observed anywhere, Clark and Lauriol (1992) were able to retrieve crude ^{14}C dates, 0.7–<2.5 ka for Caverne '85 and ~6 ka for GCG. This approach to dating ice, which is similar to ^{14}C dating of groundwater but with corrections for fractionations—along with stable isotopes—promises to be very useful in ice cave paleoclimatology where no trapped organics are available.

FIG. 15.10

Grande Caverne Glacée.

Photograph: Bernard Lauriol.

Tsi Tse Han Cave, 50 m long, has two chambers. From the main chamber (18 m long, 6 m wide, and 3 m high), two passages lead off: a short passage obstructed by ice, and a longer passage (10 m long) ending in a small second chamber that was blocked by ice in 1997 but that appears to lead to other openings. The walls of the second chamber are well decorated with speleothem deposits (principally flowstone), although less massive than those in Bear Cave. There is evidence of hunters sheltering in the cave, with a low wall constructed inside plus some red lichenous paint. Spruce trees were cached behind the wall, one of which was radiocarbon dated to about 700 year BP. Also recovered at the cave were porcupine fecal remains dating to 4270 year BP (Lauriol et al., 2001).

Caverne des Meandres is notable for the evidence of animal habitation; mummified mouse remains and bear fecal remains were recovered inside the cave below an ice plug that blocks the end of a passage littered by broken concretions. The fecal remains were dated from 37 to 38 ka. Pollen and plant remains within the sample revealed a diet composed primarily of juniper, which is toxic to modern bears. The fecal remains were therefore thought to belong to the extinct short-faced bear (*Arctodus simus*), which disappeared from Beringia at the end of the last ice age. It was the most common early North American bear during the Pleistocene Epoch.

15.3.2 NORTHWEST TERRITORIES (NT)

Karst landforms and groundwater systems are widely developed in the mainland portion of the Northwest Territories, a region that is similar in area to the entire British Isles (Ford, 2008a,b, 2009). It is divided into eastern and western halves by the north-flowing Mackenzie River. On the eastern side of the river, a series of small mountain ranges of limestone, dolomite, and gypsum form the Franklin Mountains; they have not been explored for caves. East of them, along the fringe of the Canadian Shield, these rocks are exposed in low plateaus or are buried as plains beneath glacial and lake deposits of the last ice age: Walk-In Cave (below) is representative of the low plateau karst. West of the river, the Mackenzie Mountains are a major chain extending between latitudes 60°N and 67°N, with the Continental Divide (border with Yukon Territory) in granitic rocks in the western sector and major limestone and dolomite mountains (the "Canyon Ranges") in the east. the South Nahanni Karstlands (latitude 62°N) have been comparatively well explored and studied, whereas for hundreds of kilometers north of them, little is known except for a few caves quickly blocked by ice or frozen silts around Norman Wells (Hamilton, 1995; Hamilton and Ford, 2002). No doubt there are many significant caves with perennial ice to be discovered.

15.3.2.1 Nahanni Karst

The South Nahanni River rises in the alpine granitic "Ragged Ranges" of the Mackenzie Mountains near the Yukon/NT border and flows 560 km southeast to the Liard River, the principal west bank tributary of the Mackenzie River. For a distance of ~100 km, the river passes through three spectacular antecedent canyons that it has entrenched to depths up to 1600 m in a variety of limestones, dolomites, and calcareous shales (Ford, 1974, 2010). The principal karst host rock is a Devonian platform limestone, the Nahanni Formation, 180 m in thickness that forms the rim of the First (furthest downstream) Canyon (Fig. 15.11). Numerous probable cave entrances can be seen in it, and exploration began with

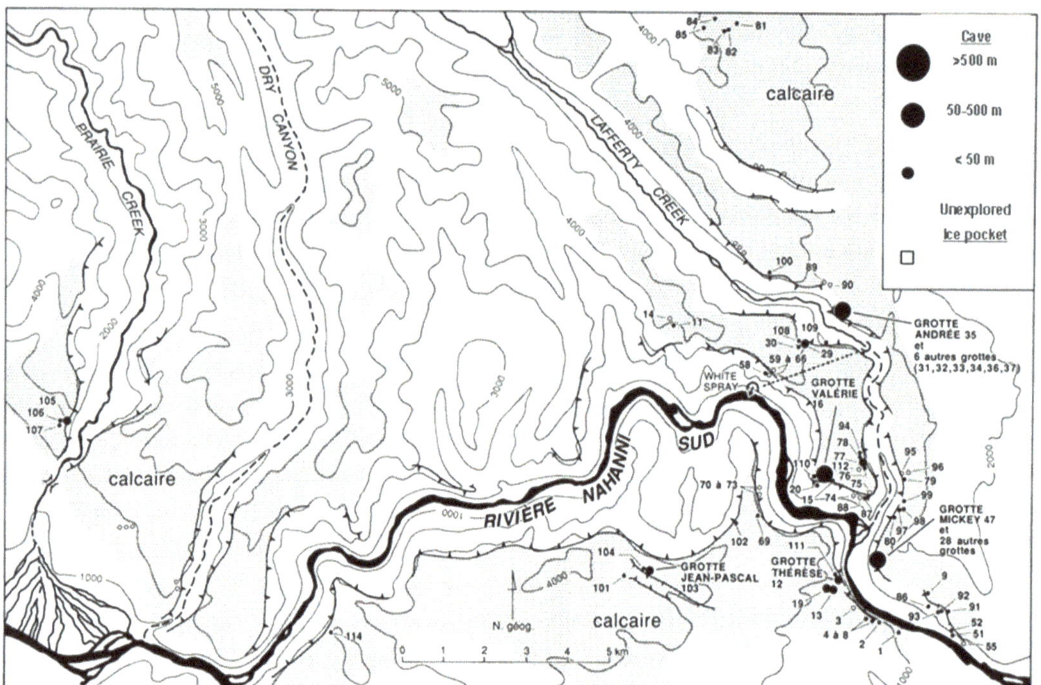

FIG. 15.11

Nahanni karst, indicating the 114 caves in the region of the South Nahanni River Gorges. The North Karst extends some 35 km north of Grotte Andree (middle right); Igloo and Arnica Caves are 18 km north in the vicinity of Death Lake.

After Schroeder, J., 1979. Le développement des grottes dans la région du Premier Canyon de la Riviére Nahanni, Sud, T.N.O. (Unpublished Ph.D. Thesis). Ottawa University, p. 265.

a Quebec party in 1970 (Marion and Poirel, 1973) when *Grotte Mickey* and some lesser caves were discovered (Fig. 15.12). The following year, South Nahanni National Park Reserve was established, in part to protect the canyons from hydroelectric development, and Parks Canada sponsored further exploration and evaluation by the Quebec group and the KRG. *Grotte Valérie* and *Grotte Andrée* (containing massive hoarfrost up to 15 cm in diameter) were discovered and mapped (Ford, 1971a, 1974; Thompson, 1976), but note that all of the hoarfrost from the 1205-m-long *Andrée* had disappeared between 1974 and 1976, except in the 10-m-long conduit at its end (Schroeder, 1979). The canyon wall entrance to *Grotte Andrée* is shown in Fig. 15.13.

From air photo analysis, Ford recognized that the karst extended 40 km north of the First Canyon as a prominent belt of giant grikes or corridors (*bogaz*), sinkholes, poljes, and springs that lay outside of the protected area. This "Nahanni North Karst" is now recognized as the most rugged, spectacular karst terrain to have been reported in any arctic or subarctic region. Exploration and study began in 1972 and continues today (Shawcross, 1973; Brook and Ford, 1974; Ford, 2008a,b, 2009; Horne, 2008). Many short caves are known, some blocked with ice barriers (as previously described) or larger flows. *Igloo Cave* (Fig. 15.14) is a relict fragment of a large phreatic cave, now ~200 m in total length and including one large room. Much of it lies below the level of the entrance and thus functions as a cold trap that has accumulated 10–100 m^3 masses of dripstone ice. *Arnica Cave* (Fig. 15.15) is a 240-m-long, intensely scalloped, vadose canyon up to 15 m in height, floored by ice in some sections (Horne, 2008). Forty kilometers west of the main North Karst area, 30 or more caves have been discovered and mapped in more recent years (Ford and Worthington, 2009). As an example, Black Ice Cave is ~120 m in length with a north-facing entrance and an abrupt drop of 5 m a short distance inside; the cave below this point is another simple cold trap with recrystallizing hoarfrost masses in the higher ceiling and terminating in a blockage of layered seepage ice that is slowly sublimating (Fig. 15.16). There have been no studies of any of these ice bodies as yet—the area is very remote, requiring floatplane or helicopter access. It was incorporated into South Nahanni National Park as part of a major extension in 2009.

The lengthiest open cave systems in the Nahanni karst are preserved along the north (updip) and east walls of the First Canyon near its mouth and in the adjoining Lafferty Canyon (Fig. 15.11). Their form is of dendritic drainage, from the sinkpoints of streams flowing from a shale cover that is now largely or entirely removed. Flow was downdip into the river channel at and just below the contemporary water tables. Development at multiple levels as the canyon deepened plus breakdown and other blockages have created ~50 m of relief between the highest and lowest points that create complexity in the distribution of seasonal (entrance zone) ice and cold trap perennial ice (Schroeder, 1977).

However, various ice types have been observed in only 18 of the 113 caves studied in the First Canyon area (Schroeder, 1979). If Grotte Valérie is still the model for understanding ice formation in caves (see next paragraph), Grotte Mickey (2270 m) and 28 other nearby caves (for a total length of 3853 m) have also favored ice formation because of preferential mechanical settings. In cliff walls, as well in these caves, joint sets seen are trending NE/SW and NW/SE and are widely opened (2–5 cm, more in some places), with or without vertical displacement (Fig. 15.12B and C). This situation suggests that it is "easier" for percolating surface water to fall down directly into dry *permafrozen* caves and feed thick ice stalagmites but no stalactites. This "mechanical decompression" process is still working in the "present" time because some dated flowstones (200 ka by ^{234}U-^{230}Th) have been broken by it (Fig. 15.12B). On the other hand, rare sponge-like ice still exists in two caves, testimony to a slow warming with melting/freezing seasonal phases.

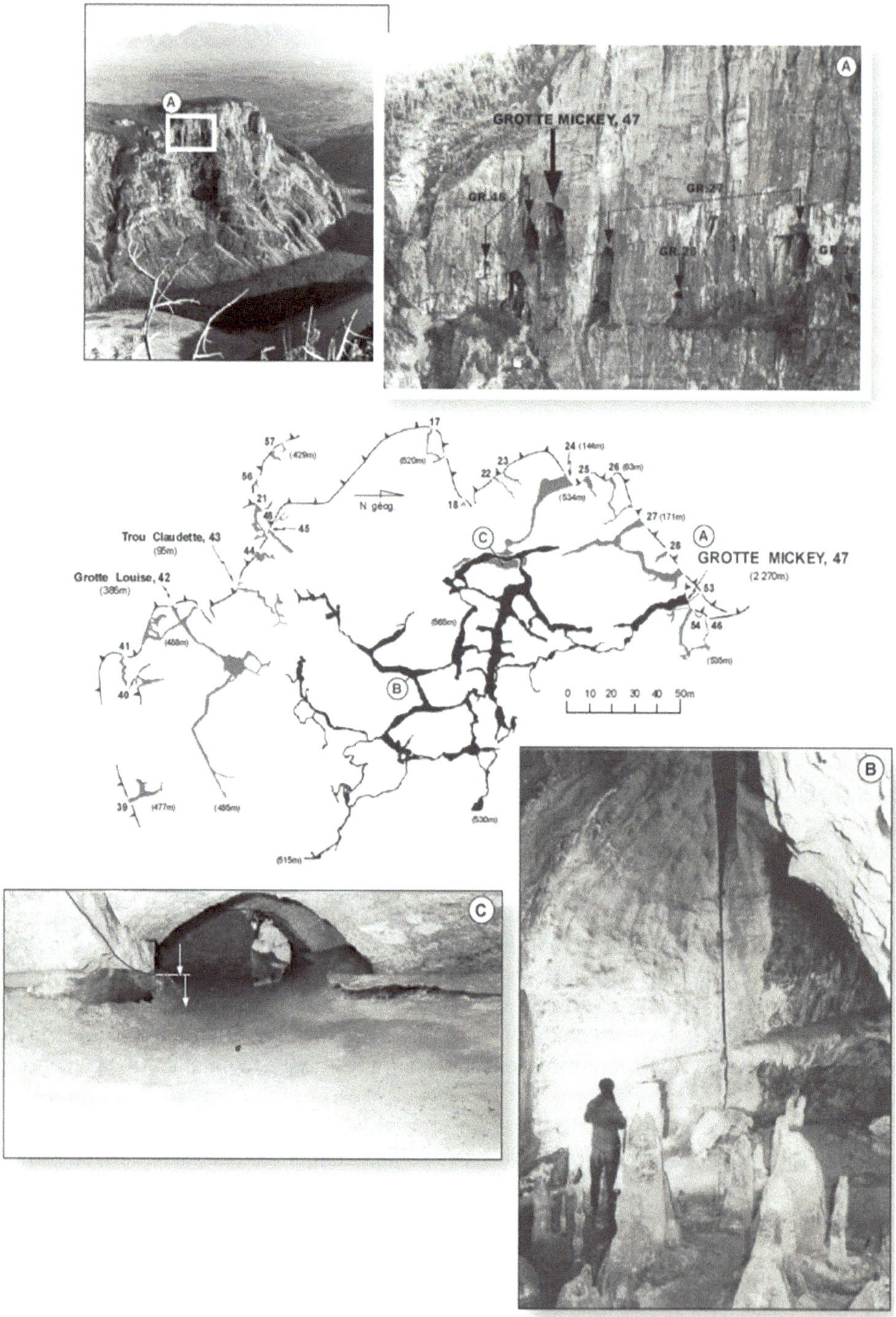

FIG. 15.12

(A) Grotte Mickey (2270 m) and associated caves (1583 m). Monique Herbeuval is seen in (B) and (C). Map, location, and features.

After Schroeder, J., 1979. Le développement des grottes dans la région du Premier Canyon de la Riviére Nahanni, Sud, T.N.O. (Unpublished Ph.D. Thesis). Ottawa University, p. 265.

FIG. 15.13

Entrance (left and inset) of Grotte Andree. The North Karst is in the backdrop.

Photographs: Greg Horne.

FIG. 15.14

Ice in Igloo Cave with Sharon Hayes (left) and Jonathan Tsetso (right), North Karst Nahanni.

Photographs: Greg Horne.

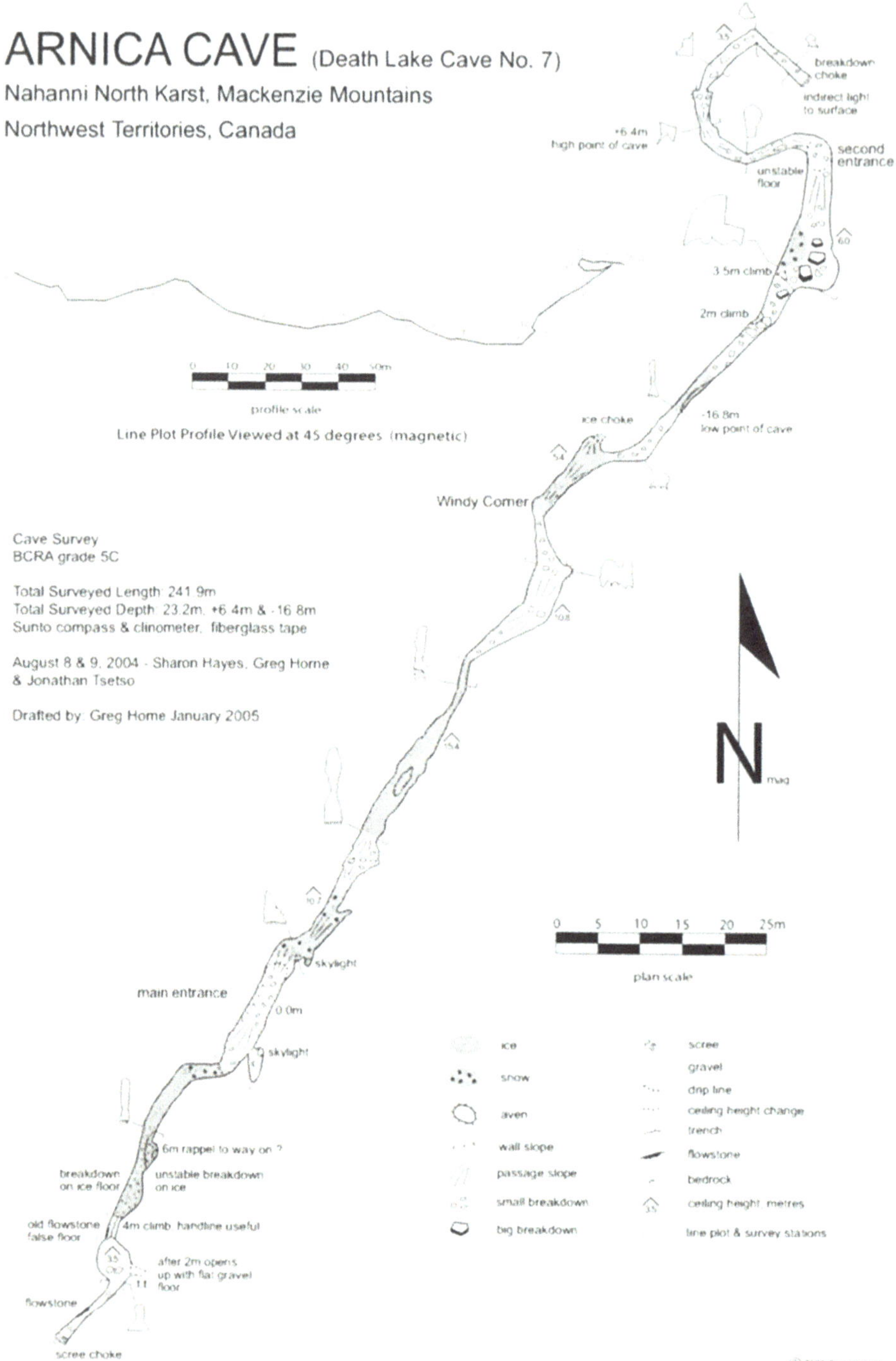

FIG. 15.15

Map of Arnica Cave.

After Horne, G., 2008. The North Karst, South Nahanni River Area. Can. Caver 68(1), 4–9.

FIG. 15.16

Jane Mulkewich, McMaster University graduate student, seen by the layered ice deposit in Black Ice Cave (left) and Sharon Hayes, Royal Roads graduate student by stratified ice in the short Blueberry Cave (right), both indicative of the conditions in the caves of the North Karst.

Photographs: Stephen Worthington (left) and Greg Horne (right).

The most studied ice cave in the canyon, Grotte Valérie, has 2 km of relict stream passages that once drained toward the river but, because of uplift and consequent canyon entrenchment, are now stranded in limestone cliffs 450 m above it (Fig. 15.17). It is an exceptionally good example of the nature of cold climate dynamics and static zones in a cave (Ford, 1971a; Ford and Williams, 2007, pp. 294–298). The external mean annual temperature of the plateau, which is 50 m above the cave, is estimated to be −7°C, and it is largely permafrozen. The first temperature measurements in the cave were made in the summers of 1971 and 1972. Five continuous temperature and humidity loggers were installed there in Aug. 2014 and recovered in Jun. 2016, for 22 months of unbroken recordkeeping (Fig. 15.18). The dynamic zonation is a consequence of the aspects and topography of the cave. The three explorers' entrances are south-facing and thus receive maximum insolation in the summer. The two in the west are 40 m higher in elevation than the eastern one. As a result, in summer cold interior air drains out through the lower eastern passage, drawing warm air in through the western entries to replace it. This creates three distinct zones. Zone 1, a "warm entrance cave" (+6–+1°C in 1971), extends from Logger 1 to the beginning of Central Gallery in which active depositions of small calcite stalactites and flowstones today indicate that the permafrost in the rock above is at least leaky, if not fully thawed. Remarkably, Logger 2 never fell below zero during the period of record, 2014–16. This sector supplies moistened air to Zone 2, a "cool exit cave" (0°C to −1.5°C in 1971, from Logger 3 to the eastern entrance; Fig. 15.17). Around Logger 3 there is still much leakage of drip water that, in the 1970s and up to 2007, was able to floor the Great

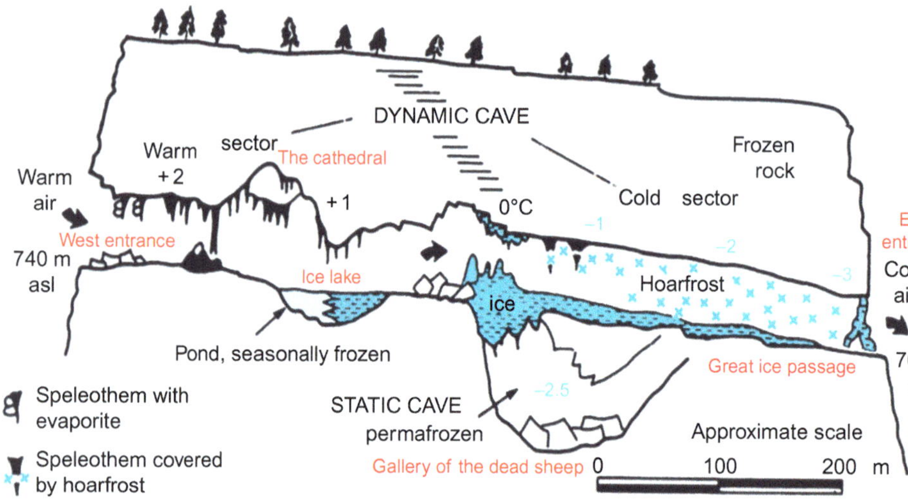

FIG. 15.17

Grotte Valérie elevation schematic.

After Ford, D.C., Williams P., 2007. Karst Hydrogeology and Geomorphology. Wiley, West Sussex, UK, p. 576.

Ice Passage with ice for a distance of ~300 m to the eastern exit. Thick hoarfrost coated the walls above the ice and in a dry tributary passage (Fig. 15.19). Much of the Gallery of the Dead Sheep, an inlet passage, is at a lower elevation that creates a cold trap that has been enhanced by build up of a thick ice dam between Loggers 3 and 4. This has created Zone 3, a static or "permafrozen" cave sitting in a lake of dense cold air in the gallery; it can be renewed only in the winter by an inflow of yet colder air from the outside (Fig. 15.17, GV #4). The passage is dry and dusty, without any overhead leakage, speleothems, or ice, but preserves the remains of 113 (as of 2016) wild mountain Dall's sheep (*Ovis Dalli*). Recent δ^2H, δ^{18}O stable isotope studies (Yonge, 2015a) largely support this mixed dynamic-static model. The data for the massive ice at the dam plot along a Local Meteoric Water Line of slope 8 (i.e., parallel to the Global Meteoric Water Line) but with an intercept of 0.8‰ (the *d*-excess value), whereas hoar plots similarly except with a *d*-excess of −4‰. The latter suggests the hoarfrost is produced via Rayleigh fractionation from incoming moist air according to the model of Yonge and Macdonald (1999) and Yonge (2014), whereas the massive ice appears to be a cold trap accumulation over a substantial period of time. Although the age of the massive ice (now undercut and accessible at its base) is not known, the disturbed skeletal remains of some individual sheep were observed to have been buried by its expansion into the Dead Sheep passage and ^{14}C dated to 2.4 ka BP. The most recent (i.e., undisturbed) skeleton lay beyond it and was dated to 600 years BP (Scotter and Simmons, 1976; Fig. 15.19).

15.3.3 WALKIN-ICE CAVE, WOOD BUFFALO NATIONAL PARK, ALBERTA, AND NORTHWEST TERRITORIES

The Walkin-Ice Cave system is representative of karst development in the southern low plateau with discontinuous permafrost and is the most easterly and lowest of the perennial ice caves currently known in Canada. It lies within Wood Buffalo National Park, which straddles the northern Alberta/Northwest

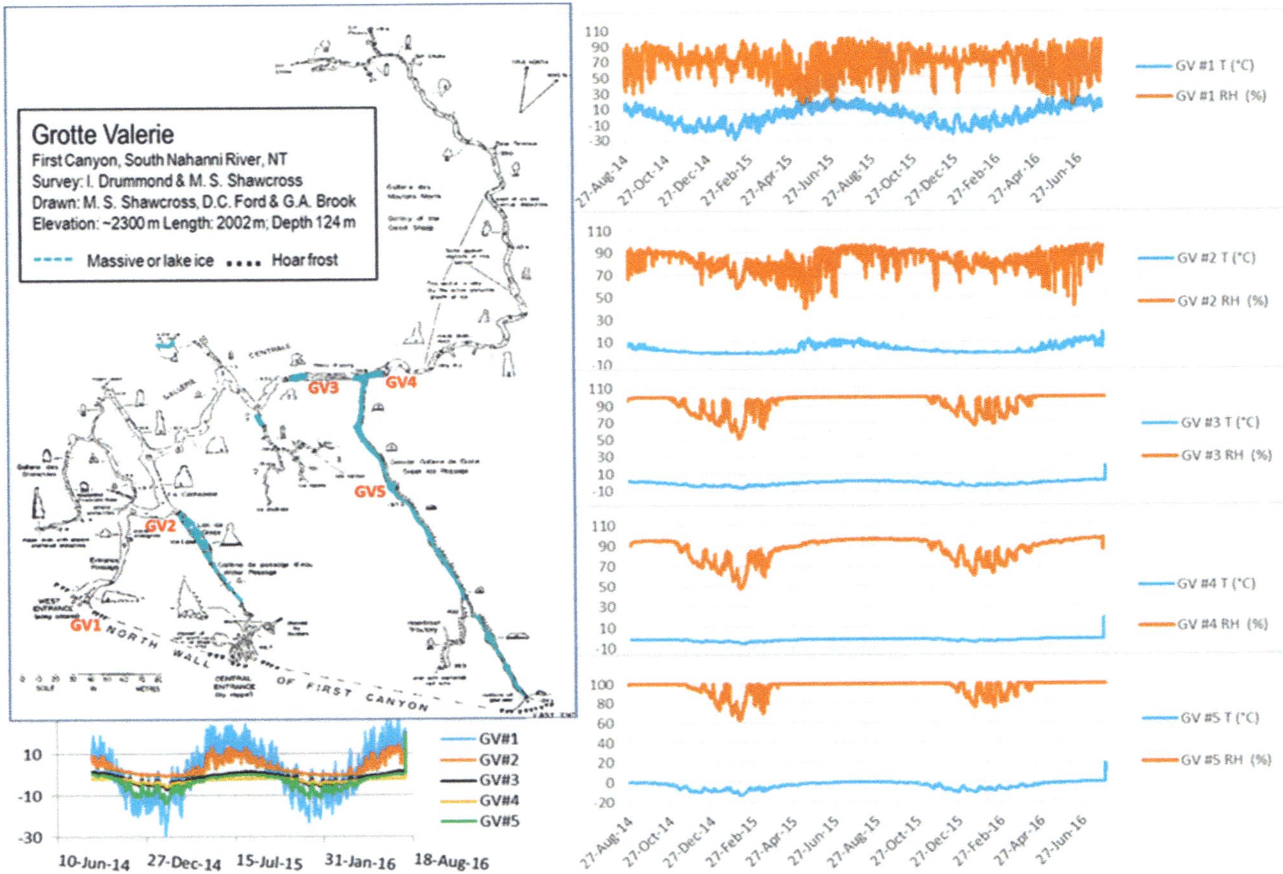

FIG. 15.18

Modified Grotte Valérie map with temperature and humidity records from five monitoring sites: GV1–GV5: compiled from data taken at 2 h intervals (from Greg Horne). All of the temperature records were plotted together in the figure below map.

FIG. 15.19

Grotte Valérie composite photograph: clockwise top left; field assistant, Lucinda views hoar and floor ice at the Junction/descending ice above the sheep remains/Greg Horne sampling the ice below previous/Dave Critchley, Melissa Carroll and Christian Stenner in the Great Ice Passage/600-year-old Dall sheep skeleton.

Photographs: Top clockwise; Derek Ford, Sylvain Foster (2), Derek Ford and Greg Horne.

Territories border (Fig. 15.20). It is the one cave known within a broad belt of karst terrain containing hundreds of large, collapsed sinkholes and blind valleys (caused by collapse of linear caves) to extend for more than 100 km behind and parallel to an escarpment 20–50 m in height along the western side of the Slave River floodplain. The plateau has a thin cap of dolomite beds (Keg Formation, Middle Devonian) overlying a thicker sequence of gypsum with interbedded detrital clastic rocks, with further clastics and salt below (Tsui and Cruden, 1984). Most of the sinkholes are between 10 and 100 m in diameter, with sides and floors of talus obstructing access to the caves between them. Many are also permanently water-filled because the cave development occurs at and below the local water table.

Walkin-Ice Cave is the exception because it is at the water table, with a flowing stream in the gypsum and collapse passages and chambers in the overlying dolomitic Keg Formation. It is currently 1.6 km long with three entrances (Walkin, Walkin 2, and Ice Cave; Fig. 15.21). Because of the rapid dissolution of gypsum, the cave is believed to be young, developing during or after retreat of the last glacial ice sheet in the region (Ford, 2009). In addition to extensive hoarfrost, the cave contains a body

FIG. 15.20

Location of Walkin-Ice Cave.

of massive stratified ice some 4 m in height close to one of the entrances, which was sampled for δ^{18}O-δ^2H (Fig. 15.22). Prominent brown bands are seen within the ice mass that are thought to be residual (cave wall) dust concentrated during periods of ablation. The dozen or so ablation horizons can be interpreted in two ways: (1) that these are modern and represent seasonal inputs of snow followed by partial melting in summer or (2) that the ice is relict and reflects secular periods of accumulation followed by partial ablation in drier periods. Against the seasonal argument, the ice contains no visible organics (pine cones, guano, etc.), which would accompany snow blowing into the entrance (plans are to try and extract pollen). Currently, accumulated snow in the entrance's doline melts, passing through breakdown and going downslope to join the streamway below. There seems to be no direct way to add it to the ice mass, which is under a substantial overhang and behind a large talus slope. Nonetheless, ice has been observed building on top of the ice mass in recent years, burying a bat roost and HOBO loggers placed at the bottom of the slope. The second possibility, that at least the lower part of the ice mass is a relict, is thought to be a possibility and appears supported by stable isotope measurements (Yonge, 2016). Fig. 15.23 compares stable isotope data for the stratified ice mass to nearby modern

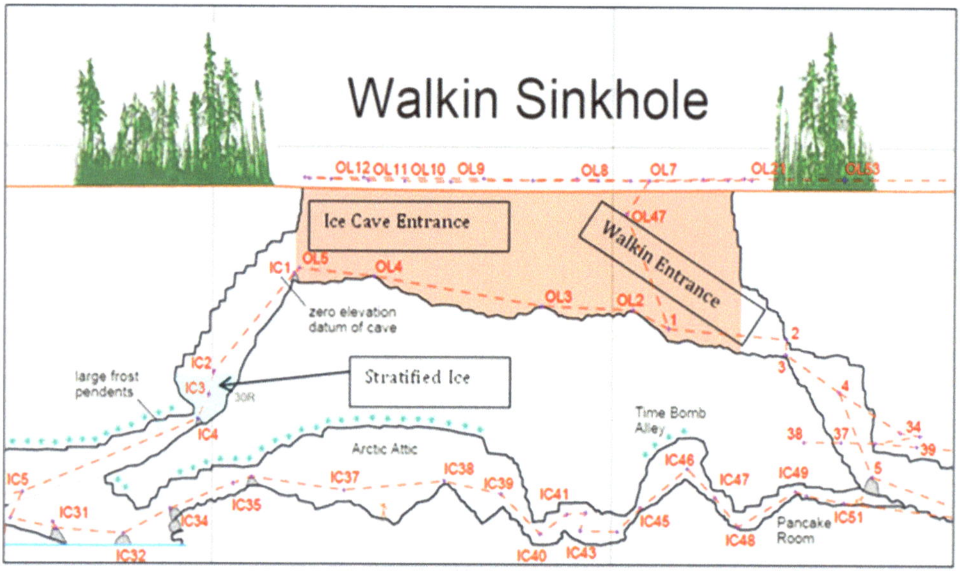

FIG. 15.21

Part of the Walkin-Ice Cave system showing the location of stratified ice.

From map prepared by Greg Horne.

precipitation at Fort Smith (Yonge, 2016). The meteoric precipitation lies somewhat off the GMWL, defining a Local Meteoric Water Line (LMWL) of a lower slope, 6.37, that is a result of the strong regional evaporative regime operating in the summer, which is kinetic in nature (Clogg-Wright, 2007). In contrast, cave ice isotope ratios scatter around the GMWL, intersecting the Fort Smith LMWL only at the most depleted (i.e., cold) values during snowfall in winter when there is no evaporation. (Hoarfrost downstream of banded ice has an obviously degraded quality during winter because of strongly inhaling cold winter air in Mar., which drastically lowers the humidity.) However, as the process is sublimation, and while it should also degrade the main ice mass, it will not lead to an evaporative isotopic signature. Nonetheless, because the ice mass is clearly derived of precipitation, why is it deviant? It is therefore possible that the ice is derived from an earlier postglacial climate. While work is ongoing to try to date the ice, a tentative matching to the climate temperature record interpreted from the Greenland Ice Core Project (GISP2; Alley, 2004; Briner et al., 2016) has been attempted (Fig. 15.24). If this is correct, then the ice accumulated during the Holocene Climatic Optimum when the climate was warmer and moister, in a high humidity-evapotranspiration regime leading to an equilibrium fraction of precipitation and a high *d*-excess of ~10‰.

15.3.4 THE ROCKY MOUNTAINS, ALBERTA-BRITISH COLUMBIA

The Canadian Rocky Mountains, straddling the Alberta/BC border in its southern reaches, extend from the US border northwesterly for about 1350 km to the low-lying Liard Plain in the Yukon. The Rockies consist largely of Paleozoic sedimentary rocks, a good proportion of which (>50%) are carbonates. Much of the area is devoid of roads and trails and is very rugged (up to 3741 m a.s.l.) requiring visits of

FIG. 15.22

Stratified ice mass in Walkin-Ice Cave with inset showing Chas Yonge measuring sampling locations *(orange tape)*. The diagram shows δ^2H of ice versus height from floor; b refers to the brown ablation bands (some minor).

Photographs: Greg Horne.

typically 2 days to a week to visit by ground travel. A good deal of exploration has been aided by helicopters, especially as caves are often found in the extensively glaciated high alpine karst. Overall, the area has been minimally explored and then mostly in the more accessible Southern Rocky Mountains where extensive cave systems have been found (up to 21 km in length); the majority of Canadian Ice Caves are found in the latter region (Fig. 15.1).

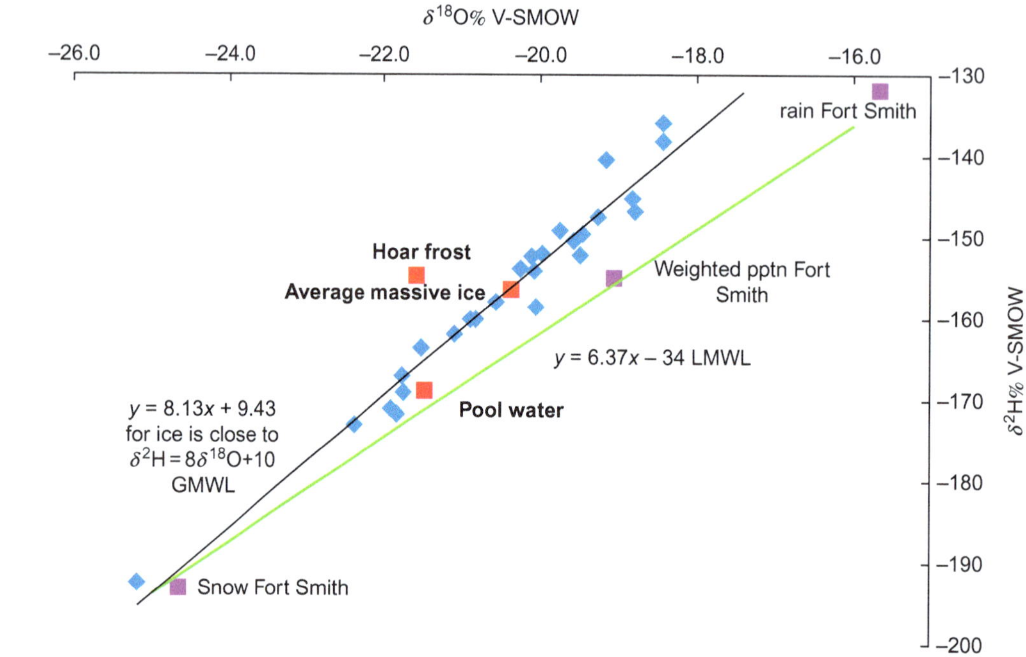

FIG. 15.23

$\delta^{18}O$-$\delta^{2}H$ of stratified ice from Walkin-Ice Cave *(blue squares)* compared to average Fort Smith precipitation.

From Yonge, C.J., 2016. The Walkin-Ice Cave System, Wood Buffalo National Park: An Archive of Ancient Ice? p. 11 (Report to Parks Canada).

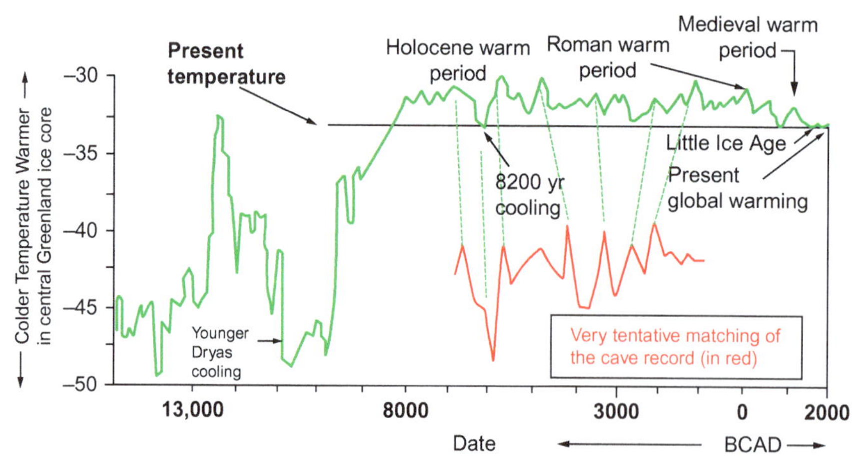

FIG. 15.24

Walkin-Ice Cave isotope record matched to an 11 ka temperature record based on GISP2 with stratified cave ice matching *(red)*.

After Alley, R.B., 2004. GISP2 Ice Core Temperature and Accumulation Data. ftp://ftp.ncdc.noaa.gov/pub/data/paleo/.../gisp2/.../gisp2_temp_accum_alley2000.txt.

15.3.4.1 Ice Trap, Jasper National Park, AB

The Ice Trap is a remarkable ice cave in a remote location up the Snaring River in Jasper National Park. Fig. 15.25 shows a rough map of the cave superimposed on a Google Earth image of the high alpine environment (entrance at 2350 m). Access to the cave is either by helicopter or a 2-day, high alpine traverse from Jasper.

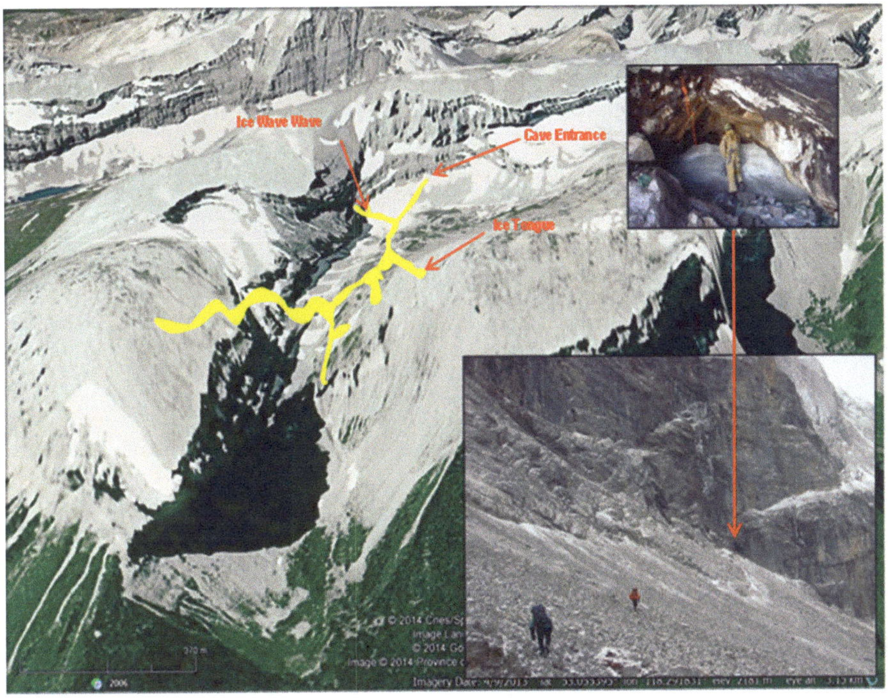

FIG. 15.25

Ice Trap Cave superimposed in *yellow* on a Google Earth image (Google Earth, 2016). Locations of the two ice masses are shown in respect to the entrance. Insets show the north-facing cliff and entrance area.

Photograph insets: Greg Horne (above) and Charles Yonge (below).

A high proportion of the cave has large, faulted passageways that vary in temperature from −2°C to −7°C, with a relative humidity ranging from 27% to 95%, well within the mountain permafrost zone. The cave is unusually dry with dusty (cryoturbated) sediments like "kitty litter," indicative of its cold and arid environment (e.g., Ford and Williams, 2007 p. 297). The cave promises interesting Holocene climatic information, as, for example, feces taken from sediments near a nest of mortified marmots was ^{14}C dated to 9560 ± 40 years.

The Ice Trap contains two areas of massive ice, the Milky Wave and the Ice Tongue. Single core samples were taken from each of the two sections of the Wave (Upper and Lower). These gave the lowest (most depleted) isotope values in the cave—the Upper less than the Lower—and almost the most depleted in all of the ice caves studied. This ordering of the low δ's suggest a *cold zone* effect (discussed

FIG. 15.26

Back view of the Ice Tongue showing sampling sites of ice for $\delta^{18}O$-$\delta^{2}H$ and packrat guano for ^{14}C age dating. Lee Hollis, proprietor of Cody Cave Provincial Park, surveys the scene.

Photographs: Greg Horne.

below), which could be associated with an unknown entrance near the Upper Wave. Unfortunately, no organic material has been recovered from either of the Waves, so the ages are unknown. Fig. 15.26 shows the back of the Ice Tongue, whose front side faceted morphology (Fig. 15.29) suggests ice retreat via sublimation. Two ^{14}C dates on packrat guano have been recovered from the backside, yielding a lower layer age of 2170 ± 30 and 1120 ± 30 years in an upper layer, suggesting a floor to ceiling age from 2.5 ka to recent times when the deposition stopped accumulating. Stable isotope studies of the visible ice layers show a slight general enrichment in δ-values over time (Fig. 15.27). One explanation for this age range is that both the Wave and Tongue formed when the cave was breached during deglaciation (at ~2.5 ka) and that a cold zone aided by permafrost was created. Those entrances are now blocked by talus. In the case of the Waves, a Rayleigh distillation of outgoing vapor would lead to progressively depleted δ-values. In the case of the Tongue, enrichment in δ-values over time might be due to the shrinking of the passage dimensions by the infilling ice. This raises the question whether cold zone ice can yield climate information (Yonge, 2015a). A tentative effort has been made by Yonge (2015b) by comparing the $\delta^{2}H$ to the *d*-excess records and documented Holocene temperatures (Fig. 15.28). While giving no quantitative information, the "hour-glass" profile of $\delta^{2}H$ and *d*-excess might be interpreted as follows.

The *d*-excess exhibits a poor but positive correlation to Holocene temperature, which might be expected because a cooler, less humid climate is associated with a higher *d*-excess and a warmer climate would exhibit the reverse (the *d*-excess relating to the humidity over the ocean source).

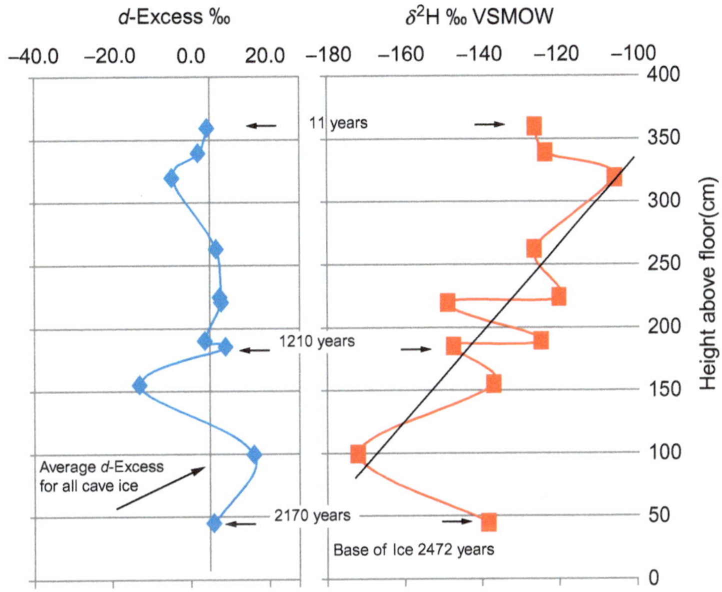

FIG. 15.27

d-Excess and δ^2H for stratified ice in Ice Trap Cave. The top and bottom dates are extrapolated. The trend line (right) shows the general increase in δ^2H with time.

From Yonge, C.J., 2015b. The fate of ice caves in northwest Canada as climate warms: what will we lose? Final Report to the Royal Canadian Geographic Society, p. 21.

The δ^2H runs counter to the *d*-excess, yielding the "hour-glass" profile. This may result from the fractionation of humid cave air as it exits through the cold zone, which is cooler in a cold climate and warmer in a warm climate. Fractionation of the sublimating vapor is greatest when temperatures are lowest and differences between the cave air and cold zone are greatest, thus a cold climate delivers the highest δ^2H—the opposite occurs in a warm climate.

Over a 9-year period, the spectacular Ice Tongue has declined substantially in mass (Fig. 15.29). The Ice Trap then is another example of a cave losing ice in a warming climate.

15.3.4.2 Disaster Point Ice Cave, Jasper National Park, Alberta

Disaster Point Cave is located at 1082 m a.s.l. in the Front Ranges of the Rockies (Table 15.1). The cave is developed in the limestone Palliser Formation (Devonian). There are two entrances at the base of a steep natural depression that act as a funnel for sediments, drifting snow, and debris. It also serves as a natural trap for many animals because of the near-vertical form of the entrances (Fig. 15.29). Radiocarbon dating suggests that the sequence of late Holocene faunal remains is relatively continuous. Two charcoal layers from massive forest fires have been excavated from sediment profile, one of which has been ^{14}C dated to 1.5 ka, and a date of 6090 ± 40 year BP on bone of *Ursus americanus* provides the current known maximum age for fauna preserved in the cave (Jass et al., 2013).

The entrances are over a 25 m ice-covered shaft that leads to the lower end of a large chamber with a steeply sloping ice floor derived from in-falling snow and seepage. Fig. 15.30 shows the location of

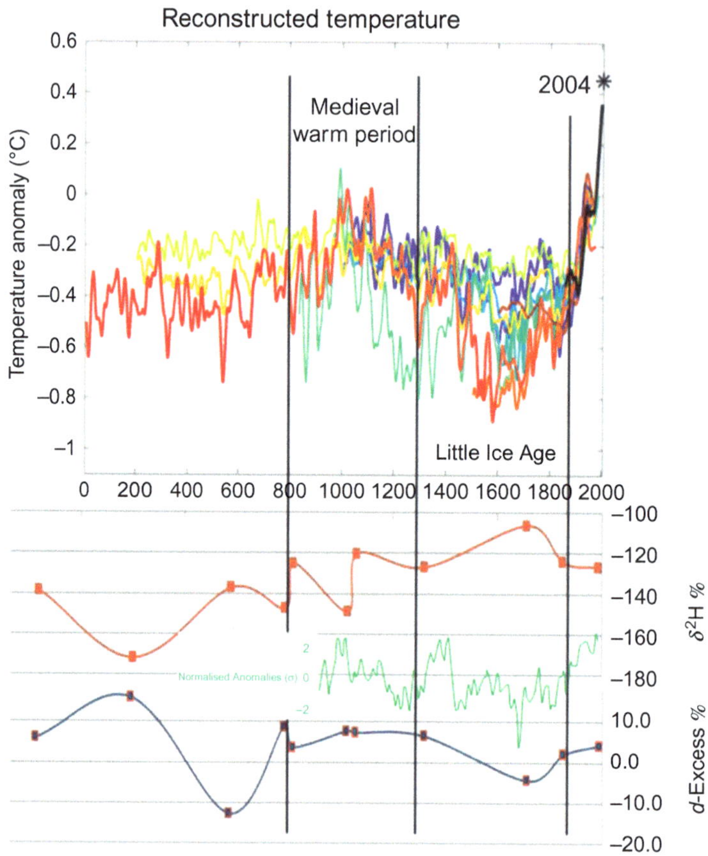

FIG. 15.28

Climate record from Ice Trap Cave compared to reconstructed global temperatures (upper plot) and the normalized temperature record for SW Canada (lower *green plot*), SERC, 2016.

Stable isotope Ice Trap data is from Yonge, C.J., 2015b. The fate of ice caves in northwest Canada as climate warms: what will we lose? Final Report to the Royal Canadian Geographic Society, p. 21.

sites where samples of ice, snow, and seepage water were collected for isotope analysis. These data define a local meteoric line (LMWL)

$$\delta^2\text{H} = 8.3\delta^{18}\text{O} - 0.8 \qquad (15.2)$$

essentially parallel to the GMWL. The low *d*-excess of ~−1‰ is suggestive of the dry continental climate in the eastern Front Ranges (see Section 15.2). Data also point to behavior as a cold trap with the ice being a mixture of seepage and snow, the dominant component being seepage similar to the river below (Fig. 15.31). This cave has potential for Holocene paleoclimatologic reconstructions because it has datable organic deposits and sequential ice essentially reflecting average precipitation.

FIG. 15.29

Ablation of the Ice Tongue over a 9-year period. Dave Critchley, biologist with the Northern Alberta Institute of Technology (top). Chas Yonge and Lee Hollis are below.

Photographs: Greg Horne.

15.3.4.3 Castleguard Cave, Columbia Icefields, Alberta-British Columbia

This cave is a simple, linear river cave created by groundwater flowing NW-to-SE from sources under the Columbia Icefield, the largest ice cap remaining in the Rocky Mountains (Fig. 15.32). The cave passes under Castleguard Mountain. The water resurges below in the Castleguard Meadows in a series of perennial and seasonal springs. It is the longest cave in Canada, currently 21,068 m, with a vertical range of 384 m. The principal passage is an alternation of vadose and phreatic parts. The cave is developed mainly in the top of Cathedral limestone formation, occasionally breaking through the Stephen Formation to come in contact with the lower Eldon Formation, all of Cambrian age (Ford, 1971b).

Most of the explorers' cave is a drained relic, with water passing down through it into lower routes that become impassably small. A number of the relict passages at the upper end of the cave, beneath

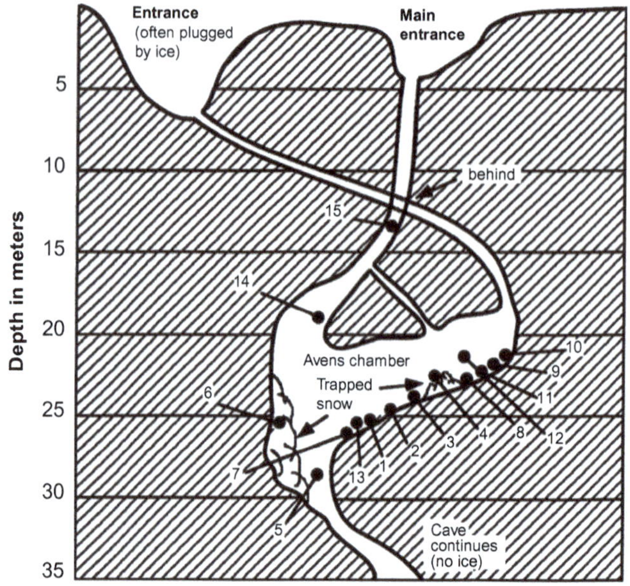

FIG. 15.30

Elevation view of Disaster Point Cave, Jasper National Park, showing ice sample collection points.

Diagram from MacDonald W.D., 1994. Stable Isotope Composition of Cave Ice in Western North America (Unpublished M.Sc. Thesis). University of Calgary, Alberta, Canada, p. 206.

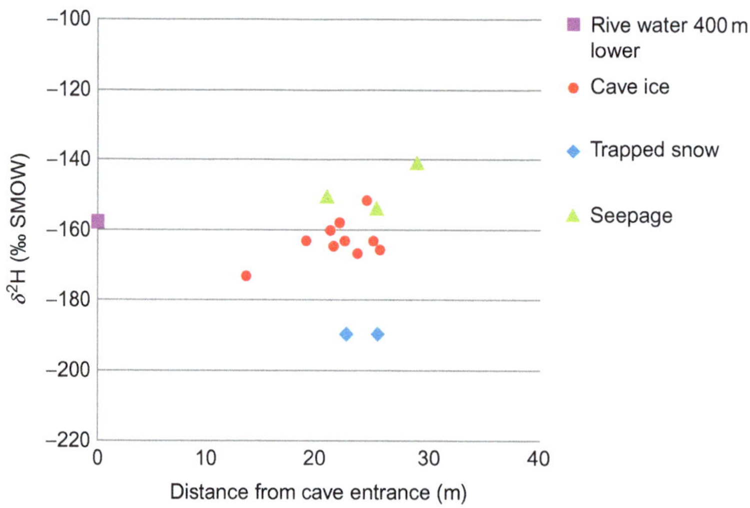

FIG. 15.31

δ^2H for seepage, trapped snow and cave ice from Disaster Point Cave. The cave ice appears to be a mixture of dominant warm season seepage and trapped snow and is similar to Athabasca River water derived from a very large catchment.

Data from MacDonald W.D., 1994. Stable Isotope Composition of Cave Ice in Western North America (Unpublished M.Sc. Thesis). University of Calgary, Alberta, Canada, p. 206.

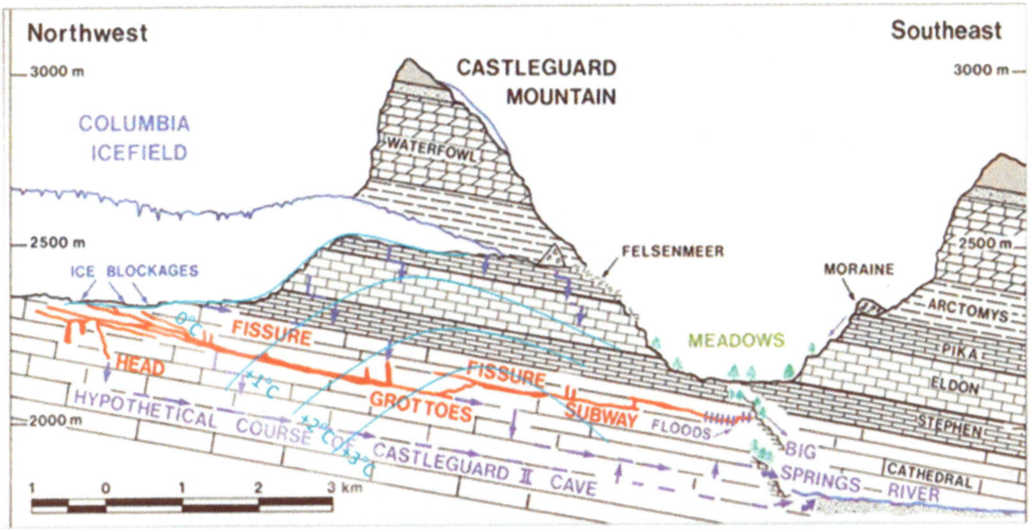

FIG. 15.32

Elevation schematic for Castleguard Cave (relict passage in *red*). Perennial ice blockages occur below the Columbia Icefield, whereas some 200 m of seasonal cold zone ice forms in the "floods" area. Diagram from Muir and Ford (1985). Approximate and partial temperature isotherms *(blue)* have been added.

Based on Ford, D.C., Harmon, R.S., Schwarcz, H.P., Wigley, T.M.L., Thompson, P., 1976. Geo-hydrologic and thermomometric observations in the vicinity of the Columbia Icefield, Alberta and British Columbia. J. Glaciol. 16(74), 219–230.

~200 m of glacier ice, terminate in sealed ice blockages. These appear to be ablating; the largest has been photographed repeatedly since 1972, revealing that the ice is not intruded (i.e., strained), that it is probably relict, and that any ablation is slow (Fig. 15.33). In warm summer weather, the downstream end of the relict cave (the only entrance for explorers) is annually flooded from a deeper overflow source. In the winter and spring, it forms a substantial cold zone in which the residual flood water freezes (Fig. 15.34). While not perennial, the ice here is an excellent example of seasonal accumulation in the entrance cold zone of a cave that continues to discharge floodwaters (Fig. 15.35). The cave functions as a climatic chimney. In the winter, interior air warmed to ~3°C by the geothermal flux flows upward into the base of the icefield, drawing cold, dense outside air in through the downstream entrance (often with strength sufficient to extinguish old-fashioned carbide lamps!); the cave walls and the residual flood water freeze. In the summer, the dynamic system reverses, the now cooler (≤3°C) interior air drains gravitationally down and outward into sunshine via the cold zone, precipitating hoarfrost on the walls and ice ponds until the first flood of the new summer melts it all. Fig. 15.32 summarizes the temperature regime around the cave system, where mean annual surface temperatures are estimated from 0°C to −14°C over the range of cave elevation (Ford et al., 1976).

15.3.4.4 Canyon Creek Ice Cave, Kananaskis Country, Alberta

Canyon Creek Ice Cave (CCIC) is located on Moose Mountain at 1775 m in the Livingstone Limestone (Mississippian). Though at a relatively low altitude, the cave supports an extensive cold zone with stratified massive ice, although this mass is now in retreat (Fig. 15.36). The cold zone appears to result

FIG. 15.33

Peter Thompson sits by the principal ice blockage in Castleguard Cave in 1972 (top). The lower photograph, taken in 2013 shows the three-decade extent of ablation. Note the coring on the lower image on right, where petrographic analysis of the ice showed that it was not intruded (i.e., unstrained).

Photographs: Tony Waltham (above) and Marek Vokac (below).

from moist cave air descending in spring through large karstified vertical fractures (the cave is 727 m long) and flowing strongly out of the entrance. Stable isotope studies support this view. The layered ice front was first sampled in 1992 when the front was 56 m from the entrance (MacDonald, 1994) and then 20 years later in 2014 when it had receded a further 94 m and the ice layers correlated (Yonge, 2015a). The cold zone ice "relaxation" length (Wigley and Brown, 1976) appears to be adjusting to warm summer melting and dry winter ablation. Fig. 15.37 displays the δ^2H of various ice types versus distance into the cave; the vertical bar shows the range of the 2014 data set. Note that when looking at averages for the stratified ice at each of those locations, shifts in δ^2H from −154‰ to −135‰ are seen. The latter averages broadly indicate the Rayleigh distillation mechanism expected across the cold zone. However, there is a great variation in the stratified ice sequences, whereas if it was purely Rayleigh distillation

FIG. 15.34

Seasonal cold zone ice close to the entrance of Castleguard Cave. Note the wind scalloping on the ice floor. (The image was taken in Apr. 1973 during the filming of Castleguard Cave. Left to right: Russ Harmon, Frank Binney Ewers, and Sid Perou.)

Photograph: Derek Ford.

FIG. 15.35

Top: Julian Carter in spring 2012 at the Ice Pingo in the entrance passage, with an inward draft and all surfaces dry. Bottom: Ralph Ewers in the same passage Jul. 1968 with a reversed draft where comparatively warm and moist air from the center of the cave is blowing out, precipitating a general hoarfrost cover. Floods later in the summer remove all of the ice.

Photographs: Julian Carter (self-portrait top) and Derek Ford (bottom).

FIG. 15.36

Entrance to Canyon Creek Ice Cave. Inset: Sampling stratified ice 152 m from the entrance in 2014.

Photographs: Nick Vieira.

from a constant vapor source (cave air), then there should be little variation. The wide variation observed in reality can be explained by the mixing of a groundwater contribution into the ice produced by sublimation and seepage in the cold zone. The data then represent the degrees of mixing: fully mixed giving the upper line, and Rayleigh Distillation the lower.

There are no dates for climate interpretation on this ice. However, there is reason to believe it is in the order of only decades old. First, fresh wood and occasional human detritus are trapped in the ice, and there are also anecdotal accounts of the ice almost completely disappearing in the recent past. It is therefore likely that variations in the layers are only of a few years and that annual variations could be observed if closer sampling was employed (visually obvious layers were sampled; Fig. 15.36).

The δ^2H amplitude is greater in the earlier sampled ice, but this was much closer to the entrance (56 m) when the ice was more extensive than at the later sampling 152 m inside the cave. The obvious implication is that external factors such as summer moisture and seepage modify the Rayleigh-generated ice more strongly closer to the entrance, where there is more interaction with outside sources.

Based on these findings, the climatic (or weather) interpretation is of more negative δ-values representing a purer Rayleigh mechanism at work. This would occur when a longer transitional period exists in the spring, securing a stable evaporitic environment along the cold zone, but one that did not melt the ice or did not allow ice formation, that is, a colder climate/weather regime. The converse (higher δ-values) implies a weakening of the Rayleigh distillation and an influence of moist summer effects (air mass and seepage). Because of the kinetic nature of the Raleigh + mixing system, only qualitative warmer and colder trends can be discerned.

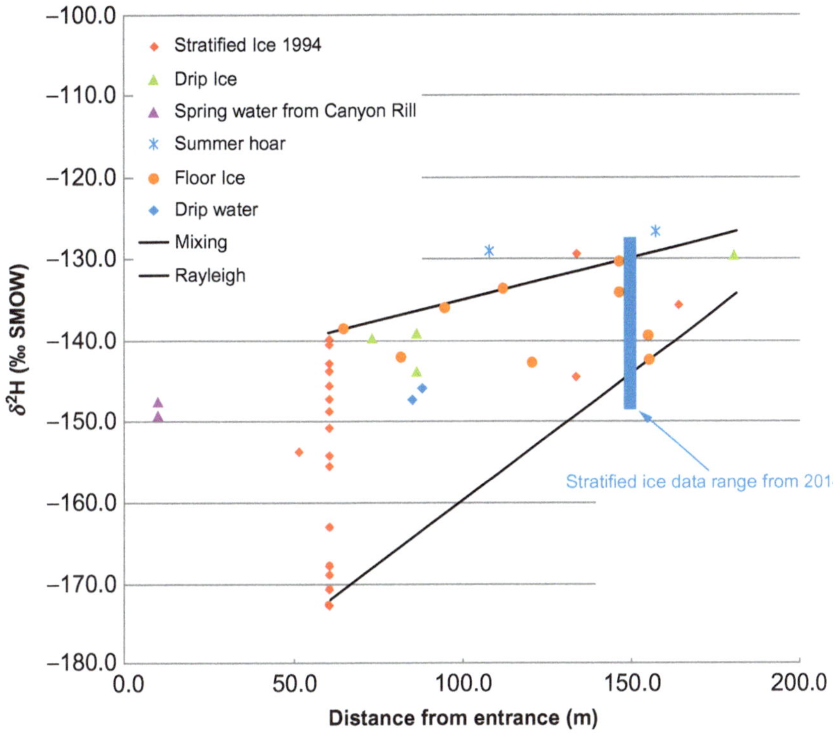

FIG. 15.37

δ^2H of ice types versus entrance distance for Canyon Creek Ice Cave.

From Yonge, C. J., MacDonald, W.D., 2014. Stable isotope composition of perennial ice in ice caves as an aid to characterizing ice cave types. In: The 6th International Workshop on Ice Caves (Procs), Idaho Falls, USA, p. 9; Yonge, C.J., 2015a. The Systematics of Perennial Ice Found in Western North American Ice Caves. Thinking Mountains, Jasper, AB, p. 13 (Abstract and Paper).

15.3.4.5 Plateau Mountain Ice Cave (PMIC), Kananaskis Country, Alberta

PMIC is in the unique ecological reserve of Plateau Mountain in the Livingstone Limestone formation (Mississippian). It is an area of relict alpine permafrost from the Late Wisconsin glaciation. The flat mountain top was not glaciated during the Late Pleistocene, and its summit permafrost has not yet reached equilibrium with current climate. At the surface, there is excellent patterned (polygonal) ground and thermokarst resulting from melting of ice wedges formed beneath stony borders during a colder climate. The slopes of Plateau Mountain show very well-developed block slopes (Fig. 15.38), in contrast to the rocky mountain sides seen elsewhere in the adjacent Rockies. At the northern end of the mountain, the >400-m-long Plateau Mountain Ice Cave (Fig. 15.39) penetrates into the relict permafrost at a shallow depth, with temperatures decreasing inward.

Public access is unusually easy because of a gas field approach road. Over the years prior to its gating in 1972, there was great concern over the disappearance of the cave ice (Fig. 15.40). First this was thought to be caused by excessive visitation, but melting continued even after the gating and implementation of a very restrictive and hence conservative access policy by Alberta Environment and

FIG. 15.38

Looking out on the blocky slopes of Plateau Mountain through the entrance gate of Plateau Mountain Ice Cave.

Photograph: Nick Vieira.

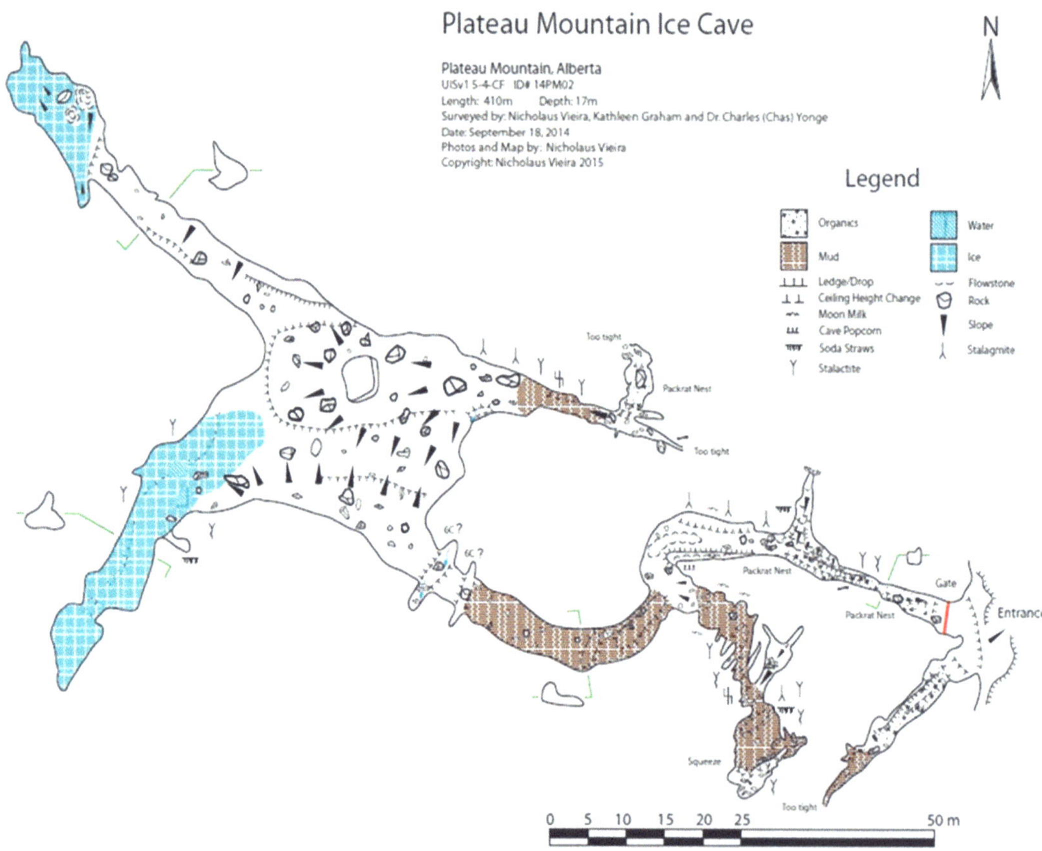

FIG. 15.39

Plan map of Plateau Mountain Ice Cave.

FIG. 15.40

Chas Yonge sampling hoar ice in the NW gallery of Plateau Mountain Ice Cave (above) and Lorraine Drummond below fresher crystals ~30 years earlier in the SW Gallery (lower). Insets show ice crystal detail from summer 1968. The upper inset shows fresh crystals, the lower show melting after too much exposure by visitors.

Photographs: Nick Vieira (upper), Ian Drummond (lower), and Derek Ford (insets).

Sustainable Development. In 2013 the management was approached for a stable isotope study of the site in order to gain an understanding of the ice formations and their fate (Yonge, 2015b). In any event, it seems likely in the current climate warming that the ice will disappear in perhaps a few decades, along with the melting permafrost on the surface.

The few data from samples collected in the two western galleries in PMIC (Fig. 15.39) show an interesting Local Meteoric Water Line (LMWL) with a slope much lower than any of the other caves studied (Yonge and MacDonald, 2014; Fig. 15.41):

$$\delta^2\text{H} = 4.2\delta^{18}\text{O} - 62.7 \tag{15.3}$$

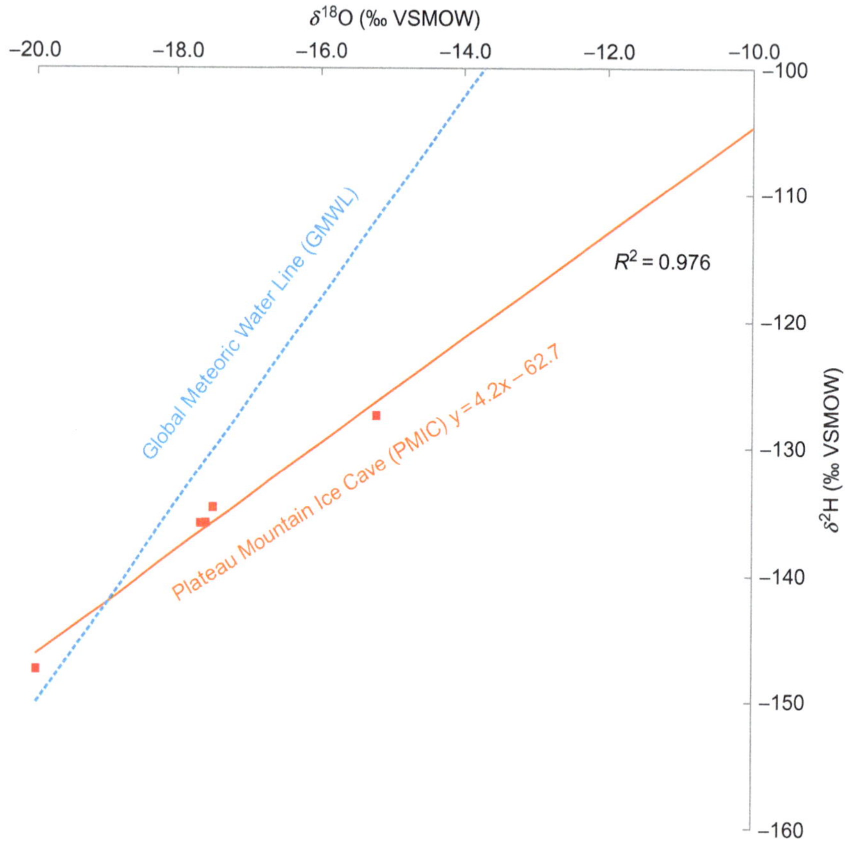

FIG. 15.41

Isotope data for Plateau Mountain Ice Cave with line slope close to a model evaporative trend of 4.

From Yonge, C.J., 2015b. The fate of ice caves in northwest Canada as climate warms: what will we lose? Final Report to the Royal Canadian Geographic Society, p. 21.

The most depleted sample—approximately lying on the GMWL—came from a pool in the ice "streamway" in the SW gallery. It appears to represent an end member and may be derived from drip water, that is, an average of precipitation at the site. All of the other samples are hoarfrost trending away from the GMWL along an apparent evaporative/mixing line of slope ~4 with increasing *d*-excess. The ice in the cave is retreating, and there is much evidence of melting and refreezing of the hoar, most likely because of this evaporative trend. For management (Alberta Environment and Sustainable Development) and climate change science, it would be interesting to core the ice to see if the cave has passed through greater or lesser evaporative periods in the past.

15.3.4.6 Serendipity, Crowsnest Pass, Alberta/British Columbia

The cave is located at 2175 m a.s.l. in the Livingstone Limestone, 525 m above the nearest perennial flowing water. The entrance, beneath a large overhang, quickly narrows to a small slot, after which the cave opens into a large ice-filled chamber with a level ice floor on the south side and a steep, stratified ice ridge on the north side. Farther into the cave the ceiling becomes decorated with masses of hexagonal ice crystals above a flat ice floor that has ablated on one side to reveal a 15 m drop, against a stratified ice mass (Fig. 15.42 main). The banding of the massive ice is very regular with bands of white, light blue, and dark ice alternating up the entire profile of the ice mass. Adjacent to the ice mass, curls of extruded ice can be found on the rock wall, mimicking gypsum flowers seen in many warmer caves (Fig. 15.42 inset). They are unusual in Canada, known only here in Ice Trap Yonge (2016), Ice Chest MacDonald (1994), and Plateau Mountain Ice Cave (Ford and Williams, 2007).

The ice appears to have built up from a combination of falling hoar and seepage. Stable isotope data on the ice ridge and ice mass support the model that the majority of the ice is derived from the hoar, which itself comes from moist summer air entering the cave and being forced to freeze at ~0°C in the permafrost zone. Fig. 15.43 shows sections of the ice masses where the δ^2H oscillates around a value of −130‰ displaced from mean annual precipitation at the site (−143‰). The higher δ-values of the ice are due to both the warm, summer vapor and the forced fractionation of that vapor at ~0°C; the low value of the *d*-excess reflects the latter (Yonge and MacDonald, 1999). Variations in the displacement (average −13‰) should allow a qualitative interpretation of climate where smaller displacements would imply cooling summers and larger displacements, warmer summers. A single radiocarbon date of 970 year BP

FIG. 15.42

ASS member Randy Spahl rappelling below massive banded ice in Serendipity. Inset: extrusion ice was found both in Serendipity and Plateau Mountain Ice Cave.

Photographs: Ian McKenzie (main) and Derek Ford (inset: example from Plateau Mountain Ice Cave).

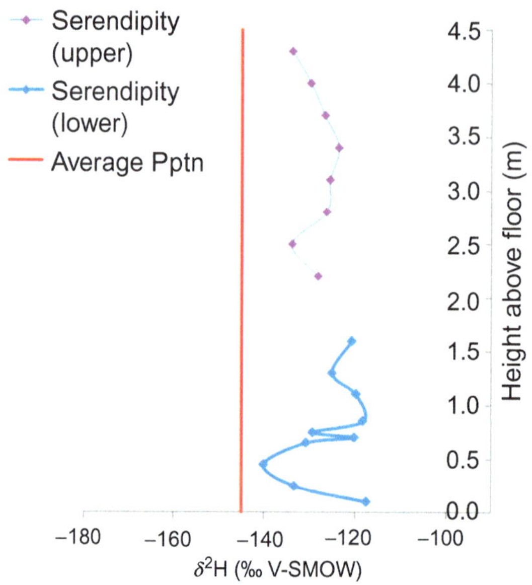

FIG. 15.43

Variation of δ^2H in stratified ice units in Serendipity (permafrost type). Note the displacement of Serendipity ice from average precipitation.

From Yonge, C.J., 2015b. The fate of ice caves in northwest Canada as climate warms: what will we lose? Final Report to the Royal Canadian Geographic Society, p. 21.

suggests that a paleoclimate record can be extracted from the ice. This example also illustrates that understanding the systematics of ice emplacement is vital if climate information is to be retrieved.

15.3.4.7 Booming Ice Chasm, Crowsnest Pass, Alberta/British Columbia

The *Booming Ice Chasm* (BIC) essentially consists of a large dipping shaft 140 m deep containing an impressive mass of ice (Fig. 15.2). With side passages, the cave is currently 961 m long and may genetically join to the Ice Palace, but is blocked by ice. It is currently the largest ice cave in Canada. The Ice Palace is in permafrost formed along an extensive fracture that also forms the main BIC shaft. Both caves are in the Livingstone Limestone of the Mississippian age.

Stable isotopes show that the BIC is essentially a cold trap cave, which receives and preserves annual snowfall Yonge (2016). Variations of δ^2H and δ^{18}O in a small ice mass give a range of composition between those expected from hoarfrost and mean annual precipitation. Any paleoclimatic interpretation would need to take into account the greater or lesser contribution of the hoar.

15.3.4.8 Ice Hall, Crowsnest Pass, Alberta/British Columbia

This large relic cave (Fig. 15.44) was entered below a 12-high headwall in a depression. It begins with a squeeze on the top of an icefall leading into a large chamber floored and sealed by ice of indeterminate depth. The cave continues into a second chamber and further on to a small passage. Although the entry has always been blocked by ice, it has been opened three times, initially by blasting and subsequently by fire. Although no formal studies have been made, it is worth noting that the entrance is blocked up by

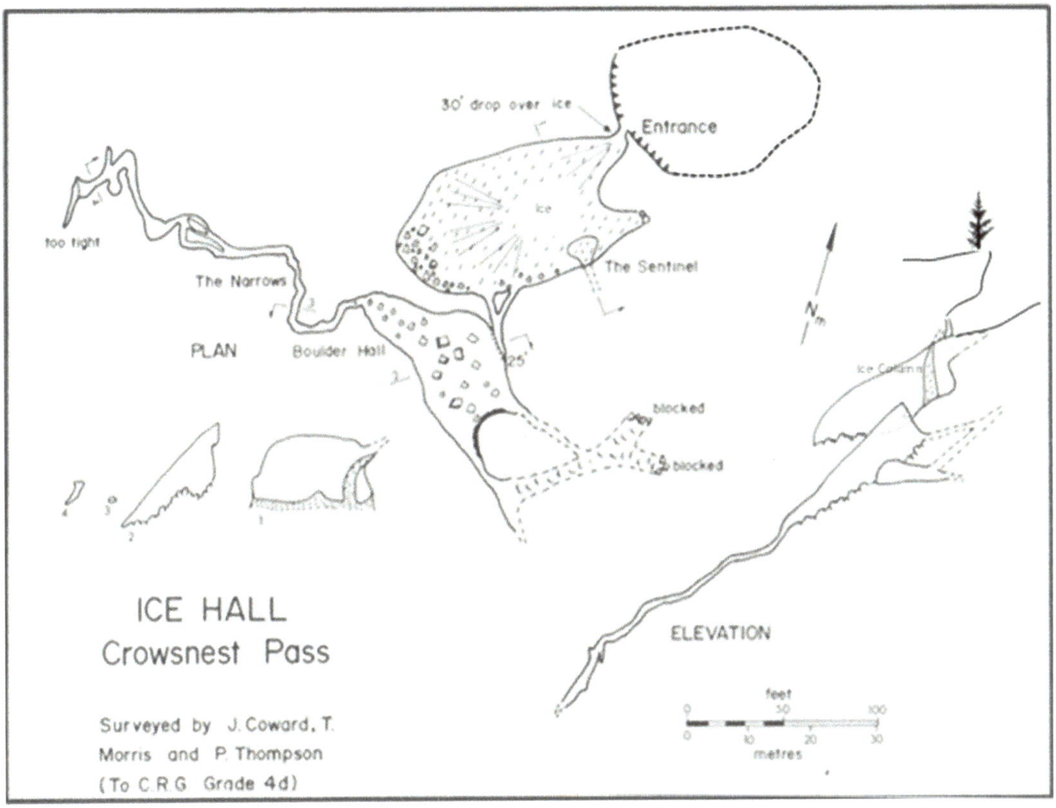

FIG. 15.44

Ice Hall.

Modified from Thompson, P. (Ed.), 1976. Cave Exploration in Canada. Canadian Caver, p. 181 (Special Issue).

the expulsion and freezing of moist cave air in summer (so much so that it almost trapped author CJY by closing up over a 2-hour period). The strong outwardly flowing air, implying a higher currently unknown entrance, appears to create a strong cold zone in the entrance constriction and shows in general that ice fillings in passages can be derived entirely from vapor.

15.3.4.9 Coulthard Cave, Crowsnest Pass, Alberta/British Columbia

Coulthard Cave is an isolated fragment of large relict cave passages in the Livingstone Limestone. The entrance of the cave faces north at 2440 m a.s.l., and all but one of the passages in the cave end in massive ice blockages. Fig. 15.45 shows the blockage in the lowest passage as it appeared in the summer of 1967; the ice was optically as clear as pure glass, with an apparent visual depth of 6+ m until the view was cut off by a bend in the passage. Marshall and Brown (1974) removed oriented ice samples and studied them by crystallographic techniques in a cold laboratory, discovering a layering (not evident to the eye) similar to that formed by the freezing of a horizontal water surface (pond). Present-day temperatures in the cave never exceed 0°C. The large scallop-like depressions in the ice indicate slow sublimation by convection currents in the cold air. An experiment obtained a sublimation rate of

FIG. 15.45

Massive ice in Coulthard Cave.

Photograph: Ian Drummond.

3 mm/year.[1] Sublimation of the ice permits entrapped sediment to reach the ice surface. Continual downward lowering of the ice surface facilitates the movement by normal trajectory of sediment toward the scalloped edges where it forms interstitial ridges. The main conclusions drawn from the ice crystallographic fabric and the total volume of the sublimation were that the ice had formed within the cave after the late Hypsithermal warm period and was not associated with any intrusion of glacier ice from the flank of the trunk glacier that filled the pass to higher elevations during the last (Wisconsinan) ice age.

[1]When mapping the cave in the summer of 1967, Ford took a Canadian penny from his jeans pocket and placed it on the tip of a cusp in a large sublimating scallop in the blockage. It would shield the ice directly underneath it. When taking their ice samples five years later, Marshall and Brown measured the amount of protection and, in the Acknowledgements for their 1974 paper, thanked him "for spending a penny" there.

Vancouver Island, British Columbia

Vancouver Island is very mountainous and has extensive limestone outcrops. The climatic regime is coastal, milder, and has much more rain and snow than in most of mainland Canada. Vancouver Island is known to have the greatest density of karst caves in the nation, extending from shallow depths beneath the sea up to 1300+ m a.s.l.

15.3.4.10 Projects Cave and Q5, White Ridges Provincial Park, Vancouver Island, British Columbia

White Ridge is an extensive karst area developed in an exposed outcrop of Quatsino Limestone (Upper Triassic) high above Gold River, a lumber milling town. It is approximately 50 km from the open Pacific Ocean and 13 km from Muchalat Inlet, a fiord. Above ~1100 m a.s.l., it displays typical alpine karst of dense karren fields with small dolines and shafts. There is much obstruction by ice and snow; Projects Cave and Q5 Shaft are two excellent examples of snow trap ice caves that contain sizable amounts of perennial ice (Griffiths, 1977; Ecock, 1984; Fig. 15.46).

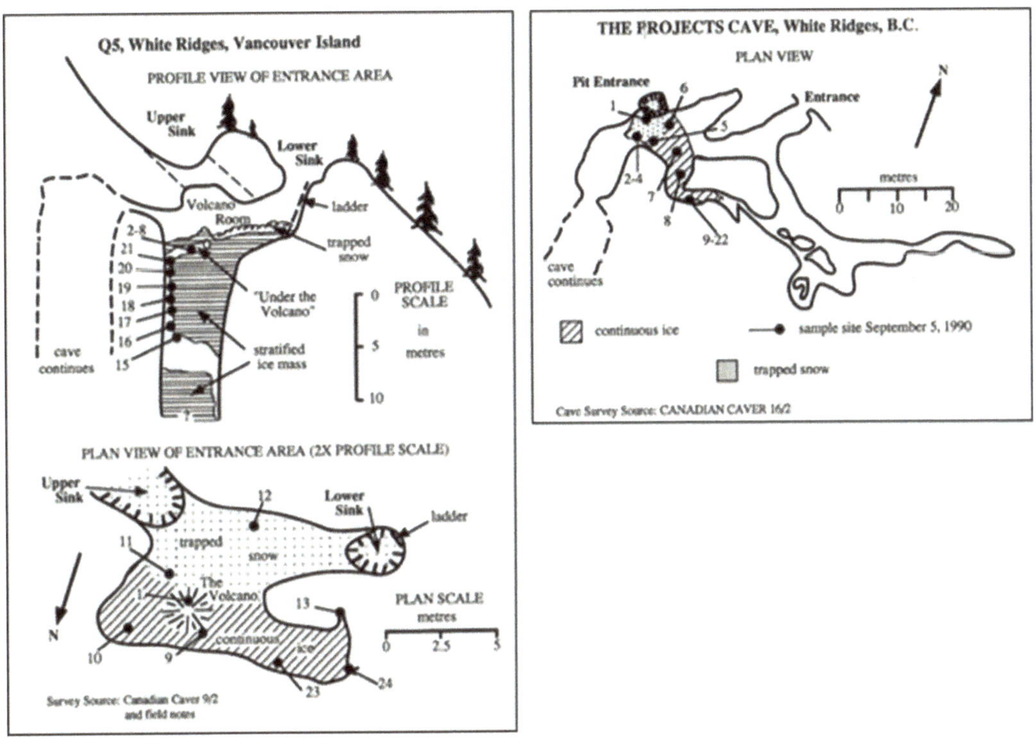

FIG. 15.46

Maps of Q5 & Projects Cave showing sample locations.

Diagrams from MacDonald W.D., 1994. Stable Isotope Composition of Cave Ice in Western North America (Unpublished M.Sc. Thesis). University of Calgary, Alberta, Canada, p. 206.

In Projects Cave, elevation 1050 m, snow is effectively captured by a 4-m-wide, funnel-shaped sinkhole that drops into the protected lower confines of the cave, so ice is restricted to the area in the immediate vicinity of the trapped snow. During the summer, the snow is under a constant barrage of dripping water so that a large proportion of it melts away, leaving the rest as a residual ice mass. As a result, this mass, which is ≥40 m in thickness and clearly stratified, is only a discontinuous record of the snowfall. Q5, in contrast, is higher (1200 m a.s.l.) and somewhat colder. It has a large sink pit acting as a funnel to direct copious snowfall into the sheltered lower parts of the cave. During the summer months, surface water draining into the cave is effectively diverted away through talus, leaving a larger proportion of the trapped snow unaffected. Seepage water melts a portion of it, which then refreezes in the <0°C cold trap. This process has created a very large, stratified ice mass >40 m thick that, being replenished from the top and melting beyond the cold trap, should rightly be called an underground glacière, comparable to, for example, the Scărişoara, Romania (Holmlund et al., 2005). Direct evidence of ice movement was detected on one occasion when it was possible to descend 15 m beneath the present floor level, in a narrow gap between the stratified ice mass and the cave walls. Along the surface of the wall, faint vertical striations were visible in a few places, an indication of rock fragments being dragged down the wall.

Fig. 15.47 (Yonge and MacDonald, 2014) shows the δ^2H variation for ice stratigraphy from the two caves. Although the ice samples oscillate around mean precipitation d-values, Q5 yields a range generally higher than those for Projects Cave. Q5 has a stream running into the entrance in the summer where the ice samples were taken, so is likely more biased toward heavier summer water. Projects Cave shows lower values, but these cluster around mean precipitation at −92‰. It has no entrance stream, and stratified ice occurs almost 20 m into the cave, a result of glacial movement perhaps. In this case, we would have more confidence in the record for Projects Cave, but would need to obtain data on its chronology as has been done at Scărişoara Cave (Holmlund et al., 2005); the figure shows possible

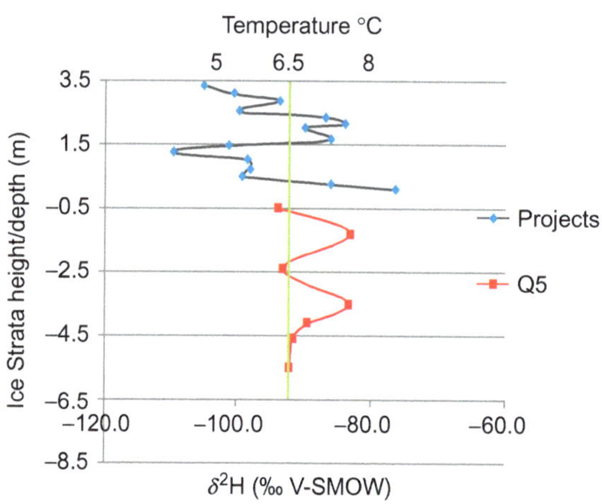

FIG. 15.47

Variation of δ^2H in stratified ice from Q5 & Projects Cave (Yonge and MacDonald, 2014). *Green line* is mean precipitation.

Temperature from isotopic values is based on Yonge, C.J., Ford, D.C., Gray, J., Schwarcz, H.P., 1985. Stable isotope studies of cave seepage water. Chem. Geol. 58, 97–105.

biannual (seasonal) variations in the δ-values. Temperatures are based on a North American curve for average cave temperatures and δ^2H values (Yonge et al., 1985; Fig. 4).

15.4 CONCLUSIONS/FUTURE WORK

Understanding these mechanisms of formation and destruction is very important if we are to discern the fate of the ice. Why worry? First, these caves are a unique resource; second, they have a significant aesthetic appeal; and third, they offer the possibility of obtaining another continental proxy climate record. For the latter, it seems that all three types of ice caves may offer climatic information: the permafrost and cold zone types may yield warming/cooling trends that are qualitative, whereas the cold trap type, especially when large enough to be considered a glacière cave, promises to yield quantitative temperatures.

Regarding the fate of the ice, we are concerned that it is disappearing at human timescale rates. For example, in Fig. 15.29 we can see the condition of the ice in 2014 along with a breach that allowed us access to the other side of the Ice Tongue. The figure shows a thinning of the Ice Tongue from 2005 to 2014. It is thus evident that this ice will disappear, perhaps in another decade, especially now that the ice plug has been breached. This suggests some urgency in studying ice caves generally. At the 6th International Ice Caves Workshop in Idaho, George Veni, (Executive Director of the National Cave and Karst Research Institute in Carlsbad, New Mexico, United States) gave a clarion call to this effect (Veni et al., 2014): "Time, money, and melting ice: Proposal for a cooperative study of the world's cave ice in a race against climate change."

ACKNOWLEDGMENTS

The first author is indebted to the following in support of writing this chapter:

Royal Canadian Geographical Society for its financial support of my ice caves research under its 2014 Independent Research Grant. William MacDonald for providing all the stable isotope data from his MSc thesis along with our years of coauthored papers and conference presentations. I thank the *Royal Alberta Museum* for its support in funding the ^{14}C dates for the ice organics. Steve Taylor of the *University of Calgary* for providing the stable isotope analyses, subsidizing the costs by 33% for the more recent analyses. *Alberta Environment and Sustainable Development* (Canmore Office) was very supportive of the ASS work in Plateau Mountain Ice Cave. The *Alberta Speleological Society* has been very active in the project, providing various volunteer field assistants, leaders, and photographers, especially Nick Vieira. Others also helped with field work, photography, and comments: Katie Graham, Christian Stenner, Ian McKenzie, and Ian Drummond. *Parks Canada*, facilitated by Greg Horne, provided helicopter transport for the work at Ice Trap; thanks for its flexibility and support.

For Greg Horne's support over several years on various ice cave projects at Castleguard Cave, Disaster Point, Ice Trap, and the Nahanni. This support includes field work, photography, facilitating stable isotope and ^{14}C analyses, facilitating National Park access and permits, edits, and baseline information.

For Bernard Lauriol's assistance in writing the Yukon section of this chapter. Thus the authors (Clark and Lauriol) of the Northern Yukon Karst Studies acknowledge the support of the people of Old Crow, the Government of Yukon Archaeological Program (Ruth Gotthardt and Gregory Hare), the National Museums of Canada (Jacques Cinq-Mars, Richard Morlan, and Richard Harington). They also acknowledge students for their theses and field assistance. Northern Yukon Karsts Studies were supported for years by the Polar Continental Shelf Project, NSERC programs, and Canadian Northern Studies Programs.

For Jacques Schroeder's assistance in writing the Nahanni section of this chapter, based on his many years of work there.

Derek Ford is deeply indebted to the Natural Sciences and Engineering Research Council of Canada for grants in aid of his research and to Parks Canada for grants and aid in kind for the South Nahanni and Castleguard Cave work.

REFERENCES

Alley, R.B., 2004. GISP2 Ice Core Temperature and Accumulation Data. In: IGBP PAGES/World Data Center for Paleoclimatology Data Contribution Series #2004-013. NOAA/NGDC Paleoclimatology Program, Boulder, CO, USA.

Atkinson, T.C., Smart, P.L., Wigley, T.M.L., 1983. Climate and natural radon levels in Castleguard Cave, Columbia Icefields, Alberta, Canada. Arct. Alp. Res. 15, 487–502.

Balch, E.S., 1900. Glacières or Freezing Caverns. Lane & Scott, Philadelphia, PA. 337 pp.

Briner, J.A., McKay, N.P., Yarrow, A., Bennike, O., Bradley, R.S., de Vernal, A., Fisher, D., Francus, P., Frechette, B., Gajewski, K., Jennings, A., Kaufman, D.S., Miller, G., Rouston, C., Wagner, B., 2016. Holocene climate change in Arctic Canada and Greenland. Quat. Sci. Rev. 147, 340–364.

Brook, G.A., Ford, D.C., 1974. The Karstlands of the South Nahanni Region, N.W.T. Contract 72-32C, Parks Canada, 501 p. (60 photo plates).

Clark, I.D., Lauriol, B., 1992. Kinetic enrichment of stable isotopes in cryogenic calcites. Chem. Geol. Isot. Geosci. 102 (1–4), 217–228.

Clogg-Wright, K., 2007. Isotope-Inferred Water Balances of Slave River Delta Lakes, NWT, Canada. Unpublished MSc thesis, University of Waterloo, p. 80.

Ecock, K.R., 1984. The Hydrology of an Alpine Karst: White Ridge, Vancouver Island. (M.Sc. Thesis) McMaster University, p. 138.

Ford, D.C., 1971a. Alpine karst in the Mount Castleguard-Columbia Icefield area, Canadian Rocky Mountains. Arct. Alp. Res. 3 (3), 239–252.

Ford, D.C., 1971a. Cavern and Associated Research in the First Canyon Area, South Nahanni River, N.W.T., Contract Report, Parks Canada, p. 115.

Ford, D.C., 1974. Geomorphology of South Nahanni National Park, N.W.T., Contract 72-32B, Parks Canada, 186 p. (7 map plates).

Ford, D.C., 2008a. Report upon the Expansion of South Nahanni River National Park Reserve. Mackenzie Mountains, Northwest Territories. p. 91 (Report to Parks Canada).

Ford, D.C., 2008a. Report upon a survey of Karst Landforms around Norman Wells, Northwest Territories. To Protected Areas Strategy Group, Ministry of Environment and Natural Resources, Government of the Northwest Territories. 233 pages plus appendices.

Ford, D.C., 2009. Mapping Known and Potential Karst Areas in the Northwest Territories, Canada. Environment and Natural Resources, Govt of the Northwest Territories. 50 p. (7 map plates).

Ford, D.C., 2010. Limestone landscapes of the South Nahanni Region, Mackenzie Mountains, Northwest Territories, Canada. In: Migon, P. (Ed.), Geomorphological Landscapes of the World. Springer Verlag, Berlin, pp. 13–20.

Ford, D.C., Williams, P., 2007. Karst Hydrogeology and Geomorphology. Wiley, West Sussex. p. 576.

Ford, D.C., Worthington, S.R.H., 2009. Final Report to Parks Canada Upon Cave and Karst Research and Evaluation in the South Nahanni and Ram River Basins, NT, Canada, vol. iii, p. 137.

Ford, D.C., Harmon, R.S., Schwarcz, H.P., Wigley, T.M.L., Thompson, P., 1976. Geo-hydrologic and thermomometric observations in the vicinity of the Columbia Icefield, Alberta and British Columbia. J. Glaciol. 16 (74), 219–230.

Google Earth, 2016. Google Earth@2016 Cnes/Spot Image, lat 53.059, lon-118.296, el 2123m, eye alt 3.5 km.

Griffiths, P., 1977. Caves of the White Ridge Mountain, Vancouver Island. Can. Caver 9 (2), 32–38.

Hamilton, J.P., 1995. Karst Geomorphology and Hydrogeology of the Northeastern Mackenzie Mountains, District of Mackenzie, NWT, Canada (Ph.D. Thesis). McMaster University p. 532.

Hamilton, J., Ford, D., 2002. Karst geomorphology and hydrogeology of the Bear Rock formation – a remarkable dolostone and gypsum megabreccia in the continuous permafrost zone of Northwest Territories, Canada. Carbonates Evaporites 17 (2), 54–56.

Harris, S.A., 1979. Ice caves and permafrost zones in Southwest Alberta. Erdkundi 33, 61–70.

Holmlund, P., Onac, B.P., Hansson, M., Holmgren, K., Morth, M., Nyman, M., Persoiu, A., 2005. Assessing the paleoclimate potential of cave glaciers: the example of Scarisoara Ice Cave (Romania). Geora. Ann. 87A (1), 193–201.

Horne, G., 2008. The North Karst, South Nahanni River Area. Can. Caver 68 (1), 4–9.

Jass, C.N., Horne, G., Critchley, D., 2013. Postglacial faunal records from cave deposits in Jasper National Park, Alberta, Canada. In: Proceedings of the 20th National Cave and Karst Management Symposium, pp. 19–22.

Lacelle, D., Lauriol, B., Clark, I.D., 2009. Formation of seasonal ice bodies and associated cryogenic carbonates in Caverne de l'Ours, Quebec, Canada: kinetic isotope effects and pseudo-biogenic crystal structures. J. Cave Karst Stud. 71 (1), 48–62.

Lauriol, B., 2016. The Ni'iinlii Njik Caves, Northern Yukon. In: Leduc, H. (Ed.), Notes on Geology and History. Government of Yukon, Yukon, ISBN: 978-1-55362-752-4.

Lauriol, B., Bertrand, P., 2016. Eaux, Glaces et Cavernes. In: MultiMondes, Montreal, QC, p. 140. ISBN: 978-2-89544-476-3.

Lauriol, B., Clark, I.D., 1993. An approach to determine the origin and age of massive ice blockages in two arctic caves. Permafr. Periglac. Process. 4, 77–85.

Lauriol, B., Carrier, L., Thibaudeau, P., 1988. Topoclimatic zones and ice dynamics in the caves of the Northern Yukon, Canada. Arctic 41, 215–220.

Lauriol, B., Ford, D.C., Cinq-Mars, J., Morris, W.A., 1997. The chronology of speleothem deposition in northern Yukon and its relationships to permafrost. Can. J. Earth Sci. 34 (7), 902–911.

Lauriol, B., Prevost, C., Deschamps, E., Cinq-Mars, J., Labrecque, S., 2001. Faunal and archaeological remains as evidence of climatic change in freezing caverns. Yukon Territory, Canada. Arctic 54 (2), 135–141.

Lauriol, B., Prevost, C., Lacelle, D., 2006. The distribution of diatom flora in ice caves of the northern Yukon Territory, Canada: relationship to air circulation and freezing. Int. J. Speleol. 35 (2), 83–92.

Luetscher, M., Jeannin, P., 2004. A process-based classification of alpine ice caves. Theor. Appl. Karst 17, 5–10.

MacDonald W.D., 1994. Stable Isotope Composition of Cave Ice in Western North America (Unpublished M.Sc. Thesis). University of Calgary, Alberta, Canada, p. 206.

Marion, J., Poirel, J., 1973. History of Nahanni Caving, 1964-1971. Can. Caver 5 (1), 30–33.

Marshall, P., 1975. Ice-blocked tubes in the Aiyansh flow, British Columbia. Arct. Alp. Res. 7 (4), 399–400.

Marshall, P., Brown, M.C., 1974. Ice in Coulthard Cave, Alberta. Can. J. Earth Sci. 11 (4), 510–518.

Muir, D., Ford, D.C., 1985. Castleguard. National Parks Centennial, ISBN: 0-660-91440-9.

Rollins, J., 2004. Caves of the Canadian Rockies and the Columbia Mountains. Rocky Mountain Books, ISBN: 0921102-94-1 p. 336.

Schroeder, J., 1977. Les formes de glaces des grottes de la Nahanni, Terretoires du nord-ouest, Canada. Can. J. Earth Sci. 14, 1179–1185.

Schroeder, J., 1979. Le dévéloppement des grottes dans la région du Premier Canyon de la Riviére Nahanni, Sud, T.N.O. (Unpublished Ph.D. Thesis). Ottawa University, p. 265.

Scotter, G.W., Simmons, N.M., 1976. The mortality of sheep within a cave. J. Mammal. 57 (2), 387–389.

SERC, 2016. Holocene Record of Climate Change. Science Education Resource Center. http://serc.carleton.edu/hazards/hurricanes/geologic_record.html.

Shawcross, M., 1973. The North Karst, South Nahanni River Area. Can. Caver 5 (1), 34.

Thompson, P. (Ed.), 1976. Cave Exploration in Canada. Canadian Caver, p. 181 (Special Issue).

Tsui, P.C., Cruden, D.M., 1984. Deformation associated with gypsum karst in the Salt River Escarpment, northeastern Alberta. Can. J. Earth Sci. 21, 949–959.

Utting, N., Lauriol, B., Mochnacz, N., Aeschbach-Hertig, W., Clark, I.D., 2013. Noble gas and isotope geochemistry in western Canadian Arctic watersheds: tracing groundwater recharge in permafrost terrain. Hydrgeol. J. 21 (1), 79–91.

Utting, N., Lauriol, B., Lacelle, D., Clark, I.D., 2016. Using noble gas ratio to determine the origin of ground ice. Quatern. Res. 83 (1), 177–184.

Veni, G., Land, L., Perşoiu, A., 2014. Time, money, and melting ice: proposal for a cooperative study of the world's cave ice in a race against climate change. In: The 6th International Workshop on Ice Caves (Procs), Idaho Falls, USA. (Special presentation).

Wigley, T.M.L., Brown, M.C., 1976. The physics of caves. In: Ford, T.D., Cullngford, C.H.D. (Eds.), The Science of Speleology. Academic Press, London, pp. 329–358.

Yonge, C.J., 2004. Ice in caves. In: Gunn, J. (Ed.), Encyclopedia of Caves and Karst. Fitzroy Dearborn, New York, pp. 437–439.

Yonge, C.J., 2015a. The Systematics of Perennial Ice Found in Western North American Ice Caves. Thinking Mountains, Jasper, AB, p. 13 (Abstract and Paper).

Yonge, C.J., 2015b. The fate of ice caves in northwest Canada as climate warms: what will we lose? Final Report to the Royal Canadian Geographic Society, p. 21.

Yonge, C.J., 2016. The Walkin-Ice Cave System, Wood Buffalo National Park: An Archive of Ancient Ice? p. 11 (Report to Parks Canada).

Yonge, C.J., MacDonald, W.D., 1999. The potential of perennial cave ice in isotope paleoclimatology. Boreas 28 ((3), 357–362.

Yonge, C.J., MacDonald, W.D., 2006. The contrast in isotopic composition of cave ice across the Divide in Western North America. In: Archives of Climate Change in Karst. Proc. Symp. Climate Change: The Karst Record IV, Baile Herculane, Romania. vol. 10. pp. 26–28. (Special publication).

Yonge, C.J., MacDonald, W.D., 2014. Stable isotope composition of perennial ice in ice caves as an aid to characterizing ice cave types. In: The 6th International Workshop on Ice Caves (Procs), Idaho Falls, USA, p. 9.

Yonge, C.J., Ford, D.C., Gray, J., Schwarcz, H.P., 1985. Stable isotope studies of cave seepage water. Chem. Geol. 58, 97–105.

FURTHER READING

Clark, I.D., Fritz, P., 1997. Environmental Isotopes in Hydrogeology. Lewis Publishers, Boca Raton, FL. p. 328.

Ford, D.C., 1966. The Nakimu Caves. Glacier National Park, BC. 84 p. (Report to Parks Canada).

Ford, D.C., 1998. Principal features of evaporite karst in Canada. In: 4th International Conference on Geomorphology, Italy, 1997. Suppl. Geogr. Fis. Dinam. Quat. III, T. 4, pp. 11–19.

Michel, F.A., 2011. Isotope characterisation of ground ice in Northern Canada. Permafr. Periglac. Process. 22, 3–12. wileyonlinelibrary.com.

Wigley, T.M.L., Brown, M.C., 1971. Geophysical applications of heat and mass transfer in turbulent pipe flow. Bound.-Lay. Meteorol. 1, 300–320.

CHAPTER

ICE CAVES IN CROATIA

16

Nenad Buzjak*, Neven Bočić*, Dalibor Paar*, Darko Bakšić*, Vinka Dubovečak[†]

University of Zagreb, Zagreb, Croatia Speleological Association Kraševski zviri, Ivanec, Croatia[†]*

CHAPTER OUTLINE

16.1 INTRODUCTION

Karst and fluviokarst in Croatia covers 43.7% (24,748 km^2) of state territory (mainland and most of the Adriatic islands, submerged karst of Adriatic Sea not included; Bognar et al., 2012). It consists of two main geomorphological units: Classical karst in the Dinaric mountain belt, and Isolated (or insular) karst in the low mountains of the Pannonian basin. The classical karst is predominantly developed in thick and well karstified Mesozoic and Paleogene carbonate rocks (limestones, dolostones, breccias, and conglomerates). It can be divided into three morphogenetic subareas: Classical karst, Fluviokarst, and Glacial and Periglacial reliefs developed on karst. While the first two types can be found in all Dinaric areas (mainland, coast, and islands, depending of geological composition, structures and geomorphological processes), the third type can be found only in higher mountain areas (~1400–1831 m a.s.l.) influenced by Pleistocene glaciation and colder, humid boreal climates during

Ice Caves. https://doi.org/10.1016/B978-0-12-811739-2.00016-4

the Holocene epoch. Though the smallest by area, this is a most promising location for ice cave research due to the colder climate.

Karst of higher Dinaric Mountains (up to 1831 m a.s.l. on Dinara Mt. highest peak) is characterized by thick, highly disturbed, and well karstified carbonate rocks (Vlahović et al., 2005; Korbar, 2009). There is a well developed deep vadose zone (>1500 m) with large vertical caves (for example, the Lukina jama-Trojama shaft system, the deepest cave in Croatia: 1431 m deep; Bakšić et al., 2013; Bočić, 2006; Stroj and Velić, 2015). There are also complex and long caves (Kita Gaćešina-Draženova puhaljka cave system: 34 km long and 737 m deep, the longest cave in Dinaric karst; Barišić et al., 2013; Croatian Speleological Server, 2017). The surface is characterized by deep and steep dolines (Pahernik, 2012). Dolines and other karst depressions (like uvalas or poljes) have specific microclimates characterized by air temperature inversion. In some cases, dolines (among other relevant factors, such as as cave morphology etc., which will be discussed later) can cause specific cave microclimates and lead to the formation of ice and snow caves (Buzjak et al., 2014). Between the higher mountain ridges and massifs of Gorski kotar, Lika, and the Dalmatian hinterland (Zagora), there are shallower karst and fluviokarst areas with poljes, highlands, hills and low mountains rich in various karst phenomena.

Isolated karst in low mountains (altitude up to 1100 m) in the Pannonian basin (mainly on Ravna gora, Ivanščica, Medvednica and Papuk Mt.) is shallow, often developed in Mesozoic and Neogene rocks in small and dissected areas even on the same massifs (Bočić, 2010; Božičević, 1974; Ozimec and Šincek, 2011).

16.2 CROATIAN ICE CAVE TERMINOLOGY

In the Dinaric Karst Mountains in Croatia, a large number of caves containing snow and ice are known. Depending on prevailing precipitate, the cave could be referred to as *sniježnica* (from Croatian *snijeg* = snow; plural *sniježnice*) if snow or firn prevails (snow cave), or *ledenica* (Croatian *led* = ice; plural *ledenice*) if ice prevails (ice cave). Both terms are specific toponyms (or speleonyms) originated as descriptive names derived from specific features important for, or noted by, the local population. In the Croatian language, *ledenica* also means freezer or icehouse (cellar, small building, or dug out hole used to store ice for preserving food or cooling drinks in the past) and these man-made objects or products are named after previously known ice caves (Barić, 1989; Dvoržak, 1989). Various synonyms can be found in literature and on maps, starting from the 17th century: *snježnica*, *snježnjača*, *snižnica*, *ledenjača*, *ledinjača*, *lednica*. According to Dvoržak (1989), the first known record of the term *ledenica* (*ledvenicza*) for an ice cave in Croatia was written by linguist and writer Juraj Habdelić in his Croatian-Latin dictionary "Dictionar," published in Graz (Austria), 1670. In his study about the Croatian-Latin dictionary published in 1710 by Fr. Matija Jakobović, Kosor (1957) recorded the first known mention of the term *snižnica* or *sniježnica* describing a cave that contains snow during the summer. These two examples prove existing knowledge of particular cave phenomena that were used for practical purposes we will discuss later.

These names are used both for caves with permanent and/or seasonal ice and snow. Locals sometimes use these names for very cold caves too, like Smrzlinka cave (Frozen cave), even though ice and snow is only seasonal in this cave, if present at all. In some cases the term *ledenica* was used for caves (mostly shafts) that served as snow and ice storages, even when they were man-made where natural ice and snow was not possible (Redenšek, 1961; Škundrica, 1954).

Since Croatian geomorphological terminology distinguishes horizontal and vertical caves or passages, (Croatian *jama*=shaft or pit; *špilja*, *spilja* or *pećina*=horizontal cave) some local names can include cave morphology in the cave name (for example Ledena jama=Ice shaft). According to some authors (Anonymous, 1913; Hirc, 1898; Malez, 1959), the term referring to a cave with snow (*snježnica*) was also used for deep, steep, and shady dolines containing deposits of snow for the entire year, or at least most of it.

16.3 HISTORICAL OVERVIEW

Scientific cave research in Croatia started in the 19th century, but even earlier there are some interesting papers dealing with physical speleology, like Nikola Vitov Gučetić's discussion about speleothems, humidity, and air currents in caves published in 1584 (Dadić, 1984). The books dealing with karst and caves, which inspired many later researchers, were those written in 18th century by Johann Weikhard Valvasor, Alberto Fortis, or Ivan Lovrić (Malez, 1984). A. Fortis (1774) was among the first authors writing about ice caves. He wrote about ice caves on Biokovo and Mosor Mt. and the use of ice and snow as a source of water for people, domestic animals, or for making ice-cream in Dubrovnik. He also recorded a belief of the inhabitants of Poljica (the area east and south-east of Split), who believed that "...removing the ice from the bottom of the mountains, where it can be maintained throughout the year, would be to challenge the North Wind that destroys their crops, therefore no one is allowed to pull it out." This is connected to the ancient widespread superstition in the Dinaric area that cold winds form underground. In the second half of the 19th century newly established institutions and organizations (faculties, museums, the Croatian Mountaineering Association) encouraged scientific karst and cave research. The first data (cave location determined by a nearby settlement) about ice cave locations (or *Höhlen mit ewigem Schnee und Eise*, as he wrote) from that time was published by Sabljar (1866). V. Klaić, a historian with an important role in the development of Croatian geography, wrote a short, but interesting, text about ice caves, where he derives ice accumulation with cave morphology and entrance exposition. He also mentions known ice caves in regions of Gorski kotar, Lika and Biokovo Mt. (Klaić, 1878). In a series of three travelogue texts about the Gorski kotar area, the famous Croatian naturalist and geographer, Hirc (1883a,b), mentioned a few ice caves and described research visits to ice caves in Mrzla draga near Mrkopalj (Aug. 9, 1883; Hirc, 1900a). Soon after that trip, he named it Pilarova ledenica (Pilar ice cave), in memory of Croatian geologist Gjuro Pilar (Hirc, 1898). He wrote about cave morphology, ice formations (ice stalactites and stalagmites) and autochthonous ice. He also measured air temperatures on the outside (16°R=20°C; R for Réaumur scale) and in the cave's hall (0.3°R=0°C). He also recorded the plant species growing there. Hirc (1893/1894) wrote an article in a newspaper completely devoted to ice caves. Besides locations of some other ice caves in Croatia, he also mentioned seven ice caves in Slovenia. He later published a very interesting discussion about the origin of ice caves with geologist Dj. Pilar (Hirc, 1898). In the book about Gorski kotar area, part of one chapter is dedicated to Snježnica pod Jožinom planom on Bitoraj Mt. (Hirc, 1898), visited on Aug. 25, 1884. From the text, it is not completely clear if it is a snow shaft or a steep rocky doline with snow at the bottom. The visitors descended on foot some 30–40 m deep. They used a thermometer to record air temperature: 11.3°R (14.1°C) at the entrance and 2.5°R (3.1°C) at the bottom. In a newspaper article where he repeats his descriptions of some ice caves, Hirc also writes that cave ice is heavier, and 10%–15% denser than regular ice. He also notes that it has a dark grey color, and that yearly growth is clearly observable. Hirc also wrote about the origin of cave ice based on the ideas of J.

Cvijić (Hirc, 1900a). Hirc also described Budina ledenica in Lika (Hirc, 1900b). In his last paper about ice caves, Hirc (1902) published the results of research in Ledena jama on Rudač in Gorski kotar. With his companions, he measured the cave length, described morphology, and measured air temperature (18°C at the surface and 5°C in the entrance passage "over the ice"). It is interesting that he recorded the existence of the ice in August. Today there is ice only during winter, and rarely in early spring; but many geoindicators preserved on cave walls and in sediments, like crioclastic debris, are traces of frost action from the past (Fig. 16.1).

FIG. 16.1

Entrance chamber in Ledena jama na Rudaču, Mar. 2006. The ice column and frozen bottom in the location of former permanent ice (according to writers from 19th century), photo N. Buzjak.

A very important year for the establishment of karst science in Croatia (then in the Austro-Hungarian Empire) was 1910, when the Geographical Section and Committee for Cave Research of the Geological Commission of Kingdom Croatia and Slavonia was founded. It was led by the world famous geologist Dragutin Gorjanović Kramberger, known for the discovery of Homo sapiens neanderthalensis in Krapina (Croatia). They started scientific cave research and the making of cave lists, which included some ice caves, too (Gorjanović-Kramberger, 1912). Girometta (1923) and Margetić (1925) published extensive speleological studies about caves in middle Dalmatia, in which they recorded some ice caves on Mosor Mt., including remarks of morphology, speleogenesis, and ice duration (Girometta, 1923; Margetić, 1925). Girometta (1927) also published results of his research on ice caves with interesting data about geographical distribution, morphology, morphometry, and microclimate. He also went into detail about air temperatures, air circulation, and water infiltration. In the paper, there is a table with monthly and mean annual air temperatures for Ledenica on Mosor Mt. (Fig. 16.2), a 50 m deep shaft. He discusses the lowest and the highest air temperatures. According to measurements on different spots (entrance, passage depth: 10 m, 20 m, 25 m, 30 m, 40 m, 45 m, 50 m) performed on Apr. 24, 1927, he noticed an unequal temperature drop comparing the upper half (slow drop) with the lower half (faster drop). He also noticed that the air temperature difference between the shaft entrance and bottom during spring, summer, and early autumn is dependent on precipitation (higher precipitation results in a smaller difference). At the end of the paper, he concluded that ice formation in that cave is caused by the accumulation of colder air connected with cave morphology.

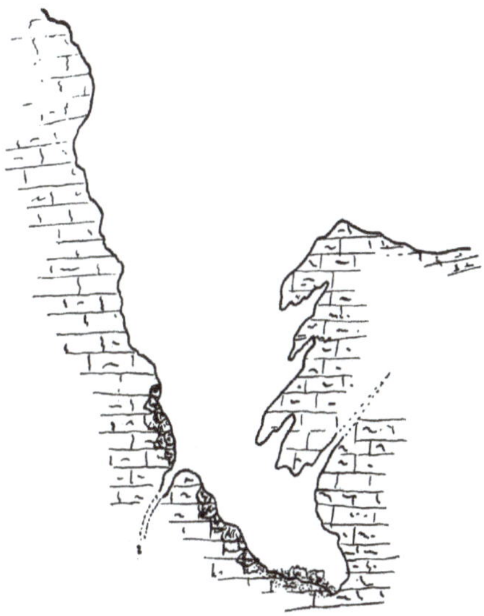

Sl. 1. »Ledenica« (Mosor). 1 : 1000.

Mjesec	Temperatura na otvoru jame:	Temperatura na podanku jame:	Razlika:
I.	**2·8**	**– 2· –**	**4·8**
II.	3· –	– 1·9	4·9
III.	5· –	– 1·2	6.2
IV.	8·4	+ 0·4	8·0
V.	11·5	1·6	9·9
VI.	15·6	2· –	13·6
VII.	**19·4**	2·8	**16·6**
VIII.	19· –	**3·1**	15·9
IX.	14· –	2· –	12· –
X.	9· –	+ 0·3	8·7
XI.	5· –	– 1· –	6· –
XII.	4· –	– 1·3	5·3
Srednja god. temp.	**9·9**	**+ 0·4**	**9·5**

FIG. 16.2

The profile of Ledenica and a table with air temperatures: Mjesec = Month, Temperatura na otvoru jame = T at shaft entrance, Temperatura na podanku jame = T at shaft bottom, Razlika = Range, Srednja god. temp. = Mean Annual T (Girometta, 1927).

In an article dedicated to cave climates and the origin of ice caves, Girometta connected this to passage morphology. According to his observations in caves on several mountains (Mosor, Biokovo, Dinara, Kamešnica, etc.), he writes that they all have one entrance, usually of "notable dimensions" and "bag shape" (Girometta, 1934). He also described the mechanism of cold traps. Girometta also described snow caves as caves containing snow fallen from outside, and which can hold snow for the whole summer. He noted that these caves have even larger entrance openings than ice caves, but the mechanism of snow preservation is the same as that of ice in ice caves. In the article he published one year later, Girometta added some new observations and data. As factors important for cave microclimate, he listed passage inclination, cave morphology, air circulation, and humidity (Girometta, 1935). Girometta described snow cave Trogrla on Kamešnica Mt. (1400 m a.s.l.), where he found deer antlers under a 3 m thick layer of snow. According to the fact that this species became extinct a "long time ago," he presumed that snow layers below the line where he found antlers "originate from ancient times." In the following, he writes about the differences in snow structure between upper layers of " rather non-compact snow" and lower layers of higher compactness, and how that means that "the lowest layers are transformed into ice." This transformation is connected with summer rains. When upper layers melt, the water drops that infiltrate to lower layers freeze and transform into ice.

A very interesting and useful article about caves in the vicinity of Perušić in Lika region was written by Rosandić (1931). He explained the cave climate and ice features of Ledenica Cave (Hirc named it Budina ledenica, a name that is still used today; Hirc, 1900b). He noted that ice is present throughout the whole year, which is not the case today. According to Rosandić, measurements of air temperature in lower parts of the cave are around 0°C with small oscillations (up to 0.5°C) and a relative humidity of 90%–96%. Year-round ice duration depends on factors including entrance exposition, cave morphology, descending passages, and vegetation (dense bush and grass) that covers the surface and prevents direct heating. Due to the accumulation of cold air in the lower parts of the cave, there is a temperature inversion that supports ice formation and preservation (Fig. 16.3).

FIG. 16.3

Reconstruction of air circulation in Ledenica Cave.

After Rosandić, Z., 1931. Iz podzemne Like (From subterranean Lika). Hrvatski planinar, 9, 239–244 (in Croatian).

In 1930, I. Krajač and his companions explored the large steep rocky doline Varnjača in Rožanski kukovi located on Velebit Mt. (today it is in a strict reserve in Northern Velebit National Park). He described the doline and several caves containing snow hidden in deeper parts, as later speleological research confirmed (Šmida, 1999).

In 1935, geomorphologist J. Roglić published an important work about Biokovo Mt. including a profile of Jarova Rupa ice cave (Fig. 16.4) and discussed morphology and ice formation in ice caves. According to his publication, upper layers of snow that fall in the caves melt, then water infiltrates deeper into the colder snow layer, where it freezes. This causes formed ice to compact under the pressure of upper layers. As an important factor for ice preservation, he emphasizes the entrance morphology and the position at the bottom of deep, steep dolines where the sun can't reach the bottom; which

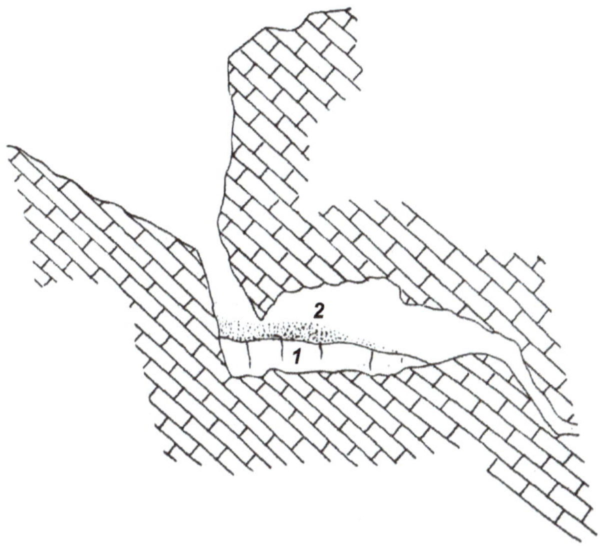

FIG. 16.4

Ice cave Jarova rupa on Biokovo Mt. Legend: 1—Ice, 2—Snow.

After Roglić, J., 1935. Biokovo—geomorfološka ispitivanja (La montagne de Biokovo—étude géomorphologique). Posebna izdanja Geografskog društva, Beograd (in Serbian).

allows snow and ice to exist year round. Roglić also wrote that a constant source of melting water has an impact on passage enlargement (Roglić, 1935).

In the period after the Second World War, there began a more intense era of speleology development and cave research. It was connected with the founding of caving groups inside mountaineering organizations, applied research for big constructing projects and militaries, mapping projects, and university research. A lot of very interesting data was collected from cave research in mountain regions which were almost completely unknown at the time. The first papers dealing with ice caves were published in the mid-fifties. Based on observations made in Ledenica u Čudinoj uvali, geologist and speleologist S. Božičević (1955) wrote about the preconditions needed for ice formation and accumulation in caves (entrance position, climate, passage morphology and inclination, the descent of cold air, and water infiltration). Božičević describes ice dynamics recorded by ice volume and ice formation changes during two research trips in 1955. In one article, he published two photographs of ice cave formations and a schematic picture of air circulation important for ice formation and preservation (Fig. 16.5). For comparison and research, two members of the "Interdisciplinary research of ice caves in Croatia" project visited the location in Aug. 2016 and noted a lack of ice in the entrance doline, where earlier researchers reported thick deposits (Božičević, 1955; Marjanac, 1956).

In 1954 geologist M. Malez explored Budina ledenica near Perušić. A cave map was made, and detailed speleological research was conducted (Malez, 1955). He also measured air temperatures, reconstructed air circulation, and observed ice formations (Fig. 16.6). Unlike his predecessors, which

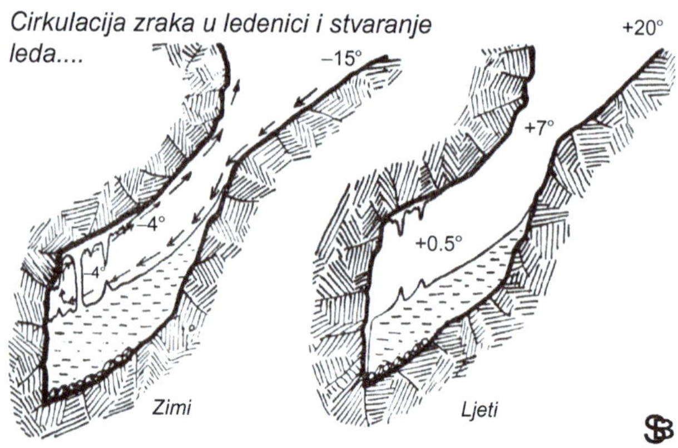

FIG. 16.5

Air circulation and temperatures important for cave ice in Ledenica u Čudinoj uvali (Božičević, 1955).

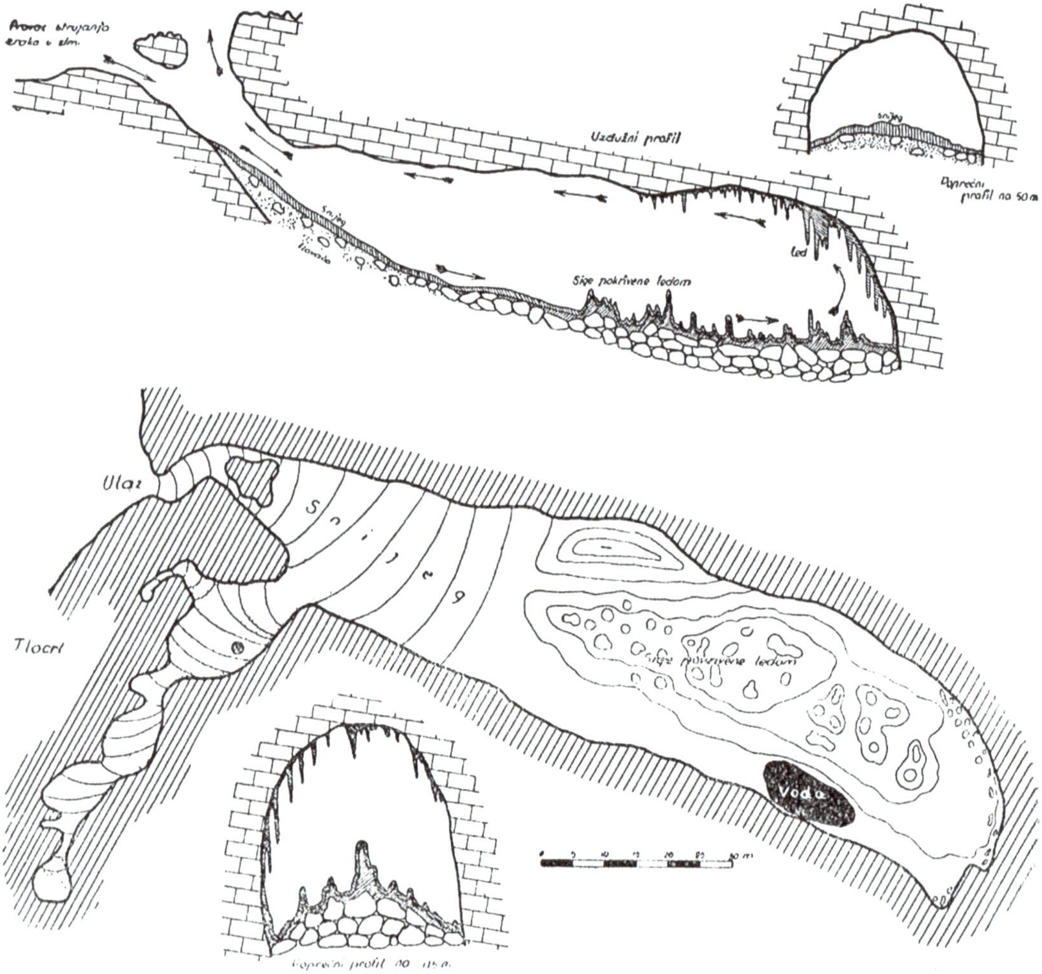

FIG. 16.6

Winter air circulation in Ledenica is important for cave cooling and ice formation. Arrows on the profile indicate air circulation in winter. Deposits (plan & profile): Snijeg=Snow, Sige pokrivene ledom=Speleothems covered with ice.

After Malez, M., 1955. Spilja Ledenica (Ledenica cave). Priroda, god. XLII, 8, 281-286 (in Croatian).

referred to it as a permanent ice cave, Malez wrote that it is a temporary ice cave because the ice is melted until winter.

During intensive research of Puhaljka cave on Velebit Mt., well known for intense air circulation (the name Puhaljka means Blow-hole), cavers collected a plethora of interesting data. They measured air temperatures and recorded freezing temperatures far from the entrance (Čepelak, 1972–1973; Malinar, 1964–1965). From one of the largest ice shafts in Croatia, Ledena jama in Lomska duliba on Velebit Mt., B. Jalžić reported ice and snow deposits, as well as interesting bluish ice formations (Jalžić, 1978–1979). Since the 1980s, owing to a growing caving community, the progress of cave education, and new research techniques, there has been an increase of caving reports, including information about ice caves. Most of the articles have general data about results from caving expeditions or single explorations, including some information about the locations and cave maps from Velebit and Biokovo Mt. (Grgasović, 1983; Kuhta, 1986–1987; Lukić, 1988–1989; Lacković, 1992; Ostojić, 1996–1997, etc.). It is important that all cave maps contain data about the position and extent of ice and snow deposits, which are useful for tracking changes. M. Garašić (1980) wrote about speleological research in a large ice cave (Ledenica kod Bunovca) on Velebit Mt. and reported an air temperature of 0.5°C in Aug. 1980. Božičević and Benček (1983), in an analysis of geological and geomorphological features of Biokovo Mt., noted the existence of large snow shafts and gave a short overview of research in the past. He also stressed the importance of ice cave research for geomorphology and palaeontology and the need for organized speleological research.

After Slovak cavers started research on northern Velebit in 1992 they, among others, discovered a few deep shafts (one of them was Manual, today known as shaft system Lukina jama-Trojama, the deepest one in Croatia and Ledena jama in Lomska duliba; Jalžić et al., 1992–1993). Since then, Croatian cavers have begun systematic explorations of that area. The result was a growing experience in exploring ice passages, an increase in the number ice and snow shafts, and an interest in collecting data relevant for ice such as air temperature (Božić, 1994; Troha, 1994; Bočić, 1997). Bakšić (1997) wrote about the microclimate zonation of Lukina jama's entrance ice zone (T_{air} = ~1°C, Jul. 1995). Due to the easier access, thick snow, and ice deposits, the entrance of Ledena jama (Ice shaft) in Lomska duliba (Jelinić, 1998–1999) is one of the most promising locations for ice cave research. Therefore, it was at this location that the first microclimate researches were performed using thermohygrographs and ice sampling for dating and isotope analysis (Horvatinčić, 1996; Horvatinčić and Božić, 1998–1999; Vrbek and Fiedler, 2000). Research in Ledena jama and Vukušić snježnica continued in recent years and were discussed later (Kern et al., 2008a,b, 2010, 2011). On the 120th anniversary of measurements conducted by Hirc (1898), members of the Croatian Meteorological Society researched snow cave microclimates and collected samples near Fužine in the Gorski kotar (Vučetić and Abramović, 2004; Rasol and Špoler Čanić, 2005). Meteorological observations from the Lukina jama-Trojama system collected during speleological expeditions gave insight into the environment in Velebit's deep shafts, with ice deposits not yet fully researched and explained (Jalžić, 2007).

In the 2000s cavers have continued systematic research in mountain areas, finding new examples of snow and ice caves on Velebit Mt. (Fig. 16.7), Biokovo, Dinara, and the Gorski kotar area. The data, with locations and cave maps, is archived with caving clubs and, more recently, in the Croatian Cave Cadastre (Bočić, 2001; Bočić, 2005; Buzjak, 1999; Kapidžić, 2004; Kuhta, 2000–2001; Kuhta, 2005; Mišćenić and Glavaš, 2006; Reš, 2006; Rubinić, 2012; Tutiš, 2007; Završki, 2012). Among them are some caves that are very useful for ice cave research due to their thick ice and snow layers (Bočić et al., 2014; Janton, 2017; Rožman et al., 2017).

FIG. 16.7

(A) Cave map of Zlatne godine shaft on Crnopac (southern Velebit), potential site for ice cave monitoring. (B) Autogenic ice in the form of a large stalagmite in the Zlatne godine shaft near bottom, photo T. Barišić. (C) Ice block in Snjeguljac cave (Velebit Mt.), photo V. Sudar. (D) Ice block profile on the depth of ~150m in Lukina jama-Trojama system (Northern Velebit National Park), photo D. Paar.

16.4 ECONOMICAL VALORIZATION OF ICE CAVES IN THE PAST

Cave ice was used in the past as a source of water for humans and cattle (Marković, 1980; Vučetić and Abramović, 2004), for preserving food, making ice cream, and cooling drinks. For inhabitants in some areas, the ice trade was an important source of income (Hirc, 1923). In some caves, it is still possible to find remnants of equipment and tools used for descending, climbing, digging, and breaking ice (axes, shovels, wedges, etc.). For descending, they used ropes and trunks (fir or beech) with trimmed branches or carved treads, which were used as ladders or stairs (Anonymous, 1913; Krajač, 1922). Fortis (1774) reports that shepherds on the mountain of Biokovo, during the summer, melted ice from caves in wooded watering places for cattle. Franić (1898) reports on ice extraction from one deep ice cave, which was then sold in the city of Gospić (Lika). Hirc (1900a,b) reports the use of ice in Gorski kotar in Croatia, Kranjska (Carniola) in Slovenia, and in Serbia; using different ways of excavation, packaging, transport before finally being exported to Trieste, Wien, and even Alexandria. Rubić (1949) and Jagačić (1957) noted conversations with local people working in ice caves on Biokovo Mt. about excavation of ice from the Velika (Crna) ledenica ice cave and some others caves (Fig. 16.8). Jagačić wrote that ice was excavated from 36 ice caves, and that the deepest one (Studenci) was 60m deep. To protect it from melting, the villagers cut the ice during the night and transport it in skins and bags, isolated with leafage and carried by mules. The main selling place was the city of Makarska on the Adriatic coast, especially since tourism developed. According to Jagačić, the daily amount of ice delivered to Makarska was 2–3 tons. Cave ice economy was of great importance for the people from the undeveloped and poor hinterland. Between the two world wars, income from the ice trade was enough to buy enough wheat to last a whole year.

In the archives of the Republic of Dubrovnik, there are many documents about the ice trade. The oldest one is from 1642 (Škundrica, 1954, 2009). There were two sources of ice storage: on Snježnica Mt. and the ice caves in Herzegovina. Due to the mild climate influenced by the Adriatic in the hilly hinterland of Snježnica Mt. on the territory of Dubrovnik, the formation of natural ice is limited or impossible. However, strong continental influence from the north causes snow precipitation and lower temperatures during the winter. The ice was made in shafts and specially adapted dolines on the northern slope of Snježnica Mt. (Vukmanović, 1980). Some villagers (called "ice men" in documents) had to fill them with snow during the winter and cover the shafts with straw and branches for isolation and shade. These ice storages were overarched by a wooden vault, with a locked door and military supervision. The other source was trade with ice from natural ice caves located in the Herzegovina mountains under Ottoman rule (today Bosnia and Herzegovina). The currencies used in trade were money, wine, gunpowder, and ammunition for guns. According to documents from the period between Jul. 20 and Aug. 20, 1783, there were 15 tons of ice delivered in Dubrovnik (Škundrica, 1954), which was consumed by the aristocracy and their guests.

16.5 GEOGRAPHICAL DISTRIBUTION OF ICE CAVES IN CROATIA

The main factors that control ice and snow cave distribution and characteristics are geographical position, elevation, climate, karst topography, cave entrance characteristics (number of entrances, location, elevation, exposition, amount of shelter), and cave morphology.

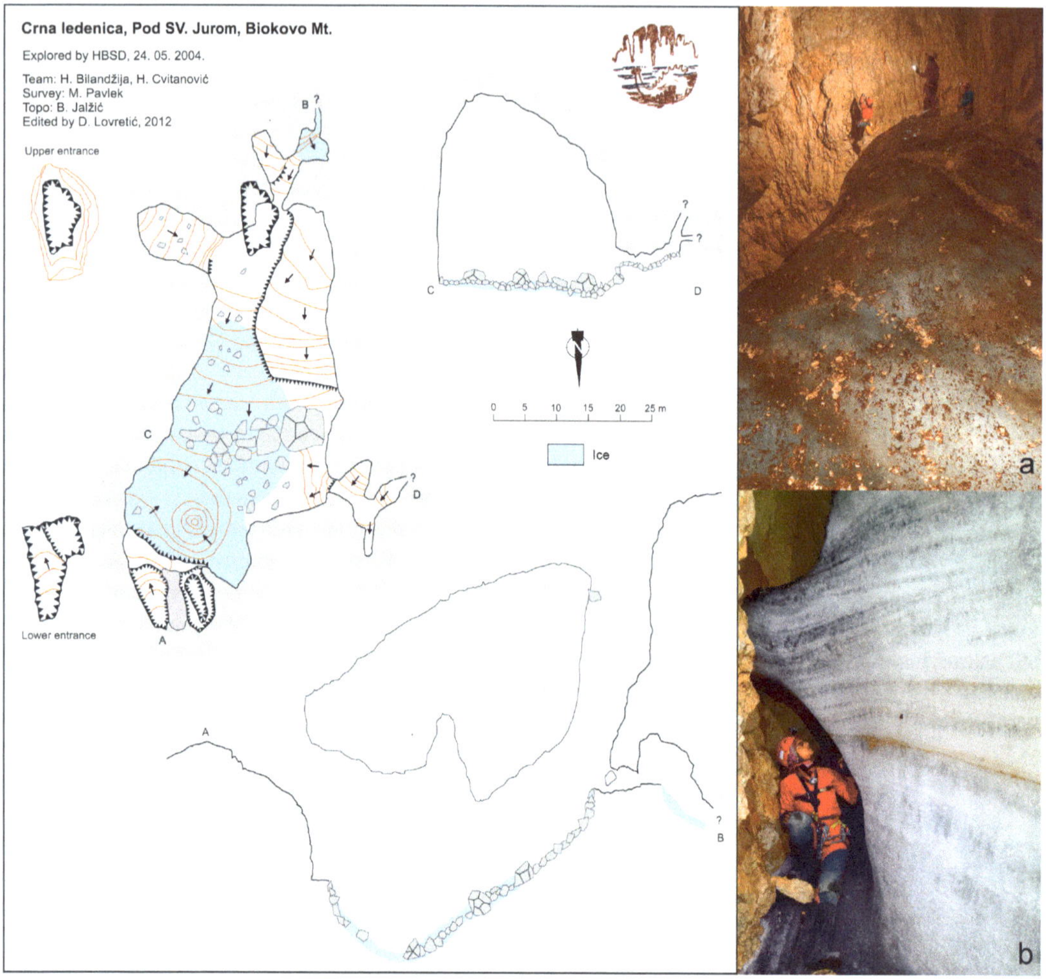

FIG. 16.8

Crna ledenica ice cave near Sv. Jure peak (1762 m a.s.l.) on Biokovo Mt. It was extensively used for ice extraction. (A) The ice pile below the lower entrance covered with cryogene debris and blocks (photo N. Buzjak, light assistant S. Buzjak), (B) up to 2.5 m thick layered ice profile in lower part (photo N. Buzjak).

According to the Köppen-Geiger climate classification (Kottek et al., 2006), most of Croatia has a temperate humid climate, with the mean monthly temperature of the coldest month higher than −3°C and lower than 18°C (Zaninović et al., 2008). The lower part of Dinaric mountain belt (Lika and Gorski kotar, and the higher parts of Istra) belongs to the climate class "Cfsbx." During the year there are no distinctly dry months, and the month with the least precipitation is in the cold part of the year. The annual precipitation has two maximums. Mean annual air temperature (MAAT) in the area around 1000 m a.s.l. is 5.5°C, and in the highest parts MAAT = 3.5°C. The coldest months are January and February (mean temperature = −2 to −5°C), and the warmest one is July (mean temperature = 12–16°C; Zaninović et al., 2008).

Only the higher mountain regions (>1200 m a.s.l.) have a snow (boreal) climate with the average temperature of the coldest month lower than −3°C (type Dfsbx). There are no dry periods, and most precipitation falls in the cold part of the year, with the maximum of rainfall occuring in autumn-winter and spring. The data from the meteorological station Zavižan (Croatian Meteorological and Hydrological Service) is illustrative for this area. It is the highest permanent meteorological station in Croatia (1594 m a.s.l.), and has collected data since 1953. The data for the standard period 1961–90 shows that the MAAT was 3.5°C. The coldest months were January and February (−4.2°C and −4.3°C), and the warmest was July (12.2°C). The range of absolute minimal air temperatures was −24.5°C (January) to 0.2°C (July), and air temperatures below 0°C are possible in every month of the warm seasons (Gajić-Čapka, 2003). During the period 2003–12, the MAAT was 4.2°C (Buzjak et al., 2014). In total there were 2590 days (71%) with air temperatures ≤0°C and only 18 (0.5%) warm days ($T_{max} \geq 25°C$). The range of absolute minimal air temperatures also increased, ranging from −23°C to 1.1°C (July). For the occurrence of very low temperatures, bura wind is an important factor (cold SE wind), and so is temperature inversion in clear and calm nights. Velebit Mt. range separates nearby Adriatic coast and hinterland, so it is an important climate border between the Mediterranean and continental areas, which have different temperature regimes. Due to the thermal influence of the warmer Adriatic, the air temperature at the western slopes decreases with elevation at a rate of about 0.7°C/100 m, and on the continental (eastern) slopes it does so with a rate of 0.5°C/100 m (Gajić-Čapka, 2003). It is therefore expected that ice caves will occur mainly on the eastern slopes.

Annual precipitation on this part of Velebit (and the Gorski kotar area) is up to 3000 mm/y (on the southern Velebit it is up to 3500 mm/y). The snow (≥10 cm) can occur from the middle of November until the middle of May. Snow cover ≥30 cm can be expected in 36% of days during a year, and in 31% of days it is not lower than 50 cm. The highest amount of snow was 320 cm (Mar. 1984), being generally between 32 and 123 cm (Gajić-Čapka, 2003). All of the provided data illustrates good conditions for snow and ice accumulation in mountain areas as illustrated with high density of ice caves in the Velebit area.

Due to the dynamics of mountain karst relief, often with an alteration of ridges and peaks with karst depressions (dolines, uvalas and karst poljes), the whole area is characterized by a climate mosaic at meso- and microscale. Even in the lower altitudes, where larger karst depression act as cold air traps, air temperature inversions often occur (Antonić et al., 1997; Buzjak et al., 2011, 2014; Horvat, 1952–1953). The area is known for abundant precipitation (2000-3900 mm/year), a high share of which is snow. In this zone, in the highest parts (>1400 m a.s.l.), periglacial environments are present with good potential for ice cave occurrences.

Atmospheric circulation from the Atlantic and the Mediterranean region, transporting humid air results in high precipitation amounts. Circulation of cold air masses from Northeastern Europe and Siberia during the winter results in low air temperatures. As a result of the combination of these two factors, the occurrence of ice caves is therefore limited to the geomorphological region of the Dinaric Mountain Belt. It is an area of karst plateaus, large karst depressions, and mountains (1.000–1.831 m a. s. l.). Its highest zone includes mountains ≥1400 m a.s.l. (Risnjak, Kapela, Plješivica, Dinara, Velebit and Biokovo; Fig. 16.9). The present knowledge on geographical distribution of ice caves is based on collected and evaluated data from caving clubs (not all included), literature, and Croatian Cave Cadastre.

According to up to now known and sorted data (321 caves), the areas of the highest frequencies of ice caves occurrences are located in the mountain areas of Gorski kotar (Šverda, Risnjak massif), Velebit, Dinara, and Biokovo mountains. Due to the numerous, and recently systematic caver's activities on Velebit Mt., this is the area with the highest frequency of known locations and data. In this

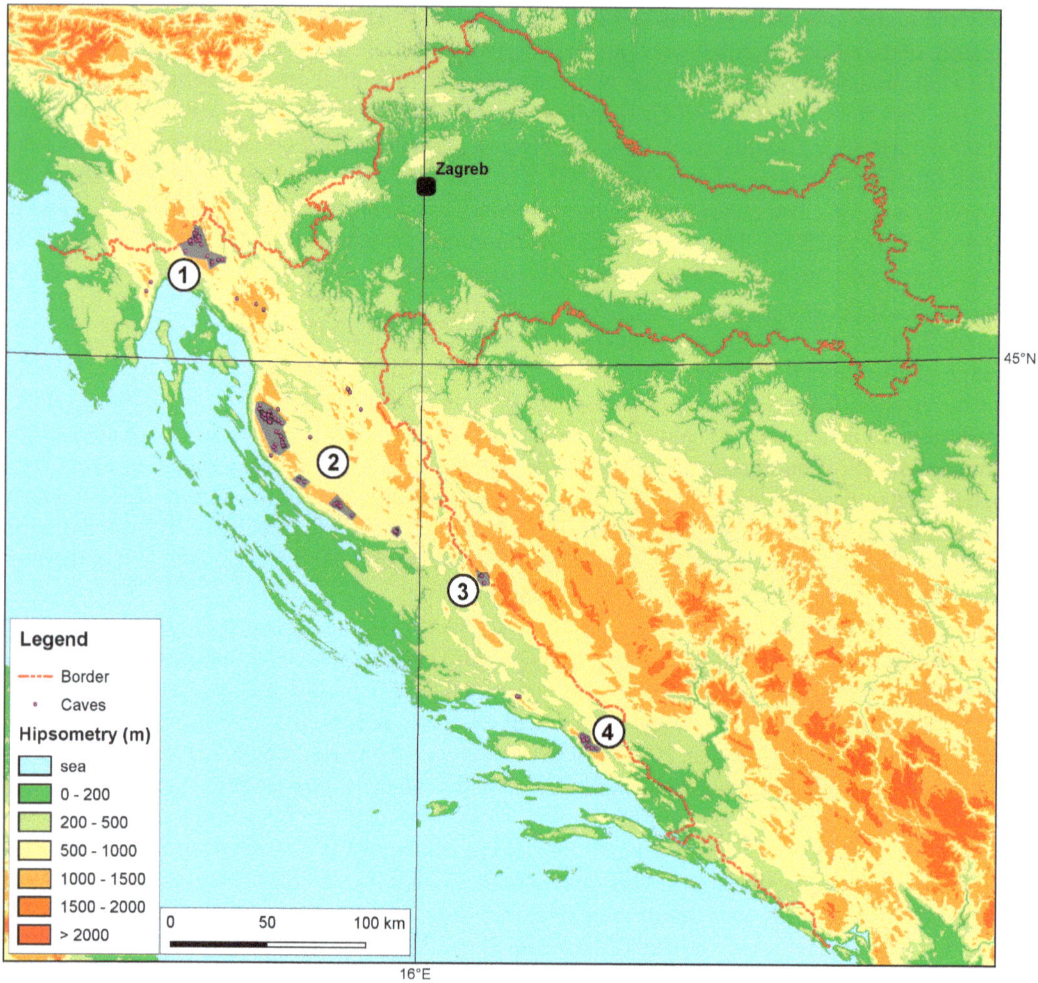

FIG. 16.9

Locations and main areas of ice and snow cave occurrence in Croatia (*N*=321 caves): 1—Gorski kotar area, 2—Velebit Mt., 3—Dinara Mt., 4—Biokovo Mt.

moment, in ice cave databases, there are 230 ice and snow caves on Velebit, which are predominantly shafts. The highest density is in the area of north Velebit (up to 25 caves/km^2; Fig. 16.10).

The frequency of ice and snow cave occurrences according to elevation (hypsometric classes) is shown on Fig. 16.11. Most of the caves are located between 1300 and 1600 m, with a Df climate type (228 or 71%). The cave's distribution is highly dependable on the state of research, but it coincides with natural conditions.

Besides "real" ice caves with permanent infillings, there are also many caves with seasonal ice and snow deposits. Further, there is a large number of caves in which these deposits no longer exist,

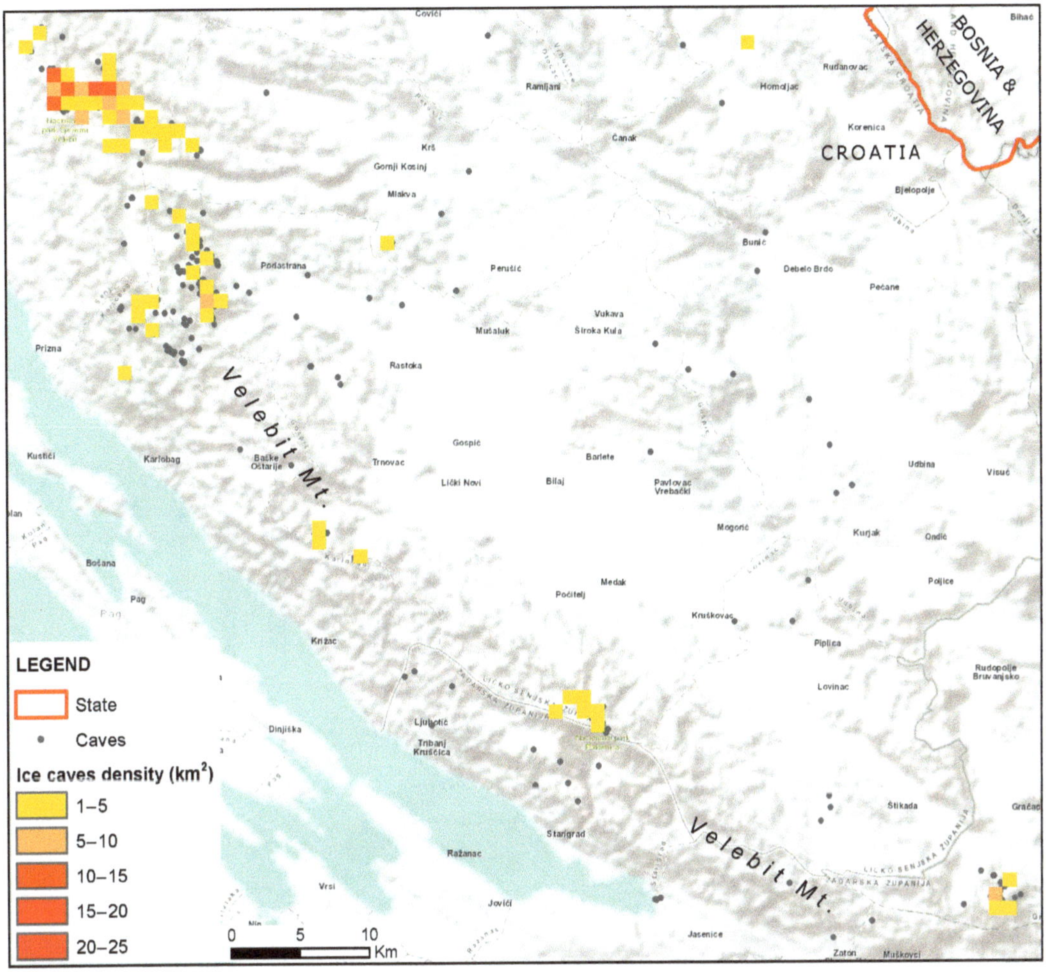

FIG. 16.10

Density map of ice and snow caves of Velebit Mt. ($N=230$ caves).

but numerous geomorphological traces of their existence and actions in the past are still present, such as the micro-relief forms and sediments that are the result of frost/defrost action. Therefore, records of previous researchers are of great importance (for example Girometta, 1923; Hirc, 1902; Rosandić, 1931). They testify that the volume of snow and ice changed (mostly reduced), or that these deposits completely melted. This is also evidenced by recent observations of cavers, especially in cases where melting opened a passage in a hitherto unknown part of the cave.

Seasonal ice and snow deposits occurs also in the cave entrances in the isolated karst of hilly areas in Pannonian basin mountains (Medvednica, Ravna gora, Papuk etc.).

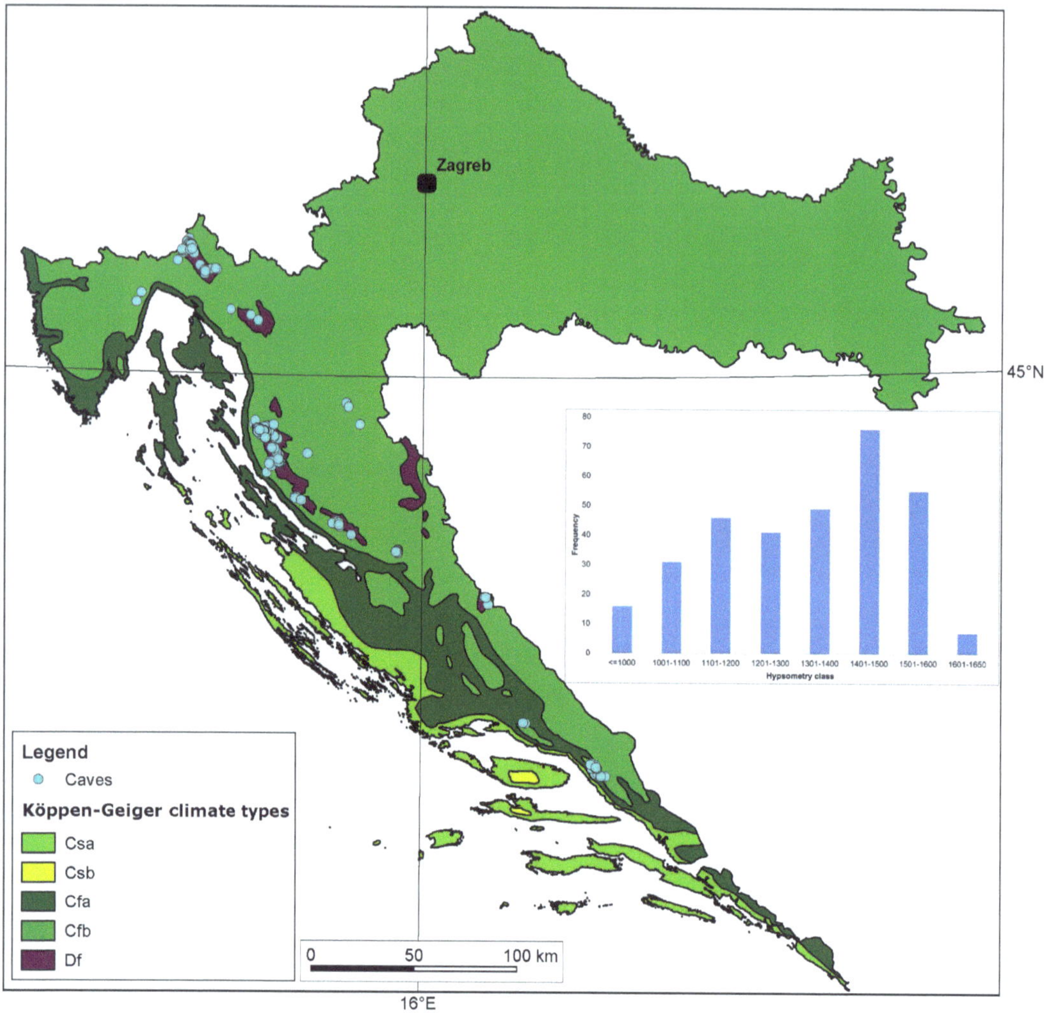

FIG. 16.11

Ice and snow cave occurrences according to climate zones and elevation (N=321 caves).

16.6 ICE CAVE MICROCLIMATES

After the oldest, up to now, known measurements taken in Ledenica on Mosor Mt. by Girometta (1927), there was a long pause. S. Fiedler did the first measurements in recent time in Ledena jama in Lomska duliba, and the entrance part of Vukušić snježnica on Velebit Mt. (Vrbek and Fiedler, 2000). Since snow and ice closed the entrance to Vukušić snježnica, the measuring point was set at the bottom of a doline in front of the entrance, so the recorded air temperatures (T_{air}) are higher than in the ice chamber, but can be used to observe the surface changes important for cave ice balance. Due to the small distance between those two locations (5 km), the conditions are similar in both features. Daily T_{air} and relative humidity (RH) were measured with a Lambrecht thermohygrograph, which was calibrated on site by

Assman psychrometer. Due to the 24-h mechanism limitations of the thermohygrograph, only daily parameters were measured.

Ledena jama is located in the southwestern part of the bottom of the Lomska duliba paleoglacier valley, so the cave's genesis and morphology is closely connected with the spreading and dynamics of the past glacier and postglacial hydrological setting (Bočić et al., 2013; Buzjak et al., 2014). It is a knee-formed vertical cave (shaft). The large (50 × 60 m) funnel-like entrance opens at the valley bottom (1235 m a.s.l.; Fig. 16.12A). Such a wide entrance (area = 3000 m^2) enables a strong microclimate

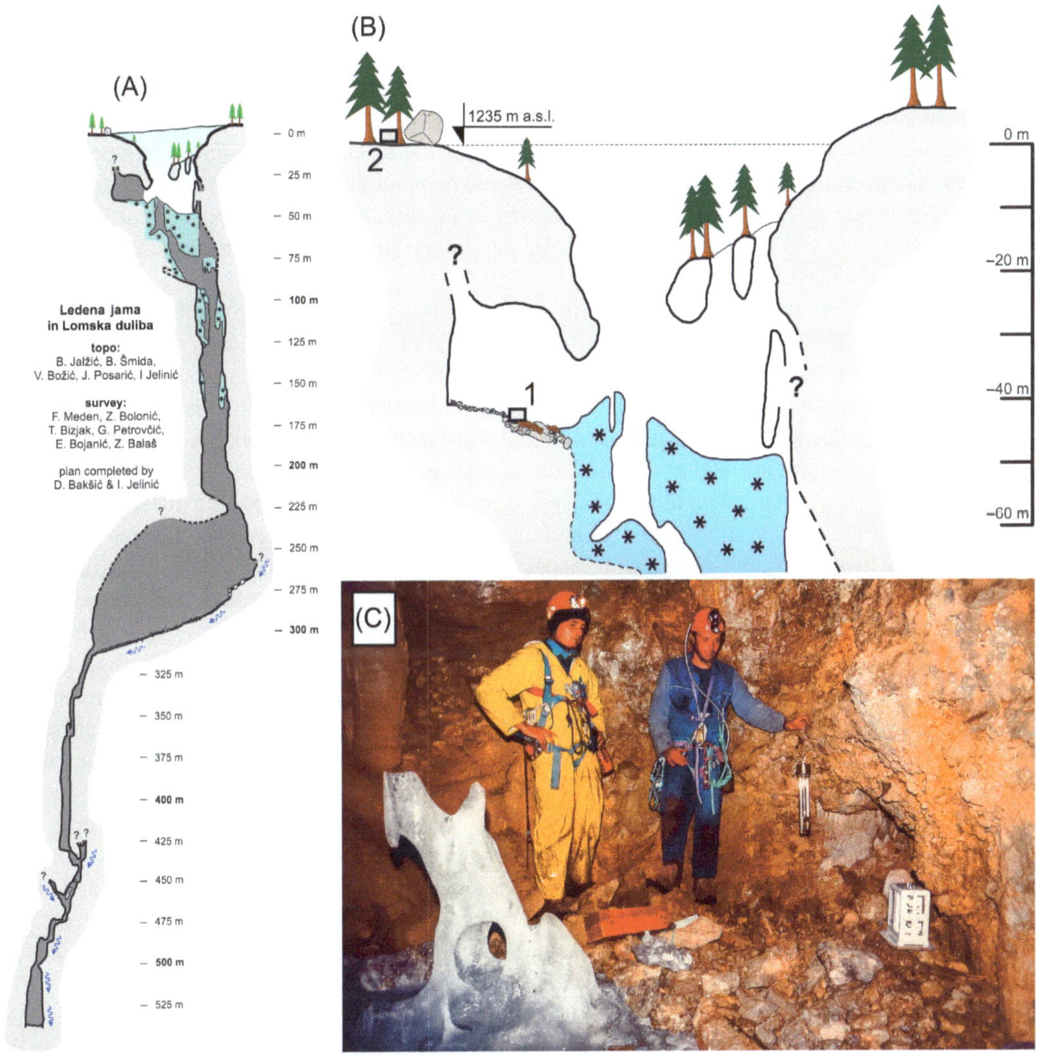

FIG. 16.12

(A) Profile of Ledena jama in Lomska duliba, (B) measuring sites in Ledena jama, in Lomska duliba (1) and on the surface (2), (C) calibration on measuring site 1995, photo B. Vrbek.

influence from the surface. Due to the steep or vertical slopes, and high ridge of Hajdučki kukovi to the south, it is, for most of the day, protected from direct sunlight. The entrance continues in a large (60×60m) chamber, with a 25m long lateral horizontal passage (Fig. 16.12B). At the depth of 50m, there is "bottom" consisting of snow, firn, and ice. It is a large mass with cooling effect. Its origin is as follows:

(1) Snow—allochthonous deposits that accumulated from outside the cave:
  (a) directly from the precipitation or
  (b) indirectly by sliding and collapsing from the steep slopes and trees around the entrance.
(2) Firn and ice—autogenic deposits forming from deposited snow (melting, refreezing, compacting) and by the freezing of percolating water.

The infilling is layered and contains various material of terrestrial origin (soil and rock particles, leaves, branches, animal bones, etc.). This infilling reaches the depth of about 90m, where the shaft continues. The total depth of Ledena jama is 536m. The first research by the Croatian Speleological Society members dates back to 1962. They explored and mapped the entrance chamber up to the depth of about 50m, where they were stopped by the ice and snow plug. Melting allowed descents into deeper passages in 1992 and 1993 by Croatian and Slovakian cavers. In 1996 cavers reached the bottom, and the cave map was finalized in 1997 (Jelinić et al., 2001).

The measuring site in Ledena jama was in a lateral passage (at that time) with ice at a depth of 40m, as well as at the surface (Fig. 16.12C). Measurements of daily T_{air} at various seasons of the year (except in winter) in the cave show negligible or no differences during the day, but there were some seasonal differences. In spring, T_{air} varied around 1°C, in summer around 0°C, and in autumn about 3.9°C (Fig. 16.13) The higher T_{air} observed in end of May than in July is probably the result of a heat influx caused be precipitation. RH at the same depth were not recorded during the day or from season to season (later measurements with more precise and sensitive equipment showed a different situation). Measurements at the surface site showed great daily variations in $T_{air,}$ as expected. The only considerable divergence from expected values, with respect to the location during the summer measurements, were recorded due to sudden night warming. Later, long-term measurements also showed greater differences due to the temperature inversions (Buzjak et al., 2014).

Long-term microclimate research using digital data loggers in Ledena jama started in 2012 (Dubovečak et al., 2016) and is still in progress (there is a plan that this location becomes permanent ice cave meteorological and monitoring site). A comparison of data from the nearby meteorological station Zavižan (6km NE from Ledena jama in the same climate region), Lomska duliba valley bottom, and in Ledena jama (Fig. 16.12, 1 and 2 on plan B) was done (Buzjak et al., 2014). The aim was to prove the influence local climate and topography had on cave environment (the effect of temperature inversions in karst depressions). According to field observations, and after Kern et al. (2008a,b), the summer period is of less importance for ice accumulation. Melting occurs during air temperature rise, and by the warm percolating water that infiltrates from the surface into the cave, as already discussed (Vrbek and Fiedler, 2000). The entering of warmer air was not observed, because it is prevented by a thick layer of denser and heavier colder air in the entrance chamber. It is marked by a sharp thermocline 20–25m below the entrance line (Buzjak, et al., 2014.). But mixing of air can be observed from results presented on Fig. 16.14. The entrance chamber acts as a cold air trap, and the lowering of air temperature on the surface causes air temperature to drop underground due to the descending of heavier, colder air into the cave. Summer oscillations are present, too, but smaller. As the surface temperature is higher than in the

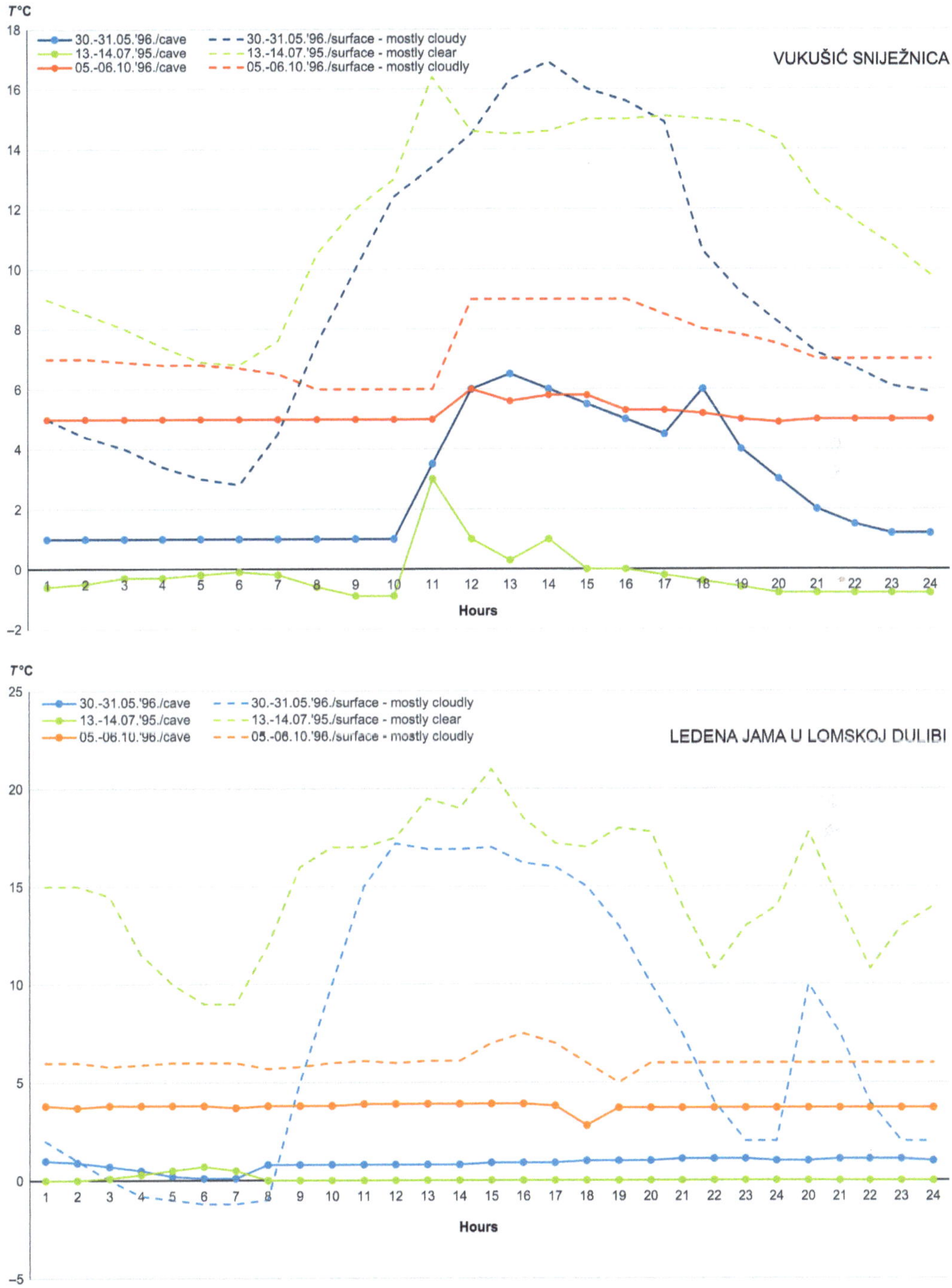

FIG. 16.13

Diurnal range of air temperatures in different seasons on the surface, in Vukušić snježnica, and in Ledena jama in Lomska duliba (Fig. 16.12, 1 and 2).

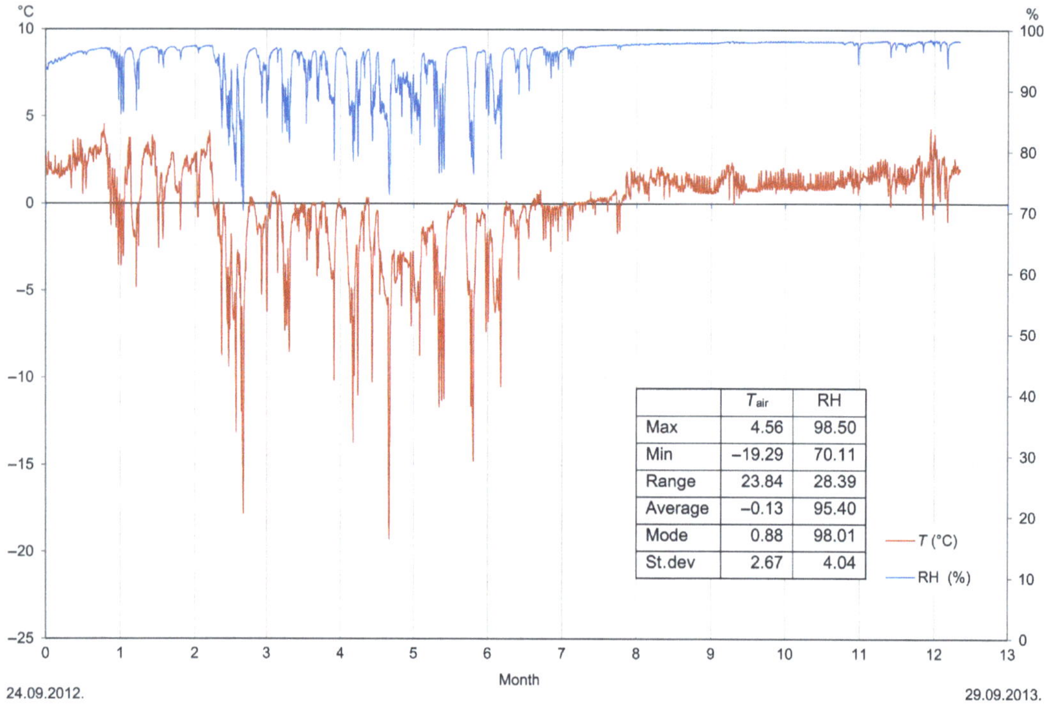

	T_{air}	RH
Max	4.56	98.50
Min	−19.29	70.11
Range	23.84	28.39
Average	−0.13	95.40
Mode	0.88	98.01
St.dev	2.67	4.04

FIG. 16.14

T_{air} and RH in Ledena jama in Lomska duliba (Velebit Mt.) Sep. 2012–13 with basic statistics.

chamber, cold air accumulates inside. In the spring and early summer, the snow melts on the surface, then temperature rises, resulting in a significant amount of drip water, which then freezes on top of ice layers at the cold bottom of the chamber (causing the growth of ice stalagmites).

According to observations and measurements, perennial ice in deep caves of northern Velebit Mt. is the result of a combination of various conditions and complex processes. Therefore it could be absent in some shafts, or spread several hundred meters below the surface. It seems that the current climate, even so high in the mountain (>1200 m a.s.l.), is not sufficient for the formation of perennial ice in all caves and shafts there by itself. For example, in the 1320 m deep Slovačka jama shaft, there is no perennial ice, even the entrance is at elevation of 1520 m a.s.l. It has a horizontal entrance that is not suitable (open) for direct snow accumulation. It was observed that air temperature in the upper part of the cave is above 0°C during the whole year. Obviously, there is not necessarily air circulation either. On the other hand, in Lukina jama-Trojama shaft system, microclimate conditions are different. The upper entrance of Trojama is at 1475 m a.s.l., and the lower entrance of Lukina jama is at 1438 m. Both are lower than in the case of Slovačka jama. Interconnections of entrances, chimneys, dozens of fissures, and small shafts at different elevations causes a chimney effect that is important for air cooling and ice and snow accumulation, especially during winter (Fig. 16.15A). This is clearly observable from continuous measurements between 2010 and 2011 (Paar et al., 2013). Even before the discovery of the higher Trojama entrance by cavers, air circulation during summer in the passages, and overflow of cold air on the lower entrance Lukina jama, was noted. Warm air causes melting, but this maintains low air

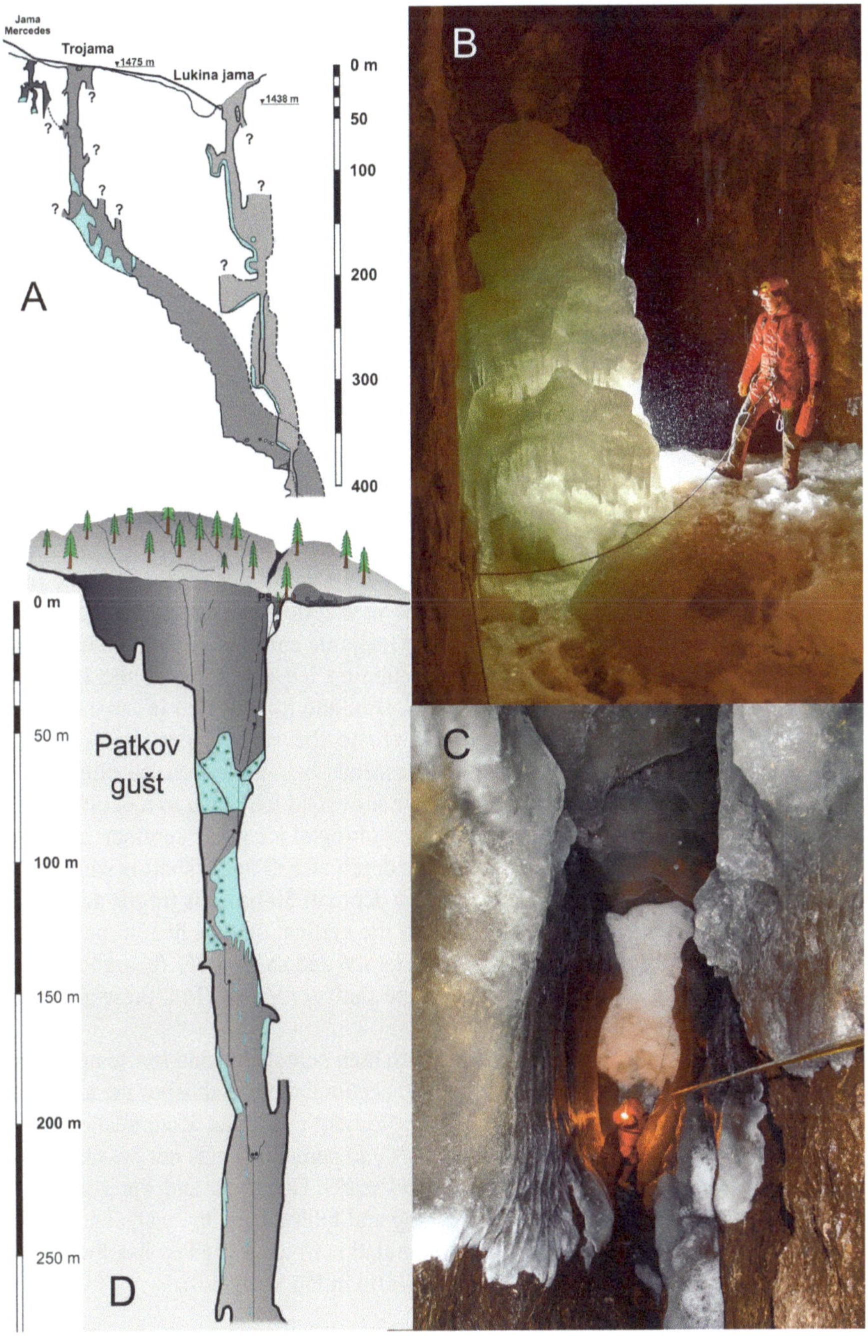

FIG. 16.15

Examples of different deep shafts morphologies on northern Velebit and ice deposits. (A) First 400 m of Lukina jama-Trojama shaft system with mapped snow, firn and ice deposits. (B) Autogenic ice in the form of stalagmite and by ice particles covered bottom of passage part at the depth of 180 m in the Trojama part of Lukina jama-Trojama system. (C) Vertical passage's walls covered by ice crust in the Trojama part of Lukina jama-Trojama system, depth ~150 m. (D) Entrance part of Patkov gušt shaft with mapped ice of the upper passage.

A: After Bakšić, D., 2015. Lukina jama. Hrvatski speleološki poslužitelj, http://speleologija.eu/lukinajama/ (accessed 15.01.17) (in Croatian); B: Photo D. Jirkal; C: Photo D. Bakšić; D: After Bakšić, D., Bakšić, A., 2008. Speleološka istraživanja jame Patkov gušt (Speleological research of Patkov gušt). Hrvatski speleološki poslužitelj, http://speleologija.eu/patkovgust/index.html (accessed 15.01.17.) (in Croatian).

Table 16.1 The Depth Range of Ice and Snow Depths in Shafts on Velebit Mt.

Shaft	Depth Range (m)
Lukina jama-Trojama	45–556
Patkov gušt	60–553
Xantipa	70–323

temperatures, and the temperature is more stable (around 0°C) than during winter (when it decrease almost up to −6°C with intense oscillations; Paar et al., 2013). In the winter, colder air, entering through the lower entrance, causes direct and evaporative cooling (Perşoiu and Onac, 2012). Strong air currents in the shaft brings snow and can cause frost, so there are snow, firn, and ice deposits. Snow is constantly present from a depth of 45–60 m to 328–556 m. The most interesting ice deposits in Trojama are between 150 and 180 m deep (Fig. 16.15B and C). In Jul. 1995 snow was recorded at the inclined shell 550 m deep below the entrance (Jalžić, 2007). Since the year 2006 (or even earlier) the system is accessible only through the Trojama entrance, since the Lukina jama entrance is clogged by snow and ice.

Another interesting example of a deep shaft containing ice and snow is the Patkov gušt shaft (Fig. 16.15D). It has an entrance at an elevation of 1450 m, and it is located in large collapsed doline. The doline entrance diameter is ~100 × 75 m. Entrance to the vertical passage is of the diameter ~65 × 30 m. At a depth of 65 m starts icy slope that descends below the first ice plug that is approximately 15 m thick. The vertical passage continues over snow and ice down to a depth of 105 m. There was a narrow passage in ice (2 × 1.5 m). Because of the melting of ice in the summer, it can be very wet. Water falls down to the shaft bottom (553 m). After the depth of 130 m the shaft is wider. The rocks are partially or completely covered with ice, coating up to a depth of 300 m. It is fragile and often breaking. The vertical passage ends with a hall 40 × 30 m. Below the vertical there is heavy snow accumulation.

Another shaft with deep ice deposits is Xantipa. The ice and snow were recorded between 70 and 323 m deep. According to data collected by cavers in the shafts of Velebit Mt., these are the deepest ice and snow deposits in the world (Table 16.1).

Strong dynamics during winter, or during periods with high percolation and low temperatures, contributes to the accumulation of perennial ice in caves. Initial accumulation establishes the temperature conditions suitable for maintaining ice during the whole year. Without this initial accumulation, temperature in Northern Velebit caves would be a few degrees above 0°C, so perennial ice is not possible. Low temperature is suitable for ice formations on the walls of vertical shafts. Due to air and water influence from the surface, especially during the summer time, ice breaking and falling from the walls into the deeper parts of the cave. This is one of the mechanisms for the accumulation of ice at big depths (down to the depth of 560 m in Lukina jama-Trojama, 553 m in Patkov Gušt, 400 m in Paž shaft etc.).

16.7 GLACIOCHEMICAL AND ICE MASS BALANCE RESEARCHES OF CAVE ICE IN NORTH VELEBIT MT.

Deep shafts of Velebit and other Dinaric mountains are complex systems that contain information about geological, geomorphological, hydrological, and ecological conditions and processes. They represent an excellent environment for study in paleo- and recent processes. The presence of perennial cave ice, as well as snow and firn in the shafts below the snow line, encourages thinking about the climate of the past, including the microclimate conditions for cave ice genesis, snow accumulation, and the impact of glacial and periglacial processes on speleogenesis. The ice in the shafts is a complex phenomenon, because ice bodies are not stable in location, shape, or volume. They are continuously changing, owing to a series of processes, including years of accumulation of snow and firn, seasonal melting and freezing, water penetration and freezing, and percolation of water until total melt. The origin and changes of ice under the influence of external and internal factors depends on its composition, microstructure, and isotopic composition, which complicates interpretation of climatic conditions during these cycles. Such research in Croatia is new and rare, but there are several promising sites that could provide interesting results.

16.7.1 RESEARCH LOCATIONS

The following researches were done in Ledena jama in Lomska duliba and Vukušić snježnica. As already stated, the upper part of Ledena jama preserves a large ice body (Fig. 16.16A). It spreads from

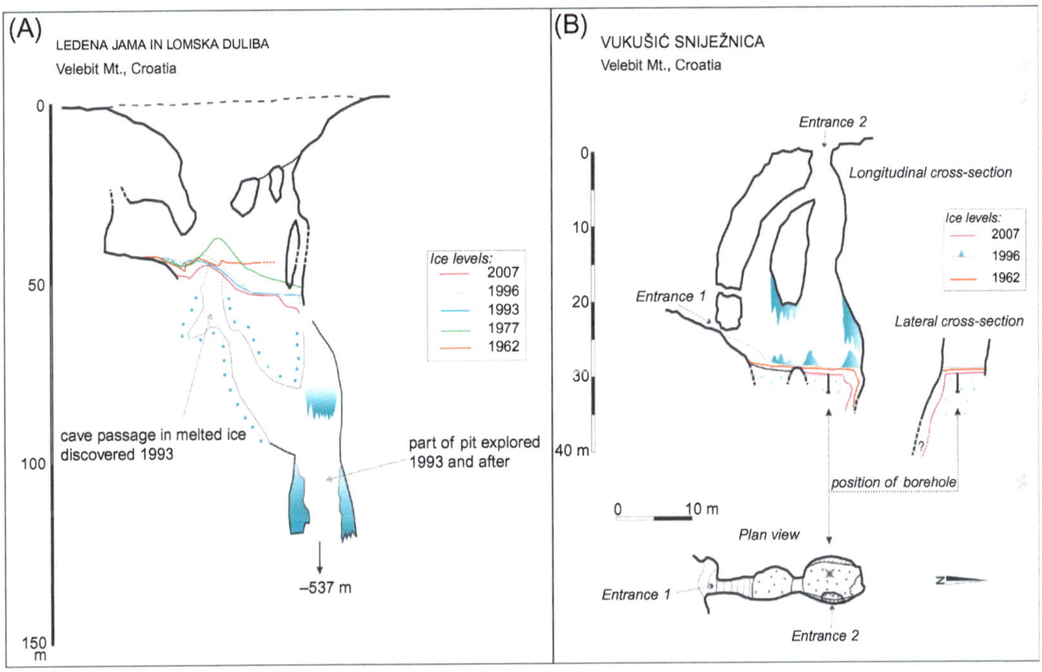

FIG. 16.16

A vertical profile of the Ledena jama in Lomska duliba entrance part (A) and Vukušić snježnica with ice core locations (B). *Color lines* represent the ice level in different years. A characteristic profile where ice level change is determined (results are in Fig. 16.17) is indicated by the vertical line.

According to Kern, Z., Bočić, N, Horvatinčić, N., Fórizs, I., Nagy, B., László, P., 2008a. Őskörnyezeti adatok a Velebit-hegység jegesbarlangjaiból (Ledena-zsomboly, Vukusic-jegesbarlang). In: Szabó, V., Orosz, Z., Nagy, R., Fazekas, I. (Eds.), IV. Magyar Földrajzi Konferencia, Debrecen, Hungary, November 14–15, 2008., Debreceni Egyetem, pp. 126–133 (in Hungarian); Kern, Z., Bočić, N, Horvatinčić, N., Fórizs, I., Nagy, B., László, P., 2008b. Palaeoenvironmental records from ice caves of Velebit Mountains - Ledena Pit and Vukušić Ice Cave, Croatia. In: Kadebskaya, O., Mavlyudov, B.R., Pyatunin, M. (Eds.), Proceedings of the 3rd International Workshop on Ice Caves. Kungur, Russia, May 12–17, 2008, Kungur, pp. 108–113; Kern, Z., Széles, E., Horvatinčić, N., Fórizs, I., Bočić, N., Nagy, B., 2011. Glaciochemical investigations of the ice deposit of Vukušić Ice Cave, Velebit Mountain, Croatia. Cryosphere 5, 485–494 (revised).

−50 m depth to −90 m depth, and takes up 20–30 m in diameter. During the speleological and morphological explorations in 1990s, isotope analysis of cave ice, dating of wood branches from ice, and speleothems were done (Jelinić et al., 2001). Vukušić snježnica is a small cave with two entrances and a length of 20 m (Fig. 16.16B). The two cave chambers are filled with permanent cave ice. Thickness of its ice deposit is estimated to be more than 15 m. Total ice area is around 50 m^2, and the volume of the ice block is estimated to be 550–750 m^3 (Kern et al., 2011).

16.7.2 CAVE ICE SAMPLING AND DRILLING

To study the isotopic composition of ice in the Ledena jama in Lomska duliba, sampling was conducted by ice profile, which is about 45 m high. For sampling vertical caving gear and technique was used. The samples were taken in Jun. and Jul. 1995, every 20 cm in depth in the first meter, and at 3, 4, 5, 30 and 40 m depth of ice layers. All depth data are referred to as depth below the 1995 ice surface. Also in two places in the ice profile (depth of 15 and 40 m), samples of the organic material were taken for radiocarbon dating (Horvatinčić, 1996; Jelinić et al., 2001).

Only in Vukušić snježnica drilling was possible (Kern et al., 2011). Two drill cores (diameter 3 cm, length 2.4 m, and 26 cm) were extracted from the ice deposit in Oct. 2008 (Fig. 16.16B). All depth data is referred to as depth below the Oct. 28, 2008 ice surface. The deepest 40 cm-long section of the 2.4 m-long core and the 26 cm-long ice core were kept as one sample and labeled as VS_lower and VS_upper, respectively. The upper 2 m long segment of the 2.4 m long ice core was cut into 36 segments. The mean sample length was ~5.3 cm. Limestone fragments up to 2 mm, and dark muddy material rich in organic matter, were observed in three consecutive samples corresponding to the 1.24–1.44 m depth range. This probably indicates a period when the ablation of the ice, and the influx of soil material into the cave occurred. Consequently, the core was subdivided into three main stratigraphic units. The first unit (0–1.24 m) is characterized by clean ice. The second unit (1.24–1.44 m) is composed of dirty ice (probably due to soil contamination). The third unit beneath 1.44 m is again characterized by clean ice (Kern et al., 2011).

16.7.3 TRITIUM CONCENTRATIONS IN CAVE ICE

In Ledena jama in Lomska duliba activity of tritium in cave ice was measured from samples taken in the ice profile at the 1st, 3rd, 5th, 30th, and 40th m depth from the surface of the ice (Horvatinčić, 1996). The maximum activity of tritium is written down in an ice sample at a depth of 3 m (2.8 Bq/L, i.e., ~23.7 TU). Due to increased activity of tritium, it is believed that this ice originates from the period 1960–65 (because of thermonuclear bombs testing). The activity of tritium (1.3 Bq/L) in samples from the surface of ice corresponds to activities of the rainfall in Zagreb at the same time as the sampling. On the basis of the measured tritium values, Horvatinčić (1996) assumes that 45 m thick layer of ice accumulated in a period of about 500 years.

In Vukušić snježnica, tritium activity was measured in the samples VS_lower (2–2.4 m depth) and VS_upper (0–0.26 m depth), and the results were 1.8 ± 0.5 TU and 9.9 ± 0.6 TU (Kern et al., 2011). For example, in VS_lower it is considered that it is the result of ice accumulation in the period between 1950 and 1955, and tritium activity is a result of mixed impact before and post the thermonuclear test period. The relatively high concentration in the near-surface sample did not match the recent activity of tritium in precipitation (2003). The tritium concentration at the 2–2.4 m depth suggests that the ice deposit was built by precipitation that fell 45 years ago, and the uppermost 0.26 m-thick ice layer formed between the early 1970s and the early 1980s, or between 1954 and 1960. (Kern et al., 2011).

The results of these studies partially coincide and indicate that the upper 2–3 m of ice was formed during the period of the second half of the 20th century. Both tritium activities were analyzed at the Laboratory for Measurements of Low-level Radioactivity, Ruđer Bošković Institute in Zagreb.

16.7.4 STABLE ISOTOPE COMPOSITIONS OF THE CAVE ICE

Stable oxygen and hydrogen isotopic composition of cave ice was researched form nine samples from Ledena jama in Lomska duliba. Stable isotopic composition were measured at the Jožef Štefan Institute, Ljubljana. Results for $\delta^{18}O$ show a range between −6.74 and −10.25‰, and for δD, −50.3 and −67.9‰. All results are expressed in relation to the VSMOW (Horvatinčić, 1996). These results are compared with MWL for precipitation in Zagreb (period 1980–95). Kern et al. (2008a,b) re-evaluated these results by comparison with the MWL of precipitation for the meteorological station Zavižan (northern Velebit) for period Sep. 2000 through Dec. 2003. This research has confirmed conclusions from Horvatinčić (1996): precipitation from the winter half of the year has a dominant role in the cave ice formation, and stable isotope composition shows stronger influence of continental climate.

16.7.5 ELEMENTAL CONCENTRATION OF THE CAVE ICE

Chemical analysis in ice cores from Vukušić snježnica was done by Kern et al. (2011) and this is still the only such research of cave ice in Croatia. The ice cores were studied in order to explore the potential of the preserved glaciochemical signal of pollution history. The upper 2 m long segment of the 2.4 m long ice core was cut into 36 segments. The mean sample length was ~5.3 cm. For each segment, a concentration of 23 elements was analyzed (metals and metalloids: Rb, Sr, Y, Ba, Pb, U, Ce, Mg, Cr, Mn, Fe, Co, Cu, Zn, Al, As, Ca, Na, K, Ni, B, Mo, Cd). These measurements were done by using the Thermo Finnigan Element2 Magnetic Sectorfield ICP-MS at the Institute of Isotopes in the Hungarian Academy of Science. Elemental concentrations and certain elemental ratios indicate a multi-source origin of the ions in ice core. Atmospheric deposition, soil water, and limestone dissolution are regarded as the three major sources of the ions. Due to the karstic setting, Ca is the most abundant ion in cave ice. So, only those chemical species not correlated with Ca possibly preserve reliable atmospheric depositional signals. The authors suggest that Cr, Cu, Pb, and Zn likely reflect past atmospheric deposition. Similar research should be carried out in the future on some other ice cave sites to determine whether the developments are of regional significance or only local (for example, because of the position of the Vukušić snježnica near an important forest road).

16.7.6 ICE MASS BALANCE ESTIMATION

Based on the reports of cavers who were visiting Ledena jama and Vukušić snježnica, levels of ice in these caves is often changing and generally lowering. In order to compare the levels of ice available, cave surveys of several generations were used (Kern et al., 2008a,b). For Ledena jama, surveys were available from 1962, 1977, 1993, 1995 and 2007, and for Vukušić Ice Cave, surveys were only available for 1962, 1996, and 2007. Data about ice levels were graphically compared on a single cave profile for each cave (Fig. 16.16). Based on this graphic display, differences of the position of the level of ice were measured (Fig. 16.17). Morphological evidences and archive data from previous surveys suggest that both ice blocks have shrunken over the past decades. The ice surface since 1962 decreased by ~0.2 m in Vukušić snježnica and by ~9 m in Ledena jama. However, the ice loss trend was not constant during this time. It seems to have become more moderate during the last decade in Ledena jama (Fig. 16.17). Soon we can expect results of new measurements that will complement the existing data and point to the development of these trends.

For a better understanding of cave ice dynamic in Velebit Mt., researches of historical data from cave maps for different localities are in progress (Bočić et al., 2014).

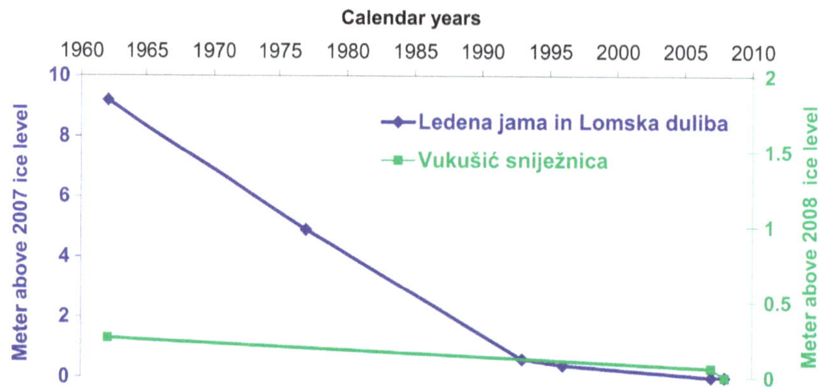

FIG. 16.17

Decrement history of ice in the two studied ice caves between 1962 and 2007. Data are presented on different *y*-axes due to the large difference in the range of ice level changes between the two caves.

From Kern, Z., Bočić, N, Horvatinčić, N., Fórizs, I., Nagy, B., László, P., 2008a. Őskörnyezeti adatok a Velebit-hegység jegesbarlangjaiból (Ledena-zsomboly, Vukusic-jegesbarlang). In: Szabó, V., Orosz, Z., Nagy, R., Fazekas, I. (Eds.), IV. Magyar Földrajzi Konferencia, Debrecen, Hungary, November 14–15, 2008, Debreceni Egyetem, pp. 126–133 (in Hungarian); Kern, Z., Bočić, N, Horvatinčić, N., Fórizs, I., Nagy, B., László, P., 2008b. Palaeoenvironmental records from ice caves of Velebit Mountains - Ledena Pit and Vukušić Ice Cave, Croatia. In: Kadebskaya, O., Mavlyudov, B.R., Pyatunin, M. (Eds.), Proceedings of the 3rd International Workshop on Ice Caves. Kungur, Russia, May 12–17, 2008, Kungur, pp. 108–113.

16.7.7 ESTIMATION OF CAVE ICE AGE

In order to estimate the cave ice age in Ledena jama, two samples of wood from the ice profile are dated by radiocarbon method (Horvatinčić, 1996; Jelinić et al., 2001). The data was run in the Laboratory for Measurements of Low-level Radioactivity, in the Ruder Bošković Institute in Zagreb. The resulting age of the analyzed material is 150 ± 100 years for the sample at a depth of 15 meters and 140 ± 90 years for the sample at 40 m depth. The deeper sample is much younger than was expected. A possible explanation is that the younger sample subsequently peaked at greater depth through a crack in the ice deposit.

Assuming that the ^{3}H-bomb-peak was at 3.75 m below the 1995 ice level, based on data given by Horvatinčić (1996), Kern et al. (2008a,b) calculated the mean ice accumulation rate ($r_{acc} = 12\,cm\,a^{-1}$). The lowermost ice layer (−40 m) can be dated to being 333 years old applying r_{acc}. Data taken from cave maps suggests the strong ice melt (8.65 m) and tritium data shows significant ice accumulation (3.75 m) in the period between 1963 and 1995. This data is in contradiction (with ice melting and ice accumulating during the same time at the same site) at first glance. However, a fast basal melting process can solve this contradiction. Based on this data, the basal melting rate is determined as $r_{bm} = 38\,cm\,a^{-1}$. And if one uses this estimation of r_{bm}, mass turnover rate of the 40 m thick ice deposit, it can be determined to only 105 years, which is a rough estimation of the age of the lowest layer of ice deposits. It is a much lower estimation of cave ice age than given by Horvatinčić (1996) based on tritium measurements, and assumes that the 45 m thick layer of ice was accumulated in a period of about 500 years. For the tritium based cave ice estimation in Vukušić ice cave (Kern et al., 2011), see the paragraph Tritium concentrations in cave ice.

Within the research in the shaft system Lukina jama-Trojama in 2011, the dating of the wood remains was performed. The wood was sampled in a vertical passage at depths of 120 and 160 m. The aim was to determine the upper age of the ice. The radiocarbon dating of samples showed an upper age of 410 ± 75 years. The dating of wood samples from Ledena jama in Lomska duliba, taken in 2012 at the depth of 50 m, under a thick layer of ice, had age of 525 ± 40 years. Those results are quite similar with results from Lukina jama-Trojama and illustrate the connection of climatic conditions on Velebit and in these shafts (Paar et al., 2013).

16.8 INFLUENCE OF PLEISTOCENE GLACIATION ON CAVES (VELEBIT MT.)

The highest parts of the Dinaric Mountains during the Pleistocene were affected by glaciation, i.e., under the influence of ice deposits. Traces of Pleistocene glaciation have been reported also in the area of northern and central Velebit in the form of valley, plateau, and cirque-like glaciers (Bognar et al., 1991; Bognar and Faivre, 2006). Their influence on the karstification and caves was studied in several locations (e.g., traces in sediments, see Rukavina, 1978–1979), and here are presented two illustrative examples from Velebit Mt.

The first example is the aforementioned Ledena jama in Lomska duliba (Bočić et al., 2013). For more about this cave, see the section on Glaciochemical and ice mass balance research on the ice cave in north Velebit Mt. This cave is located in a paleo-glacier valley of the so-called Lomski paleo-glacier (Bognar et al., 1991; Buzjak et al., 2014). Its position suggests that at least at some stage of the formation the cave was the ponor of water from melting glacial ice. This is indicated in the entrance part, on the side where the glacier and melting water was coming, and which has the form of a canyon. That area is now filled with unsorted coarse-grain carbonate sediment. Such sediment, in the form of erosion residues, is also found throughout the entrance part of the cave (Fig. 16.18). It can be assumed that a large part, or even the whole entrance to the cave at a particular time, was completely filled with the sediments of glacial and fluvial-glacial origin. Part of these sediments was later subsequently eroded and probably collapsed. Their erosion in the paleo-canyon created a new cave channel. The sample of flowstone from this channel was dated with the U/Th method in the Laboratory for Measurements of Low-level Radioactivity, at the Ruđer Bošković Institute in Zagreb. The results show that the age of this sample is 301,000 ± 55,000 years BP (Horvatinčić, 1996; Jelinić et al., 2001). On this basis, it can be assumed that this glacial and fluvioglacial sediment originated at least from the Mindel glaciation, or earlier. In the immediate vicinity of the entrance there are several larger blocks that can be interpreted as erratic blocks. However, these are only the results of the preliminary investigation. Systematic studies of the site are in progress.

Similar influence on the development of underground karst forms was studied in the 351 m long Ledenica u Štirovači (Štirovača Ice Cave) on middle Velebit (Bočić et al., 2012; 2013). The abundant erosional remnants of the fluvioglacial deposits, which fully filled much of the known part of the cave in the past, was found there. Coarse-grain clastic sediments are generally rarely present in caves, but

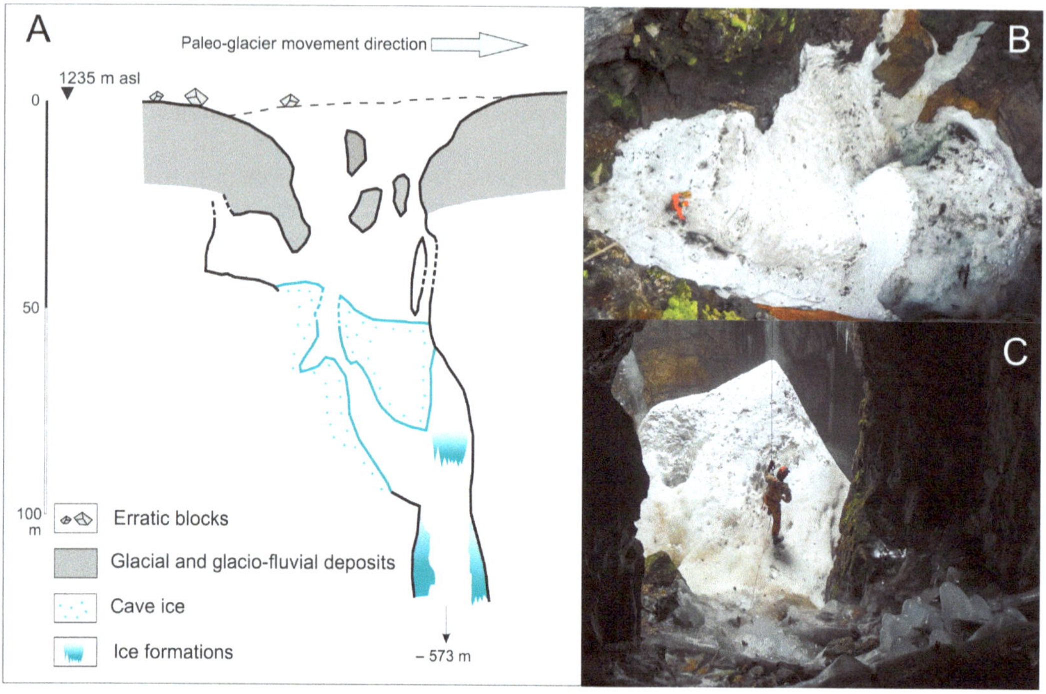

FIG. 16.18

(A) Vertical profile of entrance part Ledena jama in Lomska duliba with the position of quaternary sediment bodies of glacial and fluvioglacial origin. (B) Ice plug in entrance chamber (Sep. 2013, photo V. Dubovečak). (C) Snow pile in entrance chamber (Apr. 2014, photo D. Paar).

A: According to Bočić, N., Faivre, S., Kovačić, M., Horvatinčić, N., 2013. Influence of the Pleistocene glaciations on karst development in the Dinarides – examples from Velebit Mt. (Croatia). In: Filippi, M., Bosák, P. (Eds.), Proceedings of the 16th International Congress of Speleology, vol. 3, Brno, Czech Republic, July 21–28, 2013, Czech Speleological Society and UIS, Brno, pp. 170–172, modified.

have been discovered in this cave passage (Fig. 16.19). They occur as erosional residues and pebbles are cemented with calcitic cement. It is presumed that these sediments are of fluvioglacial origin, that is, material that was transported from the fluvioglacial fans, and which was finally sedimented in the cave. Another 20–40 cm thick sedimentary body found in this cave has been defined as limestone debris. The upper part of this sediment body is covered with a flowstone layer. This speleothem material was dated using the radiocarbon method in Laboratory for Measurements of Low-level Radioactivity in the Ruđer Bošković Institute in Zagreb. Based on the age of the flowstone crust of 8230 ± 150 years BP, as well as on its position, it may be concluded that coarse-grain clastic sediment and limestone debris are older. Based on the sedimentological properties, primarily on their cementation properties, it can be assumed that limestone debris is significantly younger then the fluvioglacial fan material. This indicates that the fluvioglacial fan cannot be related to the Würm/Holocene transition, although we may presume that the limestone debris was deposited at that time. This means that fluvioglacial sediment are deposited earlier, presumably during pre-Würm periods. During cold periods, glacial exaration produced a particularly significant quantity of sediments. Consequently, the underground karst conduits situated below the snow line during the Pleistocene were filled with high-energy sedimentary structures, as well as glacial and fluvioglacial sediments. That reduced or even stopped their conduit function, and led to difficulties in the drainage of meteoric water, lake formation within morphological depressions, and to a rise in the local water-table level. With the denudation of these deposits, and conduit reactivation a drop in the underground water level occurred (for details see Bočić et al., 2012, 2013).

FIG. 16.19

Cemented coarse-grain clastic sediment of fluvioglacial origin in the main passage of Štirovača Ice Cave (photo N. Bočić).

Besides these examples, man can find various traces of periglacial action in caves. Due to the frost action and colder local microclimate, they are present not only in high-elevation, but also in the mid-elevation caves settled in temperate climate settings across Croatia. The traces can be found in loose sediments (soil, debris, etc.), speleothems, and rocks affected by freeze-thaw cycles due to temperature fluctuations (Luetscher, 2013). The results are the movements and sorting of loose sediments and the grinding or peeling of rock and solid sediments like speleothems. This is the case in not only permanent ice caves, but also the caves in areas with low winter air temperatures. All these traces are valuable geo-indicators of environmental changes/variations in the research of historical extent of ice caves during the Pleistocene and Holocene. Systematic studies on several interesting sites in Croatia are in progress. Data is gathered from literature, results of archaeological excavations, and the sampling and mapping of cryo-clastic sediments in caves where permanent ice or snow does not occur anymore.

ACKNOWLEDGMENT

For the cooperation in collecting data and fieldwork, we thank Hrvoje Cvitanović, Roman Ozimec, Željko Marunčić, Nikolina Marić, Toni Tursić, Goran Mandić, Aida Barišić, Teo Barišić, Lovel Kukuljan, Vedran Sudar, Dalibor Reš, Goran Rnjak, Gordan Lukač, Suzana Buzjak, Branko Jalžić, Martina Pavlek, Dalibor

Jirkal, Neven Ris, Nebojša Anić, Zoran Delibašić, Željka Janjanin, Ksenija Protrka, Domagoj Tomašković, Ilija Rašić, Ljiljana Bajs, Božana Knežević, and Kazimir Miculinić. We especially want to thank all caving clubs, cavers, and cave maps authors for their contribution to the research and recording of caves all around Croatia. Their work is fundamental for scientific research and the sustainable management of the fragile karst environment in Croatia.

REFERENCES

Anonymous, 1913. Snježnice na Velebitu (Snow caves on Velebit). Planinarski vjesnik 2, 62–63 (in Croatian).

Antonić, O., Kušan, V., Hrašovec, B., 1997. Microclimatic and topoclimatic differences between the phytocenosis in the Viljska ponikva sinkhole, Mt. Risnjak, Croatia. Hrvatski meteorološki časopis 32, 37–49.

Bakšić, D., 1997. Ono što (ne)znate o Lukinoj jami (The facts you do not know about Lukina jama). Speleo'zin 7, 12–17 (in Croatian).

Bakšić, D., Paar, D., Stroj, A., Lacković, D., 2013. Northern Velebit Deep Caves. In: Filippi, M., Bosák, P. (Eds.), Proceedings of the 16th International Congress of Speleology, Vol 2, Brno, Czech Republic, July 21–28, 2013, Czech Speleological Society and UIS, Brno. pp. 24–29.

Barić, E., 1989. Terminologizirati naziv nije jednostavno (Make a name is not easy). Jezik 37 (4), 111–118 (in Croatian).

Barišić, T., Bakšić, D., Paar, D., 2013. Cave system of Kita Gaćešina - Draženova Puhaljka, the longest cave in the Dinaric karst. In: Filippi, M., Bosák, P. (Eds.), Proceedings of the 16th International Congress of Speleology, Vol 2, Brno, Czech Republic, July 21–28 2013, Czech Speleological Society and UIS, Brno. pp. 17–20.

Bočić, N., 1997. Speleološki kamp "Sjeverni Velebit '97." (Caving camp "Sjeverni Velebit '97."). Speleo'zin 7, 3–6 (in Croatian).

Bočić, N., 2001. Speleološka ekspedicija "Olimp 2000." (Caving expedition "Olimp 2000"). Speleo'zin 14, 3–6 (in Croatian).

Bočić, N., 2005. Kugina ledenica na srednjem Velebitu (Kugina ledenica ice cave on middle Velebit). Speleosfera 2, 54–58 (in Croatian).

Bočić, N., 2006. Najdublji speleološki objekti hrvatskog krša i njihove temeljne geomorfološke značajke (Deepest Caves of Croatian Karst and Their Basic Geomorphological Features). In: Međunarodni znan. skup Akademik Josip Roglić i njegovo djelo, Zbornik radova, Makarska, Croatia, 19–22 April 2006., Hrv. geogr. društvo, Zagreb, Hrvatska, pp. 161–182 (in Croatian).

Bočić, N., 2010. Geomorfologija krša Papuka - kap krša u srcu Slavonije (Karst geomorphology of Papuk). Geografija.hr. www.geografija.hr (accessed 20.01.17) (In Croatian).

Bočić, N., Faivre, S., Kovačić, M., Horvatinčić, N., 2012. Cave development under the influence of Pleistocene glaciation in the Dinarides—an example from Štirovača Ice Cave (Velebit Mt., Croatia). Z. Geomorphol. 56 (4), 409–433.

Bočić, N., Faivre, S., Kovačić, M., Horvatinčić, N., 2013. Influence of the Pleistocene glaciations on karst development in the Dinarides—examples from Velebit Mt. (Croatia). In: Filippi, M., Bosák, P. (Eds.), Proceedings of the 16th International Congress of Speleology, vol. 3, Brno, Czech Republic, 21–28 July 2013, Czech Speleological Society and UIS, Brno, pp. 170–172.

Bočić, N., Buzjak, N., Kern, Z., 2014. Some new potential subterranean glaciation research sites from Velebit Mt. (Croatia). In: Lewis, L., Kern, Z., Maggi, V., Turri, S. (Eds.), Proceedings of 6th International Workshop on Ice Caves, Idaho Falls, ID, 17–22 August 2014. National Cave and Karst Research Institute, Carlsbad, pp. 72–76.

Bognar, A., Faivre, S., 2006. Geomorphological traces of the younger pleistocene glaciation in the central part of the velebit Mt. Hrvatski geografski glasnik 68 (2), 19–30.

Bognar, A., Faivre, S., Pavelić, J., 1991. Tragovi oledbe na Sjevernom Velebitu (The traces of glatiation on northern Velebit). Geografski glasnik 53, 27–39 (in Croatian).

Bognar, A., Faivre, S., Buzjak, N., Pahernik, M., Bočić, N., 2012. Recent landform evolution in the dinaric and pannonian regions of Croatia. In: Lóczy, D., Stankoviansky, M., Kotarba, A. (Eds.), Recent Landform Evolution. Springer, Heidelberg, London, New York, pp. 313–344.

Božić, V., 1994. Speleološka ekspedicija "Lukina jama '94" i novi hrvatski rekord dubine -1392 m (Caving expedition "Lukina jama '94" and new Croatian depth record -1392 m). Spele'zin 4, 9–14 (in Croatian).

Božičević, S., 1955. Ledeno podzemlje. O ledenicama (Ice underground. About ice caves). Naše planine 3, 182–186 (in Croatian).

Božičević, S., 1974. Podzemni krški fenomeni planine Medvednice kraj Zagreba (Subterranean Karst Phenomena in Medvednica Mt. near Zagreb). Acta Carsologica 6, 95–109 (in Croatian).

Božičević, S., Benček, Đ., 1983. Tektonsko-geomorfološke specifičnosti Biokova i pojava urušnih vrtača i ledenica (Tectonic and geomorphological specificities of Biokovo Mt. and genesis of collapsed dolines and ice caves). Acta Biokovica II, 157–166 (in Croatian).

Buzjak, N., 1999. Speleološko istraživanje NP Risnjak (2. dio - 1996. godina) (Speleological research in Risnjak National Park). Speleo'zin 10, 19–21 (in Croatian).

Buzjak, N., Buzjak, S., Orešić, D., 2011. Florističke, mikroklimatske i geomorfološke značajke ponikve Japage na Žumberku (Hrvatska). Šumarski list 3–4, 127–137 (in Croatian).

Buzjak, N., Dubovečak, V., Paar, D., Bočić, N., 2014. The influence of karst topography to ice cave occurrence - example of Ledena jama in Lomska duliba (Croatia). In: Lewis, L., Kern, Z., Maggi, V., Turri, S. (Eds.), Proceedings of 6th International Workshop on Ice Caves, Idaho Falls, ID, 17–22 August 2014. National Cave and Karst Research Institute, Carlsbad, pp. 17–23.

Čepelak, R., 1972–1973. Zimsko istraživanje jame Puhaljke (Winter research in Puhaljka cave). Speleolog XX–XXI, 24 (in Croatian).

Croatian Speleological Server, 2017. List of the largest caves in Croatia. http://speleologija.hr/popis (accessed 28.01.17) (in Croatian).

Dadić, Ž., 1984. Osvrt Nikole Gučetića u 16. stoljeću na dvije špilje u okolici Dubrovnika (A review by Nikola Gučetić in the 16th century on the two caves near Dubrovnik). In: Malez, M. (Ed.), Proceedings of the 9th Yug. Congress of Speleology, Karlovac, Croatia, 17–20 October 1984. Speleological Society of Croatia, Zagreb, pp. 741–746 (in Croatian).

Dubovečak, V., Buzjak, N., Bočić, N., Paar, D., 2016. Ice cave Ledena jama in Lomska duliba (Velebit Mt., Croatia)—the promising site for ice cave monitoring. In: Mihevc, A. (Ed.), International Workshop on Ice Caves, Program guide and abstracts, Postojna, Slovenia, 16–22 May 2016. SAZU Karst Research Institute, Postojna, pp. 43–44.

Dvoržak, V., 1989. Ledenica ili… (Ice or…). Jezik 37 (4), 103–111 (in Croatian).

Fortis, A., 1774. Viaggio in Dalmazia, Venezia.

Franić, D., 1898. Uzlaz na velebitsku Visočicu (1619 m) (Ascent to Visočica on Velebit). Hrvatski planinar 4, 58–62 (in Croatian).

Gajić-Čapka, M. (Ed.), 2003. Zavižan između snijega, vjetra i Sunca (Zavižan between snow, wind and Sun). DHMZ, Zagreb (in Croatian).

Garašić, M., 1980. Ledenica u Bukovom vrhu na Velebitu (Ledenica in Bukov vrh on Velebit). Naše planine 11–12, 278–280 (in Croatian).

Girometta, U., 1923. Jame i pećine srednje Dalmacije (Shafts and caves of middle Dalmatia). Glasnik geografskog društva 9, 95–120 (in Croatian).

Girometta, U., 1927. Ledenice u srednjo-dalmatinskom kršu (Ice caves in middle-dalmatian karst). Hrvatski planinar 7, 93–97 (in Croatian).

Girometta, U., 1934. Meteorološki odnosi u pećinama i jamama (Meteorological relations in caves and shafts). Novo doba 303 (December 25), 6–7 (in Croatian).

Girometta, D., 1935. Meteorološki odnosi u pećinama i jamama (Meteorological relations in caves and shafts). Hrvatski planinar 2, 37–42 (in Croatian).

Gorjanović-Kramberger, D., 1912. Izvještaj Speleološkog odbora za godinu 1911 (Report of the Speleological Committee for 1911). Vijesti Geol. povjerenstva 2 (1911), 49–53 (in Croatian).
Grgasović, T., 1983. Speleološki logor KS PSH "Biokovo '83" (Caving camp KS PSH "Biokovo '83"). Speleon 1, 70–75 (in Croatian).
Hirc, D., 1883a. Iz hrvatske Švice. Narodne novine 224 (October 1), 4–5.
Hirc, D., 1883b. Iz hrvatske Švice. D. Hirc. Narodne novine 225 (October 2), 4–5.
Hirc, D., 1893/1894. Špilje ledenice (Ice caves). Pobratim 4 (3), 44–45 (in Croatian).
Hirc, D., 1898. Gorski kotar: slike, opisi i putopisi (Gorski kotar: images, descriptions and itineraries). L. Hartman (Kugli i Deutsch), Zagreb (in Croatian).
Hirc, D., 1900a. Nekoje znamenite pećine u Hrvatskoj (Some famous caves in Croatia). Nada 24 (December 15), 378–379 (in Croatian).
Hirc, D., 1900b. Lika i Plitvička jezera (Lika and Plitvice lakes). Tisak i naklada Lav. Hartmana (Kugli i Deutsch), Zagreb (in Croatian).
Hirc, D., 1902. Put u Ledenu jamu (The travel to Ledena jama). Prosvjeta 10 (15), 468–471 (in Croatian).
Hirc, M., 1923. Kroz Veliku Kapelu i Velebit do mora (Through Velika Kapela and Velebit to the sea). Hrvatski planinar 10, 149–160 (in Croatian).
Horvat, I., 1952–1953. Vegetacija ponikava (Vegetation of dolines). Geografski glasnik 14–15, 1–25 (in Croatian).
Horvatinčić, N., 1996. Izotopna mjerenja u ledu, jama Ledenica, Velebit (Isotopic measurements in the ice, Ledenica, Velebit). In: Zbornik radova trećeg simpozija Hrvatskog društva za zaštitu od zračenja, pp. 297–301 (in Croatian).
Horvatinčić, N., Božić, V., 1998–1999. Ledena jama na Velebitu izazov znanstvenicima (Ledena jama on velebit, challenge to scientists). Speleolog 46 (47), 47–52 (in Croatian).
Jagačić, T., 1957. Crtice s Biokova (Notes from Biokovo). Naše planine 8–12, 241–247 (in Croatian).
Jalžić, B., 1978–1979. Pojava plavog leda u Ledenoj jami u Lomskoj dulibi na sjevernom Velebitu (Blue ice in Ledena jama in Lomska duliba on north Velebit). Speleolog XXVI–XXVII, 52 (in Croatian).
Jalžić, B. (Ed.), 2007. Jamski sustav Lukina jama-Trojama (Lukina jama-Trojama shaft system). HPS, Zagreb (in Croatian).
Jalžić, B., Lacković, D., Šmida, B., 1992–1993. Lomska duliba '93. Speleolog 40 (41), 5–16 (in Croatian).
Janton, D., 2017. Speleološka ekspedicija "Srednji Velebit 2016" (Caving expedition "Srednji Velebit 2016"). Subterranea Croatica 14 (21), 29–36 (in Croatian).
Jelinić, I., 1998–1999. Ledena jama u Lomskoj dulibi. (Ledena jama in Lomska duliba). Speleolog 46/47, 17–22 (in Croatian).
Jelinić, I., Horvatinčić, N., Božić, V., 2001. Ledena jama u Lomskoj dulibi (Ledena jama in Lomska duliba). Senjski zbornik 28, 5–20 (in Croatian).
Kapidžić, A., 2004. Speleološki kamp "Paleži '04" (Caving camp Paleži '04). Subterranea Croatica 3, 8–15 (in Croatian).
Kern, Z., Bočić, N., Horvatinčić, N., Fórizs, I., Nagy, B., László, P., 2008a. Őskörnyezeti adatok a Velebit-hegység jegesbarlangjaiból (Ledena-zsomboly, Vukusic-jegesbarlang). In: Szabó, V., Orosz, Z., Nagy, R., Fazekas, I. (Eds.), IV. Magyar Földrajzi Konferencia, Debrecen, Hungary, 14–15 November 2008. Debreceni Egyetem, pp. 126–133 (in Hungarian).
Kern, Z., Bočić, N., Horvatinčić, N., Fórizs, I., Nagy, B., László, P., 2008b. Palaeoenvironmental records from ice caves of Velebit Mountains—Ledena Pit and Vukušić Ice Cave, Croatia. In: Kadebskaya, O., Mavlyudov, B.R., Pyatunin, M. (Eds.), Proceedings of the 3rd International Workshop on Ice Caves, Kungur, Russia, 12–7 May 2008, Kungur, pp. 108–113.
Kern, Z., Forizs, I., Horvatinčić, N., Széles, É., Bočić, N., Nagy, B., 2010. Glaciochemical investigations on the subterranean ice deposit of Vukušić Ice Cave, Velebit Mountain, Croatia. Cryosphere Discuss. 4 (3), 1561–1591.
Kern, Z., Széles, E., Horvatinčić, N., Fórizs, I., Bočić, N., Nagy, B., 2011. Glaciochemical investigations of the ice deposit of Vukušić Ice Cave, Velebit Mountain, Croatia. Cryosphere 5, 485–494.

Klaić, V., 1878. Prirodni zemljopis Hrvatske (Physical Geography of Croatia). C. Albrecht, Zagreb (in Croatian).
Korbar, T., 2009. Orogenic evolution of the external Dinarides in the NE Adriatic region: a model constrained by tectonostratigraphy of Upper Cretaceous to Paleogene carbonates. Earth Sci. Rev. 96 (4), 296–312.
Kosor, M., 1957. Najstariji slavonski *Dikcionar* (The oldest Slavonian *Dikcionar*). Rad JAZU, knj. 10, 5–28 (in Croatian).
Kottek, M., Grieser, J., Beck, C., Rudolf, B., Rubel, F., 2006. World Map of the Köppen-Geiger climate classification updated. Meteorol. Z. 15, 259–263.
Krajač, I., 1922. Snježnica "Špoljarova jama" pod Širokim vrhom kod Starog laza (Snow cave Špoljarova jama under Široki vrh peak near Stari laz). Hrvatski planinar 2, 32 (in Croatian).
Kuhta, M., 1986–1987. Speleološka istraživanja u Štirovači na Velebitu (Speleological research in Štirovača on Velebit). Speleolog XXXIV–XXXV, 10–16 (in Croatian).
Kuhta, M., 2000–2001. Rezultati novijih speleoloških istraživanja na Crnopcu (The results of newer speleological research on Crnopac). Speleolog 48 (49), 36–44 (in Croatian).
Kuhta, M., 2005. Speleološka istraživanja na području Crljenog kuka u NP Paklenica (Speleolgical research of Crljeni kuk in Paklenica National Park). Speleolog 53, 24–31 (in Croatian).
Lacković, D., 1992. Istraživanje na Kozjaku, Južni Velebit (Exploration on Kozjak, southern Velebit). Speleolog 38 (39), 56–57 (in Croatian).
Luetscher, M., 2013. Glacial processes in caves. In: Shroder, J., Frumkin, A. (Eds.), Treatise on Geomorphology. Karst Geomorphology, 6. Academic Press, San Diego, CA, pp. 258–266.
Lukić, O., 1988–1989. Speleološka istraživanja Crnopca na Velebitu (Speleological research of Crnopac on Velebit). Speleolog XXXVI–XXXVII, 14–26 (in Croatian).
Malez, M., 1955. Spilja Ledenica (Ledenica cave). Priroda XLII (8), 281–286 (in Croatian).
Malez, M., 1959. Speleološka istraživanja krša u 1956. god. (Speleological research in karst in 1959). Ljetopis JAZU 63, 340–354 (in Croatian).
Malez, M., 1984. Povijest speleoloških istraživanja u Hrvatskoj (Geschichte der Höhlenforschungen in Kroatien). In: Malez, M. (Ed.), Proceedings of the 9th Yug. Congress of Speleology, Karlovac, Croatia, 17–20 October. Speleological Society of Croatia, Zagreb, pp. 73–102 (in Croatian).
Malinar, H., 1964–1965. Puhaljka na južnom Velebitu (Puhaljka cave on southern Velebit). Speleolog XII–XIII, 17–20 (in Croatian).
Margetić, M., 1925. Nove jame i pećine srednje Dalmacije (New shafts and caves of middle Dalmatia). Glasnik geografskog društva 11, 44–59 (in Serbian).
Marjanac, S., 1956. Izvještaj o speleološkim istraživanjima na području Velikog Javornika, Mala Kapela (Report on speleological research in the area of the Veliki Javornik, Mala Kapela Mt.). Speleolog 3–4, 38–48.
Marković, M., 1980. Narodni život i običaji sezonskih stočara na Velebitu (Folk life and customs of seasonal herders on Velebit). Zbornik za narodni život i običaje 48, 5–139 (in Croatian).
Mišćenić, G., Glavaš, I., 2006. Međunarodna speleološka ekspedicija "Srednji Velebit '06" (Int. caving expedition "Srednji Velebit '06"). Subterranea Croatica 7, 3–8 (in Croatian).
Ostojić, J., 1996–1997. Speleološka istraživanja na sjevernom Velebitu između 1980. i 1984. godine (Speleological research on northern Velebit 1980-1984). Speleolog 44 (45), 21–23 (in Croatian).
Ozimec, R., Šincek, D., 2011. Speleološki objekti planinskih masiva SZ Hrvatske (Caves of mountainous massifs in NW Croatia). Radovi Zavoda za znanstveni rad HAZU Varaždin 22, 201–232 (in Croatian).
Paar, D., Buzjak, N., Sironić, A., Horvatinčić, N., 2013. Paleoklimatske arhive dubokih jama Velebita (Palaeoclimate archives from Velebit deep shafts). In: Marjanac, Lj (Ed.), Knjiga sažetaka, 3. znanstveni skup Geologija kvartara u Hrvatskoj, Zagreb, Hrvatska, 21–23 March, Hrvatska akademija znanosti i umjetnosti & Geološki zavod Slovenije. pp. 39–40 (in Croatian).
Pahernik, M., 2012. Prostorna gustoća ponikava na području Republike Hrvatske (Spatial Density of Dolines in the Croatian Territory). Hrvatski geografski glasnik 74 (2), 5–26 (in Croatian).

Perşoiu, A., Onac, B., 2012. Ice in caves. In: White, W.B., Culver, D.C. (Eds.), Encyclopedia of Caves. Elsevier, Amsterdam, pp. 399–404.

Rasol, D., Špoler Čanić, K., 2005. Meteorological characteristics of a snow pit in Gorski kotar region (Croatia). Hrvatski meteorološki časopis 40, 638–640.

Redenšek, V., 1961. Popis špilja i ponora u Hrvatskoj (The list of caves in Croatia). Naše planine 5–6, 147 (in Croatian).

Reš, D., 2006. Jama Snježna kraljica (-237 m), Japage, srednji Velebit (Snježna kraljica shaft). Subterranea Croatica 6, 8–11 (in Croatian).

Roglić, J., 1935. Biokovo—geomorfološka ispitivanja (La montagne de Biokovo—étude géomorphologique). Posebna izdanja Geografskog društva, Beograd (in Serbian).

Rosandić, Z., 1931. Iz podzemne Like (From subterranean Lika). Hrvatski planinar 9, 239–244 (in Croatian).

Rožman, M., Cvitanović, H., Ris, N., Resanović, A., Kukuljan, L., Rožman, T., 2017. Međunarodna speleološka ekspedicija "Japage 2016"—Srednji Velebit (int. caving expedition "Japage 2016"—middle Velebit). Subterranea Croatica 14 (21), 17–25 (in Croatian).

Rubić, I., 1949. Sa Vošca (1421 m) (From Vošac, 1421 m). Naše planine 3–4, 65–72 (in Croatian).

Rubinić, A., 2012. Speleološki kamp "Šverda 2011" (Caving camp "Šverda 2011"). Subterranea Croatica 14, 7–12 (in Croatian).

Rukavina, D., 1978–1979. Iskopavanje spilja i osobitosti koje susrećemo (The excavations in the caves and peculiarities that we meet). Speleolog XXVI–XXVII, 26–29 (in Croatian).

Sabljar, V., 1866. Miestopisni riečnik kraljevinah Dalmacije, Hervatske i Slavonije (Orts-Lexicon der Königreiche Dalmatien, Kroatien und Slavonien). Nakladom i berzotiskom A. Jakića, Zagreb.

Škundrica, Z., 1954. U starom se Dubrovniku prodavao led (In the old Dubrovnik the ice was sold). Otkrića 1 (5), 381–385 (in Croatian).

Škundrica, Z., 2009. Tajna kutija dubrovačkog arhiva II (the secret box of Dubrovnik archive II). Posebna izdanja, knjiga 28/II, 147–152 (in Croatian).

Šmida, B., 1999. Velebit. Reportàž z objavovania hlbokých priepastí Chorvátska v rokoch 1990–1998. Slovenská speleologická spločnost (in Slovak).

Stroj, A., Velić, I., 2015. Geološki profil Jamskog sustava Lukina Jama–Trojama na Sjevernom Velebitu (Geologic cross-section along the Lukina jama–Trojama Cave system, North Velebit). In: Horvat, M., Wacha, L. (Eds.), Abstracts Book of 5th Croatian Geological Congress With International Participation, Osijek, 23–25 September 2015. Croatian Geological Survey, Zagreb, pp. 240–241 (in Croatian).

Troha, D., 1994. Novija speleološka istraživanja Hajdučkih kukova (Newer speleological research of Hajdučki kukovi). Spele'zin 4, 16–17 (in Croatian).

Tutiš, S., 2007. Jama malaksalih speleologa na Crnopcu. Speleolog 55, 28–34 (in Croatian).

Vlahović, I., Tišljar, J., Velić, I., Matičec, D., 2005. Evolution of the Adriatic carbonate platform: palaeogeography, main events and depositional dynamics. Palaeogeogr. Palaeoclimatol. Palaeoecol. 220 (3–4), 333–360.

Vrbek, M., Fiedler, S., 2000. The ecological and floristic characteristics of Ledena jama pit on Velebit Mountain - Croatia. Natura Croatica 9 (2), 115–131.

Vučetić, V., Abramović, B., 2004. Meteorolozi zakoračili u podzemlje. Meteorološka mjerenja u sniježnici kraj Fužina (Meteorologists stepped into the underground. Meteorological measurements in Snježnica near Fužine). Hrvatski meteorološki časopis 39, 141–144 (in Croatian).

Vukmanović, J., 1980. Konavli. Antropogeografska i etnološka ispitivanja (Konavli. Anthropogeographic and ethnological research). Posebna izdanja DXXVII(85). Odeljenje društvenih nauka, SANU, Belgrade, p. 439 (in Serbian).

Zaninović, K., Gajić-Čapka, M., Perčec Tadić, M., 2008. Klimatski atlas Hrvatske (Climate atlas of Croatia) 1961-1990, 1971-2000. Državni hidrometeorološki zavod (DHMZ), Zagreb.

Završki, Z., 2012. Istraživanje područja Crnopca u 2012. godini (Exploration of Crnopac area in 2012). Speleolog 60, 27–40 (in Croatian).

FURTHER READING

Bakšić, D., 2015. Lukina jama. Hrvatski speleološki poslužitelj, http://speleologija.eu/lukinajama/ (accessed 15.01.17) (in Croatian).

Bakšić, D., Bakšić, A., 2008. Speleološka istraživanja jame Patkov gušt (Speleological research of Patkov gušt). Hrvatski speleološki poslužitelj, http://speleologija.eu/patkovgust/index.html (accessed 15.01.17) (in Croatian).

Croatian Agency for the Environment and Nature: Nature Protection Information System—Croatian Cave Cadastre. http://natura2000.dzzp.hr/speleo/ (accessed 28.02.17).

Hirc, D., 1899. Pogled u podzemni sviet domovine, III. Pećine ledenice (The look into subterranean world of the country, III. Ice caves). Hrvatski planinar 8, 120–123 (in Croatian).

Kuhta, M., 1990–1991. Speleološka istraživanja masiva Crnopca u 1990. godini (Speleological research of Crnopac massif in 1990). Speleolog 38 (39), 25–28 (in Croatian).

Kuhta, M., Bakšić, D., 2001. Karstification dynamics and development of the deep caves on the North Velebit Mt. - Croatia. In: Proceedings of 13th International Congress of Speleology, Brasilia, Brasil, 15–22 July. Sociedade Barsiliera de Espeleologia, Brasilia, pp. 1–4.

Škuljević, P., Buzjak, S., Buzjak, N., 2013. Mikroklimatske i florističke značajke urušne ponikve i špilje Samograd kod Perušića (Lika, Hrvatska) (Microclimate and floristic characteristics of collapsed doline and Samograd cave (Lika region, Croatia). In: Alegro, A., Boršić, I. (Eds.), Knjiga sažetaka - 4. hrvatski botanički simpozij s međunarodnim sudjelovanjem, Split, Hrvatska, 27–29 September 2013, Hrvatsko botaničko društvo, Zagreb, pp. 111–112 (in Croatian).

CHAPTER

ICE CAVES IN GERMANY

17

Christiane Meyer, Andreas Pflitsch

Ruhr-University Bochum, Bochum, Germany

CHAPTER OUTLINE

17.1 INTRODUCTION

Unlike in other European countries, caves, and thereby ice caves, are not a subject of wide public interest in Germany. Only in recent time have karsts and caves become a more popular topic in nature conservation. Traditionally, caves were subject to morphological, geological and sedimentological studies in Germany, but little attention has been paid to the processes in ice caves so far. In this chapter we present an overview of the known ice cave sites in German karst areas. Starting with a short historical overview of the ice caves described in historical literature, we will present the basic facts about actually explored ice caves today and the systematic research done so far.

17.1.1 HISTORICAL NOTES ON ICE CAVES

Ice caves are mentioned for the first time in the Harz mountains near Questenberg by Bel in 1720. In 1820 an ice cave near Roth in the volcanic area Eifel is described (cp. Fugger, 1891, p. 59; Balch, 1900, p. 246). Today the Vulkaneifel is a UNESCO Global Geopark, where seasonal ice formations are still

Ice Caves. https://doi.org/10.1016/B978-0-12-811739-2.00017-6

observed in some artificial ice caves. In the course of the 19th century several other lesser known ice cave sites are mentioned, but especially the works of Lohmann (1895), Fugger (1888), and then Fugger (1891, 1892, 1893) presented ice caves to a wider audience.

Nevertheless ice caves were, in comparison to other classical ice cave countries, only known in a reduced number of places in the Central German Uplands and in the alpine karst areas until the end of the 19th century. While most of the sites described in the Central German Uplands can be characterized as caves with seasonal glaciation, the caves in the alpine karst area of Untersberg, for example, are perennial ice caves. On the German side of Untersberg, it is especially the Schellenberger ice cave which was studied, with long breaks of several decades, from the end of the 19th century until today. In recent times, ice caves were also discovered in other German alpine karst areas, which will be described in more detail in this chapter.

The first ice cave studies were conducted in the 19th century. As the development of the ice cave theory was still ongoing, the main focus at that time was to establish a theory about cave glaciation and give proof by measurements. Fugger's theory about static ice caves would prevail after he published his measurements at the ice caves of Untersberg in 1888, and his collection of ice cave and windhole sites worldwide, as well as the history of ice cave research and studies on windholes (1891–93). Single measurements were conducted in reduced numbers due to the technical means of that time, as each measurement required hikes to the specific ice caves. Nevertheless, his extensive observations about the annual course of the ice masses, temperatures, and other details gives us today's basic information about the extent and status of the ice caves at that time. Fugger (1888) also summarized some boundary conditions for the development of ice caves which are still valid in the theoretical background of static ice caves today.

17.2 GEOGRAPHY OF ICE CAVES IN GERMANY TODAY

Though ice caves have also been described in the Central German Uplands in the past, today ice caves are mainly known in the Berchtesgaden Alps, part of the Northern Limestone Alps, close to the German-Austrian border near Salzburg, with the one exception: the artificial seasonal ice caves in Vulkaneifel, as mentioned before. Already in 1797 Alexander von Humboldt tried to explore the firn ice cave Eiskapelle at the base of Mount Watzmann (example given Wolf, 2005). Since then, this region became one of the main alpine karst areas being explored by the German speleologists. Currently close to 100 caves with perennial ice deposits are described in the alpine karst of Untersberg, Reiteralm, Steinernes Meer, Hagengebirge, Hochkalter and Hoher Göll.

Beside the karst massifs in the Berchtesgaden Alps, there exists an unknown number of caves with ice deposits at the Wetterstein massif, especially at Germany's highest situated karst area, Zugspitzplatt (see Fig. 17.1). This area is explored by German cavers since 1935 with big gaps of several decades (see Wolf, 2004). Because of the altitude, where snow and ice is commonly found in many shafts, no dedicated study of the ice deposits has been performed. Furthermore, one ice cave is mentioned in the area of Ifen/Gottesackerplateau (see Rosendahl and Niggemann, 2000). Generally, one has to state that all alpine karst plateaus are in the border area to Austria, and thus all cadaster areas are explored by both German and Austrian speleological groups, and also foreign groups. In the following chapter, we focus on the Berchtesgaden Alps.

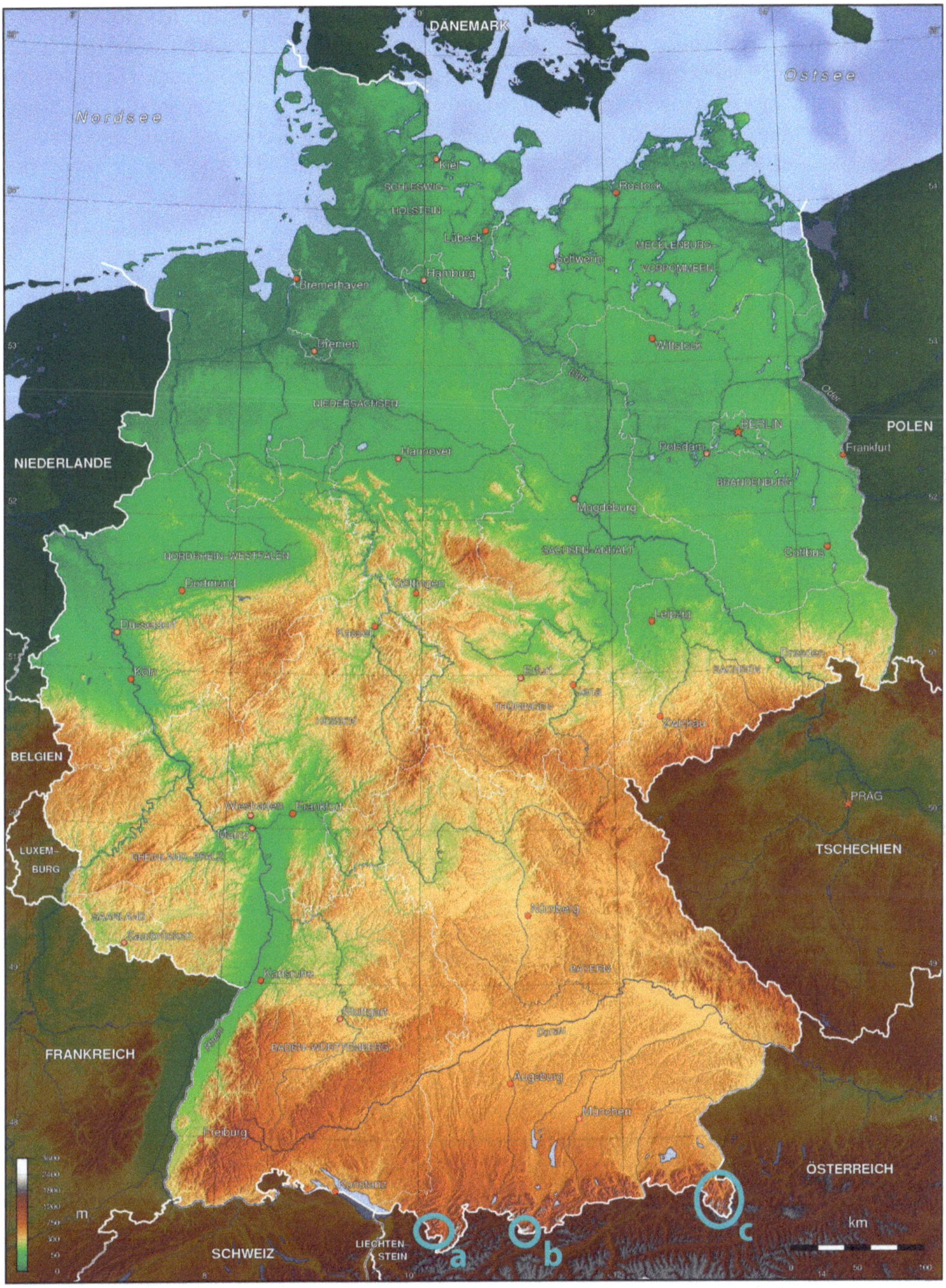

FIG. 17.1

Topographical map of Germany with location of the ice cave area (© http://www.mygeo.info, GNU Free Documentation License): Ifen/Gottesackerplateau (a), Wetterstein/Zugspitzplatt (b), Berchtesgaden Alps (c).

The sequence of rock formations in the Berchtesgaden Alps shows the development of the whole area from Upper Permian Age to today (Langenscheidt, 2005). They are part of the Northern Limestone Alps, which show over 1000 m thick carbonate sediments of Mesozoic age (Langenscheidt, 2005). High mountains with towering and rugged rock faces are determining the landscape (Konnert, 2004, Fig. 17.2). Many massifs like Steinernes Meer, Hagengebirge, Reiteralm, and Untersberg have in their central area karst plateaus, whereas Hochkalter and Hoher Göll show pronounced ridges (Langenscheidt, 1994).

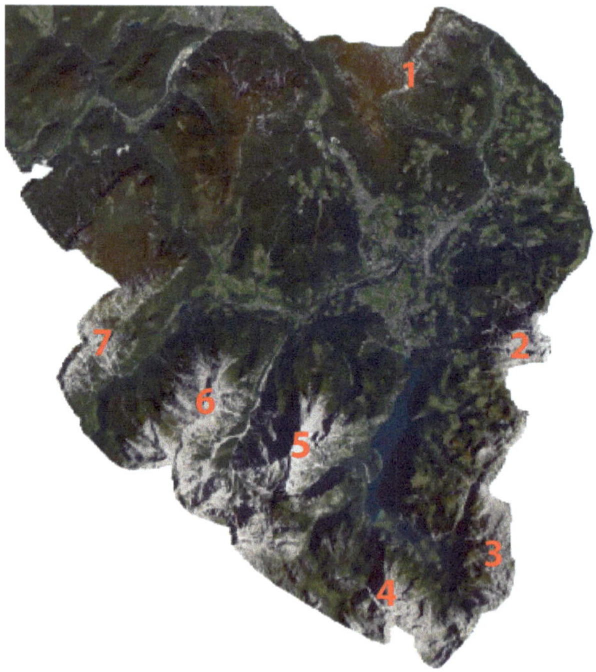

FIG. 17.2

Orthophoto of Berchtesgaden Alps with the massifs mentioned in the chapter: Untersberg (1), Hoher Göll (2), Hagengebirge (3), Steinernes Meer (4), Watzmann (5), Hochkalter (6) and Reiteralm (7).

17.2.1 CLIMATE

From the climatological point of view, the region of the Berchtesgaden Alps is located in a transition zone between maritime influences of the Atlantic and the continental climatic influences of the central Alps, which the vegetation, from ilex aquifolium to pinus cembra, also shows (Konnert, 2004). Due to the large difference in altitude of over 2000 m, the region is dominated by a typical climate of high mountains, which show highly varying climate conditions due to the strong influences of the relief. The geographical location at the Northern edge of the Alps results in high amounts of precipitation. With annual amounts of precipitation around 1200–2500 mm per year, the Northern Limestone Alps are the area with the highest amount of precipitation in Central Europe (Konnert, 2004). These high amounts of precipitation are a consequence of convective rainfalls during low pressure weather conditions, mostly during the summer, and advective orographic rainfall, which occurs mostly during winter due to Atlantic air masses from NW-W direction (Konnert, 2004).

For the alpine National Park of Berchtesgaden, which includes most of the ice cave sites, a mean annual temperature of 2.3°C at the 1800 m a.s.l., and a mean annual precipitation of 1753 at 1740 m a.s.l. is stated (Konnert, 2004). Moreover, around 40% of the annual precipitation is recorded in the summer months between June and August, while in January and February a minimum is recorded. Depending on the altitude, the first snow fall occurs in October or November, thus the number of days with closed snow cover varies between 30 and 170 days (Konnert, 2004).

17.3 ICE CAVE SITES IN GERMANY

In the following parts the basic facts about ice caves at single massifs are described. We do not claim completeness, but we tried to give the reader a good overview of the state of art, on the basis of the heterogeneous information we got. Beyond that, one can expect an unknown number of ice caves which are either not published or are explored by speleological groups we could not contact.

17.3.1 HOCHKALTER

In the cadaster area of Hochkalter, 20 ice caves are reported situated between 1765 and 2560 m a.s.l. The majority of these ice cave are vertical caves with narrow entrances with a length less than 50 m and a height difference less than 40 m. But among the vertical ice caves, the Eisl-Schlucht is a big exception. Eisl-Schlucht, which has partially perennial ice up to −120 m, and three other neighboring ice caves, are part of one new explored cave system at around 1910 m a.s.l. In summer 2010, when the cave was explored for the first time, even in a depth of −180 m, ice formations were found. In 2016 a strong ablation of the perennial ice at −120 m was observed, and the ice was nearly completely destroyed (Menne 2016, written communication). Moreover, one very short static ice cave, Firnkegelhöhle (Menne, 2008), one ponor cave with a glaciated part, Hinterbergkarschlinger (e.g., Menne, 2010), and one dynamic ice cave with a horizontal part, Eisfliegenschacht (Menne and Eckle, 2015) are reported. At Hochkalter the highest situated ice cave in Germany, Gipfelhöhle, was discovered at 2560 m a.s.l., close to the peak of Hochkalter. According to the speleological group, who is systematically exploring at Hochkalter, more ice caves can be expected close to the last remnants of the glacier at Hinterbergkar (Menne, 2016, written communication). Furthermore a strong ablation of the perennial ice bodies in the lowest situated ice caves was observed by them in the last years of exploration, while in those caves above 2000 m a.s.l. ablation is not obvious.

17.3.2 REITERALM

At the Western edge of the Berchtesgaden Alps the karstic plateau of Reiteralm is situated, which hosts perhaps the most impressive ice caves of Germany. Ten ice caves are registered here in the cadaster in altitudes between 1560 and 2144 m a.s.l. The deepest and longest among those are Yetischacht (Fig. 17.3), where a 50 m high ice mountain with an approximate volume of 53,000 m^3 is to be found at Gletscherhalle (Wisshak and Jantschke, 2005), and Eisrohrhöhle-Bammelschacht-System.

FIG. 17.3

"Gletscherhalle" in Yetischacht.

Photo by Andi Kücha.

The entrance of Yetischacht looks like many of the small entrance pits on the plateau, but is not blocked by stones in the bottom (Wisshak and Jantschke, 2005). Instead, it directly leads through a series of shafts to the ceiling of the 80 m × 50 m big "Gletscherhalle," which hosts the giant ice mountain mentioned above (see Fig. 17.3). Yetischacht can be characterized as a static ice cave (Wisshak and Jantschke, 2005). Similar features to Yetischacht can also be found in Schlundeisbergschacht, which was discovered only in 2009. A 45 m high massive ice mountain exists in the cave, which is 15 m × 4 m (Wisshak et al., 2013).

Another highly interesting and well explored ice cave at Reiteralm is the Eisrohrhöhle-Bammelschacht-System, which is situated at the foot of the NW-wall of Kleiner Weitschartenkopf. Among the caves in this part of Reiteralm, the horizontal cave Eisrohrhöhle (see Fig. 17.4) has the biggest ice volume, since it is iced during winter time as far as 300 m from the entrance (Wisshak et al., 2013). During summer the ice is periodically melting until a certain point, but in the rear part massive perennial ice is documented (Wisshak et al., 2013). In the parts Glitzerhalle and Eisdom, two interesting ice lakes exist, whose water level increases due to melting water in summer time, while during winter both lakes start freezing from the bottom to the top of the lakes (Wisshak et al., 2013). Against the general trend, in many ice caves here the ice level increases! The oldest known ice cave at Reiteralm is the Zeller Eishöhle, which was already discovered in 1909. Until 1974 the ice level of the entrance was too high, and only after that, for the first time, could speleologists enter new parts of the cave. Only in 1990 cavers could enter successfully the cave passages behind the ice, so that a detailed map could be prepared for the first time (Wisshak and Jantschke, 2005).

FIG. 17.4

Eisrohrhöhle-Bammelschacht-System.

Photo by Max Wisshak.

17.3.3 UNTERSBERG

The karst plateau of Untersberg was surely the best known massif in the Berchtesgaden Alps concerning ice caves in the past. Since the second half of the 19th century ice caves are documented, explored, and published at this mountain. One of the best known historical ice cave publications is Fugger's monograph about the ice caves at Untersberg from 1888. But there are also numerous other publications which also include information about the ice caves at Untersberg, such as Czoernig-Czernhausen (1926), Klappacher and Mais (1975) and Klappacher (1996). Here one can find older literature about the ice caves. In Klappacher and Mais (1975), 10 ice caves are described at Untersberg, of these two are on the German side of the massif, Schellenberger Eishöhle and Naturfreundehöhle. In Naturfreundehöhle, situated at 1800m a.s.l., in former times perennial ground ice in the entrance hall was reported as well as massive ice blocking in the final shaft of the cave. In spring 2002 only some little rest of these ice deposits were found (Kösling, 1988; Meyer, 2004). Another vertical ice cave on the plateau is Cannstatter Schacht at 1870m a.s.l., where at −32m in Gletscherhalle a massive ice floor exists, which is built by snow entering the cave through an open shaft (Kösling, 1988; Meyer, 2004). Another vertical ice cave is Eisbläser, in which one descends completely in the ice. The steep-walled entrance doline to the 380m deep cave is located in the center of the plateau at 1787m a.s.l. (Kösling, 1988). The ice part of the cave is limited to the entrance zone up to a depth of 80m. Currently, it is impossible to enter the deeper parts of the cave, as these parts are completely blocked by the ice (Meyer, 2016, personal communication). Also Rauchkopfschacht at 1675m a.s.l. showed in the past massive glaciation (Kösling, 1988; Meyer, 2004), but since the discovery of the cave in 1986, the ice decreased a lot. Also at Untersberg, like at the other massifs, smaller caves do have perennial ice, example given Nordstörungstopf at 1630m a.s.l. The main room of the cave has two openings, one 20m high chimney and one doline, through which the snow enters the cave. While in 2002, during the discovery, a tunnel was open to the debris below the ice, this was closed again in 2010.

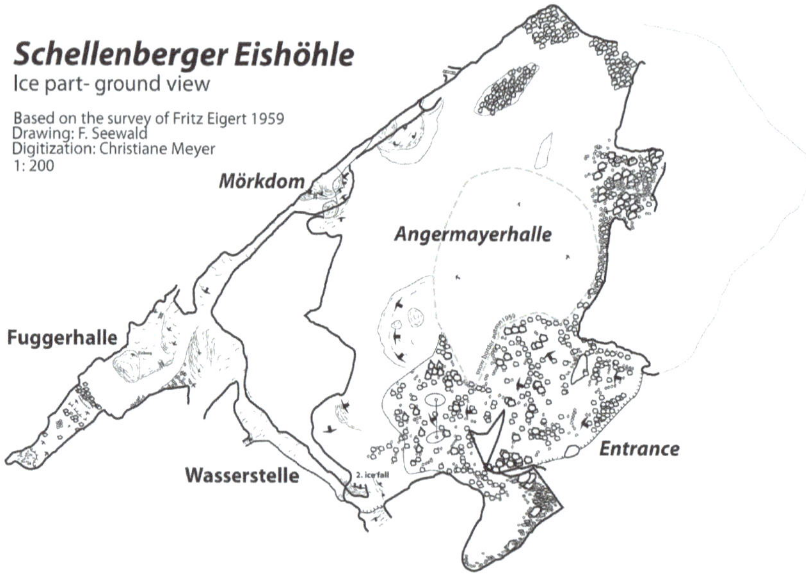

FIG. 17.5

Ground map of Schellenberger Eishöhle—ice cave part.

Digitized by Christiane Meyer.

The best known ice cave at Untersberg is Schellenberger Eishöhle (3621 m, −260 m), which has been run as a show cave since 1925. The Schellenberger ice cave has probably been known for centuries in the local population because of the big entrance portal and was first mentioned in 1826 in the Bavarian ordnance map as "Schellenberger Eisloch" (Vonderthann, 2005). It is situated at 1570 m a.s.l. on the foot of the NE-walls of Untersberg (Verein für Höhlenkunde Schellenberg e.V., 2001). The access to the cave is marked by a 4 m high and 20 wide portal, which leads to the largest room in the cave "Josef-Ritter-von-Angermayer-Halle" (see Fig. 17.5) with a length of 70 m and a width of 40 m. The floor 17 m below the entrance level of this hall completely consists of a massive ice body (Verein für Höhlenkunde Schellenberg e.V., 2001), which is surrounded by the show cave trail. The two connecting parts, Wasserstelle and Mörkdom, lead to the deepest point of the ice cave part, Fuggerhalle (see Fig. 17.6), 41 m below entrance level. From here only small fissures between the ice block and the rock give access to the deeper passages. Apart from the approximately 500 m ice cave part there is one major non-ice part, which leads from the Northeastern end through several deep shafts to the deepest point of the cave at −221 m. More literature about Schellenberger Eishöhle can be found in Klappacher (1996).

17.3.4 STEINERNES MEER AND HAGENGEBIRGE

Hagengebirge and Steinernes Meer are together with Watzmann the massifs which border the famous Königssee close to the main town of the region, Berchtesgaden. The Bavarian side of these big karst plateaus are part of the alpine National Park Berchtesgaden. Diverse speleological clubs are exploring this part of the Berchtesgaden Alps, and thus we try to give a basic overview of the most important

FIG. 17.6

Angermayerhalle at Schellenberger Eishöhle.

Photo by Lars Bohg.

ice caves. At Steinernes Meer, 24 ice caves are reported so far. The first 11 are situated in the area of Lawand, Schlosskopf, and Neuhütter, in an altitude of 1920–2095 m a.s.l. The main type of ice caves in this part of the plateau are ice caves with a horizontal part, but also vertical ice caves are listed. But the most interesting object is the cave named Sockenschleuder (example given Menne, 1999), which is situated at 1958 m a.s.l. The entrance part of the cave is a horizontal passage, which was closed with ice many times in the past. This part is followed by a little room with a massive 6 m thick ice body (Menne, 2016, personal communication), below this room also another hall with an ice body on the ground is documented (Menne, 2016, personal communication). In the area of Leiterkopf, another 13 ice caves are reported for the cadaster, located between 2037 and 2290 m a.s.l. Three ice caves out of this list show massive ground ice: Leiterkopfhöhle, Hocheggerhöhle, and Kleiner Eisboden (Peinelt, 2016, personal communication).

At the neighboring Hagengebirge, 22 ice caves are listed, which are situated between 1944 and 2210 m a.s.l. At this point we would like to especially mention Enttäuschungsloch (example given Menne, 2005b), Teeologenschacht (example given Menne, 1996), Rutschbahn (Menne, 2005a), and Regenhöhle (Menne and Eckle, 2015). In these caves massive ground ice is reported, but they are subject to strong melting.

17.3.5 HOHER GÖLL

The ice cave we have more detailed information about at the German part of Hoher Göll is Tabellenführer, which is a vertical cave with periodical active canyons and big break down domes. The cave is situated in the bare karst, since this area was covered by big snowfields all year until 2003. The most impressing part of this −295 m deep cave is the 90 m deep entrance shaft, on the bottom of which spelologist stop on top of a 12 m high ice tower (Matthalm, 2005). This massive ice formation is situated in the biggest room of the cave, called Kristallpalast (Fig. 17.7). From here a canyon with a massive ice floor continues deeper in the cave until Verzweigungshalle, which is mainly characterized by ice covered blocks (Matthalm, 2005). In all continuing cave passages, until a depth of −280 m, ice formations are observed. Until autumn 2016 a continuous increase of the ice was recorded.

FIG. 17.7

Kristallpalast at Tabellenführer.

Photo by Jens Beisswenger.

17.4 SYSTEMATIC ICE CAVE RESEARCH IN GERMANY

Though Germany has a longer research history in ice caves, systematic ice cave research is in the early stages. But the best studied ice cave in Germany, and the one with the longest research history, is, for sure, Schellenberger Eishöhle. After 1874 the cave was explored by several speleologists of the

"Landesverein für Höhlenkunde Salzburg" and later on studied by Eberhard Fugger. He started in 1869 with a study on the ice formations of the Kolowrathöhle at Untersberg (Fugger, 1888) and worked later also in the Schellenberger Eishöhle, where he carried out ice mass measurements from 1876 to 1882. The next decades, from the 1940s to the 1980s, were determined by the speleologist and long-term cave guide Fritz Eigert, who carried out ice-level measurements and temperature monitoring till the end of his activities in the cave (Ringeis et al., 2008). Since 2007 various long-term and mobile measurements (Grebe et al., 2008) have been conducted. Long-term measurements include air temperature measurements at various sites in the Schellenberger ice cave (compare Meyer et al., 2014) and ice temperature measurements in the entrance hall. Mobile measurements include, among others measurements with a thermal camera, mobile temperature measurements and a study on ice mass changes. In the summer of 2013 there were 32 new points for ice level measurements, as a prolongation of the study of Fritz Eigert, installed (Meyer et al., 2014). Since the energy supply at alpine ice caves sites is difficult, a new method of calculating airflow speed by using the given air temperature database recently started to develop (Meyer et al., 2016). This method is still under development for further applications.

Other German ice caves are also subject to scientific studies, mostly privately financed by the speleological groups themselves. For example, at Eisrohrhöhle-Bammelschacht-System, from 2006 to 2008, six autonomous data loggers were installed to analyze the dynamic airflow system of the cave in more detail (Wisshak et al., 2013). Diverse measurements also exist for the caves of Hagengebirge (Menne and Eckle, 2015), which are more related to the biological studies being carried out here, intensely concerning cave Lepidoptera (Menne and Meyer, 2005), but observations also exist concerning the ice mass changes in the ice caves of this area. The potential for future systematic research in ice caves is very good, but it is connected to bigger efforts concerning logistics and energy supply of electronic equipment, since most of the ice caves are in very remote places.

ACKNOWLEDGMENTS

We would like to thank the speleological groups exploring the caves of the Berchtesgaden Alps, who supported this chapter with information, maps, photos and personal explanations. This chapter could only be written because of your devoutness to the Berchtesgaden Alps during the last decades. We could only summarize your work. In particular we thank Benjamin Menne (Hagengebirge, Steinernes Meer, Hochkalter), Dirk Peinelt (Steinernes Meer), Max Wisshak (Reiteralm), Thomas Matthalm (Hoher Göll), and Ulrich Meyer (Untersberg) for your personal assistance in the preparation of this chapter.

REFERENCES

Balch, E.S., 1900. Glacières or Freezing Cavernes. Allen Lane and Scot, Philadelphia, PA.

Czoernig-Czernhausen, 1926. Die Höhlen des Landes Salzburg und seiner Grenzgebiete. Verein für Höhlenkunde Salzburg, Speläologische Monografien, 10.

Fugger, E., 1888. Beobachtungen in den Eishöhlen des Untersberges bei Salzburg, Salzburg.

Fugger, E., 1891. Eishöhlen und Windröhren. Separat-Abdruck XXIV. Jahresberichte der K.K. Ober-Realschule in Salzburg.

Fugger, E., 1892. Eishöhlen und Windröhren. Zweiter Theil, Separat-Abdruck XXV. Jahresberichte der K.K. Ober-Realschule in Salzburg.

Fugger, E., 1893. Eishöhlen und Windröhren. Dritter Theil (Schluss). In: Sechsundzwanzigster Jahres-Bericht der K.K. Ober-Realschule in Salzburg.
Grebe, C., Ringeis, J., Pflitsch, A., 2008. Study of temperature and airflow in the Schellenberger Ice Cave (Berchtesgadener, Limestone Alps, Germany). In: Proceedings of 3rd International Workshop on Ice Caves (IWIC-III), Kungur Ice Cave, Perm Region, Russia, 12–17.05.2008, pp. 26–29.
Klappacher, W., 1996. Salzburger Höhlenbuch Band 6 (Ergänzungsband zu den Bänden 1–5). Landesverein für Höhlenkunde in Salzburg.
Klappacher, W., Mais, K., (Gesamtredaktion) 1975. Salzburger Höhlenbuch Band 1. Landesverein für Höhlenkunde in Salzburg, 335 S., Salzburg.
Konnert, V., 2004. Standortkarte Nationalpark Berchtesgaden. Forschungsberichte 49. Herausgegeben von der Nationalparkverwaltung Berchtesgaden.
Kösling, F., 1988. Der Lehmpfuhl, Sonderheft 1. ARGE für Höhlenforschung Bad Cannstatt e.V.
Langenscheidt, E., 1994. Geologie der Berchtesgadener Berge: eine Einführung in Stratigraphie, Fazies und Tektonik. Berchtesgadener Anzeiger, Berchtesgaden.
Langenscheidt, E., 2005. Die Geologie der Berchtesgadener Alpen. In: Menne, B., Wisshak, M., Wolf, A. (Eds.), Berchtesgadener Alpen, Höhle und Karst 2004/2005. Verband der Deutscher Höhlen- und Karstforscher e.V., München.
Lohmann, H., 1895. Das Höhleneis unter besonderer Berücksichtigung einiger Eishöhlen des Erzgebirges, Diss., Univ. Leipzig.
Matthalm, T., 2005. Die Höhlen des Hohen Göll. In: Menne, B., Wisshak, M., Wolf, A. (Eds.), Berchtesgadener Alpen. Höhle und Karst 2004/2005. Verband Deutscher Höhlen- und Karstforscher e.V., München.
Menne, B., 1996. Das Wildpalfensystem. In: Klappacher, W. (Ed.), Salzburger Höhlenbuch Bd.6: Bereiche südlich und östlich des Königssees, Wildpalfensystem.
Menne, B., 1999. Forschungen der HFG Mühlacker im Hagengebirge und Steinernen Meer 1991–1998. Nationalpark Berchtesgaden, Deutschland. Forschungsbericht.
Menne, B., 2005a. Eisgrabenforschung der HFG Mühlacker 2001–2004, Nationalpark Berchtesgaden. Forschungsbericht.
Menne, B., 2005b. Das Wildpalfensystem (1335/195) im Hagengebirge. In: Menne, B., Wisshak, M., Wolf, A. (Eds.), Berchtesgadener Alpen. Höhle und Karst 2004/2005. Verband Deutscher Höhlen- und Karstforscher e.V., München.
Menne, B., 2008. Höhlenforschung am Hochkalter im Nationalpark Berchtesgaden 2002 bis 2007. Forschungsberichte der HFG Mühlacker, Bd. 1.
Menne, B., 2010. Höhlenforschung am Hochkalter. In: Tagungsband 2010, 50. Jahrestagung in Ennepetal, Verband der deutschen Höhlen- und Karstforscher.
Menne, B., Eckle, G., 2015. Berchtesgaden 2005 bis 2014, Forschungsberichte der HFG Mühlacker, Bd. 2.
Menne, B., Meyer, U., 2005. Hochkalter und Watzmann. In: Menne, B., Wisshak, M., Wolf, A. (Eds.), Berchtesgadener Alpen. Höhle und Karst 2004/2005. Verband Deutscher Höhlen- und Karstforscher e.V., München.
Meyer, U., 2004. Neuer Höhlen am Untersberg. In: In Münchner Höhlengeschichte II. Verein für Höhlenkunde in München e.V.
Meyer, C., Pflitsch, A., Holmgren, D., Maggi, V., 2014. Schellenberger Ice Cave (Germany): a conceptual model of temperature and airflow. In: Proceedings of the Sixth International Workshop on Ice Caves, 17–22 August, Idaho Falls, Idaho, USA.
Meyer, C., Meyer, U., Pflitsch, A., Maggi, V., 2016. Analyzing airflow in static ice caves by using the calcFLOW method. Cryosphere 10, 879–894. https://doi.org/10.5194/tc-10-879-2016.
Ringeis, J., Pflitsch, A., Grebe, C., 2008. Analiysis of ice level measurements in the Schelenberger ice cave in the Greman Limestone Alps. In: Proceedings of 3rd International Workshop on Ice Caves (IWIC-III), Kungur Ice Cave, Perm Region, Russia, 12–17.05.2008.
Rosendahl, W., Niggemann, S. (Eds.), 2000. Hochifen und Gottesacker. Höhle und Karst 2000/2001. Verband Deutscher Höhlen- und Karstforscher e.V., München.

Vonderthann, H., 2005. Die Schellenberger Eishöhle 1339/26 - Eine touristische Besonderheit des Berchtesgadener Landes. In: Menne, B., Wisshak, M., Wolf, A. (Eds.), Berchtesgadener Alpen, Höhle und Karst 2004/2005. Verband der Deutscher Höhlen- und Karstforscher e.V, München.

Wisshak, M., Jantschke, H., 2005. Die Höhlen der Reiteralm. In: Menne, B., Wisshak, M., Wolf, A. (Eds.), Berchtesgadener Alpen. Höhle und Karst 2004/2005. Verband Deutscher Höhlen- und Karstforscher e.V., München.

Wisshak, M., Straub, R., Jantschke, H., López Correa, M., 2013. Das Eisrohrhöhle-Bammelschacht-System(Reiteralm, Berchtesgadener Alpen). Materialhefte zur Karst- und Höhlenkunde 21, 1–110.

Wolf, A., 2004. Forschungen auf dem Zugspitzplatt. In: In Münchner Höhlengeschichte II. Verein für Höhlenkunde in München e.V.

Wolf, A., 2005. Die Eiskapelle am Königssee (1334/5) – 20 Jahre Höhlenforschung im Gletschereis des Nationalpark Berchtesgaden. In: Menne, B., Wisshak, M., Wolf, A. (Eds.), Berchtesgadener Alpen. Höhle und Karst 2004/2005. Verband Deutscher Höhlen- und Karstforscher e.V., München.

FURTHER READING

Bel, M., 1739. Philosophical Transactions, vol. XLI. The Royal Society, London.

Clemens, T., 1990. Untersuchungen an Höhlen des Roßkars. In: Materialhefte zur Karst- und Höhlenkunde (MKH), 9.

Jantschke, H., 1990. Die Geschichte der Höhlenforschung auf der Reiteralm. In: Materialhefte zur Karst- und Höhlenkunde (MKH), 9.

Klappacher, W., 1974. Die dritte Großhöhle in zwei Wochen: Schrecksatteleishöhle (1337/2) – 530 m. In: Vereinsmitteilungen des Landesvereins für Höhlenkunde in Salzburg Jg. 1974 (3), 10.

Matthalm, T., 2004. Der Tabellenführer – Eine neue Großhöhle am Hohen Göll. In: Münchner Höhlengeschichte Bd. 2.

Menne, B., 1985. Forschungsübersicht Südliches Hagengebirge/Lafelder (einschließlich: Materialsammlung zur Forschungsübersicht). In: Der Schlaz 46.

Menne, B., 1986a. Hagen 85. Beiträge zur Karst- und Höhlenkunde des Hagengebirges 3 (1).

Menne, B., 1986b. Forschungsübersicht Hagengebirge 1985: Südwestliche Randabstürze. In: Der Schlaz 48.

Menne, B., 1987. Hagen 1986. Beiträge zur Karst und Höhlenkunde des Hagengebirges 4 (1).

Menne, B., 1988a. Hagen 1987. Beiträge zur Karst- und Höhlenkunde des Hagengebirges 5 (1).

Menne, B., 1988b. Atlas des Wildpalfensystems. Beiträge zur Karst- und Höhlenkunde des Hagengebirges 5 (2).

Menne, B., 1988c. Eisgraben/Wildpalfen 1987: Eine Übersicht über die Forschungsergebnisse der HFG Mühlacker im Hagengebirge während des Jahres 1987. In: Der Schlaz 54.

Menne, B., 1989a. Hagen 1988. Beiträge zur Karst- und Höhlenkunde des Hagengebirges 6 (1).

Menne, B., 1989b. Wildpalfensystem - Die Forschungen des Jahres 1988. In: Der Schlaz 58.

Menne, B., 1989c. Wildpalfensystem/Hagengebirge. In: Mitteilungen des Verbandes der deutschen Höhlen- und Karstforscher 35 (1/2).

Menne, B., 1990. Neuhütter 1990. Höhlenkundliche Beiträge der HFG Mühlacker 1.

Menne, B., 1993. HAGEN 1990–92. Beiträge zur Karst- und Höhlenkunde des Hagengebirges 8 (1).

Menne, B., 1994. Forschung Neuhütter 1993. Der Schlaz 72: 19–24. München.

Menne, B., 1997a. Höhlen am Königssee (Berchtesgadener Alpen, Nationalpark). In: Proceedings of the 12th Int. Congress of Speleology 1997, vol. 4.

Menne, B., 1997b. Das Wildpalfensystem (Hagengebirge, Nationalpark Berchtesgaden). In: Der Schlaz 82 (Begleitheft z. 37. Jahrestagung des Verb. d. dt. Höhlen-u. Karstforscher).

Menne, B., 2001. Eisgrabenforschung der HFG Mühlacker 1999/2000. Forschungsbericht.

Menne, B., 2009. Aktuelle Forschungsaktivitäten im Katastergebiet 1333 Hochkalter. In: Atlantis. Höhlenforschung aktuell 2009 (1–2).

Menne, B., 2011. Eislschlucht 1333/46 wird tiefste Höhle am Hochkalter. In: Mitt. Verb. dt. Höhlen- u. Karstforscher, 57 (2).

Menne, B., Schmidt, F., 1984. Lafelder-Forschung der HFG Mühlacker. Beiträge zur Karst und Höhlenkunde des Hagengebirges 2 (1).

Menne, B., Wagner, W., 1983. Eisgrabenkundfahrt 1983 der HFG Mühlacker. Beiträge zur Karst und Höhlenkunde des Hagengebirges 1 (1).

Menne, B., Wagner, W., Schmidt, F., Wagner, H., Rapp, P., 1984. Eisgraben 1984. Forschungsbericht der HFG Mühlacker. Beiträge zur Karst und Höhlenkunde des Hagengebirges 2 (2).

Menne, B., Jacobi, K., Lammerer, P., Rapp, P., Schmidt, F., Wagner, W., 1990. Hagen 1989. Beiträge zur Karst und Höhlenkunde des Hagengebirges 7 (1).

Menne, B., Fischer, W., Brunner, W., 2005. Stand der Höhlenforschung im Hagengebirge und im Steinernen Meer. In: Menne, B., Wisshak, M., Wolf, A. (Eds.), Berchtesgadener Alpen. Höhle und Karst 2004/2005. Verband Deutscher Höhlen- und Karstforscher e.V., München.

Oedl, F., 1922. Die Höhlen der Ostalpen. Die Höhle in Sport, Wissenschaft und Kunst. Alpenfreund GmbH, München.

Schulzki, T., 1983. Die Forschungen der HFG Kirchheim auf der Reiteralm. In: Atlantis - Vereinsmitteilungen des Landesvereins für Höhlenkunde in Salzburg 3.

Wisshak, M., Jantschke, H., 2010. Im Höhlenruinenniveau der Reiteralm (Berchtesgadener Alpen)? Beibelkareishöhle (1337/42) und Prünzlkopfhöhle (1337/57). In: Die Höhle 2010.

Wisshak, M., Straub, R., López Correa, M., 2005. Das Eisrohrhöhle-Bammelschacht-System (1337/118) im Kleinen Weitschartenkopf (Reiteralm). In: Menne, B., Wisshak, M., Wolf, A. (Eds.), Berchtesgadener Alpen. Höhle und Karst 2004/2005. Verband Deutscher Höhlen- und Karstforscher e.V., München.

Zagler, W., 2004. Die Eisrohrhöhle am Kleinen Weitschartenkopf (1337/118). Münchener Höhlengeschichte II, VHM München e.V.

Zeller, M., 1910. Die Reiteralpe. Zeitschrift des Deutschen und Österreichischen Alpen-Vereins, 41.

CHAPTER

ICE CAVES IN GREECE

18

Christos Pennos*, Michael Styllas[†], Yorgos Sotiriadis[‡], Markos Vaxevanopoulos*,[§]

Aristotle University of Thessaloniki, Thessaloniki, Greece Geoservice LTD, Thessaloniki, Greece[†] Technological Institute of Eastern Macedonia and Thrace, Kavala, Greece[‡] Natural History Museum of Volos, Volos, Greece[§]*

CHAPTER OUTLINE

18.1 INTRODUCTION

Over 70% of Greece's territory consists of carbonate rocks (limestone, dolomite and marble). This almost uniform geological composition of the bedrock, in combination with the very active tectonic regime and the climatic setting of the broader region of the east Mediterranean, favors cave development (Pennos and Lauritzen, 2013). The exact number of caves is still unknown since there is no official archive. However, there is an estimate of more than 10,000 caves and rock shelters, and almost 10% of them occur only in a small part of the Lefka Ori mountains on Crete(Adamopoulos, 2013).

Despite the common belief that Greece lacks permanent ice deposits, a plethora of ice bearing caves have been reported from different parts of Greece (e.g., SPELEO, 2017). Most of these caves are located in high altitudes (>1500 m) and are vertically developed. This setting (i.e., high altitude, verticality) favors the seasonal accumulation of snow, since solar radiation does not reach the deeper parts of the vertical shafts. These perennial ice bodies are preserved inside the host caves when the cold winter air sinks in, displacing the warm air that is pushed out through the pit as described by Perşoiu and Onac (2012). These ice deposits depict a characteristic layering that reflects their depositional history and therefore are ideal for palaeoclimatic reconstructions (Perşoiu and Onac, 2012).

In this chapter, we introduce four high-altitude ice caves that experience different patterns of precipitation and temperature related to the major atmospheric circulation patterns and trajectories of the Eastern Mediterranean. We present the survey of each cave and characteristic photos from the ice deposits.

Ice Caves. https://doi.org/10.1016/B978-0-12-811739-2.00018-8

Furthermore, we host a literature review of the climate of Greece in relation to the large-scale atmospheric circulation in the area.

We believe that this short chapter about the *Greek Ice Caves* will introduce the existence of the ice bodies to the scientific community and will motivate researchers from different disciplines to work on them.

18.2 SETTING

18.2.1 GEOLOGICAL SETTING

The Greek peninsula shows a complex geological history, but a relatively homogenous lithological composition, as it consists mainly of carbonate rocks (marble, limestone and dolomite). In general, the northern and northeast part of the country consists of Paleozoic marbles and metamorphic rocks that are part of the old continental crust, and in the periphery of which the closing of the Tethys Sea took place during the Alpine orogenesis. In contrast, the rest of the Greek mainland, as well as most of the islands (apart from those that belong to the Hellenic arc and demonstrate considerable volcanic activity), are mainly built up of Mesozoic limestones and igneous rock formations. The carboniferous sediments represent shallow and deep marine deposits from the Tethys Sea. Overall, the general structure of Greece is characterized by a series of stacked nappes, with a composite thickness of ~5–10 km, consisting of the upper crust that was detached from the present subducted continental and oceanic lithosphere of the Adriatic-African Plate (Jolivet and Brun, 2010; van Hinsbergen et al., 2010; van Hinsbergen et al., 2005). These nappes (or "mega-units") were thrusted and stacked in an east-to-west direction since the Cretaceous (Faccenna et al., 2003; Jolivet and Brun, 2010; van Hinsbergen et al., 2005) and form a shortened representation of the paleogeographical distribution of continental ribbons and deep basins that existed in the western Neo-Tethys (e.g., Dercourt et al., 1986; Mountrakis, 2010; Stampfli and Hochard, 2009).

The Aegean Sea and its surrounding areas belong to the active continental boundary of the Alpine-Himalayan belt (Fig. 18.1) and are subjected to large-scale active deformation that stems from the subduction of the eastern Mediterranean lithosphere under the Aegean Sea, along the Hellenic Arc (Papazachos and Comninakis, 1971). Consequently, the majority of continental and coastal parts of Greece share common characteristics of back-arc extensional tectonics, expressed by the presence of a strong deformational pattern, volcanic activity, and the development of fault bounded grabens, lying in accordance with the dominant N-S extensional stress field. However, as illustrated in Fig. 18.1, the northern part of Greece is additionally influenced by a subsidiary right-lateral shear because of its co-existence with the North Aegean Trough, a dextral strike-slip structure (McKenzie, 1972).

18.2.2 THE CLIMATE OF GREECE

Located on the southern part of Balkan Peninsula, the general climatic characteristics of Greek climate correspond to "Mediterranean Climate," with wet mild winters and dry warm summers. However, this simplistic description holds truth only for a few low-lying coastal areas and for the majority of Aegean islands. Here a specific focus is given on the basic statistics of reconstructed precipitation and temperature, since they are the most important climatic variables for the formation of glacial and periglacial features, such as ice formation in caves.

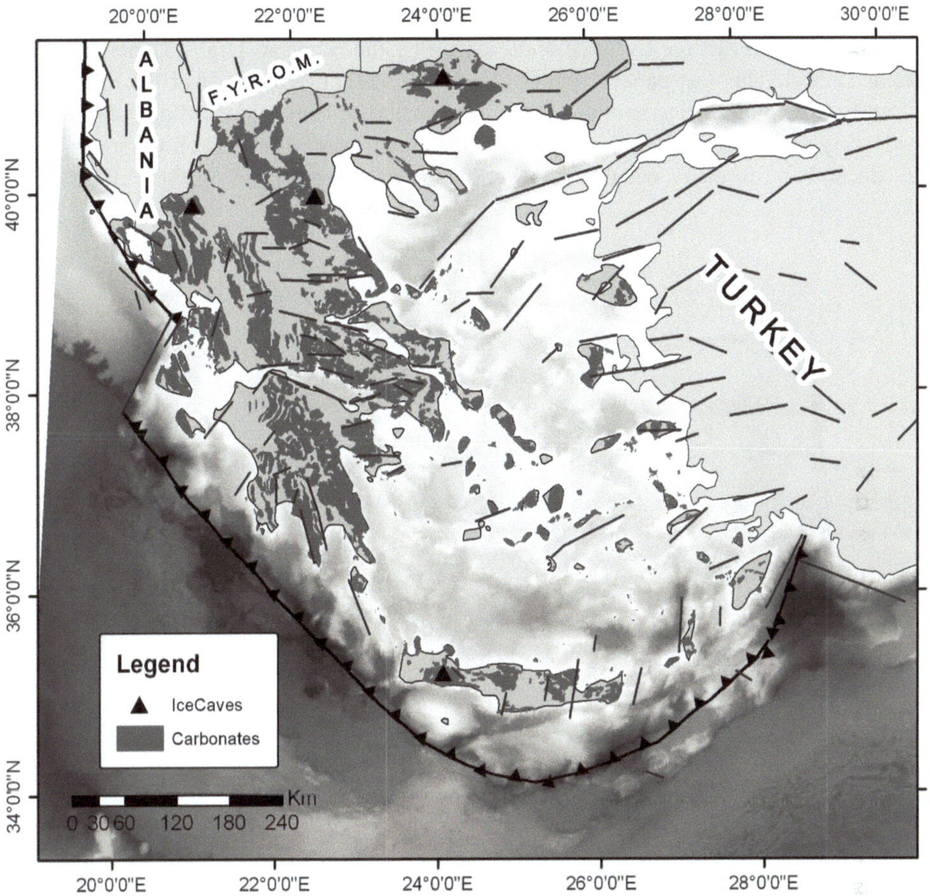

FIG. 18.1

Reference map of the broader Aegean region where the carbonate rocks, the selected Ice caves, the main tectonic lines, and the bathymetry are shown. Fault lines are extracted from Papazachos et al. (2001).

Most of the precipitation in Greece occurs during the winter and early spring (December–March), while summer precipitation occurs as convective, and often intense, rainfalls (Feidas et al., 2007; Xoplaki et al., 2000). The patterns of winter precipitation over Greece are influenced by its complicated coastline and mountainous relief. Despite the large spatio-temporal variability of winter precipitation, a significant fraction (30%) is explained by large-scale atmospheric circulation (Xoplaki et al., 2004).

18.2.2.1 The climate of Greece in relation to large-scale atmospheric circulation

Xoplaki et al. (2000) and Quadrelli et al. (2001) defined the main 500 hPa patterns connected to winter precipitation over Greece and concluded that the North Atlantic Oscillation (NAO) plays a major role in controlling precipitation over the Mediterranean region including Greece. Winters with mild temperatures and high amounts of precipitation in Greece are linked to negative phases of NAO (Bartzokas et al., 2003), whereas low winter temperatures, and a lack of precipitation over the north, northwestern parts of Greece, are associated with an enhanced frequency of northern and/or northeast continental dry and cold airflow over the area, driven by the Siberian High (Maheras et al., 1999; Xoplaki et al., 2000). This last pattern (pattern C in Fig. 18.2) originates over the Siberian Plateau and often causes precipitation and snowfall in the eastern part of Greece, including Falakro Mountain, with considerably higher amounts accumulating in the down-water direction on the south-east shores of the Aegean Sea and in many occasions on the island of Crete, including the Lefka Ori in Crete (Fig. 18.2).

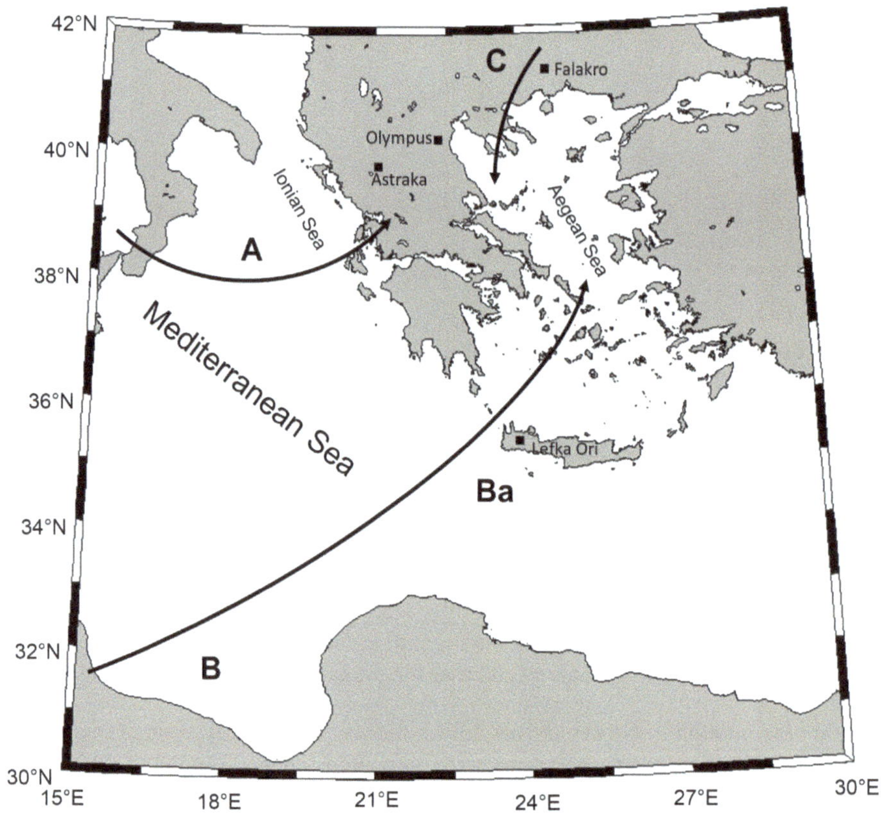

FIG. 18.2

Location of the selected mountain ranges/massifs with respective ice-caves presented in this chapter, in relation to the major patterns and trajectories of the dominant large-scale atmospheric circulation patterns that exert major controls on winter precipitation and temperature along north Aegean Sea. The movement of North Atlantic Oscillation (NAO) induced cyclonic depressions over the northwestern Mediterranean (A), cyclonic depressions generated along the lee side of the Atlas Mountains (B), and offshore western Crete (Ba), while the general direction of cold wind outbreaks of Arctic origin due to intensification of Siberian High (C) are related to low air, sea surface temperatures, and reduced precipitation in north and west Greece.

Winter NAO-driven depressions originate in the Atlantic, near the Straits of Gibraltar, and track northeastwards, barely crossing the Adriatic, with high rainfall limited to the western Greece, including the Tymfi Mountain, but with significant influence extending further east, reaching the west ramparts of Mount Olympus (Styllas et al., 2015) and, to a lesser extent, at the north-eastern part (Fig. 18.2). As these frontal depressions approach Greece from the west, they cause southwest winds (pattern A, Fig. 18.2) over the Ionian and Aegean Seas and force maritime air to move eastwards. Precipitation totals decrease south of this mean track, partly because of surface retardation due to land-air effects, and partly because the depressions decay rapidly by the time they reach eastern Greece (Flocas and Giles, 1991). In the above mechanisms of winter precipitation, the influence of the warm Aegean Sea should also be mentioned. The variability in winter cyclonic routes over Greece may be found in the complex topography of the region that yields significant monthly variations (Alpert et al., 1990).

The complex land-sea interactions, orographic precipitation variations, and frequent cold invasions of polar origin along the north coast of Mediterranean Basin, are the main causes of local cyclogenesis in specific regions, the most important of which are the Gulf of Genoa, the Atlas Mountains lee area, Southern Italy, the Central Mediterranean/southern Ionian Sea, the Cyprus region, and the Black Sea (e.g., Bartzokas et al., 2003; Feidas et al., 2007; Fotiadi et al., 1999; Prezerakos and Flocas, 1997; Trigo et al., 1999). Central Mediterranean cyclogenesis has considerable effects on the overall wetness of western Crete (Lefka Ori, pattern Ba in Fig. 18.2), but mild temperatures do not always ensue nivation and the formation and/or preservation of ice in Lefka Ori.

18.2.3 CAVES IN GREECE

More than 10,000 caves are known in Greece (SPELEO, 2017). The majority of them are vertical pits with depths ranging from a few meters up to 1200 m (Adamopoulos, 2005). Pennos and Lauritzen (2013) point out that this fact is related to the intense tectonic regime of the area that results in the fracturing of the bedrock in the accentuated topography. The high tectonic activity is responsible for the continuous lowering of the phreatic zone, creating a high vadose zone in which the vertical caves occurred (most vertically developed caves of the world are essentially vadose). Here, we present four characteristic shafts that are influenced from the different atmospheric circulation patterns and present thick ice deposits (Fig. 18.2).

18.2.4 SELECTED ICE CAVES

18.2.4.1 Chionotrypa cave, Falakro Mountain

The northernmost cave in Greece is *Chionotrypa* (Fig. 18.3) in Falakro Mountain. This ice cave is located 18.2 km northwest from Drama city in Eastern Macedonia, N. Greece, in the vicinity of the Falakro Ski Center, at an elevation of 2080 m. The cave presents a 111 m vertical expansion, and it is formed in the compact dolomitized marbles of the Rhodope Massif by vadose water circulation. The entrance of the cave is a spectacular collapsed doline 65 m in diameter.

The ice inside the cave is formed at the bottom of the collapsed doline (Fig. 18.3A) and penetrates further inside the cave. The vertical height of the ice sheet is almost 30 m (Fig. 18.3C). The formation of the ice is due to snow accumulation and compaction during winter months (Fig. 18.3B). Ice movement is suspected by Lazaridis and Makrostergios (2014) due to shearing in some speleothems at the lowest part of the cave.

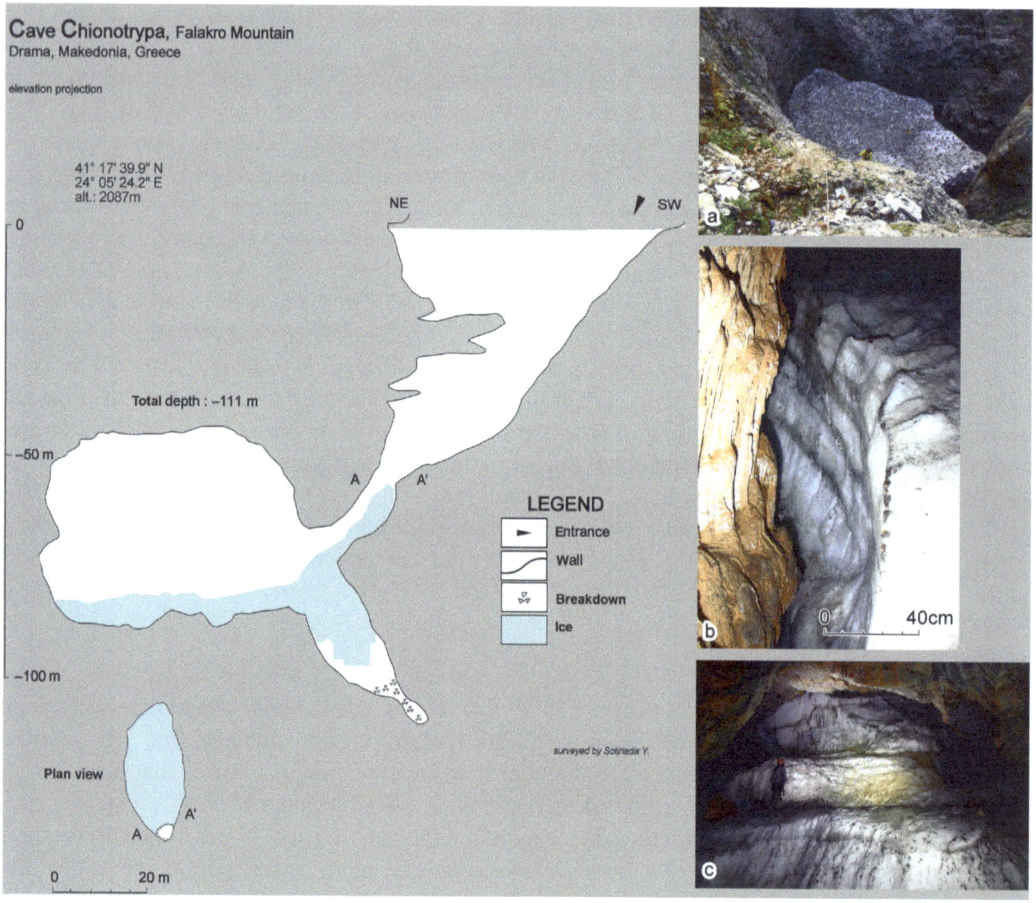

FIG. 18.3

Chionotrypa cave survey. Inlet (A) view of the ice from the surface, note that caver is 10m above the ice. Inlet (B) shows organic layers inside the ice deposit where in (C) the ice thickness is shown.

18.2.4.2 Chionotrypa cave, Mount Olympus

On Mount Olympus is found another "*Chionotrypa*" cave. The name "*Chionotrypa*" means "snow hole," and it is the name that locals have given to these ice-bearing pits all over the country. The cave has developed along tectonic discontinuities in Jurassic bedded limestones of the Gavrovo Zone in the Olympus tectonic window (Godfriaux, 1968). Its entrance lies in a slope with eastern orientation facing the Aegean Sea at 2560m elevation and has an elliptical shape of 6.6×8m, the total depth of the cave is 26m (Fig. 18.4A). The cave holds a large ice deposit that is generated due to the high amounts of snow that cover the area during the winter time, resulting in characteristic layering (Fig. 18.4B). The ice thickness fluctuates due to rising air temperature during the summer months. The minimum thickness is observed during the early autumn months (~20m), where during winter it expands up to 25m (Vaxevanopoulos, 2011). Smaller ice bearing pits in the adjacent area of Chionotrypa cave are also known (Vaxevanopoulos, 2011).

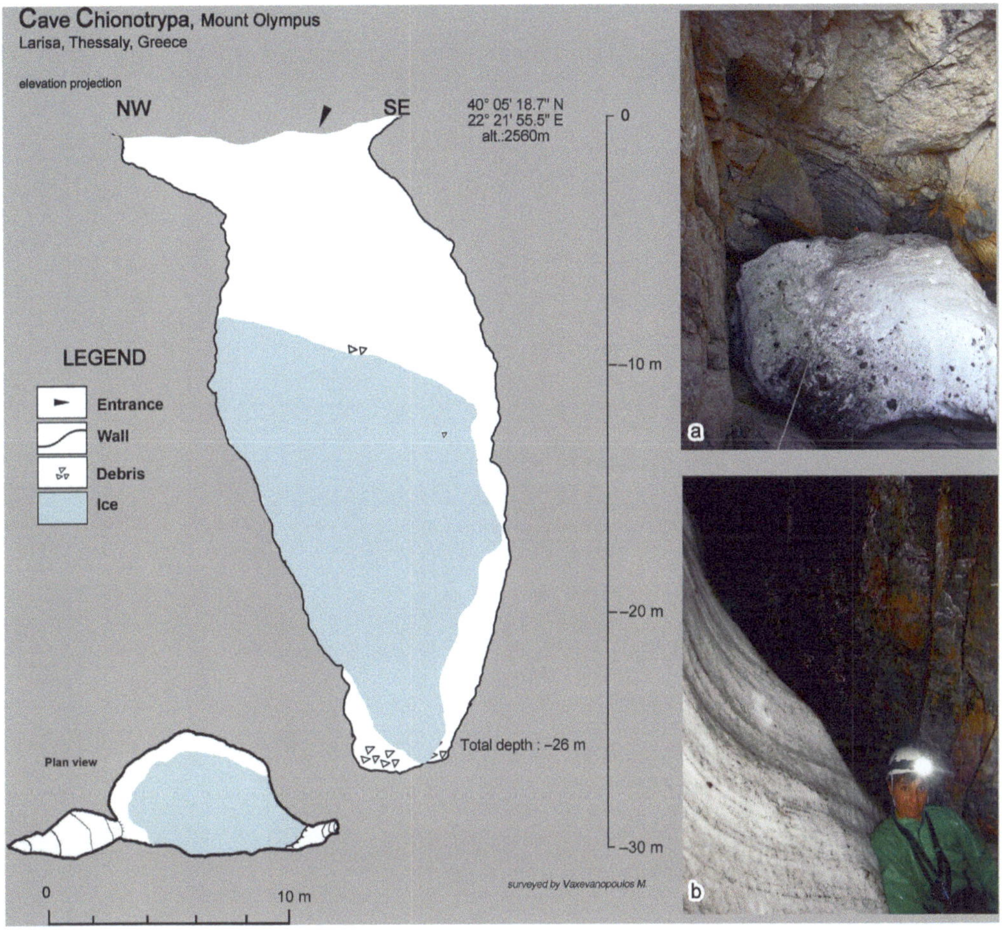

FIG. 18.4

Mount Olympus Chionotrypa cave survey. Inlet (A) view of the ice body as seen from above, and (B) organic layering at the bottom of the ice.

18.2.4.3 Provatina cave, Tymfi Mountain

Provatina cave, probably the most iconic Greek cave, constitutes a vertical pit with total depth of 407 m (Fig. 18.5). The cave is located 35.5 km north of Ioannina city at the Tymfi Mountain, W. Greece. It was first explored in 1968 by a British army expedition (Adamopoulos, 2005). The cave has developed on the edge of Tymfi Mountain's limestone bedding plane. It is formed along big vertical tectonic discontinuities of Jurassic to Upper Cretaceous limestone. The ice deposit is located at the middle part of the cave at a depth of 180 m (Fig. 18.5B) and has a mean thickness of 15 m. The ice is a mixture of snow, during winter months, and vadose waters that infiltrate the cave. The cave temperature, in combination with the position of the ice that makes it unreachable by the sun's rays, is the key factor for its preservation.

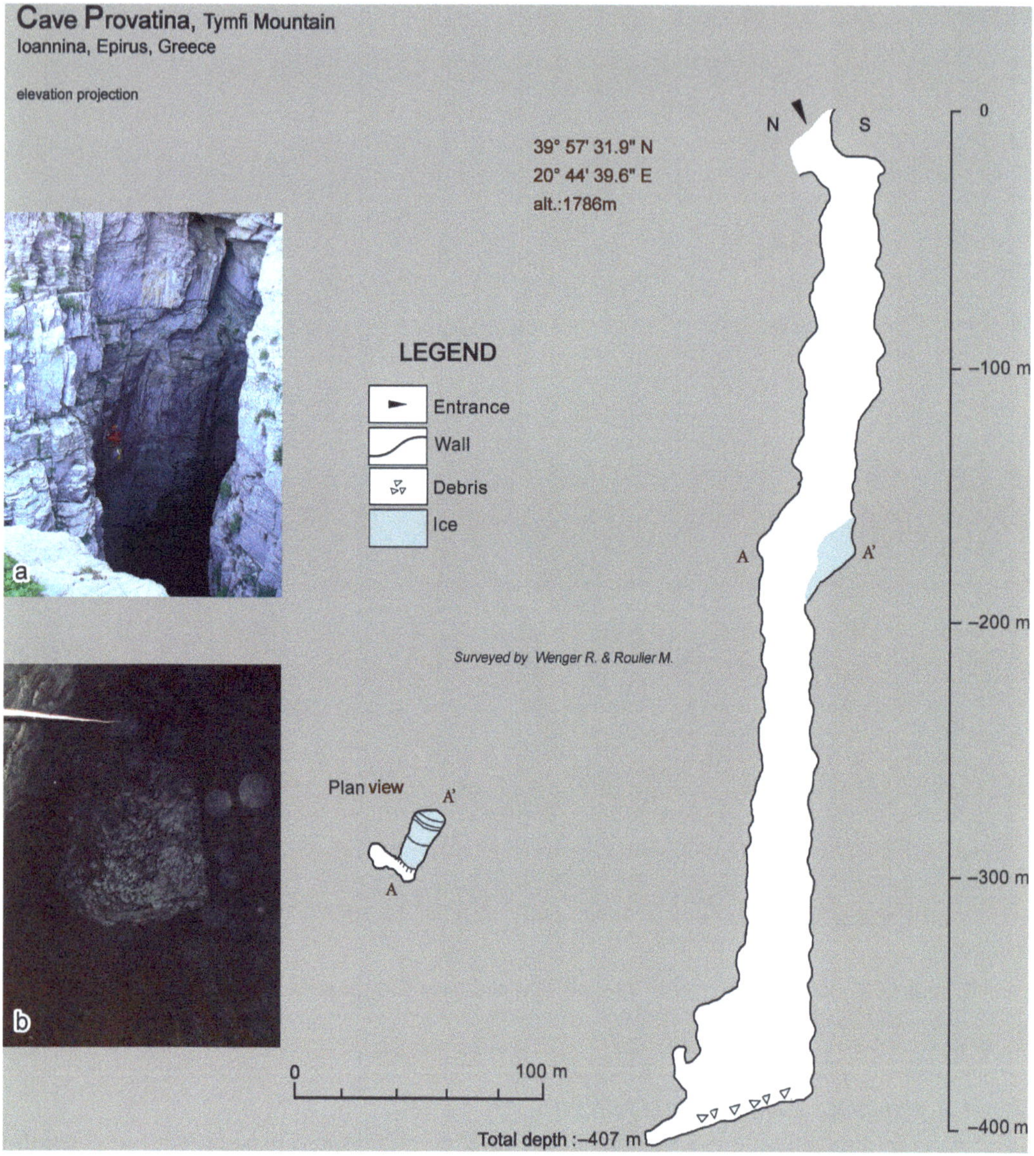

FIG. 18.5

Provatina cave survey (Graglia et al., 1982). Inlet (A) the entrance of the cave (see caver for scale) and (B) the ice deposit as seen from 80 m above (rope for scale).

18.2.4.4 Skud cave, Lefka Ori mountain range

The southernmost cave presented here is *Skud* cave. The cave is located at the Lefka Ori mountain range in the western part of Crete. The area presents the highest density of caves per square meter globally (Adamopoulos, 2005). According to Adamopoulos (*pers. comm.*) more than 100 caves and rock-shelters in the area host ice deposits throughout the year, potentially ranking them as the southernmost European Ice Caves. *Skud* (SPELEO, 2017) cave is a 46 m vertical pit that hosts a 15 m ice deposit (Fig. 18.6). The cave entrance is found at a 1915 m elevation facing toward the north. The ice is formed due to snow accumulation and compaction during the winter months, and it is preserved throughout the year because of the cave depth and entrance orientation which prevent the sun's rays from reaching the ice body (Fig. 18.6B).

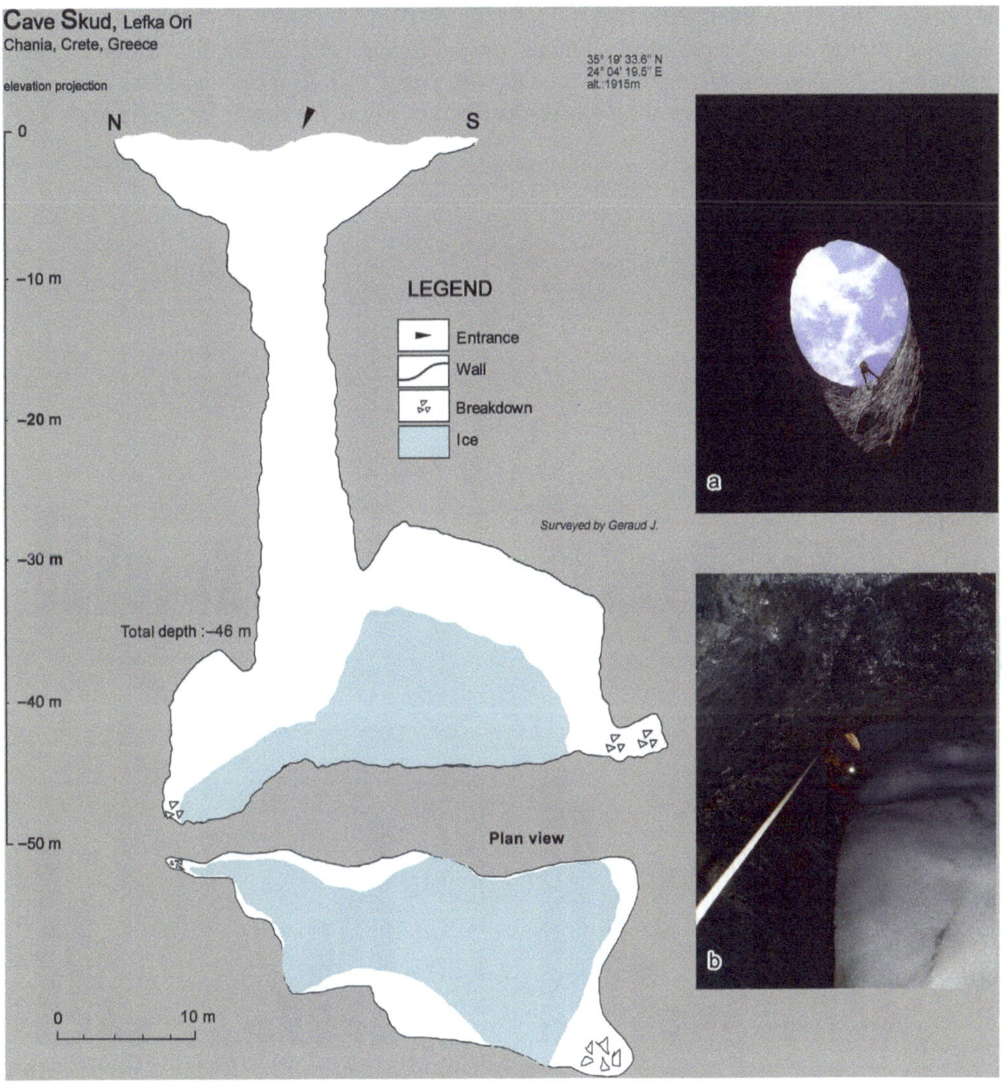

FIG. 18.6

Skud cave survey (Maire, 1982). Inlet (A) view of the entrance from the bottom of the cave, (B) the lower parts of the ice.

18.2.5 CLIMATIC CONDITIONS IN THE VICINITY OF FALAKRO, OLYMPUS, TYMFI, AND LEFKA ORI MOUNTAINS

To depict the general climatic characteristics of each separate area, data from the Climate Research Unit (CRU) gridded climatology (precipitation and temperature) are utilized. More specific, the CRU TS3.23 grid cells have been utilized (Harris et al., 2014), centered at 41.25°N, 24.25°E for Falakro Mountain, at 40.25°N, 22.25°E for Mount Olympus, at 39.75°N, 20.75°E for Tymfi Mountain, and at 35.25°N, 24.25°E for Lefka Ori, respectively. The reconstructed data spans 114 years of observations from 1901 to 2014 (Harris et al., 2014).

Average monthly variations of precipitation show a different pattern. The distribution of monthly precipitation in Falakro and Olympus is bimodal, with peaks in December and May, in contrast to Tymfi and Lefka Ori, which show a unimodal distribution where maximum precipitation occurs in December–January (Fig. 18.7).

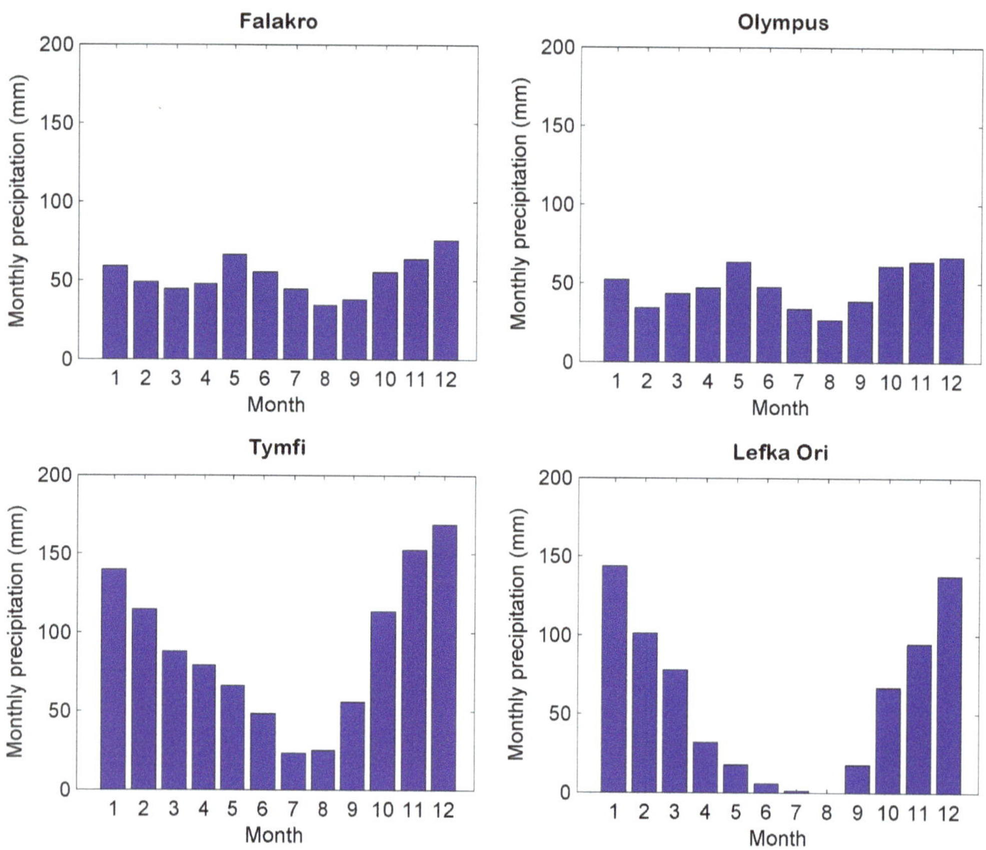

FIG. 18.7

Monthly distribution charts of precipitation for the four selected locations.

Summer precipitation in Falakro and Olympus is higher in comparison to Tymfi and Lefka Ori, the latter exhibiting precipitation values close to zero for July and August. Overall, Tymfi and Lefka Ori are characterized by wetter conditions, as average annual precipitation for the selected grid cells is approximately 1100 mm for Tymfi and 700 mm for Lefka Ori, while Falakro and Olympus are 630 mm and 580 mm, respectively. It has to be noted that the actual values in the locations of the ice caves are expected to be higher due to orographic effects.

In terms of temperature, Falakro and Olympus are characterized by colder winters. Tymfi, on the other hand, is slightly warmer, while Lefka Ori in Crete demonstrates mild temperatures (Fig. 18.8), with considerably warmer summers, so that the formation and preservation of ice in Lefka Ori requires exceptionally colder conditions in relation to the last century.

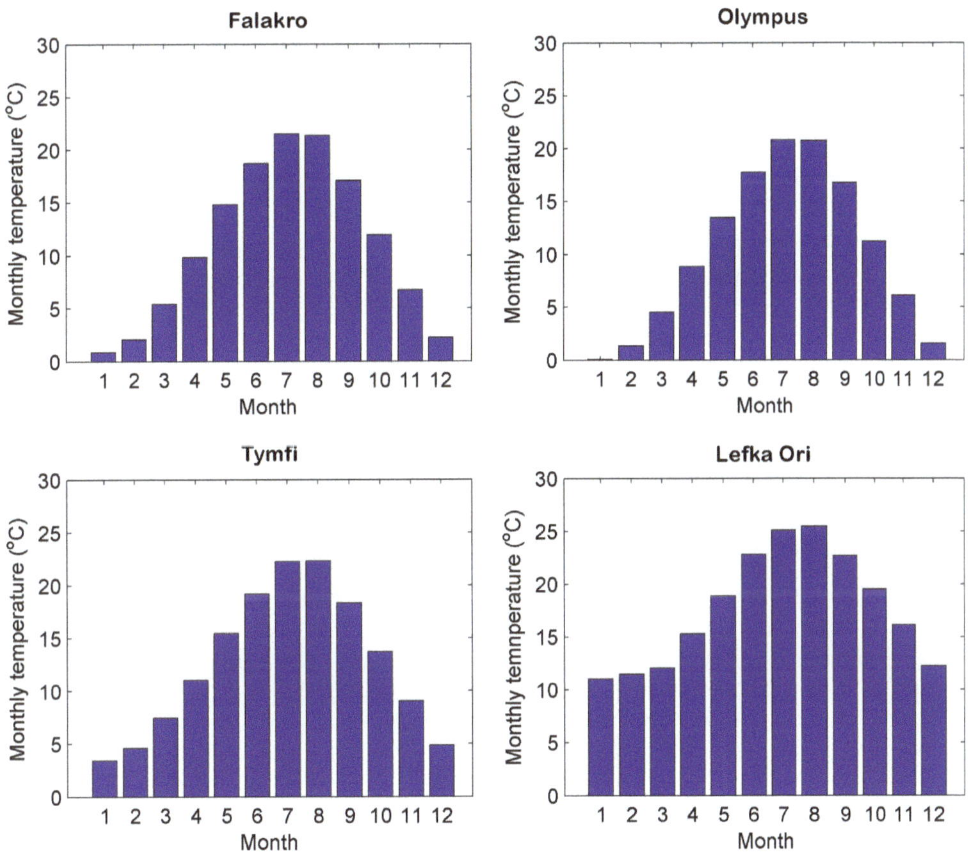

FIG. 18.8

Monthly distribution charts of temperature for the four selected locations.

The variable climatological conditions of the ice caves presented here are important in terms of their past evolution. Unraveling the information stored in these underground ice bodies may be very useful in reconstructing the contrasting patterns of previous large-scale atmospheric circulation patterns over the southern Balkans. We believe that the knowledge of the recent climatic conditions, and their relation to large-scale atmospheric variability along the selected key-sites where ice caves exist (Falakro Mountain, Mount Olympus, Tymfi Mountain and Lefka Ori), has important implications on the past (Holocene) evolution of these localities, as abrupt changes in atmospheric teleconnection patterns recorded from other proxies (i.e., pollen records, fluvial deposits and glacial variability) are expected to exert strong controls on the evolution of ice deposits as well.

ACKNOWLEDGEMENTS

Kostas Adamopoulos is highly acknowledged for the discussions he had on the topic with C.P. and Y.S., as well as for providing us unpublished data from his personal archive about the caves of Lefka Ori. Thierry Monges from the French caving club Catamaran is also thanked for providing us survey data and pictures from Skud cave. Dr. Sofia Pechlivanidou is highly acknowledged for the discussions we had on the topic. Last but not least, we would like to express our gratitude to Aris Zacharis for providing us pictures from the Chionotrypa cave at Falakro Mountain.

REFERENCES

Adamopoulos, K., 2005. The deepest and the longest caves in Greece. In: Paper Presented at the 14th International Congress of Speleology, Athens, Kalamos, Greece.

Adamopoulos, K., 2013. 1000 and 1 caves in "lefka Ori" massif, on Crete, Greece. In: Paper Presented at the 16th International Congress of Speleology, Brno, Czech Republic.

Alpert, P., Neeman, B.U., Shayel, Y., 1990. Intermonthly variability of cyclone tracks in the Mediterranean. J. Clim. 3 (12), 1474–1478. https://doi.org/10.1175/1520-0442(1990)003<1474:Ivocti>2.0.Co;2.

Bartzokas, A., Lolis, C.J., Metaxas, D.A., 2003. The 850 HPA relative vorticity centres of action for winter precipitation in the Greek area. Int. J. Climatol. 23 (7), 813–828. https://doi.org/10.1002/joc.909.

Dercourt, J., Zonenshain, L.P., Ricou, L.E., Kazmin, V.G., Le Pichon, X., Knipper, A.L., Biju-Duval, B., 1986. Geological evolution of the Tethys belt from the Atlantic to the pamirs since the LIAS. Tectonophysics 123 (1–4), 241–315. https://doi.org/10.1016/0040-1951(86)90199-x.

Faccenna, C., Jolivet, L., Piromallo, C., Morelli, A., 2003. Subduction and the depth of convection in the Mediterranean mantle. J. Geophys. Res. Solid Earth 108 (B2). https://doi.org/10.1029/2001jb001690.

Feidas, H., Noulopoulou, C., Makrogiannis, T., Bora-Senta, E., 2007. Trend analysis of precipitation time series in Greece and their relationship with circulation using surface and satellite data: 1955–2001. Theor. Appl. Climatol. 87 (1–4), 155–177. https://doi.org/10.1007/s00704-006-0200-5.

Flocas, A.A., Giles, B.D., 1991. Distribution and intensity of frontal rainfall over Greece. Int. J. Climatol. 11 (4), 429–442.

Fotiadi, A.K., Metaxas, D.A., Bartzokas, A., 1999. A statistical study of precipitation in northwest Greece. Int. J. Climatol. 19 (11), 1221–1232. https://doi.org/10.1002/(Sici)1097-0088(199909)19:11<1221::Aid-Joc436>3.0.Co;2-H.

Godfriaux, I., 1968. Etude géologique de la région de l'Olympe (Grèce). Faculte des Sciences, Universite de Lille.

Graglia, C., Recchioni, R., Ghiglia, M., 1982. ASTRAKA '81. Speleologia. Societa Speleologica Italiana.

Harris, I., Jones, P.D., Osborn, T.J., Lister, D.H., 2014. Updated high-resolution grids of monthly climatic observations - the CRU TS3.10 dataset. Int. J. Climatol. 34 (3), 623–642. https://doi.org/10.1002/joc.3711.

Jolivet, L., Brun, J.-P., 2010. Cenozoic geodynamic evolution of the Aegean. Int. J. Earth Sci. 99 (1), 109–138. https://doi.org/10.1007/s00531-008-0366-4.

Lazaridis, G., Makrostergios, L., 2014. Geology and morphology of the Chionotrypa cave (Falakro Mt, Macedonia, Greece). In: Paper Presented at the Balkan Speleological Conference "Sofia 2014" Sofia, Bulgaria.

Maheras, P., Xoplaki, E., Kutiel, H., 1999. Wet and dry monthly anomalies across the Mediterranean basin and their relationship with circulation, 1860–1990. Theor. Appl. Climatol. 64 (3–4), 189–199. https://doi.org/10.1007/s007040050122.

Maire, R., 1982. Expedition Speleo en Grece, 1981. Les Echoes Tenebres, Chalabre, France.

McKenzie, D., 1972. Active tectonics of the Mediterranean region. Geophys. J. Int. 30 (2), 109–185. https://doi.org/10.1111/j.1365-246X.1972.tb02351.x.

Mountrakis, D., 2010. Geology and Geotectonic Evolution of Greece. University Studio Press, Thessaloniki (in Greek).

Papazachos, B., Comninakis, P., 1971. Geophysical and tectonic features of the Aegean arc. J. Geophys. Res. 76 (35), 8517–8533.

Papazachos, B., Mountrakis, D., Papazachos, C., Tranos, M., Karakaisis, G., Savvaidis, A., 2001. The faults which have caused the known major earthquakes in Greece and surrounding region between the 5th century BC and today. In: Paper Presented at the Proceedings of 2nd National Conference Anti-Seismic Engineering and Technical Seismology.

Pennos, C., Lauritzen, S.-E., 2013. A conceptual model of speleogenesis in Greece. In: Paper Presented at the 16th International Congress of Speleology, Czech Republic, Brno, 21–28 July.

Perşoiu, A., Onac, B.P., 2012. Ice in caves A2. In: White, W.B., Culver, D.C. (Eds.), Encyclopedia of Caves, second ed. Academic Press, Amsterdam, pp. 399–404.

Prezerakos, N.G., Flocas, H.A., 1997. The role of a developing upper diffluent trough in surface cyclogenesis over central Mediterranean. Meteorol. Z. 6 (3), 108–119.

Quadrelli, R., Pavan, V., Molteni, F., 2001. Wintertime variability of Mediterranean precipitation and its links with large-scale circulation anomalies. Clim. Dyn. 17 (5–6), 457–466. https://doi.org/10.1007/s003820000121.

SPELEO, 2017. Archive of the Greek caves. Retrieved from http://speleo.gr/el/archive/.

Stampfli, G.M., Hochard, C., 2009. Plate tectonics of the Alpine realm. Geol. Soc. Lond. Spec. Publ. 327 (1), 89–111. https://doi.org/10.1144/sp327.6.

Styllas, M.N., Schimmelpfennig, I., Ghilardi, M., Benedetti, L., 2015. Geomorphologic and paleoclimatic evidence of Holocene glaciation on Mount Olympus, Greece. The Holocene 26 (5), 709–721. https://doi.org/10.1177/0959683615618259.

Trigo, I.F., Davies, T.D., Bigg, G.R., 1999. Objective climatology of cyclones in the Mediterranean region. J. Clim. 12 (6), 1685–1696. https://doi.org/10.1175/1520-0442(1999)012<1685:Ococit>2.0.Co;2.

van Hinsbergen, D.J.J., Zachariasse, W.J., Wortel, M.J.R., Meulenkamp, J.E., 2005. Underthrusting and exhumation: a comparison between the external hellenides and the "hot" Cycladic and "cold" South Aegean core complexes (Greece). Tectonics 24 (2). https://doi.org/10.1029/2004tc001692.

van Hinsbergen, D.J.J., Dekkers, M.J., Bozkurt, E., Koopman, M., 2010. Exhumation with a twist: paleomagnetic constraints on the evolution of the Menderes metamorphic core complex, western Turkey. Tectonics 29 (3). https://doi.org/10.1029/2009tc002596.

Vaxevanopoulos, M., 2011. Olympus Expedition. Internal Report of Proteas Caving Club, Thessaloniki (in Greek with English abstract).

Xoplaki, E., Luterbacher, J., Burkard, R., Patrikas, I., Maheras, P., 2000. Connection between the large-scale 500 hPa geopotential height fields and precipitation over Greece during wintertime. Clim. Res. 14 (2), 129–146. https://doi.org/10.3354/cr014129.

Xoplaki, E., Gonzalez-Rouco, J.F., Luterbacher, J., Wanner, H., 2004. Wet season Mediterranean precipitation variability: influence of large-scale dynamics and trends. Clim. Dyn. 23 (1), 63–78. https://doi.org/10.1007/s00382-004-0422-0.

CHAPTER

ICE CAVES IN ITALY

19

Valter Maggi*[,†], Renato R. Colucci[‡,§§], Federico Scoto[§,¶], Gaetano Giudice[,††], Luca Randazzo[††]**

University of Milano-Bicocca, Milano, Italy[*] Institute of Geosciences and Earth Resource (CNR), Pisa, Italy[†] ISMAR-CNR, Trieste, Italy[‡] Ca' Foscari University of Venice, Venezia Mestre, Italy[§] Fondazione CMCC Centro euro-Mediterraneo sui Cambiamenti Climatici, Bologna, Italy[¶] Istituto Nazionale di Geofisica e Vulcanologia sezione di Palermo – Geochimica (INGV), Palermo, Italy[**] Centro Speleologico Etneo (CSE), Catania, Italy[††] Commissione Grotte E. Boegan, Società Alpina delle Giulie, Trieste, Italy[§§]

CHAPTER OUTLINE

19.1 INTRODUCTION

Italy presents one of the largest variabilities of karstic features in the world. There are limestone outcroppings all over the country, from the Alps to Sicily, as well as in the Pantelleria Island, located in the center of the Mediterranean Sea. Karstic features are also present in the evaporates in the Northern Apennines, and in the marbles in the Apuane mountains. However, lava tube systems are also present on Etna Volcano in Sicily, and because it is active, their formation is still ongoing. Officially, 34,669 caves are included in the national speleological database (WISH, www.speleo.it), with development up to 50 km, such as the Corchia System (Apuane Mountains, Tuscany), and depth up to 1313 m, such as the Releccio Alfredo Bini system (Grigna Settentrionale, Lombardy) (Ferrario and Tognini, 2016).

Ice caves are distributed along the entire karstic area, mainly in the Central-Eastern Alps (Lombardy, Veneto, Trentino-Alto Adige-Südtirol, and Friuli Venezia Giulia regions) with probably more than 1600 existing cryo-caves, having within them permanent (multiyear) masses of ice, firn or permanent snow. In such areas, the ice deposits recorded on occasions of speleological surveys or research studies are

Ice Caves. https://doi.org/10.1016/B978-0-12-811739-2.00019-X

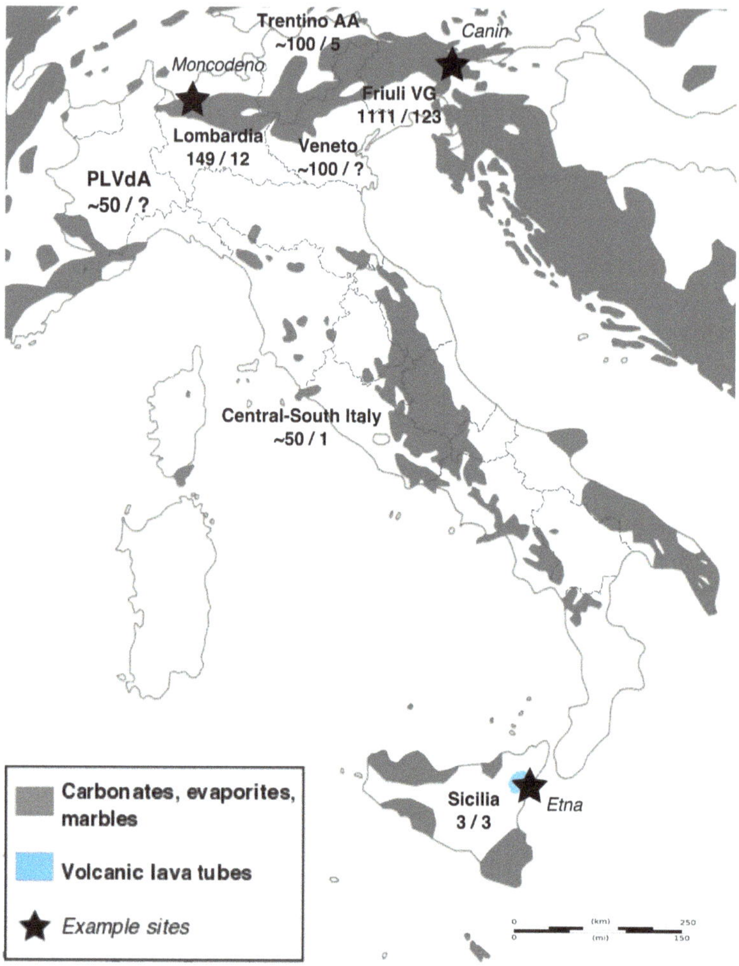

FIG. 19.1

Map of distribution of ice caves in Italy. The number below the region name represents the total of cryo-caves/total of ice caves; the star and the name in italic are the location of the example caves described in the text.

around 150. This represents 10% of all the caves with cryogenic deposits. In Fig. 19.1, the reported distribution of known cryo-caves and ice caves from different Italian regions is reported.

The consistency of ice caves both in the Alpine area, and in other Italian regions, is, however, uncertain due to: (i) the lack of widespread studies on ice caves (often the presence of an ice deposit in a cave under exploration has been seen as a negative issue by the speleologists, and for this reason they see it as something to dig and destroy), and (ii) recent climate change, which is severely reducing the ice masses almost everywhere and, in many cases, has already melted all the deposit.

In some of these ice caves, the presence of ice was recursively attested in the 1960s, 1970s, and the 1980s (i.e., LOSO1996 last survey 1956, or LOLC1897 last survey 1986 both in Lombardy), and in the last 20 years, there was a gradual ice body reduction. For this reason, in view of the fact that many caves have not been recently re-surveyed, we report here the latest information provided by the the Società Speleologica Italiana (Italian Speleological Society), and we are unable to guarantee whether or not such ice deposits are still present today (2017).

Nevertheless, the interest for ice caves, mainly in the northern Italy, dates back several centuries. One of the most discussed ice caves was the *Ghiacciaia del Moncodeno*, located in the *Grigna Settentrionale*, (not far from the ice cave LOLC1650 "Abisso sul margine dell'Alto Bregai", described in the Chapter 3), and in the past used as a private ice reserve by a wealthy family of Milano and surrounding descent.

By the end of the 1400, Leonardo da Vinci made some observations, took note of the events and places visited, in his personal diary, or as a report to Ludovico il Moro. During this period, Leonardo was nominated as a commissioner of Valsassina (North Italy) (Brevini, 2004). These were short notes that described the characteristics and spectacular nature of these places, which he wrote by himself during his journey along the foots of the mountains of Grigne. Most probably he had climbed the *Grigna Settentrionale* (North Italy), and after his visit to an ice cave he wrote the following:

> ...Queste gite son da fare nel mese di maggio. E i maggiori sassi scoperti che vi si trovano in detto paese son le montagne di Mandello, visine alle montagne di Lecche e di Gravidonia, inverso Bellinzona a 30 miglia di Lecco, e quelle di valle di Chiavenna; ma la maggiore è quella di Mandello, la quale ha nella sua basa una busa di verso il lago, la quale va sotto 200 scalini; e qui d'ogni tempo è diaccio e vento.
>
> ...These excursions are to be made in the month of May. And the largest bare rocks that are to be found in this part of the country are the mountains of Mandello near to those of Lecco, and of Gravidona towards Bellinzona, 30 miles from Lecco, and those of the valley of Chiavenna; but the greatest of all is that of Mandello, which has at its base an opening towards the lake, which goes down 200 steps, and there at all times is ice and wind.
>
> **(Translated by Richter, 1888)**

Today there is a dispute about naming the ice cave that was visited by Leonardo da Vinci at Grigna. Brevini (2004) wrote that Leonardo in person had been there to admire the unique speleological aspects of the Grigne, Fiumelatte, and the Ice Cave of Moncodeno (Ghiacciaia del Moncodeno). Through current times, other authors describe this ice cave, especially during the Little Ice Age, when, probably, the ice increase and maintainence of the deposit, created exploitable conditions for ice quarrying. Today, the Ghiacciaia del Moncodeno is practically free of ice, and only during the winters do some ice stalactites form seasonally (Bini and Pellegrini, 1998).

19.2 DISTRIBUTION OF ICE CAVES IN ITALY

As described in the previous paragraph, the difficulty encountered in realizing surveys of the cryocaves, or ice caves, does not permit a clear view of the Italian hypogean ice deposits, and, in this sense, only the Alpine area provides some distribution information (Fig. 19.1)

In the Piedmont-Ligurian area, representing the Italian western part of the carbonates outcrops, partially made by marbles, along terrain from Cuneo to Savona provinces (south-west of Piedmont region), crossing the Marguareis Altipiano and the Castelmagno area, there seems to be present around 50 caves with cryogenic deposits, but it is still not clear if some of them present permanent ice nowadays.

In effect, a structured database of caves does not exist, but distinct information can be found on the free newsletter of the Gruppo Speleologico Piemontese CAI-UGET at www.gsptorino.it, where it is reported that the main area with cryo-caves (mainly snow but it is not clear if permanent) is the Becai area, and on the Marguareis Altipiano, where at least 40 caves cryo-caves, mainly vertical shaft on collapse dolines or large fractures, have been surveyed (Gruppo Speleologico Piemontese, 2014).

In Lombardy, the recent cave database (Ferrario and Tognini, 2016) allows us to obtain all the data at the regional scale. In fact, 149 caves with ice, snow, and permanent snow are listed on the database. Most of them present only information on the database format, and from the survey draw, that ice and snow exist in the cave. These cryo-caves present a very large altitude distribution, between 1340 and 2840 m a.s.l., mainly moving from south to north, and that represents also the main change of mountain altitude in that Alpine sector. Only 12 ice caves present well known ice deposits, and two of them were studied in the last 10 years. In Table 19.1, there is a reported list of the Lombardy ice caves along with the main data. In that case, the altitude range reduced slightly from 1820 to 2780, with the main concentration from 2000 to 2150 m. In effect, 11 of the 12 ice caves are located in the Moncodeno plateau, in the northern flank of the Grigna Settentrionale (in the central Lombardy) (Table 19.1). As for the normal caves, Lombardy ice caves (but similar for all the Lombardy cryo-caves), are located in vertical shafts or collapse dolines, with a surface opening of vertical system (i.e., LOLC1650, described in the next chapter), and for this reason can be reached only by descending by rope. The ice cave distribution, both in location than altitude, is surely related to the different meteo-climatological condition along the Lombardy region, but studies do not exist that put in relationship the topographic asset of cryo-caves location and the local climatic conditions. However, because the vertical development of the cave systems, hypogean air circulation could affect cave negatives temperatures.

In the Trentino Alto Adige region, despite it being a karstic mountain area, there exists information only for the Dolomiti del Brenta, located in the western part, close to the Lombardy boundary. In that area, around 100 caves present cryogenic deposits, located between 1900 and 2750 m a.s.l., with more than 50% in the range between 2300 and 2500 m a.s.l., which are generally single vertical shaft with snow and firn on the bottom (Ischia and Borsato, 2005). Only five present permanent ice deposits, with thicknesses from a few meters to 22 m. The ice bodies are located generally in the inner part, from −10 m to −160 m depth. A continuous survey from the 80s shows a general trend of the reducing of ice mass, just like many other ice caves in Italy. In the Dolomiti del Brenta, two ice caves were exhaustively studied. Although related to different thermodynamic processes, such as wind-tube air circulation for Grotta dello Specchio (entrance elevation 1930 m a.s.l.), and cold air trap for Grotta del Castelletto di Mezzo (entrance elevation 2435 m a.s.l.), the two deposits experienced a similar evolution. In the Castelletto di Mezzo cave, the ice began to deposit during the first half of the '500, and continuously accumulated at 4.56 ± 0.61 cm year^{-1} rate until the end of '80s (maximum thickness 20.5 m, volume 2000 m^3) when, following the rising of the summer temperatures, net ablation prevailed. It is likely also that at the Specchio cave the ice began to form at the same time, but in this case the accumulation was discontinuous, with years/decades of prevailing accumulation alternated with periods when net ablation prevailed. The accumulation/ablation phases have been alternated regularly in the last 500 years, ruled by the summer temperatures, until the '80s (maximum thickness 4 m, volume 150 m^3) when the ablation rate accelerated abruptly. The lowering of the ice surface was 7.5 ± 4.1 cm year^{-1} during the period between 1997 and 2005, with the maximum value of −16.1 cm occurring during 2003 as a consequence of the extremely hot summer. At present, the evolution of the two ice deposits is controlled by the ablation, which is directly correlated with the summer

Table 19.1 The 12 Ice Caves Known in Lombardy and Related Data

Cave	Number	Coord. N	Coord. E	Name	Mountain System	Altitude	Aspect	Date
LOLC	1586	5089707	1529605	VORAGINE DI OLTRE M 40 PRESSO L'OMETTO DEL BREGAI	Moncodeno-Grigna Settentrionale	2090	N	2015
LOLC	1600	5090236	1529245	ABISSO DI VAL LAGHETTO	Moncodeno-Grigna Settentrionale	1823	N	2015
LOLC	1640	5090205	1529531	ABISSO NEL BREGAI MEDIO	Moncodeno-Grigna Settentrionale	1930	NW	2011
LOLC	1648	5089911	1529547	ABISSO DELLE SPADE	Moncodeno-Grigna Settentrionale	2047	N	2016
LOLC	1650	5089928	1529485	ABISSO SUL MARGINE DELL'ALTO BREGAI	Moncodeno-Grigna Settentrionale	2030	N	2004
LOLC	1739	5089382	1529907	GROTTA DELLE CONDOTTE FREATICHE	Moncodeno-Grigna Settentrionale	2125	N	2016
LOLC	1809	5089713	1530127	ABISSO DEI MARONS GLACES	Moncodeno-Grigna Settentrionale	2075	N	1985
LOLC	1842	5089453	1530223	GROTTA DEL PIFFERAIO	Moncodeno-Grigna Settentrionale	2140	N	2014
LOLC	1887	5089569	1530030	INFERMI NEL GHIACCIO	Moncodeno-Grigna Settentrionale	2065	N	2016
LOLC	1904	5089408	1530112	INFERNO DI GHIACCIO	Moncodeno-Grigna Settentrionale	2120	N	2015
LOLC	1916	5089419	1530202	HOLIDAY ON ICE	Moncodeno-Grigna Settentrionale	2144	N	2014
LOSO	3056	5152267	1610014	GROTTA DI CRISTALLO	Piano delle Platigliole-Monte Scorluzzo	2780	S	1998

The Cave-Number is the regional code from the Lombadian cave database, the coordinates are in kilometric Gauss-Boaga Italian system used in the regional cartography, the altitude is related to the main cave entrance, the date is related to the last cave survey.

temperatures, while the annual accumulation is minimal. Therefore, if the summer temperature will not diminish, the fast ablation process will lead to the complete extinction of the ice deposits within few decades (Borsato et al., 2006).

The Veneto region represents another karstic area of Italy, with mountain composed mainly by carbonates and dolomites rocks. Despite the karstic vocation, the information about ice caves are very scarce. Surely, cryogenic deposits are present in most of the caves, especially on the Dolomite areas and the Asiago Plateau, but a clear database of these caves does not exist. In a general survey on the Veneto speleological clubs, the information defines the existence of at least 50 caves with ice deposits. As or the rest of Italy, most of them were explored during the 60s and for this reason it is not clear if the deposits are still in the caves or not. The mountain area of the Veneto region is climatologically rainier than the Lombardy Alpine area, with precipitation form 1300–2300 mm water per year, that during the winter season is mainly snow, but less rainy/snowy in respect to the eastern Alpine area of Friuli Venezia Giulia (Colucci et al., 2016b). There, the temperature is more related to the altitude gradient and, because the west-east mountain chain orientation, there is a decrease of mean annual temperature (MAAT) northward (Costantini et al., 2013). Despite the lack in information, it is possible that the probability of cryo-cave formation is higher than the Lombardy region, and less (or equal) to the Friuli Venezia Giulia Region.

In the Friuli Venezia Giulia Region (FVG), ice caves are widespread in high elevated karstic environments. Recently, 1111 caves containing evidence of perennial cryogenic material (snow, firn and ice) have been identified by Colucci et al. (2016a), out of a total of 3068 known caves located above 1000 m a.s.l. in the area, and collectively indicated as cryo-caves.

Cryo-cave distribution was extracted by the regional inventory of FVG. Being aware that the inventory was based on reports given by speleologists over a large time interval, it's reasonable that it should be affected by a certain degree of uncertainty, despite being a useful starting tool. 123 Ice caves (11% of the total cryo-caves) were separately inventoried, taking into account those caves where firnification processes were not present, owing to the lack of direct connections with the outside, and where only the presence of ice was reported. Ice caves in FVG are generally located between 1500 and 2200 m. The highest ice cave percentage is reported in the Canin area, a high altitude karstic massif where widespread glaciokarst features (Žebre and Stepišnik, 2015) are present, and some of the lowest glacial relicts of the alpine chain still survive (Triglav Čekada et al., 2014; Colucci, 2016; Colucci and Žebre, 2016). Here, 79 (11%) of the 751 cryo-caves occur, with a mean altitude of 1928 m a.s.l. In general, cryo-cave distribution is well correlated with MAAT and altitude, being significantly more frequent where MAAT is lower than 2°C (ice caves) and 5°C (cryo-caves), therefore falling in the periglacial domain of FVG characterized by high precipitation (Colucci et al., 2016b). Caves having an entrance lower than 1000 m a.s.l. didn't report any permanent cryotic mass. The present (1981–2010) mean annual 0°C isotherm is roughly 200 m higher than the highest classified cave entrance, 100 m lower than the ELA of the LIA, and about 1000 m higher than the ELA of the LGM (Colucci et al, 2014).

Although mean summer air temperature (MSAT) does not seem to be significant in the existence of ice caves in FVG, it has been seen as the percentage of cryo-caves drops down with MSAT > 13°C. On the other hand, ice caves and cryo-caves are not present where mean winter air temperature (MWAT) is above 0°C, while their presence significantly increases with MWAT < −2°C.

Considering only the 123 ice caves, the majority of ice is present in the first 20 m of depth, while between 20 and 60 m the occurrence of ice is less frequent, and below 80 m it is only occasional.

The lithology does not drive the distribution pattern of the ice caves that are almost homogenously distributed within all the different lithologies occurring above 1000 m a.s.l. No ice has been found below about150 m from the surface.

Northern Apennines and central-southern Italy caves with cryogenic deposits are very rare. Normally, winter snow fills open cavities (as collapse dolines or large fractures), but it melts away during the spring-summer season. In Tuscany there is present only one cryo-cave, the Abisso Enrico Revel (Vergemoli, Lucca, N.102 T/LU, at 1453 m asl), an elliptical (55x9 m) and vertical system with permanent snow at the bottom (-300 m) (Sivelli and Vianelli, 1982).

On Mt. Etna, other caves with ice deposits inside were explored. Grotta del Gelo is the most renowned and is the subject of the last part of this chapter, but two others deserve a short description: the Grotta del Lago (literally "Cave of the Lake"), formed in the same lava flow of Grotta del Gelo, in which during summer the sheep use to go at watering, and the Abisso del Ghiaccio (Ice's Abyss), formed underneath the eruptive fissure of 1947, which is prevalently vertical near the entrance, reaching a depth of around 70 meters, with the presence of a great part of the ice deposit, and then intercepts the fissure bottom for a horizontal length of around 800 m. At present the bottom part of the cave is not reachable because of rock collapse and snow accumulation.

19.3 SOME EXAMPLES OF ICE CAVE STUDIES IN ITALY

Recent studies have shown a great variety of ice caves present in Italy. In this section, we describe three different ice cave system that differs in term of cave typology, hypogean climatic conditions, and ice formation (Fig. 19.1). The first study cases that we observed were vertical shafts. They have no direct connection with the surface where the snow (snowfalls or avalanches) can accumulate in the inner part. Thus in absence of direct accumulation, the only source that can generate an ice body is dripping water. Despite there being, in literature, no data about this initial stage, the ice formation process must be related to the possibility of freezing the dripping water due to cold air circulation. After the formation of the ice deposit, this will affect the microclimatic conditions inside the shaft that, generally, will be maintained and stable for a long time. An example of this typology of ice cave is the Abisso sul margine dell'alto Bregai, Mnt. Grigna Settentrionale, Lecco, where the ice deposit lies at 80 meters below the surface.

Less frequent study cases in Italy come from ice caves formed by direct snow accumulation. In fact, the probability to accumulate snow close to the cave entrances is not only related to the internal and external air temperature, but also to the amount of seasonal snow and the dynamic of snow accumulation. Examples of this are the Leupa and Vaso Caves, both in the Canin Mountain (Friuli Venezia Giulia), at the border with Slovenia, at 2200 m a.s.l. and 2300 a.s.l., with an entrance on the flank of the vertical walls. Here both snowfalls (and avalanches) and water drippings accumulate ice on the entrance halls in a complex connection.

In addition to carbonates and karstic areas, Italy also exhibits a large number of volcanic outcrops spread homogenously from the Alps to Sicily. Despite most of its volcanic activity being related to explosive eruptions, the recent history of Mt. Etna is characterized by a basic volcanism which led to fissural eruptions with lava flows and lava tube formation, as occurs nowadays in Hawaii or, back in geological time, in Yellowstone.

The last ice cave study reported in this chapter is the "Grotta del Gelo," frequently included among the most famous ice caves formed from lava tubes of the world. Although its location is not typical for an ice cave in terms both of latitude and geological setting, peculiar conditions of thermal equilibrium (for example, temperature fluctuation around zero, cold-air trap) started to develop supposedly after about twenty years since the end of the 1614–24 eruption, leading to the gradual process of subterranean freezing inside the cave. Today, ice occurs in several forms within the cave, namely seasonal lake ice, stalactites, stalagmites, and columns close to the entrance, as well as perennial ground ice in the deepest zone. The surface covered by the glacier reaches about 150 m^2, with an estimated ice volume of 350–400 m^3.

As the reader can easily understand, different typologies of ice caves and ice deposits have to be investigated by following different scientific methodologies and, therefore, also the data interpretation must be different from one case to another. Here we present the results from three different ice cave study examples based on three different scientific approaches.

19.3.1 ABISSO SUL MARGINE DELL'ALTO BREGAI, MNT. GRIGNA SETTENTRIONALE, LOMBARDY

The ice deposit LOLC 1650 "Abisso sul margine dell'alto Bregai", is located in Moncodeno (Grigna Settentrionale Regional Park, Lecco, Italy) at an elevation of 2030 m (Fig. 19.2) with an ice deposit at about −80 m below the surface. The deposit is not in direct connection with the entrance, which is at about −30 m, because there is a meander that obstructs the entry of snowfall, avalanches, and other glacial flows (Citterio et al., 2004).

An ice core obtained from the top of the ice deposit was been drilled in 2001 with an electrical powered SIPRE ice corer provided by Italian Antarctic Scientific Programme (PNRA), recovering the first 1.20 m of the ice deposit. The visual stratigraphy of the ice core, using thin sections with 10 cm resolution, observed using cross-polarised universal stage at Eurocold Lab of Earth and Environmental Sciences Dept. (University of Milano-Bicocca), show crystal structures characterized by large columnar ice with the same optical c-axis orientation. These crystals were normally found in lakes and sea ice, derived by slow freezing at the ice-water interface. The vertical alignment of bubbles along crystals boundaries indicate that the ice formation was slow enough to bound the over-sutured liquid results from these crystal grow processes. Small crystals, with more scatter c-axis orientation, were found at the top of columnar ice, related to a speedier freezing.

Using ionic chromatography in the Laboratory of Analytical Chemistry Department of the University of Florence, 5 cm resolution samples were measured, including the concentrations of major anions (NO^{3-}, SO_4^{2-}, Cl^-) and cations (Ca^{2+}, Mg^+, Na^+, K^+e NH_4^+).

The chemistry records, measured along the ice core, show that the freezing took place, from the surface (air-water interface) to the bottom, by progressive ice growing, and this process tends to concentrate the solute in the remaining water phase. Therefore, the bottom part of the ice deposit remains highly enriched in salt (Citterio et al., 2004). A general view of the stratigraphy of the whole deposit seems to show it to be composed by different ice layers, which implied the freezing of successive ponds one over another. These processes will be observed, especially in the SO_4^{2-}, no sea-salt sulfates, and NO^{3-}, where the maximum concentrations are observed at the bottom of the record.

Stable isotope ratios of oxygen were determined at the Stable Isotopes Laboratory of Geological, Environmental, and Marine Sciences Department (University of Trieste, Italy), and were calculated in reference to the international standard V-SMOW (Vienna-Standard Marine Oceanic Water). The record of $\delta^{18}O$ value in the ice core shows a decrease of heavy oxygen isotopes towards the bottom

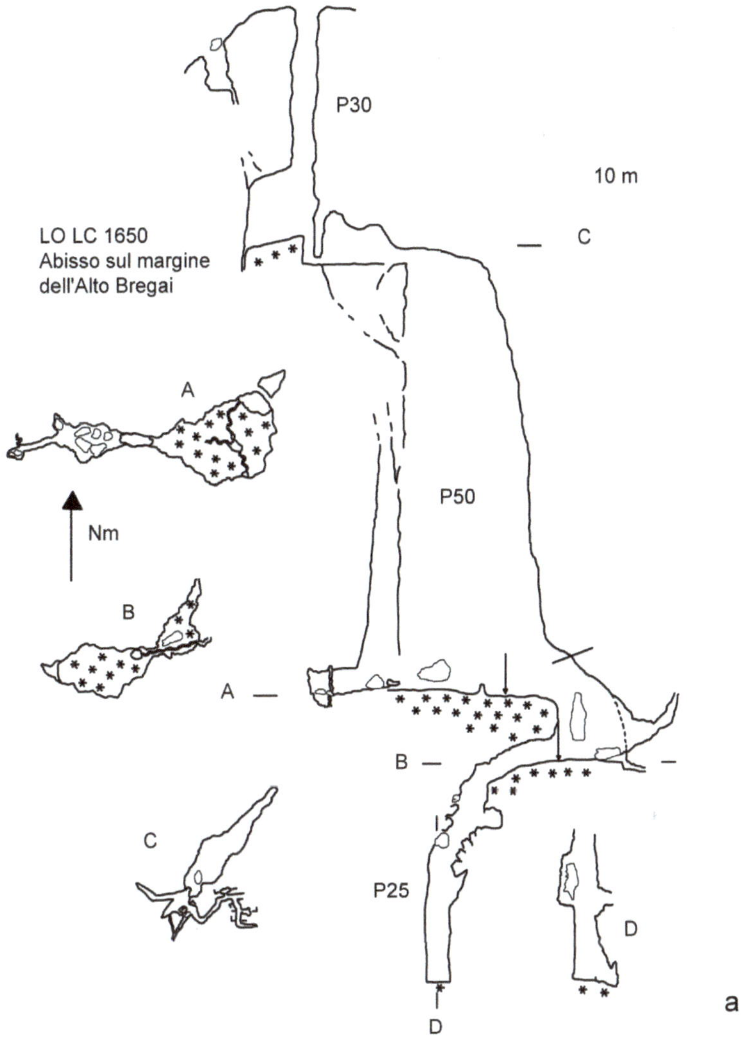

FIG. 19.2

Map of vertical profile of the LOLC1650 Abisso sul Margine dell'Alto Bregai, in the Moncondeno plateau, Grigna Settentrionale, Lombardy. The ice deposit is at −80 m depth at the bottom of P50 shaft, without any direct connection with the surface (Meander at the bottom of P30 shaft) Citterio et al (2004).

of the deposit (more negative $\delta^{18}O$ values). This decrease can be related to the oxygen isotope fractionation between water and ice forming gradually. In fact, the progressive water depletion of heavy isotopes (^{18}O) during the freezing processes produces ice more negative in terms of $\delta^{18}O$ (Fig. 19.3). Theoretically, we can assume that, after forming the first thin layer of the ice on surface, the water below the ice is insulated from the cave dripping, and the balance of stable isotopes, and the chemical species, does not change until the freezing is complete. Observing the records in Fig. 19.3, it is not possible to exclude contamination from external water. In fact, despite the general decrease of $\delta^{18}O$ from top to bottom, in the central part of the record some spikes and drops occurred, probably related to water infiltration from the pond. The same structure seems to be observed on the NH^{4+} and NO^{3-} records, but not in the sulphates records (Fig. 19.4).

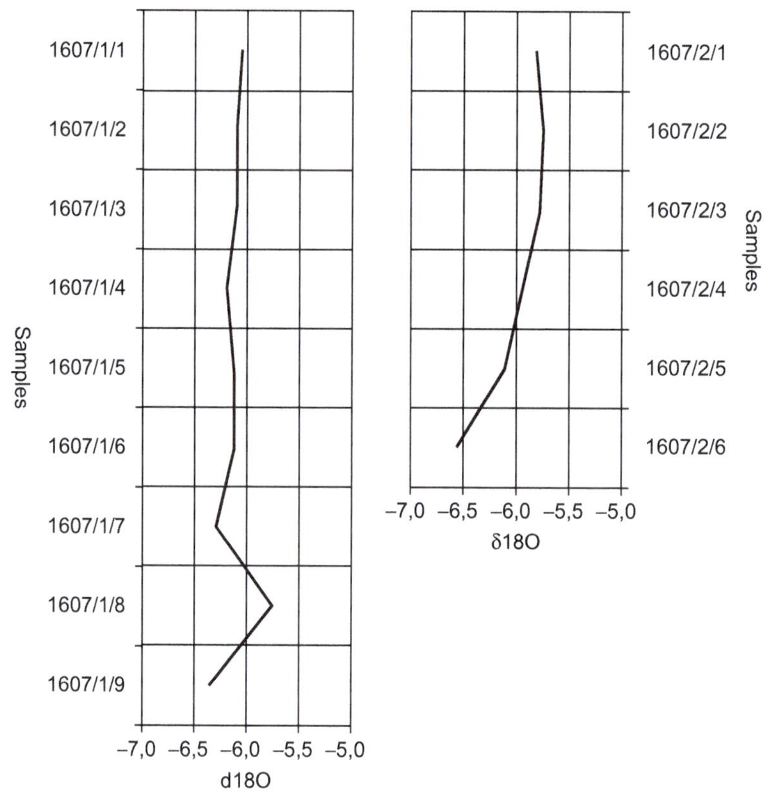

FIG. 19.3

Stable isotope measurements done in two ice layers at the LOLC1650 ice cave from ice core drilling sampling (Citterio et al., 2004).

Other measurements were done on the ice core, such as the crystal habits, preferential crystallographic orientation, and deposition of fluid inclusions, and these are in agreement with the literature descriptions of frozen lakes (Chambers et al., 1986; Tison and Haren, 1989). The crystal habit of the upper part of the ice deposit core is equi-dimensional, and the crystals are centimetric in size. Whereas

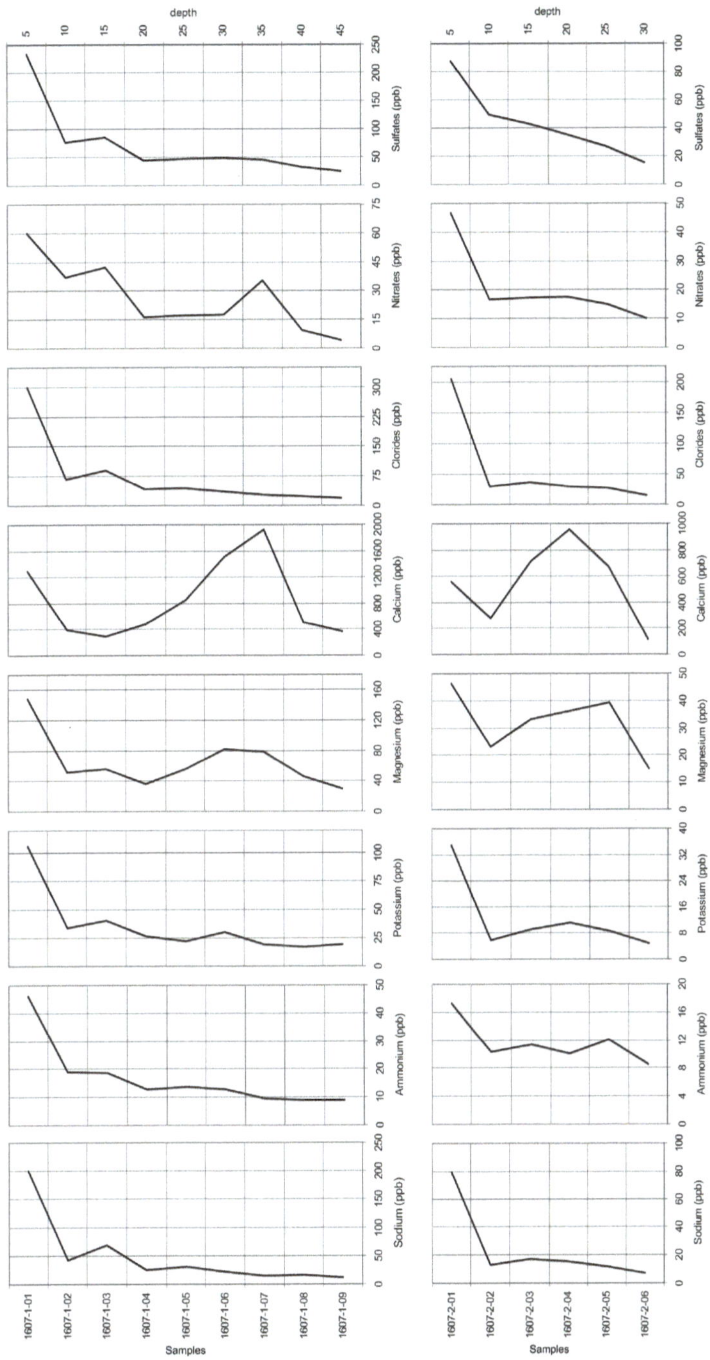

FIG. 19.4

Chemical records from the same layers measured for Fig. 19.3 at LOLC1650 ice cave. The records show the changes in chemical species concentration, which represent the relationship between ice in formation and water in ponds, was formed by water dripping and ice melting (Citterio et al., 2004).

in the deeper part crystals become columnar, vertically elongated, and vary in size from a few centimeters to many decimeters.

These differences can be related to the changing freezing regime which occurred in the pond. The centimetric crystals, with scatter c-axis orientations, in the top of ice core, can be related to the developing of ice crystals floating in water, and with the creation of the first layer sealing the pond. The columnar ice was generally formed by continuous and very slow freezing of the water molecules, using the same optical orientation of the crystal lattice.

Stefan (1891) was the first to suggest a physical model of freezing lakes; an analytical solution was developed by Turcotte and Schubert (2002) and named the "Stefan Problem". Here we present the possible application of the Stefan Problem to the ice cave deposit from ponds:

$$\frac{L\sqrt{\pi}}{c(T_m - T_0)} = \frac{e^{-\lambda_1^2}}{\lambda_1 erf \lambda_1} \tag{19.1}$$

where L = latent heat, c = specific heat, T_0 = surface temperature, T_m = interface temperature.

The geometric shape of the problem is shown on Fig. 19.3.

Water freezes at depth of solidification $y = y_m(t)$, assuming T_m as the water temperature with a uniform temperature under the ice-water interface. The depth of the ice-water interface increases with time (t), in direct proportion with the square root of time.

$$y_m = 2\lambda_1\sqrt{\kappa t} \tag{19.2}$$

κ is the thermal diffusion expressed by the equation:

$$\kappa = \frac{k}{\rho \cdot c} \tag{19.3}$$

where k is the thermal conductivity, ρ is the density, and c is the specific heat.

Applying the Stefan problem to the LOLC 1650 ice deposit, we have two unknown parameters: the superficial temperature T_0,and the freezing time t. Determination of T_0 is important for the hypogean glaciology, since it represents the original temperature that was necessary for the ice growth. Freezing time is also important, since it allows us to determine the freezing cycles with clear indications of the evolutional history of the deposit.

To preliminarily estimate these two values, the Stefan Problem was applied using values from the literature (Turcotte and Schubert, 2002). T_m (temperature of freezing interface) is considered to be constantly equal to 0°C.

The parameters are applied assuming two plausible time intervals. In the first case the modelling calculation was done between one month and 10 years, whereas in the second case the modelling calculation was done between 3 months and one year. Solving the equation shows that the complete freezing, in 1 month at 1.05 m, of the ice layer of LOLC 1650 needs a surface temperature $T_m = -38.7$°C. Whereas in 10 years the temperature would rise to $T_m = -0.3$°C. Both values appear to be improbable in that environment. It is impossible (i) to maintain very low temperatures (−38.7°C) in the caves at these latitudes, and (ii) to maintain temperatures below the freezing point for 10 years all throughout the year (Turri, 2005). Applying the boundary conditions of 3 months to 1 year modelling, using 6 months of freezing as representative of one single cold season, and considering that the water was provided (generally by dripping) during the summer season, the temperature needed to freeze hypogean water is $T_m = -5.7$°C. This result was supported by temperatures measured on the ice cave from

AWS (Automatic Weather Station installed along the LOLC 1650 cave), with few degrees colder than the current situation (Turri et al., 2005).

Based on these assumptions, the Stefan Problem could be a useful instrument of modeling the hypogean ice deposits formed by successively frozen lakes. In addition, when it is well set-up, it gives a key to decipher the story of a deposit and is a good tool to calculate the hypogean paleotemperature.

19.3.2 VASTO AND LEUPA, MNT. CANIN, FRIULI VENEZIA GIULIA

In two selected ice caves, the Leupa ice cave (LIC) and the Vasto ice cave (VIC), detailed geophysical surveys in order to describe the ice and snow within have been performed in the last years. Both caves are open on the north side of the Canin massif with entrances at about 2300 m and 2200 m a.s.l., respectively (Colucci et al., 2016a) (Fig. 19.5). The LIC is a dynamic ice cave owing to its air flow system, while the air circulation of the VIC is much more complicated and strongly influenced by the presence and amount of winter snow at the entrances. Therefore, the latter behaves alternatively as a Static or Dynamic cave. At the entrance of the VIC, steep rock walls border a narrow and deep karstic gorge opening to the north side, where the topographic characteristics favor the accumulation of wind-drifted snow. For this reason, a perennial firn/ice cone is here, partially filled also by small avalanches (accumulation season) and debris (ablation season).

FIG. 19.5

Part of the main ice deposit in the Leupa ice cave (left) and the large entrance of the Vasto ice-cave (right).

The LIC opens on the north side of Mt. Leupa, and its entrance is characterized by a 13 m wide and 4 m high opening (Fig. 19.6).

In the LIC, refreezing of seepage water (congelation ice) controls the accretion of the ice deposit. In the VIC, the metamorphism of snow (firnification processes) is likely to represent the main cause of ice accumulation. The firnification is also favored by the presence of liquid water percolating from the snow surface, allowing intensive regelation processes.

In both ice caves, besides continuous long-term temperature recordings, several ground penetrating radar (GPR) surveys have been performed. The nine GPR profiles acquired within the LIC revealed several structures in the ice body, which are interpreted as: (i) Single centimetric to decimetric clasts entrapped within the ice, found also on the actual ice surface; (ii) an air-filled cavity within the ice mass, reached during the summer period, and verified by visual inspection; (iii) the basal horizon interpreted as the ice bottom (the contact between ice and rocks); and (iv) a debris zone going from the cave entrance up to about 8 m of lateral distance. The maximum ice thickness was measured at about 4.5 m.

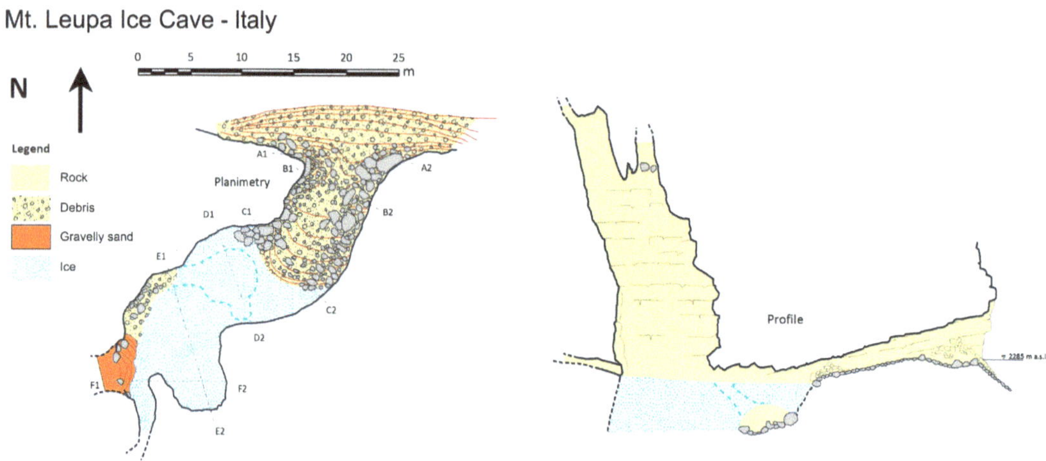

FIG. 19.6

New Survey of Mt.Leupa Ice Cave done in 2013.

At the VIC, 11 GPR profiles revealed a maximum ice thickness of about 8.5 m, with a high debris concentration within the first 2 m of the ice deposit, as well as close to the bottom of the ice deposit. In the deepest part of the VIC, weak reflection coefficients have been interpreted as possible evidence of residual permafrost that could therefore be expected in other high-elevated ice caves which exhibit a similar behavior.

The two fixed benchmarks located at the internal rock wall of the LIC represent the long-term monitoring of the elevation of the congelation ice floor since 2011, in order to collect information about surface ablation. This system was preferred to the common insertion of stakes at the ice surface to avoid a possible bias caused by eventual basal ice melting. Benchmarks are measured normally at least twice a year, generally in late autumn/early winter and late spring/early summer. The most internal benchmark is showing a rather constant annual cycle characterized by autumn minima and summer maxima, with variation in the order of 2–3 cm. A general slight increase trend was observed until 2014, when a dramatic ice surface decrease of 9 cm was recorded between Jul. 22 and Nov. 20. In summer 2015 a further ice surface decrease of 1.4 cm was measured, despite the usual positive mass balance observed at that date.

The most external benchmark showed a similar pattern compared to the others, but a general decrease trend had already been observed since 2013. Between Jul. 22 and Nov. 20, 2014 the same decrease was measured (9 cm), but 4.2 cm of lowering were also detected between Jul. 12 and Oct. 30, 2013. Such abrupt ice surface decreases have been shown to be correlated with a series of extreme precipitation events, which are able to also create about a 0.8-m-deep and around a 2-m-wide melting hole over the ice floor with an outlet bedières around 0.4 m deep and 0.2 m wide. The melting hole dramatically deepened by about 1.5 m in Nov. 2014, and the bedières were enlarged by about 15 cm on average and deepened about 20 cm while the ice floor was covered by a large amount of centimetric to decametric clasts.

A close influence of global and local climate change in the evolution of the ice deposits has been highlighted in the DIC, especially with regard to extreme weather events. The natural feedback of the ice cave's mass balance in a warming climate, and the forecasted increase of extreme weather events in the following decades, especially in regards to warmer and more intense rainfalls caused by higher 0°C isotherm, will be crucial in the future mass balance evolution of permanent ice cave deposits in this alpine area (Colucci, 2016). In the fall 2013 a 7.8 m long ice core was extracted from the VIC. The ice core has been cut and analyzed in terms of: (a) oxygen and hydrogen isotope composition; (b); black carbon and dust concentrations; (c) water conductivity; (d) mineralogical analyses via X-ray powder diffraction. The ice- core is the longest ever extracted in an ice cave in the Italian Alps.

The ice deposit, on the basis of the preliminary results on 14C radiocarbon dating, which should be regarded as the upper age limits, covers the last 2000 years. The δ18O shows rather stable values in the deeper parts of the record, while a decreasing trend, characterized by a higher variability, is observed in the upper part of the core. This change in the isotopic variability corresponds with an increase in the black carbon and dust content, as well as in the water conductivity. Such behavior may reveal a past change in the deposition dynamic. Clay minerals trapped in the ice core show the presence of soil chlorite, kaolinite, interstratified chlorite/illite layers and illite/muscovite layers. A complete characterization of this paleoclimatic record is still in progress, especially in regards to pollen analysis and further dating.

In the fall 2016, in the same area, a set of 1.0 m long horizontal ice cores was extracted from the LIC, intercepting a preserved layer of coarse cryogenic cave carbonates (CCC_{coarse}; Colucci et al., 2017, sub.) (Fig. 19.7). Such original finding represents the first alpine evidence of in situ CCC_{coarse} and the first occurrence from the southern side of the Alps. A unique opportunity to better understand the processes associated with the formation of CCCcoarse, and the well-preserved status of samples, allows for the planning of, besides U/Th datings, several different analyses, which may be associated with the precipitation of CCC. From a paleoclimatic point of view, such data will possibly contribute to the reconstruction of permafrost degradation and variability in the southern Alps during the Holocene.

FIG. 19.7

A calcite crystal as seen under SEM-EDX. See the typical euhedral (rhombohedral) crystals; with various intergrowths of scalenohedral crystals showing a chevron-type habits of the crystals surface.

19.3.3 GROTTA DEL GELO, MNT. ETNA, SICILY

The name of the cave refers both to its cold temperature and to its inner ice stack. In fact, "Gelo" is the Italian word for "bitter cold," and, *sensu lato* also means "ice".

In the past the *Grotta del Gelo* was known by shepherds who drove their flocks of sheep to water, but there is no proven evidence of other forms of economical exploitation of its ability to maintain ice in its interior, as happened for the other more easily accessible of Etna's lava caves used as "niviere," such as *Grotta della Neve*. From the 1970s it became an obliged destination for many excursionists, who considered it a goal to achieve at least once during their life.

It is likely that the formation of the stack of ice began in the second half of the 17th century, because, as it has been ascertained (Bullard, 1978), a lava flow of such a vast size as this (50 m average thickness) would take more than 10 years to cool down completely. Nevertheless, according to Marino (1998), the first known reference to the cave dates back to the late 19th century, when Sartorius von Walterhausen (1880) mentioned it as "Bocche del Gelo". About a century later, Brunelli and Scammacca (1975) cited the name of the cave in their list of Etna's caves. In the late 1970s Biffo and Cucuzza-Silvestri (1977) were induced by the growing number of hikers reaching the cave to raise the question of whether it should be visited only by scientific teams to avoid its deterioration. Bella et al. (1982) is the first known report including the access route, a description of the cave's morphologies, and a free-hand sketch. For the first time a decrease in volume of the ice was noticed, and it was linked to the 1981 eruption, whose vents opened in the western vicinity of the cave. Marino (1992) describes the glacial phenomenon identifying the cave as a cold trap and suggests that scientific institutions should focus on monitoring and protecting the ice cave.

However, the observations around the cave remained heuristic until the end of nineties, and they were limited to postulate a reduction of the ice body that was due to the 1981 eruption or an excess of visitors, but with no scientific approach.

The first scientific report about the inner climate was from 1997 to 2000: a project carried out by Centro Speleologico Etneo (CSE) and Parco dell'Etna monitored the cave's atmospheric humidity and temperature (Caffo and Marino, 1999). Finally, a new and improved network of sensors was been installed in 2013 by CSE and INGV-PA (Istituto Nazionale di Geofisica e Vulcanologia - Palermo) in agreement with Parco dell'Etna.

The cave is located in the northern flank of Etna (Sicily, Italy), one of the most active volcanoes in Europe, and, since 2013 it has been included by UNESCO in the World Heritage list.

Its entrance (37.804339°N, 14.984228°E) is located in the "A" zone of Etna Park, which is an integral reserve in the territory of Randazzo (CT) at an altitude of 2043 m above sea level, in the area called "Sciara del Follone."

As said before, the Grotta del Gelo is a lava tube resulting from volcanic activity. It is part of "Lave dei Dammusi," a pahoehoe lava field generated by the eruption of 1614–24, widely known to be one of the most protracted eruptions of Etna in historic time.

The streams of lava flow, during their evolution, were overlaid, superimposing and obstructing each other in their moving towards lower altitudes. Exposed to air, the top portion of the lava flows solidified and insulated the underlying molten lava, which continued to flow inside until the eruptive activity gradually reduced. When the supply of new lava ended, the fluid lava inside the streams drained out, forming several cavities different in size and shape. They were superficial, deep or laminar, and often surmounted by a few centimeters of rocks that resounded when somebody walked over them. This is the reason why they were called "Dammusi," which means roofs in Arabic.

As previously reported, a first morphological description, including a hand drawn sketch, was published by Bella et al. (1982), and a first account of the glaciological phenomenon was presented in Marino (1992). The latter also contains a detailed instrumental survey of the cave, realized by the speleologists of CSE which provides, with good approximation, the cave morphology and the hosted ice deposits.

The cave entrance is a large depression of more than 10 m in diameter, located in the upper part of the lava tube. A debris' inclined plane of around 30 degrees connects the entrance to the inner part, and is the place in which the snow deposit lays during winter, and it can stay there until May. There follows a large gallery, which is almost flat with different rocks' collapses, in which seasonal ice lakes usually lay, together with the ice's stalactites and stalagmites, which are sometimes really abundant. These ponds are deep, from few a centimeters to up to 30–40 cm, but sometimes (especially the first one) they are completely empty. The last part of the lava tube is occupied by the perennial ices deposits, which are almost flat and start from the end of the second lake, then incline with a slope up to 45 degrees. In this final part the ice body occupies the greater part of the gallery, joining the vault in the actual bottom of the cave (Fig. 19.8).

It is worth noting that the survey reported above describes a final part of the 15–20 m still visitable at that time, and the presence of a short gallery inside the body of ice, a condition that no longer exists at present. In fact, during the nineties this tunnel inside the ice was filled and completely included in the ice body. This was the principal evidence of the changes in the morphology, and the changes probably in the mass of the ice body, of the cave. Speculations around the origin of this phenomena stimulated the interest of the speleologists of CSE to perform a detailed study inside and around the cave.

Starting from 2013 the network of sensors consists of internal temperature and relative humidity sensors, and a weather station that was installed outside the cave from spring to autumn, with only T and RH during winter.

A detailed measurement from the summer-autumn of 2014 shows the evident correlation between the internal temperature inside the cave (cave_bottom) and the wind speed component in the direction of the axis of the cave. The correlation with the rain was also made evident. A heat water-ice mass balance has been calculated in the same period (autumn 2014), and a good agreement has been found (rain at external temperature versus ice mass decrement at cave temperature), considering only the control surface, described in the following paragraph, in which the volume variation has been measured (Fig. 19.9).

From the analysis and the observations carried out, we think that the Grotta del Gelo follows the Static Ice Cave Model, in which 3 alternating phases could be recognized (Fig. 19.10).

- A winter cooling phase in which external temperature is lower than internal, so air flows are present and the internal temperature decreases (Fig. 19.10A).
- A spring freezing phase in which melting snow and infiltration water is able to add mass to the ice body. (Fig. 19.10B).
- A summer-autumn melting phase in which no air circulation is present, even if it is sometimes disturbed by strong winds coming from the south, and the rain at relatively high temperature is able to melt down part of the ice body. (Fig. 19.10C).

Since Jul. 2013 a detailed study has been carried out, which has been driven by a strong interest in deepening the understanding of the ice body dynamics. A new topographical survey of the cave, and repeated surveys of the ice surface, quote at several points what has been conducted in a time span of

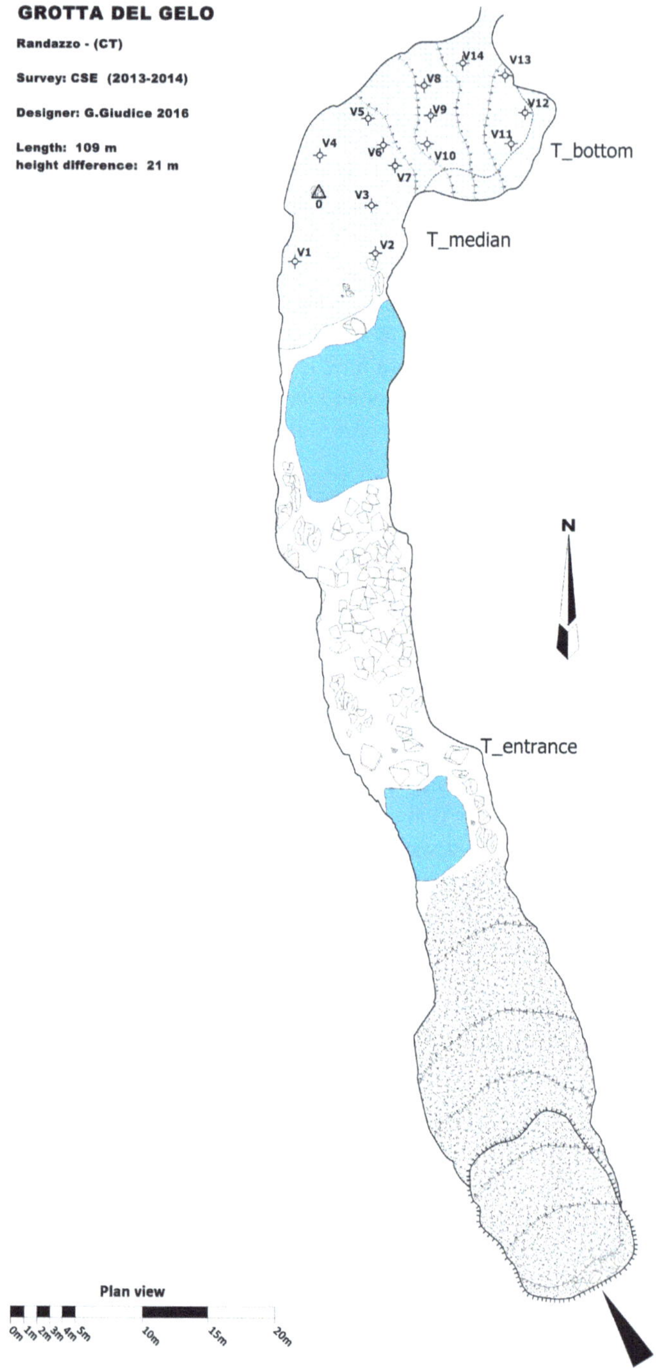

FIG. 19.8

New survey of GDG realized in 2014 by CSE. A grid of roof control points (RCP) has been identified on the ceiling of the bottom part of the cave where the perennial ice lies. From each point of the grid, manual measures are recursively performed to monitor the ice body elevation and volume change.

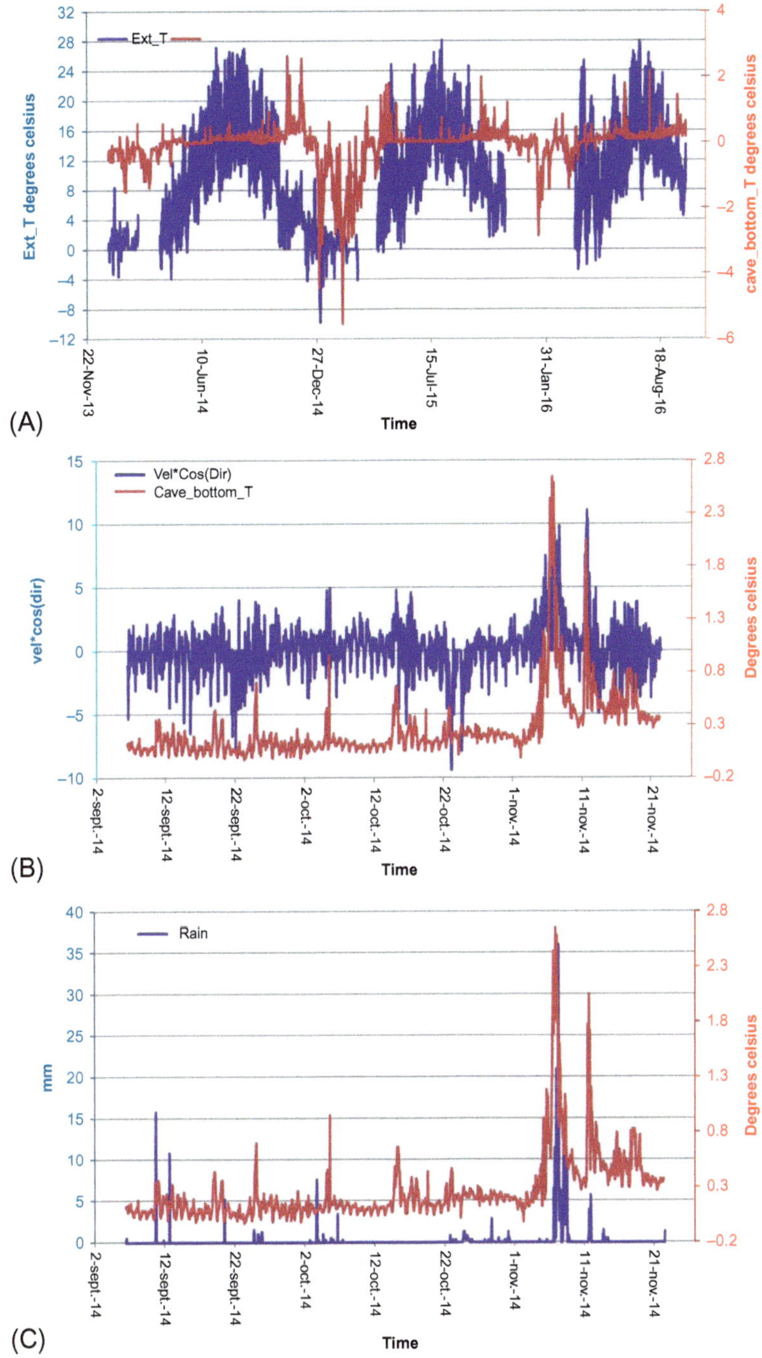

FIG. 19.9

(A) External and bottom temperature of Grotta del Gelo recorded by data loggers in 2013–16. Despite during the survey period the external temperature in the vicinity of the cave raised up to 28°C in summer, the deepest zone was globally isolated, with a temperature that fluctuated below or around the freezing point (0°C), a part of some warmer peaks that raised the internal *T.* up to +2.3°C for a short time. They can be attributed to wind gusts oriented toward the direction of the entrance (B) and precipitation fallen during high rainfall events with more than 40 mm/day (C). In winter instead, when the outside *T* was some degrees below zero (up to −7°C) the cold spills propagated trough the cave and reached the bottom by decreasing the *T* and triggering the winter phase and the spring freezing phase able to add mass to the ice body.

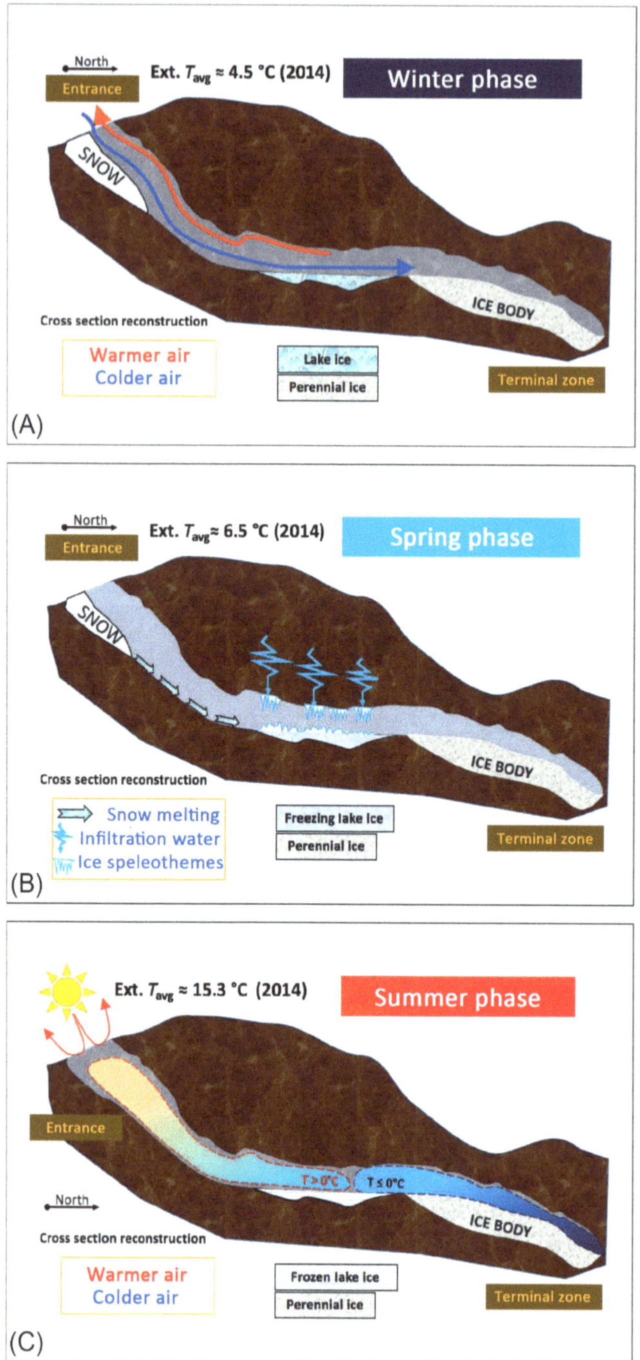

FIG. 19.10

Cross section of the GDG and its air circulation model according to the temperatures recorded. In winter (A), colder and denser air sinks through the cave and reaches the bottom where a cold air trap forms. Warmer air, less dense, is pushed out from the cave. In spring (B) melting snow and dripping are the main sources of water supply of the lake ice. Because the average temperature is fluctuating at the freezing point (or below during cold spills event) the water gradually freezes from the top to bottom and new ice can be accumulated upon the perennial ice body. This process insulates the deepest zone from the rest of the cave, creating a cold air trap, delimited by the 0°C isotherm and the underground glacier can survive for the rest of the year, even in summer when outside temperature can reach around 30°C (C).

about 3 months using a high precision total station (±2″), which was placed in a flat area of the Central Zone (identified as the origin for distance measuring) (Fig. 19.8).

Furthermore, the distance between the floor (corresponding with the glacier's surface) and a grid of 14 roof control points (spikes anchored on the cave's ceiling) has been measured every couple of months when possible, with a plumb line and a Leica Disto Laser. The main goal is therefore the mapping of the exact seasonal glacier movements and, over the years, the developing of a digital cartography that is easy to manage with a generic CAD-or GIS-based platforms. The reference surface (Jun. 2013) in the control area (137 m^2), the last surveyed surface (Oct. 2016) and the relative time variations of the normalized volume (arbitrarily considered equal to 100 m^3 in Jun. 2013) are reported in Fig. 19.11.

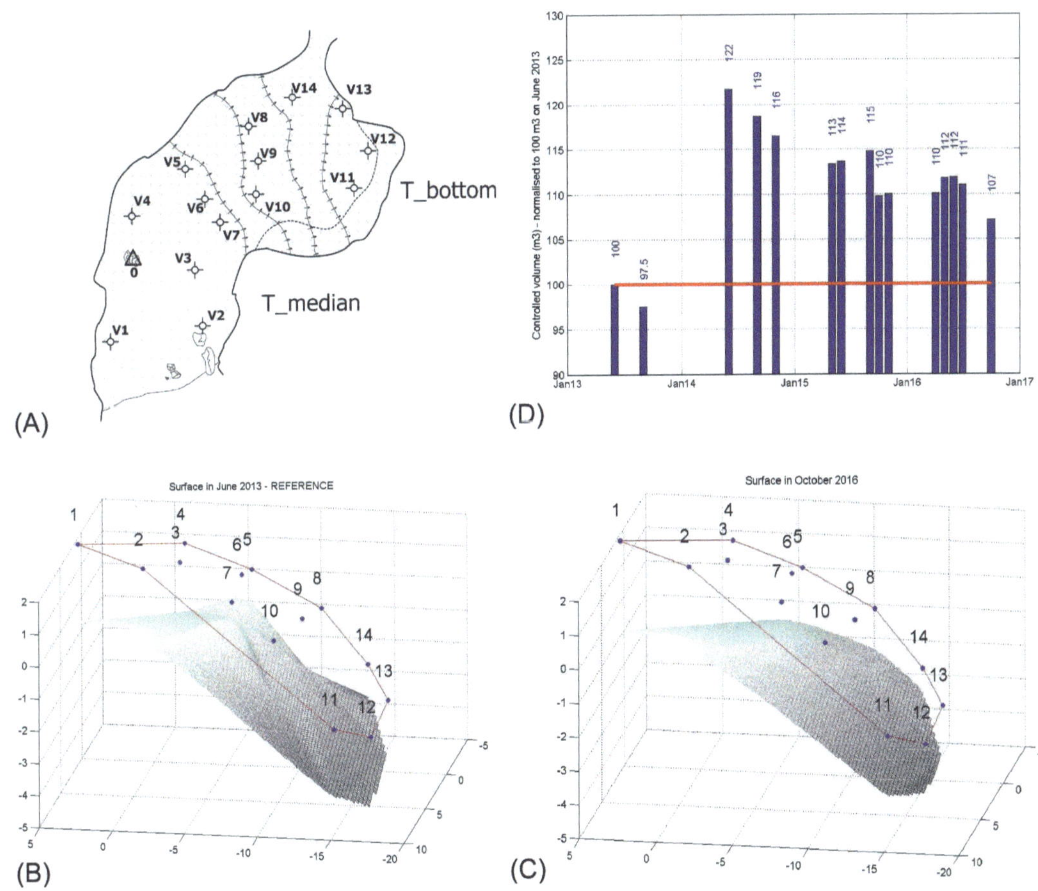

FIG. 19.11

(A) The images show the location of the Roof Control Points from which vertical distance from the ceiling to the ice body have been measured. Subsequently, the data obtained has been interpolated in a 3D plot showing the elevation and distribution change of the ice body (B and C). The graph shows the periodical volume variation by assuming a normalized ice body volume of 100 m^3 at the begin of the survey in 2013 (D).

From the analysis of these first years of surveys, it seems that the deep ice body balance (positive or negative) is strongly influenced by impulsive input of water which reached the deepest zone of the cave in the form of snow, melting water, infiltration water, or run off after precipitation.

A notable increase of the observed ice body volume occurred in 2014 (probably due to prominent previous snowfalls) and since that, apart from some tiny positive peaks, it has generally decreased until almost the value recorded at the begin of the survey. This behavior seems to be coherent also with the non-instrumental observations reported in the last decades. (Fig. 19.11)

Further aims are devoted to the realization of a 3D model of the cave and its inner ice body, to follow with more detail the volumetric variations over the years.

19.4 CONCLUSIONS

Cave deposits, together with peat bogs, swamps, lakes, and tree rings, represent an important archive of past climatic and environmental information for continental areas, and proper archives from medium to low altitude mountains, like cave sediments and speleothems, can be found well below the limit of the glaciers (Turri et al., 2003; Haeberli and Alean, 1985). Depending on their nature and formation mechanism, the underground ice deposits can contain elements that allow the study of past interactions with the climatic and environmental history of a region (Turri et al., 2003; Luetscher et al., 2005; Citterio et al., 2004; Kern et al., 2011; May et al., 2011; Perşoiu and Pazdur, 2011).

Furthermore, several Italian ice caves are located below the permanent snow line that in the Alps reaches an altitude between 3200 and 3400 m a.s.l., hence well below the glacier's altitude. Studying these caves by implementing the glaciological techniques both of sampling (ice cores) and measurements (chemistry, stable isotopes, mineral dust) allows for consideration of the ice deposits as a particular typology of speleothems, which are able to store climatic and environmental information.

As described in the previous chapter, the complexity of the Italian karst systems, and the strong differences in climate conditions, provides a large variability in ice cave behaviours. Knowing the origin of the ice caves is another important issue that can be resolved only by performing an integrated study of the air, the water, and the ice dynamics. Some caves seem to have been formed during the Holocene, with ice dated before 2000 y (Colucci et al., 2016a), while other cave present a more younger ice accumulated probably during the Little Ice Age (from AD 1300 to AD 1800) as for the LOLC1650. Despite that, a clear trend regarding the age of ice formation in the Alpine range does not exist. It seems that the older deposits are located on the eastern part of the chain. This information fit well with the distribution of the biggest ice deposits located in the karstic areas of Austria, Slovenia, Slovakia, and Romania (Kern and Persoiu, 2013).

Lava tubes in Sicily represent another important resource to improve the knowledge of ice formation mechanisms. In general, because they are easy to access, their micrometeorology and ice body dynamics can be monitored, and the underground ice analyzed, giving back the exact date of the of the cave formation.

However, due to the regular volcanic activity that is characterizing in historical time, we cannot exclude that in the next future we could observe new lava tube formation, and, where particular conditions will occur, also underground ice accumulation from the very early stages of the process.

We conclude by affirming that all that has been achieved in the past, and all that will be added in the future to reach a more global knowledge of the ice caves, needs a strict collaboration with cavers and speleological groups, because they are essential to fulfil the important function of leading survey and exploration, providing the history of the past surveys, and giving the direct support to scientists during in situ studies and sampling.

ACKNOWLEDGMENTS

Part of this work was funded by the project of strategic interest NEXTDATA (PNR National Research Programme 2011–13; project coordinator A. Provenzale CNR-IGG, WP leader V. Maggi DISAT-UNIMIB). Ice cave surveys in Friuli Venezia Giulia are conducted in the frame of the project "C3-Cave's Cryosphere and Climate" (www.c3project.net) financed by the Commissione Grotte "E.Boegan (CGEB) of the Società Alpina delle Giulie (SAG-CAI) and by the University of Trieste through the "Finanziamento di Ateneo per progetti di ricerca scientifica, "FRA 2012 and 2014" Grants, projects MONICA (MONitoring of Ice within CAves) and MACRY-FVG (Characterization and monitoring of MAritime CRYosphere in FVG). We also kindly acknowledge Francesca Vita from the Department of Life Sciences, University of Trieste (Italy) for the CCCcoarse picture taken with the Scanning Electron Microscope and the Parco Naturale delle Prealpi Giulie. Thanks to the speleologists of the Centro Speleologico Etneo and Federazione Speleologica Lombarda. Thanks to Guido Raciti, Marianna Messina, Francesco Leone, Elvira Finocchiaro and Salvatore Petralia from CSE. Special thanks also to Salvatore Caffo, the volcanologist of Parco dell'Etna, and Prof. Marco Viccaro (University of Catania) for their valuable and constructive suggestions provided during the planning and the development of Mt. Etna work.

REFERENCES

Bella, V., Brunelli, F., Cariola, A., Scammacca, B., 1982. Grotte Vulcaniche di Sicilia, notizie catastali: secondo contributo (da Si CT 26 a Si CT 50). Boll. Acc. Gioenia Sc. Nat. 15 (320), 229–292. Catania.

Biffo, L., Cucuzza-Silvestri, S., 1977. Relazione preliminare alla Direzione dell'Istituto di Vulcanologia dell'Università di Catania sull'importanza scientifica della Grotta del Gelo (Etna). In: Appendice alla comunicazione: Cucuzza-Silvestri S. (1977) – Le Grotte Vulcaniche dell'Etna e il loro studio, Atti del IV Simposio Internazionale di Vulcanospeleologia e Seminario sulle Grotte Laviche, Catania 1975, pp. 215–229.

Il Carsismo del Moncodeno (Ricerche sugli aspetti del fenomeno carsico profondo nel Gruppo delle Grigne (Lombardia): VII - il carsismo del Moncodeno). In: Bini, A., Pellegrini, A. (Eds.), Geologia Insubrica - Rivista di Scienze della Terra 3 (2), 1998. 296.

Borsato, A., Miorandi, R., Flora, O., 2006. I depositi di ghiaccio ipogei della Grotta dello Specchio e del Castelletto di Mezzo (Dolomiti di Brenta, Trentino): morfologia, età ed evoluzione recente. Studi Trent. Sci. Nat., Acta Geol. 81 (2004), 53–74.

Brevini, F., 2004. Rocce. Mondadori, 252pp.

Brunelli, F., Scammacca, B., 1975. Grotte Vulcaniche di Sicilia, notizie catastali: primo contributo. Gruppo Grotte Catania del CAI, Catania. p. 61.

Bullard, F.M., 1978. I vulcani della Terra. Paperbacks ricerca, Newton Compton, Roma. pp. 86–88.

Caffo, S., Marino, A., 1999. Il monitoraggio della Grotta del Gelo. In: Atti del IX Simposio Internazionale di Vulcanospeleologia, Catania 10-19/09/1999.

Chambers, J.L.C., Wilson, C.J.L., Adamson, D.A., 1986. A crystallographic study of the perennially frozen ice surface of patterned lake, Framnes Mountains, East Antarctica. J. Glaciol. 112 (32), 520–526.

Citterio, M., Turri, S., Bini, A., Maggi, V., Pelfini, M., Pini, R., Ravazzi, C., Santilli, M., Stenni, B., Udisti, R., 2004. Multidisciplinary approach to the study of the Lo Lc 1650 "Abisso sul Margine dell'Alto Bregai" ice cave (Lecco, Italy). Theor. Appl. Karstol. 17, 27–44 (Special Issue on Ice Caves).

Colucci, R.R., 2016. Geomorphic influence on small glacier response to post Little Ice Age climate warming: Julian Alps, Europe. Earth Surf. Process. Landf. 41, 1227–1240. https://doi.org/10.1002/esp.3908.
Colucci, R.R., Žebre, M., 2016. Late Holocene evolution of glaciers in the southeastern Alps. J. Maps 12, 289–299. https://doi.org/10.1080/17445647.2016.1203216.
Colucci, R.R., Monegato, G., Žebre, M., 2014. Glacial and proglacial deposits of the Resia Valley (NE Italy): new insights on the onset and decay of the last Alpine Glacial Maximum in the Julian Alps. AMQ 27 (2), 85–104.
Colucci, R.R., Fontana, D., Forte, E., Potleca, M., Guglielmin, M., 2016a. Response of ice caves to weather extremes in the Southeastern Alps, Europe. Geomorphology 261, 1–11.
Colucci, R.R., Boccali, C., Žebre, M., Guglielmin, M., 2016b. Rock glaciers, protalus ramparts and pronival ramparts in the south-eastern Alps. Geomorphology 269, 112–121. https://doi.org/10.1016/j.geomorph.2016.06.039.
Colucci, R.R., Luetscher, M., Forte, E., Guglielmin, M., Lanza, D., Principalle, F., Vita, F., 2017. First alpine evidence of in situ coarse cryogenic cave carbonates (CCCcoarse). Geogr. Fis. Din. Quat 40 (1) (in press).
Costantini, E.A.C., Fantappié, M., L'Abate, G., 2013. Climate and pedoclimate of Italy. In: Costantini, E.A.C., Dazzi, C. (Eds.), The Soils of Italy. Springer, Netherlands, Dordrecht, pp. 19–37.
Ferrario, A., Tognini, A. (Eds.), 2016. Catasto Speleologico Lombardo (Progetto Tu.Pa.Ca.). Federazione Speleologica Lombarda, 448 pp.
Gruppo Speleologico Piemontese, 2014. Il Becai. Grotte 161, 20–60.
Haeberli, W., Alean, J., 1985. Temperature and accumulation of high altitude firn in the Alps. Ann. Glaciol. 6, 161–163.
Ischia, M., Borsato, A., 2005. Depositi di ghiaccio nelle cavità delle Dolomiti di Brenta. In: Atti XIII Conv. Reg. Spel. Trent. - A. Adige, Villazzano, ottobre 2005.
Kern, Z., Persoiu, A., 2013. Cave ice - the imminent loss of untapped mid-latitude cryospheric palaeoenvironmental archives. Quat. Sci. Rev. 67, 1–7.
Kern, Z., Fórizs, I., Pavuza, R., Molnár, M., Nagy, B., 2011. Isotope hydrological studies of the perennial ice deposit of Saarhalle, Mammuthöhle. Dachstein Mts, Austria. Cryosphere 5, 291–298. https://doi.org/10.5194/tc-5-291-2011.
Luetscher, M., Jeannin, P.Y., Haeberli, W., 2005. Ice caves as an indicator of winter climate evolution: a case study from the Jura Mountains. Holocene 15 (7), 982–993.
Marino, A., 1992. Nota preliminare sul fenomeno glaciologico della Grotta del Gelo (Monte Etna). Geog. Fis. Dinam. Quat. 15, 127–132.
Marino, A., 1998. Indagine sul fenomeno glaciale della Grotta del Gelo (prime conclusioni dopo la campagna di raccolta dati climatologici 1997/98). In: Lavoro inedito consegnato all'Ente Parco dell'Etna nel Dicembre 1998 e reperibile presso il Centro Speleologico Etneo, Via Cagliari 15, Catania. p. 10.
May, B., Spötl, C., Wagenbach, D., Dublyansky, Y., Liebl, J., 2011. First investigations ofan ice core from Eisriesenwelt cave (Austria). Cryosphere 5, 81–93. https://doi.org/10.5194/tc-5-81-2011.
Perşoiu, A., Pazdur, A., 2011. Ice genesis and its long-term dynamics in Scărişoara IceCave, Romania. Cryosphere 5, 45–53.
Sartorius von Walterhausen, W., 1880. Der Aetna, v. II, Lipsia, pp. 373–374.
Sivelli, M., Vianelli, M., 1982. In: Società Speleologica Italiana (Eds.), Abissi delle Alpi Apuane. Società Speleologica Italiana, Bologna, pp. 225–228.
Stefan, J., 1891. Über die Theorie der Eisbildung, insbesondere über die Eisbildung im Polarmeere. Ann. Phys. Chem. 42, 269–286.
Tison, J.L., Haren, J., 1989. Isotopic, chemical and crystallographic characteristics of first-year sea ice from Breid Bay (Princess Ragnhild Coast—Antarctica). Antarct. Sci. 1 (3), 216–268.
Triglav Čekada, M., Zorn, M., Colucci, R.R., 2014. Area changes on Canin (Italy) and Triglav glaciers (Slovenia) from 1893 on based on archive imagery and Lidar. Geodetski vestnik 58 (2), 257–277. https://doi.org/10.15292/geodetski-vestnik.2014.02.274-313.
Turcotte, D.L., Schubert, G., 2002. Geodynamics, second ed. Cambridge University Press, Cambridge. pp. 162–166.

Turri, S., 2005. Dinamica ed evoluzione climatologica delle grotte ghiacciaie. Ph.D. Thesis, University of Milan, Milan 222 p.

Turri, S., Citterio, M., Bini, A., Maggi, V., Udisti, R., Stenni, B., 2003. Etude glaciologique et climatologique des cavités glacées du Moncodeno. (Grigna septentrionale, province de Lecco, Lombardie). Karstologia 42, 37–44.

Turri, S., Citterio, M., Bini, A., Maggi, V., Favaron, M., Fraternali, D., Alberici, A., Borghi, S., Colombo, M., Gottardi, R., Zappalà, D., 2005. Preliminary data recorded by a monitoring station to study the hypogean climate in an ice cave: the LO LC 1650 Ice Cave "Abisso sul margine dell'Alto Bregai" (Grigna Settentrionale, Lecco – Italy). In: 14th International Congress of Speleology, Athens, 21–28 August 2005, Congress Proceedings. pp. O–14.

Žebre, M., Stepišnik, U., 2015. Glaciokarst landforms and processes of the southern Dinaric Alps. Earth Surf. Process. Landf. 40 (11), 1493–1505. https://doi.org/10.1002/esp.3731.

FURTHER READING

Centro Speleologico Etneo, 1999. "Dentro il Vulcano: le grotte dell'Etna" Catania, 1999, pp. 252–257 - Bonaccorso R., Santi G., Marino.

Ford, D., Williams, P.W., 1989. Karst geomorphology and hydrogeology. Capman and Hall, London. 601 p.

Holmund, P., Onac, B.P., Hansson, M., Holmgren, K., Mörth, M., Nyman, M., Perşoiu, A., 2005. Assessing the paleoclimate potential of cave glaciers: the example of the Scărişoara Ice cave (Romania). Geogr. Ann. 87 (1), 193–201.

Maggi, V., Villa, S., Finizio, A., Delmonte, B., Casati, P., Marino, F., 2006. Variability of anthropogenic and natural compounds in high altitude-high accumulation alpine glaciers. Hydrobiologia 562, 43–56.

Romano, R., Sturiale, C., 1982. The historical eruptions of Mt. Etna (volcanological data). Mem. Soc. Geol. It. 23, 75–97. Roma.

Scoto, F., 2013. Nuovo Rilievo Cartografico Della Grotta Del Gelo (Etna, Versante Settentrionale) E Monitoraggio Stagionale Della Temperatura E Della Circolazione Dell'aria Al Suo Interno: Implicazioni Sulle Consistenti Fasi Di Disgelo. Unpublished BSc Thesis, University of Catania, Catania.

CHAPTER

ICE CAVES IN IRAN

20

Mojgan Soleymani*, Derek Ford†, Mohammad Nakhaei*, Alireza Nadimi‡

Kharazmi University, Tehran, Iran McMaster University, Hamilton, ON, Canada† University of Isfahan, Isfahan, Iran‡*

CHAPTER OUTLINE

20.1 INTRODUCTION

Iran is a large and mountainous country, tectonically a part of the active Alpine-Himalayan orogenic belt. The full range of the standard soluble rocks (the carbonates and the evaporites) can be found in the area (Fig. 20.1). Outcrops of salt in the Hormoz region (the southern Zagros Mountains) are especially well-known because of the occurrence of namakiers (glaciers of salt being extruded from diapirs). However, carbonate rocks are predominant in the outcroppings, covering about 185,000 km^2, or approximately 11%, of Iran's land area (Raeisi and Laumanns, 2003).

Organized amateur caving began in 1945 (Javanshad, 1995). British cavers discovered and explored a deep shaft-and-drain vadose cave, Ghar Parau, in the 1960s. Czech cavers, and the Geological Survey of Iran, organized more systematic surveying campaigns in the 1990s and early years of the present century, focusing on the Hormoz region, where they discovered remarkably long caves in salt; one of them, Tri Nahacu, is currently the longest known salt cave at 6580 m (Bosak et al., 1999; Bruthans et al., 2006). Jujar Cave was explored to a depth exceeding 1100 m in 2016 and is expected to be much deeper. With a depth of almost 560 m, the Ghala single-shaft cave was explored in 2015. The first "Iran Cave Directory" in English, published by Raeisi and Laumanns (2003), provided details on some 550 caves. A 3rd edition has increased this to more than 2000 caves (Raeisi et al., 2012).

There have been significant scientific studies of the physical nature, origin, morphology, and deposits of caves in Iran for the past 30 years or more. Because of the intense tectonic and volcanic activity

Ice Caves. https://doi.org/10.1016/B978-0-12-811739-2.00020-6

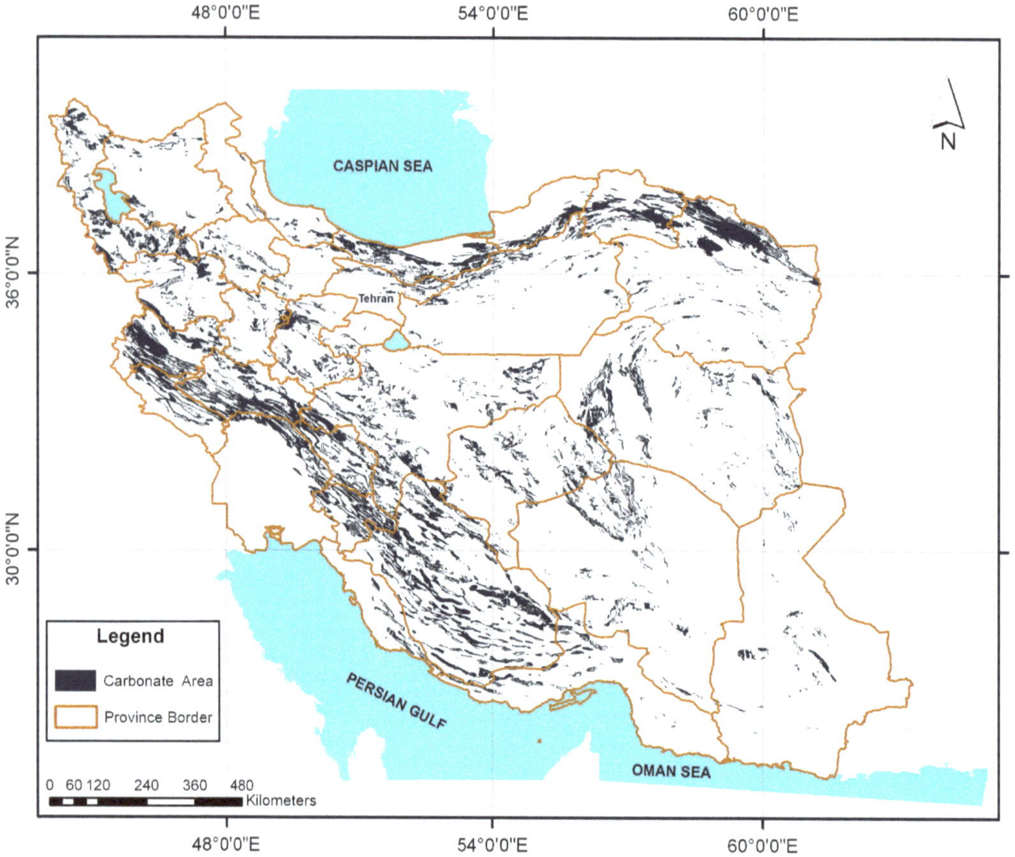

FIG. 20.1

The distribution of karst rocks in Iran.

below the surface, combined with the general aridity above it, it is not surprising that, throughout the land, many individual investigators have concluded that their caves are relicts of hypogene origin (Kowsar, 1992; Kaufmann, 2002; Yousefi et al., 2009; Sabokkhiz et al., 2012; Soleymani, 2015; Bahadorinia et al., 2016; Vardanjani et al., 2017). However, some other studies of caves near modern rivers or high in the mountains have found that they are of the globally more common meteoric origin (Raeisi and Kowsar, 1997; Rezaei and Nakhaei, 2008). Only Soleymani (2015) has reported on ice deposits in Iranian caves.

Physically, Iran is best known to the world for its mid-to low-latitude, warm-to-hot, semiarid-to-arid, climatic conditions. It is true that heat and aridity are dominant features of the lowlands, but the Cenozoic tectonic forces have created high mountain ranges and massifs. In the Quaternary Ice Ages, these hosted quite a large number of alpine glaciers, and they retain some smaller glaciers, as well as snow and firn fields, today. The potential for future discoveries of significant ice caves appears to be quite good. The two caves described below are believed to be the only examples that have received any detailed investigation up to the present time (Soleymani, 2015). They are the Yakh-Morad Ice Cave (Ghar Yakh-Morad) in the Alborz Mountains, and the Dena Ice Cave (Ghar Yakh Dena) in the northern Zagros Mountains (Fig. 20.2).

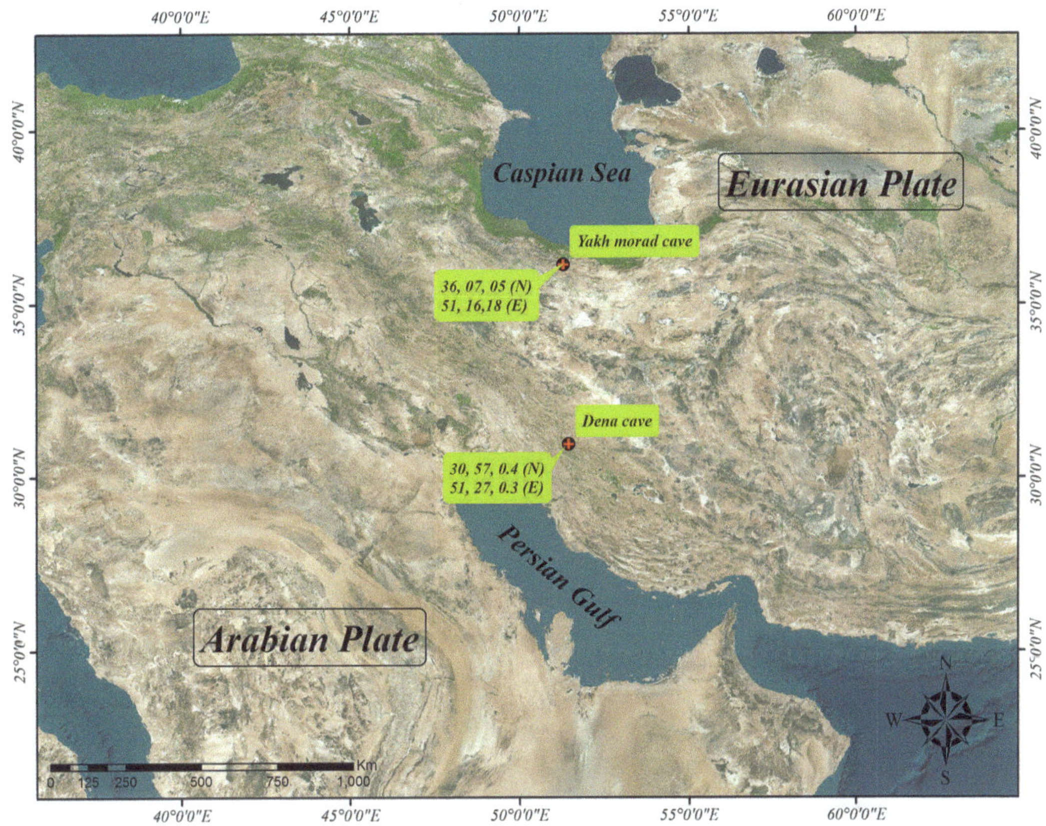

FIG. 20.2

Regional relief and tectonic, showing the location of the Dena and Yakh-Morad Ice Caves in Iran.

20.2 GHAR YAKH-MORAD, AN ICE CAVE IN THE ALBORZ MOUNTAINS

Geologically, Iran is divided into five major zones, the Zagros Mountains, Sanandaj-Sirjan Ranges, Central Iran, the East and Southeast, and the Alborz Mountains. The Alborz, lying between the Central Zone and the south end of the Caspian Sea basin, is a narrow (100 km), but elevated (3000 m), mountain belt built up to accommodate the differential plate motion between these two zones. The mountains display evidence of strong tectonic activity, with several destructive earthquakes occurring in the past. There are about 50 known caves in the central Alborz, most of which are strongly fault directed. The host rocks are Jurassic and Cretaceous limestones, with early Cretaceous redbeds and other clastic rocks in between them, and an overlying volcanic sequence succeeding them.

20.2.1 GLACIATION IN ALBORZ MOUNTAINS

The higher parts of the Alborz were heavily glaciated during the cold periods of the Quaternary. At present, active glaciers are limited to the summit zones, chiefly in an ice cap on the northern slope of Damavand Mountain (5610m), and a lesser body on Alam Kuh (4848 m). Between the crests, the regional trunk valleys (Shahhrud, Nur, Lar) have glacial U-shaped cross-sections and run in an east-west direction, with narrow youthful gorges linking them north-south (Peguy, 1959). Ghohroudi Tali (2012) showed that in the last glacial period the permanent snowline was at 1800 m a.s.l., placing more than 80% of the region under the domination of cirque and valley glacial erosion. The snowline has now receded to about 3200 m a.s.l. The mean annual temperature was ~6°C lower during the glacial periods.

20.2.2 GHAR YAKH-MORAD

Yakh-Morad Cave is located at 36°07′5.94 (N), 51°16′18.24 (E). The entrance is at 2493 m a.s.l. Fig. 20.3 shows its location in the rocky spur of a mountain overlooking a northward-draining gorge. The host rock is the Lar Formation (mid-to-late Jurassic), thick-to massively bedded gray limestone rich in ammonite fossils and chert nodules. It is overlain by thinner marly limestones, marl (Delichai Fm.), and the Tizkuh Fm. (early Cretaceous), an orbitolina-rich limestone (Soleymani, 2015).

FIG. 20.3

A Google Earth oblique view of the terrain around Yakh-Morad Cave, showing the location of the cave entrance. The hamlet of Kohnedeh is in the foreground.

According to 30-year climate normals from the meteorological station at Kohnedeh (only 550 m from the cave, at 2365 m a.s.l.), the local conditions are a steppe climate, type BSk in the Köppen-Geiger classification. Mean annual temperature at Kohnedeh is 8.9°C, and the precipitation is 205 mm. In Fig. 20.4 it is seen that the mean temperature range is substantial (27.2°C) with a significant Jan. low value of −5.6°C, rising to 21.6°C in Jul. There is pronounced summer drought. Most precipitation falls in the winter and early spring, April being the wettest month (39 mm) at the end of the winter chilling period.

A 1:500 topographic map of the cave (CRG Grade 4) was prepared by Brooks et al. (2008). Soleymani (2015) studied the lithology, geologic structure, tectonic history, and morphology of the cave, and noted the location of all deposits, including ice, and made a temperature and relative humidity traverse using a Digital thermometer (LX8011) instrument in the month of Aug. 2015.

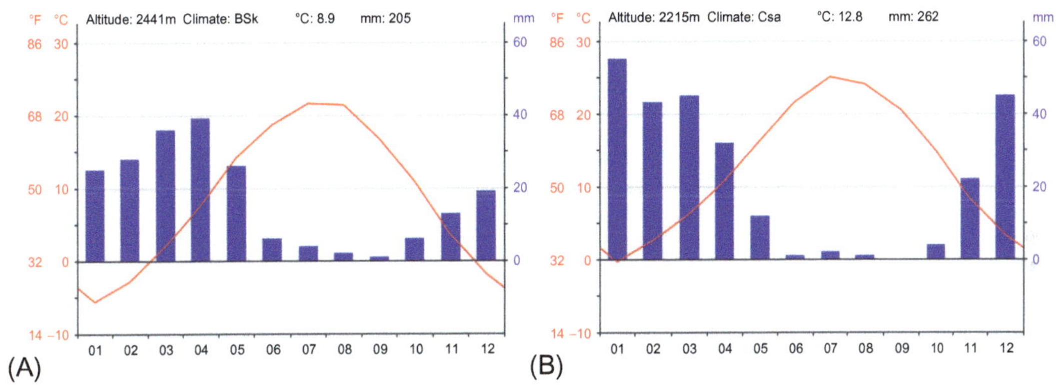

FIG. 20.4

(A) *On left*, the monthly mean temperature and precipitation climate normals for Kohnedeh, 1982–2012.
(B) *On right*, the monthly mean temperature and precipitation climate normals for Khafr, 1982–2012.

Fig. 20.5 is a schematic long section of the cave, showing the principal locations of its seasonal and perennial ice deposits, and the air temperatures and relative humidity measured in Aug. 2015. The cave may be said to consist of five levels, displaying two different morphologies (Soleymani, 2015). The two upper levels are breakdown passages and rooms (Fig. 20.6A), connected by an 11-m shaft. The walls and ceilings are densely fractured surfaces and floors that are entirely made up of fallen blocks. Almost all evidence of the previous form is lost, except for a few small solutional features and limited amounts of inactive calcite popcorn precipitates. The three lower levels appear to be younger in origin; the proportion of breakdown is reduced, and many passages display well-rounded phreatic form, plus some deep solution pockets in the walls and ceilings (Fig. 20.6B). There are also a few remnants of substantial calcite raft or shelfstone deposits, indicating that there were some long-lived pondings in the lower cave in the past (Fig. 20.6C). Many other phreatic features (pockets and impenetrably small tubes) are seen in the cliff faces outside the cave. Given its location in a steeply dipping strata in a narrow spur of limestone between mountain streams that are vigorously entrenching their valleys today, it is considered that the cave is possibly of hypogene origin, but was enlarged by stream invasions during the valley entrenchment to create the levels.

The air temperatures and relative humidity were recorded during a 1-day visit in the summer period of maximum outside temperatures. The cave displayed three thermal zones. Temperatures ranged from +4°C to +6°C in the upper levels that are older and higher in the limestone spur, and thus better ventilated; relative humidity there ranged from 75% to 85%. The middle zone appears more stable at +2.5°C to +3.5°C, RH 85%–95%. The two distinct lower levels were functioning as an effective cold trap in Aug. 2015, with temperatures between +0.5°C and +1.5°C and RH 90%–100%.

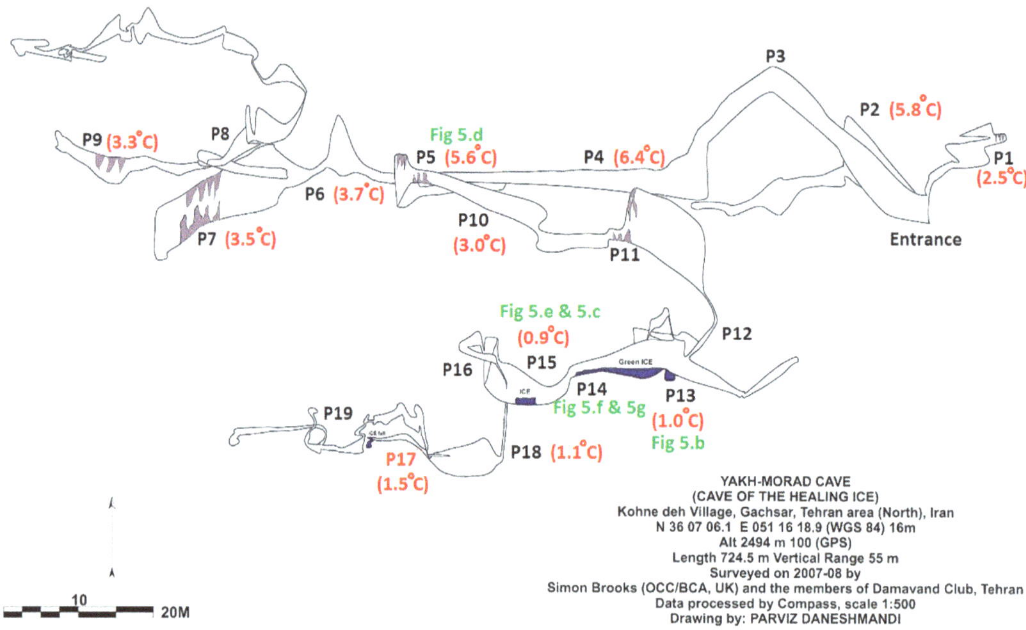

FIG. 20.5

Long section of Yakh-Morad Ice Cave showing the location of the ice forms and air temperatures measured in Aug. 2015. The *light blue* refers to seasonals, and the *dark* ones are perennials.

The known ice deposits are varieties of dripstone and flowstone formed from infiltrating waters. They are quite widely distributed throughout the cave due to the cave's broad extension within the comparatively narrow bedrock spur. The location of the cave entrance is not suitable for trapping snow and building a glacier. No hoarfrost was found in the Aug. visit, but it may well have occurred earlier in the spring-summer period. In winter, there are some thermo-indicator stalagmites where the ice is translucent, and the diameter is large during warmer spells of rapid growth, which then narrows and becomes opaque due to trapped air bubbles when it is colder. In Fig. 20.5D a few examples are asymmetric due to ablation on the upwind side, but this is minor in extent. In lower levels, it is possible that some represent several successive years where accumulation exceeded melting, which is a subject for future research.

The known perennial ices masses ("ice blocks"; Persoiu and Onac, 2012; Colucci et al., 2016) are limited to the lower cave (Figs. 20.5 and 20.6E–G). It has been noted that they are layered seasonal accumulations of ice flowstone from seepage water, with dirt bands and other discontinuities representing episodes of melting. Their total mass is difficult to estimate, because some accumulations nearly fill comparatively large and active inlets that may extend for many meters. It is evident that some of the blocks have suffered large net losses due to sublimation.

Between stations P13 and P15 there is a ponding of frozen water measuring approximately 10×3 m (an "ice lake," Ford and Williams, 2007). The depth is uncertain but may exceed 1–2 m, and it is believed to be solid to the bottom. There are some small, ice-supported, limestone clasts within the ponding, which suggests that it may have a complicated history of partial melting and refreezing.

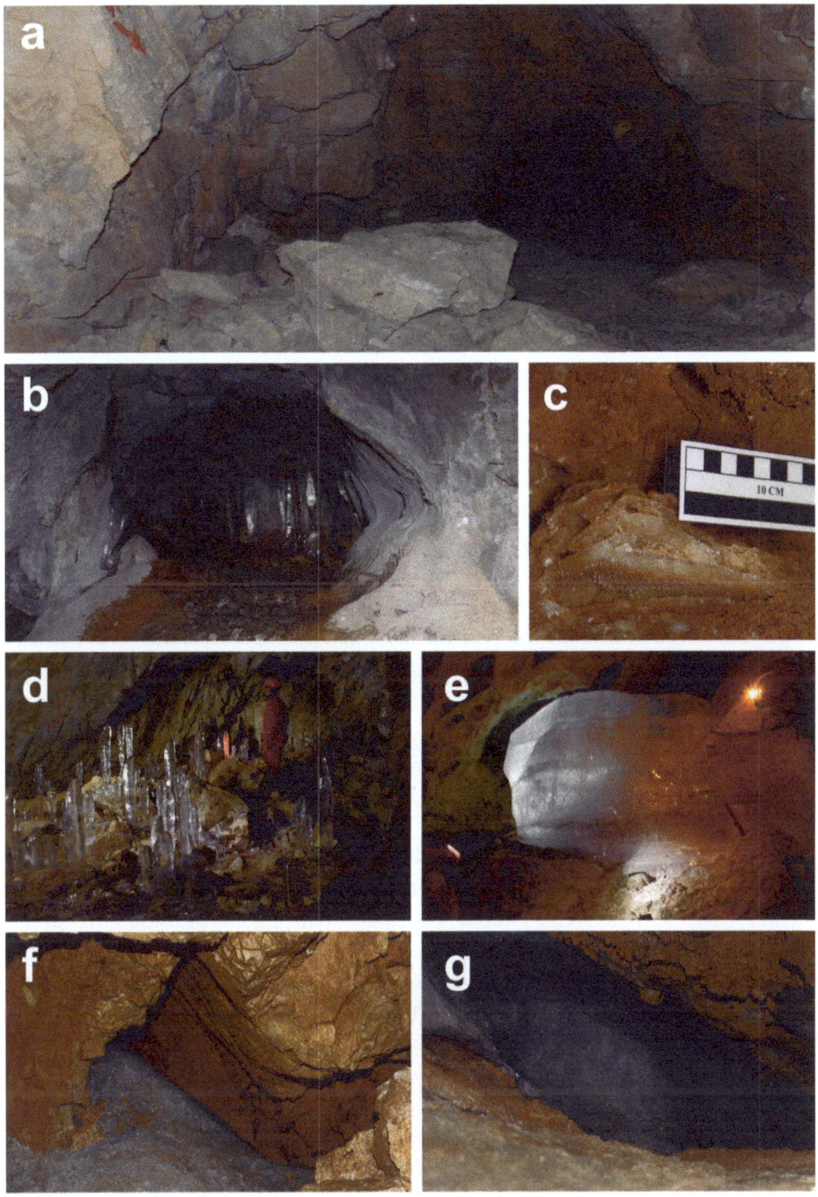

FIG. 20.6

Yakh-Morad Ice Cave. (A) A typical scene in the upper breakdown levels. (B) A well-formed phreatic tube in the lower levels. The dripstone and flowstone ice now occupying the tube is presumed to be seasonal and derived from snowmelt in epikarst a few tens of meters overhead; it was fresh in appearance and displayed no features of melting when photographed. (C) A thick calcite shelfstone or raft deposit with many successive layers attests to the existence of long-lasting ponding events in the lower cave in the past. (D) Ice stalagmites up to 1 m or more in height at P5 in the upper cave. There are some examples of "thermo-indicator growth" as defined above, but the majority have uniform cross-sections, implying rather constant temperatures and drip rates. Many displayed some loss of ice by sublimation (air flow towards the rear of the scene) but were still largely intact when photographed. (E)–(G) Three examples of "perennial" layered ice from seepage and regelation in the lower cave. (E) is approximately 1.5 m in thickness and particularly well-banded; it and (G) are now much ablated by sublimation.

It displayed no melting when visited in Aug. 2015, and it may be in a net accumulation state, perhaps capturing some meltwater trickling down from the ice deposits above it.

In summary, the Yakh-Morad Ice Cave contains small perennial ice bodies in a cold trap setting at ~2450 m a.s.l. in the Alborz Mountains, where the mean annual temperature outside is around 9°C, but it is negative in the winter months. Much of the ice appears to be diminishing rapidly; if its records of past environments and environmental changes are to be recovered, studies should be made as soon as possible.

20.2.3 GHAR YAKH DENA, AN ICE CAVE IN THE ZAGROS MOUNTAINS

The Zagros geologic zone is the largest belt of mountain ranges in Iran, and it also extends into Iraq and southeastern Turkey. The multiple ranges are oriented broadly northwest by southeast (Fig. 20.1). They were formed by the collision of the Arabian Plate with the Iranian Plate and have a total length >1500 km. Recent GPS measurements show that the collision is still active, with current rates of shortening ~10 mm/year in the southeast and ~5 mm/year in the northwest (Nilforoushan et al., 2003; Hesami et al., 2006). As a consequence of this uplift, much of the terrain lies above 2000 m a.s.l., with higher ridges and some plateaus rising to 3500–4400 m. The majority of Iranian oil resources are found in the sedimentary rocks of the Zagros, especially in the limestone Asmari Formation (Late Oligocene to Early Miocene).

20.2.4 GLACIATION IN THE ZAGROS MOUNTAINS

There are vestigial modern glaciers and firn fields at Zard Kuh (4221 m a.s.l.), and on Mount Dena (4350 m a.s.l.), in Dena National Park, which is in the more westerly high Zagros. Comparably high mountains further east, such as Kuh-i-Jupar (4135 m), Kuh-i-Lalezar (4374 m), and Kuh-i-Hezar (4469 m) are ice-free today. However, during the last glaciation, glaciers that had built up on these high grounds were able to extend down to elevations below 2000 m a.s.l. For example, a glacier up to 20 km wide, and with a thickness of 350–550 m, flowed for 17 km along a valley on the north side of Kuh-i-Jupar, descending approximately 1500 m. Under precipitation conditions comparable to those of today, such a glacier could be expected to form where the annual average temperature at sea level lies between 10.5°C and 11.2°C, but since the climate is expected to have been drier during the glaciation, the temperature must have been lower (Kuhle, 1974, 1976).

20.2.5 DENA ICE CAVE

Dena Ice Cave and its surroundings are shown in Fig. 20.7. It is an arid region, but it has residual snow and firn fields. There is one small remnant glacier (4200–3800 m a.s.l.) with prominent recessional moraines that suggests that it may have been active as recently as the Little Ice Age (AD 1550–1800).

The cave is developed in the Darian Formation (Early Cretaceous), a thick- to massively-bedded gray limestone with gray marl, that is overlain by further marls and argillaceous limestones and shales extending to the Upper Cretaceous. The limestones are dipping steeply at the cave, which is ~200 m in length and oriented close to the strike of the rock. The entrance is at 3895 m a.s.l., the highest limestone cave currently known in Iran. The cave is a relict, with mostly frost-shatter (small clast) breakdown morphology; some remnants of solution pockets and scalloping suggests that it has been a stream channel cave at times in the past. From the setting (Fig. 20.7) it could have lain behind a small cirque glacier

FIG. 20.7

Google Earth view from the west, showing the location of Dena Cave, Heram Summit, and the village of Khafr in the back ground. Note the residual glacier or firn field with prominent moraines to the right (south) of the cave and a few hundred meters above it in elevation.

during the ice ages. The entrance is now blocked by a snow bank for most of the year, but it may open in the mid-or later summer when most snow at this altitude has melted (Kosar, 2014). The passage descends approximately 10 m below the entrance level to terminate in breakdown (Fars Speleological Society, 2007). Closure of the entrance for most of the year, and the descent of the cave away from it inside, indicates that it probably functions as an effective cold trap.

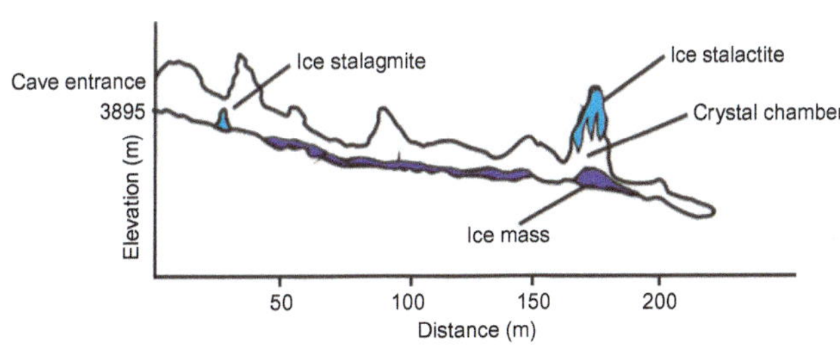

FIG. 20.8

Long section of Dena Ice Cave, showing the distribution of the ice in *blue*. The *light blue* refers to seasonals, and the *dark* ones are perennials. Topographic survey is CRG Grade 2.

There is a meteorological station in the village of Khafr, at 2250 m a.s.l. and only six km from the cave. From the 30-year climate normals (Fig. 20.4B), the mean annual temperature is 12.8°C

(Jan. −3°C; Jul. 25.1°C) at Khafr. Mean annual precipitation is 262 mm, with Jan. averaging 55 mm and a pronounced summer drought (Sep. −0 mm). The Köppen-Geiger climate type is BSk. The cave is nearly 1700 m higher in elevation than the met station. Applying the standard global temperature lapse rate of 6°C/1000 m, the mean annual temperature at the cave will be ~3°C, with the Jan. mean around −10°C, and negative mean temperatures for at least 6 months of the year. Annual precipitation at the cave will be higher than at Khafr, perhaps by twice as much (500–600 mm).

There is a thin and variable ice cover on the breakdown floor throughout the cave, partly supplied by melt from the entrance snowbank. At the "Ice Stalagmite," Points 9C and D, and the "Crystal Chamber" in Fig. 20.8, tributary seepage inlets high in the passage walls have supplied the water to build columns of ice 4 or 5 m in height, and decimeters to 1 or 2 m in diameter (e.g., Fig. 20.9D) and they then overspilled to build bigger, layered ice masses across the floor of the main cave beneath them (Fig. 20.9B and C). In Aug. 2016 it was observed that some ropes and bottles left by previous visitors were now buried to depths of 5–30 cm beneath the uppermost layers. Fig. 20.9D shows that sublimation is occurring in typical cold air convection cells in the cave today.

FIG. 20.9

Dena Ice Cave. (A) The cave entrance, seen wide open for access in Aug. 2016. For much of the year it is blocked by a snow bank. (B) In the Crystal Chamber. A large dark ice mass has accumulated here. (C) The first author descending a steep icefall in the middle of the cave. (D) A tributary icefall "medusa dripstone" on the left side in scene (C) is ~4 m in height, apparently perennial and with large (slow convection) sublimation scalloping.

This is a very preliminary study of the conditions in Dena Cave. In contrast to Yakh-Morad Cave, it appears that there is a substantial net accumulation of seepage ice below two or three tributary inlets today. If that is correct, it is most likely to be attributable to regional warming melting ground ice obstructions in the tiny tributary tubes.

20.3 CONCLUSIONS

Cave exploration and science is at an early and exciting stage of development in Iran. There are probably many other caves with perennial ice deposits to be found at high elevations in the Alborz and Zagros Mountains. This chapter is the first study of them to be published internationally.

Yakh-Morad Cave is a relict multilevel system at ~2450 m a.s.l. in the northerly Alborz Mountains, where the mean annual temperature is around 9°C, and precipitation is only ~200 mm. However, Jan. mean temperature is significantly negative (−5.6°C), and the cave is able to maintain a climatic cold trap structure. A temperature record taken at the height of the summer (Aug.) found weakly positive temperatures in the lowest trap points. At least some of the perennial ice masses there are being severely ablated, and they are believed to be unlikely to survive for many more years.

At 3950 m a.s.l., Dena Cave is 1500 m higher than Yakh-Morad. Along much of its 200 m passage net accumulation of ice may be occurring on the floor. Monitoring at annual intervals is recommended, including placement of marker pins or stakes, and coring of the ice where it is believed to be thickest.

ACKNOWLEDGMENTS

Drs. A. Nadimi, A. Persoui A.N. Palmer are thanked for professional guidance. Special thanks to geologist friends, Mansure Sani, Mohsen Asgari, Hossein Shayegan & Arash Sohrabi for assisting during the field work and other studies. Cavers Ehsan Naji, Mehdi Mohammadi, Farzad Ahmadzade, Yones Shariatmadari and Mohammad Chavoshi are thanked for their company in the caves.

REFERENCES

Bahadorinia, Sh., Hejazi, H., Nadimi, A., Ford, D., Wainer, K., 2016. The morphology and development of Kalahroud Cave. Iran. Int. J. Speleol. 45 (3), 243–257.

Bosak, P., Bruthans, J., Filippi, M., 1999. Karst and caves in the salt diapirs, SE Zagros Mts., Iran. Acta Carsol. 28 (2), 41–75.

Brooks, S., the members of Damavand Club, Tehran, 2008. Yakh-Morad Plan section, 1:500.

Bruthans, J., Filippi, M., Geršl, M., Zare, M., Melková, J., Pazdur, A., Bosák, P., 2006. Holocene marine terraces on two salt diapirs in Persian Gulf (Iran): age, depositional history and uplift rates. J. Quat. Sci. 21 (8), 843–857.

Colucci, R.R., Fontana, D., Forte, E., Potleca, M., Guglielmin, M., 2016. Response of ice caves to weather extremes in the southeastern Alps, Europe. Geomorphology 261, 1–11.

Fars Speleological Society, 2017. Profile section of Dena cave.

Ford, D.C., Williams, P.W., 2007. Karst Geomorphology and Hydrology, second ed. John Wiley & Sons, Chichester 562 p.

Ghohroudi Tali, M., Purmusa, S.M., Khosravi, S., 2012. Potential of seismicity applying multi-indicator models cave study: district 1 of Tehran City. Quantetive Geomorphology Researches: No 3, pp. 57–68 (in Persian).

Hesami, S., Yefremova, Z., Ebrahimi, E., Ostovan, H., 2006. Little known and new species of Eulophidae (Hymenoptera: Chalcidoidea) from Iran. Entomofauna 27 (28), 393–402.

Javanshad, A., 1995. Caves and Cavers. Ali Javanshad, Tehran 286 p. (in Persian).
Kaufmann, G., 2002. Ghar Alisadr, Hamadan, Iran: first results on dating calcite shelfstones. Cave Karst Sci. 29 (3), 129–133.
Kosar, N., 2014. Dena Yakh cave, the highest limestone cave in Iran. In: 1st International Congress of Geoscience, Tehran.
Kowsar, N., 1992. Speleology and Hydrogeology of Shapur Cave. (M.Sc. Thesis). University of Shiraz, Shiraz.
Kuhle, M., 1974. Vorläufige Ausführungen morpholog. Feldarbeitsergebn. a. d. SE-Iranischen Hochgebirgen am Beispiel des Kuh-i-Jupar. Z.f. Geomorph. N.F. 18 (4), 472–483.
Kuhle, M., 1976. Beiträge z. Quartärmorphologie SE-Iranischer Hochgebirge. Die quartäre Vergletscherung d. Kuh-i-Jupar. Bd. 1, Text: 1-209; Bd. 2, Abb.: 1-105. Göttinger Geogr. Abh. 67.
Nilfouroushan, F., Masson, F., Vernant, Ph., Vigny, Ch., Martinood, J., Abbasi, M., Nakhlaki, H.R., Hatzlfeld, D., Ashtiani, A., Doerflinger, E., Daignieres, M., Collard, P., Chery, J., 2003. GPS network monitors the Arabia-Eurasia collision deformation in Iran. J. Geodesy 77 (7), 411–422.
Peguy, C.-P., Blanghet, G., 1959. Les conditions meteorologiques de l'ablation su r le glacier de St. Sorlin (ete 1957). Berichte des Deutschen Wetterdiensles, Bd. 8, Tr. 54, pp. 14–17. [Studies of albedo and ablation in [957 and compa rison with conditions in 1905.].
Persoiu, A., Onac, B.P., 2012. Ice in caves. In: White, W., Culver, D.C. (Eds.), Encyclopedia of Caves. Elsevier Inc., Oxford, pp. 399–404.
Raeisi, E., Kowsar, N., 1997. Development of shapour cave, Southern Iran. Cave Karst Sci. 24 (1), 27–34.
Raeisi, E., Laumanns, M., 2003. Iran cave directory. Berliner Höhlenkundliche Berichte. vol. 10. Speläoclub, Berlin. 101 p.
Raeisi, E., Ghazi, S., Laumanns, M., 2012. Iran cave directory. Berliner Höhlenkundliche Berichte. vol. 10. Speläoclub, Berlin. 315 p.
Rezaei, M., Nakhaei, M., 2008. Speleogenesis of Katale-khour Cave, Zanjan Province. Faslname ye Zaminshenasi ye Iran 2 (6), 11–19 (in Farsi).
Sabokkhiz, F., Hejazi, S., Moghadasin, M., 2012. Analysis of geotourism of Khas e Tarash Cave by the Pralongn method. Geogr. Environ. Plann. 23 (2), 69–86.
Soleymani, M., 2015. Hydrogeology and Morphology of Yakhmorad Cave. (M.Sc. Thesis). Kharazmi University, Tehran.
Vardanjani, H.K., Bahadorinia, S., Ford, D.C., 2017. An introduction to the hypogene caves and karst of Iran. In: Klimchouk, A., Palmer, A.N., Audra, P., De Waele, J., Auler, A. (Eds.), Selected Hypogene Karst Regions and Caves of the World. Springer, Dordreicht, Heidelberg.
Yousefi, H., Noorollahi, Y., Ehara, S., Itoi, R., Yousefi, A., Fujimitsu, Y., Nishijima, J., Sasaki, K., 2009. Developing the geothermal resources map of Iran. Geothermics 39, pp. 140–151.

FURTHER READING

Soleymani, M., Nakaei, M., Nadimi, A., 2016. Hydrogeology and Morphology of Yakhmorad cave. In: 2nd International Congress of Geoscience, Tehran.

CHAPTER

ICE CAVES IN ASIA

21

Olga I. Kadebskaya*, Bulat R. Mavlyudov†

Mining Institute of the Ural Branch of the Russian Academy of Sciences, Perm, Russia Institute of Geography of the Russian Academy of Sciences, Moscow, Russia†*

CHAPTER OUTLINE

Ice Caves. https://doi.org/10.1016/B978-0-12-811739-2.00021-8

21.1 ASIA

As many areas in Asia are located in very high mountains with a cold climate (e.g., Pamir, Tien-Shan, Himalaya), these areas present the possibility to form cave glaciation in natural and artificial cavities. But in many cases the high latitude areas of Asia are more favorable for cave glaciation origin.

Permanent glaciation is known in the following countries: Kazakhstan, Uzbekistan, Tajikistan, Turkey, Iran, India, China, Japan, and Mongolia. Cave glaciation can also originate in mountain areas in Pakistan, Nepal, and Bhutan. In Asia caves ice accumulation is mainly congelation ice (different kinds of icings) and sublimation ice.

21.2 THE COMMONWEALTH OF INDEPENDENT STATES (CIS) COUNTRIES

Cave glaciation in Asia is extensive, but almost completely unstudied. There are only some places where ice in caves have been reported. It is connected not so much with impossibility of the occurrence of cave glaciation, but with weak levels of investigations in many mountain areas. We will consider those few examples which have been revealed in the extensive area of Asia (without the Asian part of Russia; Chapter 26).

21.3 KAZAKHSTAN

Cave glaciation was found only in a northwest part of Kazakhstan around Inder Lake. Karst of Inder Lake area belongs to Inder-Embensk karst district of the West Pre-Caspian karst province of the Nizhnevolzhsko-Ural karst area of the East European karst country (Golovachyov, 2012). There is sulphatic karst on northern and northeast coasts of lake.

In vicinities of Inder Lake during a cold season there is also an accumulation of cryogenic cave sediments. During expeditions in May, 2011 cave «cold bags», in which all walls are plentifully covered by ice crystals, were discovered. The floor was covered by a thick ice cover, and under entrance apertures snow-ice cones were situated. At this time the temperature on a surface exceeded +25°C. According to (Yatskevich (1937), at the bottom of some dolines on the northern coast of Inder Lake snow accumulations remain during the entire year. At that time, when the surface air temperature exceeds +35°C, at the bottom of some karst cavities and dolines air temperature remains at about 0°C (by data for Jul. 22, 1936). On the basis of the measurements (Yatskevich, 1937) it has been established that on depth from 5 to 11 m from the Earth's surface there is a sharp air temperature drop from +29.5°C to +1.5°C (i.e., on 6 m of vertical distance, temperature amplitude variation consists of 28°C!)

21.3.1 LEDYANOJ PAPORTNIK CAVE

21.3.1.1 Site and situation around the cave

The Paportnik Cave is located in an area of the Inderborskij district of the Atyrauskaya region (Kazakhstan). The cavity is formed in gypsum of the Kungur stage of the Lower Permian. The gypsum is overlain by a thin layer of old Caspian sediments. MAT = 9.3°C, $T_{Jan} = -8.6$°C.

21.3.1.2 Cavity form, cavity maps

The cave has a length about 210 m, and a depth of about 25 m. It is the largest cave in the area and a unique cave in the Northern Pre-Caspian with permanent icing (Golovachyov, 2013). The cave entrance represents a pit with a depth of about 14 m, which is located at the bottom of an asymmetric dual karst doline (Fig. 21.1).

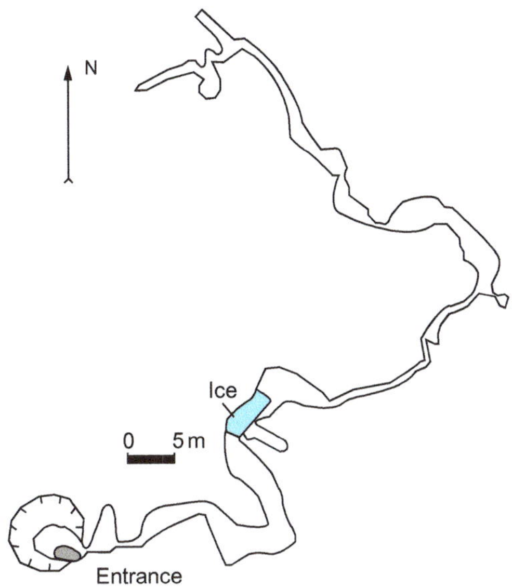

FIG. 21.1

Plan of Ledyanoj Paportnik Cave (Golovachyov, 2016).

21.3.1.3 Research history

The cave was found on the northeast coast of Inder Lake in May, 2015 during the expedition of members of the Speleology and Karstology Commission of the Astrakhan branch of the Russian Geographical Society (Golovachyov, 2016).

21.3.1.4 Ice description

At about 46 m from the entrance permanent layered ice is present with a thickness of about 2.7 m, a width of about 1.5 m, and a length of about 4.7 m (Golovachyov, 2015). During winter in a transitive zone the glaciation increases at the expense of seasonal ice (ice stalagmites and stalactites).

Permanent icing has a layered structure (Plate 21.1). There are eight horizontal layers of the white crystal ice in the central part of the icing. Layers are divided by dark ground layers (with a thickness from 0.5 to 1 cm) with vegetative debris frozen in ice. The average thickness of ice layers was about 10 cm. The thinnest ice layer was 3 cm, and the widest one was 30.0 cm (upper layer). Ice mineralization was equal to 894 mg/dm^3, and the ice belongs to the sulfat-calcium type.

Air temperature in the zone near icing at the beginning of May 2015 was −0.3°C, and at the beginning of October it was +0.4°C (Golovachyov, 2016). In May, 2016 air temperature near icing was +1.0°C.

21.4 UZBEKISTAN

21.4.1 BAYSUN TAU RIDGE

The Baysun Tau Ridge is the southwest branch of Gissarskij Ridge, located in the southeastern boundary of Uzbekistan (Surkhan-Dar'inskaya Region), close to Afghanistan and Tajikistan, near 38°41′46″N, 67°29′25″E. Baysun Tau Ridge is elongated 50 km from South-West to North-East. Ridge elevations change from 3500 to 3900 m a.s.l. The ridge is formed of Jurassic limestone. There are two main massifs on the Baysun Tau Ridge: Khodja Gur Gur Ata, and Ketmen Chapty. The climate of Southern Uzbekistan is characterized by Köppen-Geiger classifications (Peel et al., 2007) from dry hot summer (Csa) to arid cold steppe (BSk). Most of the precipitation falls between autumn and spring (Aizen et al., 1996; Sorg et al., 2012), while the summer season receives only very limited rainfall from local thunderstorms.

21.4.2 KHODJA GUR GUR ATA KARST MASSIF

The Khodja Gur Gur Ata karst massif of the Baysun Tau Ridge is elongated to 40 km. The geologic strata are presented by Cretaceous sandstone and gypsum (upper layer), and Jurassic limestone (lower layer). As a whole, the massif presents a monocline that lowers to the northwest with an angle about 10–25 degrees. The slopes looks like an incline plateau toward the South-East, and breaks off by a 400 m high vertical limestone wall that contained many caves entrances. Elevations of cave entrances range from 3200 to 3800 m a.s.l.

Permanent glaciation in Khodja Gur Gur Ata karst massif origin exists in the following caves: Ledopadnaya, Dark Star, Ulug-Bek and Morning Star. The biggest ice thickness among massif caves was found in Ledopadnaya and Dark Star caves.

21.4.2.1 Dark Star Cave

Site and situation around a cave

The Dark Star Cave is a part of the karst system Hodzha-Gur-Gur-Ata. The cave system has a set of entrances (more than 20), all from which are located on the vertical cliff wall. Gallery length of the karst system is about 16 km, and the amplitude is more than 900 m (Topography of Sverdlovsk City speleosection, 2016). Actually, Dark Star Cave has six entrances on the area of the wall with a length of about 2 km. All entrances in the cave are oriented toward the southeast. MAT = −1.2°C, T_{Jan} = −13.6°C.

Cavity form, cavity maps

The main entrance in Dark Star Cave is located at a height of about 160 m from the cliff wall base, and it has a height of 60 m, and a of width 7 m. The entrance elevation is about 3500 m a.s.l. (adjacent are the

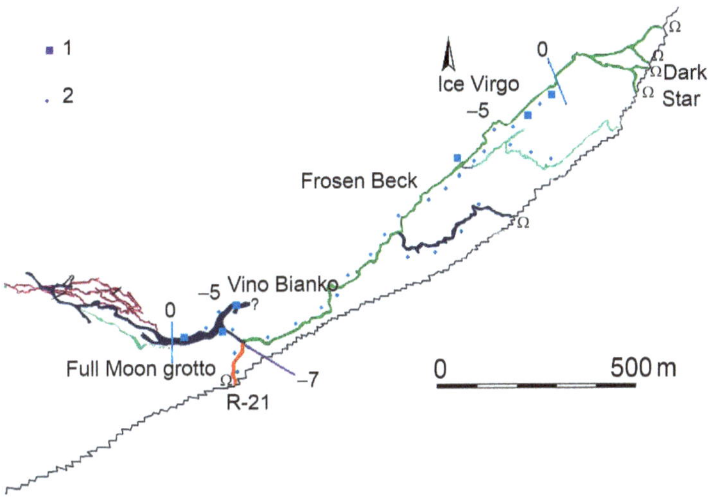

FIG. 21.2

Plan of Dark Star Cave (тopography Sverdlovsk city speleosection, 2016).

entrances Rak, Kozerog, Krasnyj Karlik, Orenburgskij, Vino Rosso). Entrance R-21, which has permanent ice close by, is located on a steep cliff wall 150m up from the base, at a height about 3380m a.s.l. and represents a meander with height more than 40m (Fig. 21.2).

In Aug. 2015 air temperature in the cave was from −7 to +4°C. Temperatures in the Full Moon Chamber, where the main quantity of permanent ice is located, is −5°C, and the humidity is 90%. Obviously, winter cold air penetrates into the cave through the lower entrances (mainly through R-21, and possibly through R-20 at 3450m a.s.l, and R-22, R-23 3400m a.s.l., which are not yet connected with the cave system). Further air stream flow is divided into two parts, one of which moves into the deep part of the massif and cools the Full Moon Chamber, and the second stream moves through Frozen Beck and Ice Virgo galleries toward the main entrance. Return draughts of warm air in the summer through main entrance move into the cave, but after 150m at the galleries, narrowing air velocity and temperature decrease in the Metro Chamber, and air does not have enough energy to warm the cave walls. In Full Moon Chamber the air temperature is always negative.

Research history

Research of caves of Khodzha Gur-Gur-Ata Ridge began in 1984, when in area of cliff wall at Katta-Tash Mountain many big caves were found at once: Berloga, Yubilejnaya, Sifonnaya, and in the next years, Festivalnaya, Ledopadnaya, Tonnel'naya, and Uchitelskaya. It was clear that additional caves are on other places of the 35-km long steep cliff wall. It was even not necessary to search caves especially, as many entrances were visible from a distance. But the question was remained: how to reach them? The entrance of Dark Star Cave was reached in 1990 by an expedition of English cave explorers (Aspex 90) by ascension from the cliff wall base for 3 days. Cave explorers from club SCS (Sverdlovsk city speleoclub) and Italian club «La Venta,» have started to work in caves Dark Star and Ulug-Bek "from above."

Ice description

First research done in 1990 established that the cave walls are almost all covered by the big ice crystals, and in the main gallery there was a set of frozen lakes, and air temperature during summer in cave galleries was between 0 and −7°C.

Directly at the cave entrance exists a snow mass, which, already in 10m from entrance walls, is covered by sublimation ice crystals (up to Full Moon Chamber and also almost in all old cave part). Permanent icings are found in two places in the cave: in Full Moon Chamber and in Ice Virgo Gallery. In Ice Virgo Gallery, the ice cover floor is sometimes flat (frozen lakes), and sometimes inclined (frozen streams). Ice thickness is up to half of a meter. Sometimes after cold winters, lakes in Frozen Beck Gallery freeze, but in some years the ice is thin, or the lakes are not frozen.

Ice originates from infiltration, condensation waters, and from entrance snow melting. When ice melts in upper parts of the galleries, water flows down and freezes in icings. In the entrance into Full Moon Chamber, there is powerful icing 20×30m, with a thickness up to 7m (Plate 21.1). Icing origin is not clear. From 2011 to 2016 icing thickness has decreased by 0.1–0.5m. In the right part of Vino Bianko Chamber, there are massive stalagmites which are 5m in height, in the left part there are also stalactites and columns (to 7m in height), frozen waterfalls, and stalagmites. A traverse of the chamber demonstrates initially an air temperature increase to 0°C, then a drop in temperature with an ice crust appearing on the floor. Further away, ice in the cave is not observed.

PLATE 21.1

Ice in some caves of Asia. (A, B) Permanent ice in Ledyanoj Paportnik Cave; (C) sublimation scaling in permanent ice in Full Moon Chamber, Dark Star Cave (photo: Tsurikhin E.A.); (D) sublimation crystals and icings in central part of Full Moon Chamber, Dark Star Cave (photo: Romeo A.).

21.4.3 PLATEAU KYZYL-SHAVAR

The plateau is located in Gissarskij National Park (Hissarskij Ridge) in the Yakkabagskaya area of the Kashkadarinskaya region of Uzbekistan. The co-ordinates are 38°55′N, 67°24′E. On the plateau, 40 caves are now known. Entrances of the majority of the investigated caves are located at an elevation between 3500 and 3600 m a.s.l. The caves are formed in Jurassic limestone. For the area the following are characteristic: dryness and sharp climate continentality. In winter the snow drops out for long periods without precipitation, in the summer it is typical that the pure sky means air dryness. There is cave glaciation in some plateau caves. Entrances in caves are located in opened cracks or at the bottom of dolines. Vertical cavities often can be blocked by icy or snow-stone plugs, and in inclined descending caves the cover can be icings, stalactites, stalagmites, columns, and sublimation crystals that develop. SIF are found in the largest caves: Gissarskaya, Sosul'ka, U109, Ledyanoj Conus/U27, etc. Apparently, caves with ice can be found on two other karst plateaus next to Kyzyl-Shevarom.

21.4.3.1 Gissarskaya Cave

Site and situation around the cave

The cave is located on the Kyzyl-Shavar karst plateau (Hissarskij Ridge). The entrance elevation is about 3530 m a.s.l. MAT = −3.3°C, T_{Jan} = −16.2°C.

Cavity form, cavity maps

Cave length is about 1761 m, with a depth of 204 m. The cave entrance is in a small doline in a thalweg of a seasonal stream in an average part of the plateau. The cave represents the cascade of vertical iced pits to a depth of about 60 m, which can change in a system of inclined meanders (Fig. 21.3).

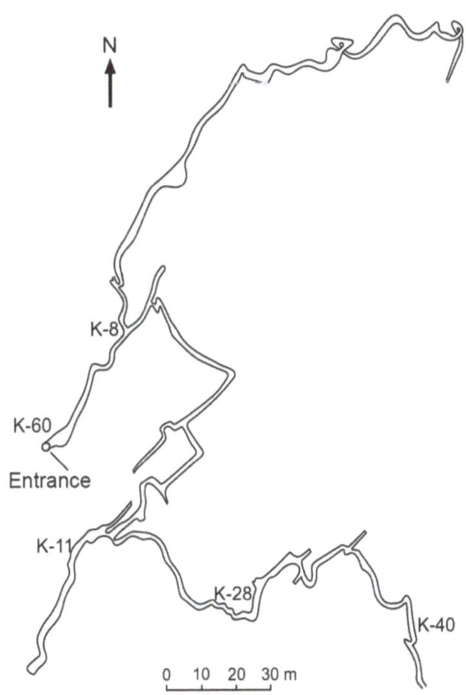

FIG. 21.3

Plan of Gissarskaya Cave (part of cave containing ice) (Sachko, 2012).

Research history

The cave was been found by a joint expedition of the Ternopol and Kiev speleoclubs in 1981, and it has received code T21. In Aug. 1982 by Kiev speleoclub "Podolia," a depth increase up to 150m was recorded. Other continuations have been found during expeditions of the Chertkovskij speleoclub "Crystal" (Ternopol, Chernovtsy) in 2011.

Ice description

The entrance part of the cave has an abundance of sublimation crystals, stalactites and stalagmites, the meander at the bottom is covered by a thick ice layer. Icings in the cave reach a depth of about 140m. In the summer in the glaciation zone, air temperature is about 0.0°C, and in a distant part in the meanders it is about 1.5°C.

21.4.4 PLATEAU KYRK-TAU

Plateau Kyrk-Tau is located 10km to the southwest from the city Urgut, and 40km to the south from Samarkand (Urgutskij area of Samarkand district). The plateau represents a spur of the Chakytkalyan Ridge and of the Zeravshanskij Ridge. The plateau is extended from the northwest to the southeast for 12km and is limited to the south by the Kashkadarya River valley, and from the north by the gorge Zaranzasaj, and continues to the Zeravshan River valley. The plateau is formed by slates and sandstones from the Precambrian and Lower Paleozoic ages, and limestone and dolomite from the Silurian and Devonian, which is broken through in the east part by an intrusion of granodiorite and granite-porphyries. The co-ordinates are: 39°20′44″N, 67°07′41″E. Elevations are about 2000–2400m a.s.l. Here there are about sixty karst cavities noted and studied. Snow in karst dolines and in separate pits remains during the whole year. MAT=6.6°C, $T_{Jan}=-6.2$°C.

21.5 TAJIKISTAN

Cave glaciation in Tajikistan is known only in the Zeravshanskij Ridge and on East Pamir. Cave glaciation is probably possible also in other high-mountainous areas of Tajikistan.

21.5.1 ZERAVSHANSKIJ RIDGE

21.5.1.1 Nuriddina Cave

Site and situation around the cave

The cave is located in the eastern part of the Zeravshanskij Ridge near to the city of Pendzhikent, in the Pendzhikent district of the Sogdijskaya region, western Tajikistan, at an elevation of about 2500m a.s.l. The cave is located in the riverhead of Kamartoshsaj in the Chakylkalyap Mountains. The cave is situated on slopes of northern exposure in vertical standing layers of strong disunited blue and gray limestone from the Late Silurian age.The cave originated on the basis of a tectonic fissure widened by karst processes. MAT=2.8°C, $T_{Jan}=-10.1$°C.

Cavity form, cavity maps

The cave entrance is oriented on the east north-east and has the form of an irregular tetragon, with a width 2.5m and a height of 4m. The cave is oriented along azimuth 270 degrees. It is an inclined descending cavity. Directly from the entrance, the cave bottom is down 5 m. The cave length is about 40 m. The cave has a second entrance, a vertical slot with a length of about 15 m.

Research history

The data of local people shows that snow and ice in the cave was always permanent. Up until 1960 snow and ice was transported to the city of Pendzhikent, where they were used it to cool ice-cream and drinks.

The ice description

Snow penetrates into the cave through an entrance in the ceiling. Ice forms as the result of snow melting, then melt water freezing at night or at the beginning of winter. Snow layer thickness changes from 1 to 4 m. The snow lies on ice, which has a thickness of about 0.5 m. The area of snow and ice is about 50 m^2. There is no ice or snow at the entrance or at the end of the cavity (Abduzhabarov, 1963). The quantity of snow depends on winter precipitation and changes from year to year by a significant amount. The biggest snow quantity in the cave exists in spring (May), and the lowest exists in autumn (September).

21.5.2 PAMIR

On Pamir, caves are developed in Jurassic limestones, which are found mainly in East Pamir at elevations above 4200 m a.s.l. Rocky karren are mainly developed at these heights, where on the flat areas rare tundra vegetation has developed, and on rocky areas vegetation is absent completely. The climate of East Pamir is characterized by cold summers and severe winters with little snow. For these areas, an almost full absence of a frostless season is typical. Frosts on soil occur all throughout the summer. MAT is equal to −5 to −7°C.

Caves of Pamir have a mainly hydrothermal origin, but they are also modeled by frosty weathering and later karst processes. Pamir caves have been very poorly investigated. Mainly, with the rare exception, these are small cavities with a length measured in tens of meters, horizontal or gently inclined. The majority of known cavities in high mountains are exposed to frost, and any water that penetrates into them freezes. Usually it happens in spring, when cover and pendent icings grow in caves. As negative air temperatures remain in caves during the summer, in combination with very low air humidity (about 30%–40%), there is ice sublimation in the caves. As measurements have shown, ice sublimation in the caves of southeast Pamir has a well expressed daily course, and it is possible to illustrate this by an example of observations for ice sublimation in cave № 17, from Aug. 1983 (Fig. 21.4) (Mavlyudov, 1987, 2008). However, in the absence of wind in caves, the intensity of ice sublimation is insignificant, and therefore in many cases even small ice accumulations in caves are permanent. Sublimation crystals grow in caves during spring time, which as a rule are seasonal and melt at the beginning of summer.

The potential possibility of natural cave glaciation can be shown in an example of artificial tunnels in rocks. For example, it can be shown in tunnels in upper courses of the Vanch River, at an elevation of about 4700 m a.s.l. in permafrost conditions, where water penetration formed permanent icings.

21.5.2.1 Putnikov Cave (Rangkulskaya Cave, Syjkyrduu)

Site and situation around the cave

The cave is located in a Jurassic limestone massif framing from the south depression of Rangkul Lake, East Pamir (Tajikistan). The cave entrance is located in the right side of the Salak-Tash gorge, at the

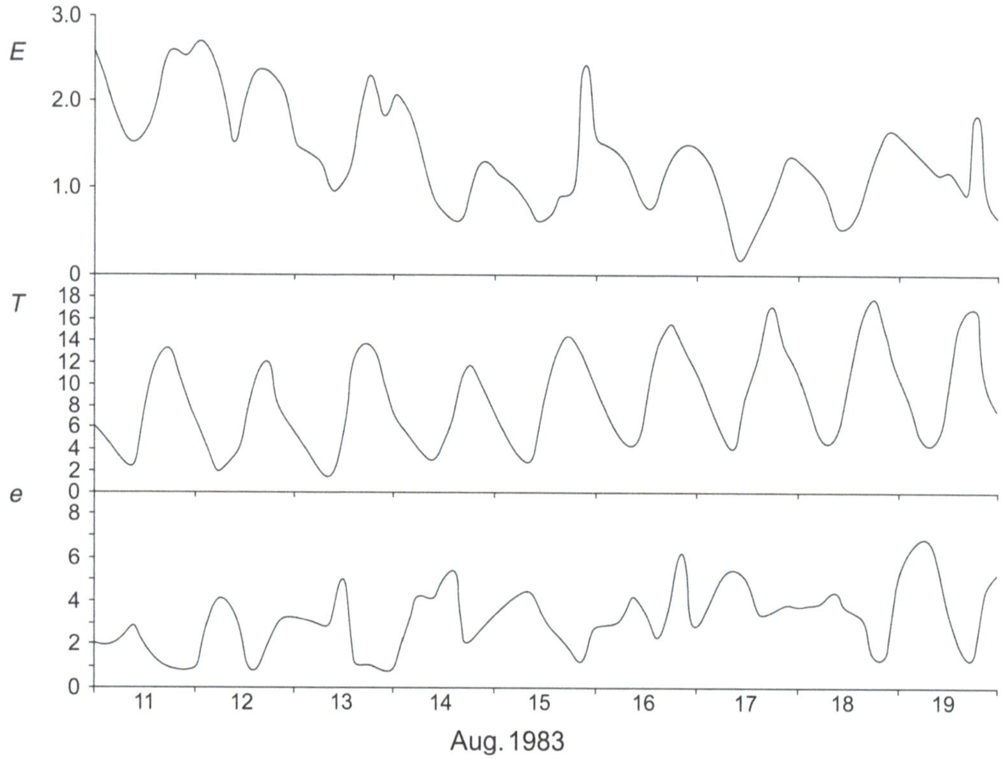

FIG. 21.4

Changing of ice sublimation intensity (*E*, 10^{-3} g cm^{-2} h), cave air temperature (*T*, °C) and humidity (*e*, mb) in cave № 17, southeast Pamir (Mavlyudov, 1987).

top of a long rubble talus, with a distance of about 4 km from the gorge mouth. The entrance elevation is about 4600 m a.s.l., and the relative height over an adjoining valley is 600 m, over a linear distance of 1200 m. The entrance is oriented to the south. MAT = −8°C, T_{Jan} = −25°C.

Cavity form, cavity maps

The entrance has a height of about 3.5 m, and a width of 5.0 m. A wide gallery (width to 25 m, height up to 25 m) begins downwards on the northeast. The cave represents the branched out system of chambers and galleries located at different levels and connected by network of fissure or cylindrical pits. Cave depth is about 240 m, and the elevation above datum is 28 m. The cave gallery length is about 2050 m (Fig. 21.5).

Research history

For local residents the cave has been known for a long time. The first publication about the cave was in 1898 (L.O., 1898). Bletshunov (1957) mentions the ice presence in the left lateral passage of the first

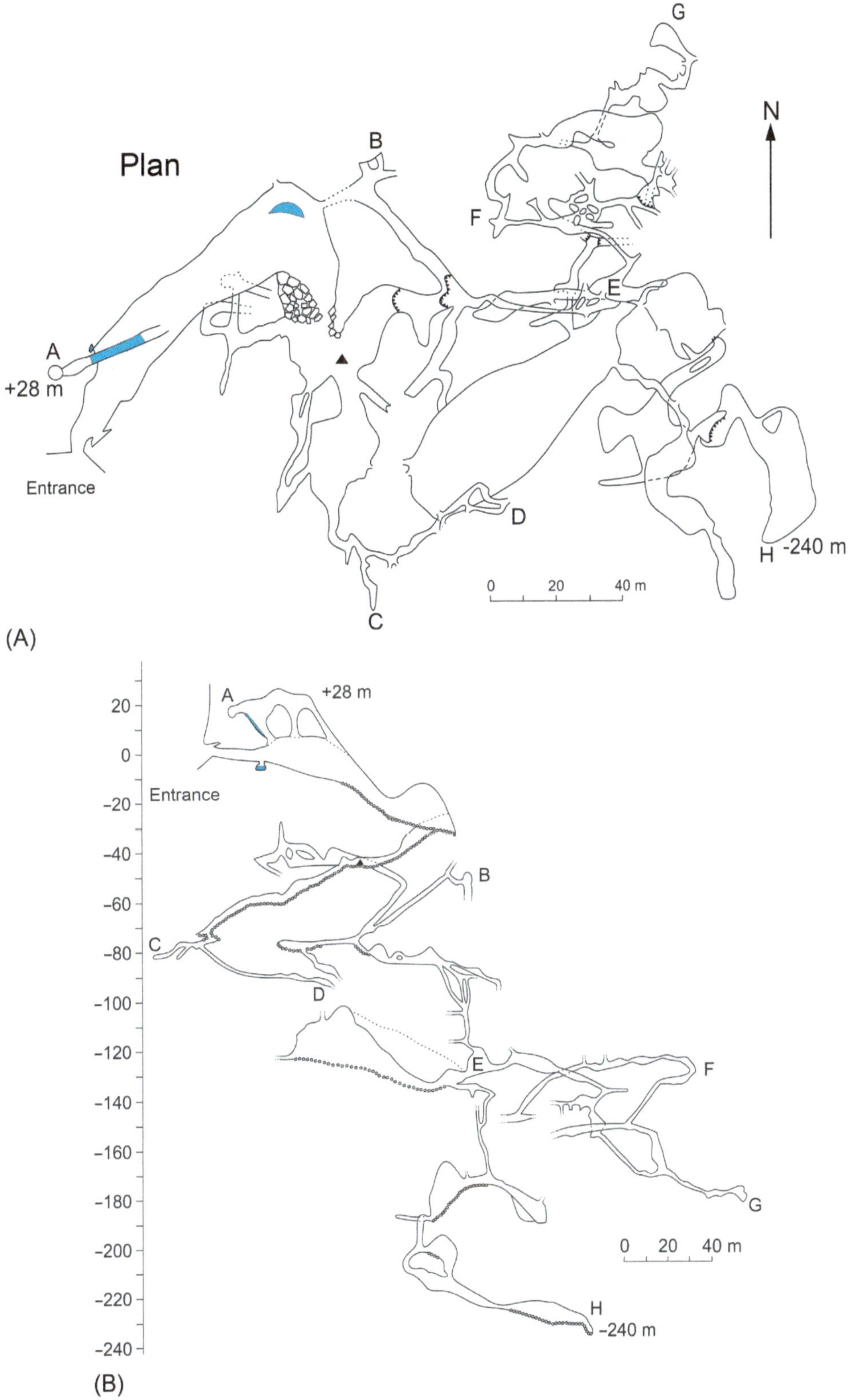

FIG. 21.5

(A) Plan and (B) longitudinal section of Putnikov Cave (Ridush, 1993).

cave chamber. In 1983, in the upper part of the cave, glaciation was observed (Mavlyudov, 1987). Ice presence in the cave was specified during subsequent speleological research (Ridush, 1993).

Forms and dynamics of ice

In caves there are two types of ice: icings and sublimation crystals. Icing originates by melt water freezing during spring. Ice is formed mainly in the entrance gallery. Because of the considerable dryness of the air (about 30%), and the negative air temperatures up to the end of summer (in Sep. 1983 was about −1 to −2°C), ice practically completely evaporates, remaining only in lateral branches and in the galleries of the remote upper level.

21.6 TURKEY

21.6.1 ALADAGLAR MASSIF

Aladaglar massif is situated in Central Taurus, in the Adana-Kayseri-Nigpe provinces of Turkey. The massif is composed mainly of Triassic, Jurassic, and Cretaceous limestone and rises to 3750 m a.s.l in elevation. Morphologically, northern (Black Aladag), central, and southern (White Aladag) sectors can be distinguished, with the local relief increasing from north to south. The high-altitude parts of the Aladaglar massif have been severely glaciated during the Quaternary period. Glacial erosion was the dominant factor in the overall surface morphology development, resulting in the formation of numerous glacial trough valleys, cirques, arêtes (narrow jagged ridges) and horn (or pyramidal) peaks. There are perennial snow accumulations in dolines and pits (Yedigoller Plateau in White Aladag). Some large shafts at the tops of ridges (higher than 3400 m a.s.l.) are blocked by ice, which contains numerous bands of frost debris within (e.g., there is about 100 m in the Ice Cave). MAT = −1.7°C, T_{Jan} = −16.6°C.

21.6.1.1 Ice Cave in Ağri (Buz Mağarasi)

Site and situation around the cave

The cave is located on the plain at the end of the foot of the small Ararat Mountain (3925 m a.s.l.) close to the main Ararat Mountain (5137 m a.s.l.), to the northeast from Doğubayazıt, near the village of *Hallaç*, Ağri district, near the boundary of Turkey and Iran. The entrance height is about 2000 m a.s.l. MAT = 5.5°C, T_{Jan} = −10.1°C.

Cavity form, cavity maps

The cave is an inclined descending cave, and the entrance is located at the bottom of a doline with a depth of about 7–8 m. From the doline bottom the inclined course begins. The cave length is about 100 m, with a width to 50 m.

Forms and dynamics of ice

In the cave there are many ice stalagmites and stalactites of the different sizes.

21.6.1.2 Buzluk Cave (Ice Cave of Harput)

Site and situation around the cave

The cave is located 11 km to the northeast from Elazig city center, Harput district. The cave was formed among limestone blocks. The entrance elevation is about 1653 m a.s.l.

MAT = 9.0°C. T_{Jan} = −4.4°C.

Cavity form, cavity maps

The ice cave is located among the rocks that are named «Icy stones»; it has a length of about 150–200 m, and a depth of about 30 m. As a result of geomorphological features, cave form (incline descending cave), and climatic conditions of the district in summer months, natural ice inside the cave is formed, in winter months ice in the cave is absent.

Research history

History of the cave is as old as the Harput district and goes back to the Urartian period (c.1500–585 BC). From ancient times, with approach of summer, the local population brought to the market in Harput ice collected in the cave to sell.

Forms and dynamics of ice

Natural ice stalactites, stalagmites, columns and ice layers are in some sections. Ice melts towards the end of summer.

21.7 INDIA

Because of a hot climate, ice in the caves of India can be found only in high mountains.

21.7.1 AMARNATH CAVE (HOLY CAVE)

21.7.1.1 Site and situation around the cave

The cave is located about 141 km from Srinagar, the capital of Kashmir (India), and 45 km from the city of Pahalgam. The cave is one of the most well-known and most sacred temples of the Hindus, and is devoted to the god Shiva. The cave is named in honor of Amarnah Mountain (5486 m a.s.l.). Cave entrance co-ordinates: 34.2149°N, 75.5008°E. MAT = −0.1°C, T_{Jan} = −11.4°C.

21.7.1.2 Cavity form, cavity maps

The cave is located in a narrow gorge in the distant end of Lidder Valley, at an elevation of about 3888 m a.s.l. Most parts of year the cave entrance is blocked by snow. The cave entrance width is about 40 m, the height is 40 m. The cave length is about 39 m.

21.7.1.3 Research history

The cave, as a temple, has been known for about 5000 years, and it has been mentioned in ancient Hindu texts. The cave is a popular place of pilgrimage for approximately 400,000 Hindus during the 45 day festival Shravani Mela in July–August, coinciding with the Hindu sacred month Shravan. According to Hindu mythology, it is a cave where the god Shiva told about the secret of an eternal Life to the divine spouse Parvati. Therefore this relic has special value for Hindus. Adherents make the 45 km pilgrimage on foot from city of Pahalgam, located approximately 96 km from Srinagar.

21.7.1.4 Forms and dynamics of ice

In the far part of the big cave, the chamber has an inclined floor, and there are three stalagmites of ice, a big one (ice Phallus of Shiva), and two lesser stalagmites (Parvati and Ganeshi). The height of the big stalagmite is about 3 m. The stalagmites melt from May until August. The big stalagmite's peak grows, and it is compressed according to moon phases, reaching the maximum height during the summer festival.

21.8 CHINA

In China, more than ten ice caves have been described so far, but only the Ningwu Ice Cave has been studied in detail (Gao et al., 2005; Meng et al., 2006; Shi and Yang, 2014; Yang and Shi, 2015), and hence more data about that cave is available. Undoubtedly, there should be caves with ice in Tibet and in the north of China on the border with Mongolia.

21.8.1 CAVE NINGWU

21.8.1.1 Site and situation around the cave

The cave is located in the shady slope of Guancen Mountain in Ningwu County, Shanxi Province, China. The entrance co-ordinates are: 38°57′N, 112°10′E. The entrance elevation is about 2121 m a.s.l. MAT = 2.3°C, $T_{Jan} = -16.5$°C.

21.8.1.2 Cavity form, cavity maps

The cave is bowl-like, with only one entrance in the upper part, therefore, it is a typical «cold bag». It extends downward from the surface to a depth of about 85 m, and has a spiral staircase close to the walls from the surface to the bottom. The widest part is in the middle of the cavity, with a width of about 20 m. The cave is visited every day by 1000 visitors (from May till October it is opened for travelers). Tourists spend practically an hour inside. At this time the cave is illuminated, and there are about 200 electric bulbs established in it.

21.8.1.3 Research history

Ningwu Ice Cave was found in 1998 (Yang and Shi, 2015) and very soon it became show cave.

21.8.1.4 Forms and dynamics of ice

The air temperature in the cave falls during the winter to −15°C. Above the depth of 40 m there is only layered ice, and there are lots of ice bodies along the wall below the 40 m mark. Ice in the cave is partially artificial. It is formed by the freezing of water, which is supplied through a hosepipe in the cave from the surface every winter. In the cave, different icings are developed (there are many ice stalactites). Ice volume reaches impressive sizes.

21.8.2 ICE CAVE VUDALYANCHI (LAVA ICE CAVE OF HEIHE)

In the Heilungjiang (China) province, the lava ice cave Vudalyanchi is located in Wudalianchi (Dazhanhe) National Forest Park, Heilongjiang province about 280 km to the north of Harbin. To preserve it, a metal door, which protects the cave from warm summer air, has been established. There are two caverns, the lava ice cavern and the lava snow cavern (Bing Dong), and a year-round colored light show. There are artificial ice figures inside the cave.

21.8.3 UNDERGROUND ICE CAVE

The cave is located in the East Jiaodebushan volcano in Wudalianchi (Dazhanhe) National Forest Park, Heilongjiang province. This cave has a total length of 515 m, and may be the longest lava cave so far known in China. The lava is dated as 0.512 Ma. The entrance of the cave is 1.4–1.8 m wide and 6–7 m high, although the ceiling becomes higher inside the cave. There is a hall 206 m beyond the entrance, with a width of 26.8 m, broken by two rock pillars, each roughly 3.2 m high, with a diameter of 4.5 m. In the main part of the cave ice sheet is about a meter thick, and covers the floor to a general gradient of 2–5 degrees. There are ice stalactites on the roof and small amounts of breakdown. This is a very cold cave with a temperature between 0°C and 5°C. The coordinates are: 48°44′11″N, 126°05′58.68″E, elevation 300 m a.s.l. MAT = 0.4°C, $T_{Jan} = -23.2$°C.

21.8.4 ICE CAVE

The cave is located in East Jiaodebushan volcano in Wudalianchi (Dazhanhe) National Forest Park, Heilongjiang province. This cave is more than 150 m long and has a vertical range of 23 m. The entrance is 0.6 m in height and 1 m wide. An initial steep slope becomes one of about 12 degrees inside the cave. Twenty-five meter from the entrance is a hall 8 m high and 12.4 m wide. The walls are covered by ice crystals, and decorated with lava drapes and stalactites.

21.8.5 GUBINGDONG CAVE (ANCIENT ICE CAVE)

The cave is located in the Jingpo Hu (Jingpo Lake) protected area in the upper-middle reaches of the Mudanjiang River, in southeast Heilongjiang Province. It is 110 km south from the city of Mudanjiang. Access to this cave lies about 15 km from the entrance gate of the crater forest park highway. There are three collapsed pits overlying a cave that branches in two, and the northern branch is known as Gubingdong. This cave has a passage diameter of about 8 m. In the summer, the surface water seeps into the cave through ceiling cracks and runs to the low-lying places in the cave, where it freezes in winter. Ice remains frozen during summer. The coordinates are: 43°50′55.7″N, 129°50′55.7″E, elevation 800 m a.s.l. MAT = 3°C, $T_{Jan} = -19$°C.

21.9 JAPAN

In Japan, nine caves hosting perennial ice have been described on Mt. Fuji (Wakimizu, 1936; Ohata et al., 1994a,b). Ice in these caves is accumulated by the freezing of water inside lava tubes, which are located from 950 to 2000 m a.s.l. (six of them between 950 and 1179 m a.s.l.).

The forest Aokigahara Jukai extends for 16 km, from Lakes Sai and Shoji to Lake Motosu, covering lava fields and slopes at the foot of the Fujiyama Mountain. Near Sai Lake there is the ice cave Narusawa, and the wind lava cave Fugaku. At the foot of the Fujiyama Mountain there are a lot of caves. Among them is Fugaku Fuketsu, a wind cave of 200 m in length, which has been formed by volcanic activity of Mount Fuji. Narusawa Hyoketsu is also a volcanic ice cave with a length of about 150 m. Water is frozen in it even during summer. Both caves are located in Aokigahara Jukai. To keep ice in some excursion caves in the summer, transparent blocks of artificial ice are put in with them, which are placed in the walls at the beginning of the caves. This process promotes less warming up of cavities in the summer. Only the Fuji Ice Cave has been investigated in detail.

21.9.1 FUJI ICE CAVE (FUJI FUKETSU)

21.9.1.1 Site and situation around the cave

The cave is located in a dense forest area, at an elevation of about 1120 m a.s.l. at the foot of Mt. Fuji, in central Japan. The cave was generated during an eruption in AD 864. MAT = 8.4°C, $T_{Jan} = -0.9$°C.

21.9.1.2 Cavity form, cavity maps

The cave is a 150 m long lava tube. Depth of the cavity is about 30 m. The cave entrance formed by the partial collapse of the roof of a roughly circular, 180 m long lava tube. Inside the cave, air temperature variations closely follow external ones, dropping to −11.2°C in winter, and leveling out at slightly above 0°C, in summer (Ohata et al., 1994a,b).

21.9.1.3 Research history

The ice cave was formed in the 864 year at eruption of volcano. During the Edo period (1603–1867) ice from cave was sent to government officials at the Edo Castle.

21.9.1.4 Forms and dynamics of ice

The cave has a perennial area of floor ice about 1100 m^2, with a volume of 3000 m^3, and a mean thickness of 2.8 m, and a maximum thickness of 4 m (Ohata et al., 1994a,b). On top of the ice block, in winter, ice columns and stalagmites develop through the freezing of infiltrating water, while in summer, ice sublimates on the wall to form hoarfrost. Mean ice level in the cave showed a 15 cm increase from 1984 to 1989, and then it suddenly started to decrease from 1989 to 1992. In the increasing stage, annual net balance (from December to November) was similar at various points, but in the decreasing stage the lowering of the level near the entrance was very large due to intense melting.

Water infiltrates inside the cave between early spring and summer, and, depending on air temperature, could either lead to ice accumulation (in spring and early summer) or melting (in late summer and autumn). Freezing and melting doesn't occur simultaneously throughout the cave, being delayed by a couple of months in its inner parts. Further, the amplitude of ice level changes is greater closer to the entrance due to the larger variability of air temperature and water availability. On the long term, the net balance of ice in the cave is strongly correlated with the average winter air temperature, and the preceding four years were an anomaly (Ohata et al., 1994b).

21.10 MONGOLIA

In Mongolia, caves are investigated very poorly. Glaciation is mentioned only in one cavity: the cave Uguul (Uguul aguj). The cave is located in an area of the Hubsugulskij National Park, located on southern coast of Khubsugul Lake, in somon Alag Erdene of Hubsugulskij ajmak. About the cave it is known only that it is a vertical cave with depth about 26 m, and the walls of its entrance pit are covered by ice. MAT = −3.9°C, T_{Jan} = −23.8°C.

REFERENCES

Abduzhabarov, M.A., 1963. Nuriddina Ice Cave. Priroda (Nature) 4, 115–116.

Aizen, V., Aizen, E., Melack, J., Martma, T., 1996. Isotopic measurements of precipitation on central Asian glaciers (southeastern Tibet, northern Himalayas, central Tien Shan). J. Geophys. Res. 101. https://doi.org/10.1029/96JD00061.

Bletshunov, A.V., 1957. Rangkulskaya cave. In: Pobezhdennye vershiny (Won tops). Yearbook of Soviet Mountaineering for 1954. Geografgiz, Moscow, pp. 237–251 (in Russian).

Gao, L., Wang, X., Wan, X., 2005. Analysis of ice cave formation in Ningwu Shanxi. J. Taiyuan Univ. Technol. 36, 455–458.

Golovachyov, I.V., 2012. Karst of Inder Lake vicinities. Geol. Geogr. Global Energy 2 (45), 7–16 (in Russian).

Golovachyov, I.V., 2013. Result of speleological researches in Inder Lake vicinities. In: Speleology and Spelestopogy. NICPTR, Naberezhnye Chelny, pp. 13–17 (in Russian).

Golovachyov, I.V., 2015. Karst and caves of Northern Pri-Caspian area. Publishing House (Astrakhan University). Astrakhan, 215 p. (in Russian).

Golovachyov, I.V., 2016. Speleological discoveries in Inder Lake vicinities. In: Kurapova, A.A., Tarkovoj, R.A., Gerasimidi, E.I. (Eds.), Astrakhan Meeting of Regional Studies. vol. VIII. Open Company "CCTEP" Publishing house, Astrakhan, pp. 35–42. (in Russian).

L.O., 1898. Rangkulskaya Cave and Charaktash. Turkestanskij Bull. (in Russian).

Mavlyudov, B.R., 1987. Caves glaciation of Pamir. Materialy glatsiologicheskikh issledovanij (Data Glaciol. Stud.) 59, 173–179 (in Russian).

Mavlyudov, B.R., 2008. Caves glaciation. Moscow, Institute of Geography RAS. 290 p. (in Russian).

Meng, X., Zhu, D., Shao, Z., 2006. A discussion on the formation mechanism of the "Ten-Thousand-Year-Old Ice Cave" in Shanxi Province. Acta Geosci. Sin. 2, 163–168.

Ohata, T., Furukawa, T., Higuchi, K., 1994a. Glacioclimatological study of perennial ice in the Fuji Ice Cave, Japan. Part 1. Seasonal variation and mechanism of maintenance. Arct. Alp. Res. 26 (3), 227–237.

Ohata, T., Furukawa, T., Osada, K., 1994b. Glacioclimatological study of perennial ice in the Fuji Ice Cave, Part 2. Interannual variation and relation to climate. Arct. Alp. Res. 26 (3), 238–244.

Peel, M.C., Finlayson, B.L., McMahon, T.A., 2007. Updated world map of the Ktsppen-Geiger climate classification. Hydrology and Earth System Sciences Discussions. European Geosciences Union 11 (5), 1633–1644.

Ridush, B.T., 1993. Syjkyrduu Cave on East Pamir. Svet (Light) 1–2 (7-8), 5–9 (in Russian).

Sachko, A.V., 2012. Results of speleological researches of Kyzyl-Shevar karst plateau (Republic Uzbekistan). Speleol. Karstol. 9, 22–28. (in Russian). Simferopol.

Shi, Ya, Yang, S., 2014. Numerical modeling of formation of a static ice cave – Ningwu Ice Cave, Shanxi, China. In: International Workshop on Ice Caves VI. NCKRI Symposium. vol. 4, pp. 7–11.

Sorg, A., Bolsh, T., Stoffel, M., Solomina, O., Beniston, M., 2012. Climate change impacts on glaciers and runoff in Tien Shan (Central Asia). Nat. Clim. Chang. 2, 725–731. https://doi.org/10.1038/nclimate1592.

Wakimizu, T., 1936. Geological and mineralogical natural monuments in Yamanashi and Hyogo Prefectures. Research Report on Natural Monuments. Ministry of Education, Science and Culture. (in Japanese)

Yang, S., Shi, Y., 2015. Numerical simulation of formation and preservation of Ningwu Ice Cave, Shanxi, China. Cryosphere Discuss. 9, 2367–2395.

Yatskevich, Z.V., 1937. Materials to studying of karst of the Indersky highland. News of the All-Union Geographical Society 69 (6), 937–955 (in Russian).

FURTHER READING

Breitenbach, S.F.M., Kwiecien, O., Sauro, F.V., Ya, Lu, Tsurikhin, E., Votintseva, A., 2013. Element and stable isotope aqueous geochemistry from Baysun Tau, Uzbekistan—tracing the source of the dripwater. In: Proceedings 16th International Congress of Speleology, Czech Republic, Brno, July 21–28. vol. 3, pp. 485–488.

Golovnina, Yu.D., 1902. On Pamirs. Notes of the Russian Traveller. Printing House of Association of I.N Kurnishev & K, Moscow. 244 p. (in Russian).

http://howinturkey.com/5-places-to-visit-in-agri/.

http://wikitravel.org/en/Wudalianchi_National_Forest_Park.

http://www.centralasia-adventures.com/ru/sights/gissarskaya_cave.html.

http://www.fazturkey.com/detail/151/buzluk-cave-.aspx.

http://www.iran.farsfoundation.nl/en/iran-knowing/tourism-in-iran/iranian-tourist-attractions/662-alborz10/5792-yakh-morad-cave.html.

https://en.wikipedia.org/wiki/Amarnath_Temple.

https://www.rutraveller.ru/place/97712.

https://www.tripadvisor.com/Attraction_Review-g1152408-d1852057-Reviews-Lava_Ice_Cave_of_Heihe-Wudalianchi_Heilongjiang.html.

İnceöz, M., Çelebi, H., 1997. Features, geology and formation of Harput Buzluk Cave (Elaziğ). Temmuz-Agustos-Eylül 24, 13–17. (in Turkish). Sayi.

Klimchouk, A., Bayari, S., Nazik, L., Törk, K., 2006. Glacial destruction of cave systems in high mountains, with a special reference to the Aladaglar Massif, Central Taurus, Turkey. Acta Carsol. 35 (2), 111–121. Ljubljana.

Klimchuk, A.B., Rogozhnikov, V.Ya, Lomaev, A.A., 1981. Karst of Kyrktau massif (Zeravshanskij Ridge, Tien Shan). Pre-print of Institute of Geological Sciences, Kiev. 53 p. (in Russian).

Shao, Z., Meng, X., Zhu, D., Yu, J., Han, J., Meng, Q., Lv, R., 2007. Detection for the spatial distribution of "Ten Thousand Ice Cave" in Ningwu, Shanxi Province. J. Jilin Univ. (Earth Sci. Ed.) 5, 961–966.

Tsurikhin, E.A., Loginov, V.L., Sauro, F., 2012. New speleological investigations on Khodzha Gur-Gur-Ata Ridge (Uzbekistan). In: Speleologiya i spelestologiya (Speleology and speletology). vol. 3. NISPTR, Naberehznye Chelny, pp. 73–76 (in Russian).

Tsurikhin, E.A., Sauro, F., Breitenbach, S., 2013. Exploration of high altitude caves in the Baisun-Tau Mountain Range, Uzbekistan. In: Proceedings of the 16th International Congress of Speleology, Brno, Czech Republic, 21–28 July 2013. vol. 2, pp. 142–147.

Wood, C., Zhang, H., 2010. Volcanic centres and lava caves in China. In: Proceedings 14th International Symposium on Vulcanospeleology, pp. 123–129.

Yalgouz-Agaj, A.G., Ardebil, L., Karimdoust, S., 2010. Identification of Some of the Geotourism Sites in Iran. World Appl. Sci. J. 11 (11), 1342–1347.

CHAPTER

ICE CAVES IN FYR OF MACEDONIA

22

Marjan Temovski

Institute for Nuclear Research, Hungarian Academy of Sciences, Debrecen, Hungary

CHAPTER OUTLINE

22.1 INTRODUCTION

Ice caves have been defined as rock-hosted caves containing perennial ice or snow, or both (Luetscher and Jeannin, 2004). In their definition, Ford and Williams (2007) considered only ice deposits, but also included seasonal deposition. Recently, Perşoiu and Onac (2012) defined ice caves as only those with perennial ice deposits. Because no scientific studies have been conducted on the ice caves in Macedonia, in order to give the widest possible base for future studies, ice caves are considered here in their broadest definition, that is, caves with perennial (ice or snow) and/or seasonal ice deposits. This chapter thus aims to summarize the known information about ice caves in Macedonia in accordance with the goal set up during the meeting of the UIS Glacier, Firn and Ice Caves Commission at the 16th International Speleological Congress in Brno in 2013 to recover information on geographic and topographic dispersion of ice caves and glacier caves (Maggi, 2013). At present, this information includes only basic data, such as the location of ice caves and the extent of ice/snow deposits within the caves.

Macedonia is a mostly mountainous country situated in the central part of the Balkan Peninsula, Southeastern Europe (Fig. 22.1), with karst areas covering 12% of the country's surfaces and distributed in more or less isolated, generally smaller areas, most of which are located in the western and

Ice Caves. https://doi.org/10.1016/B978-0-12-811739-2.00022-X

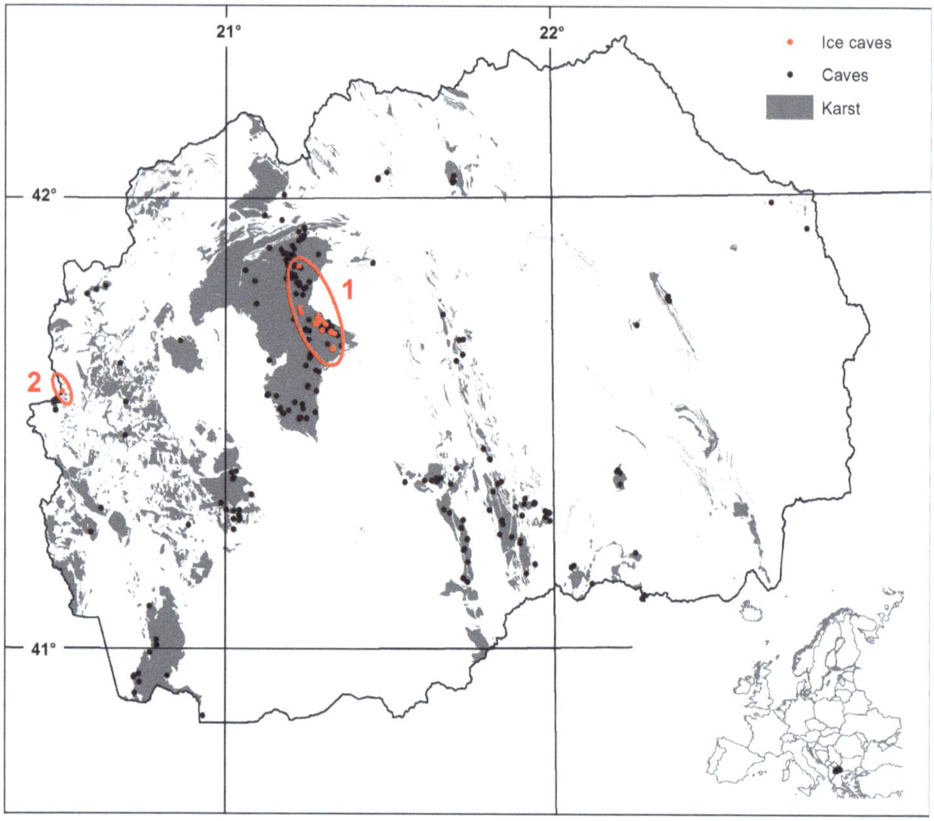

FIG. 22.1

Location of ice caves in Macedonia: (1) Jakupica Massif and (2) Krčin Mountain.

central parts of the country (Temovski, 2012). Karst rocks are mostly of pre-Cenozoic age, although ranging from Precambrian to recent times. Most of the karst systems are epigene in origin, although hypogene karst is also registered (Temovski, 2017), and are located in mid- to high-mountainous areas, with thick vadose zones, and also deep phreatic zones in many areas. Although no official cave register exists, there are less than 500 known caves, with Slovačka Jama (−650 m) being the deepest explored cave system (Carlin et al., 2011; Holúbek et al., 2013) and Slatinski Izvor (4 km) the longest explored cave (Carlin et al., 2011). Other important caves include the 230-m-deep phreatic cave Matka Vrelo (Vandermeulen, 2015), the hypogene Provalata Cave, and Karši Podot Cave (Temovski et al., 2013; Temovski, 2016a).

Ice caves (as previously defined) have been reported in two mountain areas in Macedonia: Jakupica Massif in the central part of the country and in Krčin Mountain along the western border (Fig. 22.1). Although no detailed studies have been made, a general decline in ice volume has been noted (Carlin et al., 2011), which further urges the necessity for future detailed characterization and research of ice deposits in caves in Macedonia.

Based on the size (depth and length) of the caves and the distribution of their ice/snow deposits, they have been classified as four types (Temovski, 2016b):

1. Deep caves with large ice/snow deposits and explored passages below/beside the ice/snow deposits
2. Caves (<60 m deep) with explored passages ending in snow/ice plugs
3. Smaller caves with small snow/ice deposits and explored passages below/beside the snow/ice deposits
4. Smaller caves with seasonal ice deposits

22.2 HISTORY OF EXPLORATION AND SOURCES OF INFORMATION

The data for this database was collected mostly from reports of caving explorations conducted by Macedonian, Slovene, and Slovak cavers during the past 35 years. Caves with ice or snow plugs at the bottom were reported (although not explored) during the first karst studies of Jakupica Massif (Jovanović, 1928a,b), as this is characteristic for many of the pits with large entrances on Jakupica Massif, and some were also reported later (Manakovik, 1966; Kolčakovski, 1988a,b). The exploration of ice caves in Macedonia began with the first explorations of Solunska Glava 5 in 1979 and 1980 by Macedonian cavers from SD Peoni (Carlin et al., 2011). Although the exploration of the cave was finished in 1980, detailed information (description, cave maps, and photos) was published only recently (Šmoll and Szunyog, 2005; Carlin et al., 2011). In 1995 Slovene cavers from JK Speleos - Velenje explored 15 pits on Krčin Mountain, 9 of which had snow or ice deposits, and provided a report (unpublished) with descriptions and cave maps, although coordinates and elevation of entrances were not included (JK Speleos - Velenje, 1995). Further exploration of caves on Jakupica was spurred by Slovak cavers who have organized expeditions, in collaboration with Macedonian cavers from SD Peoni, almost every year since 2004. They have explored numerous caves, including the deepest one (Slovačka Jama, −650 m). Most of their results are published (Šmoll and Szunyog, 2005, 2006; Šmoll, 2006; Szunyog, 2006; Šmoll and Sluka, 2007; Majerníčková and Imrich, 2009; Psotka et al., 2009; Sluka and Szunyog, 2011; Carlin et al., 2011; Holúbek and Šmoll, 2012; Šmoll et al., 2012; Holúbek et al., 2013) or are available on their websites, http://jakupica.speleo.sk and www.cervenevrchy-speleo.sk. Other data, that is, coordinates of entrances, were kindly provided by Martin Sluka and Jan Šmoll.

22.3 JAKUPICA MASSIF

Jakupica Mountain Massif, also known as Mokra Mountain, is located in the central part of the Republic of Macedonia. It is a complex star-shaped morphological unit consisting of several mountains that are radially dispersed from the Solunska Glava mountain peak (2540 m): Dautica to the southwest, Karadžica and Suva Planina to the north, Golešnica to the east, and Jakupica (sensu stricto) mostly in the central part around Solunska Glava (Fig. 22.2). Jakupica Massif has the highest elevation (above 1000 m) of all the mountains in Macedonia (Kolčakovski, 2006).

Jakupica Massif is a horst structure, bounded on the south and west by the Pelagonian and Poreče basins, respectively, with a well-expressed fault border at the latter and with a more gradual morphological change to the east and north, where Jakupica Massif connects to the Veles and Skopje basins

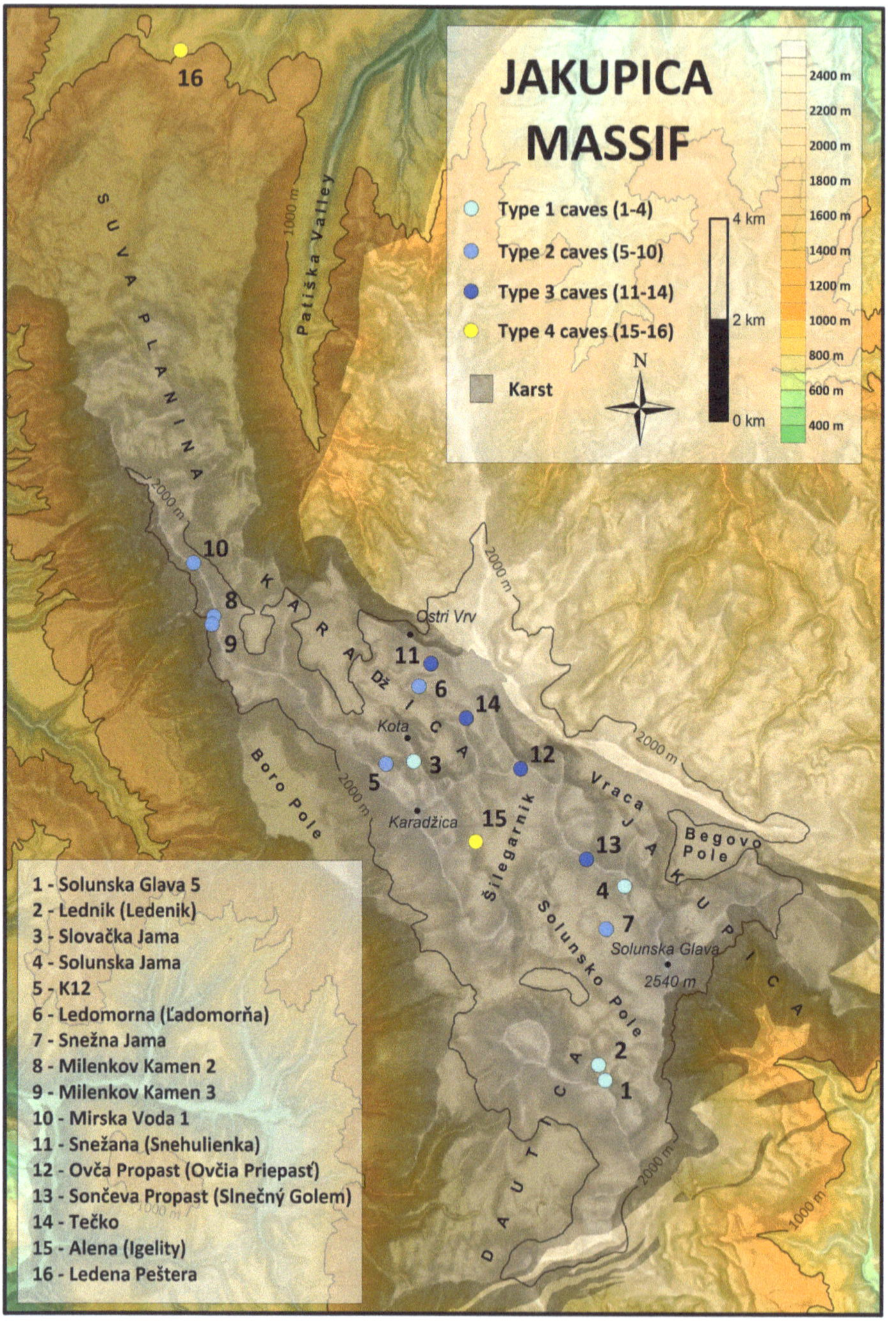

FIG. 22.2

General morphology of the central part of the Jakupica Massif with distribution of karst and ice caves.

(Dumurdžanov et al., 2005). It is part of the Pelagonian Massif, a major pre-Cenozoic tectonic unit, and it almost completely consists of Precambrian rocks, with various types of gneiss in the lower parts. It continues upward into a mixed zone of gneiss, micashist, and cipolin that is topped by a thick sequence of marble (Arsovski, 1997). The western part of the massif is developed entirely in the marble sequence, whereas the eastern part is almost entirely in Precambrian gneiss rocks.

The central part of Jakupica Massif has a plateau-like morphology, perforated by a number of glaciokarstic depressions, and the deep north-oriented Patiška glacial valley, which was scoured during at least two Pleistocene glaciations (Jovanović, 1928a). Four depressions (Solunsko Pole, Šilegarnik, Begovo Pole, and Vraca) are polygenetic karst depressions, with Solunsko Pole and Šilegarnik developed completely inside the karst terrains and with Vraca and Begovo developed at the contact with underlying cipolin and gneiss rocks to the east (Jovanović, 1928a,b; Manakovik, 1966). Begovo Pole can be temporarily partly flooded and has a number of ponors located on the southwestern border at the margin of the deluvial-proluvial deposits.

Carbonate rocks of the Jakupica Massif are highly karstified, with a deep vadose zone developed in the southern and central parts and a deep phreatic zone to the north (Fig. 22.3). The karst area continues to the west in the Poreče Basin and to the north to a hilly mountainous area cut by the deep Matka Canyon, and it is partly buried to the northeast by Neogene deposits in the Skopje Basin.

Most of the caves in the central parts are pits, among which the deepest and some of the longest explored caves in Macedonia can be found, such as Slovačka Jama (3163 m, −650 m), Solunska Glava 5 (400 m, −370 m), Solunska Jama (398 m, −274 m), Lednik (240 m, −240 m), and Babuna Spring Cave (2500 m, +95 m).

The southern parts of the karst system are mainly drained by the Krapa-Pešna/Asanoec system, the central parts are drained by the Babuna springs in the east and the Belica springs in the west, and the northern parts are drained mainly by the deep vauclusian spring, Matka Vrelo (Koritište), which was explored to the depth of 230 m (Fig. 22.3). The total karstified zone is more than 2500 m thick—taken between the highest mountain peak (Solunska Glava, 2540 m)—and the deepest karstification registered by boreholes at 0 m absolute elevation in the northeastern parts, at the border of Skopje Basin (Stračkov, 1968).

Mean annual temperatures at the Solunska Glava meteorological station (located at Solunska Glava peak) for the period 1981–2000 were −0.3°C, with an absolute minimum of −27.9°C in Mar. and an absolute maximum of 30.1°C in Jul. (Petreska, 2008). Monthly mean temperatures are below zero from Nov. to May (Fig. 22.4).

Mean annual precipitation for the same period was 820 mm (Petreska, 2008). Data from previous periods presented higher values, with 869.1 mm for 1973–1980 (as reported in Kolčakovski, 1988b) and 891 mm for 1981–1990 (as reported in Andonovski, 1997). Kolčakovski (1988b) points out that annual precipitation amounts on Solunska Glava can vary greatly, giving as examples 1977 with only 443.8 mm compared with 1982 with 1268 mm. For the period 1981–2000, the highest monthly average precipitations were in May and Dec. and the lowest in Jan. and Aug. Seasonally, during autumn this area receives the highest amounts (236.2 mm), while during the other seasons, the precipitation amounts are similar, with summer receiving the least at 187.8 mm. Snow cover remains until mid-Jun. on flat areas, whereas dolines can keep snow cover until the end of Aug. (Kolčakovski, 1988b).

On Jakupica Massif, ice or snow deposits have been registered in 16 caves, 14 of which have perennial ice and/or snow deposits and 2 have seasonal ice deposits (Table 22.1). Four of them are deep caves with large ice/snow deposits (type 1), 6 are pits finishing with ice/snow plugs (type 2), 4 are smaller caves

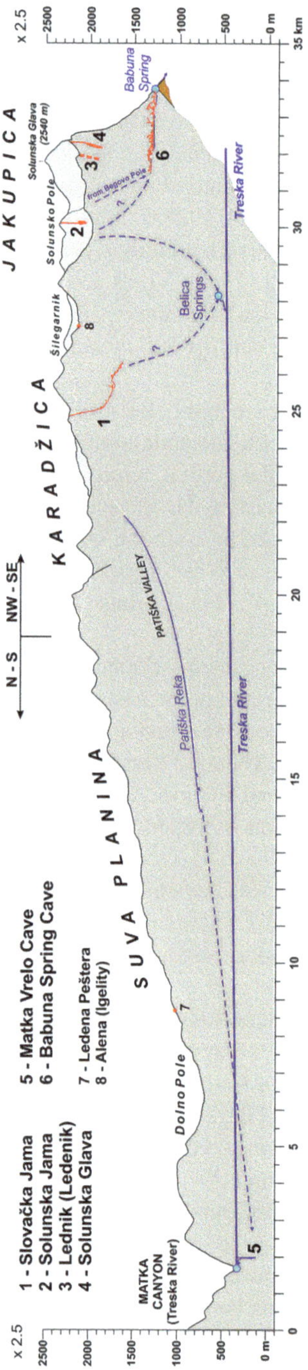

FIG. 22.3

Cross-section of the karst systems in the central and northern parts of Jakupica Massif.

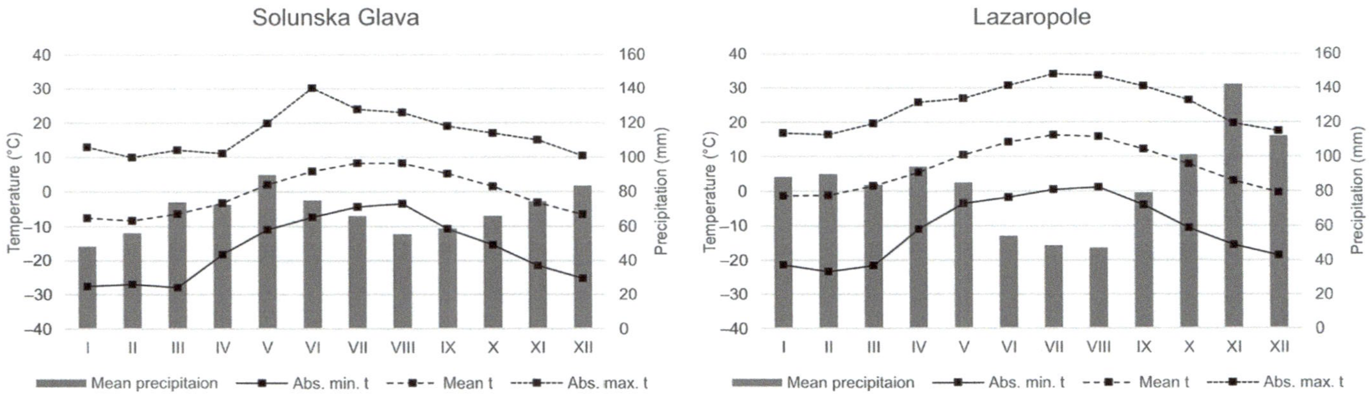

FIG. 22.4

Climate data for Solunska Glava and Lazaropole meteorological stations.

Data for Solunska Glava station (period 1981–2000) from Petreska, B., 2008. Podzemni karstni formi vo Porečkiot Basen i nivna valorizacija za potrebite na prostornoto planiranje (Underground karst forms in the Poreče Basin and their valorisation for the need of spacial planing) (Master thesis). Faculty of Natural Science and Mathematics, Skopje (in Macedonian), and data for Lazaropole station (period 1971–2000); from Study for the Revalorization of Mavrovo Protected Area, 2011. Climate Data. http://www.moepp.gov.mk/wp-content/uploads/2015/01/Study-Mavrovo-Final-7.pdf (Accessed 24 January 2017).

Table 22.1 Basic Data for Ice Caves in Macedonia

			Entrance					
No.	Location	Cave Name (Synonym)	Location (degrees[a])	Elevation (m a.s.l.)	Length (m)	Depth (m)	Deposits	Type[b]
1	Jakupica Mt.	Solunska Glava 5	21.38879 41.68301	2240	400	370	Ice, snow	1
2		Lednik (Ledenik)	21.38723 41.68577	2135	240+	240	Ice, snow	1
3		Slovačka Jama	21.34207 41.74017	2241	2620	610	Ice, snow	1
4		Solunska Jama	21.39363 41.71779	2340	398	274	Snow, ice	1
5		K12	21.33535 41.73977	2240	60	60	Ice, snow	2
6		Ledomorna (Ľadomorňa)	21.34340 41.75370	2153	27	27	Ice, snow	2
7		Snežna Jama	21.38912 41.71014	2160	50	50	Ice, snow	2
8		Mirska Voda 1	21.28765 41.77585	2107	59	51	Snow	2
9		Milenkov Kamen 2	21.29257 41.76630	2107	48	47	Snow	2
10		Milenkov Kamen 3	21.29208 41.76483	2071	64	53	Snow	2
11		Snežana (Snehulienka)	21.34636 41.75771	2164	194	58	Snow, ice	3
12		Ovča Propast (Ovčia Priepasť)	21.36837 41.73887	2293	47	41	Snow	3
13		Sončeva Propast (Slnečný Golem)	21.38436 41.72268	2325	45	15	Snow	3
14		Tečko	21.35497 41.74797	2189	18	14	Snow	3
15		Alena (Igelity)	21.35726 41.72582	2131	114	17	Ice	4
16		Ledena Peštera	21.28486 41.86772	991	–	–	Ice	4

17	Krčin Mt.	SLO-K1	20.55888[c]	1900–2100?	89	59	Snow	3
18		SLO-K4	41.58838[c]		73	53	Snow	3
19		SLO-K6			44	38	Snow	2
20		SLO-K7			37	37	Snow	2
21		SLO-K8			41	37	Snow	2
22		SLO-K9			71	32	Snow	2
23		SLO-K10			15	15	Snow	2
24		SLO-K11-12			99	53	Snow, ice	2
25		SLO-K13			28	24	Snow	2

[a]*Eastern longitude (2x.xxxxx) and northern latitude (41.xxxxx), WGS-84.*
[b]Explanation in text.
[c]Approximate location for the whole explored area on Krčin Mt. (no data for exact cave locations).

with passages partly covered by small ice/snow deposits (type 3), and 2 are small caves with seasonal ice formations (type 4). The number of caves that end with snow/ice plugs (type 2) is probably much higher than registered. All, except one cave with seasonal ice deposits (Ledena Peštera), are located in the central plateau area, with entrances above 2000 m a.s.l. Perennial ice deposits can be found from an elevation of 2328 m in the upper parts of Solunska Jama down to 1870 m at the bottom of Solunska Glava 5.

In the first group are the four most important ice caves in Macedonia: Slovačka Jama, Solunska Glava 5, Lednik, and Solunska Jama. Except for Solunska Jama, which mostly has firn deposits, the others have thick ice deposits, mostly along the cave passages, down to a depth of ~200 m, with Solunska Glava 5, as an exceptional case, having a huge ice glacier (ice cone) down to a depth of 370 m.

22.3.1 SLOVAČKA JAMA

Slovačka Jama is the deepest explored cave in Macedonia. It was previously published as having a depth of 610 m (Holúbek et al., 2013), but that depth was extended by the latest expedition in 2016, which explored 500 m of new passages down to a depth of 650 m (Pokrievka ml, 2016). The entrance is located at an elevation of 2240 m in a NNW-SSE-oriented depression on the ridge between the Kota (2309 m) and Karadžica (2472 m) peaks (Fig. 22.2).

Slovačka Jama was first reported in the literature by Kolčakovski (1988b) as Propast Kota. Kolčakovski described it as a 19-m-deep shaft with a large circular entrance with a snow-covered bottom. It was rediscovered in 2006 by Slovak cavers who, together with cavers from SD Peoni, passed the entrance's ice plug. They first explored the cave to a depth of 80 m (2006) and later to −200 m (2007) and −524 (2008), naming it Slovačka Jama na Karadžici (earlier it was known as K11) (Psotka et al., 2009; Carlin et al., 2011). The following years (2009, 2010), the entrance's passages were completely blocked by ice and were inaccessible (Sluka and Szunyog, 2011), so explorations halted. In 2011 the explorations continued, and a depth of 598 m was reached (Šmoll et al., 2012), which the following year was extended to −610 m (Holúbek et al., 2013).

Slovačka Jama is a vadose cave, with steeper passages down to a depth of 400 m, continuing to a less inclined network of interchanging phreatic and canyon passages with several sumps (Fig. 22.5).

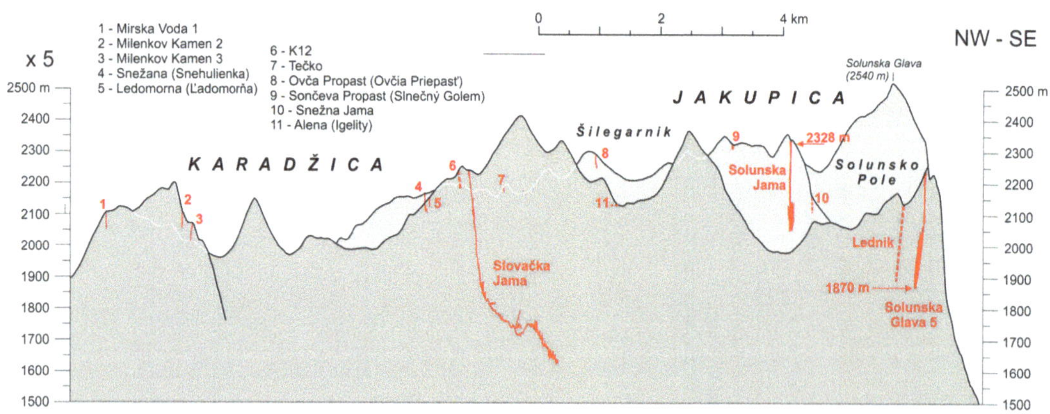

FIG. 22.5

Cross-section of the central parts of Jakupica Massif with the location of the ice caves.

The cave is located north of the Belica karst springs to which it is likely draining. The cave system has a potential vertical of 1700 m (Fig. 22.3).

Ice/snow deposits are found in the upper parts of the cave, starting from the entrance down to a depth of ~200 m (Fig. 22.6). The ice deposits are found along the passages, with some of the explored passages carved in ice (Fig. 22.7). The main ice deposits stop at the top of a 60 m shaft (Šesťdiesatka) at a depth of 200 m, but often detached ice blocks can be found at the bottom of the shaft and even down to a depth of ~300 m (Holúbek et al., 2013). The amount of ice at the entrance passages varies, which likely depends on the amount of snow in the winter. The ice can either completely clog the entrance passages or leave them open, as was the situation in 2011 (Šmoll et al., 2012).

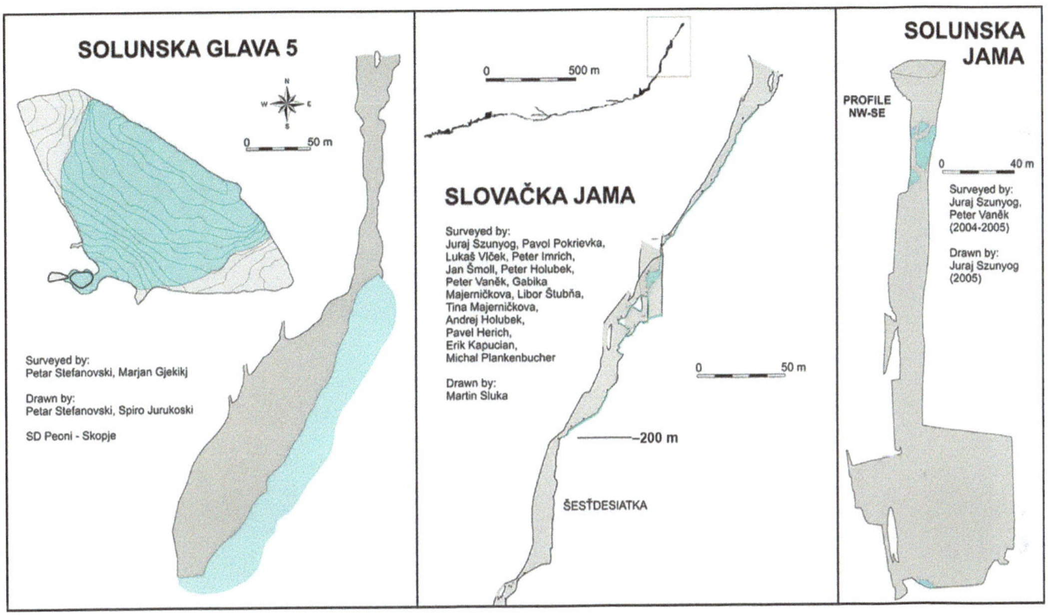

FIG. 22.6

Maps and distribution of ice/snow deposits in type 1 caves on Jakupica Massif. Ice and snow deposits are colored in *cyan*.

Maps redrawn from originals provided by Martin Sluka and Juraj Szunyog, and published in Šmoll, J., Szunyog, J., 2005. Priepast' Solunska jama. Spravodaj SSS, 1/2005, 51–56 (in Slovak) and on http://jakupica.speleo.sk/.

22.3.2 SOLUNSKA JAMA

Solunska Glava 5, with a total depth of 370 m, is the second deepest cave in Macedonia (Carlin et al., 2011). It was first explored in 1979 and 1980 by SD Peoni cavers, who reported its depth as 450 m. It stood as the deepest explored cave in Macedonia for a long time, until recent discoveries in Slovačka Jama (Kolčakovski, 2001). Although it was explored in the 1980s, detailed information on the cave was published only recently (Šmoll and Szunyog, 2005; Carlin et al., 2011), at which time a corrected depth value of 370 m was reported.

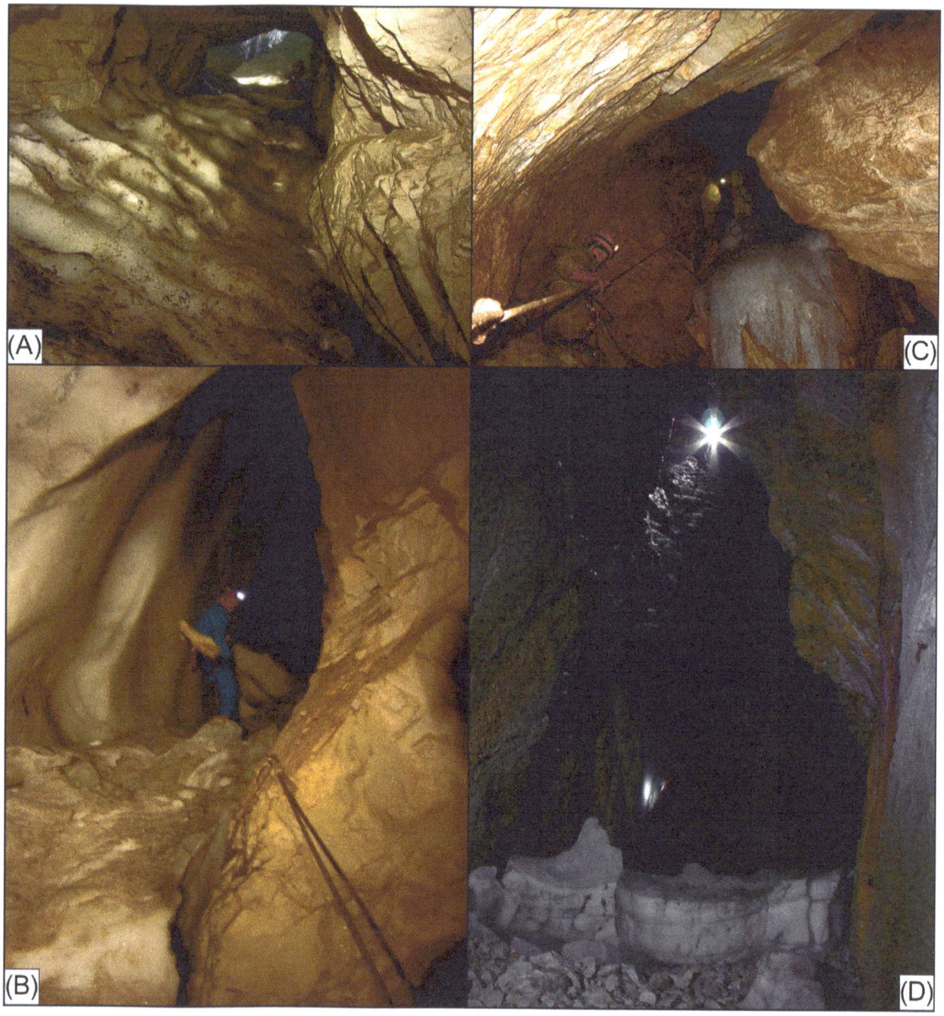

FIG. 22.7

Ice and snow deposits in Slovačka Jama: (A) entrance, (B) at depth of 150m, (C) at top of the Šesťdiesatka, at 200m depth, and (D) at bottom of the Šesťdiesatka.

Photographs courtesy of Erik Kapucian (A and B), and Libor Štubňa (C and D).

The entrance is located at 2235m elevation on a north-exposed slope in the southwestern part of the large polygenetic depression, Solunsko Pole (Fig. 22.2).

The cave has a somewhat simple morphology (Fig. 22.6) with an entrance shaft ~190m in depth that leads to a large room with a northwest-southeast direction and has dimensions of 200×250×170m (Carlin et al., 2011). The floor slopes northeast and is covered by broken rock blocks, debris, and ice deposits that start with a snow cone at the bottom of the entrance shaft, continuing downward into a large ice body. Most of the floor is covered with ice deposits, with ice stalagmites several meters high in the lower parts and an ice lake at the bottom (Fig. 22.8). The ice deposits in the large room cover an estimated area of 15,000 m^2 (based on the distribution given in the cave's map in Carlin et al., 2011); the deposits are of an unknown thickness, though in the lower parts, they are at least several meters thick. They are arguably the largest perennial ice deposits in Macedonia. Based on the morphology of the cave and the distribution of the ice deposits, Solunska Glava 5 appears to be a simple ice cold trap, with snow deposits from the entrance shaft and drip water from the ceiling contributing to the mass of the ice deposits.

FIG. 22.8

Ice and snow deposits in Solunska Glava 5: (A) at the bottom of entrance shaft, (B) view of the large ice cone, (C) closer view of the large ice stalagmites in B, and (D) ice block at the bottom of the ice cone.

Photographs courtesy of Libor Štubňa.

22.3.3 LEDNIK

Not much information is available for Lednik (Ledenik) cave, which is located in the southwestern part of Solunsko Pole. Its entrance is located at 2135 m a.s.l. in the northern part of a doline, 340 m to the northeast of Solunska Glava 5 (22.2). The cave was first explored by SD Peoni cavers, with only its depth reported in the literature. Recently, it was explored by Slovak cavers, who in 2011 first explored ice-filled passages down to a depth of 150m (Šmoll et al., 2012) and then in 2013 explored new passages below the ice deposits, down to a depth of 240m. Snow and ice deposits completely cover the floor of the large cave passages down to a depth of ~120m, where large ice stalagmites, stalactites, and cave walls covered in ice are found (Fig. 22.8). The cave continues downward with some passages developed in ice deposits, which stop below a depth of ~200m, at which point a vertical vadose passage continues downward.

22.3.4 SOLUNSKA JAMA

Solunska Jama is another deep vertical cave on Jakupica Massif. Its entrance is located at 2340m a.s.l. on the ridge between Solunsko Pole and Begovo Pole (Fig. 22.2), 1.7km northwest of Solunska Glava (2540m). The first explorations were done in the 1980s by SD Peoni cavers, who reached a depth of only 20m, after which the cave finished with a snow/ice plug (Carlin et al., 2011). In 2004 a joint Macedonian-Slovak expedition passed the entrance's ice plug and reached a depth of 150m (Šmoll and Szunyog, 2005). The following year, as part of the "Jakupica 2005" expedition, the cave was explored down to the bottom at a depth of 274m (Šmoll and Szunyog, 2005; Carlin et al., 2011). Solunska Jama has a large entrance with dimensions of 25×30m, with a vertical passage of the same dimensions that descends to a depth of 30m where it is almost completely filled with snow deposits, except for a small 1.5-m-wide passage that continues downward along the contact with the snow/ice deposits to a depth of 60m (Fig. 22.6). Below the ice plug, the passage continues with a 20-m-wide shaft down to a depth of ~190m where it joins an 80-m-high hall with dimensions of 15×25m and a bottom at a ~260m depth, from which a small passage continues and then finishes with a small pool. Snow and ice deposits are found in the upper part of the entrance shaft in the form of a 40-m-thick ice plug and on the bottom of the big hall just below the entrance shaft as a large cone formed from blocks that fell from the entrance ice plug (Fig. 22.9).

22.3.5 OTHER ICE CAVES ON JAKUPICA MASSIF

Six caves belong to the second type of caves, and they have been explored to depths of less than 60m, with explored passages ending in snow/ice plugs. These caves include K12, Ledomorna (Ľadomorňa), Milenkov Kamen 2 and 3, Mirska Voda 1, and Snežna Jama (Fig. 22.2). Snežna Jama is the only one located in the southern part of the plateau, in the large Solunsko Pole karst depression (Šmoll and Szunyog, 2005). Three of these caves are located northwest of the central plateau area: Milenkov Kamen 2 and 3 are located in the northern part of Karadžica Mountain on the western slope toward Poreče Basin, and Mirska Voda is located a little farther to the north on the mountain pass between Karadžica Mountain and Suva Planina (Sluka and Szunyog, 2011). K12 and Ledomorna are located in the central part of Karadžica Mountain on the slopes of what Kolčakovski (1988b) described as a glaciated karst valley and follow a general northwestly direction. Their entrances are located at elevations between 2071 (Milenkov Kamen 3) and 2160m (Snežna Jama), with K12 located at the highest elevation at 2240m. Morphologically, they are vertical or slightly less inclined shafts, with entrance and entrance passage diameters of 5–10m (Fig. 22.10). The passages are filled with firn and ice deposits (Fig. 22.11) and are penetrable only between the ice deposits and the wall down to depths of 50–60m (K12, Milenkov Kamen 2 and 3, Mirska Voda 1, Snežna Jama), or 27m in the case of the Ledomorna cave. Considering its proximity to the large shaft next to Peoni Hall in Slovačka Jama (at a

FIG. 22.9

Ice and snow deposits in Lednik and Solunska Jama: (A) ice and snow deposits in the upper parts of Lednik cave, (B) shaft below the ice deposits in Lednik cave, and (C and D) snow cones at the bottom of Solunska Jama.

Photographs courtesy of Libor Štubňa (A and B), and Juraj Szunyog (C and D).

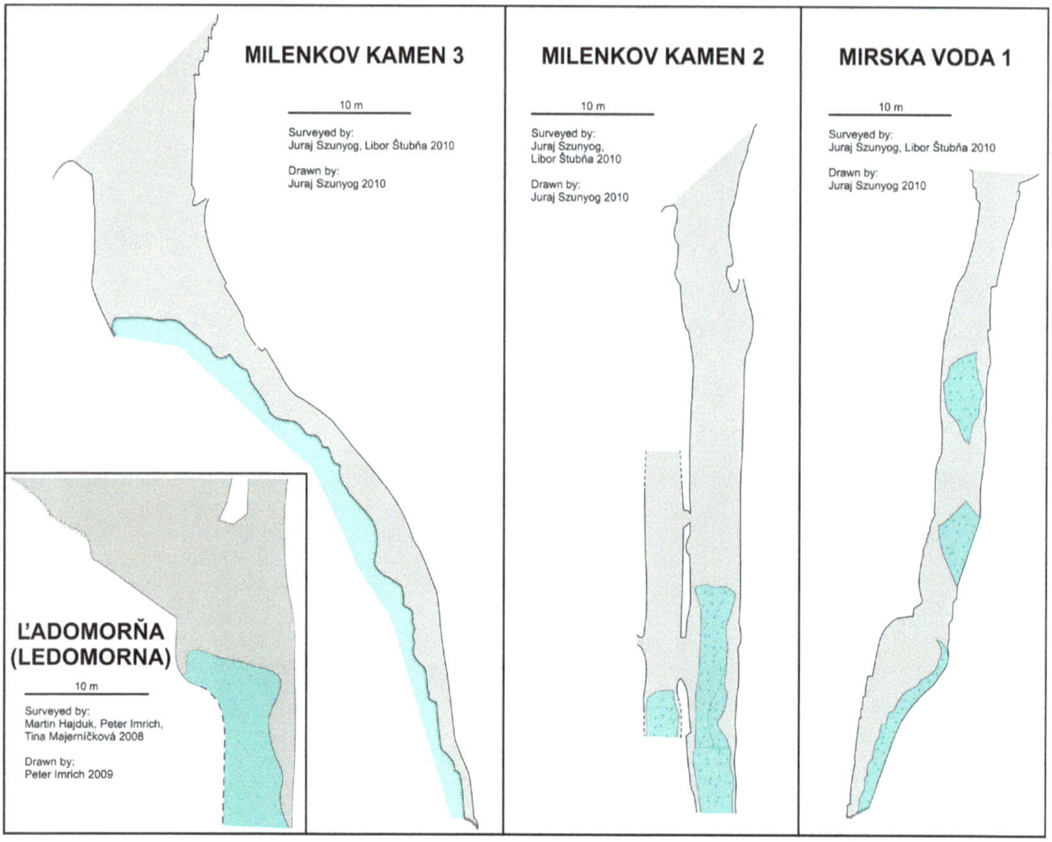

FIG. 22.10

Maps and distribution of ice/snow deposits in type 2 caves on Jakupica Massif. Ice and snow deposits are colored in *cyan*.

Maps redrawn from originals provided by Juraj Szunyog and Peter Imrich, and published in Majerníčková, F., Imrich, P., 2009. Snehulienka (Snow-white). Spravodaj SSS 1/2009, 62–65 (in Slovak) and on http://jakupica.speleo.sk.

depth of 400 m) and the organization of the passages in Slovačka Jama, K12 likely connects with Peoni Hall. These caves are morphologically similar to the entrances of Slovačka Jama and Solunska Jama, although with smaller passage diameters.

Four caves are smaller caves with passages partly filled or covered by small and mostly perenial snow deposits (type 3 caves). Snežana (Snehilienka) and Tečko are located in the northeastern part of Karadžica Mountain, Ovča Propast (Ovčia Priepasť) is located on the northern edge of Šilegarnik, and Sončeva Propast (Slnečný Golem) is located on the northern edge of Solunsko Pole (Fig. 22.2). Snežana is the largest, almost 200 m long and 60 m deep, and is located on the northern edge of the high mountain plateau, with an entrance at 2164 m (Majerníčková and Imrich, 2009). The others are located along the same aproximate diagonal to the southeast and have entrances at progressively higher elevations, with

FIG. 22.11

Ice and snow deposits in caves from types 2, 3, and 4: (A and B) snow and ice deposits in K12, (C) Milenkov Kamen 3, (D) Milenkov Kamen 2, (E) Sončeva Propast (Slnečný Golem), and (F) Alena (Igelity).

Photographs courtesy of Martin Miškov (A and B), Juraj Szunyog (C and D), http://jakupica.speleo.sk (E), and Marjan Temovski (F).

Tečko at 2189 m, Ovča Propast at 2293 m, and the southmost Sončeva Propast at 2325 m. All, except Tečko, are located on south-to-soutwest sloping terrains. They have vertical entrance shafts with large diameters (10–20 m), which in their lower parts (starting at depths from 15 to 20 m) have thick snow cones (Fig. 22.11), along which the passages continue sidewise (Šmoll and Szunyog, 2005; Majerníčková and Imrich, 2009). Tečko is the smallest cave in this group, with a northeast-exposed entrance leading to a small inclined passage, along which a snow cone of only a few meters can be found (Fig. 22.12).

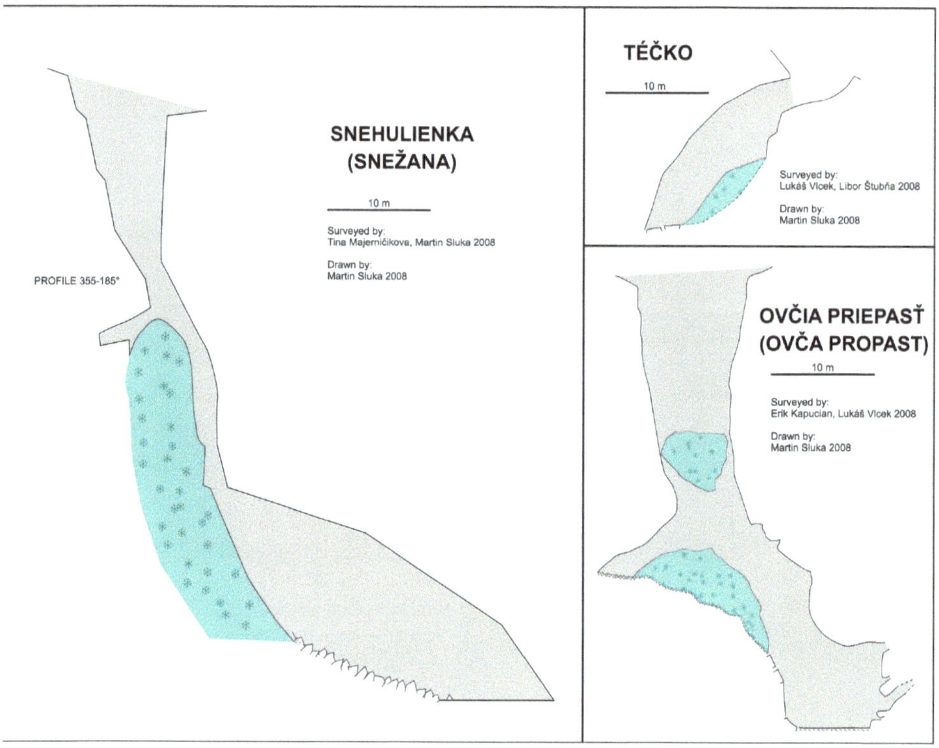

FIG. 22.12

Maps and distribution of ice/snow deposits in type 3 caves on Jakupica Massif. Ice and snow deposits are colored in *cyan*.

Maps redrawn from originals provided by Martin Sluka and published in Majerníčková, F., Imrich, P., 2009. Snehulienka (Snow-white). Spravodaj SSS 1/2009, 62–65 (in Slovak) and on http://jakupica.speleo.sk.

In addition to the caves with perennial ice or snow deposits, two caves that have only seasonal ice formations (type 4 caves) are included: Alena (Igelity) and Ledena Peštera. Alena is located in the western part of the Šilegarnik depression, with the north-exposed entrance located at 2131 m. It was first reported by Kolčakovski (1988a), who visited the cave in late Aug. of 1988. He didn't register ice formations, but he reported a cave air temperature of 7°C. Ice formations in the form of ice dripstones and flowstones (Fig. 22.12) are present in Alena until late May or early Jun. As part of the cave expeditions on Jakupica (mentioned earlier), Alena was explored in 2008 also by Slovak cavers, who named it Igelity. Ledena Peštera is located at a much lower elevation than all the other caves, with its entrance at 991 m on the northern part of Suva Planina. The cave was explored and reported by SD Peoni cavers. It has a vertical entrance passage that leads to a large room where ice formations can be found as ice stalactites and stalagmites during the winter.

22.4 KRČIN MOUNTAIN

Krčin Mountain is located in the western part of Macedonia, along its border with Albania (Figs. 22.1 and 22.13). It represents the southernmost part of the Korab horst, which is part of the Cukali-Krasta tectonic zone (Dumurdžanov et al., 2005), and is separated by faults from the neighboring Šar Planina and Bistra mountains to the north and east and the Debar Basin to the south. The main ridge is oriented in a NNW-SSE direction, with Golem Krčin (2341 m) and Mal Krčin (or Rudina, 2238 m) being the highest peaks, and continues northward to the Dešat Mountain and the highest mountain in Macedonia, Korab (Golem Korab, 2764 m). To the east and south, Krčin Mountain is delineated by the deep valley of Radika River (and the artifical Debar Lake), the main tributary to the Crn Drim (Black Drin) River, which represents the Adriatic drainage basin in Macedonia. Krčin Mountain consists almost completely of Upper Cretaceous sediments, which on the margins of the mountain are partly covered by Eocene, Pliocene, and Quaternary sediments. Most of the area consists of flysch sediments, among which carbonate rocks are found as smaller or larger bodies, layers, and lenses of platy limestones. Thermal springs, found on the southwestern and southern foothills, discharge hypogene karst systems developed in limestone and evaporite rocks (Temovski, 2013).

During the Pleistocene, the Krčin-Dešat-Korab mountain range was glaciated, as were most of the mountains in the western part of Macedonia, with glacial features much better expressed on Korab and Dešat mountains (Kolčakovski, 2006). Cirques and morains are registered only on the northeastern slopes of Krčin Mountain, with moraine deposits affected by intensive denudation.

This area receives the highest amount of precipitation in Macedonia, with values of 1400 mm reported in some areas farther north in Radika Valley (Lazarevski, 1993). Based on the climate data (1971–2000) from the weather station in Lazaropole (located at 1335 m elevation on the nearby Bistra Mountain; Fig. 22.4) and the elevation gradients for the mean annual precipitation and temperature given by Lazarevski (1993), for the areas at 1900–2100 m elevation, the mean annual precipitation is approximated to be between 1040 and 1060 mm, and the mean annual temperature to be between 2°C and 3.5°C. This area, in comparison with Jakupica Massif, receives more precipitation as expected, while the mean annual temperatures are comparable to the ones at the same elevations on Jakupica Massif.

Very little is known about karst and cave development on the higher parts of Krčin Mountain. During an expedition in 1995, Slovene cavers explored caves on the main ridge in the mountain pass between Mal and Golem Krčin peaks (JK Speleos - Velenje, 1995). Out of the 15 explored and documented caves, 9 had snow and ice deposits (Figs. 22.14 and 22.15). The caves appear to be mostly simple pits, with explored depths ranging between 15 and 60 m. Seven of them (SLO-K6, SLO-K7, SLO-K8, SLO-K9, SLO-K10, SLO-K11-12, and SLO-K13) had passages ending with snow plugs (type 2 caves), and two (SLO-K1 and SLO-K4) had small patches of snow along explored passages that continued well below the snow deposits (type 3 caves). Their exact location is unknown, with only approximate descriptions given in the expedition's report (JK Speleos - Velenje, 1995), according to which the elevations of their entrances are approximated to be somewhere between 1900 and 2100 m (Fig. 22.13).

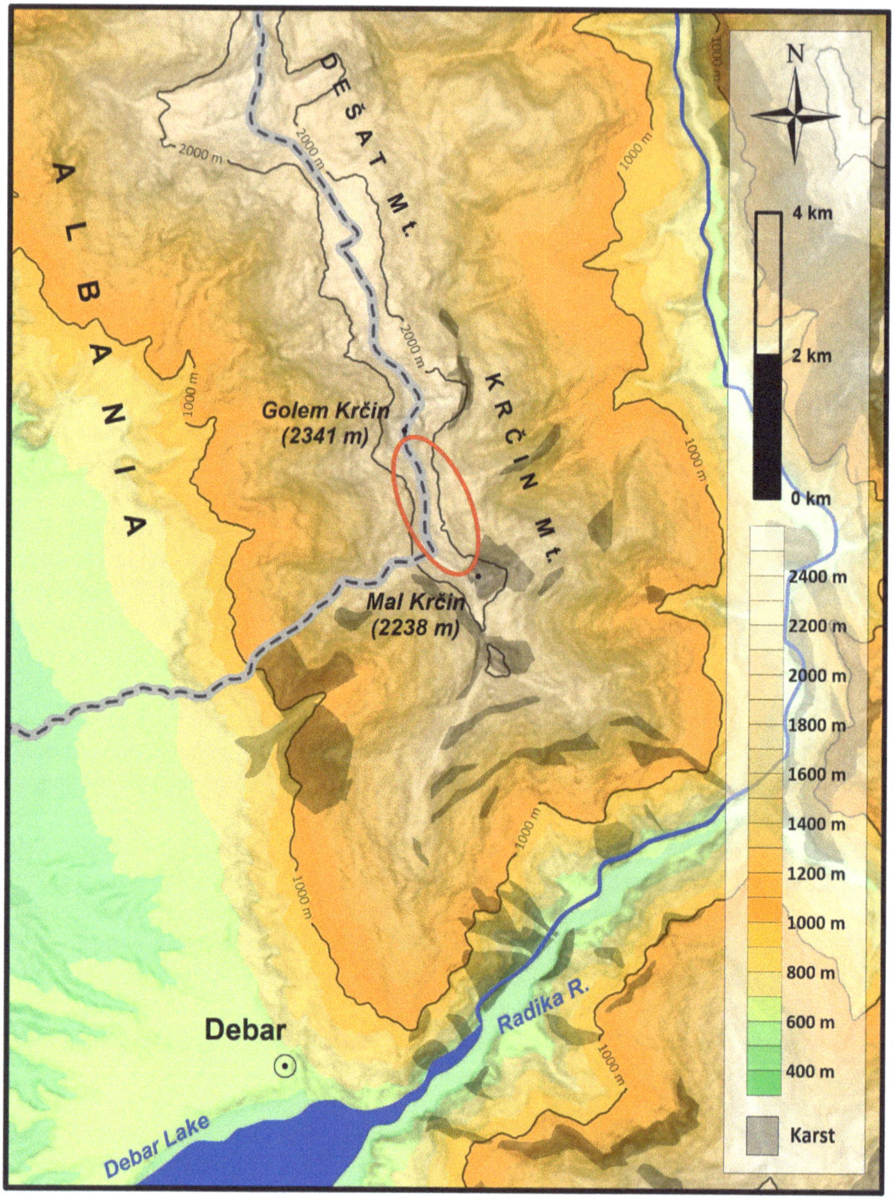

FIG. 22.13

General situation of Krčin Mountain with the approximate location *(red ellipse)* of the ice caves.

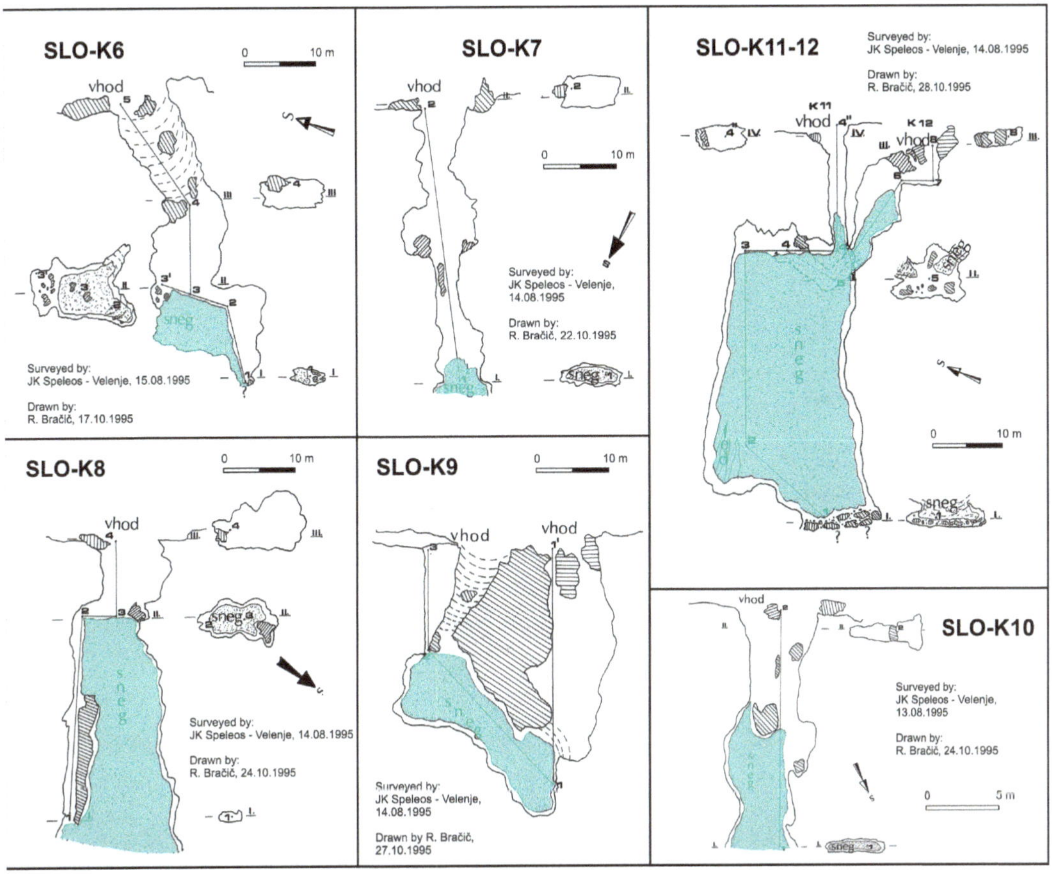

FIG. 22.14

Maps and distribution of ice/snow deposits in caves on Krčin Mountain. Ice and snow deposits are colored in *cyan*.

Maps modified from JK Speleos - Velenje, 1995. Slovenska jamarska odprava "Makedonija 95" (Slovene caving expedition "Makedonija 95"), pp. 1–67 (in Slovene).

22.5 CONCLUSION

With no previous studies conducted on the ice caves in Macedonia, this contribution aims to summarize the current information on distribution and characteristics of ice caves in Macedonia. In order to provide a better base for future studies, ice caves here are considered as caves that contain perennial ice or snow deposits and/or seasonal ice deposits. They have been further grouped based on cave size and distribution and characteristics of the ice/snow deposits. In total, 25 ice caves have been found in two mountain areas in Macedonia, Jakupica Massif and Krčin Mountain. Four of these caves are large caves with large perennial ice and snow deposits, 13 are snow/ice-plugged pits with explored passages less than 60 m deep, 6 caves have passages that are

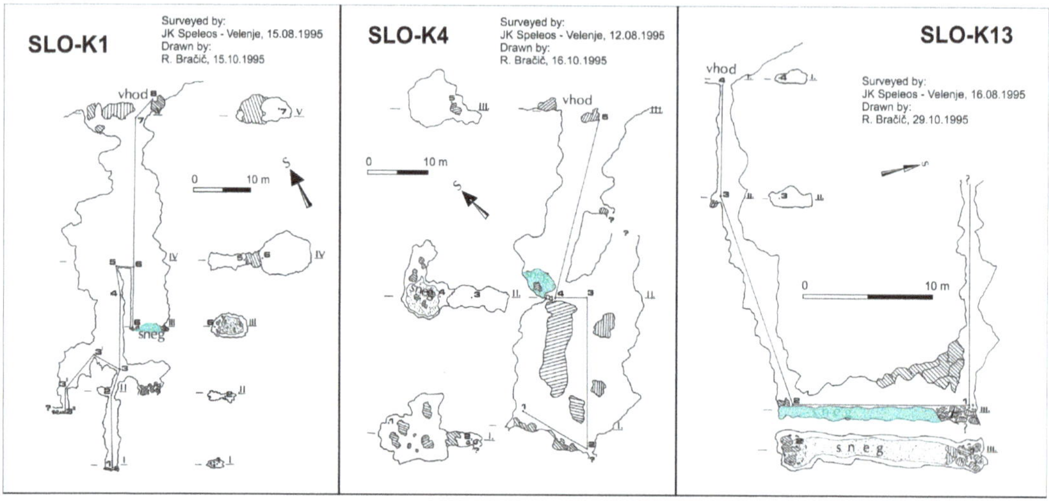

FIG. 22.15

Maps and distribution of ice/snow deposits in caves on Krčin Mountain. Ice and snow deposits are colored in *cyan*.

Maps modified from JK Speleos - Velenje, 1995. Slovenska jamarska odprava "Makedonija 95" (Slovene caving expedition "Makedonija 95"), pp. 1–67 (in Slovene).

partly covered by smaller snow/ice deposits, and 2 are small caves with seasonal ice formations. Most of them are located on Jakupica Massif, with the most important ones being Solunska Jama, Lednik, Slovačka Jama, and Solunska Glava 5, all deep vertical caves with large ice deposits, and with Solunska Glava 5 having a huge room filled with a large ice body several meters thick. Considering some not explored, snow plugged, large entrance pits on Jakupica, as well as the poorly known karst terrains in the western high mountain areas of Macedonia, the actual number of ice caves in Macedonia can be expected to be higher. Although no studies have been done, a general decline in ice volume on Jakupica Massif has been noted, and with the information on ice caves on Krčin Mountain now being more than 20 years old, detailed studies of ice caves in Macedonia are greatly needed.

ACKNOWLEDGMENTS

The information presented in this chapter is a summary of results from a number of caving expeditions and explorations over the course of more than 35 years, in which many Macedonian, Slovene, Slovak, Czech, and other foreign cavers have participated, and their efforts are greatly appreciated. I would also like to thank Martin Sluka, Jan Šmoll, Jožef Psotka, Juraj Szunyog, Peter Imrich, Martin Miškov, and Libor Štubňa for providing me with some unpublished data, cave maps, and photos from their caving explorations on Jakupica, as well as the late Rajko Bračič (JK Speleos - Velenje), who several years ago provided me with the report of their caving expedition on Krčin Mountain in 1995.

Part of the preparation of this chapter was supported by MTA Postdoctoral Fellowship Programme, No. PP-030/2015.

REFERENCES

Andonovski, T., 1997. Vlijanie na reljefot vrz klimata vo planinskite podračja vo Republika Makedonija povisoki od 1000 m (Influence of the relief on the climate in the mountain areas in Republic of Macedonia higher than 1000 m). In: Zikov (Ed.), Vlijanie na Sredozemnoto More vrz klimata na Republika Makedonija. Makedonska Riznica, Kumanovo, pp. 161–170 (in Macedonian).

Arsovski, M., 1997. Tektonika na Makedonija (Tectonics of Macedonia). Rudarsko-geološki fakultet, Štip (in Macedonian).

Carlin, R., Angelov, N., Oktavian, M., Grozdanovski, H., Žežovska, D., Žežovski, I., 2011. Unknown natural rarities in Macedonia. Speleological Society "Peoni", Skopje.

Dumurdžanov, N., Serafimovski, T., Burchfiel, B.C., 2005. Cenozoic tectonics of Macedonia and its relation to the South Balkan extensional regime. Geosphere 1, 1–22.

Ford, D., Williams, P., 2007. Karst Hydrogeology and Geomorphology. Wiley, Chichester.

Holúbek, P., Šmoll, J., 2012. Macedónsko máj 2012 (návrat po 20 rokoch) (Macedonia, May 2012 (Comeback after 20 Years)). Spravodaj SSS 2/2012, 50–52 (in Slovak).

Holúbek, P., Pokrievka, P., Sluka, M., Šmoll, J., Votoupalet, S., 2013. Balkánske speleologické výpravy v roku 2012 (Speleological Expeditions on Balkan in 2012). Spravodaj SSS 2/2013, 36–46 (in Slovak)

JK Speleos - Velenje, 1995. Slovenska jamarska odprava "Makedonija 95" (Slovene caving expedition "Makedonija 95"). pp. 1–67 (in Slovene).

Jovanović, P.S., 1928a. Glacijacija Jakupice (Glaciation of Jakupica). Posebna izdanja Srpskog Geografskog Društva 4, 1–86 (in Serbian).

Jovanović, P.S., 1928b. Karsne pojave u Poreču (Karst forms in Poreče). Glasnik Skopskog Naučnog Društva 4, 1–46 (in Serbian).

Kolčakovski, D., 1988a. Pećine visokoplaninskog krasa Jakupice (Caves of the highmountain karst of Jakupica). Speleobih 1–2, 49–57 (in Serbo-Croatian).

Kolčakovski, D., 1988b. Visokoplaninski karst na planinata Karadžica (High mountain karst of Karadžica Mountain). Geogr. Rev. 26, 95–113 (in Macedonian).

Kolčakovski, D., 2001. Speleomorfološki karakteristiki na karstniot reljef vo Republika Makedonija (Speleomorphological characteristics of the karst relief in the Republic of Macedonia). Geogr. Rev. 36, 23–34 (in Macedonian).

Kolčakovski, D., 2006. Fizička geografija na Republika Makedonija (Physical geography of the Republic of Macedonia). Institut za geografija, PMF, Skopje.

Lazarevski, A., 1993. Klimata vo Makedonija (Climate in Macedonia). Kultura, Skopje (in Macedonian).

Luetscher, M., Jeannin, P.-Y., 2004. A process-based classification of alpine ice caves. Theor. Appl. Karstol. 17, 5–10.

Maggi, V., 2013. Activity Report 2009-2013, Commission on Glacier, Firn and Ice Caves. UIS Bull. 56 (1), 24–26.

Majerníčková, F., Imrich, P., 2009. Snehulienka (Snow-white). Spravodaj SSS 1/2009, 62–65 (in Slovak).

Manakovik, D., 1966. Nekoi reljefni elementi vo porečieto na Babuna i Topolka so Titoveleškata kotlina (Some elements of the relief in the drainage basins of Babuna and Topolka and the Veles Basin). Godišen Zbornik, Geografski Institut Skopje 15, 87–139 (in Macedonian).

Perşoiu, A., Onac, B., 2012. Ice in caves. In: White, W., Culver, D.C. (Eds.), Encyclopedia of Caves. second ed.. Academic Press, Elsevier, Amsterdam, pp. 399–404.

Petreska, B., 2008. Podzemni karstni formi vo Porečkiot Basen i nivna valorizacija za potrebite na prostornoto planiranje (Underground karst forms in the Poreče Basin and their valorisation for the need of spacial planing) (Master Thesis) Faculty of Natural Science and Mathematics, Skopje (in Macedonian).

Pokrievka ml, P., 2016. Macedónsko 2016. http://www.cervenevrchy-speleo.sk/ (Accessed January 24 2017).

Psotka, J., Sluka, M., Šmoll, J., 2009. Slovačka Jama in Karadžici -524 m (Slovačka jama in Karadžica -524 m). Spravodaj SSS 1/2009, 56-61 (in Slovak).

Sluka, M., Szunyog, J., 2011. Jakupica 2010 alebo 1100km iba kvôli kopaniu (Expedition Jakupica 2010 or 1100km just because of digging). Spravodaj SSS 1/2011, 74–75 (in Slovak).
Šmoll, J., 2006. Stručná správa: Kosovo a Macedónsko 2/2006 (Brief report: Kosovo and Macedonia 2/2006). Spravodaj SSS 1/2006, 70–71 (in Slovak).
Šmoll, J., Sluka, M., 2007. Slovenska jama na Karadžici (Slovačka jama, Ceki 1) (Slovačka jama na Karadjici (Ceki 1)). Spravodaj SSS 4/2007, 50–54 (in Slovak).
Šmoll, J., Szunyog, J., 2005. Priepast' Solunska jama. Spravodaj SSS, 1/2005, 51–56 (in Slovak).
Šmoll, J. Szunyog, J., 2006. Jakupica 2005 – Macedónsko (Jakupica 2005, Macedonia). Spravodaj SSS 1/2006, 66–69 (in Slovak).
Šmoll, J., Štubňa, L., Imrich, P., Holúbek, P., Sluka, M., Pokrievka ml, P., Obuch J., 2012. Slovačka jama na Karadžici – Ceki 1/Expedícia Jakupica – Macedónsko 2011 (Slovačka Jama Cave on Karadžica Mt. – CEKI 1/ Expedition Jakupica – Macedonia 2011). Spravodaj SSS 1/2012, 66–73 (in Slovak).
Stračkov, M., 1968. Inženersko-geološki odliki na terenot od desnata strana na idnata akumulacija na rekata Treska na potegot pomegju Matka i Sušica (Engineering-geological characteristics of the terrain on the right side of the future reservoir on Treska River, on the part between Matka and Sušica). Trudovi na geološkiot zavod na SR Makedonija 13, 141–164 (in Macedonian).
Szunyog, J., 2006. Macedónsko – Albánsko 12. až 22. 10. 2006 (Macedonia – Albania 12 – 22 October 2006). Spravodaj SSS 4/2006, 46–48 (in Slovak).
Temovski, M., 2012. Površinska rasprostranetost na karstnite karpi vo Republika Makedonija (Extension of karst rock outcrops in Republic of Macedonia). Geogr. Rev. 46, 21–35 (in Macedonian).
Temovski, M., 2013. Karakteristiki na hipogenata karstifikacija i pojava na hipogen karst vo Republika Makedonija (Characteristics of hypogenic karstification and occurrence of hypogenic karst in Republic of Macedonia). Geogr. Rev. 47, 11–29 (in Macedonian).
Temovski, M., 2016a. Evolution of Karst in the Lower Part of Crna Reka River Basin (Springer Theses). Springer International Publishing, Heildelberg.
Temovski, M., 2016b. Ice caves in Macedonia. In: Mihevc, A., Zupan Hajna, N., Gostinčar, P. (Eds.), 7th International Workshop on Ice Caves, Program Guide and Abstracts. Karst Research Institute ZRC SAZU, Postojna, pp. 74–75.
Temovski, M., 2017. Hypogene karst in Macedonia. In: Klimchouk, A., Audra, Ph., Palmer, A., De Waele, J., Auler, A. (Eds.), Hypogene Karst Regions and Caves of the World. Springer International Publishing, https://doi.org/10.1007/978-3-319-53348-3.
Temovski, M., Audra, P., Mihevc, A., Spangenberg, J., Polyak, V., McIntosh, W., Bigot, J.-Y., 2013. Hypogenic origin of Provalata Cave, Republic of Macedonia: a distinct case of successive thermal carbonic and sulfuric acid speleogenesis. Int. J. Speleol. 42, 235–264.
Vandermeulen, M., 2015. 12 ans d'exploration de Matka Vrelo en République de Macédoine (12 years exploration of Matka Vrelo in Republic of Macedonia). Regards 80, 16–33 (in French).

FURTHER READING

Study for the Revalorization of Mavrovo Protected Area, 2011. Climate Data. http://www.moepp.gov.mk/wp-content/uploads/2015/01/Study-Mavrovo-Final-7.pdf (Accessed 24 January 2017).
Cave maps and photos from caves on Jakupica Massif. http://jakupica.speleo.sk/ (Accessed 27 January 2016).

CHAPTER

ICE CAVES IN NORWAY, FENNOSCANDIA AND THE ARCTIC

23

Stein-Erik Lauritzen, Lene Baastad, Julie Engelien
University of Bergen, Bergen, Norway

CHAPTER OUTLINE

23.1 INTRODUCTION

Caves containing ice occur in local pockets of permafrost that are situated in places where the mean annual surface temperature may or may not be well above freezing. This temperature difference has a threshold. As a result, low latitude ice caves can sustain only in alpine environments, and the lower altitude limit ("ice cave ELA") drops as we move into colder regions. In the Arctic, where permafrost is continuous, all caves are in a frozen state (exceptions are active talik caves). The entrapped ground ice in permafrost areas has climatic conditions similar to those of dynamic ice caves in temperate regions.

This chapter describes ice caves known in the northern regions of Europe (Fennoscandia, Iceland, Svalbard, and Greenland; Fig. 23.1) where this effect is evident. In spite of low mean annual temperatures and low altitude threshold for sustaining ice, we will see that cave ice is recessing, possibly because of the effects of global warming.

Ice Caves. https://doi.org/10.1016/B978-0-12-811739-2.00023-1

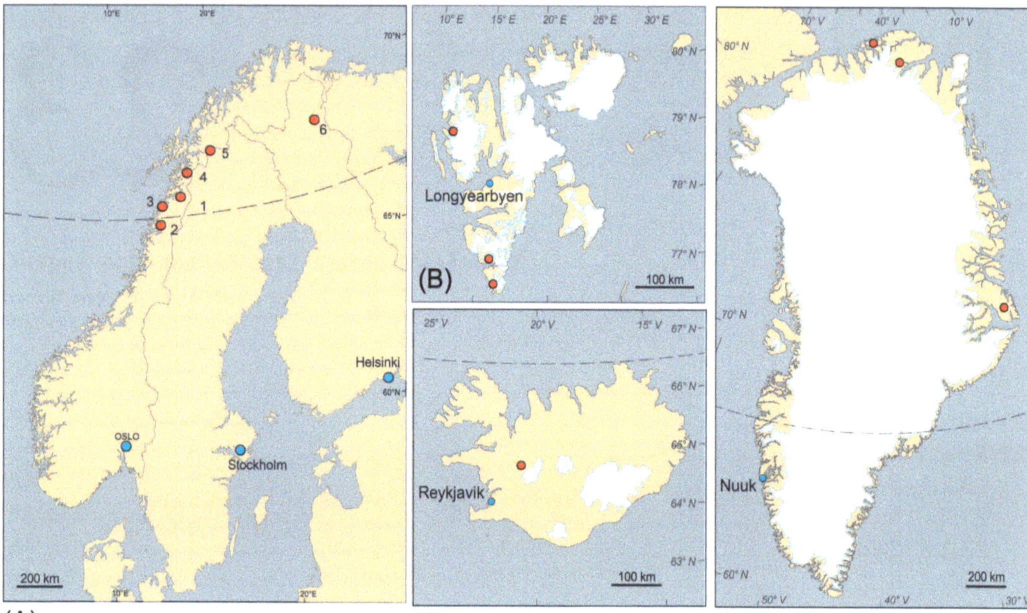

FIG. 23.1

Key maps. (A) Fennoscandia: (1) Svarthammarhola, (2) Isgrotta, (3) Greftkjelen, (4) Råggejavre Raige (RJR) and Salthølene, (5) Iskristallgrottan, and (6) Korkia-Maura cave; (B) Svalbard, *red spots* various frozen karst cave localities along the west coast; (C) Iceland *red spot* location of Hallmundahraun lava flow; and (D) Greenland, limestone areas with frozen karst.

23.2 NORWAY

Several caves in Norway bear the name "isgrotta" (ice cave), alluding to the fact that they contained ice at the time of discovery, usually in the summer months; and some explorers (e.g., Horn, 1947) have reported that passages were blocked with ice. Today many of these sites no longer contain ice. Here we discuss examples of the most important ice caves that have, or evidently had, contained perennial ice deposits.

23.2.1 SVARTHAMMARHOLA

Svarthammarhola (N67.13′E15.31′ at 295 m a.s.l.) is the largest ice cave in Fennoscandia. It is located near Fauske, north of the Arctic Circle (Fig. 23.1A). It also contains the largest cave chamber (300 × 90 × 40 m/lwh) known in this region, perhaps also in Northern Europe (Fig. 23.2). The total floor area of the upper section is >40,000 m^2. It has two major entrances with an altitude difference of 50 m; the lower entrance is at 245 m a.s.l. The cave is of the simple dynamic type (Luetscher and Jeannin, 2004), where congelation ice accumulation and ablation near the lower entrance is driven by Balch Ventilation and availability of intruding water.

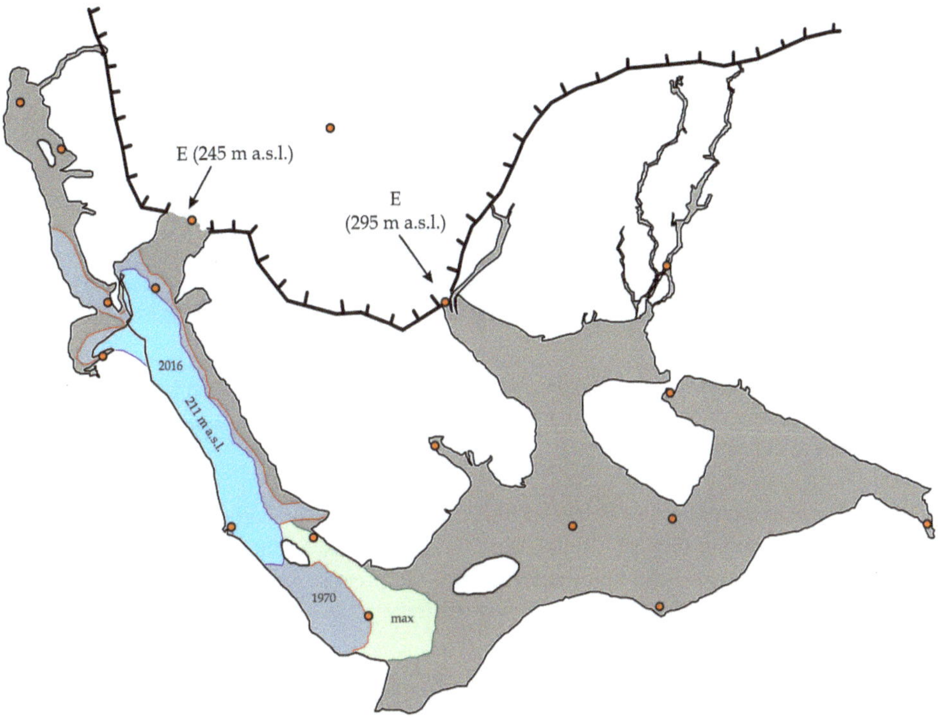

FIG. 23.2

Svarthammarhola, plan. E: Major entrances. *Red dots*, data logger stations; ice block extents: *blue*, 2016; *gray*, 1970; *green*: oldest extent (of unknown age) as traced by cryogenic precipitates.

The ice mass is situated in a horizontal tunnel near the lower entrance with a present (2016) exposed flat surface of some 3350 m^2 (Fig. 23.2). The open, horizontal surface is 160 m long and up to 25 m wide. At the downstream end, the ice is exposed at an 18-m-high ablation wall into a lower gallery (Fig. 23.3). There is also a ca 3-m-diameter ablation tunnel under the ice mass that penetrates the ablation wall. The tunnel is formed by heat brought in through an external small invasion stream. On a smaller scale, several small tunnels carrying ground air (with a relatively high radon content) appear at the ice contact with the underlying scree. Mapping suggests that the total maximum stratigraphic thickness of the ice mass is 27 m from the lowest lobe in the western ablation wall to the present ice's top surface. Upstream and to the north, because of the oblique passage cross-section, the ice tapers off and ends. With a tetrahedral shape, the ice block's volume is <15–20,000 m^3; hidden collapsed material within the ice may greatly reduce the ice's true volume.

23.2.1.1 Recent Changes in Ice Volume

Considerable changes have taken place since the cave was first discovered in 1969. The original cave survey (Heap, 1970) showed a much more extensive ice surface than today (~5450 m^2). In particular, a large ice surface at the upstream end disintegrated during the 1980s and transformed into an ice-dammed lake, which disappeared completely. The water drained underneath the ice block, presumably

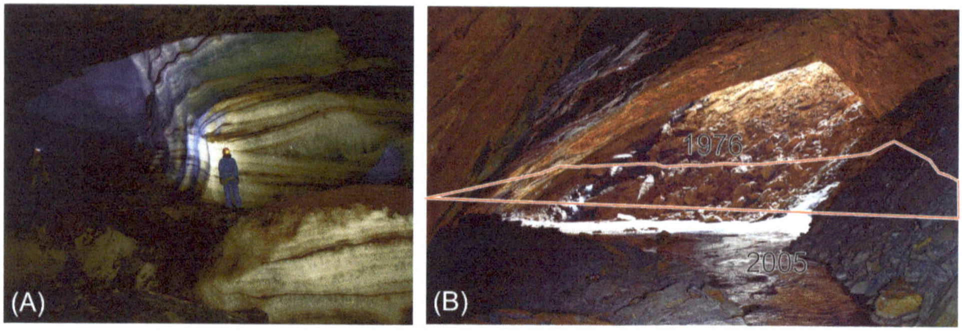

FIG. 23.3

Svarthammarhola. (A) Part of the lower ablation wall, Oct. 2016; (B) ice surface seen toward the lower entrance, 2005. The corresponding ice surface for 1976 as inferred from historical photos shown in *red line*. Height difference is about 5m.

into the lower downstream galleries. Traces of buff-colored cryogenic carbonate powder (Fig. 23.4) on boulders at higher elevations suggest a much older maximum ice surface area of about $7400\,m^2$. This may or may not be connected to the commencement of the present ice mass (<650/year, see below) or to a much older generation ice block, which is now entirely gone.

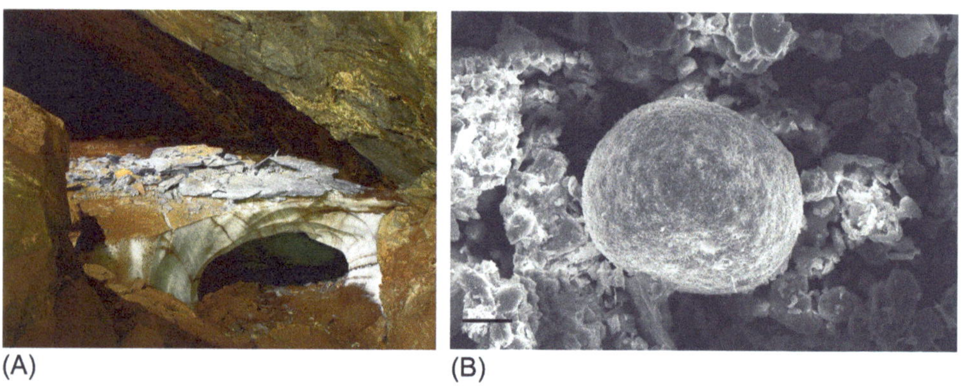

FIG. 23.4

Svarthammarhola. (A) Ice block (Oct. 2016) seen from slope under the lower entrance, showing the ablation tunnel and the recent (2015) ceiling collapse. (B) SEM image of cryogenic carbonate. Scale bar is 10μm.

Photographs taken in 1976 indicate that by 2016 the downstream ice had reduced about 5m and the upstream close to 2m. A now-deceased caver placed a spot of red paint on the rock wall before 1990, though the original position of the paint relative to the ice surface is uncertain. However, in 2005 the spot was 3.08m above the ice surface, the next year the distance increased to 3.11m, and subsequent laser measurements have revealed a progressive lowering rate, averaging 18cm $year^{-1}$. Previous measurements are not available. Linear extrapolations confirm that the ablation rate is increasing; otherwise, the red spot would have been beneath ice by about 1988. Using the historic data, combined with the direct measurements done from 2005 to 2011, it appears that the ice surface has lowered at an average rate of 0.11–0.15 m $year^{-1}$ since 1975 (Fig. 23.5). This places the paint mark at ice level somewhere between 1985 and 1990.

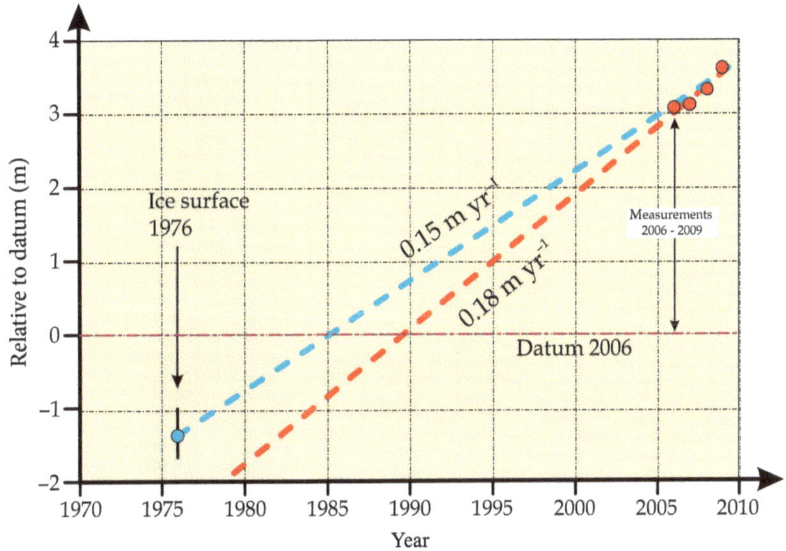

FIG. 23.5

Svarthammarhola. Ablation rate for the ice block surface. *Red dots*: direct measurements from ice surface to paint mark. *Blue dots*: ice surface as inferred from historical photos (i.e., Fig. 23.3B). *Red line*: linear extrapolation of the direct measurements, corresponding to 0.18 m $year^{-1}$. *Blue line*: same regression, but including the *blue dot*, corresponding to an average lowering of 0.15 m $year^{-1}$. See text for further discussion.

In September 2015 the roof section holding the red paint collapsed onto the ice in a previously celebrated spot where tourists gathered to take photographs (Fig. 23.4A). This is a dramatic demonstration of how modern ablation is contributing to the cave's instability. Since 1991 a deep crevasse (the "Radon Pit") has opened along the southern edge of the ice body, now exposing a small grotto below the ice mass, which is about 5 m thick at this point.

23.2.1.2 Ice Stratigraphy and Coring

From 2004 to 2005 the ice mass was core-drilled at its presumably thickest downstream end (Lauritzen et al., 2005). Rock was, however, hit in two neighboring holes at 4.5 m depth where the potential was estimated to be >20 m (Fig. 23.6). This clearly indicates that a considerable amount of roof collapse, as also seen on the present ice surface, is hidden within the ice. The drill hole was equipped with a thermistor string and capped with a polythene tube for accessing contacts. Since then the ice around the device has totally melted, suggesting between 0.5 and 1.0 m ablation since 2005, in accordance with the direct measurements previously discussed (Fig. 23.7).

A series of environmental parameters were then measured on the ice core (Engelien Bjørlien, 2006). For instance, a chemical record of various trace elements (Cu, Zn, Cd, Hg, Pb, S, Se) was made on the ice cores, with the intention of correlating the commencement (1894) of a copper smelter 25 km up-valley of the cave. An earlier, but more distant, medieval (1653–1659) silver smelter (Nasa silver mines,

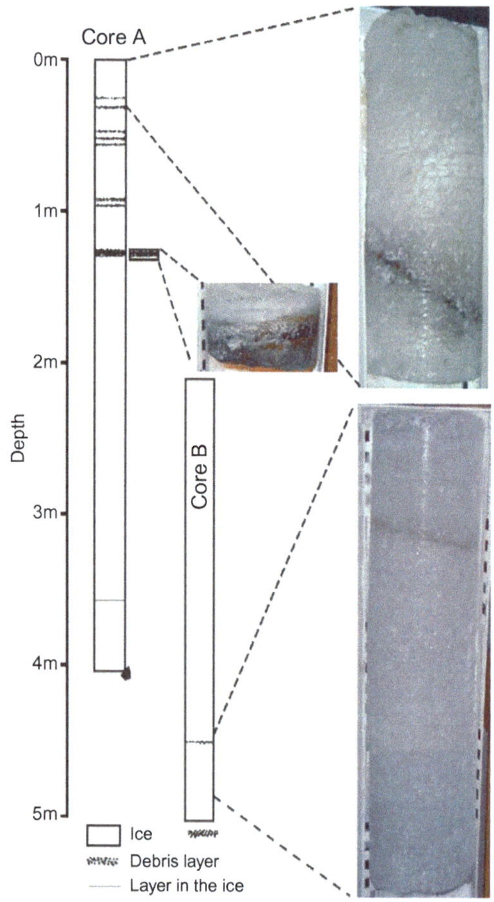

FIG. 23.6

Svarthammarhola. Details of the two ice cores with rocks and dirt-bands.

From Engelien Bjørlien, J., 2006. Utvalgte klimaproxyer fra grotteis: Svarthammarhola ved Fauske. (M.Sc. Thesis). University of Bergen.

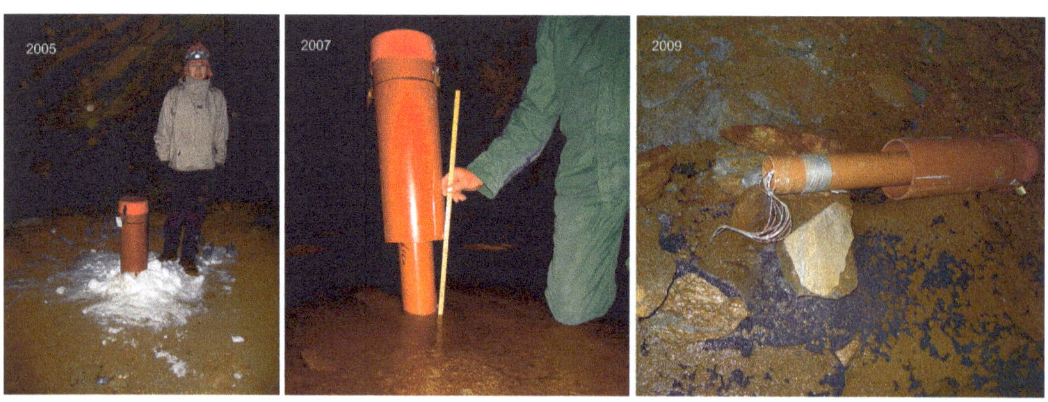

FIG. 23.7

Svarthammarhola. Ablation of ice surface around borehole casing during 4 years. Compare with Fig. 23.5.

100 km to the north) produced several hundred tons of lead oxide. Except for a weak mercury signal near the top of the sequence (which could well be a modern intrusion), no significant increase of base metals or other suspected smelter emissions (S, Se) were detected, suggesting that at least 120 years of ice accumulation is now lost (and/or that the emissions never reached the cave).

Organic matter (gyttja and plant remains) occurs at various levels within the ice mass as shown in the ablation wall. A ^{14}C date of plant fragments at the very base of the ice block dated to AD 1365 ± 75; that is, the ice is <650 calendar years old, and thus the ice accumulation seems coincident with the medieval climatic deterioration and the Little Ice Age.

23.2.1.3 Ice Stratigraphy and Stable Isotopes

Oxygen isotopes ($\delta^{18}O_w$) were measured in the ice core (Engelien Bjørlien, 2006) and in samples drilled from the ablation wall (Lauritzen, 1996). Because of the rocks encountered in the boreholes, the two records are not overlapping stratigraphically (Fig. 23.8). Since a more accurate isotope stratigraphy is in progress, only a brief discussion of these old data is pertinent. The record is supposed to cover the last 650 years, minus at least 120 years lost in recent ablation. The isotope composition represents snowmelt congelation—mainly winter precipitation—and modern (1990 congelation ice) isotopic values were only encountered twice in the record.

23.2.1.4 Ventilation Dynamics and Energy Budget

In 2005 a Campbell CR200 logger station was put in the cave to record wind speed and direction, air temperature, humidity, and ground temperature at the upper entrance; and a Campbell CR200 weather station (precipitation, air temperature, humidity, and atmospheric pressure) was placed on the plateau above the cave (Baastad, 2006). Furthermore, a set of 20 Tinytag loggers were placed at various critical places in the cave. Recording lasted, with some inevitable interruptions, from Mar. 2005 to Sep. 2006 and continued sporadically throughout 2007.

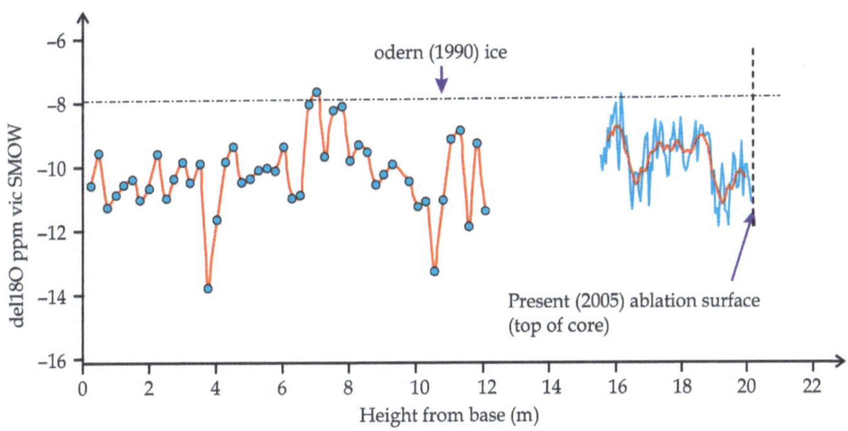

FIG. 23.8

Oxygen isotope stratigraphy of the ice block in Svarthammarhola. *Left curve*: samples from the lower ablation wall. ^{14}C dating of organic debris under the ice block indicates commencement after ~AD 1300. *Right curve*: samples from the combined ice cores; *red line* is 5-point running mean. In two instances, isotopic values attained modern (1990) values.

The cave is strongly affected by Balch-ventilation, producing winds of up to 8 (15) m s^{-1} through the upper, constricted entrance (4 m^2). Wind direction could be predicted from the difference in the virtual temperature between the cave (c) and the surface (s), $\Delta T = T_s - T_c$. $\Delta T < -3.5°C$ (cave warmer than the surface) sustained a positive (upward) chimney effect, whereas $\Delta T > +10°C$ (cave colder than the surface) sustained a downward directed wind. Between the two conditions, the ventilation displays strong oscillation, Fig. 23.9A–C.

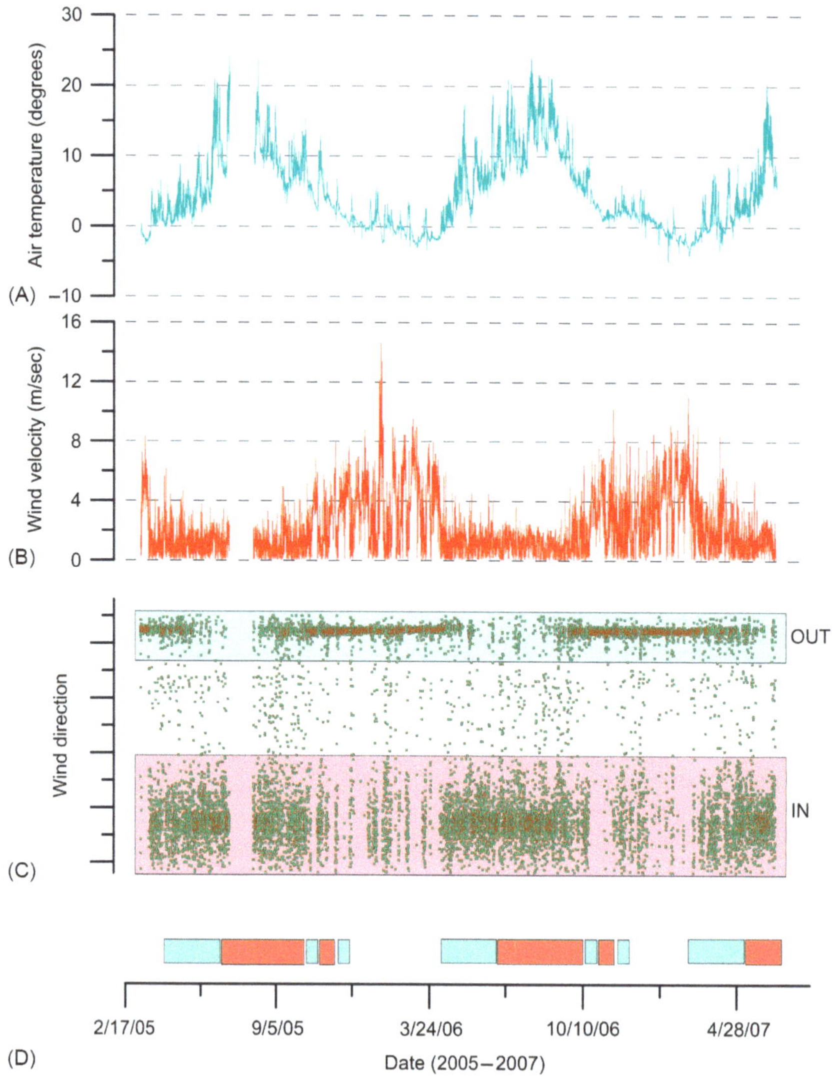

FIG. 23.9

Time-series during 2005–2007 for Svarthammarhola, upper entrance. (A) Air temperature. Note that the air comes either from the surface (inhaling) or from the cave (exhaling). (B) Wind velocity. (C) Wind direction, *blue rectangle* is exhaling or positive chimney effect, *red rectangle* represents inhalation or negative chimney effect. (D) Periods of ablation (negative energy budget) are shown in *red*. *Blue rectangles* represent periods of potential accumulation of ice. After Baastad (2006).

From the sparse logger data, Baastad (op. cit.) calculated an energy flux of −421 GJ for the years 2005 and 2006; that is, the cave consumed energy and thus the ice mass was ablating. The calculation was based on ventilation only; geothermal heat flux and heat from intruding streams were not known. Assuming that all the energy was consumed in ice melting, we can convert the calculation to 1260 m^3 ice, which when distributed over the exposed ice surface, converts to a shrinking of 33 cm $year^{-1}$ (or 4.2 cm $year^{-1}$ for sublimation). However, because of the enormous size of the upwind cave rooms, it is logical to assume that a substantial part of the heat budget was consumed by sublimation of moisture from these surfaces (2.3 and 2.6 MJ/kg for water and ice, respectively). The amount of water entering from the hillside during summer is yet unknown.

The seasonal heat budget suggests that ice accumulation could have taken place when the cave was undercooled and meltwater could enter the cave Apr. through Jun. and possibly on short occasions during the fall (Fig. 23.9D). June through October would be the period of intense ablation. Consequently, the environmental composition of the ice mass would represent winter (snow) surface precipitation.

The historic data and direct measurements suggest an average ablation of 0.11–0.14 m $year^{-1}$, corresponding to only one-third of the estimated maximum energy consumption. This may suggest that about two-thirds of the energy (280 GJ) was consumed through ice sublimation from the ice surface and by water evaporation from the large surface of the cave walls. From the historical data previously discussed, the ice mass appears to have abladed monotonically, and possibly at an increasing rate since the cave was discovered in 1969.

23.2.2 GREFTKJELEN

This 11-km-long and 350-m-deep cave system is situated on the coast south of Bodø (N66.58′E14.04′ at 350 m a.s.l.; Fig. 23.10B). The main entrance to the system consists of a huge, steeply sloping relict

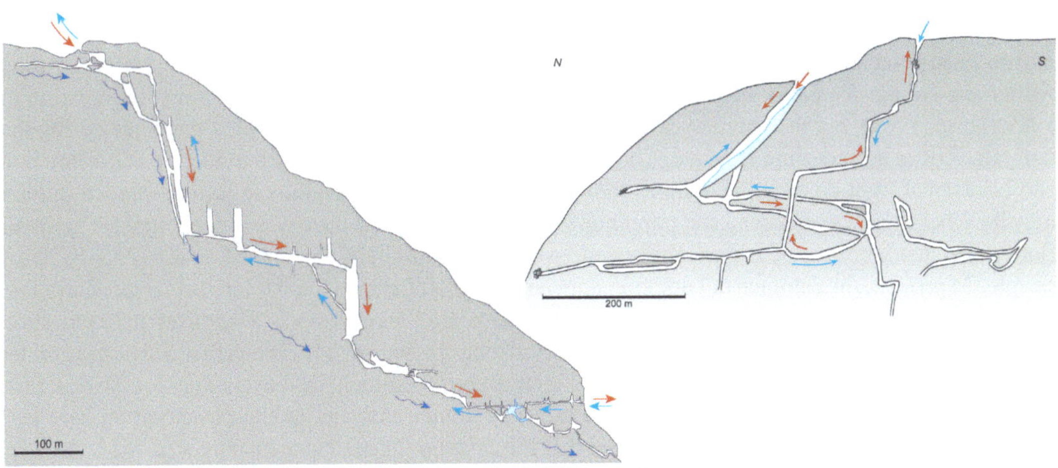

(A) (B)

FIG. 23.10

(A) Elevation map of Råggejavre Raige, the deepest cave in Northern Europe. Summer *(red)* and winter *(blue)* wind directions shown. *Curved arrows*: invasion stream movement, including the 140 m underground waterfall. From survey by Lauritzen et al. (1991). (B) Elevation map of Greftkjelen. Firn and ice mass is shown in *light blue*, winter draughts indicated by *blue arrows*, summer ventilation with *red arrows*.

Compiled from various sources (Heap, 1972; Holbye, 1983; Aasheim, 1988).

invasion canyon that contains an ice mass covered in snow. The deep cave has no known lower entrance; the cave terminates in syphons and sediment plugs, but there are smaller entrances with quite strong drafts at higher positions (425 m a.s.l.) than the main entrance. They are connected to the main cave through tortuous, narrow passages and shafts. This configuration resembles the Paradana cave system in Slovenia (see chapter 30), where an ice-filled entrance shaft carry an ice plug, and is connected to entrances at a higher altitude through complicated and in part unexplored passages, i.e., a stratodynamic cave with firn (Luetscher and Jeannin, 2004).

The cave was first explored in 1971 (Heap, 1972). Later it was greatly extended through the 1980s (Lauritzen, 1977; Holbye, 1983; Aasheim, 1988) and is still being amended. The first explorers in 1971 had little problem getting down the snow- and ice-filled vadose rift; by the following year, deep crevasses with overhanging firn and ice walls had opened. The ice mass appears to have waxed and waned in pace with annual snowfall through the past four decades (Fig. 23.10B).

23.2.3 SALTHØLENE

Situated at the headwall of the deep Tysfjord fjord complex (N68.02′E16.38′ at 643 m a.sl.; Fig. 23.1A), this more than 2000-m-long and 186-m-deep cave is developed within a thin strip of marble (Lauritzen, 2001). Deep (up to 40 m) surface shafts have snow plugs and ice bodies at the bottom. In Aug. 1985 the lowest part of the cave, a fossil vadose canyon, was filled with ice and the lowest isolated parts of the cave encrusted in hoarfrost. Recent observations are not available.

23.2.4 RJR

Råggejavre-Raige (RJR, the "Pit-tarn Cave") is situated inside the wall of another branch of the Tysfjord complex (N67.53′E16.15′ at 590 m a.s.l.; Figs. 23.1A and 23.10A). It is the deepest cave in Fennoscandia (diameter, 580 m; length, 1900+ m). It has two entrances with an elevation difference of 540 m. In an antechamber just inside the lower entrance at 80 m a.s.l., masses of ice accumulate during winter and spring. Ice masses have been observed to survive the summer since its first exploration in 1969 (Heap, 1969), but the ice masses seem to disappear along the most traveled passages during the late fall. RJR is a borderline ice cave. The lower parts of the cave bear much evidence of gelifraction and occasional light-colored precipitates that may be of cryogenic origin, suggesting that ice masses may have been more widespread and perennial in the past, before the cave was first explored. The small fjord cave, just a couple of meters above sea level, has traditionally been used as a freezer by local people. At present, the summer-fall ablation appears to just outbalance the winter-spring accumulation, suggesting that the cave is at its dynamic limit for maintaining perennial ice. The differential elevation of 540 m between the two entrances generates very strong thermal winds. The influx of warm summer water through a 140-m-high underground waterfall must be an efficient heat exchanger for the descending airflow and is probably a major heat and moisture source that can quench evaporation heat loss from cave walls. Recent observations suggest that the cave no longer supports perennial ice.

23.2.5 ISGROTTA, GLOMDAL

Isgrotta is situated in Glomdal, Rana (N66.29′E13.53′ at 225 m a.s.l.). It is a simple, static ice cave that used to contain an approximately 1-m-tall ablation wall of congelation pool ice (Fig. 23.11A). In 1977 this place contained a water table with ice rafts that, in the summer, drained leaving an open pit with the

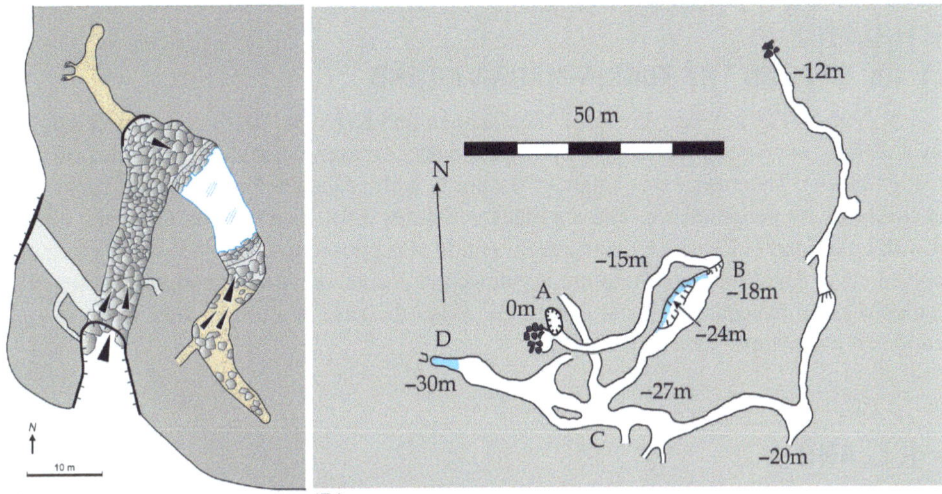

(A) (B)

Fig. 23.11(A) Plan map of Isgrotta, Glomdal, a static ice cave with congelation ice. Map from 1990, showing the pit with 1 m ice walls; the recently observed (2016) subsidence of the blocky floor not shown. (B) Iskristallgrotten, Sweden, compiled from (Engh, 1980).

ice wall, fragments of ice rafts, and proliferate hoarfrost crystals in the ceiling. Over the next 40 years, the ice wall retracted about 2 m, and recent melting left a depression in the blocky floor, where only a 40-cm-tall ice wall is exposed under collapsed blocks (observed March 2017).

23.2.5.1 Talus Caves With Ice

Talus caves (voids between rotated blocks in a rock talus) situated in slopes may, because of Balch ventilation, contain perennial ice. They may be regarded as a continuum of ice-cored rock glaciers (e.g., Humlum, 1998). There are several references to such "cold holes" in Norway, which local farmers have used as natural ice cellars, among which the talus caves in the very steep fjord walls in Western Norway are probably best known (e.g., Rekstad, 1911).

23.3 SWEDEN

23.3.1 ISKRISTALLGROTTAN

This 373-m-long and 38-m-deep cave is situated northwest of Abisko in Swedish Lapland (N68.32′E18.22′ at 894 m a.s.l.). The entrance of the now inactive cave is on top of a convex hill that restricts a potential drainage area (Engh, 1980). The cave has only one (upper) passable entrance; the ice mass is situated at the lowest section, 30 m below the surface where the passage is blocked by compact ice (Fig. 23.11B). According to Engh (op. cit.), the passage is not explorable beyond this point, and there is no appreciable draft, making the cave a static, ice cellar-type cave. Higher up in the passages, at ca −20 m below the entrance, large masses of extrusion ice fiber crystals (needle ice, "pipkrake") occurs on walls and sediments, thus the cave's name ("ice crystal cave").

23.4 FINLAND

23.4.1 ICE CAVE ON THE KORKIA-MAURA ISLAND

This cave at N68.50′E27.39′ ca 125 m a.s.l. (Kesäläinen and Kejonen, 2015) is a talus or boulder cave that developed as interspaces between displaced blocks of basement rocks (granulite) of the Baltic Shield (Fig. 23.1A). The entrance is a narrow fissure in a cliff face leading down to a room where the floor is covered with perennial ice. The ice block measures 3.5 × 10 m with an unknown thickness. It is level with Lake Inari (119 m a.s.l.), and accumulation is supposed to take place when water enters an undercooled cave. The cave may therefore be viewed as a static cave with congelation ice. The cave has been used as an ice cellar by local fishermen; today the cave is a tourist site. There is no known information on ice isotopes, chemistry, or age.

23.5 ICELAND

Among the many (500) lava tubes known in Iceland, only a few are reported to host perennial ice. The two most important ones are Surtshellir and Víðgelmir (Hróarsson and Jónsson, 1991; Stefansson and Stefansson, 2016), both within the Hallmundahraun lava flow in Western Iceland (Fig. 23.1B). Surtshellir is 3.5 km long. The surrounding lava flow is about 1100 years old (Saemundsson, 1966). The inner part of the cave was depicted in 1835 (Maiers) as having a continuous ice floor; ice was also documented by Grossmann and Lomas (1894). Surtshellir and Víðgelmir both have lost a significant amount of ice since the 1970s, and the only lava cave in Iceland with substantial ice formations is Loftshellir (A. Stefansson, p.c., 2017). There is no available information on ice isotopes, chemistry, or age.

23.6 SVALBARD AND GREENLAND

In areas of continuous permafrost, few caves are explorable, and those that are often terminate in frozen sediments, hoarfrost, and ground ice. This phenomenon is quite common in Svalbard (Lauritzen, 2006). At Fisneset, on Sørkapp Land (N76.43′16.14′ at 30 m a.s.l.), a small, horizontal karst cave in a cliff face contained an ablation wall of laminated congelation pool ice (Lauritzen 1992, unpublished). Moreover, vast areas of karstified carbonates in Eastern and Northeastern Greenland contain caves with frozen sediments and presumably ground ice (Davies, 1957; Davies and Krinsley, 1960; Loubiere, 1987).

23.7 CONCLUSIONS

The notable shrinking and often total disappearance of cave ice masses have been widely observed during the past half-century. Low-altitude ice caves are at their climatic limit of existence and are thus extremely sensitive to changes in their surroundings. Further extraction of environmental information from cave ice is therefore urgent; for the Svarthammar cave, work is in progress. The Arctic ice caves are not well explored, but they hold promise for occasional deposits of congelation ice and thus sequential information.

ACKNOWLEDGMENTS

The Bjerknes Centre for Climatic Research and The Norwegian Research Council (Project Norpast 2) are thanked for their financial support. We also thank Dr. Per Holmlund, Stockholm University, for collaboration on the ice core drilling field campaign in Svarthammarhola in 2005. We also thank Kjell Hansen for providing important historic photographs of Svarthammarhola, Roy Solbakk for kindly maintaining data loggers and Dr. Trond Ellingsen for permitting research in Svarthammarhola. We also express our gratitude to Leif Engh for providing information on Iskristallgrottan and to the many cavers who shared their observations on ice occurrence. Arni Stefansson and Kirsi Siiroinen are thanked for providing information on Iceland and Finland, respectively.

REFERENCES

Aasheim, S., 1988. Stein P'.s eventyrlige reiser. Orion Forlag.

Baastad, L., 2006. Ventilasjonseffekter og Mikroklima i Svarthammarhola ved Fauske (Ventilation Effects and Microclimate in Svarthammarhola at Fauske) (M.Sc. Thesis). University of Bergen.

Davies, W.E., 1957. Rillenstein in Northwest Greenland. Bull. Natl. Speleol. Soc. 19, 40–46.

Davies, W.E., Krinsley, D.B., 1960. Caves in Northern Greenland. Bull. Natl. Speleol. Soc. 22, 114–116.

Engelien Bjørlien, J., 2006. Utvalgte klimaproxyer fra grotteis: Svarthammarhola ved Fauske (Selected Climate Proxies From Cave Ice: Svarthammarhola at Fauske) (M.Sc. Thesis). University of Bergen.

Engh, L., 1980. Skandinavisk speleokronologi i ny belysning genom nyupptackta grottor i Vadvevagge (Scandinavian Speleochronology in a New Light From Recently Discovered Caves in the Vadvevagge Area). Grottan 15 (1), 2–22.

Grossmann, K., Lomas, J., 1894. On hollow pyramidal ice crystals. Proc. R. Soc. Lond. 55 (331–335), 113–115.

Heap, D., 1969. Report of the British Speleological Expedition to Arctic Norway 1969, Including the Work of 1968 Hulme School's Expedition. Kendal Caving Club.

Heap, D., 1970. William Hulme's Grammar School's Expedition to Arctic Nordland 1970. D. Heap, Manchester.

Heap, D., 1972. William Hulme's Grammar School, Manchester. Expeditions to Arctic Norway, 1971–72. D. Heap, Manchester.

Holbye, U., 1983. Greftkjelen og Greftsprekka, cave map. Norsk Grotteblad 11, 18–19.

Horn, G., 1947. Karsthuler i Nordland. NGU (Norwegian Geological Survey) 165, 1–77.

Hróarsson, B., Jónsson, S.S., 1991. Lava caves in the Hallmundarhraun Lava flow, Western Iceland. In: Proceedings of the 6 th International Symposium on Vulcanospeleology.

Humlum, O., 1998. The climatic significance of rock glaciers. Permafr. Periglac. Process. 9 (4), 375–395.

Kesäläinen, T., Kejonen, A., 2015. Suomen Luolat (Caves of Finland). Tallinna Raamatutrükikoda OÜ, Salakirjat.

Lauritzen, S.E., 1977. Greftdalen. Norsk Grotteblad 1 (2), 70–83.

Lauritzen, S.E., 1996. Karst Landforms and Caves of Nordland, North Norway. Guide for Excursion 2 (Climate Change: The Karst Record, Field Trip Guide). Department of Earth Science, University of Bergen, Bergen.

Lauritzen, S.E., 2001. Marble Stripe karst of the Scandinavian Caledonides: an end-member in the contact karst spectrum. Acta Cardiol. 30 (2), 47–79.

Lauritzen, S.E., 2006. Caves and speleogenesis at Blomstrandsøya, Kongsfjord, W. Spitsbergen. Int. J. Speleol. 35 (1), 37–58.

Lauritzen, S.E., Baastad, L., Bjørlien, J., Fedje, E., Holmlund, P., 2005. The Svarthamar cave research project, Fauske, north Norway. In: Proceedings of the 14th International Congress of Speleology, Athen-Kalamos, August 2005 Paper P-12, pp. 1–5.

Lauritzen, S.E., Kyselak, J., Løvlie, R., 1991. A new survey of Råggejavri-Raigi and the Hellemofjord Karst, Norway. Cave Sci. 18, 131–137.

Loubiere, J.F., 1987. Observations Preliminaires sur les Cavites de la Region du Lac Centrum (Nord-Oest Groenland). Karstologia 9, 7–16.

Luetscher, M., Jeannin, P.Y., 2004. A process-based classification of alpine ice caves. Theor. Appl. Karstol. 17, 5–10.

Rekstad, J., 1911. Om koldeholene i Hardanger (The Cold Caves of Hardanger). Naturen 35, 165–170.

Saemundsson, K., 1966. Zwei neue C14-Datierungen isländischer Vulkanausbrüche.

Stefansson, A., Stefansson, G., 2016. Surtshellir in Hallmundarhraun: historical overview, exploration, memories, damage, an attempt to reconstruct its glorious past. In: 17th International Symposium on Vulcanospeleology. Hawaii, February 6th–13th, pp. 1–11.

CHAPTER

ICE CAVES IN POLAND

24

Michał Gradziński*, Jakub Nowak[†], Grzegorz Rachlewicz[‡], Wiesław Siarzewki[§], Krzysztof Strug[¶], Witold Szczuciński[‡]

Jagiellonian University, Kraków, Poland Cracow Caving Club, Kraków, Poland[†] Adam Mickiewicz University in Poznan, Poznań, Poland[‡] Tatra National Park, Zakopane, Poland[§] The Institute of Meteorology and Water Management—National Research Institute, Wrocław, Poland[¶]*

CHAPTER OUTLINE

24.1 INTRODUCTION

Poland is mainly a lowland country. Mountain chains occupy only its southernmost part and are bordered on the north by an upland zone. Thus the majority of caves are located in the southern part of Poland. All caves in Poland in which perennial ice was noted in the second half of the 20th century are clustered in the Tatra Mountains (Western Carpathians), precisely in the western part of the mountains. The Tatra Mountains are the only mountains with alpine-type characteristics in Poland. The karst rock crops out there at altitudes between 950 and 2100 m; therefore cave entrances are found in this altitude range.

Ice Caves. https://doi.org/10.1016/B978-0-12-811739-2.00024-3

24.2 HISTORY OF DISCOVERY, EXPLORATION, AND RESEARCH OF ICE CAVES: AN OUTLINE

Several ice caves were certainly known to highlanders, namely hunters and shepherds who pastured sheep in high-mountain meadows until the middle of the last century. The first information about an ice cave in the Tatra Mountains in Poland was published by Ossowski (1883), an archeologist who was searching for a potential place for future excavations. He did not visit the cave personally, but highlanders told him about a cave with perennial ice, that is, Jaskinia Lodowa w Ciemniaku (subsequently referred to here as Ice Cave in Ciemniak). The first, although imprecise, description of this cave was provided by Pawlikowski (1887), who was one of the pioneer climbers in the Tatra Mountains. He visited the cave a few times, probably guided by highlanders. The cave became especially popular after Zwoliński's (1923) article. This cave drew the attention of many tourists, cave explorers, and scientists because it was easily accessible despite its relatively high altitude (see, e.g., Gadomski, 1926).

Many caves have been discovered, or at least mentioned in publications, since the middle of the last century, when the systematic exploration of caves in the Tatra Mountains started (see, e.g., Wójcik, 1962). The written data were included mainly in short exploration reports. More detailed information about ice caves and other caves in the Tatra Mountains is published in an inventory of Polish caves, which presently is also accessible on the web (Jaskinie Polski, 2017).

Scientific observations in Polish ice caves have been almost exclusively limited to Ice Cave in Ciemniak, where microclimatic measurements have been carried out intermittently since the 1920s. The observation of cave microclimates (Kwaśniewski, 1924; Zwoliński, 1951; Kowalski, 1953, 1955; Rygielski et al., 1995; Rachlewicz and Szczuciński, 2004); cave ice dynamics (Zwoliński, 1951; Kowalski, 1953; Siarzewski, 1994a; Rachlewicz and Szczuciński, 2004; Szukała, 2010); studies on origin, isotopic composition, and age of cave ice (Celejowska et al., 2007; Hercman et al., 2010); as well as studies on subterranean organisms (Kowalski, 1955) or their subfossil remains (Kowalski, 1953; Wójcik, 1968; Nowosad et al., 1992) have been conducted in this cave as well. Studies in other ice caves have not been practically undertaken; exceptions are the microclimatic measurements made in Jaskinia Śnieżna as a part of Jaskinia Wielka Śnieżna (Pulina, 1967, 1974) and winter bat censuses conducted in several caves, including ice caves.

Knowledge available on ice caves in the Tatra Mountains in Poland was reviewed by Koisar and Parma (1971). Subsequently, the information was presented in several unpublished reports by Siarzewski and finally summarized in review articles (Siarzewski, 1994b, 1996a). All these ice caves are located in the Tatra National Park. Access to them is strictly limited, and special permission is needed to visit them.

24.3 DISTRIBUTION, DIMENSIONS, AND TYPES OF ICE CAVES

Forty-three ice caves are listed as being in the Polish part of the Tatra Mountains (Fig. 24.1). All the caves where perennial ice was detected in the final years of the 20th century are regarded as ice caves, though some of them lack ice presently (see later section, "Ice Dynamics"). Many of them are small voids that can be classified as a shelter. The caves and their basic morphometric data are listed in Table 24.1. Some small caves in which congelation ice or recrystallized snow were reported only once in summer or autumn seasons are not included because the occurrence was likely only incidental.

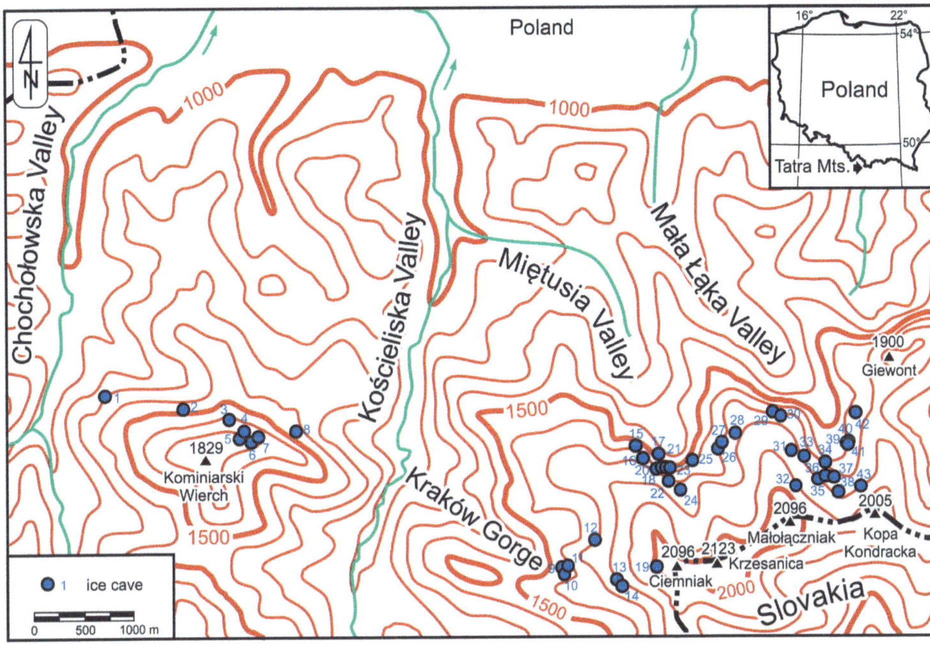

FIG. 24.1

Location of ice caves in the Polish Tatra Mountains; cave numbers are explained in Table 24.1.

Table 24.1 Ice Caves in Poland, Morphometric Data After Web Site Jaskinie Polski (2017)

	Name of Ice Cave	Altitude of Entrance m	Length m	Vertical Extent m	Dominating Type of Ice
1	Jaskinia Dmuchawa	1300	130	−18	Congelation ice
2	Dudowa Studnia	1488	80	35	Recrystallized snow
3	Jaskinia Lodowa nad Kufą	1670	20	−4	Congelation ice
4	Lejowa Szczelina	1730	8	4	Recrystallized snow
5	Szczelina ze Śniegiem	1745	24	8	Recrystallized snow
6	Tylkowa Szczelina	1750	98	25	Recrystallized snow
7	Suchy Biwak	1695	70	18	Congelation ice
8	Jaskinia Bańdzioch Kominiarski	1456 (lower entrance)	9550	562	Congelation ice
9	Śnieżna Szczelina	1613	6	0	Congelation ice
10	Szczelina pod Lodową	1591	6	2	Congelation ice
11	Jaskinia Lodowa Krakowska	1570	30	6	Congelation ice
12	Jaskinia Lodowa w Ciemniaku	1695	390	42	Congelation ice

Continued

Table 24.1 Ice Caves in Poland, Morphometric Data After Web Site Jaskinie Polski (2017)—cont'd

	Name of Ice Cave	Altitude of Entrance m	Length m	Vertical Extent m	Dominating Type of Ice
13	Szczelina w Tomanowym Grzbiecie III	1788	8	−3	Recrystallized snow
14	Szczelina w Tomanowym Grzbiecie I	1777	20	−5	Recrystallized snow
15	Studnia w Dziurawem	1543	30	−30	Recrystallized snow
16	Zimny Aven	1650	4	−1	Recrystallized snow
17	Jaskinia Lodowa w Twardych Spadach	1480	152	40	Congelation ice
18	Studnia za Murem	1555	122	−43	Recrystallized snow
19	Szczelina w Ciemniaku	2047	27	−19	Recrystallized snow
20	Studnia w Kazalnicy	1545	830	235	Congelation ice
21	Jaskinia Lodowa Miętusia	1590	85	25	Congelation ice
22	Jaskinia Lodowa Mułowa	1680	400	69	Congelation ice
23	Śnieżny Aven	1595	10	7	Recrystallized snow
24	Jaskinia Lodowa Ponorowa	1780	8	−5	Congelation ice
25	Jaskinia Lodowa Litworowa (part of Ptasia Studnia)	1576	460	−150	Congelation ice
26	Dziurka w Trawce	1763	179	28	Congelation ice
27	Jaskinia za Ratuszem Śnieżna	1720	11	−3	Recrystallized snow
28	Szara Studnia	1750	125	−50	Recrystallized snow
29	Aven z Korkiem Śnieżnym	1644	12	−10	Recrystallized snow
30	Jaskinia Strzelista	1701	83	18	Recrystallized snow
31	Ukryta Dziura	1812	32	−8	Recrystallized snow
32	Średni Lej	1880	30	−16	Recrystallized snow
33	Śnieżna Studnia	1756	12,700	763	Recrystallized snow Congelation ice
34	Świstowa Studnia	1670	28	−16	Recrystallized snow
35	Dolny Lej	1763	6	−6	Recrystallized snow
36	Bliźniacza Studnia	1795	80	−45	Recrystallized snow
37	Jaskinia Śnieżna (part of Jaskinia Wielka Śnieżna)	1700	14,954	−622	Recrystallized snow Congelation ice
38	Zaspałkowa Szczelina	1743	10	−5	Recrystallized snow
39	Dziura przy Lodowej Małołąckiej I	1641	6	4	Congelation ice
40	Jaskinia Lodowa Małołącka	1634	360	53	Congelation ice
41	Mnichowa Studnia	1640	50	22	Recrystallized snow
42	Chuda Mnichowa Studnia	1649	35	18	Recrystallized snow
43	Koprowa Studnia	1808	90	−47	Recrystallized snow

Ice caves in the Western Tatra Mountains are grouped in two extensive karst massifs, namely Kominiaski Wierch (1829 m a.s.l.) and Czerwone Wierchy, the latter consisting of four independent summits: Kopa Kondracka (2005 m a.s.l.), Małołączniak (2096 m a.s.l.), Krzesanica (2123 m a.s.l.), and Ciemniak (2096 m a.s.l.). The massifs consist of Mesozoic rocks, predominantly carbonates. The geological structure of this area is highly complicated, mainly as an effect of faulting and thrusting in mid-Cretaceous time. Various crystalline rocks occur subordinately there as well. The Tatra Mountains were subjected to long-lasting denudation after the Miocene uplift. Pleistocene glaciations exerted a substantial influence on their present relief forming U-shaped valleys, hanging valleys, cirques, and moraines.

The majority of Tatra caves are genetically associated with geomorphic evolution of this karst region in Neogene and Pleistocene time (for review, see Gradziński et al., 2009). Some extensive karst systems are connected with preglacial periods, that is, Neogene and early Pleistocene denudation. Proglacial caves are the most significant group of caves developed during the Pleistocene. They were formed by huge amounts of water released by the melting glaciers. Many caves were dissected by glacial erosion or blocked by collapses. Thus the presently accessible caves also comprise small objects that are fragments of former, more extensive ones.

The Tatra Mountains are characterized by a typical mountain climate. The mean annual temperature markedly varies with altitude, ranging between +6°C at the foot of the mountains and −4°C in the region of the highest summits (Konček and Orlicz, 1974; Hess, 1996). Similarly, the mean precipitation per year increases from c.1138 to 1876 mm. The number of days with snow cover increase in altitudes from 100 to 290. The vertical diversification of climate results in a vertical zonation of vegetation cover (Piękoś-Mirkowa and Mirek, 1996). The forest belts—lower and upper montane belt—occupy the lower part of the mountains. The timberline is located at an altitude of c.1500 m. A subalpine belt of high mountain grasslands with mountain pine extends over the timberline, that is, at an altitude of about 1500–1800 m. This is followed by an alpine belt dominated by grasslands reaching an altitude of about 2300 m, which is followed by a subnival belt above it.

The ice caves' entrances are distributed predominantly above an altitude of 1400 m (Table 24.1, Fig. 24.2). Dmuchawa cave, the entrance of which is located at an altitude of 1300 m, represents an exception (Nowak, 2008). The majority of entrances are clustered between 1500 and 1800 m a.s.l.

Perennial cave ice occurs in caves of various origins and sizes. It fills the bottom parts of vertical shafts with or without known downward continuations. Some of them are typical karst pitches with a circular outline, whereas others represent fissures, that is, tectonic fractures or bedding planes that were widened because of the gravitational instability of steep mountain slopes. The shafts have entrances in their upper parts. They reach depths between a few meters and tens of dozens of meters. Ice forms plugs at the bottom parts of the shafts. Bliźniacza Studnia (depth −45 m; Luty, 2000) and Studnia w Dziurawem (depth −30 m; Luty, 1996) are examples of blind shafts. Śnieżna Studnia (length 12,700, vertical 763; Albrzykowski, 2002) is an example of more extensive caves with ice in their entrances.

Ice also forms in more topographically complicated caves, especially the ones with multiple entries. The ice occupies spaces near the lowermost parts or at one of the lower entrances of such caves, such as Jaskinia Wielka Śnieżna (Jaskinia Śnieżna ice cave is a part of this cave; Kardaś, 2002), Bańdzioch Kominiarski (Kotarba and Wiśniewski, 1995) and System Ptasiej Studni (Jaskinia Lodowa Litworowa ice cave is a part of this system; Antkiewicz, 1999), which are among the deepest caves in Poland. Some caves included in this category are single-entrance ones, but their interiors have contact with the surface through fissures that are not traversable. The most famous ice cave in Poland—Ice Cave in Ciemniak—falls into this group (Rygielski et al., 1995). Dmuchawa and Jaskinia Lodowa Krakowska are more likely to represent this group either (Nowak, 2008 and Luty, 1994, respectively).

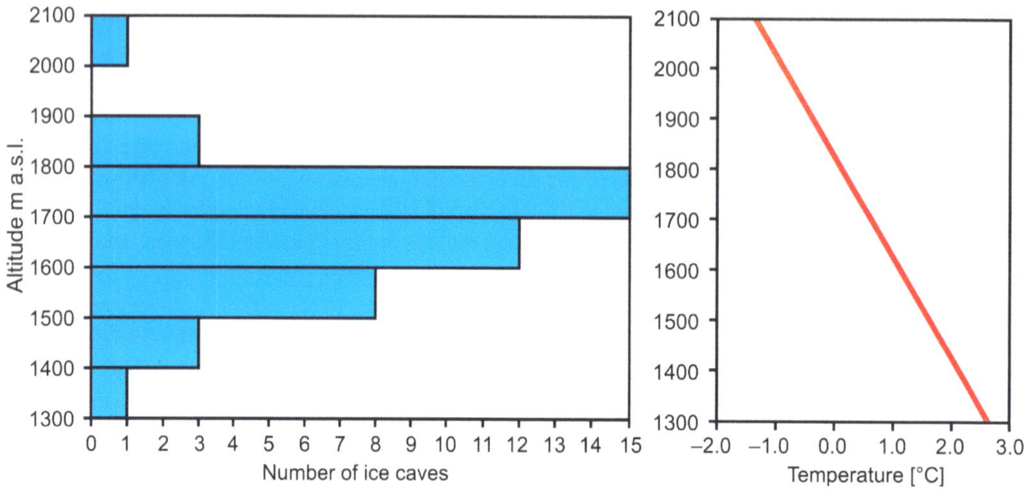

FIG. 24.2

Altitude distribution of ice caves in the Polish Tatra Mountains; vertical changes of mean annual temperature *(red line)*.

Based on data by Konček, M., Orlicz, M., 1974. Teplotné pomery. In: Konček, M. (Ed.), Klimat Tatier. Veda, Bratislava, pp. 89–179 (in Slovak).

24.4 MICROCLIMATIC CONDITIONS IN POLISH ICE CAVES

Ice can accumulate in caves or their parts only when the inside temperature drops below zero at least for some part of the year. This condition is met in some Tatra caves; however, the circulation of air inside a particular cave, which is strictly associated with cave topography, as well as outside climatic conditions are of crucial importance for the existence of perennial cave ice (Siarzewski, 1994b, 1996a,b). Two microclimatic types of caves are routinely distinguished all over the world: static caves and dynamic caves. Ice caves in the Tatra Mountains constitute both types.

Many ice caves in Poland represent a single-entrance, static-type cave. A blind passage descends from their entrance, most commonly as a vertical shaft or shafts. They have specific microclimate conditions because of sinking of cold, dense air in winter. If the outside temperature is not too high in summer, conditions favorable for the existence of perennial ice can be preserved. The persistence of cold air within such caves is facilitated by the accumulation of copious amounts of winter snow that enter the caves through their spacious entrances. Such caves are located above c. 1500 m a.s.l. in the Tatra Mountains (Siarzewski, 1994b), that is, in the climatic zone where the mean annual temperature falls below 2°C (data for the last decades of the 20th century, cf. Konček and Orlicz, 1974; Hess, 1996). Temperature measurements have not been taken in static-type caves in the Tatra Mountains.

Perennial ice is also found in some parts of Tatra caves that display dynamic air circulation resulting from the "chimney effect" (see, e.g., Wigley and Brown, 1976). Cold, dense winter air is sucked in through the lower entrance, whereas less dense, warmer air escapes through the upper entrance or non-traversable fissures that have contact with the surface. Cold air also draws in substantial amounts of snow. This mechanism causes effective freezing of cave sections, even several dozen meters in length, near the lowermost entrances of these caves. The circulation is reversed in summer. Lower entrances emit air that is colder than atmospheric air. However, in favorable conditions, the accumulated ice does not melt away completely before the next winter. Such circulation patterns have been detected in several extensive caves in the Tatra Mountains in Poland, namely in Jaskinia Wielka Śnieżna (Kardaś, 2002), Bańdzioch Kominiarski (Kotarba and Wiśniewski, 1995), and System Ptasiej Studni (Antkiewicz, 1999). All the preceding caves represent multi-entrance caves with ice accumulated near their lower entrances. The most famous ice cave in Poland—Ice Cave in Ciemniak—also belongs in this category. Although it is a single-entrance cave, the circulation is possible through chimneys and fissures in its ceiling (Fig. 24.3; Rygielski et al., 1995).

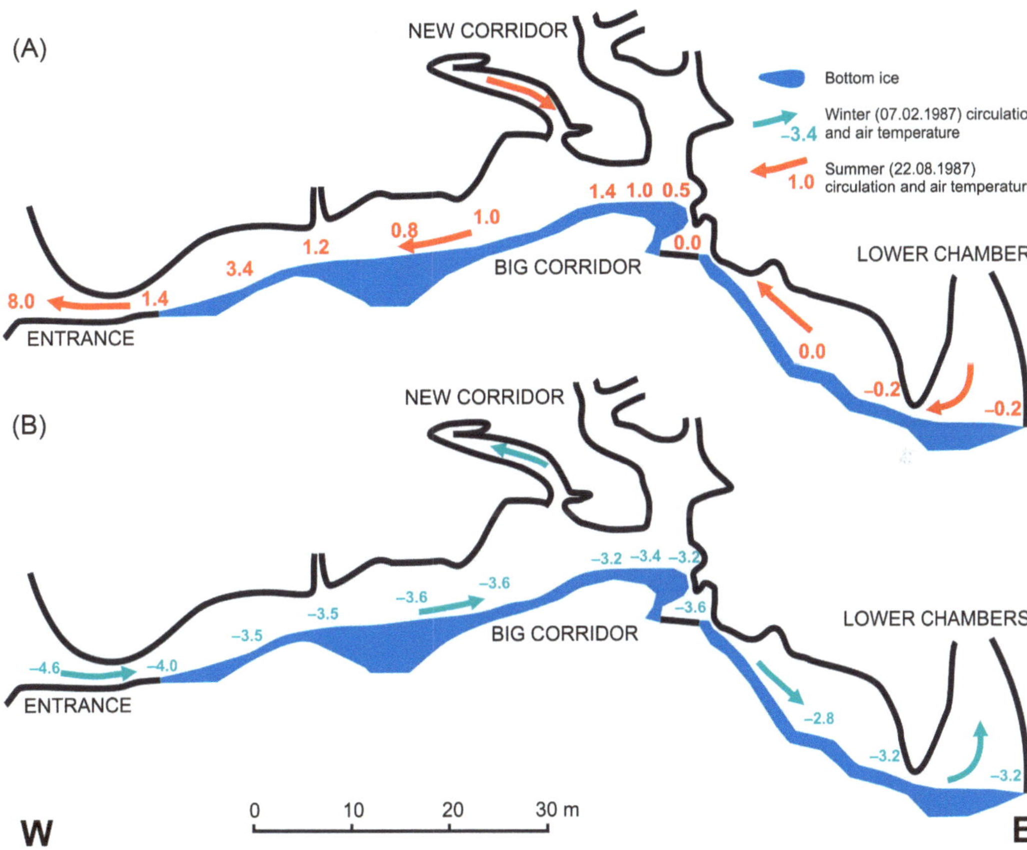

FIG. 24.3

Air temperature and air circulation in Ice Cave in Ciemniak in summer (A) and winter (B).

Based on Rygielski, W., Siarzewski, W., Wieliczko, P., 1995. Variability of the ice deposits in Ice Cave on Mount Ciemniak in the West Tatra Mountains. Questiones Geographicae 17/18, 55–64.

Microclimatic studies were carried out in Ice Cave in Ciemniak; the internal temperature was measured most frequently (Zwoliński, 1923; Kowalski, 1953, 1955; Rygielski et al., 1995; Rachlewicz and Szczuciński, 2004). The studies found that generally the temperature fell below 0°C throughout the whole cave in winter and was very well correlated to air temperature outside the cave. The minimal temperature noted in this cave was −17.7°C (Kwaśniewski, 1924). In summer the temperature slightly exceeded 0°C; however, colder air mantled the ice surface. The temperature measured over 10cm of the cave's ice bottom was 0.8°C, whereas near the ceiling, which was c. 5m higher, it was measured at 2.8°C (Rygielski et al., 1995).

Microclimatic studies were conducted in Jaskinia Śnieżna (a lower part of Jaskinia Wielka Śnieżna with the lowermost entrance of this cave system) only in the summer seasons of 1960 and 1961 (Pulina, 1967, 1974). At that time the air blew out from the entrance at a velocity of 1.5m/s, which indicated an air discharge of about 3 m^3/s (Pulina, 1967). The mean temperature in the cave ranged between 0.7°C and 3.1°C; however, it attained the lowest values in an ice-filled passage roughly 20m below the entrance (Pulina, 1967).

24.5 ICE TYPES

Three types of subterranean ice have been determined in the ice caves in Poland: (i) congelation ice, (ii) partly recrystallized snow, and (iii) hoarfrost (Pulina, 1971; Kozik, 1983; Siarzewski, 1994b). Congelation ice appears also in entrance zones of almost all Tatra caves in winter and spring seasons. Similarly, long-lasting snow can be found below vertical entrances of some caves. However, they cannot be regarded as perennial forms of snow and ice. Ground ice was also detected in caves that do not contain perennial ice covers (Pulina, 1971), which resulted in the formation of polygonal structures in clastic sediments of some caves (Pulina, 1971).

24.5.1 CONGELATION ICE

Congelation ice forms massive covers of different thicknesses that mantle bottoms of caves. Ice of this type occurs in caves with dynamic microclimates. Such ice created a distinctive ice monolith on the bottom of Ice Cave in Ciemniak, where the ice thickness reached c. 9m in the first half of the 20th century (Zwoliński, 1951). The ice volume in the middle of the 20th century was estimated at 1500 m^3 (Gradziński and Wójcik, 1961). The ice is markedly layered (Fig. 24.4; Chrobak, 1925), and two types of ice occur: opaque ice, whitish in color; and transparent layers, deep-blue in color (Celejowska et al., 2007; Hercman et al., 2010). Some laminae, pale yellow in color, have been found as well. They contain a silt-sized carbonate admixture, most probably cryocalcites. Two generations of ice are divided by a distinctive thermo-erosion surface dipping at an angle between 15° and 90° (Fig. 24.5; Celejowska et al., 2007; Hercman et al., 2010). The older generation consists predominantly of transparent ice, whereas the younger one contains layers of opaque ice alternating with transparent ice. The former ice type predominates in the younger generation. The older generation contains a group of ice stalagmites, which is visible through the transparent ice (Hercman et al., 2010). The ice from Ice Cave in Ciemniak was also studied by Jaworowski (1966, 1968), who assumed the cave ice layers to represent annual layers and sampled them in order to reconstruct history of lead pollution. He counted over 40 layers with maximum thickness of 10cm. The ice layers interpreted to form during the first half of the 20th century contained more lead than older layers (Jaworowski, 1968). The same material was subjected also to pioneering application of radioisotope ^{210}Pb by Jaworowski (1966). The ^{210}Pb activities

FIG. 24.4

Section of layered ice, Ice Cave in Ciemniak.

Photo by M. Gradziński.

FIG. 24.5

Two generation of ice divided by thermo-erosion surface, Ice Cave in Ciemniak.

Photo by M. Gradziński.

revealed general decrease with depth, which was discussed by Jaworowski (1966) in context of nuclear explosions. Meanwhile, ^{210}Pb become commonly applied dating technique and the results of Jaworowski (1966) may be also interpreted as a marker of modern age of the upper part of the ice mass and/or contamination by water percolating through the ice.

Massive covers consisting of congelation ice were noted in other Tatra caves (see Table 24.1). They attained a thickness of a few meters, for example, in Dziurka w Trawce, Jaskinia Lodowa Litworowa, Jaskinia Lodowa Mułowa, Jaskinia Lodowa Małołącka, Jaskinia Śnieżna, Śnieżna Studnia, Jaskinia Lodowa w Twardych Spadach, Jaskinia Śnieżna, and Jaskinia Lodowa Miętusia. The ice in Jaskinia Lodowa Litworowa and in Jaskinia Lodowa Małołącka was distinctively layered (Antkiewicz, 1999). No data about the internal structure of ice in other caves are known.

In addition to massive covers, congelation ice forms flowstones and cascades on cave walls, as well as dripstones, that is columns, stalactites (icicles), and stalagmites (Fig. 24.6). Such forms occur in caves with dynamic microclimates and co-occur with recrystallized snow in some caves with static microclimates. These forms are most commonly seasonal. Even in caves with perennial ice, many such forms melt completely during warm seasons. However, both written reports and published photographs testify that in the first half of the 20th century, such forms were more durable in Ice Cave in Ciemniak (Fig. 24.7; Zwoliński, 1951, 1953).

Pulina (1971) studied the internal structures of congelation cave ice in some caves in Poland, including ice caves. He found that ice crystals represent hexagonal prisms and elongated pyramids.

FIG. 24.6

Ice column in Jaskinia Lodowa w Twardych Spadach.

Photo by J. Nowak.

FIG. 24.7

Ice column in Ice Cave in Ciemniak, property of Institute of Geological Sciences, Jagiellonian University in Kraków.

Photo taken by Ludwik Chrobak in 1922.

24.5.2 RECRYSTALLIZED SNOW

Perennial ice of this type occurs in many caves with static microclimates and in some caves with dynamic microclimates as well. The key condition is the possibility of the introduction of snow into a particular cave. A spacious entrance, especially a vertical one, can provide such conditions (Fig. 24.8). Several caves represent this case, for example, Bliźniacza Studnia, Studnia w Dziurawem, Mnichowa Studnia, Śnieżna Studnia, and Świstowa Studnia. Snow falls not only directly into a cave, but it is also transported in the form of avalanches sliding down a slope and accidentally entering a cave. Another possibility is sucking of snow into a cave's interior by intense air circulation. This happens in multi-entrance caves with dynamic microclimates. Such a process operates, for example, in Jaskinia Śnieżna.

Partly recrystallized snow forms isolated lobes that mound up to a few meters in height (e.g., Studnia w Dziurawem, Świstowa Studnia) or even extensive plugs that seal the bottom parts of

FIG. 24.8

Small cave called Dolny Lej almost completely filled with partly recrystallized snow.

Photo taken in July 2012 by M. Gradziński.

vertical shafts and block possible downward passages (e.g., in Bliźniacza Studnia). The plugs exceed 10 m in thickness, as shown via observation of rimayes that periodically opened between the ice and rocky walls of a shaft. Recrystallized snow is commonly accompanied by, or even overgrown with, congelation ice.

24.5.3 HOARFROST

Hoarfrost is not commonly reported in the Tatra ice caves. Such ice was noted in Ice Cave in Ciemniak, Jaskinia Lodowa Małołącka and Jaskinia Lodowa Mułowa (Siarzewski, 1994b). Because it is very unstable, it occurs only seasonally. Therefore, it seems possible that this kind of ice occasionally forms in other ice caves, but has not been observed so far.

24.6 AGE OF CAVE ICE

Almost no precise data on the age of ice in Poland's ice caves exist. Only the age of ice from Ice Cave in Ciemniak has been studied. Hercman et al. (2010), using radiocarbon dating of moths entrapped in ice, proved that younger ice generations (see Section 24.5.1) crystallized during the Little Ice Age, which agreed with the earlier results obtained by Rygielski et al. (1995), who found cereal pollen in ice samples. Taking these studies into account, the thermo-erosion surface (Fig. 24.5) that divides both ice generations might have been formed during the Medieval Warm Period and that, consequently, the older generation of ice crystallized during the preceding cold period, that is, the Dark Ages Cold Period (Hercman et al., 2010). Wójcik (1968) found subfossil bones of *Myotis bechsteini* in Ice Cave in Ciemniak, which suggests that, in the earlier part of the Holocene, the climatic conditions in the cave were suitable for this species, which is a thermophilous one. Thus the cave probably did not have perennial ice cover then. The preceding notion indicates that the ice is late Holocene in age (see also Jaworowski, 1966). It seems reasonable to extrapolate this view to other ice caves in Poland.

24.7 ORGANISMS DWELLING IN ICE CAVES

Ice caves in the Tatra Mountains are probably inhabited by the same organisms as the other Tatra caves. However, studies were conducted irregularly and did not cover many Tatra caves. Ice Cave in Ciemniak is practically the only ice cave where more detailed studies were carried out.

Bacteria belonging to genus *Bacillus* were found in samples of water from melting ice in Ice Cave in Ciemniak (Chardez and Delhez, 1981). The same samples contained the protists *Bodo minimus*, *Bodo saltans*, *Vahlkampfia guttula*, and *Trinema lineare*. Kowalski (1955) detected various invertebrates in Ice Cave in Ciemniak, namely springtails (*Onychiurus armatus*, *O. fimetarius*), beetles (*Nebria tatrica*, *Omalium excavatum*), moths (*Triphosa dubitata*), harvestmen (*Ischyropsalis milleri*), caddisflies (*Stenophylax permistus*), and flies (*Amoebaleria caesia*, *A. spectabilis*, *Eccoptomera emarginata*, *Crumomyia* (*Stratioborbus*) *nitida*). It is not clear which part of the cave they derived from. However, Kowalski (1953) mentioned in a cave inventory that some invertebrates were found in the ice-filled passage and others in the ice-free passage.

Bats observed in ice caves in Poland belong to the cold-adapted eco group (Piksa and Nowak, 2013). The species *M. mystacinus*, *Eptesicus nillsonii*, and *Plecotus auritus* are the ones most often observed in the Tatra ice caves. These species are able to hibernate in temperatures below zero and occur in caves with perennial or periodic ice. Other bat species hibernate in ice-free parts of ice caves. They are as follows: *M. myotis*, *M. nattererii*, and *M. daubentonii*. Jaskinia Lodowa w Twardych Spadach is the largest known bat wintering site among the Tatra ice caves. More than 100 specimens have been detected there, with *M. mystacinus* being the most common (Jakub Nowak—unpublished data). Ice Cave in Ciemniak is the highest ice cave that acts as a hibernaculum (Piksa and Nowak, 2002). However, the majority of bats choose ice-free parts of these caves. A big hibernaculum is Bańdzioch Kominiarski (more than 200 specimens), but in that case, the ice part is less than 1% of the whole cave. Bats have also been observed in other ice caves, namely Jaskinia Dmuchawa (Nowak, 2008) and Jaskinia Lodowa Litworowa (Piksa and Nowak, 2013).

The entrance part of Ice Cave in Ciemniak was a temporary shelter used by chamois (*Rupicapra rupicapra*; Zwoliński, 1923; Kowalski, 1953).

24.8 SUBFOSSIL ORGANIC REMAINS IN ICE CAVES

Some subfossil mammal bones were reported from Ice Cave in Ciemniak. A rich assemblage of bones was found in the ice-free vertical series located over the ice-filled chambers of the cave's lower level (Nowosad et al., 1992). The bone assemblage was dominated by whiskered bat remains (both *M. mystacinus* and *M. brandtii*). It also comprised bones of other bats, such as *P. auritus*, *Barbastella barbastellus*, and *E. nilsoni* as well as pine marten (*Martes martes*). Wójcik (1968) mentioned subfossil bones of *M. bechsteini*. Kowalski (1953, 1955) noted a frozen whiskered bat, and Hercman et al. (2010) found moths entombed in the cave ice. Cereal pollen and pollen representing genera *Betula* and *Pinus* were reported in cave ice (Rygielski et al., 1995 and Jelonek, 2016, respectively).

24.9 ICE DYNAMICS

Ice Cave in Ciemniak is the only ice cave in Poland where ice mass balance was studied. The cave ice developed there has a several-meters-thick body covering the floor of two parts of the cave. The lower part was visited only occasionally because the passage leading to it is usually blocked by an ice mass, thus most of the investigations—including ice mass balance studies—were restricted to the upper part of the cave (Fig. 24.9). The published records on the cave ice mass geometry mapping date back to 1922 (Fig. 24.10; Zwoliński, 1923) and were followed by documentations in the early 1950s (Zwoliński, 1951), in 1986 (Rygielski et al., 1995), in 2002 (Rachlewicz and Szczuciński, 2004), and in 2009 (Szukała, 2010). These data allow for an assessment of decadal ice mass balance in the cave, and a monitoring study by Rachlewicz and Szczuciński (2004) revealed a seasonal ice mass balance pattern.

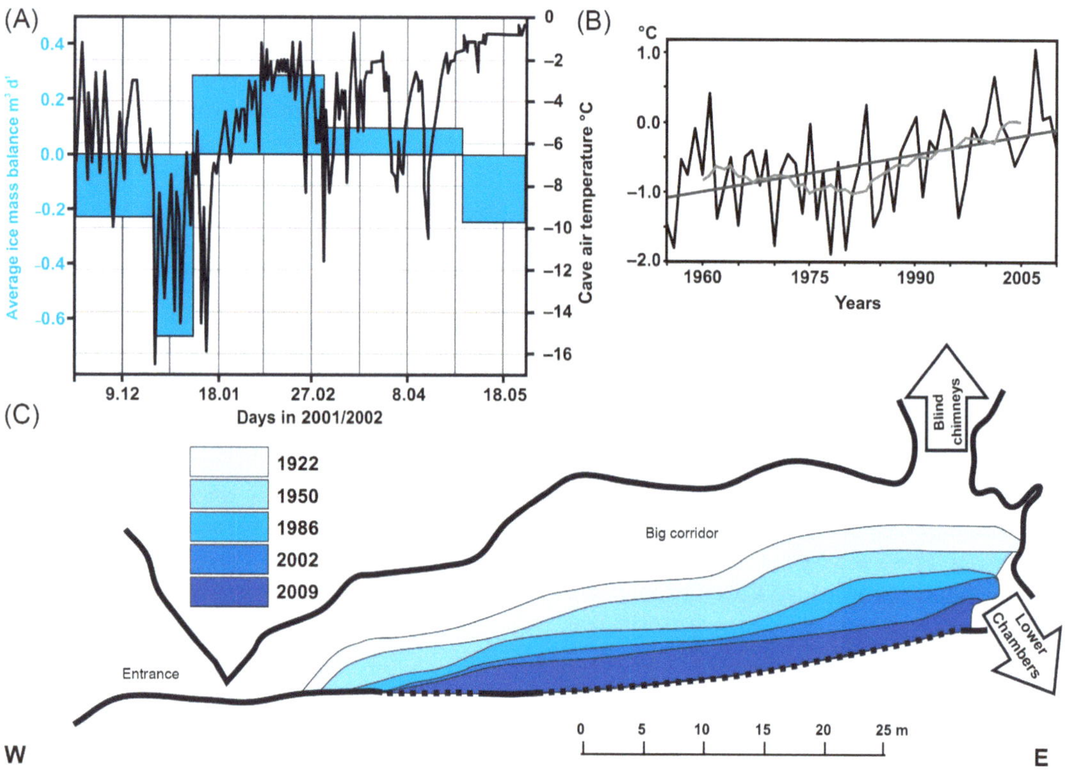

FIG. 24.9

Ice mass balance in Ice Cave in Ciemniak: (A) Seasonal ice mass balance and cave air temperature as measured during the 2001–02 season by Rachlewicz and Szczuciński (2004); (B) Mean annual temperature from Kasprowy Wierch (1985 m a.s.l.) in the Tatra Mountains showing also 11 years moving average and linear trend (after Gądek and Leszkiewicz, 2012); (C) schematic cross-section through the upper part of Ice Cave in Ciemniak with marked geometries of the ice mass in 1922 and 1950 (Zwoliński, 1951), 1985 (Rygielski et al., 1995), 2002 (Rachlewicz and Szczuciński, 2004), and 2009 (Szukała, 2010).

The seasonal mass balance of the ice mass (Fig. 24.9A) was studied in the years 2001 and 2002 along with monitoring of meteorological conditions within and outside the cave (Rachlewicz and Szczuciński, 2004). The cave's ice body was found to decrease throughout a year, except for the spring season, when the major ice mass as well as numerous small seasonal ice forms in the cave were

FIG. 24.10

Icefall in Big Corridor (Ice Cave in Ciemniak); it disappeared between 1986 and 2002 (cf. Fig. 24.9); photo taken by Ludwik Chrobak in 1922, property of Institute of Geological Sciences, Jagiellonian University in Kraków.

growing. The increase in ice mass was driven by specific conditions with temperatures below 0°C in the cave and positive temperatures outside the cave. The latter caused melting of the snow cover and drainage of the meltwater into the cave. The spring increase in the mass volume was equal to about 26% of total ice mass loss during the studied year (Fig. 24.9A). The ice was lost primarily by melting, which accounted for about 69% of the total ice mass degradation. Melting occurred during the summer and autumn, when the temperature in the cave was slightly above 0°C. However, the most intensive loss of ice mass per unit time was during winter when, in less than 2 months, about 31% of total annual ice was lost because of sublimation. The sublimation was driven by a specific winter circulation pattern when dry, heavy, and cold air blew through the cave because of the air density difference. Overall, the ice mass balance was negative, with a loss of 53.9 m^3 of water equivalent during 2001–2002.

The decadal ice mass balance was first presented by Rachlewicz and Szczuciński (2004) who used available documentation of ice mass geometry and their own new data. These calculations were later extended in time by new data from Szukała (2010). As shown by subsequent lowering of the ice mass surface (Fig. 24.9C), the cave ice mass was in continuous degradation during the 20th century. From the 1920s to 1980s, the average negative ice mass balance was on order of −20 to −30 m^3 of water equivalent per year. During the following years, the annual ice mass loss increased, reaching −60 m^3 from 2002 to 2004 (Rachlewicz and Szczuciński, 2004) and −71 m^3 from 2004 to 2009 (Szukała, 2010). The increase in negative ice mass balance is probably a reflection of an increase in mean annual temperatures (Fig. 24.9B; Gądek and Leszkiewicz, 2012). Moreover, the trend of degradation of the ice mass follows the global trend of ice cave mass degradation (Kern and Perşoiu, 2013) and may lead to the ultimate loss of perennial ice mass cover in the cave during the following decades.

Data on other caves are based on irregular and, in some caves, especially the small ones, very infrequent visits by cavers. Nonetheless, some conclusions can be reached. The tendency to decrease in ice volume in caves during past 20 years is visible in all Tatra caves. The tendency accelerated especially c.1990, which probably followed the considerable increase in the mean annual temperature at that time (Niedźwiedź, 2004). In some caves the perennial ice disappeared completely, including congelation ice (e.g., in Jaskinia Lodowa Krakowska and Jaskinia Lodowa nad Kufą) and recrystallized snow (e.g., in Studnia za Murem and Koprowa Studnia). In other caves, such as Jaskinia Lodowa Litworowa, Śnieżna Studnia, Bańdzioch Kominiarski, and Jaskinia Śnieżna, the ice level decreased substantially. This tendency likely started in the 1920s, which was indicated by observations in Ice Cave in Ciemniak. This notion is additionally supported by observations conducted outside the Tatra Mountains. For example, a credible description is illustrated by photographs and the map of a cave with perennial ice in the Beskidy Mountains (Zimna Dziura w Strzeblu, entrance coordinates: λ: 20°00′15″, φ: 49°42′09″; Leszczycki, 1931). Interestingly, the cave is located at an altitude of only 655 m. It represents a cave with static microclimatic conditions. A cover of congelation ice was confirmed there in the 1930s (Leszczycki, 1931). However, the ice cover disappeared before 1950 (Kowalski, 1954).

The decrease in ice levels resulted from human activity in particular cases. Śnieżna Studnia, where an extensive series of shaft and passages were discovered below ice-filled series, serves as an example (Albrzykowski, 2002). Most probably the discovery changed the air circulation in the cave, and the new series are frequently visited, which collectively may result in quickly vanishing cave ice. Similarly, the intentional blockage of entrances of caves characterized by dynamic air circulation prevents a cave from being frozen and filled with snow, which in turn causes a systematic decrease in the level of cave ice. On the other hand, human interaction can cause the opposite effect, that is increase the cave ice level. The discovery of an upper entrance to Bańdzioch Kominiarski, which was connected by the removal of a plug formed by slope scree, stimulated dynamic air circulation in this cave, resulting in crystallization of perennial ice near its lower entrance (Kotarba and Wiśniewski, 1995).

ACKNOWLEDGMENTS

We thank the active cavers Michał Kuryłowicz, Izabella Luty, Paweł Ramatowski, Piotr Sienkiewicz, and Zbigniew Tabaczyński, who kindly provided information on the present ice level in some caves. Renata Jach is thanked for her help with the figures.

REFERENCES

Albrzykowski, G., 2002. Jaskinia Śnieżna Studnia. In: Grodzicki, J. (Ed.), Jaskinie Tatrzańskiego Parku Narodowego. Tom 9. Wielkie jaskinie Doliny Małej Łąki. Polskie Towarzystwo Przyjaciół Nauk o Ziemi, Tatrzański Park Narodowy, Warszawa, pp. 40–71 (in Polish).

Antkiewicz, A., 1999. Ptasia Studnia – Jaskinia Lodowa Litworowa – Jaskinia nad Dachem. In: Grodzicki, J. (Ed.), Jaskinie Tatrzańskiego Parku Narodowego. Tom 7. Jaskinie wschodniego zbocza Doliny Miętusiej. Polskie Towarzystwo Przyjaciół Nauk o Ziemi, Tatrzański Park Narodowy, Warszawa, pp. 46–88 (in Polish).

Celejowska, P., Głazek, J., Jędrysek, M.O., Kicińska, D., Solarczyk, S., Szynkiewicz, A., Zieliński, M., 2007. Badania składu izotopowego tlenu w warstwach lodowca Jaskini Lodowej w Ciemniaku (Tatry). In: Stefaniak, K., Szelerewicz, M., Urban, J. (Eds.), Materiały 41. Sympozjum Speleologicznego, Kletno. Sekcja Speleologiczna PTP, Kraków. pp. 40 (in Polish).

Chardez, D., Delhez, F., 1981. Contributions à l'étude des microorganismes des cavernes de Pologne. Rocznik Muzeum Okręgowego w Częstochowie, 5. Przyroda 2, 9–13 (in Polish, French summary).

Chrobak, L., 1925. Kilka spostrzeżeń poczynionych w grocie lodowej w Kamiennem. Wierchy 3, 215–216 (in Polish).

Gądek, B., Leszkiewicz, J., 2012. Impact of climate warming on the ground surface temperature in the sporadic permafrost zone of the Tatra Mountains, Poland and Slovakia. Cold Reg. Sci. Technol. 79–80, 75–83.

Gadomski, A., 1926. Morfologia glacjalna północnych stoków Wysokich Tatr. Kotuli, Cieszyn (in Polish).

Gradziński, M., Hercman, H., Kicińska, D., Barczyk, G., Bella, P., Holúbek, P., 2009. Karst in the Tatra Mountains—developments of knowledge in the last thirty years. Prz. Geol. 57, 674–684. (in Polish, English summary).

Gradziński, R., Wójcik, Z., 1961. Concretions in Polish caves. Ochrona Przyrody 27, 213–238 (in Polish, English summary).

Hercman, H., Gąsiorowski, M., Gradziński, M., Kicińska, D., 2010. The first dating of cave ice from the Tatra Mountains, Poland and its implication to paleoclimate reconstructions. Geochronometria 36, 31–38.

Hess, M.T., 1996. Climate. In: Mirek, Z., Głowaciński, Z., Klimek, K., Piękoś-Mirkowa, H. (Eds.), Przyroda Tatrzańskiego Parku Narodowego. Tatrzański Park Narodowy, Zakopane, pp. 53–69 (in Polish, English summary).

Jaskinie Polski, 2017. http://jaskiniepolski.pgi.gov.pl/ (accessed 29.01.17).

Jaworowski, Z., 1966. Temporal and geographical distribution of radium D (lead-210). Nature 212, 886–889.

Jaworowski, Z., 1968. Stable lead in fossil ice and bones. Nature 217, 152–153.

Jelonek, M., 2016. Analiza palinologiczna lodu jaskiniowego – wstępne wyniki badań. In: Urban, J. (Ed.), Materiały 50. Sympozjum Speleologicznego, Kielce – Chęciny 20–23.10.2016. Sekcja Speleologiczna PTP, Kraków. pp. 113 (in Polish).

Kardaś, R.M., 2002. Jaskinia Wielka Śnieżna. In: Grodzicki, J. (Ed.), Jaskinie Tatrzańskiego Parku Narodowego. Tom 9. Wielkie jaskinie Doliny Małej Łąki. Polskie Towarzystwo Przyjaciół Nauk o Ziemi, Tatrzański Park Narodowy, Warszawa, pp. 71–125 (in Polish).

Kern, Z., Perşoiu, A., 2013. Cave ice—the imminent loss of untapped mid-latitude cryospheric palaeoenvironmental archives. Quat. Sci. Rev. 67, 1–7.

Koisar, B., Parma, Ch., 1971. Lodowe jaskinie w Tatrach. Taternik 47, 68–71 (in Polish).

Konček, M., Orlicz, M., 1974. Teplotné pomery. In: Konček, M. (Ed.), Klimat Tatier. Veda, Bratislava, pp. 89–179 (in Slovak).

Kotarba, S., Wiśniewski, W.W., 1995. Jaskinia Bańdzioch Kominiarski. In: Grodzicki, J. (Ed.), Jaskinie Tatrzańskiego Parku Narodowego. Tom 4. Wielkie jaskinie Doliny Kościeliskiej. Polskie Towarzystwo Przyjaciół Nauk o Ziemi, Tatrzański Park Narodowy, Warszawa, pp. 34–77 (in Polish).

Kowalski, K., 1953. Les cavernes de la Pologne, II. Państwowe Muzeum Archeologiczne, Warszawa (in Polish, French summary).

Kowalski, K., 1954. Les cavernes de la Pologne, III. Państwowe Wydawnictwo Naukowe, Warszawa (in Polish, French summary).

Kowalski, K., 1955. The cave fauna of the Polish Tatra Mountains. Ochrona Przyrody 23, 283–333 (in Polish, English summary).

Kozik, A., 1983. Strefa "peryglacjalna" w jaskiniach. Prace Naukowe Uniwersytetu Śląskiego 614. Geologia 7, 99–101 (in Polish).

Kwaśniewski, K., 1924. Badania Groty Lodowej. Głos Zakopiański 2 (18), 2 (in Polish).

Leszczycki, S., 1931. Grota lodowa na Strzeblu, w Beskidzie Wyspowym. Wierchy 9, 183–185 (in Polish).

Luty, I., 1994. Jaskinia Lodowa Krakowska. In: Grodzicki, J. (Ed.), Jaskinie Tatrzańskiego Parku Narodowego, Tom 5. Jaskinie Wąwozu Kraków. Polskie Towarzystwo Przyjaciół Nauk o Ziemi, Tatrzański Park Narodowy, Warszawa, pp. 154–158 (in Polish).

Luty, I., 1996. Studnia w Dziurawem. In: Grodzicki, J. (Ed.), Jaskinie Tatrzańskiego Parku Narodowego. Tom 6. Jaskinie zachodniego zbocza Doliny Miętusiej. Polskie Towarzystwo Przyjaciół Nauk o Ziemi, Tatrzański Park Narodowy, Warszawa, pp. 98–101 (in Polish).

Luty, I., 2000. Bliźniacza Studnia. In: Grodzicki, J. (Ed.), Jaskinie Tatrzańskiego Parku Narodowego. Tom 8. Jaskinie Doliny Małej Łąki. Polskie Towarzystwo Przyjaciół Nauk o Ziemi, Tatrzański Park Narodowy, Warszawa, pp. 142–145 (in Polish).

Niedźwiedź, T., 2004. Reconstruction of summer temperature in the Tatra Mountains since 1550. Prace Geograficzne 197, 57–88 (in Polish, English summary).

Nowak, J., 2008. Dmuchawa – koniec mitu. Jaskinie 53, 21–22.

Nowosad, A., Jagielska, A., Strzelczyk, J., 1992. Bone remains of bats and marten from Cave Lodowa in Ciemniak (Western tatras, Poland). Przegląd Zoologiczny 36, 199–205 (in Polish, English summary).

Ossowski, G., 1883. Czwarte sprawozdanie z badań antropologiczno-archeologicznych w jaskiniach okolic Krakowa dokonanych w 1882 r. oraz rozpoznanie przygotowawcze do badań jaskiń tatrzańskich. Zbiór Wiadomości do Antropologii Krajowej 7, 66–88 (in Polish).

Pawlikowski, J.G., 1887. Podziemne Kościelisko. Pamiętnik Towarzystwa Tatrzańskiego 11, 33–48 (in Polish).

Piękoś-Mirkowa, H., Mirek, Z., 1996. Plant communities. In: Mirek, Z., Głowaciński, Z., Klimek, K., Piękoś-Mirkowa, H. (Eds.), Przyroda Tatrzańskiego Parku Narodowego. Tatrzański Park Narodowy, Zakopane, pp. 237–274 (in Polish, English summary).

Piksa, K., Nowak, J., 2002. Noteworthy records of northern bat *Eptesicus nilssonii* (Chiroptera: Vespertilionidae) in the Tatra Mountains. Acta Zool. Cracov. 45, 321–324.

Piksa, K., Nowak, J., 2013. The bat fauna hibernating in the caves of the Polish Tatra Mountains, and its long-term changes. Cent. Eur. J. Biol. 8, 448–460.

Pulina, M., 1967. Les observations concernantes le microclimat de Gouffre Śnieżna. Speleologia 3 (1), 21–30 (in Polish, French summary).

Pulina, M., 1971. Typy ladu v tatranských jaskyniach. Slovenský Kras 9, 57–74 (in Slovak).

Pulina, M., 1974. Niektoré mikroklimatické charakteristiky jaskýň v Poľských Tatrách. In: Konček, M. (Ed.), Klimat Tatier. Veda, Bratislava, pp. 702–707 (in Slovak).

Rachlewicz, G., Szczuciński, W., 2004. Seasonal, annual and decadal ice mass balance changes in the ice cave Jaskinia Lodowa w Ciemniaku, the Tatra Mountains, Poland. Theor. Appl. Karstol. 17, 11–18.

Rygielski, W., Siarzewski, W., Wieliczko, P., 1995. Variability of the ice deposits in Ice Cave on Mount Ciemniak in the West Tatra Mountains. Questiones Geographicae 17 (18), 55–64.

Siarzewski, W., 1994a. Jaskinia Lodowa w Ciemniaku. In: Grodzicki, J. (Ed.), Jaskinie Tatrzańskiego Parku Narodowego, Tom 5. Jaskinie Wąwozu Kraków. Polskie Towarzystwo Przyjaciół Nauk o Ziemi, Tatrzański Park Narodowy, Warszawa, pp. 142–153 (in Polish).

Siarzewski, W., 1994b. Jaskinie Lodowe w Tatrach. In: Grodzicki, J. (Ed.), Jaskinie Tatrzańskiego Parku Narodowego, Tom 5. Jaskinie Wąwozu Kraków. Polskie Towarzystwo Przyjaciół Nauk o Ziemi. Polskie Towarzystwo Przyjaciół Nauk o Ziemi, Tatrzański Park Narodowy, Warszawa, pp. 11–47 (in Polish).

Siarzewski, W., 1996a. Ice caves in the Polish Tatra Mountains. In: Kotarba, A. (Ed.), Przyroda Tatrzańskiego Parku Narodowego a człowiek. Tom I. Nauki o Ziemi. Tatrzański Park Narodowy, Zakopane, pp. 98–101 (in Polish, English summary).

Siarzewski, W., 1996b. Warunki mikroklimatyczne jaskiń tatrzańskich. In: Grodzicki, J. (Ed.), Jaskinie Tatrzańskiego Parku Narodowego, Tom 6. Jaskinie zachodniego zbocza Doliny Miętusiej. Polskie Towarzystwo Przyjaciół Nauk o Ziemi, Tatrzański Park Narodowy, Warszawa, pp. 13–29 (in Polish).

Szukała, P., 2010. Zmiany bilansu masy lodu w Jaskini Lodowej w Ciemniaku (Tatry Zachodnie, Polska) z zastosowaniem technik trójwymiarowego modelowania jaskiń. In: Nauka a zarządzanie obszarem Tatr i ich otoczeniem, tom I. Zakopane. pp. 145–149 (in Polish).

Wigley, T.M.L., Brown, M.C., 1976. The physics of caves. In: Ford, T.D., Cullingford, C.H.D. (Eds.), The Science of Speleology. Academic Press, London, pp. 329–358.

Wójcik, S., 1962. Jaskinie lodowe w Tatrach Polskich. Wierchy 30, 227–233 (in Polish).

Wójcik, Z., 1968. Geomorphological development of the limstone areas of the Tatra Mts. and other karst massifs in the Western Carpathians. Pr. Muz. Ziemi 13, 3–169 (in Polish, English summary).

Zwoliński, T., 1923. Nieznane groty Doliny Kościeliskiej. Wierchy 1, 26–43 (in Polish).

Zwoliński, S., 1951. Le grotte de glace dans la Tatra. Zabytki Przyrody Nieożywionej 4, 19–23 (in Polish, French summary).

Zwoliński, S., 1953. Jaskinie lodowe. Wszechświat 54, 55–61 (in Polish).

CHAPTER

ICE CAVES IN ROMANIA

25

Traian Brad*, Carmen-Andreea Bădăluţă†, Aurel Perşoiu*

Emil Racoviţă Institute of Speleology, Cluj-Napoca, Romania Ştefan cel Mare University, Suceava, Romania†*

CHAPTER CONTENTS

25.1 INTRODUCTION

There are carbonate rock outcroppings in Romania (Fig. 1) over 5000 km^2, mainly as Mesozoic limestone and dolomites in the Carpathian Mountains and chalk in SE Romania (Dobrogea). However, the area prone to cave development is far larger, as carbonate rocks are often covered by both younger (Neozoic) and older (pre Mesozoic) deposits. Exploration of caves started in the early 20th century and intensified in the 1970s and 1980s, leading to the discovery of c. 12,500 caves. Most limestone areas are located between 200 and 1500 m above sea level (asl), in the forested zone, with only a few caves (mostly vertical shafts) occurring at high altitudes (above the tree line) in the Southern Carpathians. The longest cave is Vântului Cave (>54 km), and the deepest cave is V5 Cave (>600 m), both located in NW Romania. The climate is continental temperate, with mean annual temperatures above 0°C in most karst areas (except for the high altitude karst in the Southern Carpathians (Fig. 1B). At lower altitudes, snowpack lasts for c. 5 months in the NW mountainous areas and for c. 2 months in SW. The combination of low latitude, mainly horizontal cave entrances, and warm climate does not allow for extensive permanent cave ice formation. However, there are a few notable exceptions.

25.2 ICE CAVE RESEARCH

Our knowledge of the presence of ice in caves in Romania and associated processes dates back to 1927, when Emil Racoviţă published a monograph of Scărişoara Ice Cave (SIC), which was based on two years of investigation (Racoviţă, 1927). In his study, Racoviţă investigated the origin and morphology

Ice Caves. https://doi.org/10.1016/B978-0-12-811739-2.00025-5

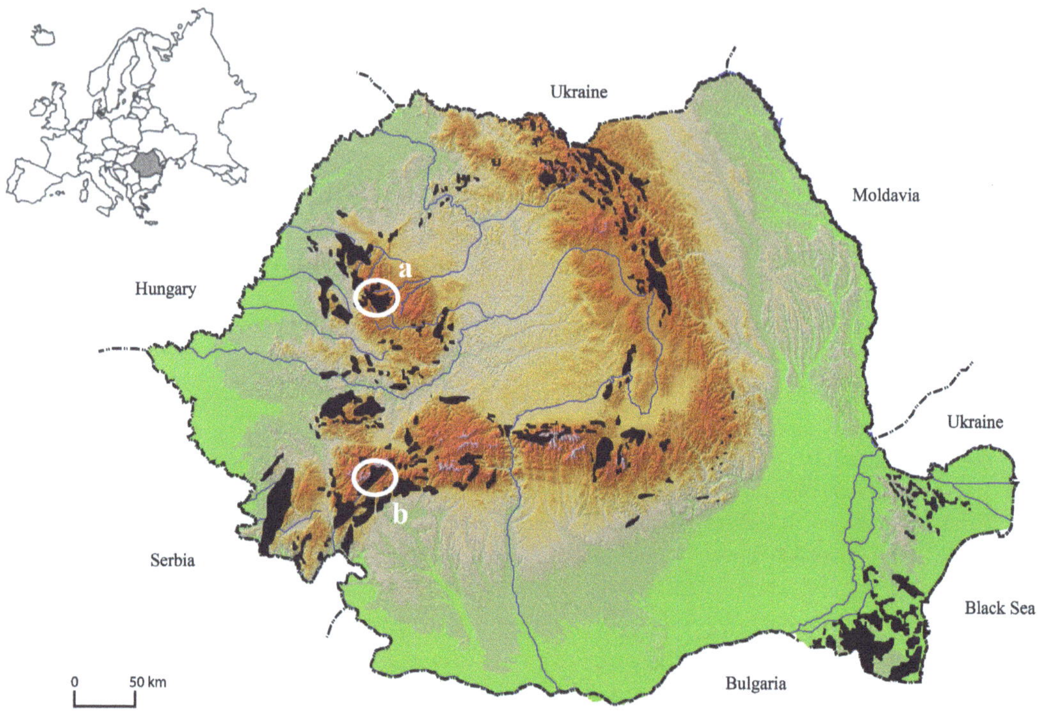

FIG. 1

Map of Romania showing the locations of ice caves: (A) Apuseni Mountains, (B) Retezat Mountains.

of ice and the cave biota and famously concluded: "*I think I have shown the great scientific interest presented by the Scărișoara Glacier Cave. Not taking into account the solution of the enigmas presented by the history of the glacier, numerous problems within all the branches of natural history may be thoroughly considered upon this occasion, with the help of periodical or continuous observations and by conceiving experiments.*" This was the start of almost a century of investigation, which resulted in perhaps the best understood ice cave in the world. Having SIC as a study site somewhat overshadowed other ice caves in the Apuseni Mountains, and their systematic research has been initiated only after 2000. Studies in SIC resumed in 1946 (Șerban et al., 1948), when a (semi)continuous monitoring program was initiated, aiming to understand the cave climate and the dynamics of ice and fauna. This program had two major periods of intensive research, between 1963 and 1968, and between 1982 and 1992, when monthly measurements of various climatic (air, rock, and ice temperature), glaciologic (ice dynamics), and biologic measurements were made. Two of the most interesting studies were published by Pop and Ciobanu (1950) and Șerban et al. (1967). The first of these attempted to reconstruct the vegetation history of the Apuseni Mountains by analyzing pollen entrapped in the ice, while the second tried to obtain palaeoclimate information by analyzing the stable isotopic composition of cave ice, which was just a couple of years after the first such study was carried out on polar ice cores. The pollen study set the age of ice in Scărișoara to c. 3500 years, an age, which was corrected to ~1000 years half a century later (Perșoiu and Pazdur, 2011). The lack of proper equipment and possibilities to date the ice lowered the scientific value of the 1967 study, but it set a goal for researchers in Romania that was only recently achieved (Perşoiu et al., 2017). The studies of the 1960s through 1990s formed the basis (Racoviţă and Onac, 2000) for a renewed scientific assault on SIC after 2000, when the previously accumulated data, combined with new data (chemical and isotopic composition of cave ice and calcitic speleothems, precise age determinations), led to the development of comprehensive studies on the genesis and long-term dynamics of the ice block and the deciphering of the palaeoclimatic information it hosts.

25.3 ICE CAVES IN ROMANIA

Ice caves in Romania are located in two regions (Fig. 1). In the Retezat Mountains (Fig. 1B), two perennial ice caves are found, but semipermanent snow deposits can also be found at the bottom of vertical pits, which are limited in extent and thickness and prone to melting in warm and wet summers. In the Apuseni Mountains (Fig. 1A), ice in caves forms by the freezing of infiltrating water, resulting in massive ice bodies. Research carried out over the past 100 years identified six caves with permanent ice, of which two (Gheţarul de la Barsa and Gheţarul de la Vârtop) have lost these deposits since AD ~2000. Presently only four caves have permanent ice in the Apuseni Mountains: Scărişoara Ice Cave (SIC), Focul Viu Ice Cave (FV), Borţig Ice Cave (BIC), Gheţarul din Poiana Vârtop Cave, and in all of those caves, ice is rapidly receding.

25.3.1 ICE CAVES IN RETEZAT MOUNTAINS

High altitude (>2000 m als) karst is very limited in extent in Romania, occurring in small patches in the Southern Carpathians only (Fig. 1). Of these, the most representative karst region is located in the Retezat Mountains, where several shafts deeper than 50 m are located above the treeline (~1800 m asl), which gives the region favorable morphology for snow accumulation. At the surface, snow persists for c. 5 months/year, reaching up to 3 m in thickness. Semiperennial snow accumulations are found in most pits. However, these deposits generally melt towards the end of summer and in early autumn. In the early 1980s, two shafts with perennial snow were described: Avenul Mare cu Zăpadă din Albele-Găuroane (AZA) and Avenul cu Gheaţă din Dâlma Brazii cei Vineţi (ABV).

AZA (Fig. 2) has an approximately 60 m thick snow deposit, which occupies the bottom of a c. 8 m large shaft. At the time of discovery, speleologists were able to descend 84 m in a rimaye between the rock and the snow pack; however, this rimaye was subsequently closed by fresh snow, and no further descents were possible. The deposit was described as transitioning from snow to crystalline ice, with numerous layers of embedded organic matter, thus offering the promise of a possibly unique environmental archive of winter climate variability.

25.3.2 ICE CAVES IN APUSENI MOUNTAINS

As mentioned above, today only four caves in the Apuseni Mountains host perennial ice. Gheţarul din Poiana Vârtop has been described by Feier et al. (2001). In this cave, ice occurs on an inclined slope leading from the entrance downwards. It formed by the freezing of inflowing water, most likely as thin sheets of ice, on the inclined slope. The thickness of the ice is unknown, but a rimaye between the ice and the rock descends for 2–3 m before closing off, putting an upper limit of 12,000 m^3 on ice in the cave. The other three ice caves—Borţig, Focul Viu, and Scărişoara—share similar morphologies (Fig. 3), ice blocks dynamics, and climate.

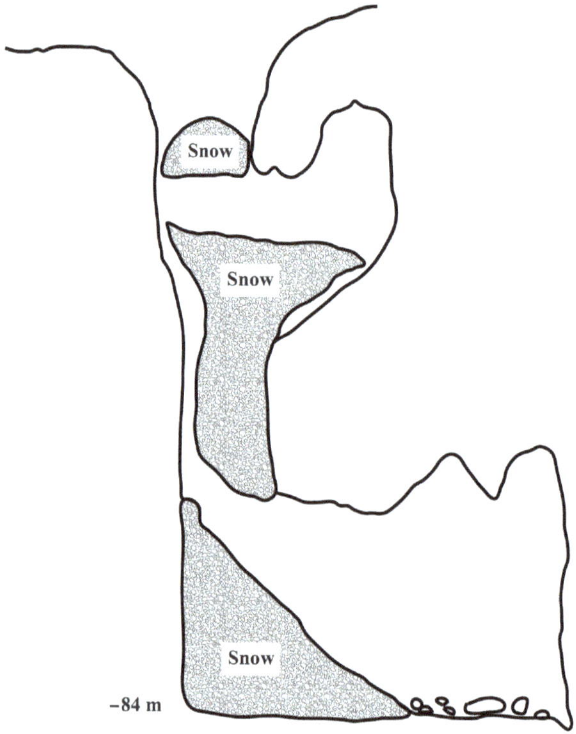

FIG. 2

Cross-section through Avenul Mare cu Zăpadă din Albele-Găuroane.

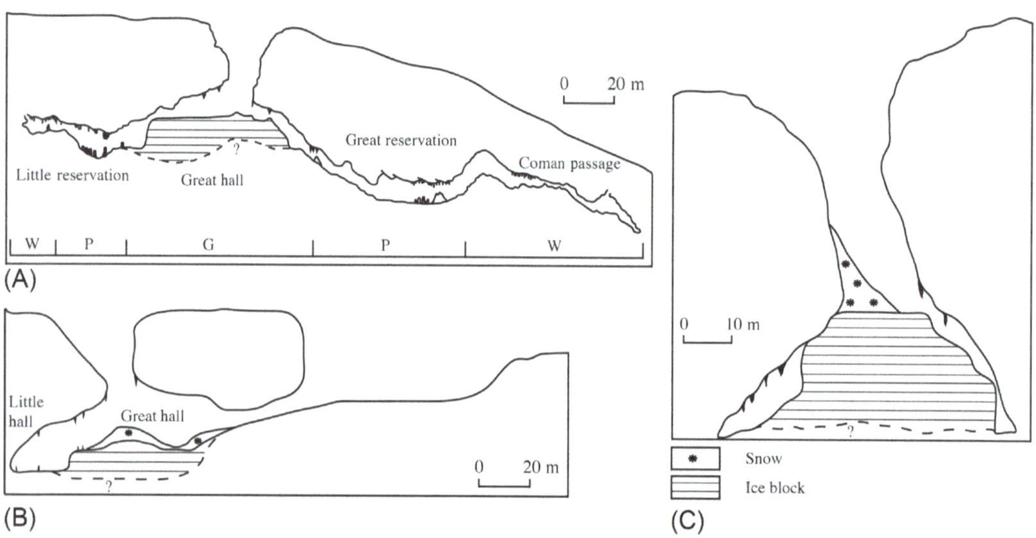

FIG. 3

Cross-sections through Scărișoara (A), Focul Viu (B), and Borțig (C) ice caves.

These three caves are located between 1150 and 1165 m asl in the transition zone from birch to spruce forests. The morphology of the caves is similar, each being a descending cave with entrances located at the upper parts only, a morphology, which favored the inflow of cold air in winter and subsequent freezing of water to form large ice bodies—100,000 m^3 in SIC, 25,000 m^3 in FV, and 30,000 m^3 in BIC. In all three caves, small ice speleothem develop in winter and spring, when inflowing dripping water freezes to form stalagmites, stalactites, and ice columns. In FV and BIC, these formations usually melt towards the end of summer, but in Scărişoara, large fields of perennial ice speleothems develop (Fig. 4) due to a larger ice volume and the peculiar cave morphology.

FIG. 4

Field of ice speleothems in Scărişoara Ice Cave, Romania.

The peculiar morphologies of the caves lead to specific climatic conditions. Air temperature variations in the caves are controlled by variations of the same parameter outside in winter (defined as the time interval when external air temperature is below 0°C), and by the absorption of the heat (transferred by conductive processes through the rocks and air in the entrances) that accompanies the melting of ice in summer (Fig. 5).

As a direct consequence of the peculiar climate inside the cave, the infiltrating water freezes, and layers of ice accumulate to build up large deposits of perennial ice. The ice deposits in SIC, FV, and BIC consists of regularly spaced sequences of annually laminated layers, each formed by a couplet of clear ice and sediment (organic matter, calcite, vegetal remains, etc.), visible on the lateral, ablation walls of the ice blocks (Fig. 6).

In summer, high dripping rates bring large quantities of warm water (5–6°C) into the caves, partly melting the uppermost layers of ice. Temporary lakes develop on top of the ice blocks, but these will eventually freeze, top to bottom, in autumn. In winter, temperatures below 0°C lead to the undercooling and freezing of the walls, and thus the dripping of water stops. The inflow of cold and dry air leads to intensive ablation of ice, while on the overcooled walls large hoarfrost deposits emerge, generated by the warm air exiting the lower parts of the caves. The growing process of the ice block slows down after autumn, taking place only in periods of mild weather, when the infiltrating water is rapidly

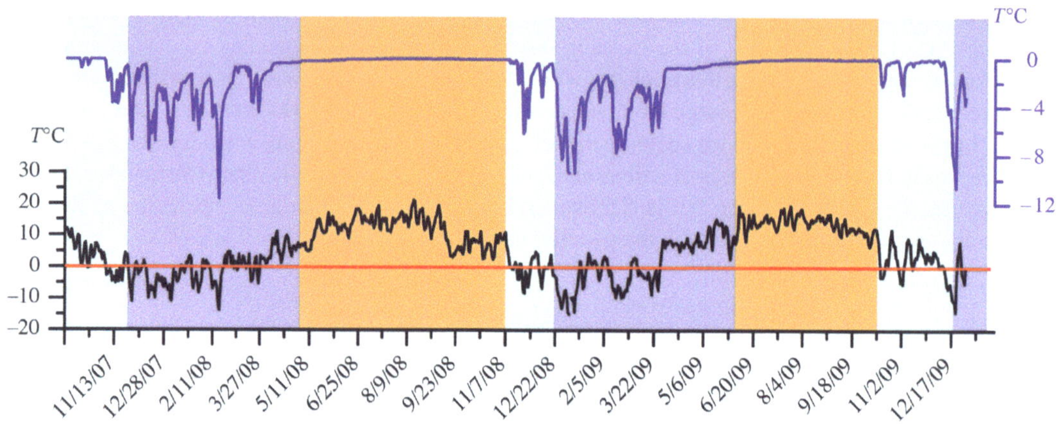

FIG. 5

Air temperature variations outside and inside an ice cave. All descending caves have a similar behavior in terms of air temperature variability (see other chapters in this volume), hence this variability applies to any such cave.

FIG. 6

Ablation wall in Scărișoara Ice Cave. Organic and inorganic-rich strata are visible on the left and right ends of the wall. In the middle, recently frozen water covered the layering.

freezing, forming either thin layers of ice on top of the autumn layers, or ice stalagmites and stalactites. In spring, negative temperatures are preserved in the cave, while outside, rising air temperatures and increased precipitation causes the melting of snow and the infiltration of larger amounts of water into the cave. The dripping water arrives in a cool cave environment, thus leading to a rapid accumulation of ice. When temperature in the cave rises above 0°C the heat induced by dripping water is higher than the cooling effect of ice, and the melting of ice begins. Snow that reached the bottom of the caves also partly melts (Fig. 7), further adding to the water (and ice) balance of the caves.

FIG. 7

Early summer in Focul Viu Ice Cave. Direct sunlight leads to melting of snow and refreezing of water as ice, but also to the melting of the ice, in spots where it directly shines on it.

During melting, more than one annual layer could be melted away, leaving a combined, thicker layer of sediments. Accumulation patterns differ, forming both organic-rich and calcite-rich layers (Fig. 6). Organic, impurity-rich layers are generated by the influx of material from the outside, which was washed inside the cave during heavy or prolonged rains. These layers are indicators of humid periods with during wet summers. Calcitic layers form during autumn and winter as a result of rapid CO^2 degassing during water freezing. Cryogenic calcite with high $\delta^{13}C$ values is deposited on the surface of ice and later incorporated in the forming ice blocks.

Observations of ice layering and structure, its isotopic composition, and contents in SIC, FV, and BIC (Holmlund et al., 2005; Kern et al., 2009; Perşoiu et al., 2017) have shown that these forming processes have acted similarly during the past couple thousand years, building up the ice blocks as successive layers of lake-ice deposits. However similar in structure, the ages of ice deposits differ in the three caves: the oldest ice is preserved in Scărişoara Ice Cave (10,500 years, Perşoiu et al., 2017), and in Focul Viu and Borţig the ice is much younger: 1800 years in FV (Citterio et al., 2005), and ~1200 years in BIC (Kern et al., 2007).

25.3.2.1 Focul Viu Ice Cave

Focul Viu Ice Cave is a small (107 m long), descending cave in the central part of the Apuseni Mountains, Romania (46.27°N, 22.68°E, Fig. 8). It is located at an altitude of 1165 m (Orghidan et al., 1984), in the upper part of a limestone ridge. A small, descending entrance is giving access to a large chamber ("Great Hall", 68 × 46 m), followed by a smaller chamber ("Little Hall", 20 × 5 m). The ceiling of the Great Hall is open to the sky through a large shaft, through which snow and tree trunks have fallen inside the cave to form a large, cone-shaped pile (Perşoiu et al., 2007). The floor of the Great Hall is occupied by a massif, layered, ice block (20 m thick, 30,000 m^3). Between the northern side of the ice block and the host rock, a narrow rimaye gives access to the lower, ice-free sector of the cave.

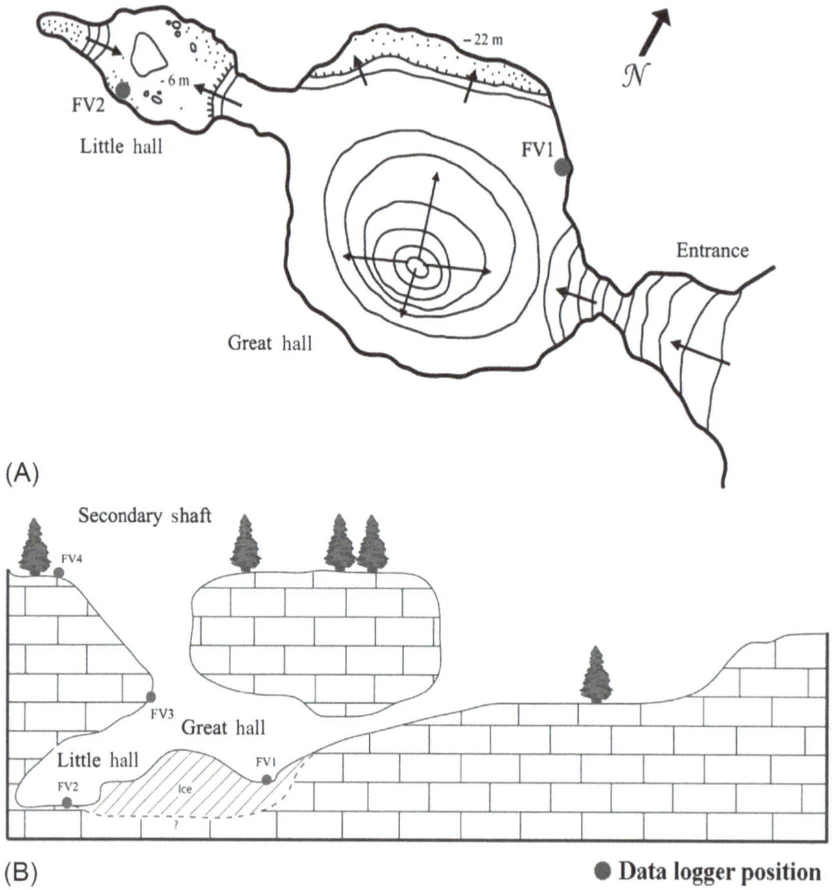

FIG. 8

Map and cross-section of Focul Viu Ice Cave, with the location of air measurements station.

Perşoiu et al. (2007).

Air temperature variations inside the cave (Fig. 9) result in a peculiar type of circulation (Perşoiu et al., 2007). As long as external air temperature is above 0°C, there is no air inflow inside the cave (due to differences in air density), thus the warming trend (Fig. 9) in the cave is due only to conductive heat transfer through air and rock. However, due to differences in air temperature and density, a convective cell develops, transporting air between the two rooms of the cave (summer type circulation). As soon as the external temperature drops below internal ones, a rapid inflow of cold air occurs through the lower entrance, leading to a general drop in temperature throughout the cave. During these periods, internal air temperature variations are clearly following the external ones.

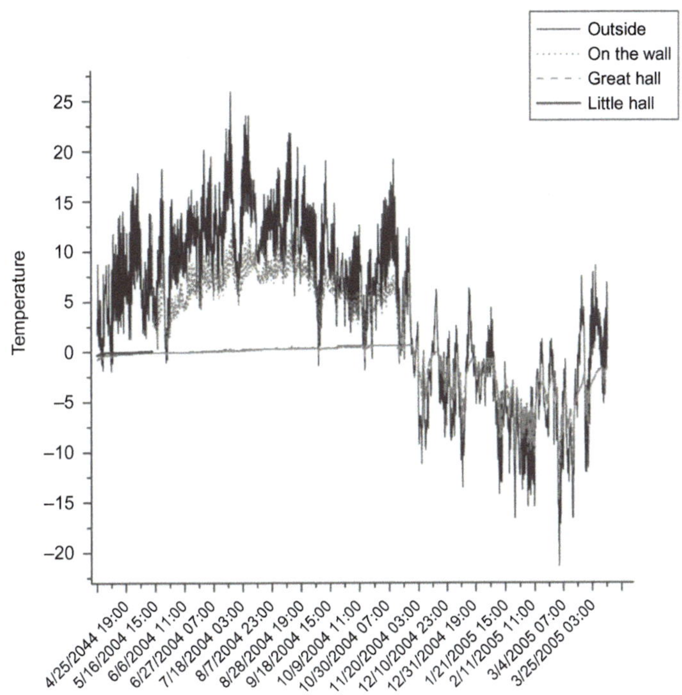

FIG. 9

Air temperature variations in Focul Viu Ice Cave.

Perşoiu et al. (2007).

In the secondary shaft, air temperature variations are following the external ones, but they are always lower than these due to the influence of the ice block (cooling by conductive transfer of heat). When the temperature outside drops below the internal one, the cold air that is entering through the lower entrance sweeps through the entire cave and pushes out the warm air from the inner parts of it through the secondary shaft.

The laminated structure of the ice block hinted that it might preserve past environmental changes, thus leading to intensive drilling and analytic efforts in the cave. Fórizs et al. (2004), Kern et al. (2004), Citterio et al. (2005), and Bădăluță et al. (2017) dated the ice and analyzed O and H stable isotopes in water, as well as O and C stable isotopes in cryogenic cave calcite, in an effort to reconstruct the climate history of the past ~1000 years.

A positive correlation between air temperature and the isotopic composition of precipitation, as well as drip water in the cave, has been found. The water isotopic values in the ice core show low values up until AD 900, higher values between AD 900 and ~1300 (Medieval Warm Period, MWP), and again lower values after AD 1300 (Little Ice Age, LIA), reaching their minimum after AD 1800. The isotopic composition of cryogenic cave calcite shows slightly higher values in the MWP and lower values in LIA, possibly suggesting a climatic influence. Modern observations are too short to be able to calibrate this putative signal, but there is work in progress (Bădăluță et al., 2017) addressing this issue.

25.3.2.2 Borțig Ice Cave

The Borțig Ice Cave (46.56°N, 22.69°E; 1236m asl) is the third largest ice cave in Romania. The ice block (25,000 m^3) has accumulated by the freezing of water at the bottom of a 45m deep shaft (Fig. 3C). Its maximum measured thickness reaches 22m, although the morphology of the ice block is still unknown. The water that feeds the cave originates from snow melt at the bottom of the shaft, dripping water, and direct rain on the top of the ice. Like in other ice caves in the Apuseni Mountains, a shallow lake develops on top of the ice block, the lake freezes in winter to form a new layer of ice. The morphology of the cave and ice block is rather simple: a deep, roughly circular shaft has its bottom filled with ice and snow. Geothermal heat partly melted the ice at the contact with the rock, opening a narrow rimaye, which can be explored down to a depth of 23m, where ice and rock are in contact. Dripping and infiltrating water led to the formation of ice stalagmites on the upper surface of the ice block, but these usually melt every summer.

Radiocarbon dating of wood remains in the ice gave an age of ~1075cal. BP for the ice at 10m below the present surface (Kern et al., 2007). A recent study by Kern et al. (2009) assessed the suitability of the ice for palaeoclimatic studies and found (based on tritium measurements) that, contrary to Scărișoara Ice Cave, where melting led to the disappearance of more than 1.5m of ice in the past decades (Perșoiu and Pazdur, 2011; Perșoiu et al., 2017), in BIC, ice has continuously accumulated since at least 1950, with no indication of melting associated to recent warming (Kern and Perșoiu, 2013).

25.3.2.3 Scărișoara Ice Cave

The Scărișoara Ice Cave has been continuously studied for almost a century, with the results of the earlier studies summarized in a book by Racoviță and Onac (2000), available online at http://www.kartsportal.org.

The cave is located in the Bihor Mountains (46°29′23″N, 22°48′35″E), NW Romania, at an altitude of 1165m above sea level. The local people visited the cave well before the first mentioning in literature. Apparently, they have collected ice from the cave and used it to preserve their food, or they have melted the ice and used it as drinking water in the dry summer periods. The first documents mentioning the cave date back to the mid 19th century (Vass, 1857), while the first scientific reports present the results obtained after an Austrian expedition that was organized in Bihor Mountains. One of these publications (Peters, 1861) presents a brief description of the cave and the ice stalagmites, together with information on rock age and tectonics. Another publication (Schmidl, 1863) provides a map and describes the cave in more detail. Later, Scărișoara Ice Cave was visited several times by Emil Racoviță and René Jeannel (Fig. 10), and their results on the genesis and dynamics of the ice block and ice stalagmites were gathered together in an important monograph (Racoviță, 1927).

The cave ice age (3500 years) was estimated for the first time on basis of the analysis of pollen extracted from the ice layers (Pop and Ciobanu, 1950). Later research, performed especially by scientists of the Emil Racoviță Institute of Speleology, brought more thorough and elaborate results on the age of ice, cave climate and topography, genesis and evolution of the ice block, the formation and dynamics of ice stalagmites, paleoclimate reconstructions on the basis of various proxies present in the sedimentary ice, alongside detailed biospeleological and microbiological information.

Cave morphology

Scărișoara Ice Cave is part of a larger karst system developed on three levels and stages of karstification (Fig. 11), and it is carved in Upper-Jurassic limestone (Bucur and Onac, 2000). Together with Pojarul

FIG. 10

Early 20th century visit to Scărișoara Ice Cave.

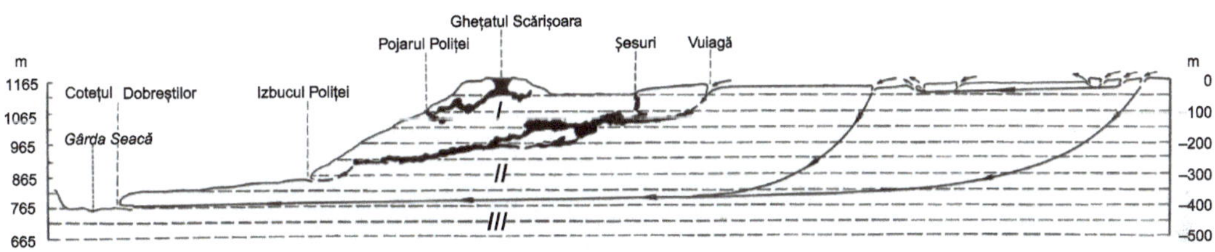

FIG. 11

Karstification levels below Scărișoara Ice Cave (Rusu et al., 1970).

Poliței Cave, Scărișoara represents the oldest level of this system, a fossil level, while the other two levels located underneath are longer and still active.

The access into the cave is made on a series of stairways through a 60 m wide and 48 m deep shaft (Fig. 12). Large and permanent deposits of snow are present at the bottom of this shaft, in front of an imposing portal, 24 m high and 17 m wide (Fig. 12).

Passing through this portal, the floor of the Great Hall (Sala Mare) is represented by the top of the great fossil glacier located mainly in this room. It extends for over 3000 m^2, and is a perfectly horizontal ice surface with four conic ice formations, one to the left, towards the Great Reservation, and the other three joined together at the distal end of the ice block, just before the Church (Fig. 13).

Temporary ice stalactites may form in the cave ceilings during winter. The Church is a smaller room located at the northwestern end of the ice block. Here the horizontal ice floor of the cave descends for over 8 m on a steep slope, leading to a room fully occupied by ice stalagmites (Fig. 4).

FIG. 12

Entrance of Scărișoara Ice Cave in winter.

FIG. 13

The Great Hall in Scărișoara Ice Cave.

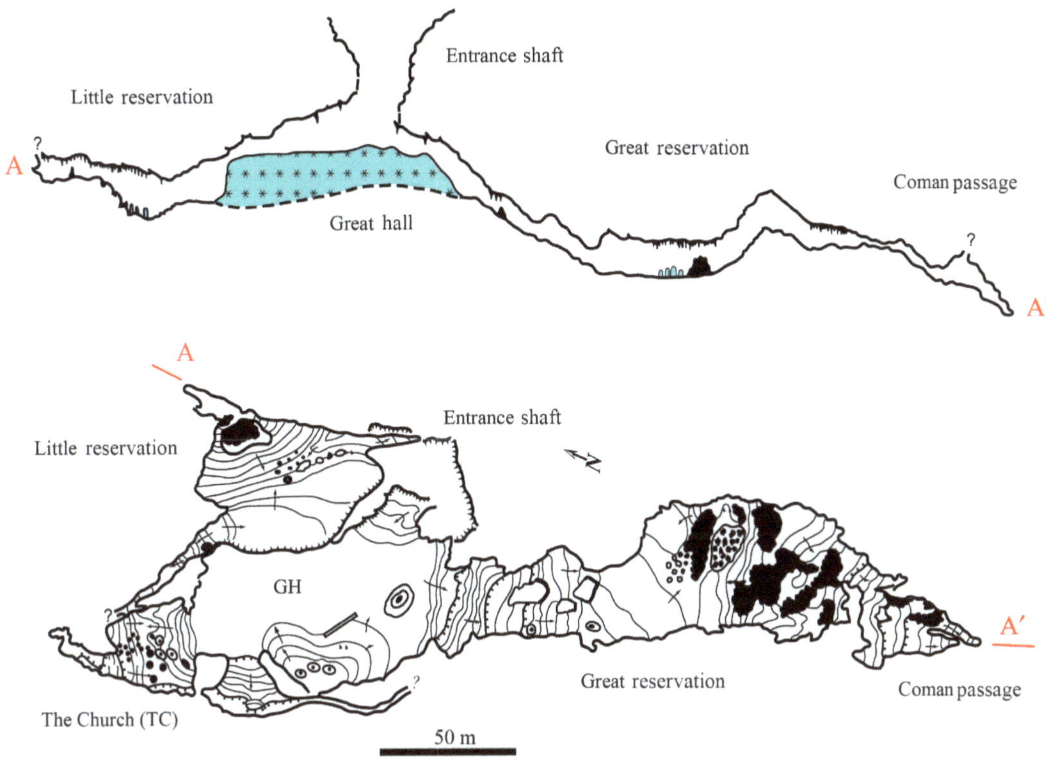

FIG. 14

Map and cross-section of Scărișoara Ice Cave (line AA′ indicates the profile of the cross-section).

The Great Hall and the Church represent the touristic area of the cave, which can be visited on a horizontal wooden pathway installed above the ice floor. On the north and south sides of the Great Hall, the ice block withdraws and leaves access to the two profound sectors of the cave, the Little Reservation and the Great Reservation, respectively (Fig. 14).

The Little Reservation can be accessed by descending a 15 m high vertical ice face, which represents the northern flank of the underground glacier, where the annual stratification of ice can be easily observed (Fig. 6). The Little Reservation is formed of a large room that contains ice stalagmites in its first descending part. The floor of its second half ascends abruptly, and it is covered with calcite crust and ends up in a narrower corridor called Palatul Sânzienii (or the Palace of the Gentle Fairy), which is scattered with calcite speleothems.

The Great Reservation is incomparably larger than the Little Reservation, accounting for one third of the total length of the cave. It opens on the southern flank of the ice block, after descending a steep slope on ice. The Great Reservation starts with the Maxim Pop Gallery, a heavily descending sector covered with rocks, boulders, and tree trunks fallen from the surface through the entrance shaft. This gallery leads into a more horizontal and wider area, with numerous ice stalagmites similar but larger than those present in the Little Reservation. Following this gallery the visitor climbs up a large agglomeration of rocks that have fallen from the cave ceilings. On top of this deposit, calcite stalagmites and stalagmitic domes have formed, suggesting the name of this sector: the Cathedral. The Great Reservation ends up onto a descending gallery, called Coman Gallery, ending at 105 m below surface.

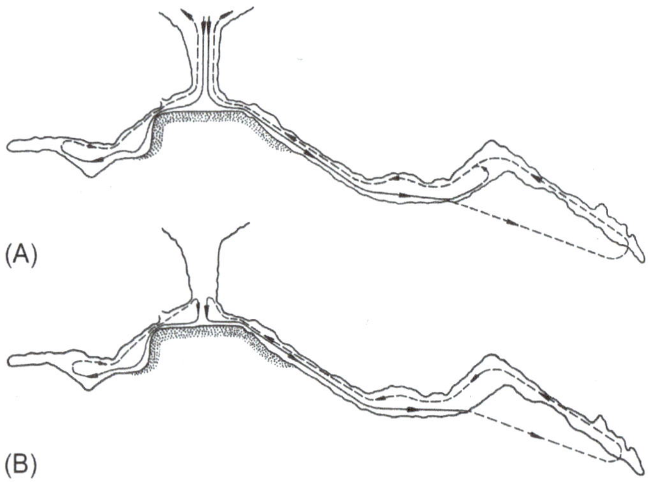

FIG. 15

Air circulation in Scărişoara Ice Cave (A: winter; B: summer).

Cave climate

Scărişoara Ice Cave is a descending cavity, connecting with the surface only in its upper part, through the large shaft at the entrance. In this cave the air circulation is governed primarily by the differences between air temperature and density at the surface, compared to those present underground. During winter the descending cold air from the surface reaches all three rooms neighboring the ice block until those sectors where ice stalagmites form (Figs. 3 and 15).

In the Great Reservation the cold air may reach the profound parts of the cave through the openings between the large deposit on the Cathedral floor. In sectors where the air becomes warmer the air circulations reverses, and the air current moves along the cave ceilings, through the entrance shaft, towards the surface (Fig. 15A). In summer the air exchange with the surface ceases, and the air circulation follows the same pattern as in winter, but only underground (Fig. 15B). When descending the entrance shaft during summer, one can feel this temperature threshold somewhere in the second half of the descent.

The air circulation creates the separation of four climate zones of the cave (Racoviţă, 1984): a transition zone located into the entrance shaft, a glacial zone in the Great Hall, a periglacial climate in all three rooms bordering the ice block, and a relatively warmer zone located in the profound sectors of the cave. The largest temperature variations occur within the glacial zone in the Great Hall, where the air temperature may drop during winter to $-14.5°C$, while during summer the temperature in this area comes up to $0°C$ or slightly above.

Glaciology

In Scărişoara Ice Cave the ice is represented by the ice block in the glacial zone, and ice stalagmites in the periglacial areas. Having a volume of more than $100{,}000\,m^3$, and an average thickness of 22 m, the ice block is the largest of this kind in the world. The ice block forms by the annual freezing of water that percolates from the surface through the entrance shaft and the rock pores and fissures. During summer a 3–5 cm deep lake forms on the ice block surface as a result of ice melting and seepage water infiltration. The water that infiltrates carries debris, which contains soil, vegetal rests, invertebrates, and microorganisms, which decant down onto the lake bottom. During winter the lake freezes, trapping the organic debris at the bottom, resulting in the formation of one annual layer of ice formed on top of the stratified ice block, which can be nicely seen after descending into the Little Reservation (Fig. 6). This continuous annual construction does not lead to the gaining in height and filling up the Great Hall by the ice block, as the oldest ice layers melt at the bottom of the ice block at a rate of 1.5 cm/year (Perşoiu, 2005). The age of this older ice was estimated to be around 10,500 years, based on radiocarbon dating of the containing debris (Perşoiu et al., 2017), but the ice block obviously started to form earlier then this age, as its bottom melts continuously, so the first layers were gone long before the research has started. In periglacial areas of the cave, the influence of the massive ice block located nearby favors the formation of various ice stalagmites on the floor of the three rooms bordering the block. In the Church, the icy floor is covered with more than 100 perennial ice stalagmites (Perşoiu and Pazdur, 2011) of various sizes and shapes. The size and shape of these ice stalagmites may differ from one year to another on the basis of the local climate regime. The dimension of these stalagmites is directly proportional with the amount of water percolating from the above surface, which feeds these stalagmites and the temperature that favors their formation or partial melting. The ice stalagmites are formed from successive layers of hexagonal monocrystals, which are disposed perpendicularly on the stalagmite growth axis and radial on its extremity. In an ideal homogenous environment, with a constant temperature and no ventilation, the ice stalagmites grow as cylindrical

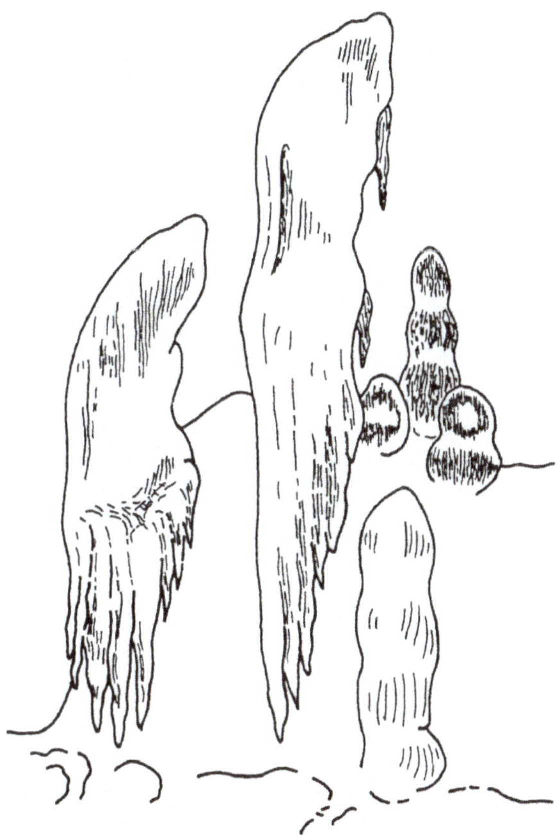

FIG. 16

Ice stalagmites with bilateral symmetry (original drawing by Emil Racoviță).

vertical columns with the upper end thickened. When the air circulation and the thermal gradient increase, this ideal radial symmetry is replaced by a bilateral symmetry, and the ice stalagmites grow preferentially on the opposite side of the air current, and they are now called "spectral ice stalagmites" (Fig. 16).

The influence of temperature on the ice stalagmites is considerable in several parts of the periglacial zone, and these ice columns are formed of thick and transparent zones, which alternate with thinner and opaque zones (Fig. 17).

FIG. 17

Thermoindicator stalagmites in Scărișoara Ice Cave.

These speleothems are called "thermoindicator" ice stalagmites (Viehmann and Racoviță, 1968) and are formed in conditions of large fluctuations of temperature. The thick and clear zones were formed under relatively higher temperatures, when the percolating water froze gradually and evenly, while the thin and opaque zones were formed at lower temperature values, when the water froze almost instantly, retaining air and calcium carbonate that did not have sufficient time to be removed out of the percolating water.

During summer, the dripping water forms circular excavations on the thickened top end of the ice stalagmites. The depth and width of these cavities are directly proportional with the thermal and mechanical impact of the falling water drops. This impact is sometimes as strong, and as such the cavity turns into a tub, which perforates the entire ice stalagmite down to its bottom.

The ice is absent in the profound zones of the cave, where the air temperature raises around to +4°C to 5°C (Palatul Sânzienii, Coman Gallery). In these warmer sectors of the cave, cave fauna appears, represented especially by large populations of cave beetles (*Pholeuon* (s. str.) *knirschi glaciale*), alongside nematodes, oligochetes, spiders (*Platybunus bucephalus*, *Nesticus racovitzai*, *Troglohyphantes racovitzai*), mites (*Ixodes vespertilionis*), springtails (*Onychiurus multiperforatus*, *O. armatus*, *O. sibiricus*, *O. granulosus multisetis*, *O. rectopapillatus*, *O. variotuberculatus*, *Oncopodura crassicornis*, *Tomocerus minor*), hymenoptera, amphipoda (*Niphargus laticaudatus*), and several species of bats (*Myotis myotis*, *M. oxygnathus*, *M. mystacinus*, *M. brandtii*, *M. dasycneme*, *M. daubentonii*, *Nyctalus noctula*, *Plecotus austriacus*, *Vespertilio murinus*).

Scărișoara Ice Cave has yielded a unique window into the past climatic environmental history of the region. Several studies (Perşoiu and Pazdur, 2011; Feurdean et al., 2011; Perşoiu et al., 2017) have addressed this issue, and they are all discussed in detail in the "Palaeoclimatic significance of cave ice" section in this book.

REFERENCES

Bădăluță, C., Ersek, V., Piotrowska, N., Perşoiu, A., 2017. Oxygen and carbon stable isotopes in cryogenic cave calcite (CCC)—possible proxy for past climate changes. In: European Geosciences Union General Assembly 2017, Vienna, Austria.

Bucur, I.I., Onac, B.P., 2000. New Data Concerning the Age of the Meso- Zoic Limestone From Scărișoara (Bihor Mountains). Studia Univ., Babeș-Bolyai, Geologia. XLV (2), 13–20.

Citterio, M., Turri, S., Perşoiu, A., Bini, A., Maggi, V., 2005. Radiocarbon ages from two ice caves in the Italian Alps and the Romanian Carpathians and their significance. In: Mavlyudov, B.R. (Ed.), Glacier Caves and Glacial Karst in High Mountains and Polar Regions, pp. 87–92.

Feier, N., Munteanu, G., Onac, B.P., 2001. Mineralogy and speleogenesis of the Ice-Cave from Poiana Vartop (Bihor Mountains, Romania). Theor. Appl. Karst. 13–14, 81–85.

Feurdean, A., Perşoiu, A., Pazdur, A., Onac, B.P., 2011. Evaluating the palaeoecological potential of pollen recovered from ice in caves: a case study from Scărișoara Ice Cave, Romania. Rev. Palaeobot. Palynol. 165 (1–2), 1–10.

Fórizs, I., Kern, Z., Nagy, B., Szántó, Z., Palcsu, L., Molnár, M., 2004. Environmental isotopes study on perennial ice in the Focul Viu Ice Cave, Bihor Mts., Romania. Theor. Appl. Karstol. 17, 61–69.

Holmlund, P., Onac, B.P., Hansson, M., Holmgren, K., Mörth, M., Nyman, M., Perşoiu, A., 2005. Assessing the palaeoclimate potential of cave glaciers: the example of Scărișoara Ice Cave (Romania). Geogr. Ann. 87A (1), 193–201.

Kern, Z., Perşoiu, A., 2013. Cave ice—the imminent loss of untapped mid-latitude cryospheric palaeoenvironmental archives. Quat. Sci. Rev. 67, 1–7.

Kern, Z., Fórizs, I., Kázmér, M., Nagy, B., Szántó, Z., Gál, A., Palcsu, L., Molnár, M., 2004. Late Holocene environmental changes recorded at Ghețarul de la Focul Viu, Bihor Mts, Romania. Theor. Appl. Karstol. 17, 51–60.

Kern, Z., Molnár, M., Fórizs, I., Perşoiu, A., Nagy, B., 2007. A Porcika-jégbarlang padozati jegének képződésével kapcsolatos következtetések glaciológiai megfigyelések és geokémiai jellemzők vizsgálata alapján. Karsztfejlődés 12, 315–330.

Kern, Z., Molnár, M., Svingor, É., Perşoiu, A., Nagy, B., 2009. High resolution, well preserved tritium record in the ice of Borţig Ice Cave (Bihor Mts., Romania). Holocene 19 (5), 729–736.

Orghidan, T., Neagra, S., Racoviţă, G., Lascu, C., 1984. Pesteri din Romania: ghidturistic. Sport-Turism, Bucuresti, pp. 85–87.

Perşoiu, A., 2005. Evidence of basal melting of the ice block from Scărişoara Ice Cave. In: Mavlyudov, B.R. (Ed.), Glacier Caves and Glacial Karst in High Mountains and Polar Regions. Institute of Geography of the Russian Academy of Sciences, Moscow, pp. 109–112.

Perşoiu, A., Pazdur, A., 2011. Ice genesis and its long-term mass balance and dynamics in Scărişoara Ice Cave, Romania. Cryosphere 5 (1), 45–53.

Perşoiu, A., Feier, I., Citterio, M., Turri, S., Maggi, V., 2007. Preliminary data on air temperature in Focul Viu Ice Cave (Bihor Mts., Romania). In: Zelinka, J. (Ed.), Proceedings of the 2nd International Workshop on Ice Cave, Demänovská dolina, Slovak Republic, pp. 62–64.

Perşoiu, A., Onac, B.P., Wynn, J.G., Blaauw, M., Ioniţă, M., Hansson, M., 2017. Holocene winter climate variability in Central and Eastern Europe. Sci. Rep. 7, 1196. https://doi.org/10.1038/s41598-017-01397-w.

Peters, K.F., 1861. Geologische und mineralogische Studien aus dem südöstilchen Ungarn, insbesondere aus der Umgegend von Rézbánya. Sitzung-Ber. Akad. Wiss. Wien, Math. Naturwiss. Classe 43 (1), 385–463.

Pop, E., Ciobanu, I., 1950. Analize de polen în ghiaţa de la Scărişoara. Anal. Acad. Rom. Seria Geol. Geogr. Biol. III (2), 23–52.

Racoviţă, G., 1984. Sur la structure méroclimatique des cavités souterraines. Theor. Appl. Karstol. 1, 123–130.

Racoviţă, G., Onac, B.P., 2000. Scarisoara Glacier Cave. Monographic Study. Carpatica, pp. 1–140.

Racoviţă, E., 1927. Observations sur la glacière naturelle dite "Gheţarul de la Scărişoara". Bull. Soc. Sci. Cluj III, 75–108.

Rusu, T., Racoviţă, G., Coman, D., 1970. Contributions à l'étude du complexe karstique de Scărişoara. Ann. Spéléol. 25 (2), 383–408.

Schmidl, A.A., 1863. Das Bihar-Gebirge an der Grenze von Ungarn und Siebenbürgen. Verl. Förster u. Bartelmus, Wien. 442 pp.

Şerban, M., Coman, D., Givulescu, R., 1948. Découvertes récentes et observations sur la glacière naturelle dite "Gheţarul de la Scărişoara". Bull. Soc. Sci. Cluj X, 174–210.

Şerban, M., Blaga, L., Blaga, L., Chifu, A1., Ciobotaru, T., 1967. Contribuţii la stratigrafia depozitelor de gheaţă din Gheţarul de la Scărişoara. Lucr. Inst. Speol. "E. Racoviţă", VI, pp. 107–140.

Vass, I., 1857. Eine Wanderung nach der Eishöhle bei Skerisora. Verh. u. Mitt. des sieb. für Naturwiss. zu Hermannstadt VIII, 162–170.

Viehmann, I., Racoviţă, G., 1968. Stalagmitele de gheaţă termoindicatoare. Dări de Seamă Comit. Geol. LIV, 353–363.

FURTHER READING

Contribuţii la stratigrafia depozitelor de gheaţă din Gheţarul de la Scărişoara. Lucr. Inst. Speol. "E. Racoviţă", VI, pp. 107–140.

Iţcuş, C., Pascu, M.-D., Brad, T., Perşoiu, A., Purcărea, C., 2016. Diversity of cultured bacteria from the perennial ice block of Scărişoara Ice Cave, Romania. Int. J. Speleol. 45 (1), 89–100.

Popa, I., Kern, Z., 2009. Long-term summer temperature reconstruction inferred from tree ring records from the Easter Carpathians. Clim. Dynam. 32, 1107–1117.

CHAPTER

ICE CAVES IN RUSSIA

26

Bulat R. Mavlyudov*, Olga I. Kadebskaya†

*Institute of Geography of the Russian Academy of Sciences, Moscow, Russia**

Mining Institute of the Ural Branch of the Russian Academy of Sciences, Perm, Russia†

CHAPTER OUTLINE

Ice Caves. https://doi.org/10.1016/B978-0-12-811739-2.00026-7

26.1 HISTORY OF RESEARCH ON CAVES WITH ICE IN RUSSIA

The study of caves with ice has a rich history. Ice caves in Russia were first mentioned in 1690 (Maximovich, 1952). Since then this natural phenomenon has drawn the attention of travelers and researchers, and at the present time, there is vast literature describing seasonal ice formations (SIF) in caves. However, if the study of SIF in the Earth's natural and artificial cavities has not as yet taken an independent scientific direction—in cave glaciology or speleoglaciology—the reasons are twofold: (1) the slow development of speleology as a science, and (2) the little attention paid to this ice from a glaciologic perspective.

The information about SIF in caves is scattered in numerous geological, geographical, glaciological, karstological, local history, and speleological publications. Analysis of the literature shows that the information is very irregular. First, in some cases, there is a brief mention of SIF presence in cavities;

in other cases, there are detailed descriptions of SIF or of caves with SIF as a whole. Second is the limited quantity of general publications. Usually, the study of a natural phenomenon involves a gradual deepening of knowledge, from simply ascertaining the phenomenon to the establishment of its causes and the mechanism behind its functioning. In the study of cave ice, we, on the one hand, already have surprisingly correct investigations and far-reaching conclusions, while on the other hand, we have only modest descriptions or mentions of SIF research, which complicates consideration of the chronological order of cave SIF research. Knowing the degree of conditional characteristics and that the steps toward cave SIF overlap in time, it is possible to distinguish the following periods (steps!) of development: (1) acquaintance with ice in caves (16th century through 19th century); (2) research on ice as a filler in caves or as underground ice (since the 19th century); and (3) research on cave glaciation (since the 20th century). The transition from one step to another occurs at points of accumulation of a certain degree of knowledge about cave SIF, and each step considers questions about the increasing degree of generalization.

Acquaintance with ice in caves (16th century through 19th century). This time was characterized by mentions of ice in caves or by descriptions of separate caves with ice. Caves were visited as a rule while other sorts of activities were being conducted—preparation of geological and geographical descriptions of a district, land management, and so on—or simply from curiosity. It is possible to find research describing caves with ice. Descriptions by F. I. Stralenberg (1730), I. G. Gmelin (1733), I. P. Falk (1768), I. Lepyokhin (1770), and others (Maximovich, 1947) are devoted to Kungur Cave (Ural Mountains). From 1735 through 1740, I. G. Gmelin visited and described an "ice cave" on the Yenisei River (Gvozdetskij, 1954a,b). B. Vahushti (1745) told about Khorkhebi Cave with ice on the Caucasus (Maruashili, 1974). In 1746 N. P. Rychkov visited a valley of the Ik River cave with ice (Stupishin, 1959). The Inderskie Caves with ice were described by P. S. Pallas in 1769 during his travels (Maximovich, 1947). M. Severgin (1809) was informed about several caves with ice in Eastern Siberia: Balaganskaya Cave on the Angara River and Udinskaya and Ledyanaya Caves on the Lena River (Gvozdetskij, 1954a,b).

Through the middle of the 19th century, publications devoted to caves with ice were almost exclusively descriptive in character. Researchers of this time considered caves with ice more often as a unique natural phenomenon, but the reasons for the formation of the ice were not investigated. Achievements of this descriptive period include the following: (1) ice can be formed and can exist in separate caves; (2) caves with ice are found in different places; and (3) it is possible to find ice in various forms in caves (frozen lakes, ice dripstone, formations, etc.).

Research of ice as filler in caves or as underground ice (since the 19th century). Active exploring pursuits, geographical research and geological surveys, and land reclamation (settling, surveying under construction of objects and roads, etc.) have led not only to the detection of new caves but also to more steadfast attention to them, their genesis, and the sediments that fill them (including ice).

Research at this step is pointed in two directions: the study of cold conditions and the origin of ice in caves, and the study of the ice in caves.

The abundance of caves with ice has encouraged researchers to look for the reasons why the ice originated in them. E. S. Feodorov (Fyedorov, 1883) thought that the main reason was connected with the gravitational cooling of inclined caves. For example, in Kungurskaya Cave, he determined that the cooling of cavities was connected with ventilation caused by cold air in the winter and to ice evaporation (Fyedorov, 1883). V. Ya. Altberg established that there was not enough ice evaporation to explain Kungurskaya Cave's cooling, because in separate cavity locations, cooling occurred but ice was absent, so there was nothing to evaporate (Altberg, 1930, 1931; Altberg and Troshin, 1931). M. P. Golovkov (Golovkov, 1938, 1939a,b) thought that the majority of authors did not solve the basic question: what is the source of the low temperatures in caves? The cold in caves can be imported from outside or can originate inside cavities. He believed that both were involved. V. S. Lukin (Lukin, 1962, 1965) showed that in Kungurskaya Cave, water and ice evaporation used up to 30% of the energy, as they were consumed by dry, cold air flowing through the cave in the winter.

Observations have shown that for horizontal caves with more than one entrance, the gravitational theory of cold accumulation is incomplete, because in these caves there are no gallery declivities, and cold air has no place to flow in. Yu. Listov (1885) found that a principal cause of similar caves' cooling is air movement between entrances (chimney effect) (Listov, 1885). These conclusions agree with later works of researchers: for example, A. A. Kruber (Kruber, 1915), V. Ya. Altberg (Altberg, 1930, 1931), M. P. Golovkov (Golovkov, 1938, 1939a,b), V. S. Lukin (Lukin, 1965), V. M. Golod (Golod, 1981; Golod and Golod, 1974a,b, 1975, 1978, 1981).

Two other possible reasons for cold accumulation in caves are heat absorption at adiabatic air expansion and baric air circulation in cavities. Theories on cooling of caves at adiabatic air expansion are held by R. A. Alexeev and V. I. Belyak (Alekseev and Belyak, 1970; Belyak, 1973), who thought that this way of cooling was the primary action behind ice formation in caves. V. S. Lukin (Lukin, 1965) showed that, of the common daily cooling, adiabatic air expansion provides no more than 3% toward the cumulative effect of cavity cooling. Researchers such as V. N. Dublyanskij (Dublyanskij and Lomaev, 1980) and L. I. Sotskova (Sotskova, 1981; Sotskova and Dublyanskij, 1982) thought that baric changes in a free atmosphere can influence air movement in cavities, and V. E. Dmitriev (Dmitriev, 1980b) thought that baric changes can be one of the reasons for cavities to cool.

The following hypotheses are also offered as reasons for ice formation in caves: ice formed during glacial age, summer heat (in warmer weather, the quantity of ice is greater), chemical processes of cooling, and waves of heat and cold. V. E. Dmitriev thought that the relict ice can be kept in some caves near the permafrost boundary (Dmitriev, 1979).

At the beginning of the 20th century, it was found that the main reason for ice accumulation in caves is water freezing in frozen cavities. Thus cooling is due to cold air moving from the surface and circulating in the cavities during the wintertime.

Research on cave ice has considered its form, composition, structure, dynamics, and movement. Ice studies have been conducted from different perspectives: karstologists and speleologists consider ice to be cave sediments, and geocryologists and glaciologists consider cave ice to be fillers as underground ice (Dmitriev, 1980a,b). Both groups have also considered ice in caves to be foreign (allogenic) formations.

Numerous forms of ice in caves have been known to exist in Russia, but E. S. Fyedorov (1883) emphasized that in essence cave ice forms do not differ from surface ice forms. He considered one possible special ice form, that is sedimentary ice generated from frost crystals that fall from the roof of caves (Fyedorov, 1883). I. D. Cherskij (1876) was the first to pay attention to the presence of permafrost in caves (Gvozdetskij, 1954a,b). Therefore, from the beginning of the 20th century and until now, with one exception, no new forms of cave ice have been found. The exception is cave glaciers, which had been assumed to exist, but their existence had not been proved. More recently, cave glaciers were found in Kuznetsk Ala Tau (Vistengauzen and Dmitriev, 1977; Dmitriev, 1972a,b, 1977, 1980a).

Study of cave ice structures began in the 20th century. M. P. Golovkov (1939a,b) investigated the structure of ice stalactites in Kungurskaya Cave and showed that such structures represent monocrystals with an axis ***C*** beginning at a nucleation site and lengthening to become an ice stalactite.

He considered the study of ice structures as key to understanding their origin (Golovkov, 1938, 1939a,b). P. A. Shumsky studied the sublimation of ice crystal structures (Shimsky, 1955). E. P. Dorofeyev (1969) studied the structure of ice sublimation crystals in Kungurskaya Cave and found the temperature conditions necessary for the growth of different morphological forms of such crystals (Dorofeyev, 1969).

Cave ice dynamics were seldom mentioned in publications up to now. Alekseev and Belyak (1970) thought that the dynamics of flowstone ice in caves were defined by the quantity of water coming in from the outside and the processes of ice evaporation and melting caused by air streams (Alekseev and Belyak, 1970).

According to the publications, researchers were guided mainly by separate caves' ice forms and the study of their structures, their dependence on certain growth conditions, and their existing characteristics and evolution. Thus ice forms were not considered as a component of the uniform phenomenon—cave glaciation.

V. E. Dmitriev (Dmitriev, 1980b) measured the velocity of the movement of cave glaciers in Kuznetsk Ala Tau at about 0.2–0.5 m/year. Yu. E. Lobanov (oral message, 1981) paid attention to ice movement on the inclined taluses of Sumgan-Kutuk Cave in the Ural Mountains. E. V. Shavrina found the same in Ledyanaya Volna Cave on the Pinego-Kuloj plateau (Shavrina, 2002).

Only the most rudimentary geochemical research of cave ice has been done, despite the fact that there are many varieties of chemical compounds. This can be explained by the fact that scientists generally use only individual ice samples for analysis, which do not reflect the conditions and dynamics of the broader geochemical effects on ice structures in a cave. Nevertheless, V. N. Dubljansky collected available analytical data on cave ice and tried to point toward the basic principles of cave ice composition (Dublyanskij et al., 1992).

Here are the general achievements in the development of our knowledge about ice in caves and caves with ice: (1) finding of large number of caves with SIF in many areas in Russia; that is, they are shown to be widespread; (2) development of principal reasons for the accumulation of cold in caves—(a) favorable cavity formation and (b) cavities' winter ventilation that leads to the accumulation of cold in rocks as a result of their heat exchange with the cold air; and (3) finding that cave ice is specific, as reflected by its structure, composition, character of accumulation, and movement.

The study of cave glaciation (since the 20th century) is still in an early stage of development. The first to speak about cave glaciation was apparently N. I. Karakash (1905), who compared descriptions of Kungurskaya Cave done in 1848, 1883, and 1905 and who concluded that "the glaciation of [the] first parts of Kungurskaya Cave increase[s]" (Karakash, 1905). In regard to cave glaciation, N. I. Karakash understood the degree of cavity filling by ice. A. A. Kruber (1915) thought that specific ice accumulation in separate caves was not an obstacle to finding the general laws of ice accumulation in cavities (Kruber, 1915). Thus we did not have a very good representation of cave glaciation.

One of the first steps in the research of cave glaciation was to separate caves in which SIF accumulated into different types. By the beginning of the 20th century, three basic types of these caves were identified: horizontal (with entrances at different elevations), inclined descending, and vertical (Kruber, 1915 and other). Later N. A. Gvozdetsky (Gvozdetskij, 1968) expanded the list of basic caves with SIF to seven, including in the list caves with permafrost and some complex types of cavities that combine the elements of two basic types (e.g., an inclined cave and a pit).

M. P. Golovkov (Golovkov, 1938, 1939a,b) paid attention to events that even the analyses of caves with ice descriptions occurring at different times did not address: increasing and decreasing glaciation.

Although these events occurred, it was not clear why they occurred. Was it because of an increasing quantity and spreading of ice in a cavity, to increasing or decreasing of the mean annual temperature (MAT) in a cave for any period, and so on? He did not think that we could answer those questions because there still were no quantitative characteristics for many of the geophysical phenomena in caves. In essence, in his publications, M. P. Golovkov was the first to present questions about cave glaciation.

In 1938 Yu. A. Bilibin (Alekseev and Belyak, 1970) proposed the scheme of air circulation in karstified limestone in the Aldan Plate. In these cavities, one entrance was situated at watershed level and another one at the bottom of the mountain. Air circulation took place in the summer from the top, down, and in the winter from below, upward. Bilibin considered that such air movements can destroy permafrost.

Alekseev and Belyak (1970) thought that the formation of ice in caves in eastern and central Siberia was caused by sudden decreases in the temperature of moving air at its adiabatic expansion after it moved through a narrow passage in a cave. This point of view was criticized by V. E. Dmitriev (1980) (Dmitriev, 1980b).

As a result of the research in Kungurskaya Cave and also on the basis of the analysis of the data from a natural model of caves with ice, V. S. Lukin (Lukin, 1962, 1965) established a structure for the thermal balance of Kungur Cave and a ratio for the processes supporting the existence of ice in the cave. Lukin also proved the interrelation and interdependence of these processes, and the possibility of permafrost formation in caves in modern time has distinguished between cave permafrost and superficial permafrost (permafrost that penetrates the Earth's crust along cave channels). However, V. S. Lukin considered glaciation of only one specific cavity.

V. M. Golod (Golod, 1981; Golod and Golod, 1974a,b, 1975, 1978, 1981) described, in the form of a mathematical model, the thermal regime of a horizontal cave with ice using observation data from caves in the Pinego-Kuloj karst plateau located north of the Russian Plain. As the basis of the mathematical model, he entered a representation of the presence of positive and negative temperature anomalies at the upper and lower entrances in a horizontal cave (Lukin, 1965). This model has allowed us to estimate the dimensions of the freezing zone in caves, the dynamics related to dependence on changes in external conditions, and the temperature conditions along the caves. An analysis of this model has allowed us to estimate temperature preconditions of cave glaciation and to approach its spatial dynamics. Golod also outlined further perfections of the model that will allow us to consider not only dimension variations of the cooling zone in a cave but also actual ice accumulation and disappearance in cave channels, that is, to characterize glaciation dynamics of each particular cavity.

The first to actually consider cave glaciation was V. E. Dmitriev (Dmitriev, 1980a,b). He thought that "all set of various types of ice in karst cavities as a single whole, united by the environment in which they develop—by the cave, should be considered as a separate group in range of modern glaciation of the Earth." For cave glaciation, V. E. Dmitriev understood all sets of ice in caves. For example, in the caves in the Altai and Sayan Mountains and Kuznetskij Alatau mountains, he examined the dynamics of cave glaciation (including fluctuations of cave glaciers), a regime of cave glaciation (temperature field, ice accumulation and ablation, freezing, and melting). By considering cave glaciation as a set of cave ice, V. E. Dmitriev actually concluded that cave glaciation existed, though separately from the caves with only an oroclimatic base of glaciation. It reflects in his offer to research ice of caves by glaciologists and caves and their climate—by karstologists. That is, in essence, V. E. Dmitriev criticizing reference of ice of caves to cave sediments itself comes back to this thesis but in the latent form.

Cave glaciation in our opinion is not only a set of cave ice features but also a set of processes that results in the existence of ice in caves (Mavlyudov, 1989a, 2008). These processes (rocks cooling, air circulation, ice accumulation and ablation, etc.) are caused both by structure and by all stages of evolution of karst cavities. Thus glaciation is not simply imposed on karstic cavities but is also the major stage of their evolution, which by the way, was mentioned by V. E. Dmitriev (Dmitriev, 1980b) as he considered the origin and development of nival-corrosive cavities. From this perspective on cave glaciation, the common laws of cave glaciation arose (Mavlyudov, 1989a,b, 2008).

Last year, a popular investigative direction became the study of cryogenic minerals in caves (Andrejchuk and Galuskin, 2001; Andrejchuk et al., 2013 and others).

26.1.1 THE GENERAL DESCRIPTION OF CAVES WITH ICE IN REGION

In Russia there are now more than 10,000 caves. Approximately 10% of all karstic cavities known in the former USSR (including caves in countries described in Chapter 21 of this book) are subject to permanent glaciation (in separate areas of the country—the Caucasus—it can reach about 50%; Dmitriev and Chujkov, 1982). According to V. E. Dmitriev (Vistengatzen, Dmitriev, 1977) in the former USSR, there were more than 600 caves with permanent ice (data for 1980).

26.1.1.1 Spatial distribution

On the basis of modeling and research of caves with ice in the area of the former USSR, a map of cave glaciation was constructed (Mavlyudov, 1989a,b) (Fig. 26.1).

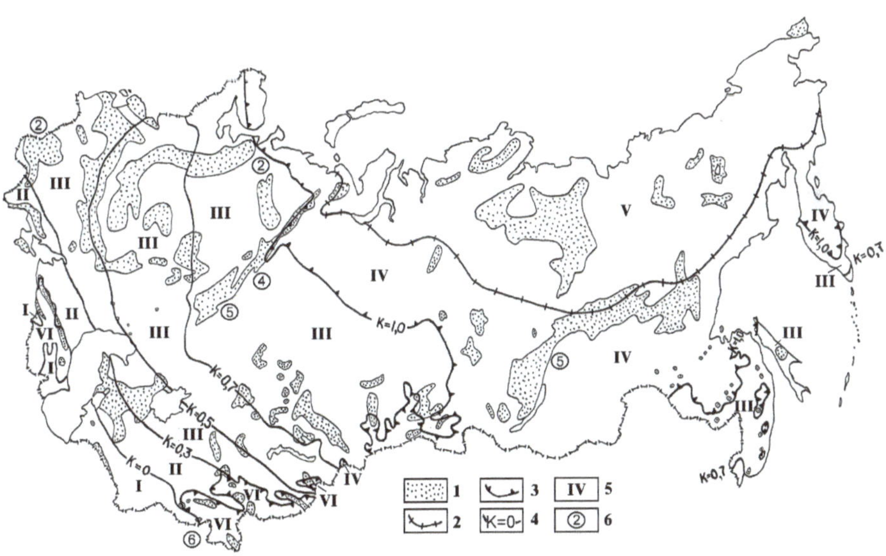

FIG. 26.1

Map of cave glaciation in area of the former USSR (Mavlyudov, 1989a,b, 1997): (1) karst rocks. Boundaries of permafrost distribution: (2) continuous, (3) sporadic and discontinuous. (4) Isolines of caves glaciation index and their values, (5) number of a zone of cave glaciation (V, every; IV, majority; III, separate; II–III, seasonal cave glaciation; I, cave glaciation is absent), and (6) difference (***a***) of MAT of district and (neutral zone) of caves.

As we can see, conditions for the occurrence of cave glaciation are available in most parts of the former USSR. Areas of karst rocks where the occurrence of natural caves is possible are shown on the map. Points corresponding to the most significant areas for caves with ice distribution are also shown on the map (Fig. 26.2). The most significant areas with caves with ice in Russia (and in the former USSR) are mountain areas—Crimea, Caucasus, Ural Mountains, Tien-Shan, Pamir, Altai, Kuznetsk Ala-Tau, Sayan Mountains, and Far East region—and also the plains areas: Pinega, the Russian Plain, Priuralye, Pribaikalye, and Transbaikalie.

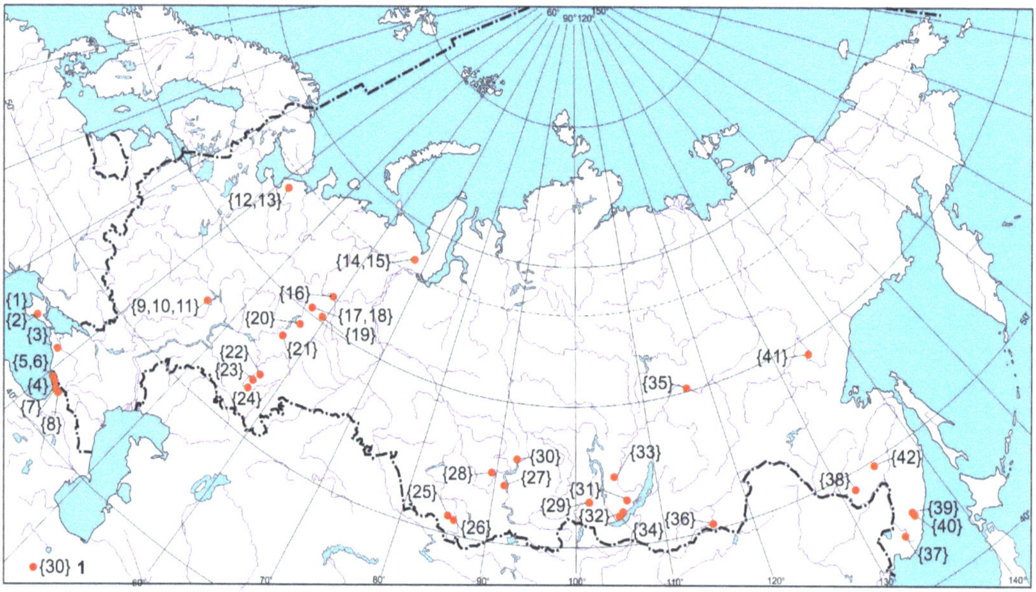

FIG. 26.2

Most significant areas in Russia with caves with ice distribution: (1) number of caves (similar to description).

26.1.1.2 Types of caves with ice

In Russia all types of caves in which glaciation development has been found include "horizontal" caves with entrances located at different elevations, inclined descending caves, and vertical pits, and also a combination of these cavities. Special types of cavities are caves in freezing rocks (in permafrost), which cover more than half of former USSR area.

26.1.1.3 Ice types in caves

All types of cave ice available in nature are found in the caves of Russia. Congelation ice (icings: integumentary and pendent, ice of reservoirs and rocks) prevails in horizontal caves, but sublimation ice occurs in a lower quantity (usually up to 5% of total ice). In inclined descending caves, the ice can be developed as congelation ice (icings: integumentary and pendent, ice of reservoirs and rocks) and as sedimentary and metamorphosed snow. Sublimation ice usually occurs seasonally, and perennial forms have a very limited distribution. Vertical cavities have mainly sedimentary and metamorphic ice; congelation and sublimation ice occur in insignificant quantities.

26.1.1.4 Dynamics of ice in caves

The dynamics of ice in the caves of Russia have been poorly studied, though a few caves have received more attention. First, the Kungurskaya Ice Cave was studied by the laboratory of the Mine Institute of the Urals Branch of the Russian Academy of Sciences. Regular research on the cave's climate and features of ice accumulation and ablation were carried out, mainly concerning icings. A study of the dynamics of cave icings was conducted in Pinega (Pevcheskaya Estrada Cave (Г-1) and Bolshaya Golubinskaya Cave) by employees of the Pinezhsky Nature Reserve. Irregular observations of the dynamics of snow and ice were conducted in the Snezhnaya cave system in the Caucasus by employees of the Institute of Geography, Russian Academy of Sciences. At the end of 1970 and the beginning of 1980, the dynamics of cave glaciers in Kuznetsk Ala Tau were studied by V. E. Dmitriev (Tomsk State University). In some other caves, ice dynamics have been observed, but only for their level of quality.

Research on the ice mass balance in caves is a labor-intensive and time-consuming business; therefore data about ice mass balance has been received only for individual caves. In this regard, there is a question about the reception of the information about the ice mass balance in cavities with the use of indirect methods. One example is the study of layered thicknesses of ice in caves. There are also other reliable indirect methods of defining the change tendencies of the ice mass balance in caves: length of negative temperature anomalies (NTA) zone in horizontal caves, depth position of an ice ledge in inclined descending caves, and depth position of the snow line in pits and shafts.

Model calculations and observations in caves have shown that the length of the NTA zone in horizontal caves is very sensitive to changes in external air temperatures (especially to long, periodic oscillations) (Golod, 1981; Golod and Golod, 1974a,b; Mavlyudov, 1985). This means that an increase in air temperatures outside caves causes a reduction in the length of the NTA zone and the reduction of cavity cooling and increase of ice ablation. At decreasing external air temperatures, the length of the NTA zone increases by increasing cavity cooling and increasing the area of the ice accumulation zone. In the first case, a negative ice mass balance will exist in the caves; in the second case, there will be a positive ice mass balance. A result of the climate change's influence on the ice mass balance in caves may be the length of changes in the NTA zone in similar cavities at different locations, at movements from south to north (Mavlyudov, 1985).

For inclined descending caves, a qualitative criterion for ice mass balance changes can be the depth of their ice ledges (Mavlyudov, 1988a,b). These ledges are formed as follows. During winter, the air temperature gradually increases as it advances from a cave's entrance deeper into the cavity, which decreases the cave's cold reserve in this direction. Water flows into a cave in the spring, mainly through the entrance, which means that the quantity of growing ice will gradually decrease along with the cavity's depth. During the summer at advancements into the cavities' depth, we will see a fast air temperature drop close to the entrance, but further on, the air temperature hardly varies and remains close to zero, which provides ice melting mainly only at the entrance part of a cave cavity. At the end of a zone of intensive ice melting, an ice ledge that will be annually renewed. As the climate warms outside the cave, the ice ledge will move deeper into the cavity (which means a negative ice mass balance), and when the climate cools, the ice ledge in the cave will move toward the entrance (which means a positive ice mass balance). As a result, ice ledge depth in inclined caves in different climatic conditions can illustrate (Fig. 26.3) geographical position changes and serve as a model for changes in an external climate.

The reflection of the snow and ice mass balance in pits is their snow line (Mavlyudov and Vturin, 1988), which is stable with fluctuations of an ice mass balance close to zero, as observed in superficial snow fields (Glazyrin, 1985; Pertsiger, 1986). Interannual oscillations of pit snow lines in this case apparently do not exceed a few meters. A negative ice mass balance leads to a move of the snow line deeper into a pit, whereas a positive balance leads to the snow line being closer to the surface. A change in the depth of a snow line in similar pits, with an increase in the absolute elevation of a district, is similar to a change in their ice mass balance with a change in the quantity of atmospheric precipitation. That is, an increase in the quantity of solid precipitation and in absolute height moves a pit's snow line to a ground surface, and a reduction of solid precipitation and height removes the snow line from the ground surface. For example, there are some pits on a southern slope of the Bzybskij Ridge (Caucasus) for which the connections between the depth position of their snow line and their district's absolute height are known (Fig. 26.3).

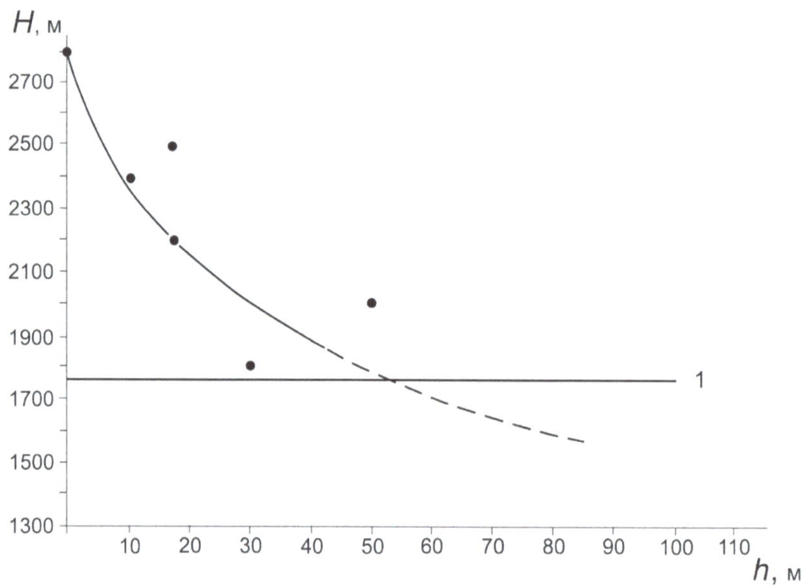

FIG. 26.3

Connection of depth of position of pits' snow line (*h*) and absolute height of district (*H*) (the Bzybskij Ridge, Caucasus, 1986). (1) Height of local forest boundary (tree line).

26.2 ICE CAVES IN CRIMEA

Most of the caves with SIF in Crimea are located on high-mountainous karst massifs in the yailas (mountain pastures), including Karabi, Ah-Petri, and Chatyr-Dag. All the cavities originate in limestone.

The cavities that are subject to glaciation are inclined descending cavities (Bolshoj Busluk) and vertical cavities (pits on the upper plateau of Chatyr-Dag). All caves with SIF are located in the high mountains where in the winter the temperature is negative and where snow remains for a good deal of time.

Special research of ice in caves of Crimea was not done up to now (no publications).

Kruber (1915) was the first person to study ice in the caves of Crimea. In addition, in Crimea the snow's role in karst formation was studied, and aspects of the formation of nival-corrosive pits in particular were studied (Dublyanskij, 1963; Dublyanskij and Shutov, 1967).

In vertical cavities, sedimentary and metamorphic ice remain year-round, and congelation ice (icings: covering and pendent) prevails in inclined descending caves.

Last tens of year the quantity of ice in caves began to decrease, including in the most significant caves with ice in Crimea, the Bolshoj Busluk and the Trekhglazka. There is no information about changes in SIF conditions in vertical cavities.

26.2.1 CAVE BOLSHOJ BUZLUK (BUZLUK-KOBA, LEDYANAYA CAVE)

26.2.1.1 Site and Situation Around the Cave

The cave is located in an average part of Karabi-jajly at an elevation of about 1045 m a.s.l. on almost unforested tableland (entrance coordinates: 44°51′32″N, 34°32′06″E. MAT = 6.4°C, $T_{\mathrm{Jan}} = -3.3$°C).

26.2.1.2 The Cavity Form, Cavity Maps

The Bolshoj Buzluk cave is a classic incline descending cave. It is a huge karst collapse with a depth of about 41 m with permanent ice on the bottom (length, 165 m; depth −81 m). The cave begins from the mouth of a big karst doline with a diameter of about 20 m in which beech trees grow. From a depth of 35–40 m from the entrance, there is a big chamber 70 m in length and with a bottom covered by permanent ice. At the lower part of the cave, there is a deep pit (19 m) with a diameter of about 0.5 m and with water flowing on thc bottom (Fig. 26.4).

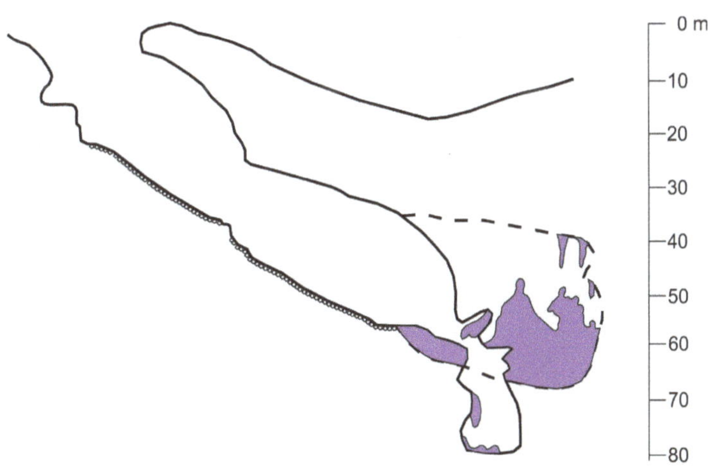

FIG. 26.4

Longitudinal section of Bolshoj Buzluk Cave (Dublyanskij, 1977).

26.2.1.3 Research History

Local residents have known about the cave for a long time. Kruber (1915) was the first person to publish information about a cave with ice in it.

26.2.1.4 Forms and Dynamics of Ice

From a depth of 35–40 m from the entrance, there is a chamber whose bottom is covered by permanent ice. At the left and right of the arch, garlands of ice stalactites up to 3 m length hang down and a rough system of ice stalagmites rises. The chamber's bottom is covered with transparent, bluish ice that is formed from snow arriving there in the winter. Every year as the doline gains snow, it almost fills the entire cave. The area of the ice covering in the cave varies from year to year, sometimes reaching 1000 m^2 and a volume up to 5000 m^3. The age of ice is noted as being up to 50 years (according to 1980 data).

In past years, winters have become warmer, and as the quantity of solid precipitation has decreased, the quantity of ice in the cave has decreased. Ice retreats have opened new pits which were covered by ice earlier, and ice columns have melted by various amounts.

A. A. Kruber (1915) wrote that the degree of cave glaciation is different in different years. In 1908 and 1910 only the lower part of the chamber was covered by ice (about a fifth of its area). In 1911 after a severe winter, two-thirds of the slopes were covered by snow and ice.

26.2.2 TRYEKHGLAZKA CAVE (LEDOVAYA, VORONTSOVSKAYA CAVE)

The cave is located 700 m from the top station of a ropeway on Ah-Petri at an elevation of about 1200 m a.s.l. (entrance coordinates: 44.453 N; 34.054 E. MAT = 5.7°C, $T_{Jan} = -3.6$°C).

Penetration into this tourist cave is possible through one of three entrances—large apertures located near each other and reminding three eyes or eye-sockets. The Tryekhglazka Shaft begins by a 30 m doline that opens into the dome of the first big chamber. The second entrance represents a double shaft. It consists of two pits divided by an enormous rock block. The second shaft leads to a wall in the chamber of the first doline. The cave's total area is about 620 $м^2$, and its total volume is 3200 m^3. At the cave's bottom, there is a huge snowdrift in the form of a cone. Sometimes this snow mass does not melt, even during the hottest summer days.

Local residents have known about the cave for a long time, but no scientific studies have been performed so far.

One sight in the cave is a snowdrift towering in the center with a height of 6–7 m, which was created by snow blowing from the surface. There are seasonal ice stalactites and ice stalagmites. Meltwater from snow spreads along the chamber floor and freezes, forming a feature similar to a skating rink. Some years this area has reached more than 200 m^2. In the east wall of the chamber is a hole, which is the entrance into a small pit completely filled by ice.

26.3 CAUCASUS

A set of high mountainous karst massifs surround the Caucasus's perimeter. Among them, the most significant ones are Fisht, Arabika, Bzybskij, Hipstinskij, Rachinskij, Khvamli, Okhachkue, and Shakh-Dag. All karst massifs were constructed from Jurassic or Cretaceous limestone and have elevations ranging from 2000 to 4200 m a.s.l. Heavy snow falls in the mountains and stays for a long time (up to half a year or more). Strong winds promote the redistribution of snow on the surface of massifs, providing snow accumulation in pits and in the entrances to inclined descending caves.

Vertical cavities and complex cave systems receive the biggest distribution of snow on the Caucasus. In them the main quantity of snow collects. Inclined descending cavities with ice are less typical in this area.

On the Northern Caucasus ice is known to be in a cave, that is, the Kholodil'nik (Ostapenko, 1991) in the Tryu-51. The cave is located on the Tryu-Yatyrgvarta karst massif between the Small Laba River and its left inflow Urushten, Mostovsky district, Krasnodarskij Land (icing with length of about 100m and thickness of about 2–3m).

There are ice accumulations in caves on the Fisht-Oshten-Lagonaki karst massif: Ledyanaya Cave (Fisht, elevation of entrance about 2167m a.s.l.), the Dneprovskaya Cave with a depth of about 130m (Lagonaki, the Stone Sea), the Paryashaya Ptitsa (Fisht), and so on.

On the Arabika massif, a significant quantity of SIFs are noted in the upper parts of the cavities: Martel (elevation about 2316m a.s.l.), Vakhushti Bagrationi (elevation about 2125m a.s.l.), Belaya Loshad' (elevation about 2350m a.s.l.), Ahtiarsktya (elevation about 2520m a.s.l.), Lednik (elevation about 2285m a.s.l.), Ledyanoj Drakon (elevation about 2290m a.s.l.), Dykhanie Korolevy, and many others.

On the Bzybskij karst massif, approximately 50% of all karst pits contain permanent SIF (Dmitriev and Chujkov, 1982).

On the Hipstinskij massif, there are many pits and caves with permanent SIF (Shaft Glukhaya, elevation about 1750m a.s.l.). The best known and biggest accumulation of SIF is located in the Snezhnaya-Mezhennogo-Illusiya-Banka (entrance Snezhnaya) cave system.

The first rather comprehensive study of the scientific characteristics of caves in some areas of Georgia (e.g., Racha and Imeretita) was published in "Description of Georgian Kingdom" by Vakhushti (1941). Episodic research of caves in high mountains of the Caucasus and thus caves with SIF began at the beginning of the 20th century. One of the first to visit the Arabika massif was the celebrated French speleologist Martel in 1903 (Martel, 1909). A. A. Kruber also worked on the Arabika massif (1912). It is especially important to note N. A. Gvozdetskij's monographic research (1954), in which he studied karst in different areas of the Caucasus. Regular research of caves on the southern slope of the Caucasus began in 1958 when the Institute of Geography Vakhusty in Tbilisi organized the Laboratory of Karstology and Speleology. In particular in 1961 and 1962, the laboratory organized expeditions on the Arabika massif where cavities with large quantities of SIF as shaft features were found. Martel and Vakhushti Bagrationi also opened and investigated this area (Kiknadze, 1972; Tintilozov, 1976). Amateur speleology started to develop at this time, and research of the caves in the Caucasus began on a grandiose scale, but SIF studies in caves were not yet being done. Since 1979 regular research of SIF in the Snezhnaya cave system, which contains the most SIF among the caves in the former USSR, have begun. Research was conducted by the Institute of Geography, USSR Academy of Sciences until 1982 (Mavlyudov, 2008). However, only single visits were made (Mavlyudov, 2014). In 1981 the Tryu-51 cave on the Tryu-Jatyrgvarta karst massif in Northern Caucasus, which contains icing with a length of about 100m and thicknesses ranging from 2 to 3m, was opened and investigated. Research on a cave's climate and ice dynamics were carried out by Plotkin L. A. (data are not published) (Lozovoj, 1984). In Kholodil'nik Cave from 1987 to 1990, Ostapenko A. A. conducted research on a microclimate at icing (Ostapenko, 1991).

Sedimentary and metamorphic ice prevails in vertical caves in the Caucasus, but other ice types appear less often. In inclined descending caves, which occur in the Caucasus much less frequently, congelation ice (ice of water reservoirs and icings) prevails, but sedimentary ice is almost absent.

The main vertical cavities with permanent SIF are situated in the range of heights limited by the forest level (from 1700 to 1800m a.s.l.) and the snow line (on a southern slope of Caucasus at 2700–2800m a.s.l.). The lower boundary is determined by the snow concentration in cavities, where it is practically absent; and the upper boundary is determined by the pits' snow line uplift to the surface, which means that all existing pits with wide entrances will be completely filled with snow; that is, they will be preserved.

The quantity of snow accumulation has been estimated only for the Snezhnaya branch of Snezhnaya-Mezhennogo-Illuziya-Banka, where it reached 3000 m^3 (or 300 g/m^2 water equivalent) in reference to the area of the pit entrance (the index of snow concentration is about 10, whereas for other cavities of this area, it is estimated as 1.4). Thus the snow thickness accumulated at different depths in Snezhnaya from year to year. On other places at the bottom of an entrance pit, it can reach 30 m/year (Mavlyudov, 2008, 2014).

In the Snezhnaya branch, snow accumulation is cyclic (the period being about 35–40 years) and is connected with both the external climate and the cavity morphology.

In general, pits with snow in the Caucasus struggle with two conditions: climate warming and changes in the quantity of solid precipitation. Climate warming over the past several decades has lowered the snow line in pits with SIF, but in multi-snow winters, these snow losses have been compensated for and sometimes exceeded.

For inclined descending caves, the past several years have tended to reduce the amount of glaciation and decrease the quantity of ice. For example, in Sakenule Cave (close to the Nikortsminda settlement, on the slopes of Rachinskij Ridge) glaciation that existed in 1960s has now disappeared.

26.3.1 LEDYANAYA CAVE

The cave (investigated by Lazovoj S.P. in 1977) is situated on a southern massif on Fisht mountain (Western Caucasus), Krasnodar Land, in a zone where most cavities are concentrated at an elevation of 2167 m a.s.l. The cave originated in Jurrasic limestone. MAT = −1.8°C.

The cave is about 50 m in length. The entrance has a vaulted aperture at the base of a rocky ledge that closes an upper part of a valley and that is directed abruptly downward. The height of the rocky wall reaches 15 m and the entrance height without snow is about 4 m. Further on, the cave leads to a corridor with an abruptly inclined snow floor. At the snow base, there is a flat area, the bottom of a periodically appearing lake and the cave branches behind the lake. The main gallery is directed to the west and the other branch to the northwest. The lake is situated almost in the middle part of the cave. The chamber with the lake is up to 8 m in height and has a width of 6 m. Five meters from the entrance, an open pit in the cave's ceiling connects with the ground's surface.

Even in the summer, snow fills the entire bottom of the valley before the cave's entrance and the entrance to the gallery. Cave SIF are presented by snow, cover icings on the walls and floor, lake ice, ice stalactites, stalagmites, and columns. Until the autumn snow, only icings, ice, and fragments of lake ice remain.

Samples of cave ice (on July 25, 1977) showed mineralization: sublimation ice, 32.6 mg/l; icings on walls, 58.3 mg/l; cover ice, 63.7 mg/l; ice stalactites, 44.9 mg/l; ice stalagmites, 70.4 mg/l (ice composition, hydrocarbonate-calcareous).

26.3.2 SNEZHNAYA-MEZHENNOGO-ILLUZIYA-BANKA CAVE SYSTEM

26.3.2.1 Site and Situation Around the Cave

The Snezhnaya-Mezhennogo-Illuziya-Banka cave system is located on a southern slope of the Razdelny ridge (a spur of the Bzybsky Ridge) within the Hipstinskij karst massif (Republic of Abkhazia). The cave has four entrances located at different elevations. The most interesting entrance is the one into the Snezhnaya branch because there are accumulations of permanent SIF. This is located at an elevation of about 1900 m a.s.l. within a subalpine zone on a slope oriented to the south. Some trees (birch and juniper) are found only on slopes of a doline entrance. MAT = 2°C, T_{Jan} = −5°C (Photo 26.1).

PHOTO 26.1

Examples of ice accumulations in Caucasus caves. Snezhnaya Cave System. A – entrance shaft at the beginning of winter, B – snow-firn-ice cone in Bol'shoj Chamber, C and D – in Gvozdetskij Chamber (photo of A. Shuvalov).

26.3.2.2 The Cavity Form, Cavity Maps

The total length of the cave system is greater than 31 km, and the depth is about 1780 m. The part of the cave system that is subject to glaciation is a partially isolated branch that begins at an entrance pit with a depth of about 40 m. Snow at the bottom of pit does not melt completely. Further on, there is a system of steeply inclined slots with widths ranging from 2 to 4 m (height more than 15 m) that are partially filled with snow and ice. At a depth of about 100 m, there is an expansion (10 m × 15 m) that was named the Ice Chamber of Gvozdetskij. In the chamber along with snow there are ice stalagmites, stalactites, and cover icings. The system of pits ends at the top part of the Bolshoj Chamber (width 100 m, length 140 m, height about 60 m), most of which is occupied by a snow-ice cone (Fig. 4.1.9).

26.3.2.3 Research History

The cave system (namely, the Snezhnaya branch) was opened in 1971 to members of the speleoclub at Moscow State University who went down to a depth of about 300 m. Almost all further expeditions sought to go further into the cave, but research of cave glaciation did not occur. B. R. Mavlyudov first went to the cave in 1971 and studied the SIF in Snezhnaya. However, direct research of SIF in the cave began in 1979 under the aegis of the Institute of Geography, USSR Academy of Scientists. At that time, the dynamics of ice and features of cave's climate were studied. Research in the cave happened regularly until 1982, after which only irregular visits were undertaken. The dynamics of a snow-ice cone were studied by A. Degtyaryov, but the data were not published.

26.3.2.4 Forms and Dynamics of Ice

SIF in Snezhnaya are presented mainly by snow, firn, and metamorphic ice in galleries and cones, but congelation ice is represented by icings formed in the northwestern and southern parts of the Bol'shoj Chamber, where there are water inflows in the cavity. In places where water drops fall on the cone's surface, small pits or stalagmites grow. The maximum pit depth on the slope of the snow-ice cone reached 23 m between 1970 and 1980. The cone has subsequently been filled by snow and continues to grow.

The SIF volume in the cone comprised about 50,000 m^3 in 1980 and 90,000 m^3 in 2000. The total SIF volume in the cavity changed from 60,000 to 100,000 m^3 during that time (Mavlyudov, 2014, 2016). The air temperature in the Bol'shoj Chamber changed from −1.5°C in the winter to 0°C in the summer. This cave branch has a constantly descending draught of air during the year.

In the winter the snow on the cone usually cannot directly penetrate, but it does penetrate into portions of the chamber in the form of avalanches when a lot of fresh snow (sedimented and blowing) collects at a pit entrance. Avalanches occur in the Bol'shoj Chamber until in the southern end of an entrance pit a snow plug forms.

The height of the snow-ice cone in the Bol'shoj Chamber was about 30 m from 1970 to 1980. Then in 1990 the cone's height began to increase, although it has not blocked an entrance aperture into the Bol'shoj Chamber. As a result, snow in the Bol'shoj Chamber has ceased to penetrate. After, at first snow filled the galleries between an entrance pit and the Bol'shoj Chamber and then also the entrance pit. As a result, this cave entrance was inaccessible for a number of years. During this time, the cone did not receive new snow. The chamber was not cooled by winter air, which led to melting in the dome and a decrease in size. In the late 1990s, the chamber had a positive ice mass balance, which then became negative when snow ceased to enter the cave. When the entrance opened again, the ice mass balance became close to zero. We estimated the ice mass balance of the cave for 1985 and 1986, as shown in Table 26.1.

Table 26.1 Ice mass Balance in Snezhnaya Branch (Caucasus) in 1985 and 1986

Ice Mass Balance Components	Ice Volume, m^3 w.e.
SIF accumulation	3000
SIF melting	−2900
SIF evaporation	−25
The sum:	75

Specific ice mass balance, g/sm²: 0.8.

As we can see, the ice mass balance's receipts and debits are approximately equal. If a year has more multiple snowfalls than expected, then a lot of snow will penetrate into the cave (as was the case in 1967 and 1968, 1986 and 1987, and 2000 and 2001), and the ice mass balance will be sharply positive, because ice melting in the lower part of a glaciation zone is less intensive than at the bottom of an entrance pit. If snow accumulates only in a vertical part of the cave (in a winter with little snow), the cave's ice mass balance will be negative. Apparently, it is necessary to look at the ice mass balance in separate places in a cavity. For example, the ice mass balance of the Bol'shoj Chamber (depth is about 200 m) throughout a number of years (from 1971 to 1985 and after 2001) was close to zero, or negative, and only in the winter of 1986 and 1987 was it positive. There the age of ice exceeds 500 years. The ice mass balance during the entire time of observation (since 1971) was negative in the part of the cave where the depth was –40 m to 140 m from the surface. The ice mass balance of the entrance pit varies from year to year. Its age, which is certified by the age and thickness of permanent snow at the bottom of the pit, does not exceed 4–7 years. Thus the ice mass balance of the lower parts of a cave is more morphologically dependent than the upper parts of a cave. This means that the influence of external climate changes on SIF accumulation in deep parts of a cave can be blocked by a snow dam in hypsometrically higher places in the cave.

Snow can penetrate the Bol'shoj Chamber two ways: by falling directly through the lower entrance, as has been observed some years, or by entering through a higher entrance. The second case is possible only when the first entrance is completely blocked by snow and ice. During the period of research, the system for the receipt of snow in the cave repeatedly varied, and eventually a situation adverse to snow accumulation in the Bol'shoj Chamber developed. This can occur two ways: (1) growth of a snow-ice cone upward can lead to corking the system of snow penetration "from below," and (2) there can be a corking by snow in the galleries from above. In 2000 the top of a snow-ice cone reached the arch of the Bol'shoj Chamber, which speaks to the first scenario. In the summer of 2006, the entrance pit in Snezhnaya was completely blocked by snow and was not opened during two summer seasons. In 2008 a route from an entrance pit was opened, and since then it has not closed. During the blockage, the height of the cone began to decrease. By 2013 the cone's height had diminished to approximately 15 m from the chamber's roof. Thus the duration of one cycle of snow accumulation in the cave was equal to about 35–40 years. Determining the uniformity of the duration of cycles of SIF accumulation in the Snezhnaya branch will require further observation. Regarding the character of ice stratification on the snow-ice cone, opened through a pit in the ice, historically, there have been longer periods when snow did not penetrate the Bol'shoj Chamber. At that time there was considerable ice melting on the cone, and non-uniformity of melting was distinctly traced in visible outcrops of ice. Thus long-term cycles of ice formation in the upper part of the Snezhnaya branch appear to be connected with the external climate (snow accumulation) and to depend on the cavity's structure.

26.3.3 SHAFT OF MARTEL

The cave entrance is located in the Ortobalagan trough valley (Arabika karst massif, Republic of Abkhazia) and represents a wide pit filled with snow. The entrance's elevation is about 2316 m a.s.l. MAT = 0.1°C, $T_{Jan} = -7.5$°C.

Depth of the shaft is about 135 m, and the length is 250 m. It has a wide entrance, and a snow cone is at the bottom (−80 m) (Fig. 26.5).

The shaft has been known since the beginning of the 20th century. It was investigated in 1960 by Georgian cave explorers. Further research has been connected with only the cave passage.

At the bottom of the first pit in this shaft, there is a snow cone that, in a warm season, reaches a height of 25 m. The bottom part of the cone is transformed into dark blue ice crystals. The mineral composition (hardness) of the ice stalactite at the lower part of the pit is about 8 mg/L.

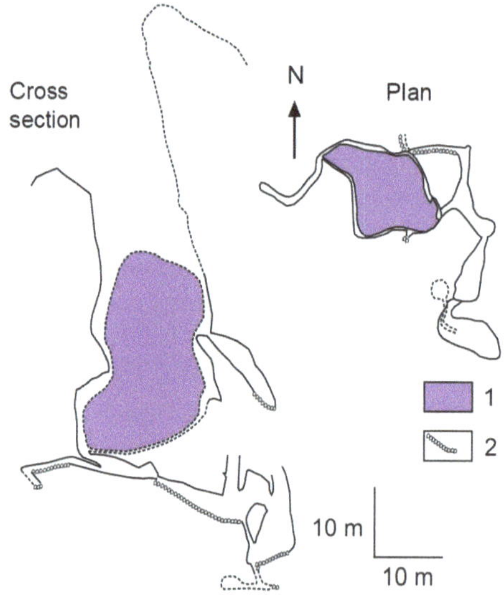

FIG. 26.5

Martel Shaft. Survey by Brukhanov A., Zubareva N., Sheremet S. (Dneprovskij, 2007).

26.3.4 SHAFT OF VAKHUSTI BAGRATIONI

The shaft entrance is located on the right abrupt branch of the left tributary in the Gelgeluk trough valley, 45 m above its bottom (Arabika karst massif located in the area of Shrattovaya Valley, the Gagrskij Ridge, Republic of Abkhazia). Entrance elevation is about 2125 m a.s.l. Entrance coordinates are 43°33′01″ N, 39°50′39″ E. Shaft originates in Jurassic limestone. MAT = 1.3°C, $T_{Jan} = -6.5$°C.

The entrance is an asymmetric collapse doline that leads to an inclined meandering gallery that at the depth of 50 m leads into a 71 m pit (with some ledges). The depth of the shaft is about 250 m, its length is 1967 m, its projective length is 1265 m, its area is greater than 560 m^2, and its volume is more than 4400 m^3 (Fig. 26.6).

Georgian cave explorers found and investigated the cave in 1960. Only the main passage was investigated further.

The snow cone at the shaft entrance is traced to a depth of 46 m; the top part is granular snow, or firn, and the lower part is blue crystallized ice.

The total amount of SIF in the cave is estimated as being 1300 m^3 with an ice volume of about 400 m^3 (Tintilozov et al., 1966). In the Snow Chamber, limestone walls are covered by icing. On the eaves are seasonal ice accumulations and sublimation ice crystals. The mineral composition of the stalagmite ice at a depth of 25 m is 85 mg/L. The dynamics of the ice are unknown.

26.3.5 CAVE SKHVAVA

26.3.5.1 Site and Situation Around the Cave

Skhvava Cave is located in beechen forest on northern slope of the Rachinskij Ridge at an elevation of about 1310 m (Shuashvavi village, Ambrolaurskij region, Georgia). MAT = 5.8°C, $T_{Jan} = -5.3$°C.

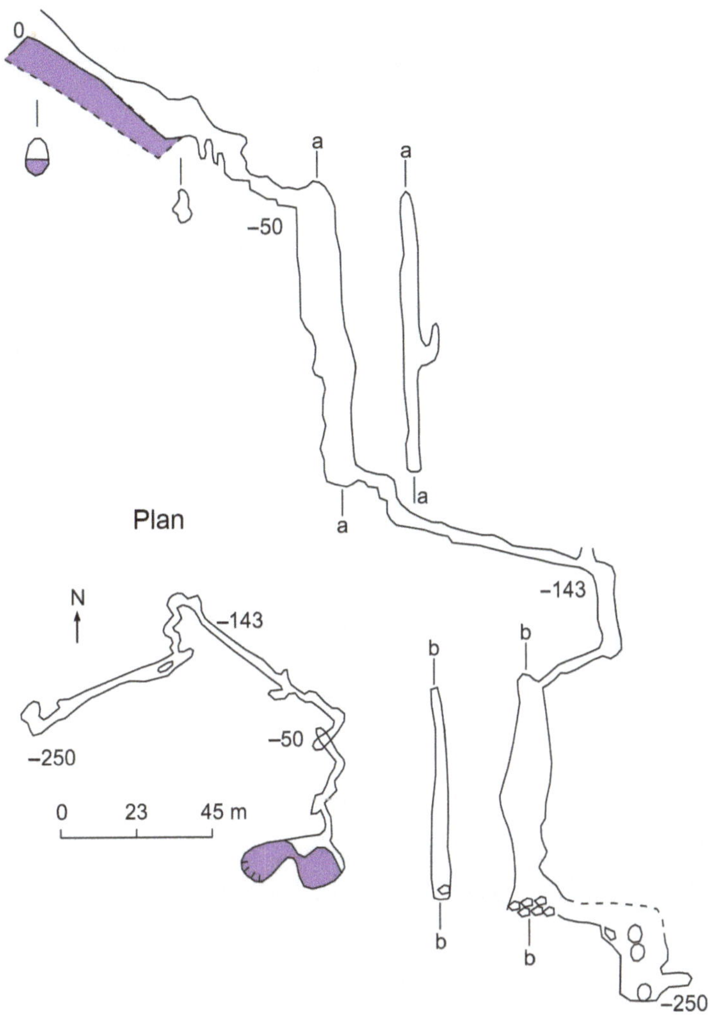

FIG. 26.6

Shaft of Vakhusti Bagrationi. Longitudinal section and plan.

26.3.5.2 The Cavity Form, Cavity Maps

The cave represents a classical descending inclined cavity, with a chamber of variable height, and an average floor inclination of about 40°, which begins at the bottom of a deep collapse doline (Fig. 26.7). Cave floor incline is WWN.

26.3.5.3 Research History

The cave is easily accessible; consequently local residents have known about it for a long time and have used it in the summer for storage of perishable food. The first scientific research in the cave was conducted in 1960 (Pirpilashvili, 1962). We did our research in the cave in 1984 (Mavlyudov, 2008).

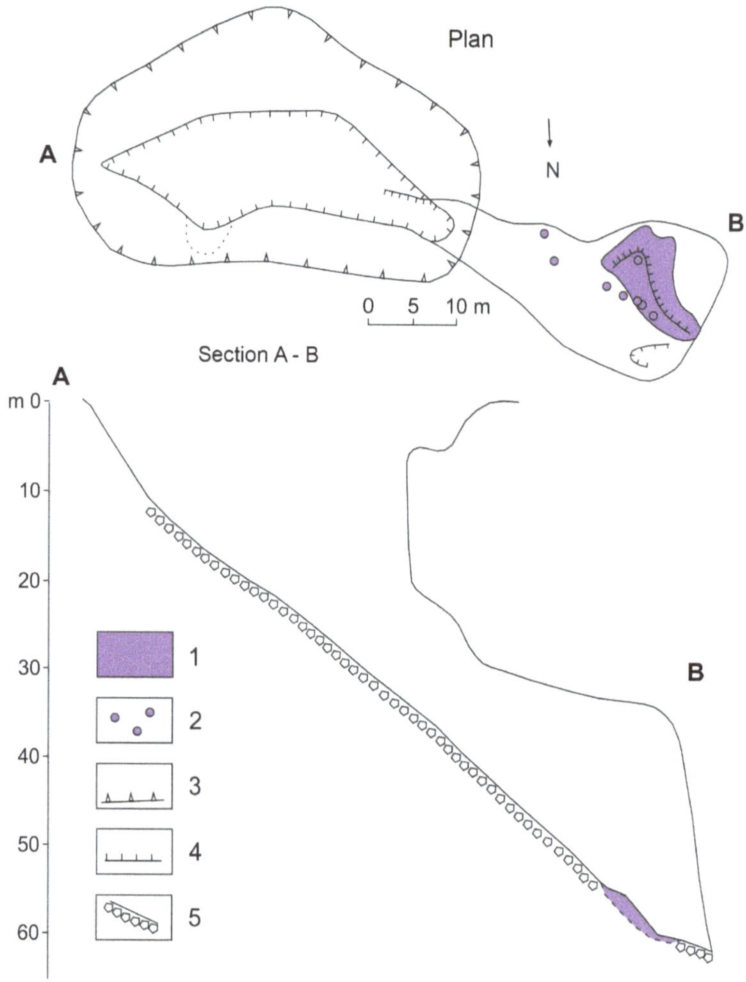

FIG. 26.7

Plan and cross-section of Skhvava Cave (Caucasus): (1) icing, (2) ice stalagmites, (3) boundary of doline, (4) scarps, and (5) talus slope (Mavlyudov, 2008).

26.3.5.4 Forms and Dynamics of Ice

Ice forms: snow masses, metamorphic ice, icings, and seasonal sublimation crystals. Permanent ice is available only at the lower part of cave and has an area about 50 m^2; the ice thickness is unknown. According to observations in the summer of 1984, the intensity of the ice melting in the cave was about 0.5 mm/day in a water equivalent that corresponded to the ice loss at the bottom part of the cave of about 4 m^3/year. According to annual ice layers (10 of them were visible), the average thickness of the annual ice remnant corresponded to approximately 40 cm, which meant that the quantity of freezing ice was about 48 cm/year. Considering that new ice is formed on about half of the area covered by ice, we can estimate the volume of cave ice freezing there as approximately 12 m^3. Thus, in a zone of permanent glaciation, the accumulation of cave ice is three times as great as the loss of ice through melting (ice evaporation does not exceed 0.3 m^3). The specific ice mass balance of this part of the cave is approximately equal to 16 g/cm^2. Common ice mass balance in the cave is difficult to estimate because the volume of seasonal snow and ice accumulation at the end of a talus slope is unknown.

Snow accumulation in the cave is seasonal in nature. The bulk of the snow accumulates in an entrance doline, but only an insignificant portion penetrates into the cave chamber, with the former melting at the end of May, and the latter melting by the end of July (Pirpilashvili, 1962). It is impossible to estimate what part of the meltwater will be transformed into ice. Despite the fact that the common ice mass balance of the cave is positive, apparently, there are years when it is zero or negative, although there are disagreements concerning the interpretation of the ice stratigraphy? Most likely years with an average or a considerable amount of solid precipitation are favorable for ice accumulation in a cave, because at those times more water will penetrate into a cold cavity and turn into ice. Thus in the summer in such a cave, only part of the winter cold reserve is used.

26.3.6 CAVE BOGA

This cave is situated on the Khvamli karst massif, in the south part of Lekhchumskij Ridge, at the watershed of the Rioni and Tskheniskali Rivers, 25 km from Kutaisi, Georgia. Entrance elevation is about 1710 m a.s.l. MAT = 3.6°C, $T_{Jan} = -7.4$°C.

The cave begins at a wide doline.

Local residents have used the cave for a long time as a natural warehouse refrigerator.

At a depth of 25 m in the cave are frozen pools and icings (cover and pendent—stalactites, stalagmites, and ice covers). At the lower part, almost all the steep walls are covered by thin icings.

26.4 ICE CAVES IN THE RUSSIAN PLAIN

Caves with ice on the Russian Plain are located in two places: the Novgorod and Arkhangelsk areas. This does not mean that permanent cave glaciation is impossible in other places on the Russian Plain, although there are no favorable morphological preconditions suitable for this purpose. Nevertheless, in cavities with favorable conditions, glaciation will occur. For example, perennial icing with a thickness of about 1 m has been found in an ancient mine located in the Sernurskij area of the Mari-El Republic.

In the territory of the Nizhniy Novgorod, inclined descending cavities are favorable for glaciation, and in the Arkhangelsk area, horizontal caves with ice prevail.

The study of this area and its caves began in the 18th century. In 1768 P. S. Pallas visited and for the first time described the Bornukovskaya Cave and took measurements of the air temperature. I. I. Lepyokhin (1771, 1772) investigated a karst in the Volga region at the east edge of the Russian Plain. He also visited the basin of the P'yana River.

In 1837 the karst formations at the basin of the rivers of Northern Dvina—the Pinega, Kuloj, and Mezen rivers—were surveyed by A. Shrenk, who described in detail the Kulogorskaya gypsum cave at the Pinega.

V. V. Dokuchayev (1883) gave a detailed description of the karst at the basins of the P'yana and Serezha rivers, including the Bornukovskaya Cave located in gypsum, with a thickness of about 24 m. His report showed that the cave began with a big entrance (height, 4 m; width, 6 m) that at 8–10 m led into a huge chamber with a length ranging from 85 to 95 m, a width of 21 m, and a height ranging from

8 to 10 m. At the end of chamber were two small lakes (depth up to 2 m) in which small fish were found. Dokuchayev also provided data on the air (11°C) and water (6°C) temperature in the cave.

The paper, "About Karstic Phenomena in Russia," by A. A. Kruber (1900) was the first regional report about karst on the Russian Plain.

N. Mazarovich (1912) described karst in valley of the P'yana River, including the Bornukovskaya Cave and two small caves in the Ichalkovskij pine forest.

A. V. Stupishin (1956) analyzed features of karst development and distribution in the Middle Volga region, studied the reconstruction of karst paleo-landscapes, and denoted karst divisions according to districts.

H. P. Torsuev considered conditions of karst formation and established some laws for distribution of a karst in the Onega-Dvina interfluve area and the Kuloj plateau. He defined karst stages of development and provided a division scheme for the karst of these areas.

From 1966 to 1974 Leningrad researchers on the Dvina-Mezens interfluve conducted about 20 speleological expeditions. During this time, they opened 130 karst caves. Later, speleological activities were continued.

In these caves, all forms of cave ice were found, but sedimentary ices were met very seldom (e.g., the Bolshaya Golubinskaya Cave in the Pinega district).

In caves of the Arkhangelsk area, in a few cavities, cave glaciation changes are strong, while in many cases, the glaciation is relatively stable. For example, in caves in the Nizhniy Novgorod area, glaciation degrades, and in the Arkhangelsk area, a reduction in ice formation in caves during the summer stage has been noted.

26.5 ICE CAVES IN THE NIZHNIJ NOVGOROD AREA

In the Nizhniy Novgorod area, caves subject to permanent glaciation are in the Perevoznenskij, Buturlinskij, and Ardatovskij districts. All the caves in this region have gypsum origins; consequently, their morphology can change quickly (within ten years).

The majority of these caves have a bag form, with entrances situated 5–6 m above the bottom. The entrance goes downward to an inclined extending channel that ends in a roundish chamber with a vaulted ceiling. The length of the caves ranges from 15 to 25 m. These caves, as well as the Bornukovskaya Cave (Buturlinskij district in the Nizhniy Novgorod area), are inclined descending caves. In the Balahoninskaya Cave (Ardatovskij district in the Nizhniy Novgorod area), some ice is perennial.

Despite the fact that conditions for cave glaciation are quite favorable in this area, few caves have permanent ice, which mainly is connected with the morphology of the caves. Available bag form cavities can have a great deal of water or an amount that is not sufficient enough to protect SIF from external influences and to preserve them during the summer.

A few caves located in the Nizhniy Novgorod area are subject to glaciation: the Studencheskaya, Kholodnaya, Teplaya, and Anonymous caves located in the Ichalkovskij pine forest on the right coast of the P'yana River (Perevoznenskif district of the Nizhniy Novgorod area) (Russkikh and Mikheev, 1971; Russkikh and Ivanov, 1992). In the summer of 1980, the temperature in the caves ranged from −0.4°C to −0.6°C.

Icings and ices of lakes are perennial mainly in this area. Sublimation ice crystals can be preserved in a number of caves until the end of summer.

It was noticed in 1980 that the air temperature was negative during the summer in caves in the Ichalkovskij pine forest and that, in all caves, some ice remained the entire year. However, in more recent years, the temperature inside the caves has risen during the summer (it can rise up to 3°C and higher). This trend means that, except for the Kholodnaya Cave, glaciation has become seasonal.

26.5.1 KHOLODNAYA CAVE (LEDYANAYA)

26.5.1.1 Site and Situation Around the Cave

This cave is located near a village in Ichalki in the Ichalkovskij pine forest on the right coast of the P'tana River (left tributary of the Sura River at right tributary of Volga River), Perevoznenskij district of the Nizhnij Novgorod area. Entrance coordinates: 55.42770°N, 44.54722°E. MAT=4.2°C, $T_{\mathrm{Jan}}=-11.2$°C.

26.5.1.2 The Cavity Form, Cavity Maps

This inclined descending cave is located on the northern slope of a broad gull. The cave consists of two chambers: inclined chamber with name Svetlyj (Light) and flat chamber with name Temnyj (Dark). The length of the Temnyj Chamber is about 15 m, the width is 12 m, and the height is 8 m. At the bottom of the Temnyj Chamber, there is a lake with cold and transparent water (Fig. 26.8).

26.5.1.3 Research History

The cave has been known from the 15th or 16th century. The ice has not been studied.

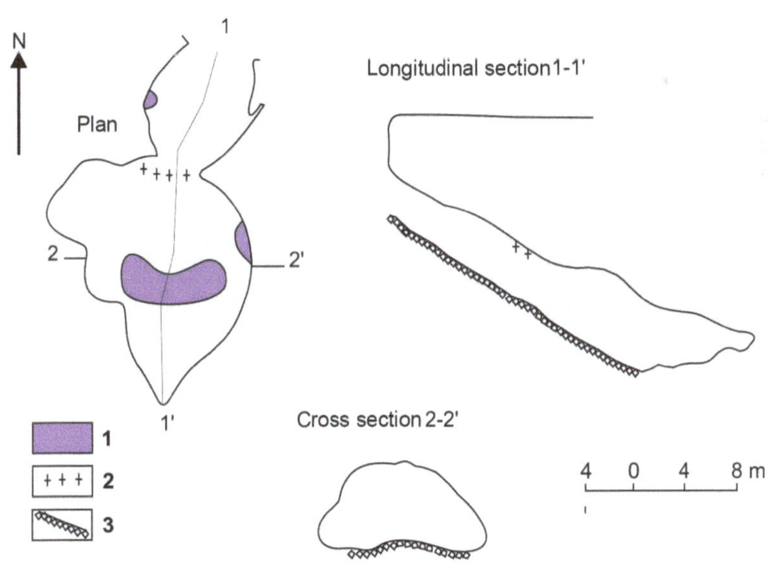

FIG. 26.8

Plan and cross-sections of Kholodnaya Cave (Russkikh and Ivanov, 1992): (1) icing ice, (2) sublimation ice, and (3) talus slopes.

26.5.1.4 Forms and Dynamics of Ice

Congelation (ice of lake, stalactites) and sublimation ice are in the cave. The temperature in the cave during the summer does not exceed 3°C, and ice and hoarfrost in the heart of the cave do not melt throughout almost all of the summer. Even in the summer, the lake is sometimes covered with ice.

Although in the summer of 1980, the air temperature in the caves of the Ichalkovsky pine forest was negative (ranging from −0.40°C to 6°C), now because of the winters' warming air temperatures, the air temperature during the summer months is positive, which has led to a decrease in the amount of permanent ice in the caves. Some years the presence of ice in the caves is seasonal.

26.5.2 BALAHONINSKAYA CAVE

26.5.2.1 Site and Situation Around the Cave

This cave is situated on a slope to the south of the watershed of the Serezha and Tesha rivers, in the area of the former Gipsovyj settlement, 2 km to the south of the Balakhonikha village (Arzamas district of the Nizhniy Novgorod area), at the east side of an old, unused gypsum open-cast mine. MAT=4.2°C, T_{Jan}=−11.2°C.

26.5.2.2 The Cavity Form, Cavity Maps

The cave origin is gypsum. The narrow entrance is in the form of a horizontal crack with a width of 2.5 m and a height of 0.75 m. The entrance leads to a chamber with a length of about 12.5 m, a width of 8.8 m, and a height of 2.8 m that then leads to a narrow gallery (Fig. 26.9). The average length of the accessible part of the cave is about 70 m. The temperature in the cave during the summer ranges from 2°C to 3°C.

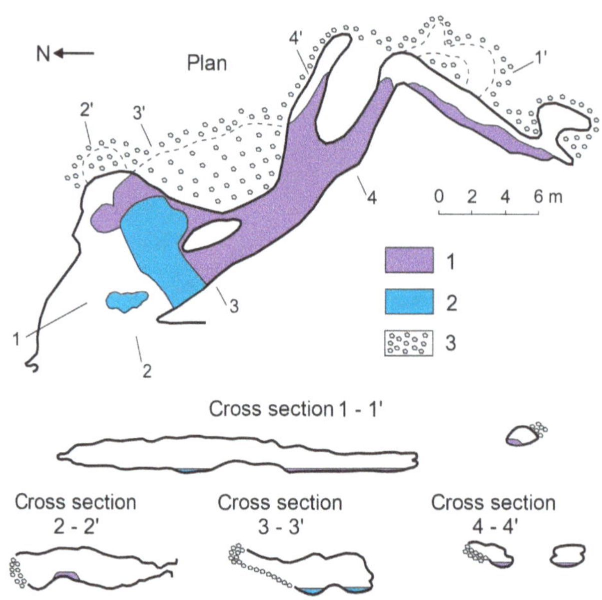

FIG. 26.9

Plan and cross-section of Balahoninskaya Cave (Russkikh and Ivanov, 1992).

26.5.2.3 Forms and Dynamics of Ice

The lake in the cave is derived from very clean water that flows into the cave practically all summer. It is long, cold and covered by ice, thawing little, forming a superficial pool under which ground ice can remain.

26.5.3 BORNUKOVSKAYA CAVE

26.5.3.1 Site and Situation Around the Cave

The cave is located on the right slope of a valley at the southern branch of the P'yana River, 1.5km west-northwest from the Bornukovo village in the Buturlinskij district of the Nizhnij Novgorod area. Entrance coordinates: 55.389849°N, 44.780784°E. MAT=4.2°C, $T_{Jan}=-11.2$°C.

26.5.3.2 The Cavity Form, Cavity Maps

In the summer of 1929, the cave's length was 73m, the width was 37m, and the height was 8.5m. In 1958 an explosion caused the gypsum, open-cast mine entrance to collapse, but it opened the way into two other cavities with lengths of about 200 and 60m.

Now the cave consists of a series of connected grottoes that, from the entrance, spiral downward and to the left. The first chamber is near the entrance, but it is not easy to penetrate. It is necessary to creep through a narrow crack to get to the first small chamber. From there, you go through a second narrow crack to get to the main chamber, which extends from the southwest to the northeast and goes downward under a corner at a 55-degree angle to the horizon. In the upper third of the cave from the main grotto to the right, there is another larger chamber. This chamber has a steep slope of about 60° and a length of about 40m (Fig. 26.10).

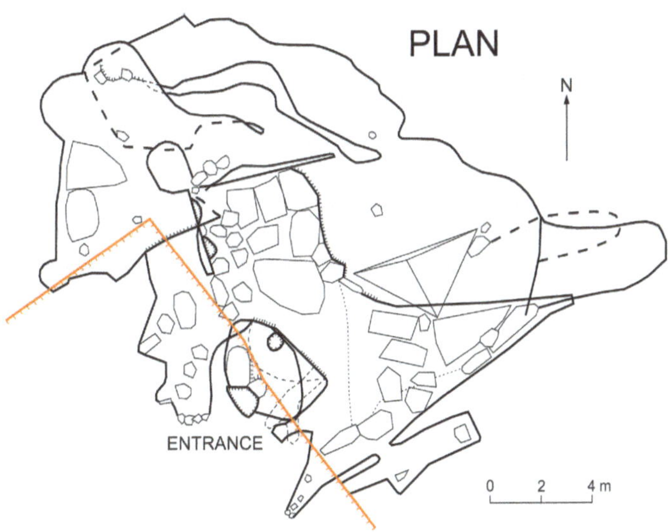

FIG. 26.10

Plan of Bornukovskaya Cave. Brown: boundary of quarry in gypsum. Survey of Lavrov I.A. and Chugaeva A.A., 2003.

26.5.3.3 Research History

P. S. Pallas visited the cave in the 18th century. The cave was also visited by R.-N. Murchison in 1842, Miller in 1875, V. V. Dokuchaev in 1884, A. N. Mazarovich in 1912, and L.V. Dal' (unknown year).

26.5.3.4 Forms and Dynamics of Ice

The dynamics of the cave normally results in icings (cover and pendent) and during the winter, sublimation crystals. In the 18th century, ice in the cave was seasonal, according to the input of water caused by river floods. Now because of the collapse of the arch (in part artificially), the cave is located above the flood level of the P'yana River that had promoted ice preservation in the cave.

26.6 ICE CAVES IN THE ARKHANGELSK AREA

In the Arkhangelsk area, 470 caves are known (in January 2014), among which 267 are situated in protected areas. The majority of the caves are in gypsum-anhydrite layers, from the Sakmarian stage of the lower series of the Permian system, with a thickness of about 50 m. Dolomite and limestone from the Assel stage (Asselian stage of lower series of the Permian system P_{1as}) are the local basis for the karst. Red-colored rocks from the Ufa stage (Ufimian stage of lower series of the Permian system P_{2u}) reside on the surface of massifs in separate places. The cave formation result from the neotectonic raisings and the concentration of powerful glacial water supplied during the epoch of development and degradation in the last glacial period (Shavrina, 2002).

The main caves are horizontal and subhorizontal systems of galleries located at different levels. The largest cave systems with sulfidic sediments are not only in the reserve area but also throughout Russia, for example, the Kulogorskaya-Troya Cave (length 17.5 km).

Most of the Pinezhskij reserve is a complex combination of karst landscapes, and for the most, the reserve is covered by forests (87%), while bogs cover about 10% of the area, and the small areas by meadows, bushes, and light forests. Forest vegetation comprises fir, pine, larch, and birch trees.

Monitoring observations were conducted in the Pevcheskaya Estrada (Г-1) and Bolshaya Golubinskaya caves for 20–25 years by the employees of the Pinezhskij reserve (E. V. Shavrina, pers. comm.) and include measurements of cave air temperatures and qualitative descriptions of the ice conditions in the caves. An analysis of the long-term observations in the Г-1 Cave showed that (1) winter air temperatures in the cave remained almost constant; (2) the maximal cave air temperatures showed a negative trend connected with decreasing summer temperatures (Fig. 26.11). However, after 2006 the maximum cave air temperatures began to increase, which has led to a growing difference between outside temperatures and cave temperatures of up to 0.3°C and 2°C in mean years; (3) the difference in amplitudes of maximum and minimum air temperatures in caves in various places has increased since 2000 by 0.5–8°C (Fig. 26.11); (4) the MAT of permanent glaciation zones decreases by 0.3–2.3°C, and for seasonal glaciation zones, it increases by 0.2°C. There are steadily decreasing trends of maximum and minimum temperatures for zones with permanent glaciation and of the stability or small growth in zones of seasonal glaciation.

Data about karst phenomena, including caves in the Arkhangelsk area, are the result of works by A. I. Shrenk (1855), R. A. Samojlovich (1909), M. B. Edemskij (1931), J. D. Zekkel (1934), Ya. T. Bogachyov (1934), N. F. Tessman (1958), V. P. Torsuev (1965), and others. In 1960 the geographer-karstologist A. G. Chikishev with students from Moscow University produced a schematic plan and description of the Kulogorskaya-1 and Kulogorskaya-2 caves. Since 1966 the Leningrad Section of Speleology (LSS,

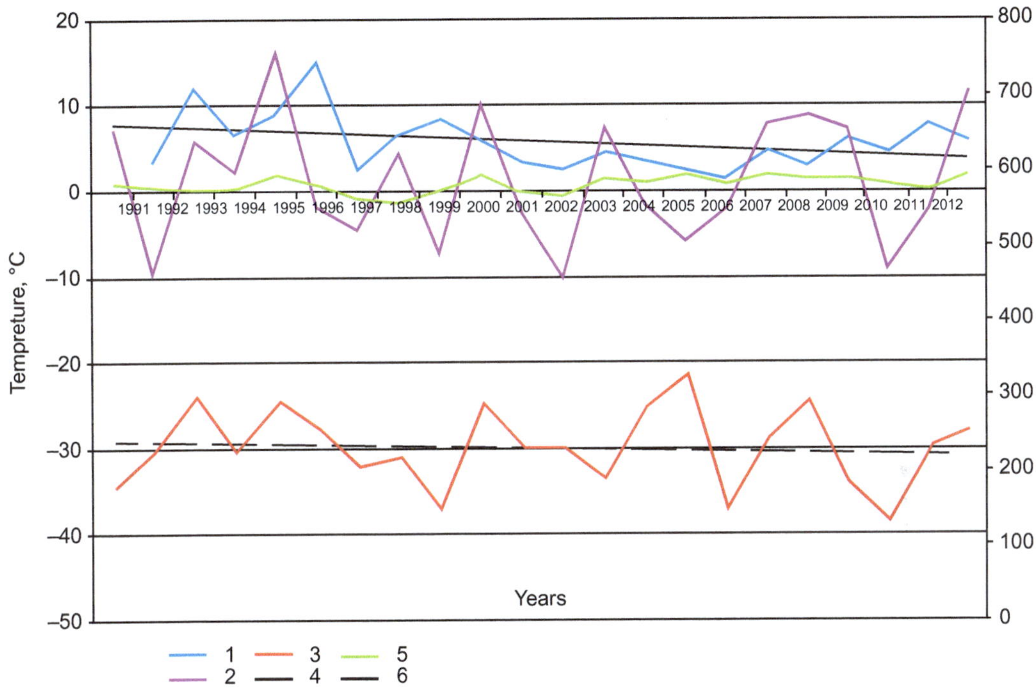

FIG. 26.11

Variations of MAT and precipitation on surfaces and maximum and minimum temperatures in NTA zone in cave Г-1 from 1991 to 2011 (Shavrina, 2012): (1) maximum air temperatures in NTA zone, (2) yearly precipitation on the surface, (3) minimum temperatures in NTA zone, (4) linear trend of maximum temperatures in NTA zone, (5) MAT on the surface, and (6) linear trend of MAT on the surface.

chief, V. M. Golod) has conducted systematic expeditions in the region. Their work included searches, topographic surveys, descriptions of caves, and scientific investigations (including the caves' climate). In 1970 some thermometric observations were done by the Karstic Group, PGO "Arkhangelskgeology." In 1980 these works were continued by cave explorers of the Arkhangelsk City Speleological Section.

The MAAT on surfaces closely to caves is equal −0.5°C, the low temperatures in the cavities (MAT = 2°C in neutral zone of caves), and the penetration of considerable volumes of water (in liquid and vaporous phases) in the caves lead to wide development of OTA zones and seasonal and permanent SIF. In caves, congelation ice (icings, cover and pendent; ice in streams; and lakes) prevail. Permanent icings exist in 25% of the caves. Sedimentary and sublimation ice is mainly seasonal.

Typical forms of ice are icings, stalactites, stalagmites, stalagnates, ice of lakes and streams, and sublimation crystals. In caves in this area, there are original forms—screens that develop in a zone of water spraying in waterfalls that create freezing ice covers in vertical streams. The mineralization of ices in caves can vary within three orders of magnitude (Fig. 26.12).

For seasonal ice, three cycles of development are allocated. The prewinter cycle is characterized by the development of ice crystals, stalactites, stalagmites, stalagnates, icings, crusts, and ice on lakes and streams. The prespring cycle typically has formations of icings and growth of sublimation crystals. In the summer cycle (after flood), ice forms from freezing of rehumidified friable sediments, ice crystals

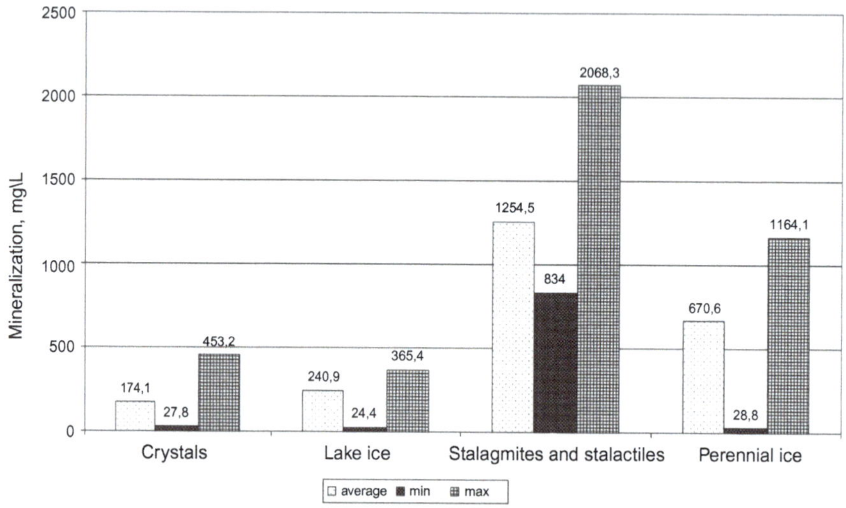

FIG. 26.12

Mineralization of cave ice in Pinega area (Shavrina, 2012).

grow, and ice flowstones develop. At the beginning of the 21st century, development of ice became an exception rather than the rule during the summer cycle.

Because of the overall reduction of seasonal ice, zones of SIF development are displaced. This new reality is connected with the migration of water streams and with changes in air temperatures on the Earth's surface and in caves. At the beginning of January 2006, ice melting on the lake in Bolshaya Golubinskaya Cave was observed; by the end of January, the lake had frozen again. In 2007 this lake was frozen only in January. In the winter of 2008–09, for the first time, the lake was not frozen, which was connected with long, flooding rains (until the end of December) that preserved high water temperatures (up to 2.5°C) until the end of February. In the colder winter of 2009–10, the lake was frozen in January.

Now in Pinega, caves occur with decreasing per-annual ice volumes. Also, a shift in the development time of seasonal glaciation occurred in the mid-1990s. Because of the short-term temperature growth from rain flooding into caves, perennial ice melting has increased. Seasonal ice formation now begins one to two months later, at the time of the long autumn flooding of underground streams.

During the 21st century, the only cave with a steady growth in the volume of ice was Ledyanaya Volna Cave, which happened because the cave icing was above the stream of water. In 2004 the cave entrance was closed because of a large collapse. In Yubilijnata Cave (C-26) from 1960 to 2005, the volume of permanent icing decreased twice. According to radio carbon analysis of remnants of wood, this ice is more than 200 years old. In the summer of 2009, icing in cave Г-1, which was observed in 1960, completely melted.

26.6.1 KULOGORSKAYA CAVE SYSTEM (CAVE KULOGORSKAYA-TROYA)

26.6.1.1 Site and Situation Around the Cave

This cave system is situated in the Pinezhskij district of the Arkhangelsk area at the watershed of the Pinega and Kuloj rivers at the base of the Kulogorskij ledge. The entrance elevation is about 20 m a.s.l. MAT = -0.5°C, T_{Jan} = −13°C.

The karst rock is gypsum layered with dolomite, forming homogeneous layers with thicknesses ranging from 0.2 to 7 m, which are covered with a layer of strongly fissured dolomite and limestone with a thickness of about 3 m (Malkov et al., 2001).

26.6.1.2 The Cavity Form, Cavity Maps

Kulogorskaya cave system has a complex morphological structure as in plan (line, square, and maze style) and is arranged in longitudinal sections (there are three levels) (Fig. 26.13).

The cave system comprises big Kulogorskaya-13 (Troya) and other small cavities, Kulogorskaya-1, Kulogorskaya-2, and Kulogorskaya-3. Not all the connected caves are accessible to people, but they all are connected in a united transport system of water and air.

At high water, the lower part of the cave system is flooded (water uplifting up to 1.5 m).

26.6.1.3 History of Research of a Cave

Kulogorskaya Cave was first described by A. I. Shrenk in 1837. Brief descriptions of this cave were presented by R. A. Samojlovich (1909), M. B. Edemskij (1931), N. V. Tessman (1958), A. G. Chikishev, and V. M. Andreev (1960). In 1962 A. G. Chikishev developed a schematic plan and description of the cave.

Since 1966 the Kulogorsraya-1 (K-1) and Kulogorsraya-2 (K-2) caves have been studied by members of the Leningrad Section of Speleology (headed by O. V. Buzunov and M. P. Golod). Since 1974 cave work has been conducted by the geologists of the Karstic Group of the Arkhangelsk Geological Service (headed by Yu. I. Nikolaev, V. N. Malkov, and V. A. Kuznetsova) and cave explorers of the Arkhangelsk Regional Section of Speleology. They performed topographic surveys and hydrogeological and microclimatic observations.

In 1980 Arkhangelsk cave explorers dug out an entrance to the Kulogorskaya-13 cave (K-13, Troya). In 1987 the total length of the cave system exceeded 14 km. In 1991 a vertical pit at a depth of 21 m was excavated and it became an artificial upper entrance into the cave. The temperature inside the cave was 2.5°C (Tokarev et al., 2015).

The ice in the cave was not studied.

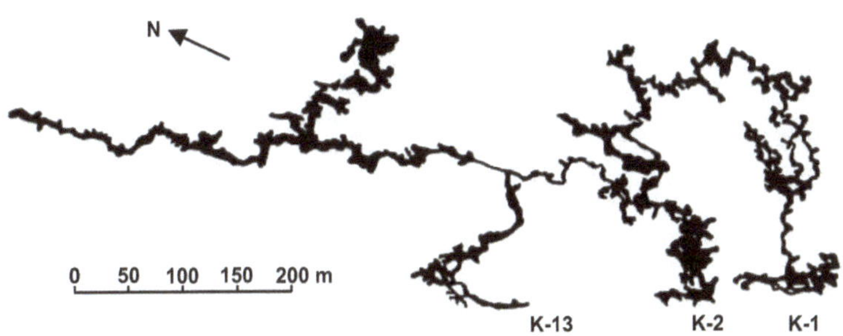

FIG. 26.13

Plan of Kulogorskaya cave system (Troya): (K-1) Kulogorskaya-1 Cave, (K-2) Kulogorskaya-2 Cave, and (K-13) Kulogorskaya-13 Cave.

26.6.1.4 Forms and Dynamics of Ice

Ice collects in the winter at the lower entrances in the Kulogorskaya cave system, but ice melts almost completely at the time of the spring floods. The greatest quantity of permanent ice is in Kulogorsraya-1 Cave, whose galleries range from 0.7 to 1.5 m in height. This cave is affected less by floods than the other caves in the system because it is the uppermost cave in the system and therefore during floods is not always completely filled with water. This state of affairs promotes the formation of an extensive NTA zone, where there are negative air temperatures all year and SIF develop. Thus stalactites, stalagmites, ice on the boundary of water streams, icings, and sublimation crystals in separate places remain in the cave all year around.

26.6.2 GOLUBINSKIJ PROVAL CAVE {13}

26.6.2.1 Site and Situation Around the Cave

This cave is located on the right coast of the Pinega River in the area of the Golubinskij geological preserve, approximately 17 km from the Pinega settlement. The entrance into the cave is situated at the mouth of a broad gull (ravine), Taraskan'ya Shel'ya. An abrupt descent into the cave begins at a small ledge on the side of it ravine; it is equipped with a wooden ladder. Cave origin is a gypsum layer from the Sakmarian stage of series of the Permian system. There are anisomerous white and gray gypsum with prolayers of blue anhydrite and varicolored clays with selenite proveins. The entrance elevation is about 34 m. MAT = −0.5°C, T_{Jan} = −13.0°C.

26.6.2.2 The Cavity Form, Cavity Maps

The cave length is about 1622 m, the area is 5267 m^2, the volume is 8255 m^3, with a vertical amplitude of 17 m. A submeridional system of tectonic cracks controls the orientation of the main galleries. The cave has three levels with a height difference ranging from 2.5 to 4 m. The main gallery is developed on the second level, and partly on the first level. Lateral galleries are on the second and third levels. The cave plan is presented in Fig. 26.14.

Between an entrance grotto and the Forum Chamber, galleries cross each other on an orthogonal grid. Here, there is a NTA zone with permanent glaciation. The Forum Chamber is elongated in a sublatitudinal direction, section arch. The chamber's length is about 24 m, its width is 6–8 m, and its height is 3–5 m.

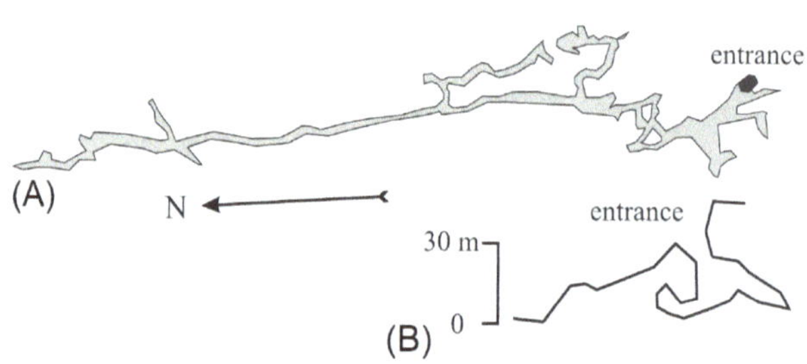

FIG. 26.14

Plan (A) and longitudinal section of entrance area (B) of Golubinskij Proval Cave.

The length of the main gallery is about 500 m, and it connects with the Forum and Kruglyj chambers. The gallery has flat-oval and rhombic cross-sections. Most of the cave is dry. There are only two short streams: northern and southern.

26.6.2.3 Research History

The cave was opened in 1964 by Leningrad cave explorers. In 2003 it was equipped for excursions (up to 200 persons per day).

26.6.2.4 Forms and Dynamics of Ice

All year in the NTA zone, sublimation crystals and icings (cover and pendent) are observed, and in the lower part of the entrance, doline sedimentary ice (firn) has accumulated. Icing fragments were opened at two points in the cave's lateral contour. An outcrop of friable sediments have accumulated in the cave with a thickness of about 6 m and an age range of about 10.2 thousand years BP to 7.8 thousand years BP. Two intervals with cryogenic concretions of radially radiant gypsum ("gypsum hedgehogs") that mark previous periods of cave glaciation have been found (Photo 26.2).

26.7 ICE CAVES IN THE PRIURALIE AND URAL

The Urals cover a large area that lies in four climatic zones and has a complex geological structure. In Fig. 26.15, the eastern and western boundaries of the Urals are defined according to Chibilyov (2011) and Shakirov (2011). The Urals region is divided into nine geographic areas, as shown in Fig. 26.15. The division into different regions (Chibilyov, 2011) is based on geology, geomorphology, structure, and tectonics.

The climate varies from tundra to semi-arid conditions. The highest annual precipitation, about 1200–1400 mm, is in the Cis-Polar Urals, whereas in the Northern and Southern Ural, it is about 500–600 mm. In the Mughodzhary, the increasing influence of the continental climate has resulted in an annual precipitation reduction of 300–450 mm.

Virtually all types of surface and underground karst features are present in the Urals. Rock formations directed longitudinally are favored for comparative analysis of processes in karst landscapes at different latitudes. The most intense karstification occurs in Paleozoic sediments. In the eastern margins of the East European Craton and in adjacent parts of the Ural foredeep, sulfate rocks (gypsum and anhydrite) interstratified with thin limestone and dolomite layers of the Irenskaya Unit are karstified, and to a lesser extent, so are limestones and dolomites of the Filippovsky Unit of the Kungurian Stage and limestones of the Artinskian Stage of the Lower Permian. Salt-containing and sulfate sediments occur mainly in the Ural foredeep. The folded zones of the western Urals and the central Ural uplift are characterized by karst development in the Devonian, Carboniferous, and Permian carbonate strata with a total thickness of about 2000 m. The most intensely karstified area is the western slope of the Southern Ural.

At this point in time, more than 3200 caves with a total length of about 244 km are documented in the Urals. Among ice caves developed in carbonate rocks, the most representative ones are the Yanganape caves (length 30–36 m), Ledyanaya near the Pechora River (160 m), Medeo (60 m), Eranka (500 m), Ledyanaya-D'yavolskij kurgan near the Sosva River (176 m), Mariinskaya (1000 m), Askinskaya (230 m), Kinderlinskaya (9113 m), Kutukskaya-1 (520 m), and Ylasyn (487 m). Ice caves

PHOTO 26.2

Ice formations in caves in Pinega area. Photos by V. Maltsev.

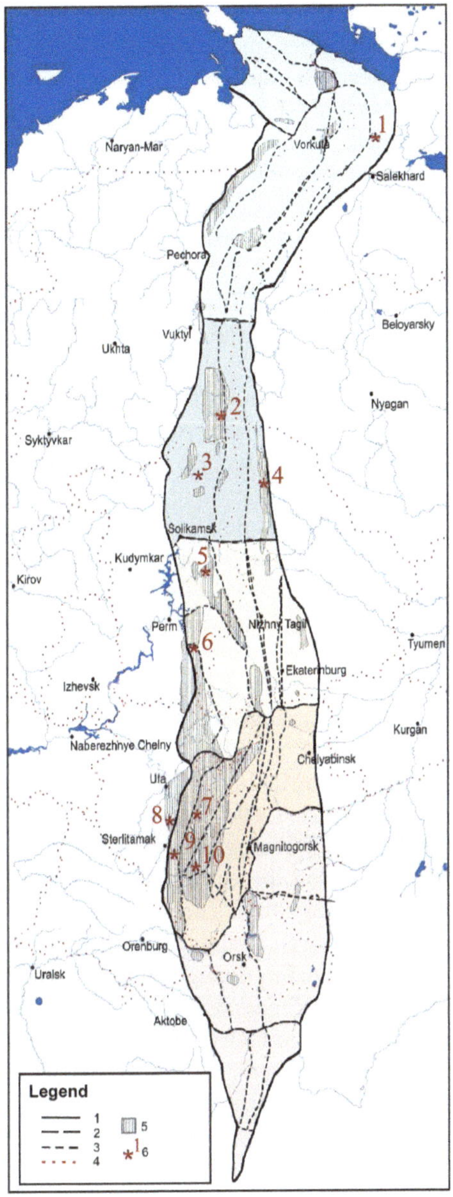

FIG. 26.15

Location of the main karst areas and ice caves in the Ural (Chibilyov, 2011). 1–4 boundaries: (1) of the Ural physiographic land, (2) of regions, (3) of subregions, (4) administrative, (5) main karst areas, and (6) object with indications of ice caves and their numbers: 1, Yanganape and Nyavape ridges region; 2, Ledyanaya cave (in Pechora River); 3, Eranka and Medeo caves; 4, Ledyanaya Cave (in Sosva river); 5, Mariinskaya and Usvinskaya Ledyanaya caves; 6, Ordinskaya, Uinskaya Ledyanaya, and Kungurskaya ice caves; 7, Askinskaya Cave; 8, Kutukskaya-1 Cave; 9, Moskovskaya, Maksyutovskie, and Bashkainskaya caves; and 10, Kinderlinskaya Cave.

in sulfate rocks include Kungurskaya Ice Cave (5700 m), Oktyabr'skaya (290 m), Ordinskaya (4900 m), Uinskaya Ledyanaya (227 m), and many other small caves.

To the best of our knowledge, the earliest mention of caves in the Northern Pri-Ural area was in the map of the world created by the Italian monk Fra Mauro in c.1459 (Materials, 1871). The original map is stored in the Doge's Palace in Venice.

In the first half of 18th century, in conjunction with the expansion of geographical research, study of the Ural caves began. In 1703, under the decree of Peter I, S. U. Remezov received the task of drawing a plan for the city of Kungur and the Kungursky district. He was the first person to develop a plan of Kungurskaya Cave.

From 1720 to 1723 the chief of the Ural and Siberian mountain factories, V. N. Tatischev, visited Kungur city many times. Having visited the cave, he drew some conclusions and founded an explanation about the cave's formation.

At the beginning of 18th century, I. G. Gmelin visited the Kungurskaya Ice Cave. He paid attention to the "dirt" on the ice surfaces, which consisted of gypsum, and determined its formation by freezing the gypsum particles out of the water. I. G. Gmelin also had been made the first temperature observations in the cave.

The end of the 18th century was the time of academic expeditions organized under M. V. Lomonosov. Reports of participants in the 1768–1774 expeditions of I. I. Lepyokhin, N. P. Rychkov, P. S. Pallas, I. P. Fal'k, I. I. George, and others contain descriptions of many caves and ice. E. S. Fyodorov described ice in the Kungursky Ice Cave in 1883 and found real "cave ice" underneath the ground's surface (Maximovich, 1947).

N. I. Karakash worked in the Kungurskaya Cave at the beginning of the 20th century. He estimated changes in cave glaciation on the basis of his analysis of historical data.

The first classification of ice in caves was made by G. A. Maximovich (1945). He published the first set of instructions for studying ice in caves (Maximovich, 1946, 1963). In 1947 G. A. Maximovich did the first review of caves with ice (Maximovich, 1947).

The best studied cave in the region is the Kuengurskaya Cave.

The Yanganape and Nyavape ridges (on the eastern slope of the Polar Ural more than 100 km north of the Arctic Circle) host several ice caves. The crests of the Yanganape and Nyavape ridges reach altitudes of 250–300 m a.s.l. and are well distinguished from the flat tundra. The surface area of the southern Yamal Peninsula has an undulating topography, with absolute surface elevations of 55–80 m a.s.l. Cryogenic landforms such as frost mounds and thermokarst depressions with shallow lakes and marshes are well represented there. The permafrost thickness in the study area reaches 100–150 m. Geologically, the crests of the Yanganape ridge consist of old reefs of the Lower Devonian (Khromykh and Belyaev, 2010) with a thickness of about 170 m. Caves are located in the margins of the reef structures surrounded by deep sea sediments with pelagic fauna and extrusive rocks at the base (Khromykh and Belyaev, 2010). Reefs vary in shape and size and are generally round or oval in plan view and in conical sections. A reef's base diameter ranges from 0.5 to 1 km. Six caves with lengths not exceeding 35 m are known within the area. All caves are situated in permafrost.

Karst development in the Yanganape ridge region (the western slope of the Urals) occurred in two stages. First, hypogene cavities were formed soon after the reef constructions. The second stage of karstification occurred in modern permafrost and frost-weathering conditions. MAT = −6.1°C, T_{Jan} = −23.8°C.

26.7.1 CAVE OF V. N. CHERNETSOV

The cave is located in western board of a broad gull in a southern part of Yanganape Ridge. The elevation of the entrance is about 146 m a.s.l.

The cave entrance is situated at the lower part of a rocky wall with a height of about 3–5 m and represents a rectangular slit-like aperture with a width of about 3.6 m and a height of 0.7 m (Fig. 26.16).

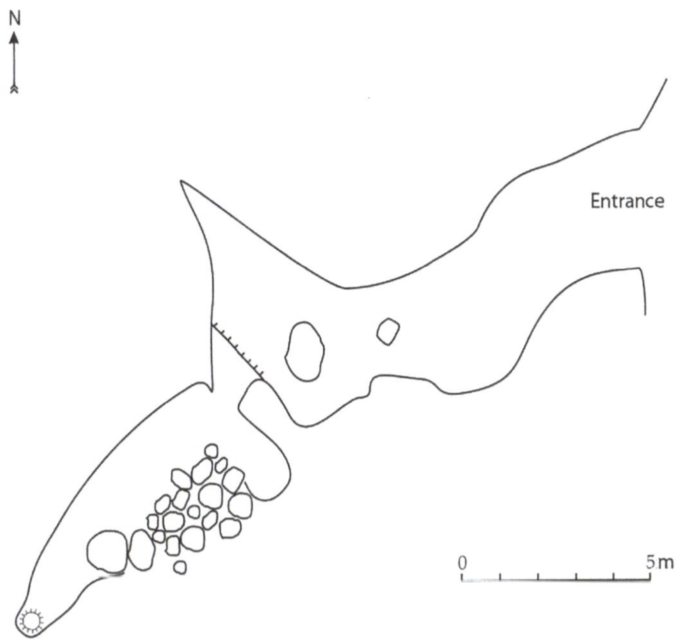

FIG. 26.16

Cave of V. N. Chernetsova (Yanganape-2), plan.

A horizontal gallery with a length of about 9.6 m and a height of about 1 m begins at the entrance. The gallery's floor is covered with frozen clay sediments with limestone debris. The gallery leads into a chamber with a height of about 3 m, a width of 1–1.5 m and a length of 6.3 m and with an orientation defined by a diagonal crack. A narrow manhole (height 0.4–0.5 m) begins in the western wall of that chamber at a height about 1.5 m from floor level. The total cave length is about 24 m.

The caves were observed in 1990 but were only described in 2012.

There is permanent icing in the middle part of the cave (width 0.5 m, length 3 m); in a distant part of the cave under the organ, there are ice stalactites and stalagmites.

26.7.2 CAVE ACADEMICHESKAYA (YANGANAPE-3)

The cave is located on the western slope of Yanganape Ridge at the lower part of a rocky ledge. Entrance height is about 180 m a.s.l.

The cave has three slit-like entrances, which with a small bias lead into a common small chamber with a width of about 3 m and a height of 0.9 m (Fig. 26.17). The height of the entrances range from 0.5 to 0.8 m, and the width of the galleries range from 2.5 to 3 m. The central gallery has a length of 6 m, extreme about 10 m. The total cave length is about 26 m.

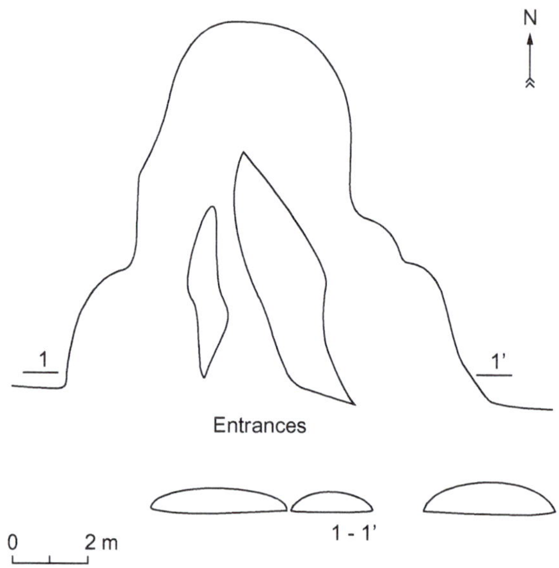

FIG. 26.17

Plan of Academicheskaya Cave.

There is icing close to the cave's entrance with a thickness of about 0.8 m, and the ice is mixed with ground on which bone rests. Datings of the bones show their age to be less than 2500 years BP. Bones from the bottom layer, according to radio carbon dating, are 4100 ± 200 years BP.

26.8 ICE CAVES IN THE NORTHERN URAL

A covered type of karst prevails in the Northern Ural. Karstified carbonates (limestones, dolomites, and dolomite limestones) from the Middle Devonian up to the Lower Permian and sometimes older (the Silurian and the Ordovician) occur within the folded zone of the western Urals. The carbonates have a total thickness of about 2000 m. This area was covered by glaciers several times, which resulted in some superficial and underground features being filled by glacial deposits. Among the 45 caves discovered in the area, one contains the largest quantity of permanent ice, Ledyanaya Cave.

26.8.1 LEDYANAYA CAVE

The cave was opened and studied by Guslitser and Kanivets in 1958 (Guslitser and Kanivets, 1965). It is located in the right board of a broad Jordanian gull in 150 km from its mouth at an elevation of about 16–17 m above water level of the Pechora River. From the cave entrance, it is possible to rise easily from a broad gull along a gentle slope. MAT = −1.9°C, T_{Jan} = −18.8°C.

Cave length is about 160 m. The cave was generated along three systems of tectonic cracks (Fig. 26.18).

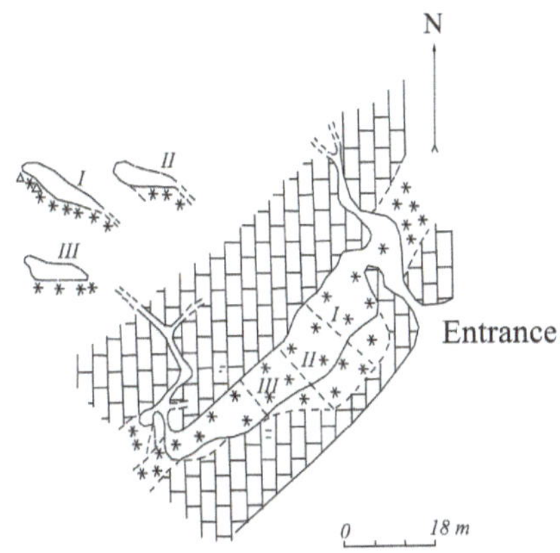

FIG. 26.18

Plan of Ledyanaya Cave on Pechora River (black dots indicate the presence of ice) (Guslitser and Kanivets, 1965).

The main wide gallery goes in a southwesterly direction. The ice floor descends at a 15–20 degree incline to a gallery, then quickly extends into a big chamber that is elongated from southwest to northeast. Its length is about 40 m, its width is about 5–15 m, and its height is about 2–4 m.

The floor in the gallery and chamber is formed by thick layered ice. The floor surface in the biggest part of the chamber is horizontal. The ice is very flat and smooth and looks like the ice of a frozen lake. From the western end of the chamber, by a system of uplifting to the north galleries, it is possible to go about 30 m. These galleries do not contain ice. Icing and ice-cemented talus of limestone blocks are found on the northern wall of the southeast gallery. The cave contains congelation ice as icings (cover and pendent), lake ice, and sublimation crystals (they sometimes cover ice stalactites). In the distant southwest end of the chamber in monolithic thick ice, deep niches were formed by winter ice evaporation. In this place, the ice outcrop is very visible. There is an increase of multilayered ice that is dense and transparent and has rare air bubbles. Each layer ends with a milky-white opaque ice covered by a very thin ice layer with cryogenic minerals. The boundary between the layers is very sharp. Lower layers have the greatest thickness. The thickness of the lowest layer exceeds 60 cm.

The ice structure shows that its formation occurred as a result of fast penetration into the cave and subsequent freezing of a large amount of water. There were several such penetrations and freezings. There are no data about the age of the ice. Ice preservation is provided now by the low position of the chamber in comparison with the cave entrance, where cold winter air can flow into the cave. The air temperature in the chamber at the end of the summer of 1958 did not rise above −4.5°C.

In the Kolva and Vishera Rivers basin area of Perm Land there are more than 90 caves. The biggest cave with permanent glaciation of this area is Yeranka Cave, and with the biggest ice thickness is Medeo Cave.

26.8.2 YERANKA CAVE

The cave is located in the Srednevisherskij area of carbonate karst in the folded zone of the western Urals. Yeranka Cave was developed in the Asselian and the Sakmarian strata of the Lower Permian. MAT = 2.9°C, T_{Jan} = −17.3°C.

The cave entrance is located at a height of 20 m on the slope of the southeastern exposition, 350 m from the mouth of the Yeranka River, which is a left-bank tributary of the Beryozovaya River. The total length of the cave is 500 m. It is an inclined descending cavity (cold bag) that produces perennial ice with a thickness up to 3.5 m in the first grotto. The entrance has a triangular form: height is about 2.5 m and width is about 6 m (Fig. 26.19).

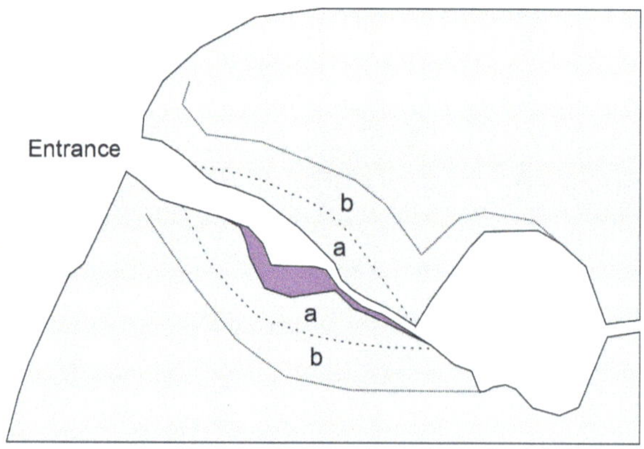

FIG. 26.19

Schematical longitudinal section of entrance part of Yeranka Cave. Estimated frozen zones in the first grotto (NTA zone) located symmetrically from icing: (a) in summer and (b) in winter.

In the first grotto, which is 15 m from the entrance, there is perennial icing and vertical steps (height 2 m) that lead to a lower (flat) part of the grotto, where there is a narrow inclined gallery (length about 20 m). This gallery leads to the big first chamber with an arch height of about 8–10 m. The temperature there during the summer of 2012 (August) was −0.5°C. In August of that summer, the air temperature in the second chamber, away from the entrance, was positive (+3.9°C). The presence of scree detached from the walls in the second chamber's entrance allows us to assume that periodic water freezing can occur there during the winter.

An chemical analysis of ice showed its low mineralization (141.76 mg/dm^3) of hydrocarbonate calcium composition. The low ice mineralization was connected with the effect of the inflow of meltwater on the icing surface during the spring.

26.8.3 MEDEO CAVE

Stone Pekhach is among the most interesting and best known karst areas in Prikamye. It is located in the Srednevisherskij area of carbonate karst of the folded zone of the western Urals on the right coast of the Bereyozovaya River at the place of confluence with the Bad'ya River.

Rock rich with fossils was formed by layered gray siliceous limestones of Sakmarian Unit of the Lower Permian.

Medeo Cave (Badinsky Ice Cave; also Bad'inskaya Ledyanaya) is located in the western part of Stone Pekhack, in a region with MAT=2.9°C, $T_{Jan}=-17.3$°C.

The cave entrance (height about 3 m and width 7 m) is located about 30 m above river level. The cave consists of two chambers with the sizes 25 m × 16 m and 17 m × 13 m. The total cave length is about 60 m (Fig. 26.20).

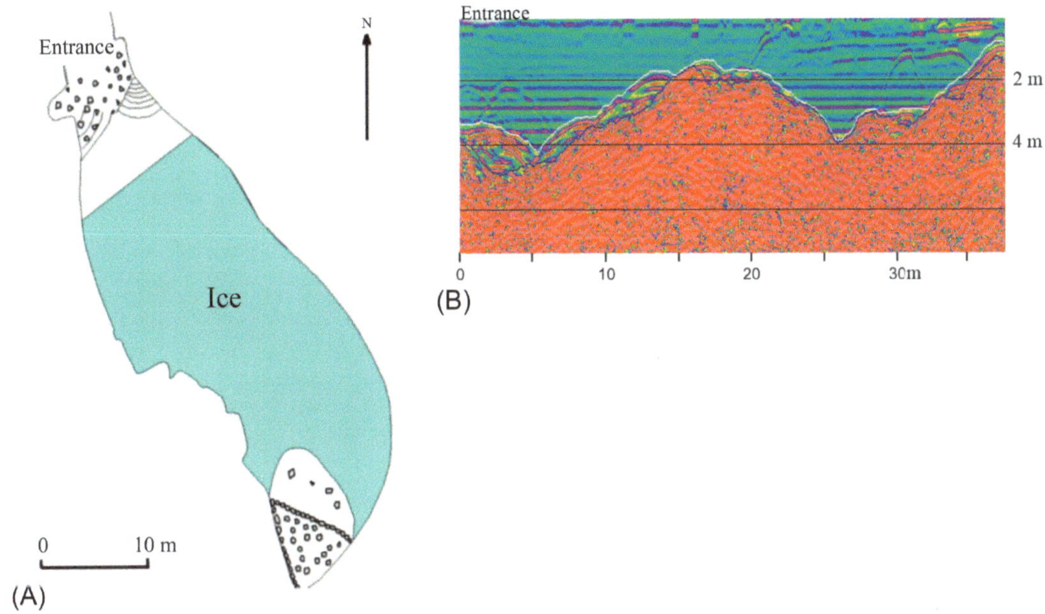

FIG. 26.20

Plan of cavity (A) and longitudinal section of ice thickness in Medeo Cave, as shown by GPR (B) (Stepanov et al., 2014a,b,c).

The ice morphology was described for the first time in works of Kadebskaya O. I. and Chajkovskij I. I. (2012) and the ice thickness in the work of (Stepanov et al., 2014a,b,c).

The floor of the chamber is covered with a layer of permanent ice, for a total area of about 600 m^2. Ice closely adjoins all the cave walls. The greatest ice thickness was measured by georadar and found to be 4.5 m (Stepanov et al., 2014a,b,c); it is located near the cave entrance. Further inside the cavity, the ice thickness gradually decreases to 0.5 m. The volume of permanent ice in 2011 (March) was about 940 m^3, which allows us to consider this underground icing as the largest one by volume in the northern Ural Mountains.

From March 29 to September 8, 2011, the ice thickness on the cave's east wall increased by 8.2 cm, and on the cave's west wall by 10.1 cm. Accumulation and ice melting in the cave are cyclic in nature. When icing grows up in the entrance to the gallery roof, ice inside the cave begins to melt.

Water on the ice surface can move in the form of streams or as a film. Ice structures, from fine-grained to gigantic crystals, influence the formation and distribution of cryogenic minerals. The ice has a mineralization of about 57.9 mg/dm^3 and belongs to the hydrocarbonate potassium-sodium type.

26.8.4 LEDYANAYA CAVE (D'YAVOLSKOE GORODISHE)

The cave (first described in 1942) is located 30 km from the town Severouralsk in the Severouralsk district of the Sverdlovsk Region on the right coast of the Sos'va River, 250 m WSW from the big collaps known as D'yavolskoe Gorodishe. Entrance elevation is about 201 m a.s.l. MAT = −0.5°C, T_{Jan} = −18.4°C.

Cave length is about 176 m, with a vertical range of about 7 m (Fig. 26.21).

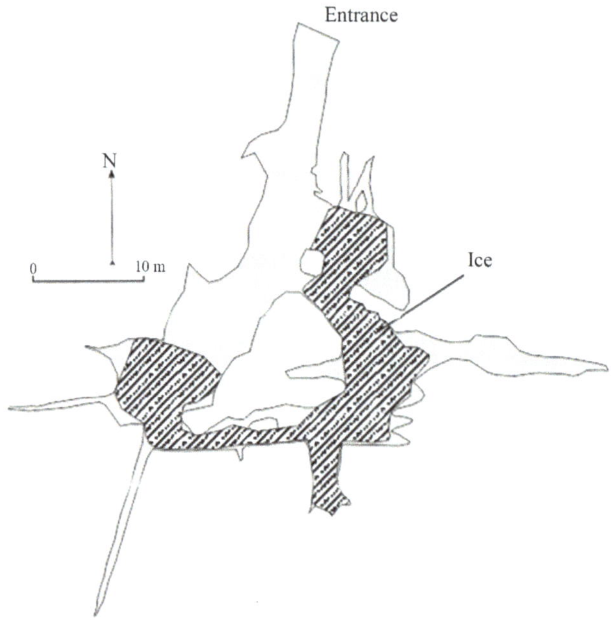

FIG. 26.21

Plan of Ledyanaya Cave (D'yavolskoe Gorodishe). Survey of Sverdlovsk speleogly, E. Zurikhin, 2011.

The cave is almost completely filled with ice; in the summer, water is on ice surface. The mineral ikaite was found on the icing surface.

26.8.5 MARIINSKAYA (GUBAKHINSKAYA) CAVE

The cave is located in the Kizelovsky area of carbonate karst in the folded zone of the Western Urals close to mouth of the left board of the Mariinskij broad gull, near the settlement Verkhnyaya Gubakha of Perm Land. MAT = 3.6°C, T_{Jan} = −16.4°C.

The cave (described for the first time by M. S. Gurevich in 1932) is formed of massive limestone of Vizey and the Serpukhov layers of the Lower Carboniferous. The cave has two entrances and is located in a long, rocky ledge with a height of about 7 m. The entrance elevation from a broad gull bottom is about 80 m; from the level of the Kos'va River, it is 120 m. Absolute entrance altitude is about 275 m a.s.l.

The length of the cave galleries is about 1000 m (Figs. 26.22 and 26.23), and the depth is 47 m (Valujskij, 2000).

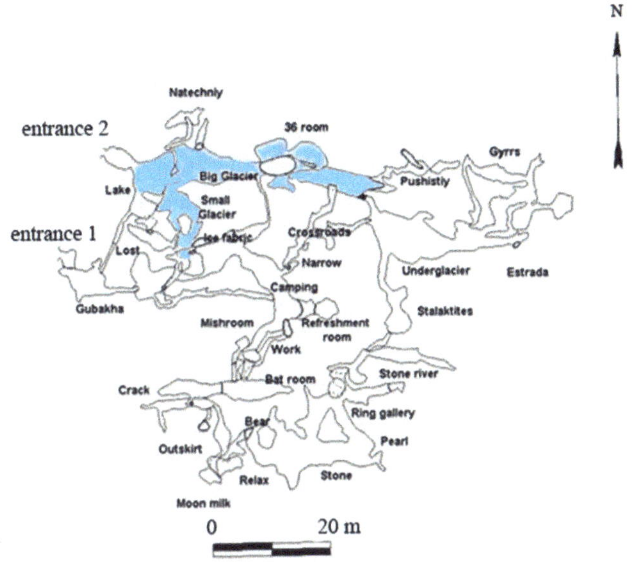

FIG. 26.22

Plan of Mariinskaya Cave (Valujskij, 2000).

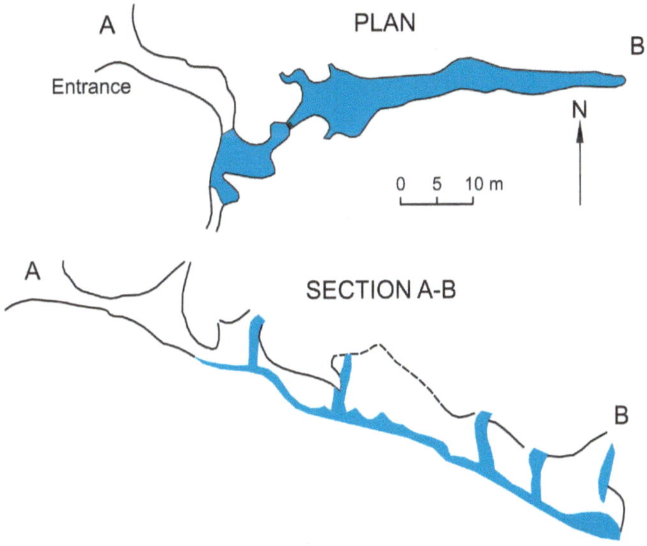

FIG. 26.23

Plan and longitudinal section of Mariinskaya Cave (area with ice) (Mavlyudov, 2008).

The cave has four levels that are connected by gradual transitions and pits. The right entrance (entrance 1) represents a rather narrow crack with a height of about 1.5 m.

Twenty meters to the left of the first entrance, there is a second entrance of triangular form with a height of about 1.2 m from which a 62-meter-length icing begins with an ice thickness of about 1.5–2 m. The total area of ice cover in the cave is about 300 m^2. Seasonal icing in the cave became perennial in 1976 after the installation of a concrete stopple at the bottom of one gallery (Mavlyudov, 2008).

The annual ice accumulation in the lower part of the cave can be estimated as 160 mm or more, and in an initial and average part of the cave, there is 15–52 mm of ice (observations 1985–1986). An average year's ice accumulation in the cave is more than 5 m^3.

From 1985 to 1994, the cave experienced blockage by ice in the second entrance, which kept cold air from penetrating into the cavity. In the winter, the cold air stopped entering the cave, and the air temperature in the cavity increased, and soon permanent icing started catastrophically decreasing and began receding from the walls. By 2007 icing at the second entrance was melting and ice accumulation in the cave was renewed. Thus accumulation and ice melting in the cave is cyclic.

Ice closely approaches all the cave walls. In the middle part of the cave, the ice is pure and transparent and is visible at all thicknesses. Close to the entrance, the ice is white and matt, and in the upper part of the cave, the ice is covered with clay sediments and rock fragments.

In the cave are developed icings (cover and pendent), and in the summer, at the end of the icings, ice breccia are formed from fallen ice stalactites.

26.8.6 KUNGURSKAYA ICE CAVE

26.8.6.1 Site and Situation Around the Cave

Kungurskaya Ice Cave is located in suburb of Kungur city in the Perm Land on the bank of the Sylva River. From a physical-geographical perspective, the cave is situated in the east part of the Russian Plain and is located at the joint of two geographical areas: the High Transvolga and the Ufa Plateau. Ledyanaya Mountain (above the cave) is a plateau of a highly karstified massif ranging over the bottoms of the Sylva and Shakva River valleys at 90–96 m above them. The slopes of the massif are abrupt with fragments of destroyed terraces.

The Kungurskaya Ice Cave is developed in sulfidic sediments of the Lower Permian (Kungurskij Stage). The general thickness of the karst rocks reaches 100 m.

Carbonate and sulfate layers have their own names. The cave consists mainly of gypsums and anhydrites of the Ledyanopeshernaya Layer ($_{ld}P_{1K}^{ir}$); in some of the cave's grottoes overlying sediments of carbonate-sulfidic Nevolinskaya Layer ($_{nv}P_{1K}^{ir}$) are also open. Above lies the Shalashinskata Layer ($_{sh}P_{1K}^{ir}$). These three layers together form the Irenskij horizon of the Kungurskij Stage.

The cave entrances now have a natural aperture at the base of a rocky escarpment (old entrance) and also two artificial tunnels; the entrance and exit are located at elevations accordingly, at 118, 120, and 129 m a.s.l. from the river side. The absolute elevation of the massif surface changes around the cave from 167 to 189 m. The rock thickness above the cave ranges from 37 to 85 m. MAT = 2.1°C, $T_{Jan} = -15.5$°C.

26.8.6.2 The Cavity Form, Cavity Maps

The cave is a horizontal labyrinth spreading from along the Sylva River valley (on the level of the first terrace above the flood plain) in Ledyanaya Mountain at 700 m. The cave length is about 5.7 km (Fig. 26.24). The total area of the cave is 65,000 m^2, and the volume is 206,000 m^3 (Kadebskaya, 2004).

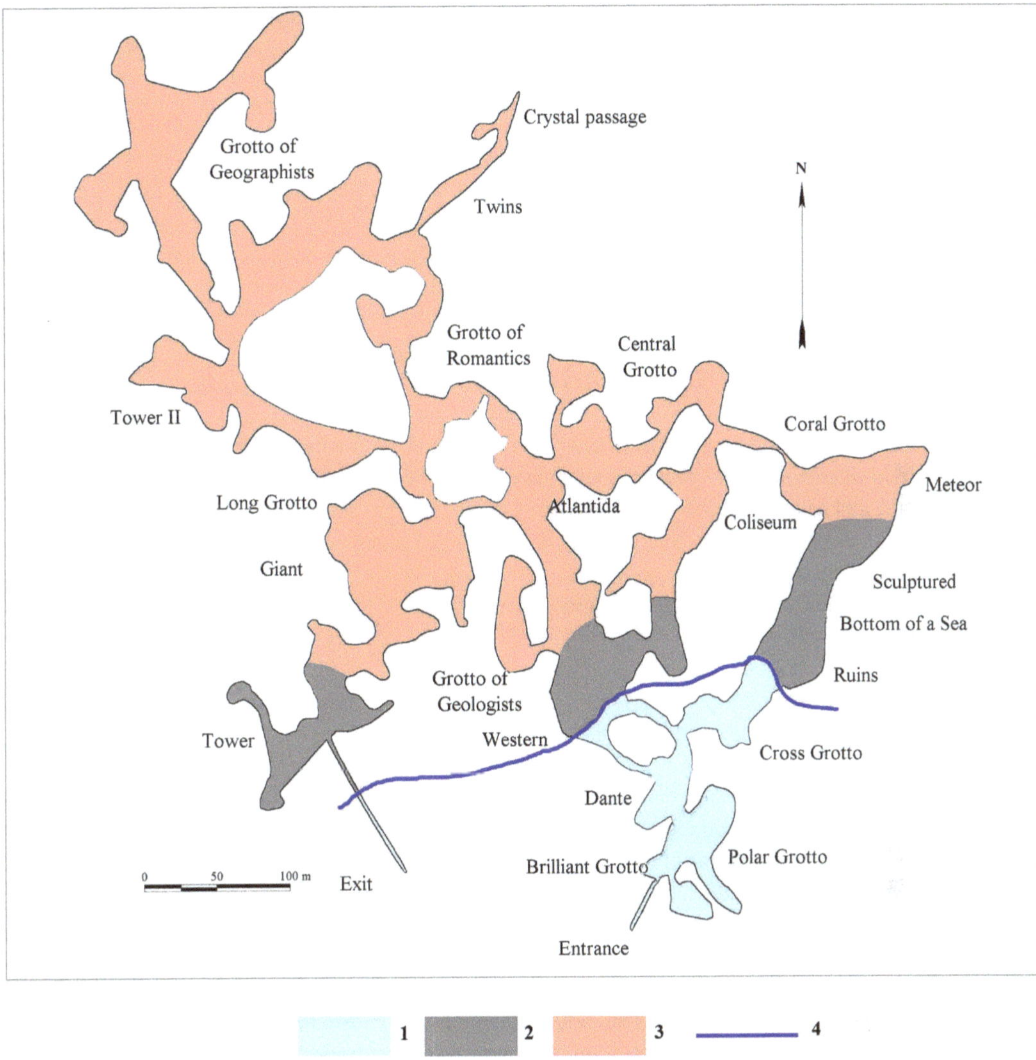

FIG. 26.24

Plan of Kungurskaya Ice Cave (Dorofeyev, 1967): (1) NTA zone, permanent ice, (2) seasonal ice, (3) area without glaciation, and (4) boundary of permanent ice. Year of survey 1964.

The main morphological cave feature is the prevalence of large collapse chambers (grottos) with lengths and widths ranging from 30 m to 200 m. The cave is one of the longest gypsum caves in Russia and the largest among them in volume. The cavities are located at the level of ground water drained from the karst massif from the Sylva River (or a little above it). Therefore, in many grottos, there are numerous underground lakes (about 70), 12 of which are quite large (from 130 to 1500 m^2). The total area of the lakes is about 7500 m^2 or 1% of the total cave area. The present period of cave development can be defined as late-Vadoz, which is accompanied by increasing rock collapses due to the chambers' large dimensions.

26.8.6.3 Research History

The history of the study of ice in Kungurskaya Cave is connected with the history of cave studies in general. The first researchers did not find ice in caves, but then when ice was first found, everyone interested in cave studies began to pay attention. The complexity of these descriptions over different years makes it possible to estimate the dynamics of cave glaciation for long periods of time. However, visitors and researchers in the mid-19th century did not study ice.

At the beginning of the 18th century, I. G. Gmelin paid attention to the "dirt" consisting of the dust of gypsum found on ice surfaces in the Kungurskaya cave and explained its formation by freeze-out of particles of gypsum from water (Maximovich, 1947). The first research on cave glaciation was done by E. S. Fyedorov (1883). He thought cave ices were only the ice that was formed by melt-out of sublimation crystals, because such ice was only in caves, whereas all other kinds of ice can occur on outside surfaces. He paid attention to the presence of gypsum dust layers in outcrops of friable sediments in Krestovyj Grotto and associated them with previous periods of ice accumulation in caves.

N. I. Karakash concluded that cave glaciation increased by comparing cave descriptions in 1848, 1883, and 1905 (Karakash, 1905).

V. J. Altberg carried out research on cave glaciation (1930, 1931). He concluded that ice evaporation cannot be a unique source of cave cooling, because in some places in caves, ice is not present, so there is nothing to evaporate, but cooling occurs (Altberg, 1930, 1931).

M. P. Golovkov investigated the structure of ice stalactites in Kungurskaya Cave and explained that by studying an ice's structure, its origins can be determined (Golovkov, 1938, 1939a,b).

After entrance tunnels were built, cave glaciation began to degrade quickly, which led to more interest in investigations. The research conducted by V. S. Lukin in 1950 not only helped to establish the reasons for the occurrence and existence of cave glaciation but also led to a number of engineering projects for cave glaciation restoration. He carried out research in Kungurskaya Cave and also studied glaciation dynamics in a specially constructed artificial mine with two entrances at different heights that was used as a natural cave model. This work led to the ability to model air movements in karst massifs (Fig. 26.25).

It has been shown that the intensity of winter air circulation in a cave almost exceeds by twice the summer circulation that is favorable for the formation and preservation of ice in caves. V. S. Lukin was first to find climatic zones in caves: zones of negative and positive temperature anomalies (Lukin, 1962, 1965). In a negative temperature anomaly (NTA) zone, the MAT is essentially below the temperature of the karst massif, and in a positive temperature anomaly (PTA) zone, the MAT is considerably above the temperature of the karstic massif. If the NTA zone temperature is negative, there is permanent cave glaciation. The research in the Kungurskaya Cave allowed the creation of a universal system for describing climatic systems in caves (Mavlyudov, 1994).

Versatile research of ice and glaciation in the cave was done from 1960 to 1995. Temperature and glaciation in a cave as well as the structure of ice and the conditions of different ice growth and cave permafrost (Dorofeyev, 1969, 1988, 1990, 1991; Dorofeev and Mavlyudov, 1993) were investigated.

From 1982 to 1986 the dynamics of glaciation in the cave and the ice mass balance based on measurements of ice melting, evaporation, and accumulation were further investigated.

Since 2000 O. I. Kadebskaya has studied the dynamics and control of cave glaciation.

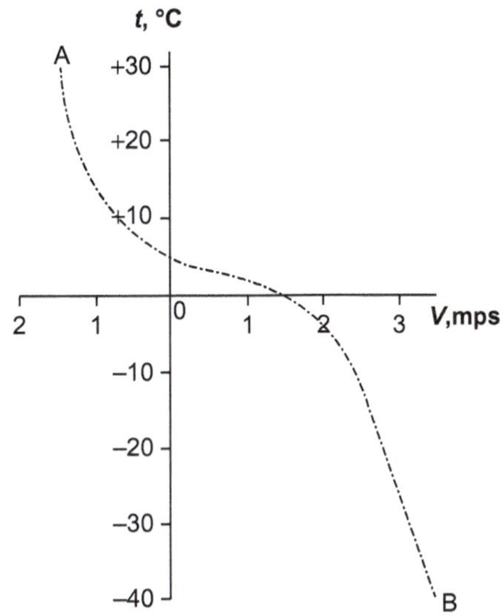

FIG. 26.25

Dependence of direction and speed of air stream at Kungurskaya Cave entrance from surface air temperature: (A) summer air draught, and (B) winter air draught (Lukin, 1962).

Throughout the past 60 years, detailed and stationary geological, speleological, biological, and other observations in the cave have formed the basis for an electronic database for the cave, and also for the publication of a monograph (Dublyanskij, 2005).

26.8.6.4 Forms and Dynamics of Ice

In the cave, the following ice formations are well developed: congelation (leak, segregation ice, and ice-cement), sublimation (crystal formations), and sedimentary metamorphic ice (Photo 26.3).

The area of ice distribution in the cave is estimated to be 1000 m^2, and the volume is estimated to be 500 m^3.

Cave observation has been conducted for about 300 years. In the first cave descriptions, about 250 years ago, the variability of ice in caves that sometimes blocked entrances was mentioned. Glaciation fluctuations were connected with outside climate changes. In plans and descriptions before 1720, ice was not mentioned (probably it is not enough ice or it absolutely is not present). Sometime between 1722 and 1734, there was collapse on Ledyanata Mountain that closed one of the galleries. The collapse probably happened in the open upper entrance that had aided the production of glaciation. Earlier, in 1770, ice had reached Meteornyj Grotto.

A cold regime existed in the cave during the first quarter of the 19th century, after which warming began. When the cave entrance was closed, ice glaciation in the cave decreased; conversely, when the entrance was opened, glaciation increased. Such natural regulation of glaciation was controlled by an external cycle of climate warmings and coolings. This process continued until 1914 when a door was established at the entrance, after which cave glaciation began to depend on artificial ventilation.

In 1937 a tunnel was constructed in Brilliantivyj Grotto that modified the natural circulation of air in the cave. By the end of 1940, glaciation in the cave had practically disappeared.

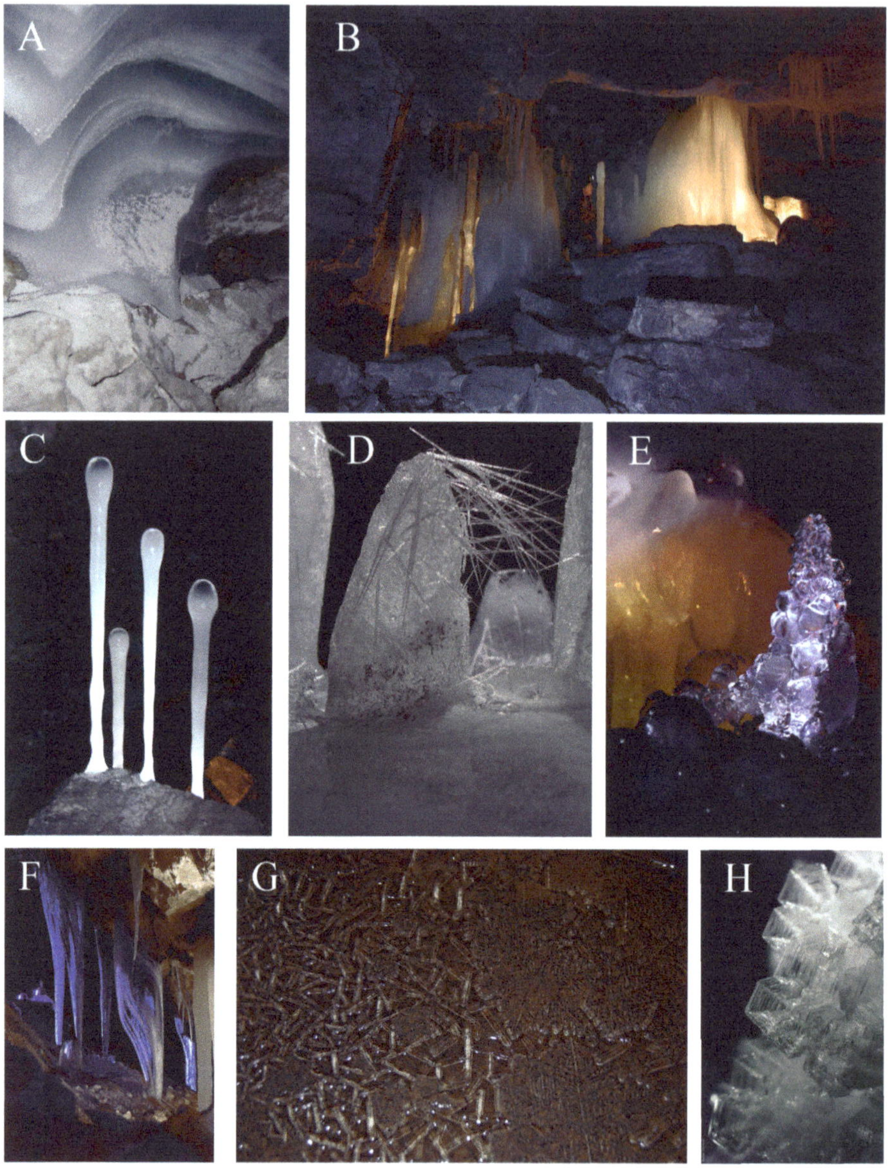

PHOTO 26.3

Types of ice in Kungurskaya Ice Cave: (A) permanent ice in Polarnyj Grotto, (B) permanent ice in Krestovyj Grotto, (C) club-like stalagmites in Ruiny Grotto, (D) sublimation crystals on stalagmites in Polarnyj Grotto, (E) coralloid stalagmites in Ruiny Grotto, (F) ice curtains-flags in Morskoe Dno Grotto, (G) crystals on lake surface in Krestovyj Grotto, and (H) sublimation crystals in Pervyj Grotto.

Scientific research and technological advances enabled the restoration of glaciation in the early 1950s. In 1969 ice formed in Meteornyj Grotto. In 1972 an exit tunnel was constructed, but was not completely enclosed until a door was constructed until 1985. During this period, draughts through the two entrances reduced glaciation at the entrance but increased it at the exit.

In 1979 a flood penetrated the cave and consumed much of the cold that had accumulated in the cave rocks (Ezhov et al., 1990). Since 1980 winter ventilation of the cave has become irregular, and cave glaciation has started to degrade. In 1983 the old entrance collapsed and was filled with ice. The entrance tunnel was opened in the winter of 1985–86.

The irregularity of cave ventilation led to the degradation of cave glaciation. The boundary of permanent glaciation at the entrance was reduced to 100 m. In the summer, the temperature in the entrance to the grottos began to rise to positive values, activating roof collapses in the chambers.

In 2000 winter cave ventilation was restored only through the entrance tunnel. Cave glaciation then recommenced, although at the cave's exit, the presence of ice became seasonal.

As we can see, Kungurskaya Cave glaciation is very sensitive to climate changes. Climate cooling extends the NTA zone in the cavity and increases the quantity of ice in it; climate warming reduces the NTA zone in the cavity and the amount of ice in it. These changes, however, are also affected by anthropogenic regulation of the cavity's ventilation, but in any case, changes can appear through artificial mechanisms as well as natural ones.

Continuous monitoring of ice accumulation and ablation in the cave from 1985 to 2016 show that the ice mass balance in different parts of a cave are distinctive from one another.

In Brilliantovyj Grotto, in First Grotto, in the gallery between the Brilliantovyj and Polarnyj grottos, in Dante Grotto and in the Gore Tolstyakam gallery the ice mass balance has been defined only by ice evaporation. From 1985 to 2016 the specific ice mass balance of these parts of the cave ranged from 0 to 0.05 g/sm^2 (or from 0 to 5 mm w.e.), and the ice evaporated layer was about 3–30 cm (Fig. 26.26A). Ice ablation in Krestovyj grotto change during time and depend from winter cave ventilation (Fig. 26.26B).

In Polarnyj Grotto from 1987 to 2000, ice evaporated with the same intensity as in Brilliantovyj Grotto. But after additional winter ventilation (by opening a door in the tunnel), from 2000 to 2005 accumulated ice isolating the grotto. As a result the MAT in the grotto began to rise and permanent ice began to melt (it melted completely in 2015). Artificial destruction of the ice barrier in the winter of 2016 restored a negative MAT, and from 2016 to 2017, ice in the central part of the grotto grew from 30 to 70 cm.

In Krestovyj Grotto in 1985 and 1986, ice grew to 150 mm in the winter. Beginning in 1987 permanent ice started to degrade and by 2001 had completely melted (Kadebskaya, 2004). At MAT = 0.5°C, the total annual ice melting was 180–198 mm. The area of permanent ice distribution from 1985 to 2001 changed from 1020 to 530 m^2. Restoration of air ventilation in 2002 and 2003 led to the displacement of the seasonal glaciation boundary in Morskoe Dno Grotto. Now in Scandinavskij and Krestovyj grottos, the MAT is negative, and the growth of icing ranges from 0.1 to 1.7 m.

Maximum sublimation ice formation occurs when low temperatures in the external air arrive in the cave (February–March). At sharp cold snaps, all cross-sections of the cave gallery are occupied by streams of dry, cold air that lead to the evaporation and falling of ice crystals. This has led to the formation of sedimentary-metamorphic? why metamorphic glacial? ice on the floor of Brilliantovyj and Polarnyj grottoes. In Brilliantovyj Grotto in 1985 and 1986 measurements showed about 0.7 m^3 (in w.e.) of sublimation ice at an average sublimation intensity of about 0.2 mm/day; however, the intensity of ice sublimation was irregular over time. In the First Grotto sublimation crystals are permanent, and the thickness of ice accumulation on the walls is more than 0.7 m.

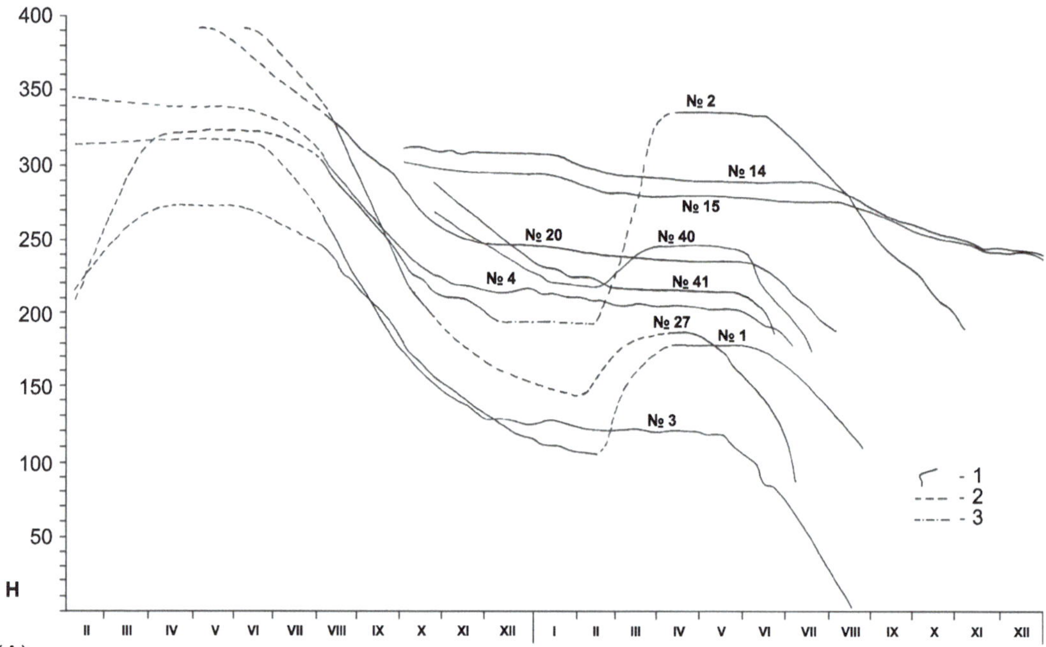

(A)

(B)

FIG. 26.26

(A) Ice ablation accumulation (A–H) in Kungurskaya Cave from 1985 to 1986 at different points (stake numbers): (1) measured data, (2) probable values, and (3) ice absent (Mavlyudov, 2008). (B) Changing of ice ablation in Krestovyj grotto during time.

Now in all cave grottoes in the NTA zone, there is a period of restoration of glaciation and increasing ice thickness.

26.8.7 ASKINSKAYA CAVE

The cave is located in the Arkhangelskij district of the Republic of Bashkortostan in the middle part of the left slope of the Karanyurt River valley in Ulutau Ridge, which is on the western slope of the Southern Urals (Kudryashov and Salikhov, 1968). The first research of the cave was done by G. V. Vakhrushev, I. K. Kudryashov, and E. D. Bogdanovich. The cave is situated on the left slope of the Askin River 1.8 km south of the Solontsy village. The cave origin is Devonian limestone, pointed in a west-northwest direction (Asimuth 280°, angle 34°). MAT = 4.0°C, $T_{Jan} = -14.7$°C.

The cave has length of about 230 m and a vertical range of about 34 m. A 9 m × 22 m, arch-like entrance is located about 60 m above water level in a board of rocky massif and is oriented north-northeast. The cavity is an inclined descending (cold bag) form. The entrance doline ends at a mouth with a height of about 0.6–0.7 m and width of 6 m. Further on, an ice ledge goes down and into a big chamber. The cave is almost completed represented by one chamber with a length of about 104 m and a width ranging from 40 to 60 m, and a height up to 22 m. The cave volume is about 50,000 m^3. The chamber leads to narrow galleries with western and southern orientations.

Askinskaya Cave is considered, by area, to have the most cave ice in the Ural Mountains.

The quantity of cave ice grew five to six times from 1930 to the middle of the 1960s. In 1924 cave ice covered the slope and one-fourth of the chamber area close to the entrance. In 1966 ice covered all the surface of the chamber floor (Kudryashov and Salikhov, 1968). This means that throughout 42 years, the annual increase of ice thickness was about 22 mm. Taking into account ice melting, we can say that the annual increase in ice thickness was about 27 mm. From 1985 to 1994 cave glaciation did not change considerably, and ice covered all the chamber's surface (Mavlyudov, 2008). Based on published data, the age of the permanent cave glaciation is more than 80 years.

In 2011 ice covered all chamber's surface, and in many places icing is covered by clay sediments mixed with a cryogenic flour, vegetative remains, and dust. The cave now has 11 large ice stalagmites with heights up to 7 m and widths at the base up to 3 m.

One georadar survey showed that the thickness of per-annual cave ice changes from 0.15 to 2.00 m and almost uniformly decreases from the entrance to distant parts of the cave (Fig. 26.27) (Stepanov et al., 2014a,b,c).

Close to the entrance there is a sharp increase in ice thickness, which deepens in rocks with widths ranging from 1.5 to 2.0 m. The ice surface in this part of the cave is flat and is at the same level as nearby places. The ice deepening is elongated from the entrance to the western wall of the chamber and continues to a narrow gallery that is blocked by stalagmites. This bed deepening is probably connected with the western gallery and is the result of water action in the past. At the same place, there are many rock fragments and other inclusions (in lower part of icing). The ice volume (excluding ice of stalagmites) was calculated to be 1480 m^3.

In summer of 2010 (July) the temperature in the NTA zone did not exceed +0.1°C; in the winter it changed from −1°C to −0.6°C (Kuzmina, et al., 2014).

Congelation ice in the cave has a mineralization of about 141–145 mg/dm^3, with hydrochemical types, hydrocarbonate (GC) and hydrocarbonate calcium sodium (GCS).

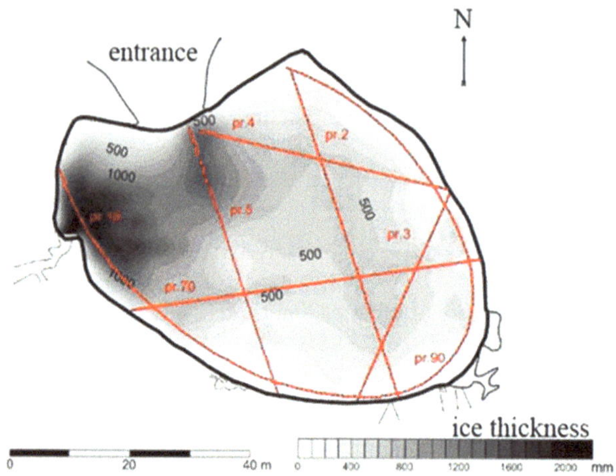

FIG. 26.27

Distribution of ice thickness in Askinskaya Cave (Stepanov et al., 2014a,b,c).

26.8.8 CAVE KINDERLINSKY (30 ANNIVERSARY OF VICTORY)

26.8.8.1 Site and Situation Around the Cave

The cave is located in the Gafuri district of Bashkortostan (South Ural) near the mouth of the Zilim River's left tributary, the Kinderlya River. Its origin is Devonian bituminous limestone on the eastern slope of the Tashastinskaya syncline. Entrance coordinates: 54°09′25″N; 56°51′16″E. MAT=4.0°C, $T_{\mathrm{Jan}}=-14.7$°C.

26.8.8.2 The Cavity Form, Cavity Maps and Sharts (the Plan, a Cut)

Kinderlinskaya Cave is the deepest and third longest cave in the Urals (length 9113 m, amplitude 215 m) (Caves, 2010). The cave entrance is located on the southern slope of a karst massif.

The cave is characterized by a maze pattern, vertical chimneys, arched ceilings, and other morphological features suggesting a hypogene origin. The cave represents an inclined system of passages developed in north, northeast, and northwest directions and forms four levels. The trapezoid entrance (12×7 m) faces south. The chambers are small in area, but their height reaches 80–90 m (Fig. 26.28). The most remarkable are the Klassicheskij, Atlantida, and Figur chambers. The cave is rich with flowstone. Along with the usual calcite forms (stalactites, stalagmites, and others), it has anthodites and gypsum crystals.

26.8.8.3 Research History

The cave was used for long time by local hunters for meat storage. In 1940 it was shown to geologist G. V. Vakhrushev, but he did not study it. Cave research began in 1974 when the hunter A. Karanaev showed the cave to Andreev A. S., head of the speleological section of Sterlitamak.

26.8.8.4 Forms and Dynamics of Ice

The cave is horizontal but with entrances at different heights. The NTA zone close to the entrance has a length greater than 150 m. This zone has permanent icing with widths up to 12 m and lengths greater than 100 m (Fig. 26.28).

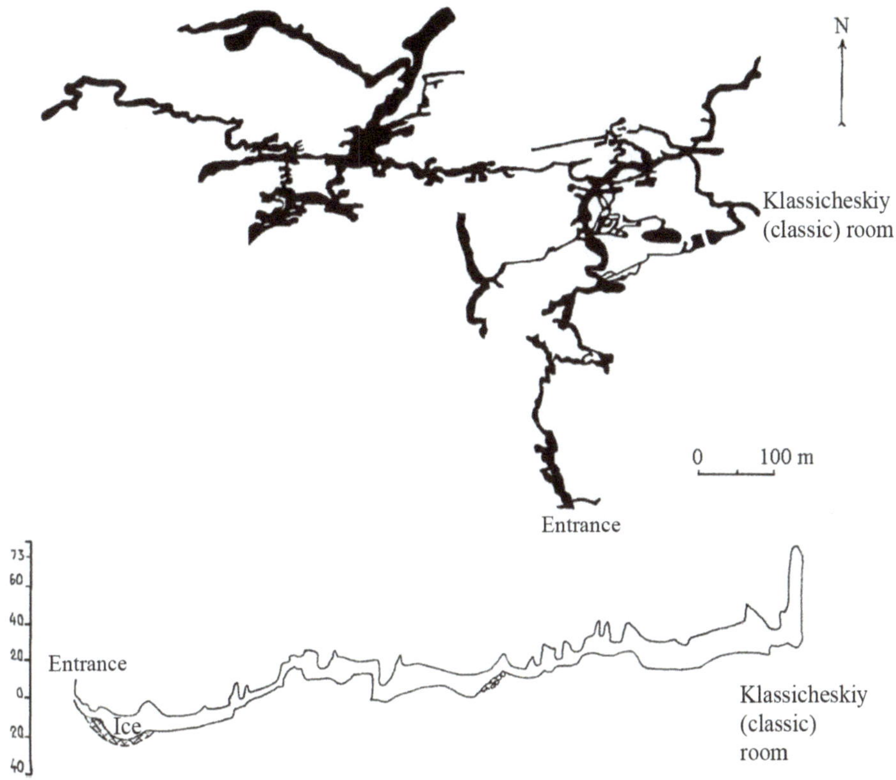

FIG. 26.28

Plan of Kinderlinskaya Cave and longitudinal section of area with glaciation. Survey of speleoclub of V. Nassonov, Ufa.

The air temperature in this zone does not exceed −0.3°C in summer and can fall to −20°C in the winter, depending on the temperature outside the cave. The air temperature is practically constant and is +6°C in the central part of cave (neutral zone).

Icing is characterized by a complex bed structure with a step ledge about 4 m high and also by sagging in the form of ship keel, where the ice mass is at its maximum thickness. The ice thickness changes from 0.5 m in the lower part of icing to 6–8 m at an entrance lattice (Stepanov et al., 2014a,b,c).

As a result, it is possible to see small degradations of cave glaciation in the cavity over time.

A chemical analysis of the ice showed some increase in ice mineralization from the entrance to the interior of the cave (from 185.0 to 224.5 mg/dm^3). Also, there was an increase in the pH but maintenance of all macrocomponents, except chlorine. Low ice mineralization was connected with icing formation from meltwater arriving through the cave's entrance during the spring. The behavior of the main chemical components reflects gradual freezing-out or flowing away. The congelation ice is of a hydrocarbonate-calcium type.

26.8.9 KUTUKSKAYA-1

The cave is located in the Kutuksky area within the Belsko-Nugushskoe interfluve between Tamantay and Kibiz ridges on the western slope of the Southern Ural Mountains. The cave entrance is located at an elevation of about 610 m a.s.l. in the Meleuzovskij district of Bashkortostan (area of National park "Bashkiriya"). The origin of the caves in the Kutuksky comprises massive gray limestone in Visay layers of Lower Carboniferous and Famenskij Layer of Upper Devon. MAT = 4.6°C, $T_{Jan} = -11.3°C$.

The cave is a cold "bag" cave. The cave length is about 520 m, and the gallery inclination is about 15–20°.

From the entrance to the central part of the biggest cave chamber (50 m × 70 m), there is extended permanent cave ice. In the chamber's center, there is a partly frozen clay talus. At the boundary of the ice, there is a clay cover with a thickness of about 0.5 m. In the western part of the cave, there is a narrow gallery with a water stream (Fig. 26.29).

The air temperatures in 2015 (July) in the central cave chamber was about −0.2°C. The results of a georadar survey showed that the maximum ice thickness was about 3.5 m (Kadebskaya and Stepanov, 2016). The greatest ice thickness was below the entrance ledge. The length of permanent icing was about 60 m (Photo 26.4).

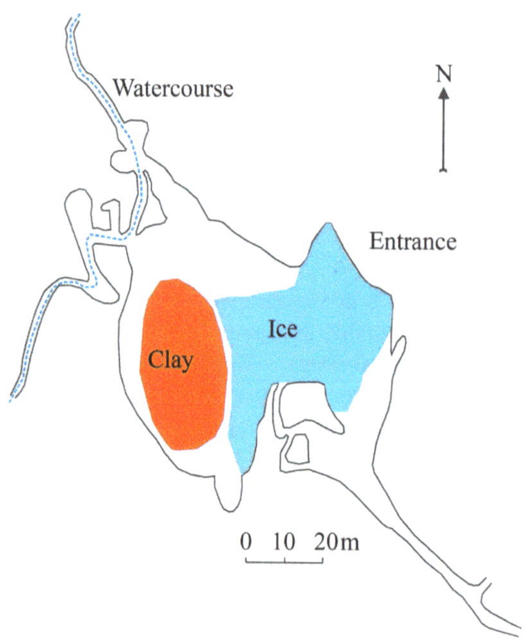

FIG. 26.29

Position of underground icing in Kutukskaya-1 Cave (cave plan by A.I. Smirnov, survey of BSU expedition, 1963).

PHOTO 26.4

Caves with permanent ice in Ural Mountains: (A) Medeo Cave (Northern Ural Mountains), (B) Mariinskaya Cave (Middle Ural Mountains), Michael Ovsejchik photo, (C) Us'vinskaya Ice Cave (Middle Ural Mountains), Michael Ovsejchik photo, (D) Bolshaya Mechkinskaya Cave (Middle Ural Mountains), Yaroslav Zanda photo, (E) Kinderlinskaya Cave (Southern Ural Mountains), and (F) Kutukskaya-1 Cave (Southern Ural Mountains).

26.9 ICE CAVES IN ALTAJ

The SIF in the caves of Altai have not been studied. Observations in the literature mainly relate to episodic inspections of the caves. Climatic observations in the caves are limited to separate measurements of air temperature. SIF are known in all karst areas in Altaj. The elevation of the caves' entrances does not seem to influence the presence of glaciation—SIF accumulations are known in caves with entrance elevations from 400 to 2000 m a.s.l. A total of 30 caves are known, with SIF of two-thirds of them being vertical. A major factor in SIF accumulation is the potential to collect a considerable mass of snow at the entrances of the caves.

By duration of the existence of these caves, it is possible to allocate seasonal, annual, and permanent SIF accumulations. Glaciers like the snow masses in vertical caves of Altai are noted with depths from 4 to 40 m from the surface. Some of them look like snow-ice stopples in tubes of karst shafts above their bottom.

Accumulations of sublimation ice during winter are marked closely to the entrance of many caves. Altaj 2 has cavities in which ice was observed on water surfaces. Ledovitaya Cave (Terektinskij Ridge) represents a shaft intersecting the ceiling of a chamber. At a depth of 30 m from the chamber's surface, there is a lake with an area of about 240 m^2. Under an aperture in the entrance pit, there is an ice island. A second similar cave is Pesochnaya Cave (the Bashchelakskij Ridge), which looks like a pit, where at a depth of 11 m, there is water and on the water's surface ice floes float. Monitoring observations on permanent ice in Tashur Cave were done by Vistingauzen V. K. in 1969. In the middle of the summer of 1969, large accumulations of ice, snow, crystals of hoarfrost, and ice stalagmites were observed, and snow penetrated at a depth of 10 m into an extensive chamber through two apertures in the ceiling. In the middle of the summers of 2005 and 2008, the cave did not contain as much ice. At the same time, SIF accumulations in the inclined Dmitrievskaya mine in the New-Chagyrskij area had completely disappeared. In 1969 ice in the mine had depths ranging from 20 to 40 m in an entrance aperture.

At the same time, in Tigirek-2 Cave, located 3 km west of Yashur Cave, the quantity of sedimentary-metamorphic ice increased. In 1959 ice of an insignificant thickness covered the bottom of an inclined gallery from the doline bottom to the mouth of the first pit. In 2008 this gallery was completely filled with snow, and ice blocked access to the depths of the cave. On an ice surface, there were small lakes. Prospective thickness of the ice stopple was several meters. Near Tigirek-2 Cave, there is Firnovaya Cave with a deep inclined crack in the rock massif, where 5 m from the entrance, accumulations of firn have been tracked to a depth of about 30 m, at which point the firn almost completely block the cavity. In Ledyanaya Katushka Cave, which is located 1 km from New-Chagyrskij, icing remains. Entrances in all these caves are situated at elevations ranging from 370 to 770 m a.s.l. In all cases, observations were made in July. For stationary observations of permanent ice in the Altaj area, more approached the Ledyanaya Katushka Cave area close by settlement Ust-Chagyrka of Krasnoshchekovskij district (Vistingauzen, 2010).

26.9.1 KUL'DYUKSKAYA CAVE

The cave is situated in the Shebalinskij district of the Republic of Altai, in the Cherginskij Ridge, near the Kul'dyuk stream, the right tributary of the Kuela River of the Cема River basin. Entrance coordinates are 51°27′17″ N, 85°28′28″ E. The cave entrance has an almost square form and is located on the right slope of the Verkh-Cherga River near the confluence of its right tributary, Munduki, at an elevation of about 300 m a.s.l. Climate is sharply continental. MAT = 1.8°C, $T_{\text{Jan}} = -16°C$.

The cave length is about 130 m, and the depth is about 38 m. The cavity consists of four parts: the area close to the entrance arch, a lateral part through the passage-cornice, the Altaj Chamber with high ceiling, and the Lustra Chamber with ice (Fig. 26.30).

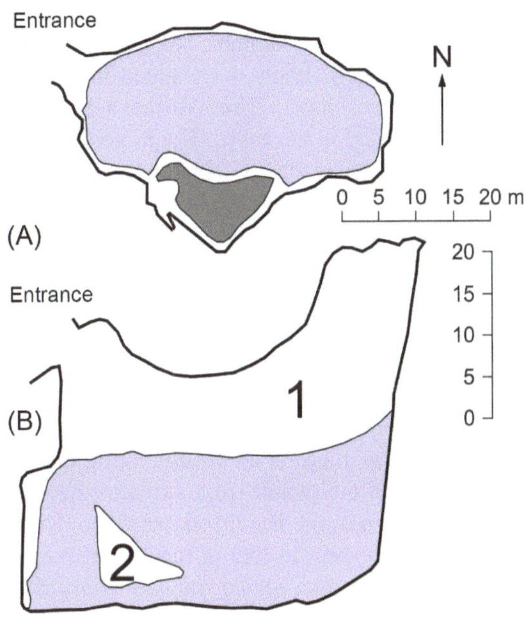

FIG. 26.30

Plan (A) and longitudinal section (B) of Kul'dyukskaya Cave: (1) Altaj Chamber and (2) Lustra Chamber (Marinin, 2001).

The cave was opened in 1976 by an expedition of cave explorers from the Mountain-Altay Pedagogical Institute.

The cave has mainly congelation ice (95%), with the remaining 5% formed by the accumulation of snow and ice breccia caused by the collapse of ice stalactites. The height of the ice columns in the cave reaches 17 m, the thickness of the layered cover ice is about 20 m, the area of ice distribution is about 510 m^2, and the ice volume is about 7200 m^3. In 1980 an increase in the cave's ice mass blocked the entrance into Lustra Grotto.

26.9.2 OROKTOJSKAYA (AJGARINSKAYA) CAVE

The cave is located near Oroktoj village on the left coast of the Oroktoj River in the Chemalskij district of the Republic of Altaj. MAT = 1.8°C, $T_{Jan} = -16$°C.

The cavity is developed in marble limestone of the Cambrian Period and in the plan looks like a zig-zag. The total cave depth is about 25 m, and the length is 53 m. The cave includes an entrance ledge (3 m), a ledge along icing, the big collapsed Mrachnyj Grotto, an inclined 15-meter tunnel, and Nizhnij Grotto.

The cave has icing with a length of about 6 m and stalagmites with heights up to 1 m and diameters of about 30–40 cm.

26.10 ICE CAVES IN KUZNETSK ALA TAU

Permanent ice and permafrost were noted in approximately half of the studied caves (Tsykin, 1981a,b). Firn-ice plugs were found in vertical caves (Efremkinskaya-6, Krest, Surpriz, etc.), and icings were found at the bottom of the entrance pits of some caves (Marinovskaya, Nakhodka, Habzasskaya). In some caves, ice completely melted during the summer, but permafrost remained in cave sediments (caves, Badzhejskaya, Torgashinskaya, etc.). Icings were noted in inclined caves (Zhemchuzhnaya, Krutaya, Kirillovskaya, Ledyanaya, Arochnaya, Efremkinskaya-1, Narvskaya, etc.). Congelation, metamorphic and sublimation ice prevailed in the caves. The largest SIF accumulations had their own movement and were referred to as cave glaciers (caves, Bidzhinskaya, Krest, Meshok) (Dmitriev, 1979). Their volume reached 2000–3000 m^3. It has been noted that sublimation ice is mainly seasonal, but it can survive more than one summer in caves with cover icings and cave glaciers (Divnogorskaya-1, Mokraya, Vladimirskaya, etc.). The greatest quantities of ice were noted in caves with large concentrations of snow in the entrance pits.

The first printed data on the area's caves appeared at the beginning of the 18th century in the publications of F. I. Strapenberg, J. G. Gmelin, and P. S. Pallas. Ice and microclimate stationary research in caves in this area was organized from 1975 to 1977. Until 1982, settlement Small Syya was the scientific base for the Institute of Merslotovedeniya SB AS USSR.

In inclined cavities, along with snow, there is an accumulation of congelation ice as a result of water inflow from the caves' arches and meltwater from surfaces that enters through the caves' entrances, for example, snow accumulation in the Bidzhinskaya Cave (Kuznetskij Ala Tau) (Dmitriev, 1977). Observations from 1970 to 1975 showed that at the end of the period of accumulation, the snow thickness in the feeding zone of the cave glacier ranged from 0.5 to 0.7 m. Because of freezing snow, meltwater formed a crust of superimposed ice with thicknesses up to 0.5 m, which was 0.1–0.3 m during spring snowfalls; after that a new layer of superimposed ice formed. Snow did not penetrate into the tongue of the cave glacier, here there was freezing of melt water from the glacier; that is, formation of congelation ice as for the most of the year air temperature in this part of cave was negative.

According to a sporo-pollen analysis, the relic thickness of ice in Nakhodka Cave in Kuznetsk Ala Tau presumably came from the Holocene (Late Quaternary) epoch (Dmitriev, 1977). That is, it had an age of about 10,000 years BP. However, this ice is relic instead of active, and this cave is located near the boundary of permafrost, which was extended greatly during the last glacial period.

26.10.1 BIDZHINSKAYA CAVE (LEDYANAYA)

The cave is located 6 km northwest of the Tolcheya settlement in spurs of Azyrtal Ridge. Entrance elevation is about 593 m a.s.l. Entrance coordinates are 54.05°N; 91.00° E. MAT = 3°C, $T_{Jan} = -17°C$.

The entrance is located at the bottom of a small doline. The mouth of this pit has a roundish form with a 2 m × 2 m section. The pit depth is about 10 m. The cave length is about 171 m, the depth is about 28 m, the area is 1300 m^2, and the volume is 8000 m^3 (Fig. 26.31). The air temperature at the end of the east gallery is +3°C, and the humidity is 95%.

At the bottom of an entrance pit, the cave glacier's thickness is up to 10 m and the area is about 150 m^2. The glacier fills the lower part of a small grotto from which twisting galleries begin in eastern and

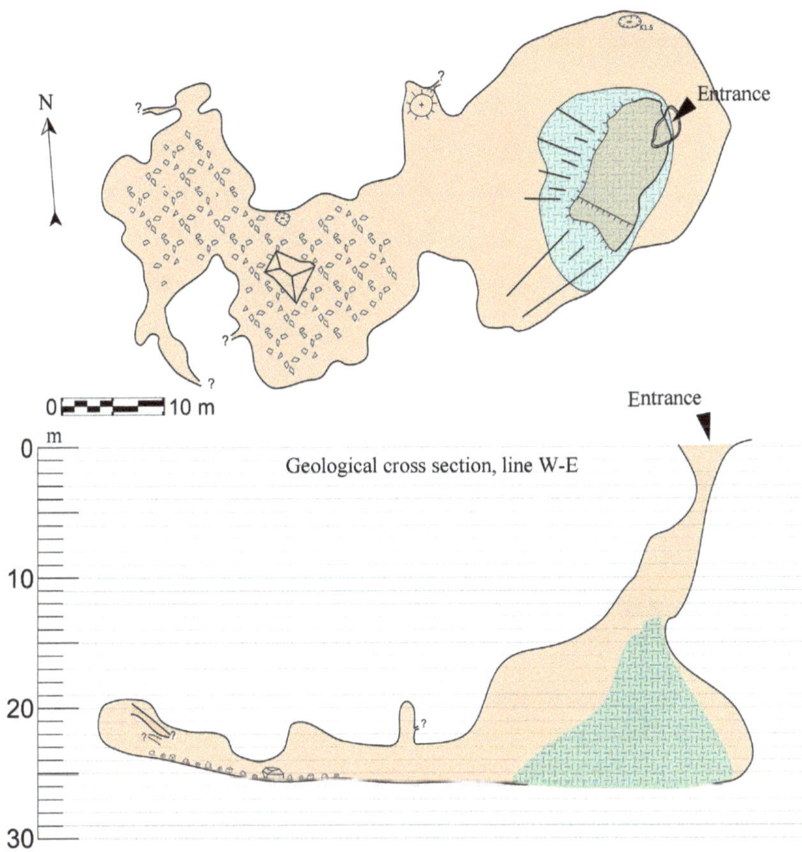

FIG. 26.31

Plan and longitudinal section of Bidzhinskaya Cave. Survey by Berezhenko D., Lankovich A., Vasil'ev A., 2013.

southern directions. At the beginning of each gallery are small tongues of the glacier. Wood fragments about 2000 years BP (V. E. Dmitriev's, pers. comm.) have been found in the ice along with the remains of animal bones.

26.10.2 CAVE KREST (SYJSKAYA)

The cave is located in Kuznetskij Ala Tau, in the Ijusskij karstic area, on the Syjsko-Efremkinskij speleological site, in the Shirinskij district of the Republic of Hakassija. The cave is situated in a rocky crest close to the Efremkino-Communar road. The nearest water source is the Malaya Syya River.

The cave is located close to the crest of the Cyjskij Ridge, at an elevation of about 340 m above the level of the Malaya Syya River, over a limy open-cast mine; it is visible from the Small Syya settlement and is 2 km to the west of the Malaya Syya River. Entrance altitude is about 750 m a.s.l. It is one of highest caves in this area. MAT = 3°C, $T_{\mathrm{Jan}} = -17$°C.

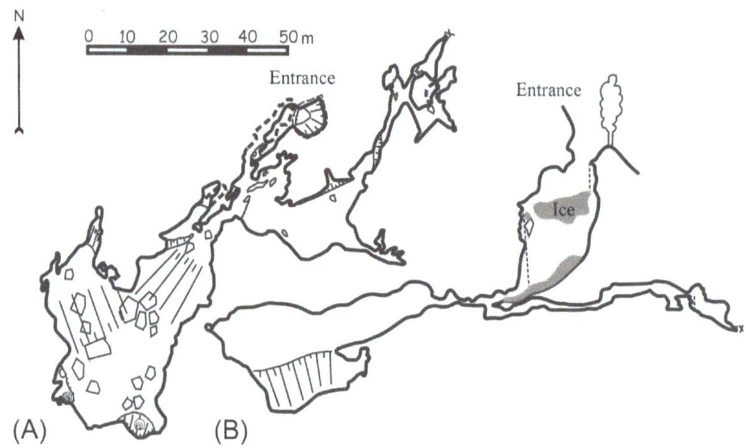

FIG. 26.32

Plan (A) and longitudinal section (B) of Krest Cave. Survey by Medvedev A., Kirichenko V., Krasonotarsk Speleo Club, 2001.

The cave entrance has crossing trapezoid diagonals, with a size of 10 m × 12 m. The depth of the entrance shaft is about 30 m. The bottom of the shaft leads to a passage into a large inclined chamber and, in an opposite direction, to a sequence of small grottos, galleries, and dead-end pits. The cave length is about 380 m, the depth 65 m, the area 2415 m^2, and the volume 12,075 m^3 (Fig. 26.32).

The first printed data about caves of the area appeared at the beginning of the 18th century and contained the works of F. I. Stralenberg, J. G. Gmelin, and P. S. Pallas. From 1975 to 1977 stationary research on ices and a microclimate were organized in the caves. In Malaya Syja settlement was until 1982 the science base of the Institute of Merzlotovedeniya (Permafrost Institute) of SB AS USSR.

A snow-ice stopple formed in the upper part of the entrance shaft. From there, ice gets to a shaft bottom, where a cave glacier up to 4 m thickness has formed.

26.11 ICE CAVES IN THE SAYAN MOUNTAINS

26.11.1 BOL'SHAYA ONOTSKAYA CAVE

The cave is located on the right coast of the Onot River in the East Sayan foothills; its origin is dolomite of the Kamchadal'skaya Layer of the Lower Proterozoic.

The cave has two entrances: the main entrance is arciform and the upper entrance is a pipe-like collapse. The cave represents a practically straight gallery, with widths from 4 to 10 m, that ends in a big grotto with a length of about 20 m and a width of 16 m (Fig. 26.33). The cave length is about 226 m, with a vertical range of about 60 m.

Seasonal congelation ice has been observed in the cave in the large central grotto in the form of stalactites, stalagmites, and cover icings on the walls and floor. Ice in the cave forms from percolating water infiltrating through cracks in the roof system and also through the upper entrance. The water

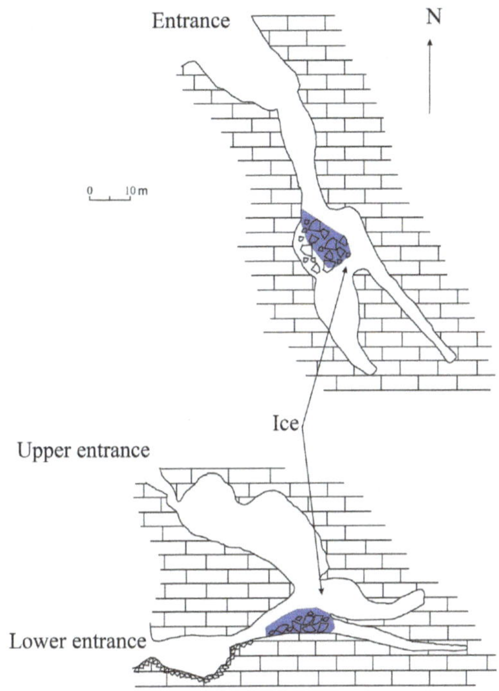

FIG. 26.33

Plan (upper) and longitudinal section (lower) of Bol'shaya Onotskaya Cave. Survey by Speleo Club Arabika, 2010.

composition in the cave consists of hydrocarbonate magnesian with a mineralization of about 148.9 g/ dm^3, which is caused by interactions between the underground water and the magnesian dolomite.

Among the cryogenic minerals is a rare mineral, lansfordite; in addition, the cryogenic flour combines with ikaite with impurities of gypsum and barite.

26.11.2 LEDOPADNAYA CAVE

The cave is located in the Berezovskij district in Krasnoyarsk Land, 21 km from highway M-54 to the west of Krasnoyarsk. In the karst area of Prienisejskaya, there is a fold-block zone in the East Sayan area. The cave entrance is situated in Stolby National Park. The entrance, in the form of an oval pit with a size of 2 m × 5 m, is located on a massif slope near the rocky crest. Entrance elevation is about 350 m a.s.l., and it is 210 m above the level of the Yenisej River. MAT = 1.4°C, $T_{Jan} = -15.7$°C.

The cave (found in 1975 by A. Baklanov) has small maze systems at different levels of depth, with the systems connected by vertical pits or by a complex of steeply inclined galleries. Cave origin is light-gray, massive, fissured limestone from the Upper Rihpean–Vend (Upper Proterozoic) period of the Ovsyankovata Layer. The cave length is about 589 m, the depth 188 m, the area 1230 m^2, and the volume 5960 m^3 (Fig. 26.34).

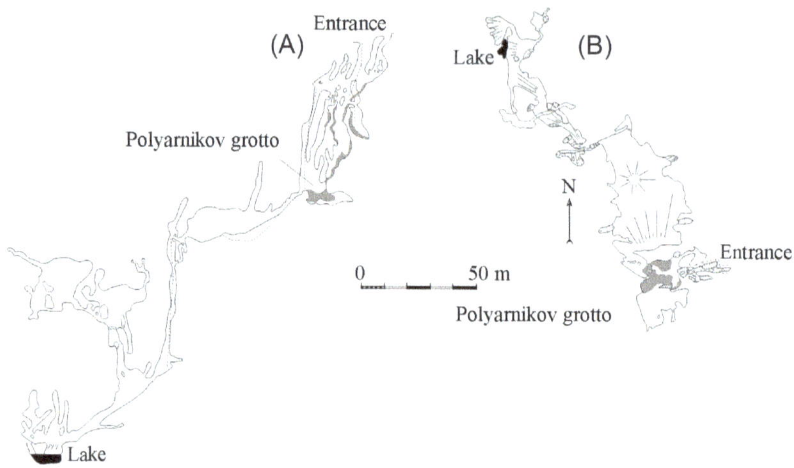

FIG. 26.34

Plan (A) and longitudinal section (B) of Ledopadnaya Cave. Survey by Medvedev A., Zaliev F., Krasonoyarsk Speleo Club, 2001.

The entrance part of the shaft up to the Pamyatnikov Grotto is covered with congelation ice. Icing decreased during the past decade.

26.12 ICE CAVES IN PRIBAIKALYE AND TRANSBAIKALIA

In the Irkutskaya area, the MAT changes from −1°C to −4°C, and the temperature of rock massifs is 2.5 to 4°C (Philippov, 1981, 1990). Perennial ice has been noted in the entrance portions of caves with a distinct warm part—Hudugunskaya, Balaganskaya, Argarakanskaya, Mechta, Nizhneudinskaya—and also in cold bag caves with wide entrances and constantly negative temperatures (−1°C to −5°C): Holopovskaya, Buryatskaya, Otonskie, and so on.

In caves in the Irkutskaya area, perennial icings are developed in the Bol'shaya Bajdinskaya (Philippov, 1989), Talovskaya, Zagadaj (Philippov, 1997), Okhotnich'ya, Khrustalnaya (Bazarova, 2010; Bazarova and Gutareva, 2011), Mechta, and Aja-Raydovaya caves.

26.12.1 BAJDINSKIE CAVES AND CAVE MECHTA

These caves are located 10 km to the south of the Olkhonskie Vorota strait, at elevations of about 250 m above the level of Baikal Lake; their origin is marble of the Olkhon series of the early Archean (Lower Proterozoic) period (Bazarova et al., 2014). MAT = 0°C, $T_{\mathrm{Jan}} = -20$°C.

Mechta Cave length is about 830 m, with a vertical amplitude of about 52 m.

Mechta Cave was opened in 1960, and in 1979 was already included in Soviet tourist booklets as a cave equipped for visiting.

It is especially interesting that Bol'shaya Bajdinskaya Cave has permanent icing (Fig. 26.35). The ice body is located 18 m from the entrance. It has steeply inclined (75–80°) ledge blocks that lead to a

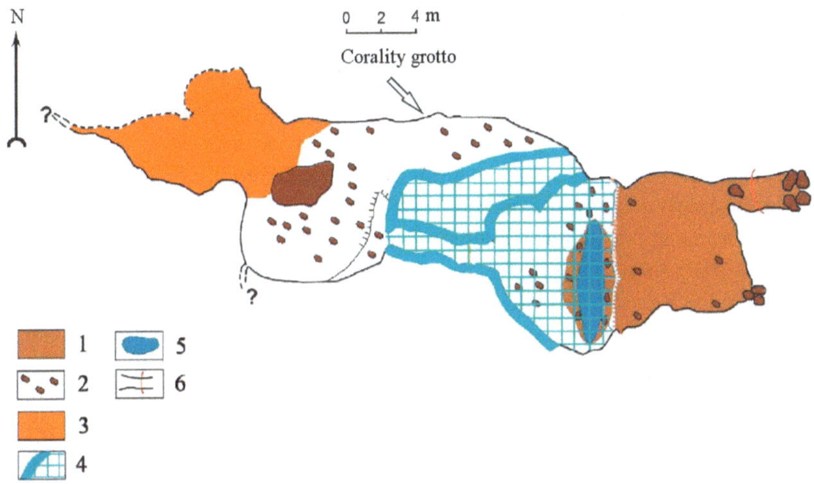

FIG. 26.35

Plan of Bol'shaya Bajdinskaya Cave: (1) soil resedimented from surface, (2) fragments and blocks, (3) sandy loam and gruss sediments, (4) ice body and its boundaries, (5) lake, and (6) boundary of snowdrift during the cold period of year. Survey by A. B. Alekseev, etc. (Speleo Club Mechta, Irkutsk).

gallery in a deeper part of the cave, the Karallitovyj Chamber, which was opened by cutting through the ice and building an artificial steeply inclined tunnel. The ice body is in the form of an ice stopple that degraded extensively over the past several decades.

Now insignificant seasonal formation of ices occurs in the grotto close to the entrance because of (1) water percolating from the surface through tectonic cracks and tubes with diameters ranging from 10 to 30 cm, (2) meltwater from the cave lake, (3) meltwater from snow drifting into grotto at a distance of 3 m, and (4) sublimation of water vapor (hoarfrost on cave ceiling and walls). In the mouth of tubes there are ice edges and ice stalactites with lengths up to 2 m. The thickness of fresh icings in different places is about 15 cm. Icings cover underlying horizontal layers of lake ice with obvious angular differences. Insignificant formation of icings with thickness of a few centimeters occur on the wall of an ice body inside the cave in the Korallotovyj Chamber.

Modern congelation ice forms at the end of autumn and beginning of winter by the freezing of infiltration water and also by melt and infiltration of water during the spring and beginning of summer.

Ice body size is as follows: length about 14 m, thickness 9.5 m, and complex internal structure. Congelation ice (icings and lake ice) forms the ice body; sedimentary-metamorphic ice has not been observed.

The upper 1.5 m of ice is typically bluish-greenish, is clean, has a high degree of transparency, and has a considerable quantity of gas bubbles. The thickness of the separate ice layers ranges from 3 to 5 cm and at times up to 18–20 cm.

In the right part of the ice outcrop, there are two intersecting ice veins with a thickness up to 20 cm that are cracked inside and that consist of transparent, colorless ice. In the left part of the ice outcrop is a clearly visible slanting lamination at the point of contact with the cave wall. It is possible to study the structure of the ice body at depths greater than 1.5 m and to penetrate into a distant part of the cave (in Korallitovyj Chamber) by using an inclined tunnel that Irkutsk cavers dug in 1984.

At the beginning, the tunnel crosses muddy horizon VIII with a complex structure. It consists of five pulled-together mud layers containing a considerable amount of marble debris, pit-run fines, various wood remains, and more rarely bones brought into the cave (apparently by people). The radio carbon age gathered from boughs at the top layer ranged from 925±25 (COAH-3050) to 320±30 years (COAH-3051).

An especially difficult history of icing development is embodied in lower 3m of ice. The presence of numerous mud horizons containing soil sediments from the surface indicates numerous periods of icing degradation. Obviously over time, icing size (thickness and area) experienced essential fluctuations. At the present time, fossil icing has retreated about 6m, based on findings in resediment soils in prospect holes dug in the Korralitovyj Chamber.

According to radio carbon dating of wood in the lower ice layers, growth of the modern ice body began about 3000 years ago [2710±30 (COAH-3047)] in late Holocene. Ice body growth was not continuous; there are numerous mud horizons corresponding to periods during which the ice body degraded and inclusions in ice were projected on younger layers.

A considerable quantity of large (up to 2cm) crystals and ikait aggregates remain in permanent icing and on its surface.

26.12.2 CAVE SYSTEM AYA-RYADOVAYA

This is the longest cave on the coast of Baikal Lake. The cave length is about 1350m, with a vertical range of about 70m (Osintsev, 2010). Entrance coordinates: 52°47.842′ N; 106°33.102′ E. MAT=0°C, $T_{Jan}=-20$°C.

The cave was discovered by a party of Irkutsk geologists looking for deposits of carbonate raw materials in Aja Bay from 1946 to 1953. Data about the cave was first presented by G. P.Vologodskij in the monography, "Karst of Irkutsk Amphitheatre" (Bazarova et al., 2014). Cave origin is calcite and dolomite-calcite graphite containing marble from the Olkhonskaya series of the Upper Arlhty (Lower Proterozoic) (Philippov, 1989).

The marble contains small layers of biotite gneisses (Gutareva, 2009) and consonant dykes of pegmatite granites (Philippov, 1989). In a distant part of the cave (Ledyanoj Grotto), permanent icing is found (Philippov, 1997).

Pollen analyses suggested that before complete melting (in 1993), the ice could have been at least 8000 years old (dating to the Boreal phase of the Holocene).

26.12.3 BOTOVSKAYA CAVE

This is the longest cave in Russia. It is located on the Verhnelenskoe karst plateau in the Central Siberian uplands. Cave entrances are located in the upper part of the left side of the Bota River (Philippov, 1994). Entrance coordinates are 55.3079° N; 105.3442°T. MAT=−3.5°C, $T_{Jan}=-26.8$°C. Botovskaya Cave origin is algal limestone of the Ust'kutskata layer of the Lower Ordovician (visible thickness about 8m) clumped among sea sandstones of the same layer (Philippov, 1999).

In 2010 the cave length was about 64,435m, with a vertical range of about 40m (Osintsev, 2010). In the plan, cave represents a meshy-polygonal one-level labyrinth (Fig. 26.36).

In cave galleries located close to the surface, there are ice formations presented as congelation (cover icings, stalactites, stalagmites) and sublimation ice. Permanent icing close to the cave entrance is referred to as "Medeo."

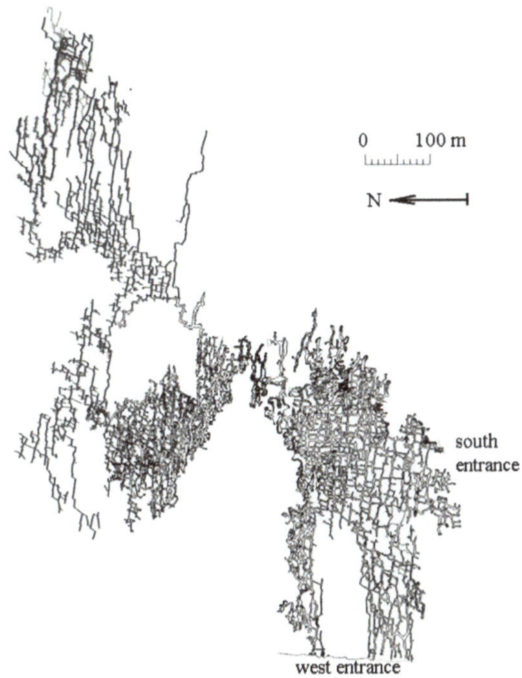

FIG. 26.36

Plan of Botovskaya Cave. Survey of Arabica speleoclub, Irkutsk, A. Osintsev, 2015.

26.12.4 OKHOTNICH'YA CAVE

The cave is located on the Primorskij ridge. It was opened in 2006 to tourists. Cave origin is oncolytic and stromatolitic limestone and dolomite of the Uluntujskaya Layer of the Upper Proterozoic. MAT=0°C, $T_{Jan}=-20$°C.

The entrance leads to a big grotto, with a width ranging from 15 to 20 m and a height ranging from 5 to 10 m, that leads to a steeply inclined talus that goes down into the cave. Okhotnich'ya Cave is third in terms of length in the Baikal region. The cave length is about 5700 m, with a vertical range of about 77 m (Osintsev, 2010). The cave orientation follows a series of subparallel faults pointing north-northeast. Galleries represent both volume passages and narrow high passages with typical slit-like form in cross-sections and heights up to 25 m.

There are seasonal snow and ice and cryogenic minerals in the cave (Bazarova et al., 2011, 2014).

26.12.5 LENSKAYA LEDYANAYA CAVE

The cave is located in the middle of the Lena River, Yakutia. The entrance to the cave is 2.5 km from Tinnaya village and is situated on rocky escarps on the left coast of the Lena River. MAT=−3.5°C, $T_{Jan}=-26.8$°C.

The uplift from the cave entrance follows a steeply inclined talus about 117 m above water level. The entrance is clearly visible from the river. Cave origin is a monocline in layered Cambrian limestone (falling to east, angle 20°). The cave length is about 134 m, and the amplitude is about 19 m. The main gallery is oriented north-northeast and has a width of about 7 m and a height ranging from 1.5 to 3 m. There are many large block collapses in the cave. Fifty meters from the entrance, after a narrow way with blockages, there is a large chamber with ice. The chamber width ranges from 10 to 15 m and has a height up to 10 m and a length of about 30 m (Fig. 26.37).

The cave was widely known in the 18th and 19th centuries, mainly because it was situated close to the Irkutsk post path, Yakutsk. The cave has been undeservedly forgotten, and in the 20th century and thus far in the 21st century has not been studied. It is mentioned in extracts based on the abstract of an article by A. I. Losev and published by V. A. Obruchev. The first cave plan (1785), which this author has not seen, is stored in state archives in the Kostromskaya area (Filippov, 2009).

The first publication about the cave was presented by A. I. Losev (1815).

The remarkable fact is that the cave is an ascending cave and an ice grotto is located 19 m above the entrance. Congelation ice has developed in the cave (Photo 26.5).

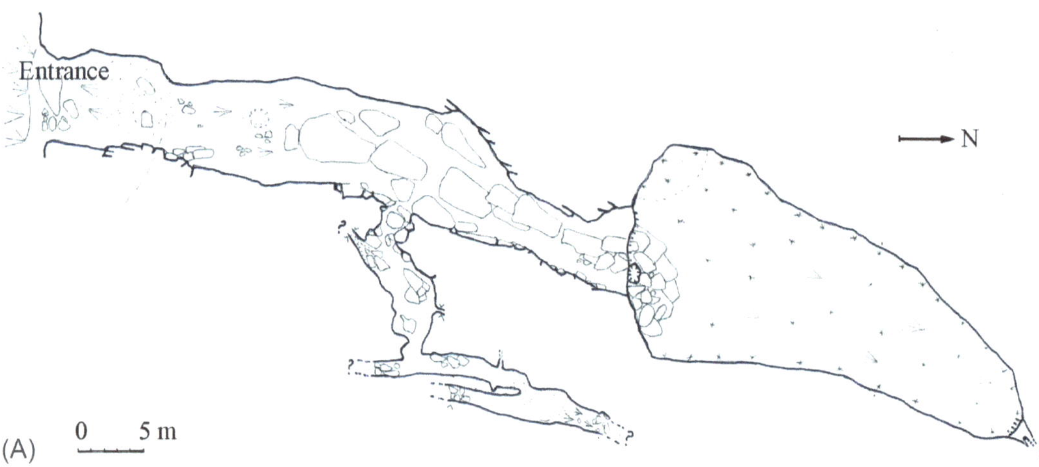

(A)

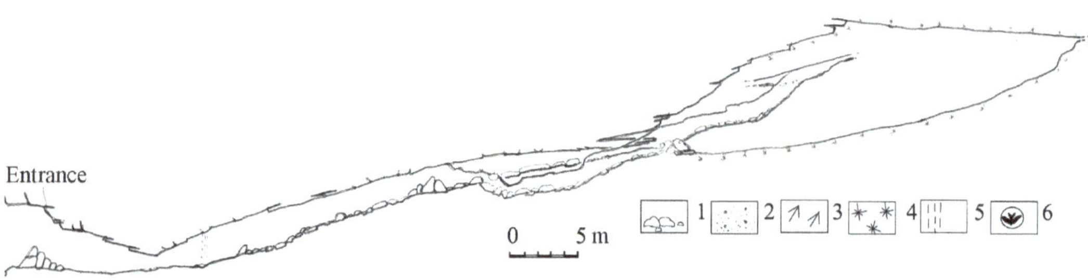

(B)

FIG. 26.37

Plan (A) and longitudinal section (B) of Lenskaya Ledyanaya Cave (Osintsev, 2009): (1) rock blocks, (2) rock debris, (3) direction of inclination, (4) ice, (5) water drops, and (6) bats.

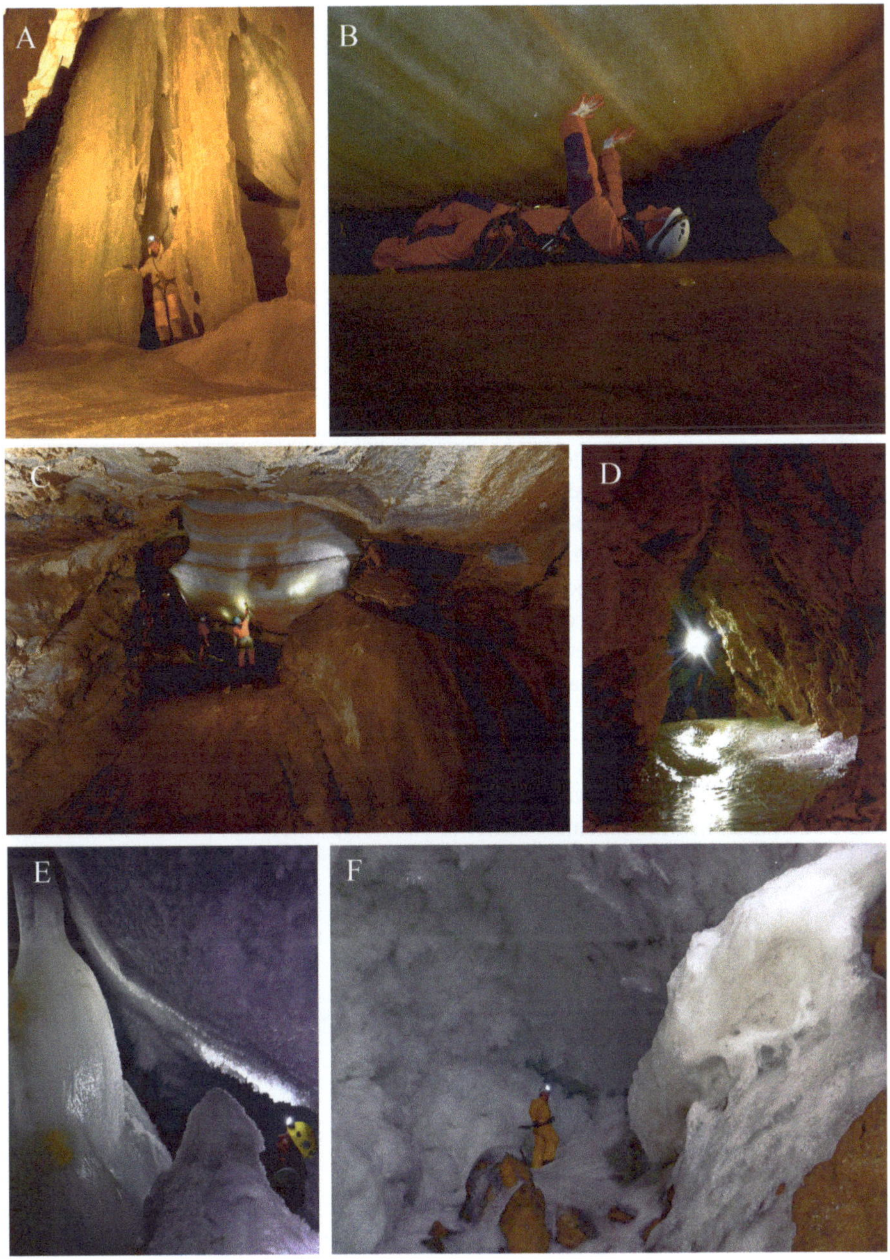

PHOTO 26.5

Caves with ice formations in Siberia: (A, B) Kuldjukskaya Cave, Altaj (photo by D. Schwarts), (C) Bol'shaya Bajdinskaya Cave, Irkutsk area (photo by Yu. Shurka), (D) Botovskaya Cave, Irkutsk area (photo by A. Osintsev), (E) Elanskaya Ice Cave, Yakutia (photo by A. Osintsev), and (F) Lenskaya Ice Cave, Yakutia (photo by A. Osintsev).

26.13 ICE CAVES IN TRANSBAIKALIA AND THE FAR EAST

26.13.1 KHEETEJ CAVE

26.13.1.1 Site and Situation Around the Cave

The Kheetej Cave is located 95 km southeast of the Mogojtuj settlement in spurs of the Ketuj-Nuru Ridge, on Onon River's right bank at the boundary of the Ononskij and Mogotujskij areas in Transbaikalian Land. The cave is about 7 km from the nearest settlement, Ust-Borzya. The cave's surroundings are mainly mountain steppe communities on graded slopes of hills (grassy and shrubby vegetation). The cave elevation is about 840 m a.s.l. Entrance coordinates are 50.630737°N, 115.759506°E. MAT = −0.3°C, T_{Jan} = −22.7°C.

26.13.1.2 The Cavity Form, Cavity Maps

The cave length is about 150–160 m. The cave has two levels and two cavities named Sukhaya (Dry) and Mokraya (Wet), which are connected to each other. The cave has four chambers: Ledyanoj, Yurta, Tupikovyj, and Kostyanoj. The first level is represented by a doline with a diameter of about 60 m and a depth of about 30–35 m. Twelve-meter ice steps descend into Bol'shoj, or Ledyanoj, Chamber. The chamber in the plan resembles a rhombus that extends in a NW direction. Its width at the bottom reaches about 52–57 m, the length about 70 m, the area about 2400 m^2, and the height about 26 m (Fig. 26.38). In the western part of the chamber, there is an ice stalagmite about 35 m in diameter. The next level of the cave, Kheetej-dry (Letnyaya), consists of doline that is larger by area and depth than the doline in Kheetej-wet. The doline at the bottom is filled with large blocks of limestone.

26.13.1.3 Research History

The first information about the cave was provided by I. G. Gmelinym in 1735. In the 19th century the cave was called Mikhachin, which means butcher or cannibal, which was probably connected with the

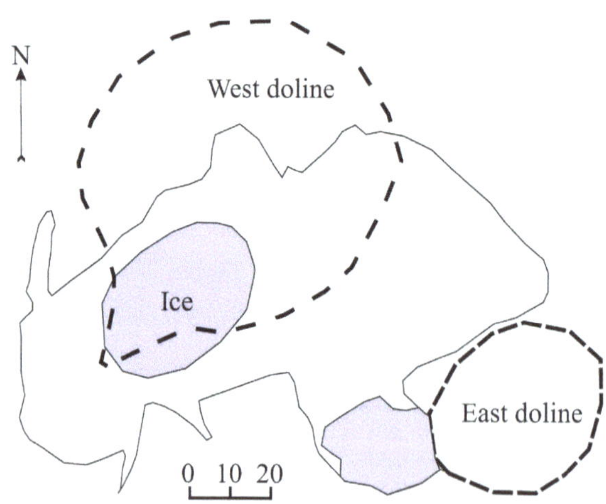

FIG. 26.38

Plan of Kheetej Cave. Survey by Yadrishenskij, 1990.

death of the people who tried to visit the cave. Complete data about the cave were acquired as a result of complex studies done by expeditions of the Transbaikalian branch of the USSR Geographical Society in 1980 and later by a group under the leadership of I. I. Zheleznyak.

26.13.1.4 Forms and Dynamics of Ice

There are congelation (icings) and sublimation ice in the cave. Within a year in the Ledyanoj Chamber, the air temperature can change from −7.1°C in the winter to about zero (−1°C to +0.8°C) in the summer; the MAT is negative. In the small-volume chambers, Yurta and Tupikovyj, ice does not form, and the temperature oscillates from +1°C to +6°C.

26.14 ICE CAVES IN THE FAR EAST

In the Far East cave glaciation is found in cavities in the Khabarovskij and Primorskij provinces (Dyomin, 1981). Permanent icings are noted in the Starogo Medvedya, Steregushee Koh'ye, Dal'giprolestransa, Ledyanaya, Priiskovaya, Ledyanaya Malutka, Komsomol'skaya, and Kholodil'nik caves. All named caves are inclined descending caves. In such caves, ice is present as icings formed by melting snow that penetrates the caves. During the winter, close to the caves' entrances, sublimation crystals grow on gallery arches and walls. In neutral zones of such caves, the air temperature remains practically constant at about +3°C to +5°C (Table 26.2).

26.14.1 PRIISKOVAYA CAVE (SINEGORSKAYA-2 OR ZOLOTAYA)

The cave is located in the Jakovlevskij district of the Primorskij province, 8.5 km NE of the lumberers' LZP-3 settlement, on the northern slope of a mountain with an elevation 522 m a.s.l., on the left side of the Peshernyj Spring valley (right tributary of Ptiiskovyj rivulet). The cave entrance's elevation is about 500 m a.s.l.

The cave is a big chamber connected to the surface by a collapse doline on one end that gradually descends and narrows and becomes a narrow passage leading into the branched part of the cave. The cave length is about 270 m, and the depth is about 69 m (Fig. 26.39). Flowstones, kassiterit as alluvial deposits, other metals, and also fragments of medieval ceramics have been found in the cave.

Table 26.2 Morphometric Data of Ice Caves (Bersenev, 1989)

Cave Name	Length, m	Area, m^2	Volume, m^3	Entrance Height, m a.s.l.	Kind of Ice	Ice Volume, m^3
Starogo Medvedya	130	440	2100	330	I, CK	180
Ledyanaya Katushka	27	33	95	400	I, CK	
Ledyanaya	325	2600	18,500	280	S,	
Steregushee Kop'ye	1100	2770	6600	220	I, S, CK	250
Ledyanaya Malyutka	98	220	570	340	I, CK	
Kholodil'nik	60	105	330	600	I, CK	160
Priiskovaya	290	760	2400	530	I, CK	90
Vitnitskaya	51	80	139	395	S	

S, snow; I, icings; SK, sublimation crystals.

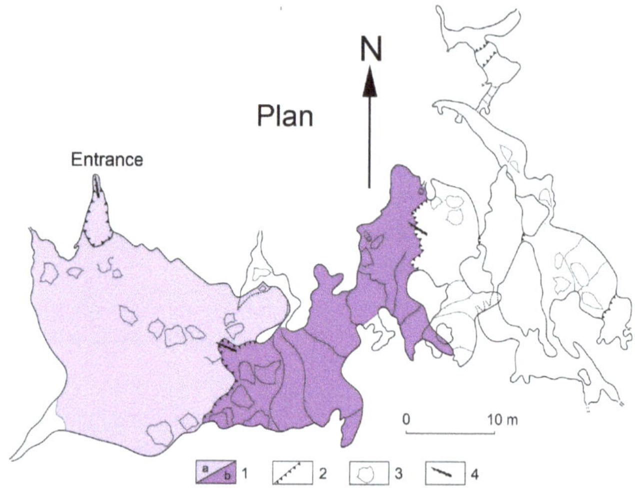

FIG. 26.39

Plan of Priiskovaya Cave: (1) ice: a – seasonal, b – permanent; (2) ledges in ice and rocks; (3) approximate isolines showing relief of cave floor; and (4) tree used as ladder. Survey of Vladivostok speleoclub.

The cave was first surveyed in 1964 by A. Ya. Barabanov, though it had been discovered a bit earlier. In 1965 E. G. Leshok visited the cave. The cave plan was made by V. I. Bersenev in 1968, and later he made some additions, including detailed descriptions.

The cave has developed permanent congelation ice (icing with length about 27 m and thickness about 1 m) and has seasonal sublimation ice close to the cave entrance.

26.14.2 STAROGO MEDVEDYA CAVE

The cave is situated 30 km from the Birakan settlement in the Evrejskij Autonomous district of Khabarovskij Land near the Bira River, on the southern slopes of the Sutarskij Ridge. The cave is known to have a permanent ice body (icing). The elevation of the cave entrance is about 330 m a.s.l. MAT = −1.6°C, T_{Jan} = −27.1°C.

The cave consists of two chambers connected by a narrow passage with a length of about 5 m (Fig. 26.40).

In the first chamber there is a dome-shaped ice accumulation with an area 60 м^2 and a thickness up to 7 m (average is about 3 m). On a 5-meter natural outcrop of ice, alternating dark and light ice layers with an average thickness of about 4–6 cm (maximum, 10 cm) are visible.

26.14.3 CAVE KHOLODIL'NIK (POLARNAYA)

The cave (opened in 1972) is located in the area of the Dal'negorsk settlement on northern slope of Sakharnaya Golova Mountain (875 m a.s.l.) at height about 300 m above stream level, 2 km from a highway. Cave origin is massif of Triassic limestone. Entrance elevation is about 600 m a.s.l. MAT = 1.8°C, T_{Jan} = −15.0°C.

The cave has a length of about 60 m, a depth of 23 m, a vertical range of 26 m, an area of 105 m^2, and a volume of 330 m^3. The cave entrance was formed at the base of a dry valley and represents an angular

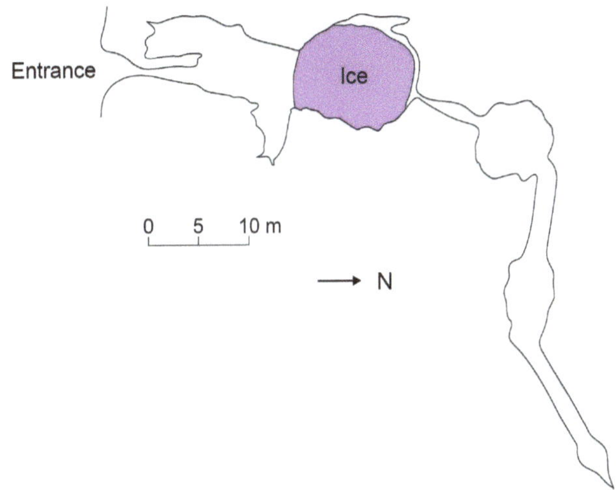

FIG. 26.40

Plan of Starogo Medvedya Cave (Demin et al., 1980).

form collapse with a width of about 7.5 m and a height of about 6 m. The second entrance aperture, which is nearly filled with fragments of limestone, is located on a terrace close to a crest 10–15 m above the previous entrance and represents a triangular aperture with a height of 0.5 and a width of 1 m. From the entrance, an inclined gallery (angle 30°) begins with a length of about 30 m, a width of 5–6 m, and a height of about 10 m. The gallery floor is filled with fragments of country rocks. Eighteen meters from the entrance, the talus debris on the gallery floor reaches a horizontal ice surface and continues under it. The maximum height of the gallery's arch is above the ice.

The cave was opened in 1972. Biologists have conducted investigations in the cave.

The ice area is about 20 m^2, and the ice thickness is up to 9 m (according to some data, it is about 11 m). The total volume of the ice is about 160 m^3. It is completely filled by ice, with certain bag-like deepening in a distant part of the cave. It seems that after the cave opened, the quantity of the ice in it increased a little. But in the past few years, the ice has begun to decrease, which is probably connected with the fact that a pit was dug at the left wall at a depth of about 4 m in 2005. By studying the annual layers, the ice's age is estimated to be 75 years.

26.14.4 LEDYANAYA MALYUTKA CAVE

The cave is located to the east of the city of Dal'negorsk (Primorskij province), in the Gorbusha River valley, on the southwest slope of Dovgalevskaya Mountain, at the watershed of the Malyutka and Isvestkovyj rivulets. The relative height above the river is about 25 m; the absolute elevation is about 340 m a.s.l. Entrance coordinates: 44.617° N; 135.67° E. MAT = 3.4°C, $T_{Jan} = -13.4$°C.

The entrance in the cave is located at the bottom of a doline that leads downward (at an angle of about 25–30°) to a gallery covered by ice. The second collapse entrance is located in the gallery arch. The cave is developed on the basis of two systems of cracks and is characterized by alternating hardly passable passages and small chambers. The cave ends where limestone comes in contact with siltstone. The cave length is about 98 m, the depth is about 20 m, the area is 220 m^2, and the volume is 570 m^3 (Fig. 26.41). Widely spread collapse sediments and flowstones are in the cave.

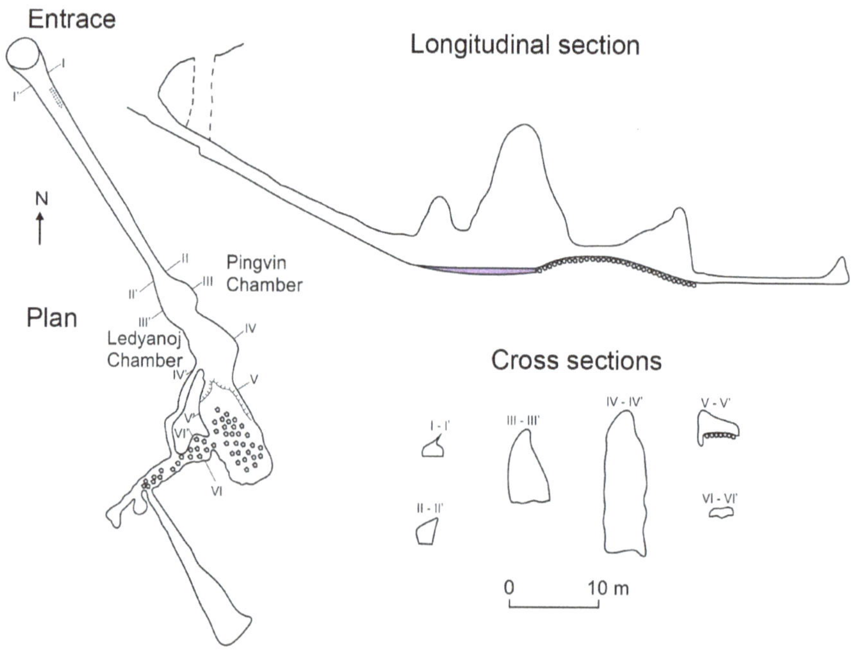

FIG. 26.41

Plan, longitudinal and cross-sections of Ledyanaya Malyutka Cave. Survey by O. I. Prikhodchenko et al., 1969 with additions of Yu. N. Bersenev, 1972.

The cave was investigated in 1969 by O. I. Prikhodchenko who developed the plan and descriptions.

It contains permanent congelation ice (icing) and seasonal stalagmites, stalactites, and sublimation crystals.

26.14.5 CAVE ABAGY-DZHE (ZHILISHCHE CHYERTA, ABAKHTY-DIETE)

The cave is located on the left coast of the middle current of the Majya River (Uchuro-Majskaya structural-formation zone) in the area of the Ajano-Majskij district in the Khabarovskij province. The entrance elevation is about 258 m a.s.l. MAT = −10.1°C, $T_{\text{Jan}} = -41.2$°C.

The cave length is about 1400 m, the depth is 16 m, the area is 5300 m^2, and the volume is 30,000 m^3. Country rocks are subhorizontal, calcinated, and brecciated cataclastic dolomite of the Tsipandiskaya layer. The cave has pronounced, northeast-spreading subvertical fractures. The cave represents a system of radially dispersing, northeast-spreading tectonic cracks that have expanded because of corrosion. Passages formed on the northwest-spreading cracks have a subordinate value. Chambers with an area up to 1400 m^2 occur in places where the galleries cross in different directions. The galleries' widths range from 0.5 to 2.5 m, and their heights range from about 5 to 10 m (Fig. 26.42). About 70% of the cave is constantly covered by water. The water's depth can reach 11 m. The level of the cave water synchronously changes with water level changes in the Majya River. During floods, many galleries become siphons, and some of them become impassable as they pinch out and upward.

During the winter water in the cavity freezes from above (first 0.05–0.20 m). In the summer the ice partially melts. Lines of ice remain on the cave walls. At water level fall on chamber floors, ice floes

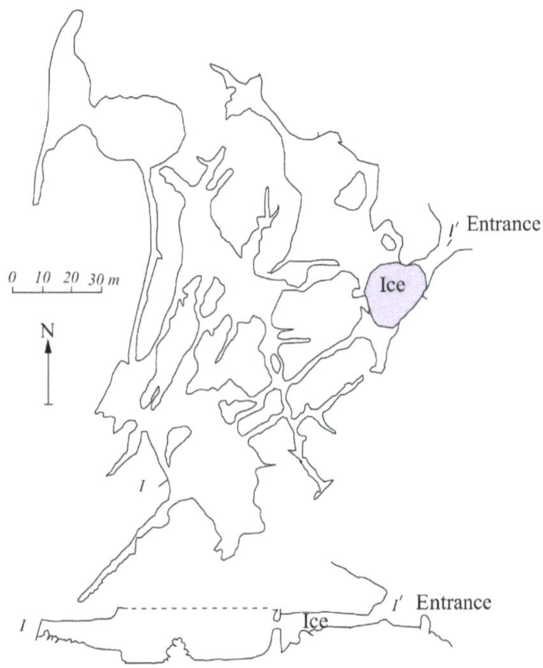

FIG. 26.42

Plan and longitudinal section of Abagy-Dzhe Cave (Bersenev, 1989).

with area to 10 м^2 can be found. Close to the entrance, ice of various origins can be observed in the summer: congelation (from water freezing in the spring during floods, forming layers with a thickness up to 5 cm with air layered between them), metamorphic (formed by snow blown into the cavity), and sublimation ice whose crystals develop mainly on the chamber's southern wall close to the entrance. There is a tube with a diameter of about 2 m on the chamber's arch close to the entrance. Now the pipe is filled with an ice stopple, and a suspended lake (during summer) exists on its surface. At the beginning of 20th century, this stopple was absent because the part of the chamber that was close to the entrance had a greater quantity of ice. The regulation of cold and ice accumulation in the cave was natural until 1940. In 1940 the cave entrance was closed with a timber blockhouse door (Afanas'ev, 1988). Thus the temperature in the cave was kept at a negative level. As a result, the cave galleries and vertical pi (organ) were blocked by ice in distant parts of the cave. In 1960 the blockhouse was disassembled, but the previous conditions for the existence of ice in cave were not restored.

26.14.6 STEREGUSHEE KOP'YE CAVE

The cave (discovered by he Khabarovsk cavers 25 years ago) is located in the vicinity of the Hail of Pobedinskoe rural settlement, 70 km from the train station, Sanboli, Khabarovskij, in the Khabarovskij province, at the middle of a Kur River watercourse. The cave elevation is about 220 m a.s.l. and is 32 m above river level. Cave origin is Permian limestone. Coordinates: 49,43972° N; 134,55238° E. MAT = 1.7°C, T_{Jan} = −21°C.

The cave length is about 1100 m, the depth is 35 m, the area is 2770 m^2, and the volume is 6600 m^3. The cave has a number of big chambers (up to 400 м^2) united in one system by numerous intertwining galleries with different dimensions. Four entrances are located at the bottom of collapse dolines (Fig. 26.43). A central entrance begins at the bottom of a large doline by a big inclined chamber. The chamber floor is made of stone talus partially covered by ice. Further on, there is a big area of permanent icing. There are three cave levels with a vertical range of about 5–10 m. In some places, the upper levels are united forming a high chamber with arches and eaves. Two pits in the cave are constantly filled with water; lakes with depths up to 2.5 m are periodically formed. During snow melting or downpours, streams penetrate into the cave and can cut "canyons" in the ice with depths up to 1 m.

Developed icings and sublimation ice are in the cave. The ice formation is directly connected with the origin of the collapse dolines and the collapse of the gallery arches: snow masses form at the bottom of dolines, and meltwater freezes in the frozen cavities forming icings. The air temperature ranges from −0.64°C to −2.20°C in the NTA zone. In the central chambers, the temperature is 1°C to −5°C. Long-term observations show that the quantity of ice in the first chamber increases annually, although the total amount of ice in the cave has decreased, which has allowed reuniting some caves—Kvadrat, Truba, Burunduk, Alyenushka, Giprolestrans—that were previously separated because the connections between the galleries were blocked by ice.

There were interesting results of radio carbon dating of wood from under icing in the cave. It appears that two samples taken in one place showed the age as 6793 ± 14 and 948 ± 74 years BP (Photo 26.6) (Bersenev, 1989).

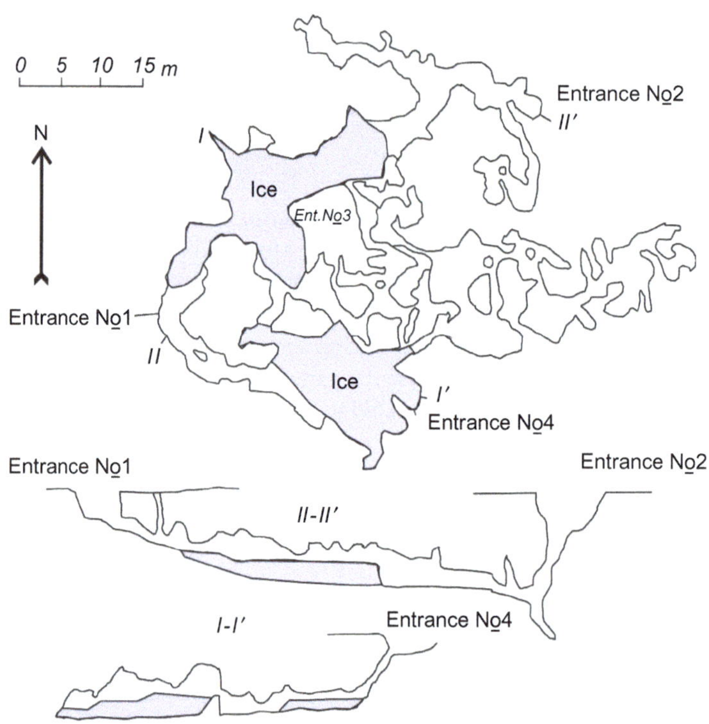

FIG. 26.43

Plan and longitudinal sections of Steregushee Kop'ye Cave (Bersenev, 1989).

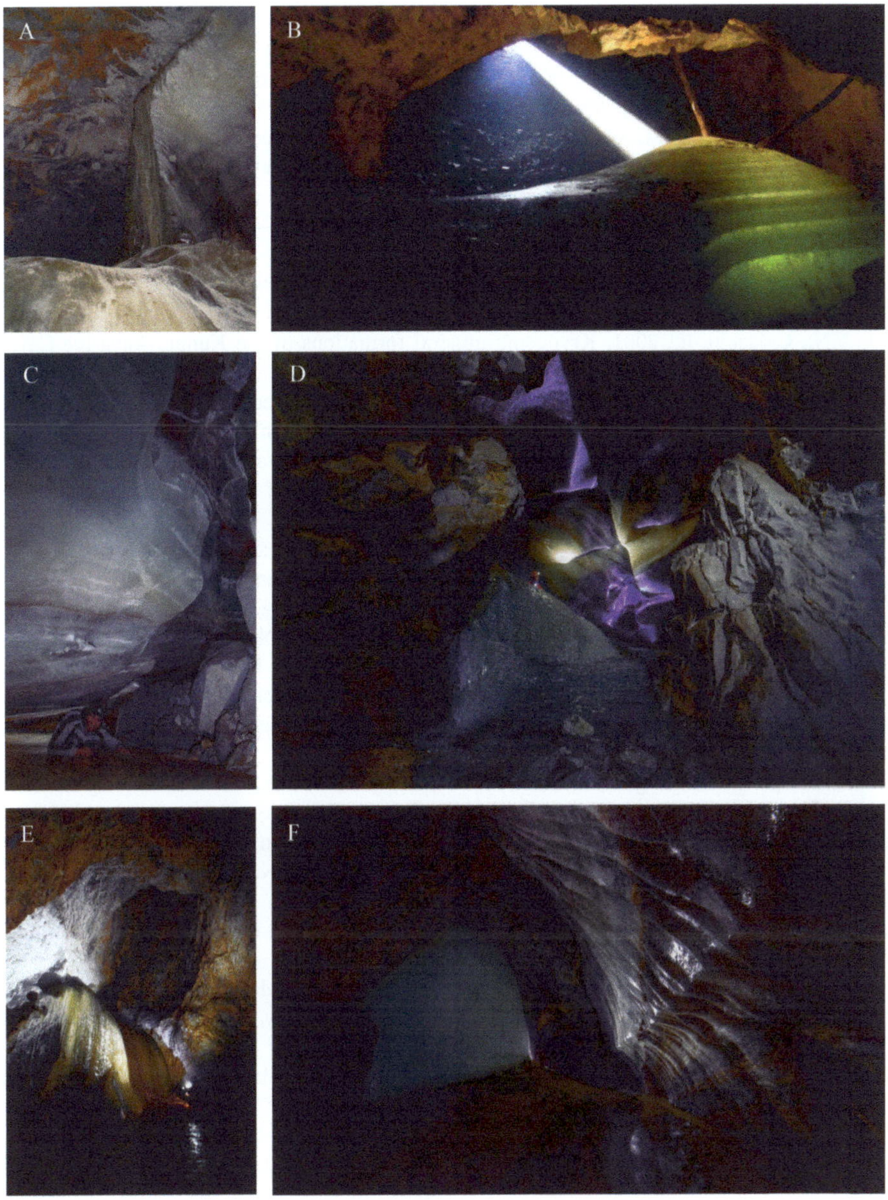

PHOTO 26.6

Caves with ice of Siberia and Far East. A – perennial ice in Oroktojskaya Cave (Republic Altaj); B – perennial ice in Starogo Medvedya cave (Jewish autonomous region); C, D, F – perennial ice in Stereguslheye Cop'yo Cave the (Khabarovsky Krai); E – perennial ice in Heyetej Cave (Transbaikalian Land).

REFERENCES

Afanas'ev, P.Yu, 1988. Microclimate change of Abagy-Dzhe Cave as a result of time economic use. In: Problems of Studying of Technogenic Karst. pp. 96–98. Abstracts of reports. Kungur (in Russian).

Alekseev, V.R., Belyak, V.I., 1970. Caves ice of Southern Siberia. Bull. Moscow State Univ. 1, 59–65 (in Russian).

Altberg, V.Ya, 1930. Observations in the Kungurskaya Ice Cave in 1928. Trans. State Hydrol. Inst. 26–27, 69–78. Leningrad (in Russian).

Altberg, V.Ya., 1931. Observations in the Kungurskaya Ice Cave in 1929–1930. Trans. State Hydrol. Inst. 32, 77–92 (in Russian). Leningrad.

Altberg, V.Ya., Troshin, V.F., 1931. About new forms of crystal ice. Trans. State Hydrol. Inst. 32, 93–107. Leningrad (in Russian).

Andrejchuk, V.N., Galuskin, E., 2001. Kryogenic mineral formations of the Kungurskaya Ice Cave. Peshery (Caves) 27–28, 108–116 Perm (in Russian).

Andrejchuk, V.N., Kadebskaya, O.I., Chajkovskij, I.I., 2013. Cryogenic Mineral Formations of Kungurskaya Cave. Silesian University-Mine institute UB RAS, Sosnovets-Perm. 128 pp. (in Russian).

Bazarova, E.P., 2010. Kryomineral formations of Irkutskaya and Chitinskaya areas. Speleology and spelestology: development and interaction of sciences. In: Collection of Materials of the International Scientific-Practical Conference. NSPI, Naberezhnye Chelny, pp. 35–38 (in Russian).

Bazarova, E.P., Gutareva, O.S., 2011. New data about mineralogy of caves of Irkutskaya area. In: Structure of Lithosphere and Geodynamics: Materials of XXIV All-Russia Youth Conference (Irkutsk, 19–24 April 2011). IEC SB of RAS, Irkutsk, pp. 62–63 (in Russian).

Bazarova, E.P., Gutareva, O.S., Kononov, A.M., Ushchapovskaya, Z.F., Nartova, N.V., Осинцев, A.V., 2011. Mineral of Okhotnich'ya Cave (Baikal region, Irkutskaya area). Speleol. Karstol. 7, 5–14 (in Russian).

Bazarova, E.P., Kononov, A.M., Gutareva, O.S., Nartova, N.V., 2014. Specific of cryogenic mineral formations in Okhotnich'ya Cave in Pribaikalie (Irkurask Region). Cryosphera Zemli 17 (3), 67–76. (in Russian with English summary).

Belyak, V.I., 1973. Spetsific of icing formation in caves of Southern Siberia. Problems of icing formation. Trans. Zabajkalkaya Branch USSR Geogr. Soc. XCII, 65–67. Chita (in Russian).

Bersenev, Yu.I., 1989. Karst of Far East. Nauka, Moscow 182 pp. (in Russian).

Caves of Volga basin, Ural and Pri-Ural, 2010. Statistical reference book. Naberezhnte Chelny: NSPI, 71 pp. (in Russian).

Chibilyov, A.A., 2011. Ural Mountains: Natural Diversity and Euro-Asian Boundary. Ekaterinburg, Institute of steppe UB RAS. 169 pp. (in Russian).

Dmitriev, V.E., 1972a. About ice formations of caves of Kuznetskij Ala Tau and the Batenevskij Range. Quest. Geogr. Kuzbass Mount. Altaj 6, 65–69 Novokuznetsk (geology, geomorphology, physiography) (in Russian).

Dmitriev, V.E., 1972b. About ice formations of some caves of Kuznetskij Ala Tau. Quest. Geogr. Kuzbas Mount. Altaj 6, 65–69. Novokuznetsk (in Russian).

Dmitriev, V.E., 1977. Ice formation on caves glaciers. Nat. Econ. Siberia 19–20 Yakutsk (in Russian).

Dmitriev, V.E., 1979. Specifics of oscilations of cave glaciers of Kuznetsk Ala Tau. Materialy Glatsiologicheskikh Issltdovanij (Data Glaciol. Stud.) 36, 181–185 (in Russian).

Dmitriev, V.E., 1980a. To a question on the complex approach to cave ice studying. Quest. Geogr. Siberia 13, 99–102. Tomsk (in Russian).

Dmitriev, V.E., 1980b. Caves glaciation as a part of the Earths glaciosphere. Karst Siberia Far East 130–145 Vladivostok (in Russian).

Dmitriev, V.E., Chujkov, V.D., 1982. Geographical regularities of caves glaciation. In: Conditions, Tasks and Methods of Studying of a Deep Karst of the USSR. p. 136 (in Russian).

Dneprovskij Speleo Club, 2007. http://www.speleo.dp.ua.

Dorofeev, E.P., Mavlyudov, B.R., 1993. Dynamics of Kungurskaya Cave glaciation. Peshery (Caves) 23–24, 131–140 Perm (in Russian).
Dorofeyev, E.P., 1967. New plan of Kungurskaya Ice Cave. Peshery (Caves) 5 (6), 110–113 (in Russian).
Dorofeyev, E.P., 1969. Ice crystals of the Kungurskaya Cave. Peshery (Caves) 7/8, 30–39. Perm (in Russian).
Dorofeyev, E.P., 1988. Evolution of glaciation in Kungurskaya Cave. Peshery (Caves) 21, 36–41. Perm (in Russian).
Dorofeyev, E.P., 1990. Permafrost and underground ice of Kungurskaya Cave. In: Problems of Geometeorology and Accumulation of a Winter Cold. pp. 18–24. Sverdlovsk (in Russian).
Dorofeyev, E.P., 1991. Icings of Kungurskaya Cave. In: Problems of Icing Formation. Institute of Geography SB AS USSR, Novosibirsk, pp. 181–188 (in Russian).
Dublyanskij, V.N., 1963. About a snow role in karst formation and a feed of karst water. Proc. USSR Acad. Sci., Sect. Geogr. 2, 69–75 (in Russian).
Dublyanskij, V.N., 1977. Karst Caves and Shafts of Mountain Crimea. Nauka, Leningrad. 182 pp. (in Russian).
Dublyanskij, V.N. (Ed.), 2005. Kungurskaya Ice Cave: Experience of Mode Absérvations. UB RAS, Ekaterinburg. 376 pp. (in Russian).
Dublyanskij, V.N., Lomaev, A.A., 1980. Karst Caves of Ukraine. Naukova Dumka, Kiev. 180 pp. (in Russian).
Dublyanskij, V.N., Shutov, V.I., 1967. Corrosion-nival karst cavities of mountain Crimea. Trans. USSR Geogr. Soc. 99 (6), 468–476 (in Russian).
Dublyanskij, V.N., Andrejchuk, V.N., Bersenev Yu, I., et al., 1992. Chemical composition of ice of karst cavities of the USSR. Eng. Geol. 5, 119–123 (in Russian).
Dyomin, L.V., 1981. Ice formations of caves of south of Far East. In: Accumulation of Winter Cold in Rocks and its Use In National Economy. pp. 50–52. Perm (in Russian).
Ezhov, Yu.A., Dorofeyev, E.P., Lukin, V.S., 1990. Floods in Area of City Kungur (Their Reasons, Dynamics, Prediction and Arrangements Struggle With Them). UB AS USSR, Sverdlovsk. 48 pp. (in Russian).
Filippov, A.G., 2009. Lenskaya Ice Cave in Yakutia. Peshery (Caves) 32, 130–138. (in Russian).
Fyedorov, E.S., 1883. Note about Kungurskie Caves. Mater. Geol. Russia XI, 217–243. SPb (in Russian).
Glazyrin, G.E., 1985. Distribution and a Mode of Mountain Glaciers. GIMIZ, Leningrad. 181 pp. (in Russian).
Golod, V.M., 1981. Mathematical model of aerothermohydro-dynamic processes in an aeration zone. In: Accumulation of Winter Cold in Rocks and Its Use in National Economy, pp. 6–8. Perm (in Russian).
Golod, V.M., Golod, M.P., 1974a. Microclimate of gypsum caves of Pinezhye. Caves of Pinego-Severodvinskoj Karst Area, 128–154. Leningrad (in Russian).
Golod, V.M., Golod, M.P., 1974b. Mikroklimat of gypsum caves of Pinezh'ya. Cave of Pinego-Severodvinsk Karst Area. pp. 128–154. Leningrad (in Russian).
Golod, V.M., Golod, M.P., 1975. Analysis of ice caves formation. Conditions and Problems of Karst-Speleological Researches 37–38 Moscow (in Russian).
Golod, V.M., Golod, M.P., 1978. Problematics and a technique of microclimatic observations in caves. In: Research of Karst Caves With a View of Their Use as Excursion Objects. pp. 194–196 Tbilisi (in Russian).
Golod, V.M., Golod, M.P., 1981. Specifics of microclimate of gypsum caves of Pinego-Severodvinskaya karst area. In: Accumulation of Winter Cold in Rocks and Its Use in National Economy. pp. 49–50 Perm (in Russian).
Golovkov, M.P., 1938. Research of ice of the Kungurskaya Cave. Trans. Sci. Res. Inst. Earth Crust LSU 1, 54–81 (in Russian).
Golovkov, M.P., 1939a. Note about structure and morphological features of ice crystals. Trans. All-Union Mineral. Soc. 68 (2), 163–170 (in Russian).
Golovkov, M.P., 1939b. Research of ice of the Kungurskaya Cave. In: Proceedings of LSU, Serie Geology-Soil Science, vol. 21 (5), 11–35. (in Russian).
Guslitser, B.I., Kanivets, V.I., 1965. Caves of Pechorian Ural. Nauka (Leningrad Division), Moscow-Leningrad. 134 pp. (in Russian).
Gutareva, O.S., 2009. The Kajnozojskij karst in Priolkhonie (southwest coast of Baikal Lake). Speleol. Karstol. 2, 77–82 (in Russian).

Gvozdetskij, N.A., 1954a. Karst. Geografgiz, Moscow. 352 pp. (in Russian).
Gvozdetskij, N.A., 1954b. Karst, second ed. Geographgis, Moscow (in Russian).
Gvozdetskij, N.A., 1968. Types of ice caves in the USSR. In: Actes du IV Congress Internationale de Speleologie en Yougoslavie, Ljubljana. vol. 3, pp. 463–467 (in Russian).
Kadebskaya, O.I., 2004. Geoecological conditions of the Kungurskaya Ice Cave and adjoining area, its protection and rational use. (Author's abstract of PhD dissertation). Perm University, Perm 28 p (in Russian).
Kadebskaya, O.I., Stepanov, Yu.I., 2016. Characteristic of icings and cryogenic minerals in caves of Kutuk area (Bashkortostan). Geol., Geogr. Glob. Energy 2, 30–39 (in Russian).
Karakash N.I., 1905. Kungurskaya "Ice" Cave in Urals Mountains. Transaction of Imperial SPb. Society of Naturalist 36 (1) (Reports of sessions for 1905). SPb. (Printing house of Merkur'ev M.). 1905–1906, 11–24 (in Russian).
Khromykh, V.G., Belyaev, S.Yu, 2010. Devonian reeves of east slope of Polar Ural Mountains. Lithosphere 2, 25–34.
Kiknadze, T.Z., 1972. Karst of Arabika massif. Metsniereba, Tbilisi. 248 pp. (in Russian).
Kruber, A.A., 1900. About karstic phenomena in Russia. Zemlevedenie IV, 1–34. (in Russian).
Kruber, A.A., 1915. Karst area of Mountain Crimea. . Moscow, 319 pp. (in Russian).
Kudryashov, I.K., Salikhov, R.A., 1968. Reasons of cold conservation and ice formation in Ackinskaya Cave. Trans. Baskirian Branch USSR Geogr. Soc. 5, 68–79 (in Russian).
Kuzmina, L.Yu, Chervyatsova, O.Ya., Gilyamzyanova, N.F., Ryabova, A.S., 2014. Aerobiological research of the Askynskaya Ice Cave (southern Ural Mountains). In: Speleology and Spelestology. vol. 5. NSPI, Naberezhnye Chelny, pp. 303–311 (in Russian).
Listov, Yu, 1885. Caves—ice-houses. Data Russ. Geol. 12, 105–280. Sankt-Petersburg (in Russian).
Losev, A.I., 1815. Description of caves in Irkutskaya province. Proc. Free Econ. Soc. LXVII, 137–157. SPb (in Russian).
Lozovoj, S.P., 1984. Lagonakskoe Upland. Publishing House, Krasnodar. 160 pp. (in Russian).
Lukin, V.S., 1962. Observations over air mode of karst massifs on example of Kungurskaya Ice Cave. Trans. Inst. Geol. UB AS USSR 62, 107–119. Sverdlovsk (in Russian).
Lukin, V.S., 1965. Temperature anomalies in caves of Preduralie and critical analysis of theories of underground cold. Peshery (Caves) 5 (6), 164–172 (in Russian).
Malkov, V.N., Gurkalo, E.I., Monakhova, L.B., Shavrina, E.V., et al., 2001. Karst and Caves of Pinezh'ya. Moscow, pp. 10, 27, 93 (in Russian).
Martel, E.A., 1909. La côte d'Azure Russe (Riviera du Caucase). Le massif de l'Arabika, Paris.
Maruashili, L.I., 1974. Volcanic caves of Georgia. Peshery (Caves) 14–15, 156–159. Perm (in Russian).
Mavlyudov, B.R., 1985. Geographical peculiarities of spreading of caves with ice. Materialy glatsiologicheskikh issledovanij (Data Glaciol. Stud.) 54, 193–200 (in Russian).
Mavlyudov, B.R., 1988a. Melting of snow and ice in karstic cavities. Materialy glatsiologicheskikh issledovanij (Data Glaciol. Stud.) 64, 123–128 (in Russian).
Mavlyudov, B.R., 1988b. Melting of snow and ice in karst cavities. Materialy glatsiologicheskikh issledovanij (Data Glaciol. Stud.) 64, 123–128 (in Russian).
Mavlyudov, B.R., 1989a. Caves glaciation of the Soviet Union (Author's abstract of PhD dissertation). Moscow. 23 pp. (in Russian).
Mavlyudov, B.R., 1989b. Snow and ice in karst cavities of the Hipstinsky massif (the Western Caucasus). In: Problems of Karst of Highlands. Metsniereba, Tbilisi, pp. 205–208 (in Russian).
Mavlyudov, B.R., 1994. Caves climatic systems. In: Questions of Physical Speleology. MFTI, Moscow, pp. 6–24 (in Russian).
Mavlyudov, B.R., 1997. Map "Caves glaciation" (scale 1:40000000). In: Atlas of Snow-Ice Resources of the World Moscow, p. 356 (in Russian).
Mavlyudov, B.R., 2008. Caves Glaciation. Institute of Geography RAS, Moscow. 290 pp. (in Russian).

Mavlyudov, B.R., 2014. Dynamics of snow and ice accumulation in the Bol'shoj chamber of cave system Snezhnaya-Mezhennogo-Illusiya (the Western Caucasus). In: Proceedings of V Regional Scientifically-Practical Conference "Karst and Caves of Caucasus: Results, Problems and Prospects of Researches", Sochi. pp. 213–216. (in Russian).

Mavlyudov, B.R., 2016. Cave system Snezhnaya-Mezhennogo-Illyuziya, the Western Caucasus. Bol. Geol. Min. 273 (1), 219–235 (in English).

Mavlyudov, B.R., Vturin, B.I., 1988. snow-ice formations of karst cavities of Khipstinskij massif. Materialy glatsiologicheskikh issledovanij (Data Glaciol. Stud.) 62, 148–153 (in Russian).

Maximovich, G.A., 1946. Short Instruction on Studying of Cave Ice and Ice Caves. Molotov, 10 pp. (In Russian).

Maximovich, G.A., 1947. Cave ice. Trans. USSR Geogr. Soc. 79 (5), 237–250. (in Russian).

Maximovich, G.A., 1952. About the first description of ice caves. Trans. USSR Geogr. Soc. 84 (1), 103–104. (in Russian).

Maximovich, G.A., 1963. The instruction on studying of cave ice and ice caves. Methodol. Karst Stud. 9, 27–35. Perm (in Russian).

Osintsev, A.V., 2009. Lenskaya Ice Cave in Yakutia—the newest history. Peshery (Caves) 32, 138–143. Perm (in Russian).

Osintsev, A.V., 2010. Large caves of Baikal region - newest researches. In: Speleology and Spelestology: Development and Interaction of Sciences. Collection of Materials of the International Scientific-Practical Conference. NSPI, Naberezhnye Chelny, pp. 99–101 (in Russian).

Ostapenko, A.A., 1991. About ice sediments of Kholodil'nik Cave. In: Study and Usage of Karst of Western Caucasus. pp. 81–82 Sochi (in Russian).

Pertsiger, F.I., 1986. Long-term mode of bloing snow field. Trans. Tp. MASRI. V. 117 (198), 117–123 (in Russian).

Philippov, A.G., 1981. Ice caves of Irkutsk area. In: Accumulation of Winter Cold in Rocks and Its Use in National Economy, pp. 58–60. Perm (in Russian).

Philippov, A.G., 1989. Geology of caves of Baikal Lake coast. In: Proceedings 10th International Congress of Speleology (13–20 August 1989), Budapest. vol. 2, pp. 583–585 (in Russian).

Philippov, A.G., 1990. Microclimate of caves of the south of Siberia. In: Problems of Geometeorology and Accumulation of Winter Cold. pp. 64–67. (In Russian). Sverdlovsk.

Philippov, A.G., 1994. Cave Botovskaya. In: Questions of Physical Speleology. pp. 142–160 M: MFTI (in Russian).

Philippov, A.G., 1997. Cave ices of Irkutskaya area. Svet (Light) 2 (17), 13–16 (In Russian). Kiev.

Philippov, A.G., 1999. Genesis of Botovskaya Cave. Geomorfologiya 1, 108–115. (in Russian).

Pirpilashvili, P.M., 1962. Natural refrigerator of village Skhvava. Inform. GSSR Acad. Sci. 29 (6), 791–798. (in Russian).

Russkikh, A.V., Ivanov, A.D., 1992. Cave and Karst Lakes Volgo-Vyatskij Land. Kirov, 87 pp. (in Russian).

Russkikh, A.V., Mikheev, A.A., 1971. Some microclimatic characteristics of underground cavities of Gor'kovskaya area and Marijskata ASSR. In: Accumulation of a Winter Cold In Rocks and Its Use in a National Economy Perm (in Russian).

Shakirov, A.V., 2011. Physiographic Zoning of Ural Mountains. UB RAS, Ekaterinburg. 617 pp. (in Russian).

Shavrina, E.V., 2002. Karst of southeast part of Belomorsko-Kulojskoe plateau, its protection and rational use. (Abstract of PhD dissertation) . Perm, 22 pp. (in Russian).

Shavrina, E.V., 2012. Result of observation in caves with using of temperature loggers. In: Long-Term Dynamics of Ecosystem Components of Natural Complex of Pinezhskij Reserve and Adjacent Areas. pp. 15–21 Arkhangelsk (in Russian).

Shimsky, P.A., 1955. Foundation of Structural Ice Researches. Publishing House of the USSR Academy of Science, Moscow. 492 pp. (in Russian).

Sotskova, L.M., 1981. Air mode of karst cavities. In: Accumulation of Winter Cold in Rocks and Its Use in National Economy. pp. 11–12 Perm (in Russian).

Sotskova, L.M., Dublyanskij, V.N., 1982. Mikroclimate of karst cavities. Zemlevedenie (Earth Sci.) 14, 79–81 (in Russian).

Stepanov, Yu.I., Kichigin, A.V., Tajnitskij, A.A., Kadebskaya, O.I., 2014a. Study of perennial ice by georadar and by mineralogical researches on an example of Medeo Cave (Northern Ural Mountains). Probl. Region. Ecol. 1, 132–138 (in Russian).

Stepanov, Yu.I., Tainitskiy, A.A., Kichigin, A.V., Kadebskaya, O.I., Mavlyudov, B.R., 2014b. Study of multiyear ice in Medeo Cave (Northern Ural). In: 6th International Workshop on Ice Caves: Proceedings Symposium 4/ NCKRI. Carlsbad. pp. 25–30 (in English).

Stepanov, Yu.I., Tajnitskij, A.A., Kichigin, A.A., 2014c. Georadar of research of underground icings in caves of Ural Mountains. In: Kadebskaya, O.I., Andrejchuk, V.A. (Eds.), Complex Use and Protection of Underground Spaces: International Science-Practical Conference on 100-Year-Old Anniversary of Scientific and Tourist-Excursion Activity in Kungurskaya Ica Cave and on 100 Anniversary Birthday of V.S.Lukin. GI UB RAS, Perm, pp. 65–68 (in Russian).

Stupishin, A.V., 1959. Cave ice of Middle Volga region and the nature of its formation. Speleol. Karstol. 53–62 Moscow (in Russian).

Tintilozov, Z.K., 1976. Karst Caves of Georgia. Metsniereba, Tbilisi. 274 pp. (in Russian).

Tintilozov, T.Z., Changashvili, G.Z., Okrodzhanashvili, A.A., 1966. Karstovo-Speleological features of limestone massif Arabika. Probl. Geogr. Georgia 60–83. (in Georgian with Russian abstract).

Tokarev, I.V., Franz, N.A., Vahhi, I.E., Krainyukova, I.A., Sorokin, S.V., 2015. Study of water regime of Pinega speleo area (Arkhangelsk region, Russia) by stable isotopes (σ_2H, σ_{18}O). Peshery (Caves) 38, 17–34 (in Russian).

Tsykin, R.A., 1981a. Cave ices of East Sayan and Kuznetskij Ala Tau. In: Accumulation of a Winter Cold in Rocks and Its Use in a National Economy. pp. 37–38. Perm (in Russian).

Tsykin, R.A., 1981b. Karst caves of vicinities of Krasnoyarsk. In: Environment of the Minusinskaya Depression (Western Part of KAFEC Area). Pedagogical institute, Krasnoyarsk (in Russian).

Vakhushti, 1941. Geography of Georgia (the Description of Georgian Kingdom). Tbilisi (in Georgian).

Valujskij, S.V., 2000. Caves of Perm Area. Bank of Cultural Information, Ekaterinburg 135 pp. (in Russian).

Vistengauzen, V.K., Dmitriev, V.E., 1977. Glaciation of USSR karst caves. In: Actions for Increase of Stability of an Subgrade in Karst Areas of Baikal-Amur Magistral and Other Questions of Karstology. pp. 66–67 Krasnoyarsk (in Russian).

Vistingauzen, V.K., 2010. Mountain Rudnichnaya with cave Rudnaya and with cave-mine "Novo-Chagyrskij"—zone of recreation of state natural complex preserve of regional value "Chinetinskij". In: URL: "Red Book of Altaj Land: Especially Protected Areas" (Electronic Resource of Local Access). Barnaul (in Russian).

FURTHER READING

Abdrakhmanov, R.F., et al., 2002. Karst in Bashkortostan. Ufa. 382 pp. (in Russian).

Bakka, S.V., Kiselyova, N.Yu., 2009. Especially Protected Natural Areas of the Nizhnij Novgorod Area. The Annotated List. Ministry for Protection of the Environment and Natural Resources of the Nizhnij Novgorod area, N.Novgorod 560 pp. (in Russian).

Bazarova, E.P., Kadebskaya, O.I., Kononov, A.M., Ushchapovskaya, Z.F., 2016. Conditions and features of cryogenic mineral formation in caves of a southern part of a folded frame of the Siberian platform (Western Pribaikalye and East Sayan Mountains). Bull. Perm Univ. Geol. 2 (31), 22–34 (in Russian).

Bogdanovich, E.D., 1971. Kutukskie Caves. Bashkiria. Guide. pp. 19–420. Ufa (in Russian).

Bolotov, V.V., Konstantinov, A.V., Rudenko, Yu.T., Shevchenko, Yu.S., 1985. Kheetejskie Caves. Nature and an Economy of Chita Area. pp. 56–59 Chita (in Russian).

Bol'shakov, P.M., 1958. About karst of Ust-Borzinskoe deposit of limestone. Trans. Irkutsk Mine-Metall. Inst. 5, 101–106 (in Russian).

Cave of Pinego-Severodvinsk Karst Area, 1974. Leningrad (in Russian).
Chajkovskij, I.I., Kadebskaya, O.I., 2014. Mineral formations of Yeranka Cave in Northern Ural Mountains. Probl. Mineral. Petrogr. Metallog. 17, 92–107 Perm (in Russian).
Chernyaeva, K.P., Sedymova, V.V., 1972. Caves-ice-houses of some areas of Mountain Altaj and Kuznetskij Ala Tau. In: Problems of Glaciology of Altaj, pp. 121–123. Tomsk (in Russian).
Chikishev, A.G., 1978. Karst Russian Plain. Nauka, Moscow 190 pp. (in Russian).
Cochrane, J.D., 1824. Narrative of a Pedestrian Journey Through Russia and Siberian Tartary From the Frontiers of China to the Frozen Sea and Kamchatka; Performing During the Years 1820, 1821, 1822, and 1823. H.C.Carey, L I.Lea, and A.Small, Philadelphia Collins AND Hannay, N.Y. 415 pp.
Demin, L.V., Bersenev Yu, I., Tatarnikov, V.A., 1980. Karst of Primorskij, Khabarovskij Lands and of Amur region. In: Karst of Far East and Siberia. Publishing House of Far East Branch of USSR Academy of Sciences, Vladivostok, pp. 5–54. Vladivostok (in Russian).
Dmitriev, V.E., 1980c. Glaciation of caves as a part of the earths glaciosphere. In: Karst Siberia Far East, pp. 130–145. Vladivostok (in Russian).
Dmitriev, V.E., Vistingauzen, V.K., 1977. Glaciation of karst caves of the USSR. In: Actions for Increase of Stability of an Earthen Cloth in Karst Areas of BAM and Other Questions of Karstology. Theses of Reports of All-Union Scientific and Technical Conference, October, 19–21th 1977. Krasnoyarsk, pp. 66–67 (in Russian).
Dublyanskij, V.N., 2000. Amusing Speleology. Ural Ltd., Chelyabinsk. 526 pp. (in Russian).
Dublyanskij, V.N., Klimchuk, A.B., Kiselyov, V.E., Vahrushev, B.A., Kovalev, Yu.N., Melnikov, V.P., Ryzhkov, A.F., Tintilozov, Z.K., Chujkov, V.D., Churubrova, M.L., 1987. Large karst cavities of the USSR. V III. In: Speleological provinces of the Big and Small Caucasus. VISITI, Moscow. 87 pp. (in Russian).
Dvukhzhil'nyj, V.V., 1981. Biological researches of ice in Kholodil'nik Cave. In: Karst of Far East: Scientific and Practical Value Of Karstological Researches, pp. 56–57. Vladivostok (in Russian).
Dyemin, L.V., Bersenev, Yu.I., Tatarnikov, V.A., 1980. Karst of Primorskij, Khabarovskij Lands and of Amur region. In: Karst of Far East and Siberia. pp. 5–54 Vladivostok (in Russian).
Erman, A., 1838. Reise um die Erde durch Nord-Asien und die beiden Oceane in den Jahren 1828, 1829 und 1830. In: Band 2: Reise von Tobolsk bis zum Ochozker Meere im Jahre 1829. Verlegt bei G.Reimer, Berlin. 444 pp.
Erman, A., 1848. Travels in Siberia: Including Excursions Northwards, Down the Obi to the Polar Circle, and Southwards, to the Chinese Frontier. vol. 2. Transl. by W.D.Cooley. L.: Longman, Brown, Green & Longmans, 536 pp.
Filippov, A.G., Alekseev, A.B., Sergienko, T.J., Goryunova, O.I., 1997. Birch bark bucket from the Bolshaya Bajdinskaya Cave (Priolkjonie). In: Dulovsky Readings of 1997 (Section Archeology and Ethnography). Materials of Reports and Messages. Listok, Irkutsk, pp. 82–88 (in Russian).
Geography of the Nizhnij Novgorod Area, 1991. The Manual. Volgo-Vyatskoe Book Publishing House, Nizhnij Novgorod, 207 pp. (in Russian).
Gurkalo, E.I., 2005. Speleological examination of the Arkhangelsk area. Northern Speleoalmanac, 41–44. Arkhangelsk (in Russian).
Gurkalo, E.I., Shavrina, E.V., 2002. Foundations of allocation of karst-speleological objects as nature monument. Ecology of northern areas of Russia. Problems, situation forecast, way of development, decision. In: Proceedings of International Conference. vol. 2. IENS UB RAS, Arkhangelsk, pp. 971–974 (in Russian).
http://www.nordspeleo.ru/cca/cca_6/icetemple_6.htm.
Ilukhin, V.V., Aronov, M.P., Sakharov, B.V., 1965. The Bornukovo Caves. Peshery (Caves) 5 (6), 128–134 (in Russian).
Kadebskaya, O.I., Kosintsev, P.A., 2012. Karst phenomena of Yanganape and Nyavape Ridges (Polar Ural Mountains). Probl. Region. Ecol. 6, 42–48 (in Rissian).
Kadebskaya, O.I., Tchaikovskij, I.I., 2014. Mineral formation of Pobeda Cave (Bashkortostan), connected with formation and melting of permanent ice. Proc. Russ. Acad. Sci. Ser. Geogr. 3, 66–72 (in Russian).

Kadebskaya, O.I., Tchajkovskij, I.I., 2012. Specific of kryo—and mineral—enesis of Medeo Cave (Northern Ural Mountains). Peshery (Caves) 35, 41–47 (in Russian).
Khorev, B.S., 1964. Gor'kij Region (the Nature, the Population, an Economy). Gorkij Book Publishing House, Gor'kij 240 p (in Russian).
Kiknadze, T.Z., 1963. Formation of cave ice in karst shaft Vakhushti Bagrationi. Inform. GSSR Acad. Sci. 31 (2), 325–331 (in Russian).
Konstantinov, A.V., Rudenko, Yu.T., 1986. Caves of Chitinskoe Transbaikalia. In: Archaeological and Ethnographic Researches in the Eastern Siberia. pp. 107–109. Irkutsk (in Russian).
Kruber, A.A., 1912. Trip on Arabica. Nat. Sci. Geogr. 1, 1–19. (in Russian).
Laxman, L.V.E., 1890. his life, travel, researches and correspondence. Translation from Swiss by E.Palander. SPb., 488 pp. (in Russian).
Losovoj, S.P., 1989. From researches history of caves and karst. In: Kubanskij kraeved. Publishing house, Krasnodar, pp. 34–46 (in Russian).
Luzan, P.I., Retinskij, K.V., 1994. Cave of Yanganape massif: interaction of physical weathering and a karst. In: Questions of Physical Speleology. MFTI, Moscow, pp. 161–166 (in Rissian).
Lyudkovskij, G.V.B., Mavlyudov, B.R., Morozov, A.I., Usikov, D.A., 1980. Shaft Snezhnaya deepest in the USSR and the third in the world. Европейска регионална конференция по спелеология. Sofia - НРБ 22–28.09, 1980. Сборник от материали. V. 2. Sofia. 1983. PP. 266–271 (in Russian).
Lyudkovskij, G.V., Mavlyudov, B.R., Morozov, A.I., et al., 1981. About research of Snezhnaya—the deepest cave of the USSR (the Western Caucasus). Rep. USSR Acad. Sci. 259 (2), 437–442 (in Russian).
Mal'chikova, I.Yu, 2000. Ice in caves of Transbaikalia. News RGS 132 (5), 38–43 (in Russian).
Marinin, A.M., 1969. Kuldyukskaya ice cave. Trans. Altaj Depart. Geogr. Soc. USSR 9, 22–33 (in Russian).
Marinin, A.M., 1972. Caves of Altaj. In: Problems of Glaciology of Altaj. pp. 24–26 (in Russian). Tomsk.
Marinin, A.M., 1990. Karst and Caves of Altaj. Novosibirsk, 148 pp. (in Russian).
Marinin, A.M., 1998a. Kuldyukskaya cave—the largest underground glacier in the north of Altaj. In: Especially Protected Natural Areas and Objects of Republic Altaj and Mountain Systems of the Centre of Eurasia: Ways and Problems of Sustainable Development. pp. 106–108. Gorno-Altajsk (in Russian).
Marinin, A.M., 1998b. Kuldyukskaya Cave—The Largest Underground Glacier in the North of Altaj. Especially Protected Natural Areas and Objects of Republic Altaj and Mountain Systems of the Centre of Eurasia: Ways and Problems of Sustainable Development. Gorno-Altajsk, pp. 106–108 (in Russian).
Marinin, A.M., 2001. Speleological pages of Red Book of Altaj Republik. Peshery (Caves) 27–28, 138–150. (in Russian).
Martin, V.I., Smirnov, A.I., Sokolov, Yu.V., 1993. Bashkiria Caves. Peshery (Caves) 23 (24), 30–59. Perm (in Russian).
Materials for Hystory-Geographical Atlas of Russia, 1871, pp. II–IV (in Russian).
Mavlyudov, B.R., 1979. Some data about karst of the Hipstinsky high-mountainous massif. In: Karst of Central Asia and Highlands. pp. 45–47. (in Russian). Tashkent.
Mavlyudov, B.R., 1980. Feature of a structure of snow-ice formations in shaft Snezhnaya in the Western Caucasus. Materialy glatsiologicheskikh issledovanij (Data Glaciol. Stud.) 40, 189–193 (in Russian).
Mavlyudov, B.R., 1981. Condition of snow and ice formation in a karst shaft Snezhnaya. In: Accumulation of Winter Cold in Rocks and Its Use in a National Economy, pp. 62–64. Perm (in Russian).
Mavlyudov, B.R., 1989c. Glaciation of Caves of the Soviet Union. (Abstract of PhD thesis). Moscow, 23 pp. (in Russian).
Mavlyudov, B.R., 1993. Snow-ice formations of caves—an indicator of a modern environment of mountain region. Materialy glatsiologicheskikh issledovanij (Data Glaciol. Stud.) 76, 53–60. Moscow (in Russian).
Mavlyudov, B.R., 1996. Snow accumulation on southern slope of the Bzybsky Ridge and a role of karst relief in its redistribution. Materialy glatsiologicheskikh issledovanij (Data Glaciol. Stud.) 80, 146–150 (in Russian).
Mavlyudov, B.R., 2001. Rhythmic processes in caves glaciation. Materialy glatsiologicheskikh issledovanij (Data Glaciol. Stud.) 90, 138–142 (in Russian).

Mavlyudov, B.R., 2005. Reflexions about ices in caves, about terminology, and not only. Northern Speleoalmanac 6, 45–62. Arkhangelsk (in Russian).
Mavlyudov, B.R., Morozov, A.I., 1984. Shaft Snezhnaya. Peshery (Caves) 19, 15–25 Perm (in Russian).
Mavlyudov, B.R., Vtyurin, B.I., 1988. Snow-ice formation of karst cavities of the Hipstinsky massif. Materialy glatsiologicheskikh issledovanij (Data Glaciol. Stud.) 62, 148–153 (in Russian).
Maximov, G.M., 1993. Karst and caves of the centre of Mountain Altaj. Svet (Light) 4 (10), 20 (in Russian).
Maximovich, G.A., 1945. Classification of ice in caves. Proc. USSR Acad. Sci. 9 (3–6), 565–570 (in Russian).
Maximovich, G.A., 1947a. Cave ice. Proc. USSR Geogr. Soc. 79 (5), 237–250 (in Russian).
Maximovich, G.A., 1947b. Speleographical sketch of the Molotov Province. Speleol. Bull. 5–42 Perm (in Russian).
Medvedeva, E., 2005. Iskat' (expedition to cave Vakhushti Bagrationi, Arabika). Svet (Light) 2 (28), 18–19 Kiev (in Russian).
Mikheyev, V.E., 1999. Karst Pri Yenisej fold-blocks zones. Stolbovskij karst area. In: Krasnoyarsk National Park "Stolby" (in Russian).
Obruchev, V.A., 1933. History of Geological Research of Siberia. Second Period (1801–1850). Publishing house of the USSR Academy of Science, Leningrad 230 pp. (in Russian).
Philippov, A.G., 1993. Cave of Irkutsk area. Peshery (Caves) 23 (24), 71–83 (in Russian).
Philippov, A.G., 2009. Ice Lenskaya Cave in Yakutia. Peshery (Caves) 32, 130–138. Perm. Вып. (in Russian).
Philppov, A.G., 1997. Cave ice of Irkutsk area. Svet (Light) 2 (17), 13–16 Kiev (In Russian).
Potapov, S.S., Parshina, N.V., Titov, A.T., et al., 2008. Cryogenic formations of caves of Pinezh'ya. In: Mineralogy of Tekhnogenez. IM UB RAS, Miass, pp. 18–43 (in Russian).
Rakviashvili, K.Sh., 1965. Some questions of studying of ice caves of Georgia. Caves Georgia 3, 48–51. Tbilisi (in Russian).
Rakviasvili, K.Sh, 1965. Some questions of ice caves studying in Georgia. Caves Georgia 3, 48–51. Tbilisi.
Russkikh, A.V., 1990. Microclimatic characteristic of underground cavities of Central Volga region having negative temperatures. In: Problems of Geometeorology and Accumulation of Winter Cold. pp. 42–44. Sverdlovsk (in Russian).
Schukin, N.S., 1833. Trip to Yakutsk. SPb, p. 106 (in Russian).
Schukin, N.S., 1844. Trip to Yakutsk, second ed. Printing House of Department of Military Settlements, SPb 315 pp. (in Russian).
Semivskij, N., 1817. New Curious and Authentic Narrations About the Eastern Siberia. Military Printing House of Principal Headquarters, SPb. 227 pp. (in Russian).
Shavrina, E.V., Guk, E.V., 2005. Modern dynamics of ice formations in Pinega caves. In: Mavlyudov, B.R. (Ed.), Glacier Caves and Glacial Karst in High Mountains and Polar Regions. Institute of Geography RAS, Moscow, pp. 113–117 (in English).
Shavrina, E.V., Malkov, V.N., 2008. Monitoring of karst and exogenic geological processes. In: Components of Ecosystems and a Biodiversity of Karstic Areas of the European North of Russia (on Example of Pinezhskij Reserve). pp. 64–74 (in Russian).
Shomysov, N.M. Geological excursions on Gor'kij region. http:digger.h1.ru.
Site: dalnegorsk.ru.
Site: http://outdoor-khv.ru/node/255.
Site: https://vladspeleo.ru/misc/peshhera-xolodilnik-peshhera-legenda/.
Stepanov, Yu.I., Kadebskaya, O.I., Tajnitskij, A.A., Kichigin, A.V., 2013. Studying of permanent ice in Yeranka cave (Northern Ural Mountains). Peshery (Caves) 36, 30–36 (in Russian).
Stupishin, A.V., 1967. Plain Karst and Regularities of Its Development on an Example of the Central Volga Region. Publishing house of Kazan University, Kazan. 292 pp. (in Russian).
Taskin, G., 1829. Geognostic description of Onon-Borzinskaya valley. Mine J. 3 (7–8), 167–208 (in Russian).

Torsuev, I.P., 1964. Karst of Interfluve of Onega and Severnaya Dvina Rivers: Experience of Physiographic Characteristics of Karst of North of Russian Plain. Publishing House of Kazan University, Kazan 101 pp. (in Russian).

Tsykin, R.A., 1993. Caves of Altaj-Sayan mountain area. Peshery (Caves) 23 (24), 59–70 Perm (in Russian).

Tsykin, R.A., Tsykina, Z.L., Tchernyaev, K.P., 1979. Caves Altaj-Sayan Mountain Area. The Manuscript. Krasnoyarsk, VINITI. Dep - № 1875-79, 163 pp. (in Russian).

Vistingauzen, V.K., 2009. Cave Yashur. In: Red Book of Altaj Land: Especially Protected Natural Areas. pp. 206–207. Barnaul (in Russian).

Vistingauzen, V.K., 2011. Snow-ice formations in caves of Altaj. In: Karst Systems of the North in Changing Environment. pp. 151–152. Moscow (in Russian).

Vistingauzen, V.K., Platonova, C.G., Bondarevskaya, S.A., 2009. Broad gull Strashnyj. In: Red Book of Altaj Land: Especially Protected Natural Areas. pp. 199–201 Barnaul (in Russian).

World Longest Gypsum Caves, 2008. Compiled by: Bob Gulden, November 05. 2008. http://www.caverbob.com/home.htm.

Zheleznyak, I.I., Mal'chikova, I.Yu., 2005. Caves Kheetej. Express Publishing House, Chita. 114 pp. (in Russian).

CHAPTER

ICE CAVES IN SERBIA

27

Dragan Nešić*, Jelena Ćalić†

Institute for Nature Conservation of Serbia, Niš, Serbia Geographical Institute "Jovan Cvijić" of the Serbian Academy of Sciences and Arts, Belgrade, Serbia†*

CHAPTER OUTLINE

27.1 INTRODUCTION

In this chapter, the term "ice caves" refers to both perennial and seasonal ice caves, following the classification of Ford and Williams (2007), as opposed to the stricter classification by Luetscher and Jeannin (2004), which would, taking into account only perennial ice caves, leave Serbia with only four established sites. Considering the fact that in five of the listed ice caves, the ice sometimes lasts the whole year while in some years melts completely, we had to conditionally introduce the term "occasionally perennial," which in this case lies in-between perennial and seasonal. In the Serbian geomorphology literature, the caves with ice and/or snow are called *ledenice* (ice caves) and *snežnice* (snow caves); these words are from the language of the native people, but are accepted in national scientific terminology as well (Gavrilović, 1974).

Ice caves at temperate latitudes are usually expected to be located in the zones of relatively cold periglacial alpine environments in high mountain areas above the tree line. However, there are few such areas in Serbia, and most of them do not have an underlying carbonate geology. On the other hand, there are examples where ice and snow tend to accumulate in caves below the tree line for longer than an usual winter season. Such accumulations of subterranean ice are considered to be azonal processes, which are not typical for the given elevation range (zone). Exceptions of this kind are located mainly in the medium-high mountains of the Carpatho-Balkan and Dinaric ranges (Fig. 27.1) in Serbia. Locations of the ice caves fall between 670 m a.s.l. and 1776 m a.s.l.

Ice Caves. https://doi.org/10.1016/B978-0-12-811739-2.00027-9

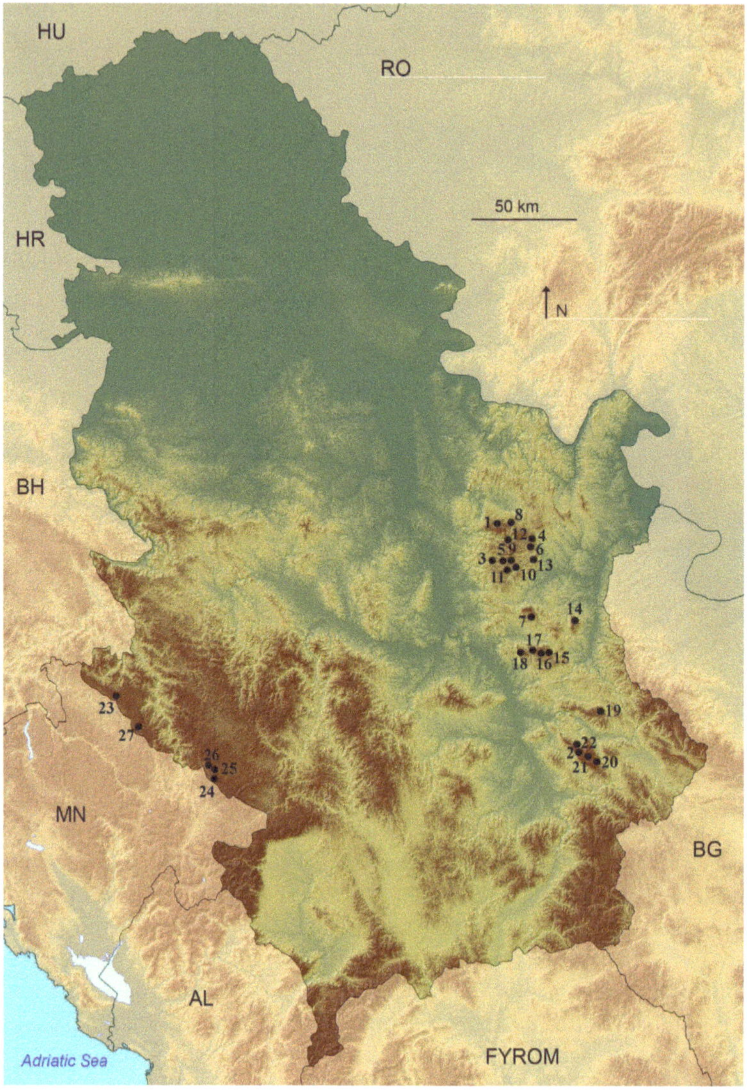

FIG. 27.1

Spatial distribution of ice caves in Serbia (base map generated from SRTM data).

In general, it can be assumed, based on theory and the majority of literature references, that the ice in the perennial ice caves in Serbia is mostly of relict and endogenic character (formed through congelation of cave waters). Some seasonal ice caves host recent and/or exogenic ice, as well as snow accumulations at the entrances. Long-term observations, despite being very irregular in time, have indicated that ice in the static ice caves in Serbia is mostly formed by freezing of percolation waters and very rarely by concentrated input. Gradual transformation of snow to ice is mostly lacking, except in the case of a unique perennial ice cave in SW Serbia (Lednica, in the village of Krnjača on Ožalj Mountain). High amount of precipitation in the colder parts of the year, and the absolute number of frosty and snowy days, has a significant impact on the quantity of underground ice (Nešić, 2002b).

27.2 HISTORY OF RESEARCH

The first observation of an ice cave in Serbia dates back to the first half of the 19th century. French/Austrian geologist Ami Boué (1840, 1891), during his geological explorations of the "European Turkey," visited several caves in Serbia. One of them was the Rtanjska Ledenica ice cave on Mt. Rtanj in Eastern Serbia, where he measured the air temperature at the bottom. Mačaj (1866) mentioned the Tupižnička Ledenica ice cave in his general "topographic" overview of the Knjaževac county.

The end of the 19th century in Serbian ice cave research was marked with explorations done by the founder of Serbian speleology and karstology, Jovan Cvijić (Cvijić, 1893, 1895, 1896). He gave a rather high hierarchical place to ice caves, within his general classification of caves: (1) "proper" caves (river caves and dry caves), (2) rock shelters, and (3) ice caves.

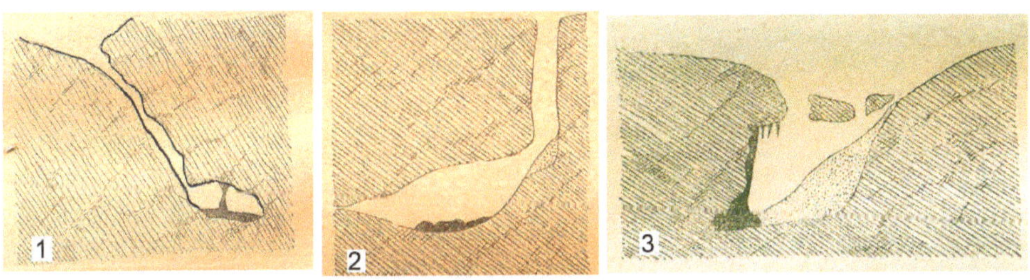

FIG. 27.2

Jovan Cvijić's sketches of ice caves in Eastern Serbia: (1) Rtanjska Ledenica, on Rtanj Mt.; (2) Stojkova Ledenica, on Kučaj Mt.; and (3) Ledenica u Ždrebici, on Suva Planina Mt.

After Cvijić, J., 1895. Pećine i podzemna hidrografija u istočnoj Srbiji (Caves and underground hydrography in Eastern Serbia). Glas Srpske kraljevske akademije nauka, 46, 1–101 (in Serbian).

Cvijić compiled the first list of ice caves in Serbia (there were 11 of them at that time), with small-scale maps (Fig. 27.2) and air temperature measurements. Apart from stressing the prerequisites for the existence of ice caves—specific morphology (cold air traps), drip waters or snow—he suggested three classifications of ice caves: static and dynamic (although he stated that there are no dynamic examples in Serbia), "permanent" (perennial), and "periodic" (seasonal). Morphologically, Cvijić differentiated three types of ice caves: (a) short ice caves resembling rock shelters (mostly of N exposure, thick outside vegetation, mildly inclined single passage), (b) ice caves of "aven" (shaft) type (term borrowed from French; entrance at the bottom of a doline, inclined passage called "stromor," and an ending small chamber), and (c) ice caves with shaft entrance (straight vertical channel of uniform cross-section descending to a large chamber). Cvijić also mentioned the local inhabitants' perceptions of the ice caves, either as small but valuable water reserves for shepherds on the waterless high karst plateaus or as a potential means for selling ice during the summer, for festivals and fairs. This is how he mentions the visit of Ami Boué to the Rtanjska Ledenica ice cave:

> Ami Boué visited Rtanjska Ledenica 58 years ago and it is of great importance to present his observations. Accompanied by the Serbian prince Miloš, he visited this ice cave in August 1836 (Die europäische Turkei von A. Boué I Band. Wien 1889, p. 88 – Recueil d'itinéraires dans la Turquie d'Europe par A. Boué, Tome second. Vienne 1854, p. 322). He found a lot of snow and ice in it, and measured the temperature of −2°C in the cave, while the air temperature outside of the cave was 23–24°C. At that time, there were thick woods around the Rtanjska Ledenica, while now it is surrounded mostly by meadows.
>
> **(Cvijić, 1895)**

In the second half of the 20th century, several geomorphologists (Petrović, 1964; Milić, 1968; Petrović, 1976; Lazarević, 1998) studied ice caves to a certain extent, but mostly within the broader analyses of general karst morphology. The beginning of the 21st century brought several new papers on ice caves, based on a series of temperature/humidity measurements, carried out mostly within seasonal dynamics during several years (Nešić, 2002a,b,c, 2017; Nešić et al. 2008a,b). There have been no data logging and no absolute dating of ice so far. Published papers on snow and ice in the underground karst of Serbia have focused mostly on particular scattered examples, but both detailed analytical research and summarizing, synthesis studies are generally lacking.

27.3 TYPES OF ICE CAVES IN SERBIA

Based on the length of the extended duration of ice and snow in Serbian caves, the results of explorations and research and the previously mentioned classification criteria, we have identified 27 ice caves in Serbia. Four of them are perennial, five are occasionally perennial, and 18 are seasonal (Tables 27.1 and 27.2). At the time of Cvijić (1895), as many as 9 out of 11 known ice caves were perennial. He wrote, "In the remaining nine caves, the ice is present in July and even later, during the whole summer; they are therefore in the group of permanent ice caves."

27.3.1 PERENNIAL ICE CAVES

Perennial ice and snow occur in only four caves in Serbia, highlighting the rarity and irregularity of underground ice in azonal conditions below the tree line in the medium-high mountains. Three caves of this kind are situated in the Carpatho-Balkan mountains of Eastern Serbia, and one in the Internal Dinarides in southwestern Serbia. In the Carpatho-Balkanides, these caves are Tisova Jama on Beljanica Mountain (1336 m), Mijajlova Jama on Kučaj Mountain, and Provalija on Suva Planina Mountain (1808 m), while the Dinarides host Lednica ice cave, near the village of Krnjača on Ožalj Mountain, close to the border with Montenegro.

Tisova Jama is situated in the central part of the high karstified plateau of Beljanica Mountain, with the entrance at 920 m a.s.l. The large cavity consists of a large doline, vertical passage, and a vast inclined chamber, 181 m long and 105 m wide (Fig. 27.3).

Table 27.1 Ice Caves in the Carpatho-Balkan Mountains of Eastern Serbia

No	Name	Location	Entrance Elevation	Entrance Dimensions	Length	Depth	Passage Inclination	Origin of Ice	Duration of Ice/Snow	Source
1	Tisova Jama	Beljanica Mt.	920	180-160	331	235	Vertical/ inclined	Snow	Perennial	Ćalić Ljubojević and Ljubojević, 2000
2	Provalija	Suva Planina Mt.	1649	10-7	53	58	Vertical	Snow	Perennial	Nešić, 2017
3	Mijajlova Jama	Kučaj Mt.	710	15-8	297	167	Vertical/ inclined	Percolation water	Perennial	Petrović, 1976
4	Dubašnica	Kučaj Mt.	905	100-30		276	Vertical	Percolation water	Occasionally perennial	cave register
5	Zla Ledenica	Kučaj Mt.	900	30-15	150	37	Vertical/ inclined	Snow	Occasionally perennial	Petrović, 1964
6	Stojkova Ledenica	Kučaj Mt.	865	30-17	150	47	Vertical/ inclined	Snow and percolation water	Occasionally perennial	Cvijić, 1895
7	Rtanjska Ledenica	Rtanj Mt.	926	24-9	63	40	Inclined	Percolation water	Occasionally perennial	Nešić, 2002a
8	Ledena Peć	Beljanica Mt.	670	9-6	108	18	Inclined	Percolation water	Seasonal	Cvijić, 1895
9	Dobra Ledenica	Kučaj Mt.	850	13-3	105	20	Inclined	Percolation water	Seasonal	Petrović, 1964
10	Ledenica u Maloj Brezovici	Kučaj Mt.		8	43		Inclined	Percolation water	Seasonal	Cvijić, 1893
11	Ledenica u Velikoj Brezovici	Kučaj Mt.								Cvijić, 1893
12	Ledena Pećina	Kučaj Mt. (Vrtačelje)	922	5-4	231	92	Inclined/ cascading	Percolation water	Seasonal	Nešić et al., 2008a

Continued

Table 27.1 Ice Caves in the Carpatho-Balkan Mountains of Eastern Serbia—cont'd

No	Name	Location	Entrance Elevation	Entrance Dimensions	Length	Depth	Passage Inclination	Origin of Ice	Duration of Ice/Snow	Source
13	Gaura Đin Šotaća	Kučaj Mt.	800		10		Inclined	Snow and percolation water	Seasonal	Previnac, 1951
14	Tupižnička Ledenica	Tupižnica Mt.	1060	13-11	63	26	Inclined/ vertical	Percolation water	Seasonal	Nešić, 2002a
15	Veliki Ledenik	Devica Mt.	1123	22-8	51	25	Inclined	Percolation water	Seasonal	Nešić, 2002a
16	Mali Ledenik	Devica Mt.	1145	8-4	42	8.5	Inclined	Percolation water	Seasonal	Nešić, 2001
17	Ledenik kod Sićermanske Livade	Devica Mt.								Nešić, 2001
18	Ledenik	Ozren Mt.	1120		43	20	Inclined	Percolation water	Seasonal	Cvijić, 1924
19	Snežnica	Svrljiške Planine Mt.								Cvijić, 1895
20	Ledenica u Ždrebici	Suva Planina	1370	18		24	Inclined	Snow and percolation water	Seasonal	Cvijić, 1895
21	Krstasta Provalija	Suva Planina								Milić, 1962
22	Ledenica Trem	Suva Planina	1776	10-6	13	2	Rock shelter	Snow and percolation water	Seasonal	Cvijić, 1895

All dimensions are in meters (m). The numbers in the first column correspond to locations in Fig. 27.1.

Table 27.2 Ice Caves in the Dinarides of Southwestern Serbia

No	Name	Location	Entrance Elevation	Entrance Dimensions	Length	Depth	Passage Inclination	Origin of Ice	Duration of Ice/Snow	Source
23	Lednica	Krnjača village	1152	8-5			Vertical/ inclined	Snow and percolation water	Perennial	Personal observation
24	Ledenica	Piskova Poljana (Pešter)	1518	7-3	69	6	Ponor cave	Percolation and inflow waters	Occasionally perennial	Nešić, 2015
25	Špela Bores	Vojselov Bunar (Pešter)	1215	2-10	43	23	Vertical/ inclined	Percolation water and snow	Seasonal	Nešić, 2015
26	Bezdan u Mezgraji	Mezgraja (Pešter)	1316	9		40	Vertical/ inclined	Percolation water and snow	Seasonal	Nešić, 2015
27	Golubova Pećina	Kamena Gora village	1346	1-14	37	11	Vertical structural cavity	Snow and percolation water	Seasonal	Personal observation

All dimensions are in meters (m). The numbers in the first column correspond to locations in Fig. 27.1.

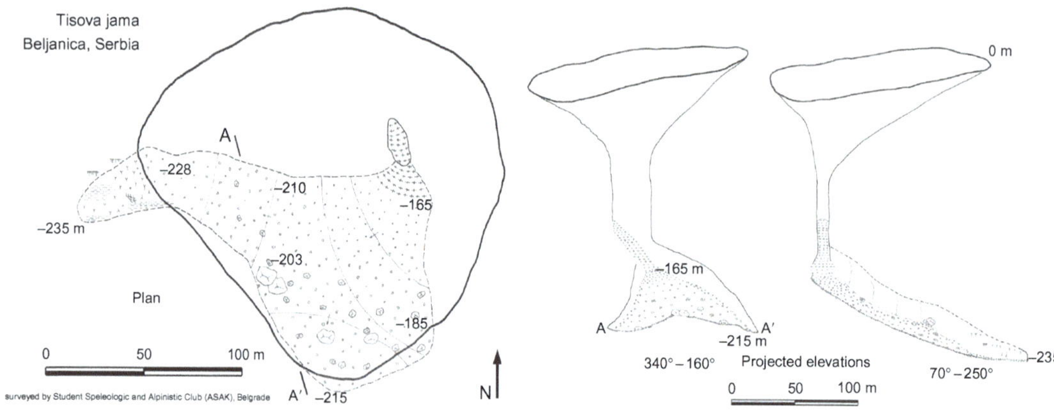

FIG. 27.3

Cave map of the Tisova Jama.

After Ćalić Ljubojević, J., Ljubojević, V., 1999. Tisova jama (Tisova Jama Cave). Zaštita prirode (Protection of Nature), 51 (2), 27–31 (in Serbian); Ćalić Ljubojević, J., Ljubojević, V., 2000. Caves below collapse dolines—case study of Tisova Jama (Eastern Serbia). Acta Carsol. 29 (2), 95–101.

The dimensions of the doline are 160 m by 180 m, with an approximate surface of 9.07 hectares. The total depth, measured from the rim of the doline, is 235 m. At -130 m, where the vertical passage enters the chamber, a large ice-snow accumulation, 50 m long and 10 m thick on average, with approximately 5000 m^3 of accumulated ice and snow (Ćalić Ljubojević and Ljubojević, 1999, 2000) is present. The vast doline is the area that collects the snow in winter, from which the snow then slides farther down, like through a funnel. At the bottom of the vertical shaft, morphological conditions are favorable for ice and snow accumulation. It is suspected that the ice from the Tisova Jama dates back to the Little Ice Age (a period of colder climate), but this assumption needs to be checked by absolute dating methods. The difference between the ice quantity in 1988 and 1999, when this cave was twice surveyed, shows a decreasing trend (Ćalić Ljubojević and Ljubojević, 1999, 2000), which can probably be explained by the recent influence of climate changes (as also discussed by Luetscher et al. (2005), who used examples from the Swiss Jura to show the decay of cave ice in the past decades).

Mijajlova Jama in the western part of the Kučaj mountains (710 m a.s.l.) is an impressive vertical cave with three entrances and a total depth of 167 m (Petrović, 1976). Through the freezing of percolation waters, ice crusts form on the walls of the shaft, further collapsing at the foot of the vertical drop to form accumulations of crushed ice and debris resembling an underground rock glacier.

Suva Planina Mountain (1808 m) is the highest limestone morphostructure within the Carpatho-Balkanides of Eastern Serbia and has a vast karst-leveled surface called Valožje (1400–1600 m). This area hosts both relict and recent high-mountain karst, as well as relict periglacial relief. Taking this into account, it is not strange that *Provalija Cave* (Fig. 27.4), located close to the mountain top (Trem, 1808 m), hosts perennial ice and snow accumulation (Nešić, 2017).

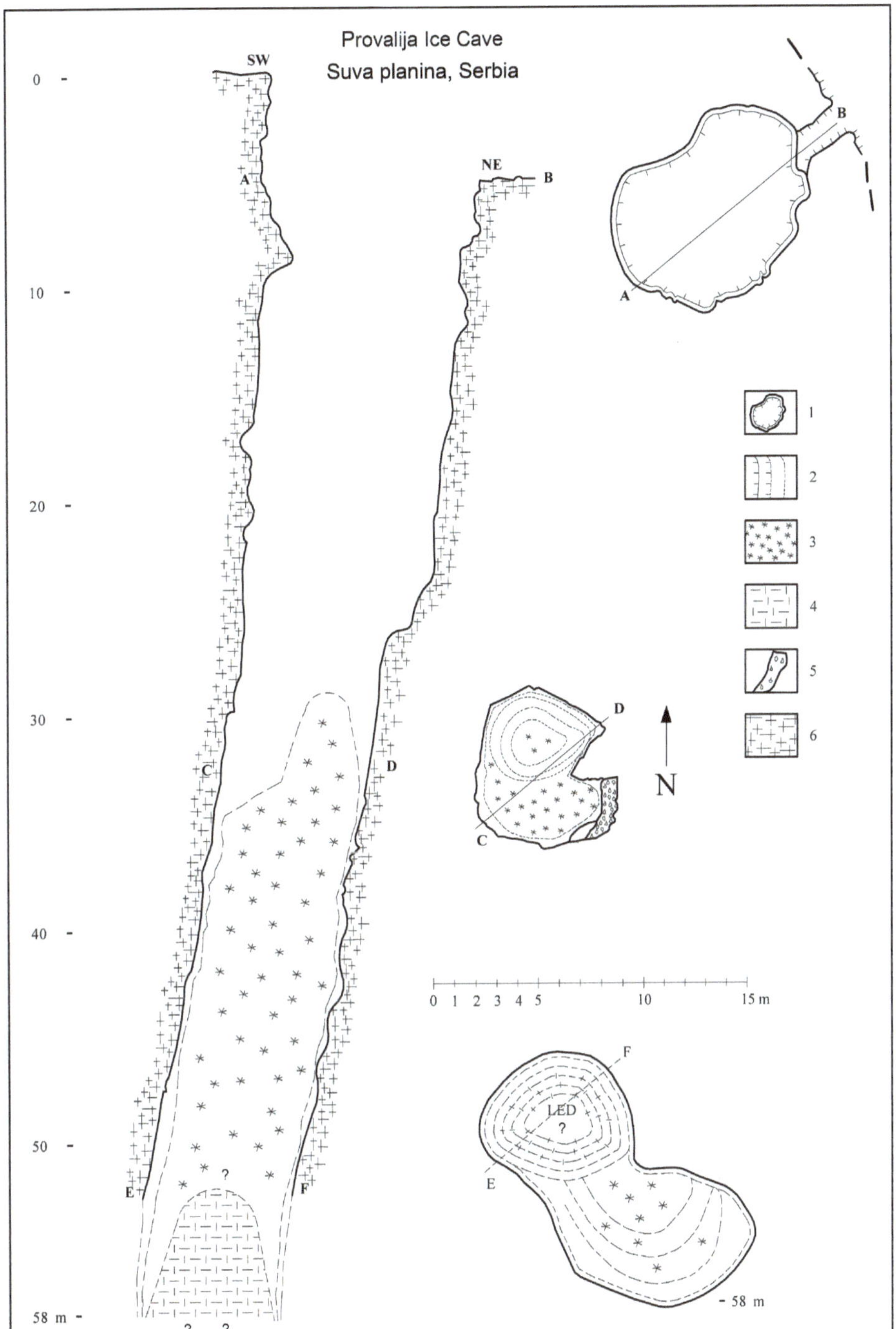

FIG. 27.4

Provalija Ice Cave (1 – Entrance; 2 – Step, pit; 3 – Firn; 4 – Ice; 5 – Terrace with rubble; 6 – Massive limestone).

After Nešić, D., 2017. Jama Provalija na Suvoj planini (istočna Srbija), primer podzemnog visokoplaninskog karsta (Provalija pit on Suva planina (eastern Serbia), an example of underground high mountain karst). In: Proceedings of the 8th Symposium on Karst Protection, Pirot (in press).

This cave is situated in the structural depression Kozija Grbina, SW of the highest peak, at 1649 m a.s.l. It consists of a sub-vertical passage 7–10 m in diameter and 58 m deep, but speleomorphological exploration is not complete yet. At the depth of 30 m, a large accumulation of compact ice is present, with a visible ice thickness of 8–10 m (volume about 385 m^3). This is permanent ice with snow cover that fluctuates in thickness with seasonal weather conditions. On Suva Planina Mountain (Valožje plateau), other caves have been detected (e.g., Krstasta Provalija, Savina Propast; Milić, 1962) with indications of prolonged presence of ice and snow, but these phenomena are still insufficiently explored.

There is only one known perennial ice cave in the Serbian part of the Dinarides: *Lednica Cave* in the village of Krnjača on Ožalj Mountain, located on the side of Tmuša sinking river, with entrances at 1152 m a.s.l. This cave is still insufficiently explored and has not been surveyed, but the first indications point out that one of the main factors of ice formation is the accumulation of snow. Below the entrances, the cave passage is inclined and covered with compact ice on a length of about 50 m. It seems that in the conditions of seasonal partial melt, the meltwater freezes again at the contact with the ice below. At the bottom of the cave a spacious chamber is present, which is the zone of cool air concentration. The climate of Lednica can be linked to its climatic context, with precipitation amounts in this part of the Dinarides exceeding 1000 mm (Ducić and Radovanović, 2005).

27.3.2 OCCASIONALLY PERENNIAL ICE CAVES

Long-term observations, although at irregular intervals, especially during the last decade of the 20th century and since, have shown that there are several caves where ice lasts the whole year or longer, sometimes interrupted by periods without summer ice. For these ice caves, we have conditionally accepted the term "occasionally perennial ice caves." Representatives of this group are Zla Ledenica, Stojkova Ledenica, and Dubašnica on Kučaj Mountain, Rtanjska (Mužinačka) Ledenica on Rtanj Mountain (Carpatho-Balkanides), as well as the Ledenica ice cave on the Pešter plateau in the Dinarides of southwestern Serbia.

The data on *Zla Ledenica* (alternative name in Vlach language: Gaura Rea) are very scarce—just one source. Petrović (1964) published a short note about it, together with the nearby seasonal ice cave Dobra Ledenica (Gaura Bună). At the time of Petrović's explorations, in September 1962, there was an accumulation of ice and snow that probably, at least that year, lasted all year. However, personal contacts with the local population and hunters who visit the area, indicate that in recent years there has been no ice in Zla Ledenica, which must be taken conditionally because it was not verified.

Stojkova Ledenica on the karst-leveled surface of Dubašnica in the eastern part of Kučaj Mountain was first explored and mapped by Cvijić (1895), then subsequently was visited by Jeannel and Racovitza (1929) within a biospeleological study, and in the second half of 20th century was remapped by Lazarević (1998). This is a large karst cavern whose entrance was formed by collapse. At the times of Cvijić and Lazarević, the cave was reported as having perennial ice, but in recent visits no ice was present in early autumn (Nešić, 2006, pers. comm.)

Dubašnica Cave, also situated on eastern Kučaj, is the second deepest cave in Serbia (−276 m). Being almost completely a vertical pit, it is occasionally visited by caving teams for sports reasons, but unfortunately measurements of temperatures were never properly organized or published. The teams reported finding ice in summer months, but there were also years without deposited ice or snow.

The entrance to *Rtanjska Ledenica* on Rtanj Mountain, lies at an elevation of 926 m a.s.l. It is an example of a static ice cave with an inclined, 63-meter-long passage that reaches a maximum depth of 40 m. Freezing of percolation waters contribute greatly to the ice balance. In some years, the ice deposits exist during the summer and are continuous throughout the year or for several consecutive years. One such case was recorded in autumn 2004, when there was about 1 m^3 of ice. Relatively detailed air temperature measurements (not data logging) were carried out in 1997 (Nešić, 2002a). Ami Boué visited this cave in 1836 and later wrote, "In Cretaceous limestones, natural shafts are also frequent, although they are rarely so nicely characterized as in the case of Ledenica, in the woods on the southern slopes of Mt. Rtanj" (Boué, 1891).

Observations of the *Ledenica Cave*, located in the Žilindar-Kruščica area on the Pešter plateau in SW Serbia, fit well with what was written about occasionally perennial ice caves. Here, ice is formed by freezing of percolation and surface inflow water, as this cave is fed by a short ponor stream. On April 26, 2008, the whole surface of the cave was covered by an ice sheet 0.5–1 m thick, with a total volume of 40–50 m^3, which by October 16, 2008, was reduced to 4.3 m^3 (Nešić, 2015). This observation places the cave in the category of occasionally perennial ice cave, considering also the fact that during the explorations in the autumn of 2007, there was no ice at all.

27.3.3 SEASONAL ICE CAVES

All other caves with ice and snow in the Serbian karst (Tables 27.1 and 27.2) are seasonal static ice caves, with annual renewal of ice and/or snow deposits, these lasting longer than snow outside the caves. The ice is primarily formed by freezing of percolation waters, and occasionally by snow firnification. In the seasonal static ice caves of Serbia, ice melts under the influence of heat advected by warm drip waters—reverse process than during formation of ice deposits—as well as the general warming of air, bedrock, and infills. The maximum recorded warming compared with the winter temperatures in these ice caves was 5.5°C (Nešić, 2002a).

It can be concluded that seasonal ice caves are the dominant type of ice caves in the Serbian karst. On average, snow and ice in these caves lasts about 6–8 months per year, depending on weather conditions.

27.4 OUTLINE OF TEMPERATURE DYNAMICS

There have been relatively few attempts to explain the climatic functioning of azonal static ice caves in Serbia (Cvijić, 1895; Nešić, 2002a). Measurements of cave climate parameters were carried out in Rtanjska Ledenica, Tupižnička Ledenica, and Veliki Ledenik caves on Devica Mountain. These measurements were done using the simple mercury thermometer (0.2°C precision), mounted on a tripod 1 m above the ground. Data loggers were not available. The results pointed to the regularity of permanent minimal negative air temperatures inside the ice caves (–0.2°C to –0.8°C) while outside temperatures were up to 6°C to 7°C, or very cold air from 0.4°C to 1.2°C, with outside temperatures of 14.8°C to 21.3°C (Nešić, 2002a; Nešić et al., 2008b). This climatic condition of permanent concentration of cold air in ice caves is called the "closed period" (e.g., Girardot and Trouillet, 1885, cited by Luetscher and Jeannin, 2004; Cvijić, 1895), which in present climatic conditions in Serbia occurs throughout most of the year (Fig. 27.5A and B). Inversion of air temperatures in the static ice caves was recorded during periods with outside negative air temperatures, while inside the ice caves (although with temperatures below zero), the temperature was several degrees warmer. This was recorded in Tupižnička Ledenica on January 9, 1997, when the temperature was −2.9°C outside the cave and −1.5°C inside the cave.

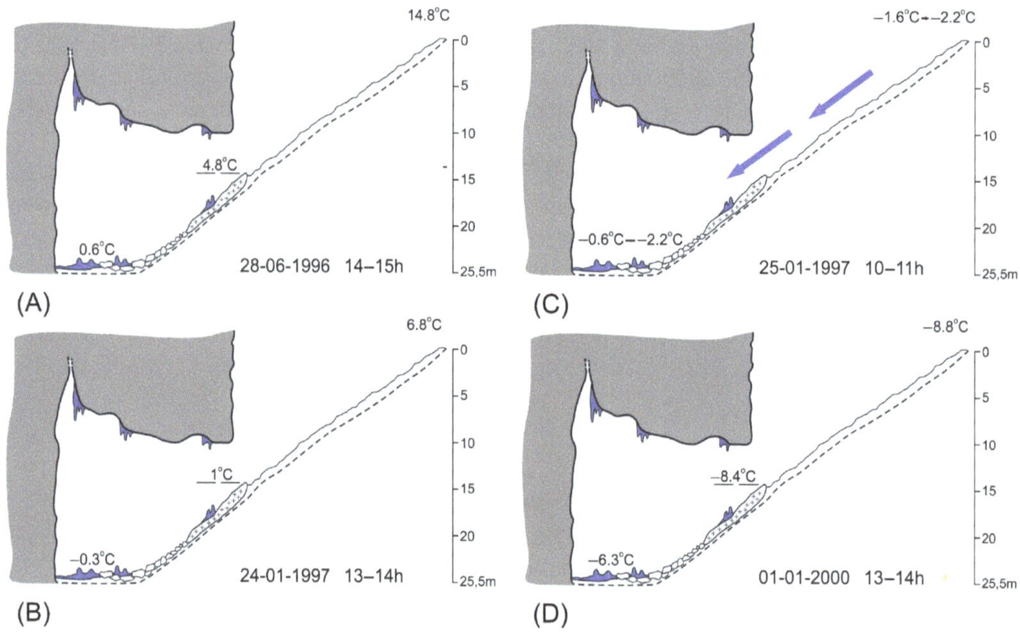

FIG. 27.5

Climatic functioning of the static ice cave Veliki Ledenik on Devica Mt.: (A, B) long-lasting static functioning ("closed period") without noticeable air circulation; (C) extremely short-lasting dynamic moment, with observed and measured sudden inflow of cold air from the outside ("open period"); and (D) static phase, without noticeable air circulation, but with expected subsequent air inflow because of the difference in temperature (the moment was not observed in the field on a given date, but suspected).

More detailed observations were made in the Veliki Ledenik ice cave on Devica Mountain. On December 31, 1999, it was −6.2°C outside and −4.4°C inside, and on January 1, 2000, it was −8.8°C outside and −6.3°C inside (Nešić, 2002a) (Fig. 27.5D). This means that azonal ice caves may have a closed period even during temperature inversions. On the other hand, the same cave showed the dynamic characteristics typical for the open period on January 25, 1997. D. Nešić was carrying out temperature measurements in the cave during the temperature inversion, and exactly as he was taking the measurement, a sudden inflow of cold air caused a wind-like effect, with a temperature drop of 1.6°C (Fig. 27.5C). Based on a number of observations of climate conditions in ice caves and making top/bottom air temperature profiles, we came to the conclusion that the short open period of static ice caves, which refers to the inflow of cold air during the winter, should be linked only to the rare conditions of a sudden drop in the outside temperature on winter nights, whereas in general the main climatic characteristics of this type of ice cave remain unchanged (Nešić, 2002a).

27.5 CONCLUSIONS

Azonal static ice caves in Serbia are caves with permanent presence of cold air, which is, apart from the general climatic and weather conditions in the mountains below the tree line, mostly conditioned by the geometry of the caves. The most usual types are shallow inclined caves where the underground chambers are positioned laterally in relation to the mostly spacious entrances. This arrangement enables the formation of cold air traps, additionally helped by northerly exposed entrances, as well as by thick vegetation in their immediate surroundings, protecting them from direct sunlight.

Ice caves in Serbia are distributed in an elevation belt between 670 m a.s.l. and 1776 m a.s.l. All of them fall in the category of static ice caves, with an almost constant duration of closed periods (in terms of air circulation) and only rare, extremely short-lasting open periods.

Using the criterion of the extended duration of ice and snow, we differentiate three types: perennial (4 caves), occasionally perennial (5 caves), and seasonal (18 caves) (Tables 27.1 and 27.2).

At the end of the 19th century, 9 out of 11 known ice caves in Eastern Serbia had perennial ice (Cvijić, 1895), and today all of them are only seasonal ice caves. At the present time, no detailed monitoring has been set up in order to systematically keep a record of these changes, but this may be part of future activities (cf. Luetscher et al., 2005).

Despite more than 150 years of research, the static ice caves of Serbia are insufficiently explored, both climatically and morphologically. It is necessary to upgrade the methods for data logging and absolute dating, as well as for a multidisciplinary approach (e.g., biospeleological studies of ice caves and their frigophyllic fauna).

In the past, ice in Serbian ice caves had economic significance, when it was taken out and used for cooling foods and drinks. With the development of artificial cooling systems, economic interest in these caves has diminished, but the issue of monitoring climate change could put these phenomena in the limelight again.

REFERENCES

Boué, A., 1840. La Turkie d'Europe. Arthus Bertrand, Paris 526. p.

Boué, A., 1891. Geološka skica evropske Turske (Geological sketch of European Turkey). Translation from French by Jovan Cvijić and Jovan Žujović Geološki anali balkanskog poluostrva (Annales géologiques de la péninsule Balkanique)3, 1–157. (Annex) (in Serbian).

Ćalić Ljubojević, J., Ljubojević, V., 1999. Tisova jama (Tisova Jama Cave). Zaštita prirode (Protect. Nat.)51 (2), 27–31 (in Serbian).

Ćalić Ljubojević, J., Ljubojević, V., 2000. Caves below collapse dolines—case study of Tisova Jama (Eastern Serbia). Acta Carsol. 29 (2), 95–101.

Cvijić, J., 1893. Geografska ispitivanja u oblasti Kučaja u istočnoj Srbiji (Geographic research in the area of Kučaj in Eastern Serbia). Geološki anali balkanskog poluostrva (Annales géologiques de la péninsule Balkanique)5, 7–172 (in Serbian).

Cvijić, J., 1895. Pećine i podzemna hidrografija u istočnoj Srbiji (Caves and underground hydrography in Eastern Serbia). Glas Srpske kraljevske akademije nauka 46, 1–101 (in Serbian).

Cvijić, J., 1896. Les glacières naturelles de Serbie. Spelunca – Bulletin de la Societé de Spéléologie, II 6–7, 64–77.

Cvijić, J., 1924. Geomorfologija. Knjiga prva (Geomorphology. Book One). Državna štamparija, Beograd1–588. (in Serbian).

Ducić, V., Radovanović, M., 2005. Klima Srbije (The climate of Serbia). Zavod za udžbenike, Beograd. pp. 5–212 (in Serbian).

Ford, D.C., Williams, P.W., 2007. Karst Hydrogeology and Geomorphology. John Wiley & Sons, Ltd., Chichester. 562 pp.

Gavrilović, D., 1974. Srpska kraška terminologija. Kraška terminologija Jugoslovenskih naroda II (Serbian Karst Terminology. Karst Terminology of Yugoslav Peoples, II). Savez geografskih institucija Jugoslavije, Beograd. pp. 1–73 (in Serbian).

Girardot, A., Trouillet, L., 1885. La Glacière de Chaux-les-Passavant. Mémoires de la Société d'Emulation du Doubs 9 (5), 449–524.

Jeannel, R., Racovitza, E., 1929. Enumération des Grottes visitées 1918-1927 (7-e série). Biospéologica 54. Archives de Zoologie expérimentale et générale, Paris, Tome 68, fasc. 2.

Lazarević, R., 1998. Kras Dubašnice, Gornjana i Majdanpeka – pećine, jame, kraška hidrografija (Karst of Dubašnica, Gornjan and Majdanpek—Caves and Karst Hydrography). Srpsko geografsko društvo (Serbian Geographical Society), Beograd. pp. 1–302 (in Serbian).

Luetscher, M., Jeannin, P.-Y., 2004. A process-based classification of alpine ice caves. Theor. Appl. Karstol. 17, 5–10.

Luetscher, M., Jeannin, P.-Y., Haeberli, W., 2005. Ice caves as an indicator of winter climate evolution: a case study from the Jura Mountains. The Holocene 15 (7), 982–993.

Mačaj, S., 1866. Građa za topografiju okruga Knjaževačkog (Materials for topography of the Knjaževac County). Glasnik Srpskog učenog društva 19, (in Serbian).

Milić, Č., 1962. Glavne odlike krasa Suve planine (The main characteristics of Mt. Suva Planina karst). Zbornik radova Geografskog instituta "Jovan Cvijić" SANU (J. Geogr. Inst. "Jovan Cvijić" SASA)18, 93–154 (in Serbian).

Milić, Č., 1968. Jame kao indikatori periglacijala u krasu istočne Srbije (Vertical caves as indicators of periglacial environment in the karst of Eastern Serbia). Cvijićev zbornik. SANU, Beograd. pp. 69–81 (in Serbian).

Nešić, D., 2001. Kraška morfologija u slivu Toponičke reke [Karst morphology in the drainage area of the Toponička Reka River]. Unpublished M.Sc. Thesis, Faculty of Geography, University of Belgrade, pp. 1–306.

Nešić, D., 2002a. Rezultati speleoloških i speleoklimatoloških istraživanja Velikog ledenika na Devici, Rtanjske i Tupižničke ledenice (Results of speleological and speleoclimatological research of the ice caves Veliki Ledenik on Devica Mt, Rtanjska Ledenica and Tupižnička Ledenica). Glasnik Srpskog geografskog društva (Bull. Serbian Geogr. Soc.)82 (2), 45–54. (in Serbian).

Nešić, D., 2002b. Led i sneg u speleološkim objektima planinskih područja Srbije i Crne Gore (Ice and snow in the caves of Serbian and Montenegrian mountains). In: Proceedings of the Conference "S planinom u novi vek", Kopaonik. pp. 89–94 (in Serbian).

Nešić, D., 2002c. Ledenice i snežnice Karpato-balkanskih planina istočne Srbije (Ice and snow caves in the Carpathian-Balkan mountains of Eastern Serbia). In: Proceedings of the Conference "Ekološka istina", Donji Milanovac. pp. 57–61 (in Serbian).

Nešić, D., 2015. Results of the speleological exploration on Pešter plateau. Bull. Serbian Geogr. Soc. 95 (4), 1–30.

Nešić, D., 2017. Jama Provalija na Suvoj planini (istočna Srbija), primer podzemnog visokoplaninskog karsta (Provalija pit on Suva planina (eastern Serbia), an example of underground high mountain karst). In: Proceedings of the 8th Symposium on Karst Protection, Pirot (in press).

Nešić, D., Pavićević, D., Ognjenović, S., 2008a. Rezultati speleomorfoloških i biospeleoloških istraživanja jame Vrtačelje (Ledena Pećina) (Results of speleomorphological and biospeleological explorations of the Vrtačelje Cave—Ledena Pećina). Zaštita prirode (Protect. Nat.)58 (1–2), 15–25. (in Serbian).

Nešić, D., Pavićević, D., Petrović, B., Zatezalo, A., 2008b. Rezultati novijih istraživanja Tupižničke ledenice (Results of recent studies of Tupižnička Ledenica). Zaštita prirode (Protect. Nat.)59 (1–2), 67–79. (in Serbian).

Petrović, D., 1964. Dve ledenice na Kučaju (Two ice caves on Mt. Kučaj). Glasnik Srpskog geografskog društva (Bull. Serbian Geogr. Soc.)44 (1), 69–73 (in Serbian).

Petrović, J., 1976. Jame i pećine SR Srbije (Caves of Serbia). Vojnoizdavački zavod, Beograd pp. 1–511 (in Serbian).

Previnac, M.M., 1951. Neke prirodne znamenitosti u karstu istočne Srbije (Several natural phenomena in the karst of Eastern Serbia). Zaštita prirode (Protect. Nat.)2–3, 359–367 (in Serbian).

CHAPTER

ICE CAVES IN SPAIN

28

Enrique Serrano*, Manue Gómez-Lende*, Ánchel Belmonte†, Carlos Sancho†, Javier Sánchez-Benítez‡, Miguel Bartolomé§, María Leunda§, Ana Moreno§, Bernard Hivert¶

Universidad de Valladolid, Valladolid, Spain Universidad de Zaragoza, Zaragoza, Spain† CES Alfa, Madrid, Spain‡ Instituto Pirenaico de Ecología, CSIC, Zaragoza, Spain§ AS Charentaise, Angouleme, France¶*

CHAPTER OUTLINE

28.1 INTRODUCTION

Ice caves are found in the Iberian Peninsula and Canary Islands in different massifs of the Pyrenees, the Cantabrian Mountains, Betic Mountains, and the Teide. They are always in calcareous areas of the highest massifs, except on the Teide, a volcano of 3718 m height, which is the highest peak in Spain, where they are located in volcanic environments. The remaining caves are found in the Paleozoic calcareous outcrops that, in the case of the Cantabrian Mountains, have Westfaliense limestones that form the highest summits, such as the Picos de Europa (2650 m), the Palentine Mountains (Espigüete, 2450 m), and to a lesser extent in the calcareous cover of the Cretaceous Age in Castro Valnera (1777 m). In the Pyrenees, ice caves are mostly found in the South Pyrenean Sierras in limestone massifs with a Cretaceous, Paleocene, and Eocene cover, stretching from the western edge of Aragón in the Sierra of Alano, with ice caves in the sierras of Aísa, Tendeñera, Ordesa-Monte Perdido (3355 m) and Cotiella, to the Catalonian Pyrenees in Montsec. In general terms, ice caves are dominant in the high mountain

Ice Caves. https://doi.org/10.1016/B978-0-12-811739-2.00028-0

massifs and are practically absent at the lower altitudes of the sub-alpine belts, where there are only caves with seasonal ice, exploited by the commercial ice industry, as in Montsec.

From the prior lack of knowledge, or the sole consideration that the caves were a resource for the supply of ice, the first studies were conducted in the 20th century, and focused on heritage or their exceptional thermal behavior. Those studies were carried out by scholars and cavers, mostly French, among whom R. Jeannel, E.G. Racovitza, N. Casteret, P. Dubois, H. Salvaire, P. Bernand and M. Van Thienen stand out. It will be in the 21st century that records of fossilized ice from Spanish ice caves become interesting as exceptional sources of environmental information and paleodata archives for the reconstruction of past climates. Ice caves are not only studied because of their high heritage value, but also as excellent proxies to understand past environmental changes.

The fact that several measurements carried out in ice caves, such as endoclimatic conditions or the isotopic values of the ice sequences, correspond to climate parameters (Luetscher et al., 2005), lead to dedicate scientific efforts to the study of ice caves before their disappearance in the near future. The study of Spanish ice caves is now just beginning, and limited by several factors. These limitations can be reduced by obtaining more detailed knowledge of the underground ice masses on the following aspects: monitoring karstic systems in order to better understand the factors driving ice variations, the use of multi-proxy methodologies in reconstructing past climatic variability, and the replication of the signal obtained by various methods to extract information on regional climatic change (Moreno et al., 2016).

28.2 HISTORY OF RESEARCH

Ice caves were known in Spain as ice stores due to their practical use, but the first references from a naturalist viewpoint were made by Alexander Von Humboldt in 1799. Previously, the Teide Ice Cave was cited in 1667, when British travelers climbed the Teide summit. Humboldt visited the Canary Islands on a trip to America and, together with the botanist Aimé Bonpland, spent 6 days traveling around the Tenerife Island. He showed great interest in the Teide (3718 m), and they proceeded to climb it, but before doing so they took some time to explore the ice cave situated in the Altavista area, also known as Los Neveros (Fig. 28.1). Their visit to this ice cave was very successful, and made its existence and singularity known in scientific and cult circles in the Canaries, Spain, and Europe. However, the cave was already known by the local population as the source of the ice and snow extracted from it and brought down by mule to the nearby villages, as it is referred by Bernardo Cologan, who published his piece 5 years before Humboldt's visit. In 1814 Humboldt published his account of the trip, and his hypothesis on ice formation, in his book "Voyage aux régions équinoxiales du nouveau continent" (1814), in which he described "la Caverne de glace, placée à 1728 toises de hauter, par conséquent au-dessous de la limite où commencent les beiges perpé uelles sous cette zone." He compared it with the "glacieres" of the Jura and the Apeninos and attributed its conservation more to the snow feed than to the external temperatures or the altitude. He also mentioned the ice cave of the Teide in letters he sent to Bonpland before his work was edited, which dealt with the details of the measurements he made on the volcano (Bourguet, 2003). Years later, the cave was studied once more by the Italian astronomer Charles Piazzi Smyth (1858), and ice caves became more common in publications, such as Browne (1865) and Balch (1900).

In the 19th century ice caves were not only considered to be simple stores, but indicators of the Earth's physical processes, and were of even greater interest than glaciers. As Thury indicated in the middle of the 19th century, ice caves not only conserve ancient ice, but also conserve large ice masses

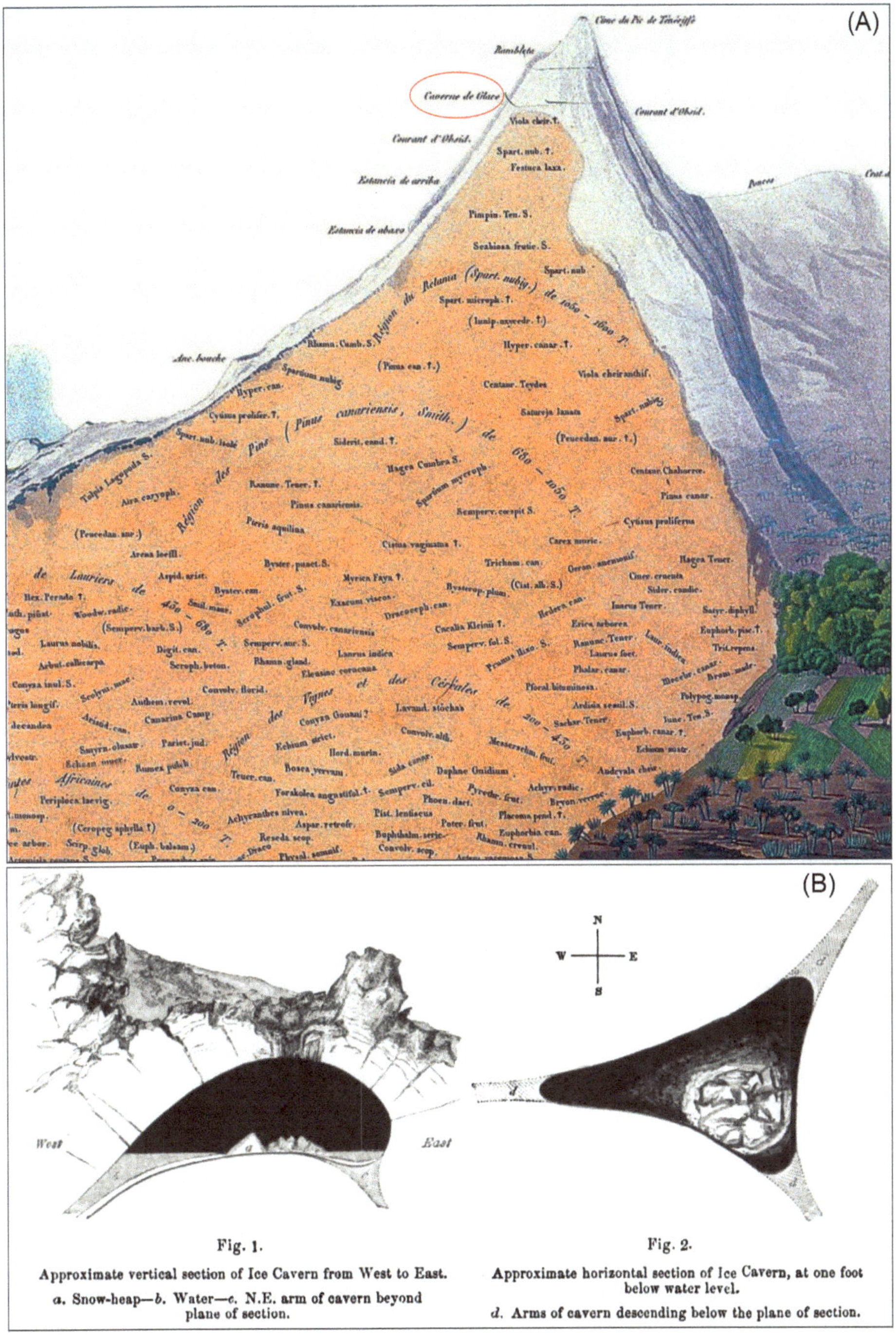

FIG. 28.1

(A) The Teide ice cave represented as "Caverne du Glace" (Caverna de Xelo in the Spanish edition) in the "Tableau Physique des Isles Canaries" by A. von Humboldt (1814). (B) Map of Teide Ice.

which were formed when water gets inside the caves. The growing importance of the study of ice caves from the 19th century helped to consider them as another element of the cryosphere, and of particular interest to climatic development. In Spain, however, since Humboldt's work, the much poorer scientific context of the 19th century lead to an almost complete absence of scientific research on ice caves until very recently. Only Puig y Larraz (1896) mentioned the "Cueva del Gel" or the Llimiana ice cave in Montsec, a very low altitude cave well known in the area, which was later studied in 1910 by Jeannel and Racovitza (1912). In that work, the presence of ice inside the cave is described together with the record of high temperatures in summer, finally considering it to be "glaciéres."

It is true that ice caves have always been known as a source of raw material for the ice market, and just as the inhabitants of Llimiana in Montsec knew of the "Cova del Gel" (Cave of the Ice), the shepherds of Sotres, in Picos de Europa knew of the "Cueva del Xelu" (Cave of the Ice). But the recent diffusion of their existence would come about following speleological explorations. The explorations of Norbert Casteret in the Pyrenees in the 1920s were outstanding, since they represented the onset of modern exploration, and the first time that the discovery of subterranean frozen landscapes in Spain was made public. Casteret Cave, probably the most famous Spanish ice cave thanks to the explorations by the renowned speleologist, was discovered in 1926, and soon became known as the highest ice cave in Europe (Casteret, 1953, 1961).

The discovery of Casteret Cave in 1926 took place while Norbert Casteret was traveling to the summit of Monte Perdido from la Brecha de Roland. The famous French speleologist explored some entrances and caves with the idea of visiting them again in the future. In one of them, then known as la Espluca Negra, and now known as Casteret Ice Cave, he discovered a new underground world similar to the one previously described in other European alpine areas, where ice caves have been studied since the 18th century, but completely new in the Pyrenees. When he entered the now world famous cave he contemplated "*one of the strangest and most rare decorations we may have seen in the world; a frozen lake and beyond, coming from the Earth's entrails, a river of ice,(...) this glacial underground world, almost unimaginable, stunning, it is extraordinary!*" (Casteret, 1933). Thus the exploration and, above all, the dissemination of the existence of ice caves in the Marboré-Monte Perdido massif began in what is now the National Park of Ordesa and Monte Perdido. Casteret would not return to the cave until 1953, when he explored and mapped the entire cave, which he considered to be the highest in the world. Casteret discovered that *"each step new and marvellous aspects, from amongst the most of our planet: an underground glacier and a natural cathedral submerged in the entrails of the Earth*" (Casteret, 1961) and he worked hard on the mapping and exploration of these caves, publishing the book "Les glaces souterraines. Les plus éleveés de Monde" in 1953 (Fig. 28.2).

In spite of the fact that the exploration of ice caves had begun, it was not until 1958 that their scientific study started, when Dubois (1958) made the first measurements of temperatures in the cave and of the size of the subterranean ice. Additionally, he made the first chronological inferences on its possible age pointing to a hypothetical Würm age.

Between 1953 and 1962, the Spéléo Club Alpin Languedocien of Montpellier made fifteen surveys and explored all the ice caves then known. Later on, the GEK of the Institut Universitaire of Technologie of Perpignan continued the work from 1968. In the following decades, considerable progress was made, and the first scientific studies were carried out. In 1982 Henri Salvayre published a scientific study in which he made the first exhaustive inventory of ice caves, including origin, formations, morphologies, and types of ice, including some geochemical analysis (Salvayre, 1982). He also unsuccessfully attempted to draw up a timeline of the ice genesis using an analogy with the Scărişoara Ice Cave (Romania), thus

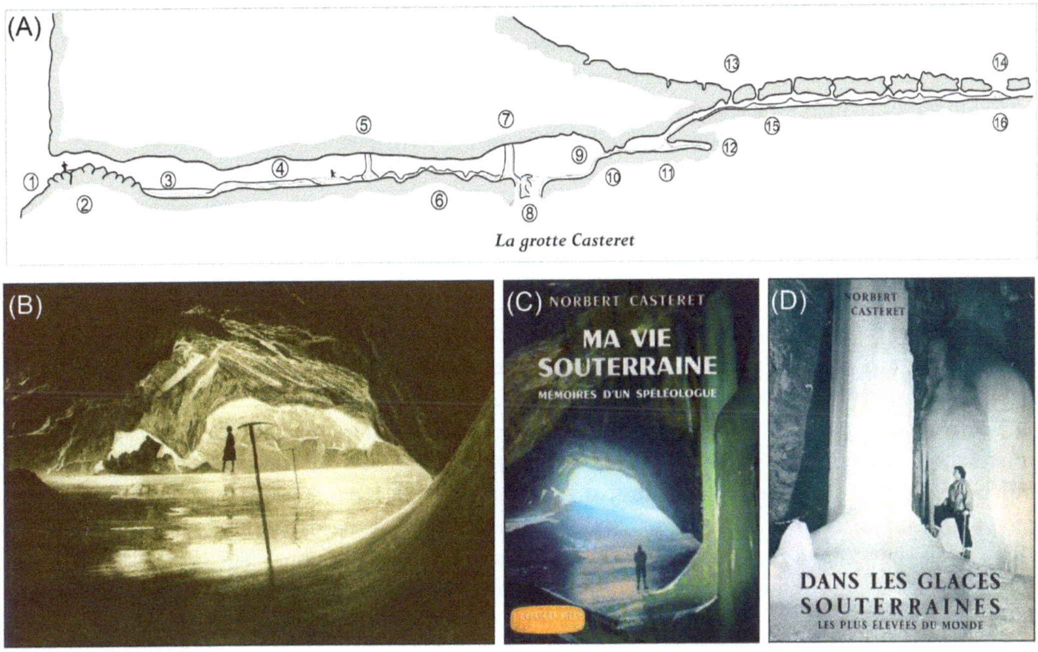

FIG. 28.2

Norbert Casteret discovered, explored, mapped, and took the first pictures in the Casteret ice cave. (A) Profile of Casteret ice cave redraw from the original map by Norbert Casteret (1961): 1, snow; 2, moraine; 3, subterranean lake of 700 m^2; 4, glaciere; 5, ice column; 6, rock and ice chaos; 7, Ice column; 8, unexplored ice pit; 10, cat hole, violent airflow; 12, dead-end; 13–14, lapies; snow shaft; 15–16, subterranean snow. (B) Picture of N. Casteret. (C and D) Book covers showing ice column and ice lake in the Casteret Ice Cave in the 50s of 20th century.

dating the ice to 6000 years (Salvayre, 1982). A first synthesis of the previous studies and new ones was carried out 5 years later. The 32 previously considered ice caves were reduced to 27, with descriptions of cryospeleothems and estimates of ice losses (Bernand and Van Thienen, 1987).

In the Pyrenees, French speleologists remained interested in the ice cave exploration of Marboré, and at the same time other massifs with ice caves were being explored. Thus from 1960 the GE of Badalona studied the area of Escuaín, in which the ice caves in the Sierra de las Tucas were catalogued. An inventory of ice caves and their altitudes became available in the 1970s (Salvador-Franch, 1977). In the Sierra de Tendeñera in 1972 the ERE of the CEC began speleological works that were continued by the Espeleo Club Gracia and the GEP of Toulouse, making possible the exploration and mapping of caves such as those of Soaso and Fenez in the Arañonera system. Recently, the ERE of CEC discovered the frozen cave of Peña Forca in the Pyrenees (2002), and the AS Charentaise discovered the HS4 ice cave in the Picos de Europa (2012), which indicates the extraordinary potential remaining to be discovered in the Iberian mountains.

In the Cantabrian Mountains the presence of ice caves has been mentioned since the 1970s, though detailed maps and the exploration of ice caves took place mainly in the 1990s. At the end of the 1980s and the start of the 1990s, the caver groups SE Geológicas, SEII, and the York University Cave and Pothole Club carried out frequent surveys in the massif of Cornión, in which the ice caves there were mapped (Llastra-B3, Pozu Vega Huerta). Also, in the massif of Urriellos, between 1971 and 1982, the AS Charentaise explored the area of Lloroza, where they mapped some ice caves such as K5. However, the systematic exploration and mapping began in the 1990s, when CES Alfa and AS Charentaise groups worked together to map 40 ice caves in the area, which meant that the massif de Urriellos in Picos de Europa, together with the Marboré-Monte Perdido in the Pyrenees, were the areas with the highest density of ice caves in the Iberian Peninsula.

The most significant contributions coming from speleological explorations were the references to the presence of ice in the 1970s and, especially, in the 1980s in the massifs of Picos de Europa and Monte Perdido-Marboré. In those areas, several ice caves and ice remnants were mentioned and catalogued, often without a full description of their characteristics. Nevertheless, the cave topographies, and the essential information about the ice caves, were provided by groups such as ERE of CEC, GE Badalona, CES Alfa, and AS Charentaise, which enabled and facilitated the posterior study of ice caves in the 21st century. As previously mentioned, these are the only research studies regarding ice caves, with the exception of some occasional mentions in relatively recent and more generic studies of geomorphology, speleology, or geoheritage (Cerdeño and Sánchez, 2000, 2005; Serrano and González-Trueba, 2005; González-Trueba, 2007; Serrano et al., 2009; Gómez-Lende et al., 2011). Therefore, the scientific studies on ice caves have been published mostly in the last 8 years. Similarly, the study of ice caves in the Pyrenees is of renewed interest in the 21st century after the inventory and cataloguing work of the 1980s (López-Martínez and Freixes, 1989). Ice caves are now included in the permafrost maps of the Pyrenees, using previous inventories (Serrano et al., 2009). The study of ice caves began in Picos de Europa in Castil (Berenguer-Sempere et al., 2014; Gómez-Lende et al., 2016), and later with the collaboration of CES Alfa and the University of Valladolid from 2010 in Verónica and Altaiz (Gómez-Lende et al., 2011; Gómez-Lende and Serrano, 2012a,b,c; Gómez-Lende et al., 2014; Gómez-Lende, 2015, 2016). At the same time, in the Pyrenees, the first studies began in the massif of Cotiella, centering on cave A294 (Belmonte and Sancho, 2010; Sancho et al., 2012; Belmonte, 2014; Belmonte et al., 2014), and in Monte Perdido, in the Casteret and Sarrios ice caves (Leunda et al., 2015; Moreno et al., 2016; Bartolomé et al., 2016).

28.3 ICE CAVES IN SPAIN

Ice caves are defined as caves with permanent ice, even when the ice remaining in autumn is of very small size. The perennial ice bodies can be accompanied by seasonal cryospeleothems and other morphologies linked to freezing processes. In this work, we consider all caves with permanent snow, metamorphic ice or speleothems, and snow shafts more than 10 m deep, which have been cited in papers, speleological reports, mapped or pictured in speleotopographies, or photographed. Although it is difficult from speleological sources to discern if the ice is permanent or seasonal, we have not considered the caves with only seasonal ice in this work, except those with permanent ice until very recent times, such as the Teide Ice Cave or LLimiana (Montsec). Shafts with perennial ice or firn in the bottom and more than 10 m in depth have been included.

In Spain, ice caves are found today in the Pyrenean and Cantabrian Mountains (Fig. 28.3) due to the altitude of those massifs. In fact, their altitude, difficult access, and exploration are the main reasons that limit their study, diffusion, and use, as described by Maire (1990, p. 508), who pointed out that high mountain and sub-arctic ice caves are less studied due to difficult access.

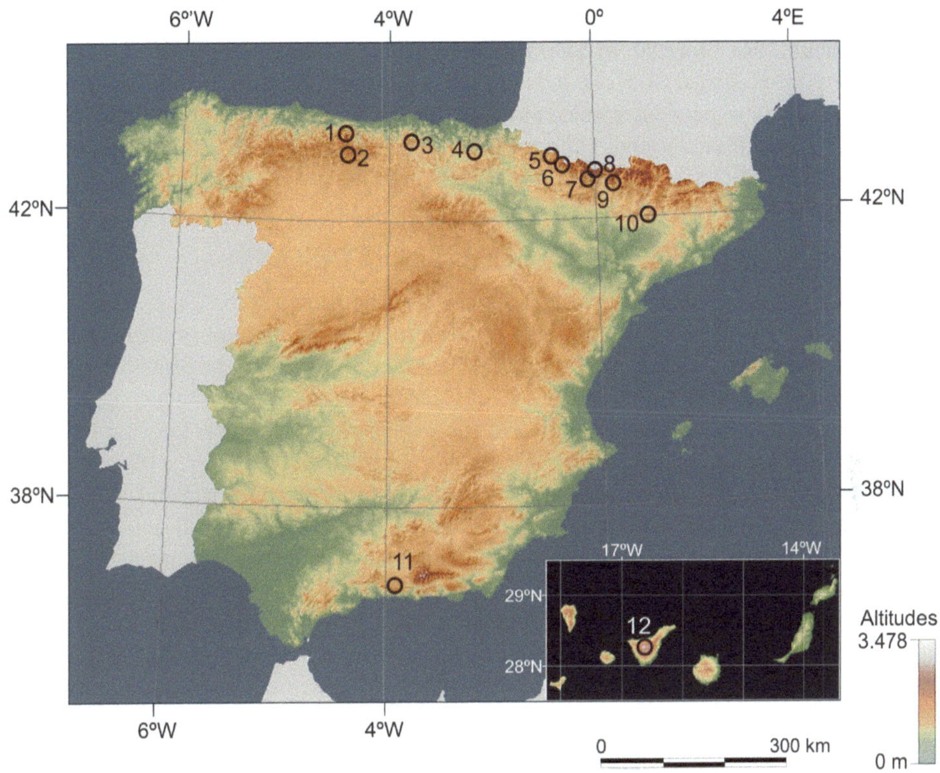

FIG. 28.3

Places where ice caves have been cited or mapped in Spain. 1, Picos de Europa; 2, Palentine Mountains, Espigüete; 3, Burgos Mountains, Valnera massif; 4, Aitsgorri; 5, Añelarra and Alano; 6, Aísa; 7, Tendeñera; 8, Marboré-Monte Perdido-Las Tucas; 9, Cotiella; 10, Montsec; 11, La Tejeda; 12, Teide.

While ice caves are very common in the middle and low alpine mountains (Glacière de Chaux-les-Passavant, at 500m; Monlés, at 1.135m, Eisriesenwelt, at 1.641m), in the Urals (Kungur Ice Cave, at 120m), the Carpatos (Scărișoara at 1.165m), and the Low Tatras (Dobšinská at 969m), in Spain only 2% are located at altitudes lower than 1600m. Only 9% are found below 2000m altitude, and 28% are above 2500m. Thus most of the caves (63% in Spain and 72% in the Iberian Peninsula) are located between 2000 and 2500m a.s.l., corresponding to the transition between the sub-alpine and alpine belts in the supraforestal high mountain, where snow plays a fundamental role (Gómez-Lende, 2015).

The latitudinal location of the Iberian Peninsula at an Atlantic-Mediterranean junction influences the development of ice caves, which are highly sensitive to climatic change. Firstly, the marginal position of the Iberian Peninsula, with respect to the overall extension of cold environments and the moderate altitudes of its summits (3479m in the Peninsula and 3718m in the Canaries), determine the ice cave locations in the highest geomorphological belts. The direct influence of the humid Atlantic climate in the Cantabrian Mountains and the western Pyrenees determines the development of most ice caves in the highest karstic massifs, such as Monte Perdido or Tendeñera.

There are caves at occasional sites at more southern latitudes or lower altitude, but this has now been placed in doubt. For example, in Sierra de la Tejeda (Granada-Málaga), under the peak of La Maroma (2065 m), the so-called Cueva de la Nieve or La Maroma (2050 m), together with other similar ones in the same area, have been interpreted as snow shafts (Durán Valsero and Molina Muñoz, 1986). Some caves have snow all year round, although only La Maroma surpasses 10 m depth, deeming it as an ice cave (Citterio et al., 2005), and they have been kept in the inventory.

28.3.1 ICE CAVES IN THE TEIDE VOLCANO (CANARY ISLAND)

The most well-known is the Teide Ice Cave, also called as the Altavista Ice Cave, which was studied by Humboldt in the 18th century. Rosales (1994) made a speleological review of the island in which he mentioned three more caves, Helada II, the Cuevas Heladas de Villar, and the Sima del Hielo, without specifying either altitudes or position. These are all volcanic caves or tubes in which outside temperatures deriving from altitude lead to the cooling of snow melt waters and their freezing in the volcanic conducts.

The ice cave of the Teide is located in the slopes close to the Altavista hut. Currently, it has seasonal cryospeleothems, but not perennial ice (Martínez de Pisón and Quirantes, 1981; Martín-Moreno, 2006, 2010), although there are snow accumulations during some summers, which makes its classification as an ice cave doubtful. The ice cave of the Teide is a lava cave in which the snow melt waters are filtered through black lava flows before freezing to form cryospelothems in the form of "*stalactites, stalagmites, verglas on the ground and parietal ice adhering to the walls* (…) *sometimes forming thin ice columns and a waterfall at the bottom*" (Martínez de Pisón and Quirantes, 1981). These authors pointed out the variations in volume and location of the cryospelothems throughout the year depending on water availability, with the most abundant cryospelothems appearing in spring. There was no mention to any ice block, only small pools that freeze, and a permanent snow cone in the entrance that contributes to a slight warming of the cavity due to the reflection of the sun's rays onto the domes and walls. Martín-Moreno (2006) points to the temporary nature of the ice in the cave, records the relative humidity ~100%, and observes different types of cryospelothems and verglas, visible until today. The cave was blown up by dynamite by the army in order to extract large quantities of ice, which may have led to endoclimatic changes and the end of the frozen masses in its interior. In spite of these modifications in the cave, the endoclimatic conditions still favored the formation of seasonal ice that was used by the army to take provisions of ice (Table 28.1) (Miranda, 2004; Martín-Moreno, 2010; Cedrés, 2015).

Table 28.1 Ice Caves in the Tenerife Island (Canary Islands)

Volcano	Cave	Altitude m.a.s.l.	References
Teide	Cueva de Hielo del Teide	3400	Martínez de Pisón and Quirantes, 1981
	Ice cave 2 Pico Viejo	–	Rosales, 1994
	Cuevas Heladas de Villar	–	Bacallado, 1995
	Sima de Hielo Montaña Rajada	–	Martín-Moreno, 2006

28.3.2 ICE CAVES IN THE PYRENEES

In the Pyrenees there are ice caves in at least seven different massifs: Anie-Añelarre, Sierra de Alano, Aísa, Tendeñera, Marboré-Monte Perdido-Las Tucas, Cotiella, and Montsec (Table 28.2). All of them belong to the South Pyrenean Sierras, formed by calcareous rocks of the Cretaceous and Tertiary Ages, which make up a group of at least 24 caves with a known ice presence (Salvador-Franch, 1977; Salvayre, 1982; Bernand and Van Thienen, 1987; López-Martínez and Freixes, 1989; Silvestru, 1999; Serrano et al., 2009; Sancho et al., 2012). The caves are small in number, and at the lower altitudes of the Inner Sierras, concentrated around Monte Perdido, the highest calcareous massif in the Pyrenean Mountains exist, where some authors point to the presence of several dozen ice caves (Fig. 28.4).

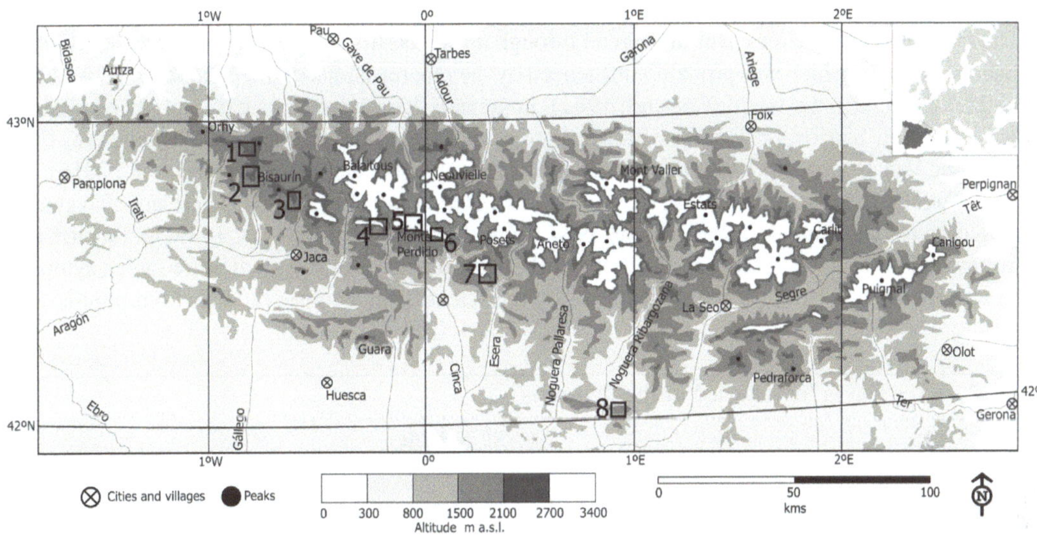

FIG. 28.4

Location of main massifs where ice caves have been cited or mapped in the Pyrenees. 1, Añelarra; 2, Peña Forca, Alano Range; 3, Lecherines, Aísa Range; 4, Arañonera system, Tendeñera Range; 5, Marboré-Monte Perdido; 6, Las Tucas Range; 7, Cotiella massif; 8, Montsec Range.

In Marboré-Monte Perdido the ice caves are found in an altitude range of 420 m, between 2640 and 3060 m, with a mean altitude of their entrances of 2750 m, which coincides with the distribution of possible permafrost (Serrano et al., 2009). In the Sierra de Las Tucas, to the east of Monte Perdido, there is also a group of half a dozen ice caves, this time in a broader range of 560 m, between 1900 and 2460 m. Their mean altitude is 2270 m, far lower than the caves of Marboré-Monte Perdido and clearly lower than the 0°C isotherm. In Tendeñera, the inventoried caves with ice extend from 2259 to 2390 m in a range of just 130 m, and have a mean altitude of the entrances of 2315 m, but the low altitude of the summits must be taken into consideration (2538 m). To the east and west the ice caves are dispersed, with one sample in each massif, at altitudes of 2238 m in Cotiella. To the west is Lecherines in the Sierra de Aísa, whose entrance, at 2080 m, is one of the lowest, and Peña Forca at 2300 m. From Lecherines a longitudinal gradation is observed, with a slow elevation of ice caves from west to east, at the same time as the altitudinal range of ice caves becomes broader, culminating in Marboré-Monte Perdido, where the caves are at higher altitudes and in a broader altitudinal range (Fig. 28.5). Structural conditions, with calcareous outcrops at a given altitude, determine the distribution of the highest ice caves. Nevertheless, the absence of these at lower altitudes is limited by the climatic conditions because of the transition to the Mediterranean climate.

Finally, at very low altitudes between 1298 and 1310m in the Montsec Range, the Cave of Llimiana is found, formerly known for having cryospelothems and sheets of seasonal ice. It was cited by Puig y Larraz (1896), and Jeannel and Racovitza (1912) studied it as early as 1910. These authors described it as a "glaciere" and recorded its temperatures, though its endoclimatic characteristics were studied by Cardona (2010). Today it is not considered an ice cave because the cave has seasonal parietal ice and cryospeleothems but not permanent ice.

The Pyrenees is where ice caves are spread through more massifs and have a greater presence in the published literature, which was probably influenced by the repercussions of the Casteret Ice Cave being the highest ice cave in the world (Casteret, 1961). In general, ice cave studies have only been made on caves in Casteret, Sarrios, or dealing with Marboré caves as a whole (Dubois, 1958; Salvayre, 1982; Bernand and Van Thienen, 1987; Leunda et al., 2015; Bartolomé et al., 2016; Moreno et al., 2016), and A294 of Cotiella (Sancho et al., 2012; Belmonte, 2014; Belmonte et al., 2014). There are also syntheses on ice caves (Bernand and Van Thienen, 1987; López-Martínez and Freixes, 1989) and their use as environmental indicators (Serrano et al., 2009). Still, dedicated cryological studies are rare, a symptom of the scarce attention paid them, until recently, by the scientific and conservationist communities.

Table 28.2 Ice Caves Inventoried in the Pyrenees

Massif	Cave	Altitude m a.s.l.	References
Monte Perdido-Marboré	Casteret	2640	Duverneuil et al., 1974
	Sarrios 1–6	2790–2810	Salvador-Franch, 1977
	Casco 1–3	2750–2830	Salvayre, 1982
	Torre 1–3	2850–2860	Bernand and Van Thienen, 1987
	Cilindro	3020–3060	López-Martínez and Freixes, 1989
	Faja Luenga	2780	
	Brecha 1–3	2790	St. Pierre, 2007
	La Roya 1–5	2715–2725	Bartolomé et al., 2015, 2016 Leunda et al., 2015
	Cristales	?	
	CS-8	?	
Las Tucas	Avenc del gel (C8) Revilla	2460	Salvador-Franch, 1977
Escuain	C33	2360	
	C40 "Avenc Sala Gelada"	2440	
	B13 "Cova Fresca"	2320	
	B17 (?)	2140	
	"Avenc de la Neu" (?)	1900	

Table 28.2 Ice Caves Inventoried in the Pyrenees—cont'd

Massif	Cave	Altitude m a.s.l.	References
Añelarre	AN 62 "Sima del Hielo"	1988	Puch, 1987
	"Sima de Linza"	–	ARSIP, 2009
	N1	2075	
	N2	2070	
	LO7-17	2076	
	LO7-3	2071	
	AN118	2077	
	AN230	2087	
	C213	–	
Sierra de Alano (Ansó)	Peña Forca Ice Cave	2300	ERE-CEC, 2003
Sierras de Aísa	Lecherines Ice Cave	2080	
Tendeñera	ST2. "Cueva del Hielo"	2200	Espeleo Club de Gràcia, 1992
(Arañonera)	Soaso Ice Cave	2259–2400	Puch, 1987
	T3 Fenez Ice Cave	2390	
	S-10	?	
	T1 Grajera de Turbón	2208	
	A 31	2473	
Cotiella	A294	2260	Sancho et al., 2012
	A69	2170	Belmonte, 2014
	A152	2191	Belmonte et al., 2014
	A405	2338	
	C27	2456	
	B94	2440	
	Helada Brecha Brujas	2700	

28.3.3 ICE CAVES IN THE CANTABRIAN MOUNTAINS

Most of the ice caves of the Cantabrian Mountains (Fig. 28.5) are concentrated at a single location, the Picos de Europa. This is a calcareous massif formed by limestones from the Carboniferous Age, thrusted out and reaching a thickness of over a thousand meters. They are in three massifs, Ándara, Cornión, and Urriellos, which reach 2648 m (Torre Cerredo), with a highly developed karstic system (Collignon, 1985; Fernandez-Gibert et al., 2000; Rosi, 2004; González-Trueba, 2007; Gómez-Lende, 2015; Ballesteros, 2016) in which large vertical caves predominate, and where 12 of the 95 caves in the world of over 1000 m depth are located. Nine of these caves are in the massif of Urriellos (FEE, 2011). In the massif of Cornión, which forms broad karstic shelves, ten ice caves, characterized by their low altitude, have been recognized. All of them are located between 1945 and 2000 m, with a mean altitude of 1985 m, and are without a doubt the lowest altitude in the Iberian Peninsula for the presence of grouped ice caves.

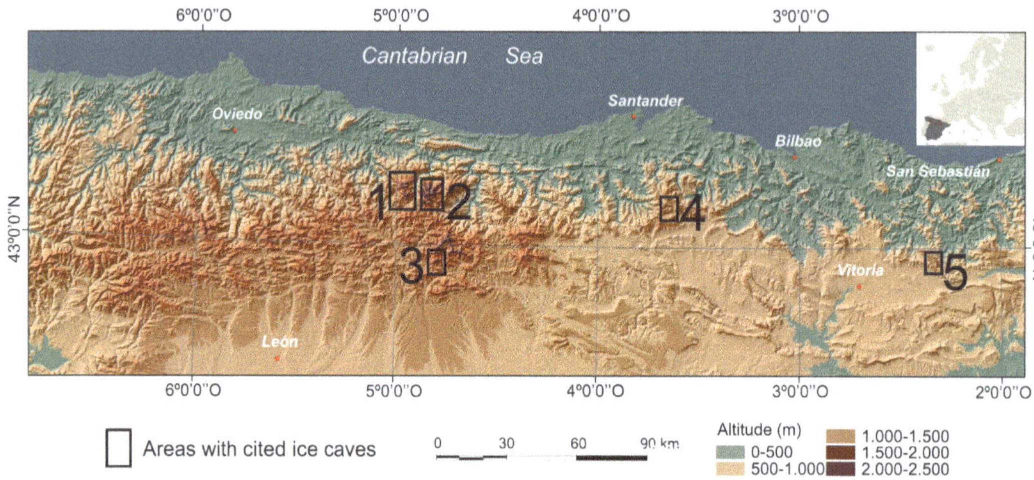

FIG. 28.5

Places where ice caves have been cited or mapped in the Cantabrian Mountains. 1, Western Massif, Cornión, Picos de Europa; 2, Central Massif, Urriellos, Picos de Europa; 3, Espigüete, Palentine Mountains; 4, Valnera, Burgos Mountains; 5, Aitzgorri.

If we take as the definition of an ice cave a cave formed on a rocky substrate in which there are perennial water accumulations in a solid state, either snow or ice (Perşoiu and Onac, 2012), 71 ice caves have been inventoried in the massif of Urriello up to this date (Table 28.3). They are found at greater altitudes than in the western massif and are more numerous, and therefore are probably not linked to the more exhaustive exploration in this sector. The inventory carried out by CES Alfa and AS Charentaise from the 1970s, in the southern part of the massif of Urriello, which mapped 290 caves, indicated that 24% of the caves in this part of the massif contained ice somewhere in their interior. This ice is distributed around the cave mouths, from 10m of depth to 260m of depth in the large ice accumulations of HS4. It is noteworthy that over 40 ice caves are located below 20m, with maximum depths of 110m (TA19) and 260m (HS4), but the remainders are between 20 and 40m deep. There are ice caves at between 1955 and 2420m of altitude, which developed in very different ways. In at least 17 caves there are great volumes of ice at these depths, although the volume and state of most of them is still not known. 40% of the caves are located between 2000 and 2100m a.s.l. To the east lies the most significant altitudinal range for the development of ice caves. The entries of many of them are in the bottoms or sides of the glaciokarstic basins, covered with thick layers of snow during winter, and affected by temperature inversion. In some sectors, a greater density of ice caves has been detected, as in the sector of Hoyo Oscuro, which is attributed to favorable topoclimatic conditions for the generation and conservation of ice in the caves. At over 2300m, 16 ice caves are located with a great variability in altitudinal distribution.

Outside the Picos de Europa, there are several caves with snow and ice deposits at low altitude. On the north face of the Espigüete, there is a calcareous summit of 2450m formed by limestones from the Carboniferous Age, and there are nine ice caves mapped by the GIS Vacceos and, since 1978, by the UE Vallisoletana (Martínez, 1977; Geoda, 2010; UEVa, 2011). These are small cavities located between 1910 and 2140m, with a mean altitude of 2035m, and all of them are situated in a northern orientation at the foot of the large calcareous face, which creates favorable topoclimatic conditions like high

Table 28.3 Ice Caves in the Cantabrian Mountains

Massif		Cave	Altitude	Cave	Altitude	References
Picos de Europa	Vega Huerta	Llastral B3	–	I32	–	OUCC, 1984
		Alfa 31	2000	K6	–	Borreguero, 1986
		Alfa 68	2000	I8 Vega Huerta	–	Kemp, 1985
		K-9010	1945			LLoyd, 1985, 1990
	Junjumia	A26	–	A1 Juriazu	–	Senior, 1987
	Ario	E9	2000	E20 La Jayada	1979	Ortiz, 1990
		E18	1962	Asopladeru P36	1369	YUCPC, 1991
		F7H	1940	Pozu la Nieve	1311	Preziosi et al., 1997
		F38	1950	F5	–	OUCC, 2007
		F45	1990	F57 C. del Arco	–	Verheyden et al., 2007
		5/4		F60	–	Greaves, 2008
		La Jayada	1750	F80	–	Rogers, 2008
		7/9	–	F90	–	Estevez, 2011
		5/9	1870	E9	2000	Fernández et al., 2011
		C13	–	E20	1979	Bottomley, 2016
		D2	1778	E25	1980	
		Jorcada	–	E26	1980	
		13/9	–	F6 Paso Doble	–	
		G3	–	F2Las Perdices	1866	
				F20L'Redondo	1937	
	Ondón	Torca 54	1580	Torca Daniel	1440	
	Urriellos	Verónica	2230	T14	2345	Fabriol, 1975
		Altaiz	2190	HR10	–	Cerdeño and Sánchez, 2000
		HS4	2350	V6	2211	
		Castil	2095	T1	2420	CES-Alfa and A.S. Charentaise, 2016
		ES9	2055	V40	2278	
		Tiro Llago	2300	P5	2025	Gómez-Lende, 2016
		Abisu l'Xelu	1819	CV11	2365	
		P9	–	HR7	2354	
		H5	2321	ES01	2032	
		ES22	2087	HS13	2291	
		ES20	2047	N24	2063	
		ES21	2045	V35	2270	
		P1	1955	P20	1981	
		N43	2143	CV27	–	
		N3	2040	N12A	–	
		CV13	2352	V4	2140	
		CV22	2352	V15	2329	
		CV5	2384	JO17	2226	
		M5	2310	N12	2088	

Continued

Table 28.3 Ice Caves in the Cantabrian Mountains—cont'd

Massif		Cave	Altitude	Cave	Altitude	References
		PV2	2313	V1	2290	
		P30	2081	ES17	2027	
		TA22	2058	N26	2044	
		N19	2092	ST01	2192	
		ES21	2045	TA19	2080	
		TA20	2051	2N	2042	
		HR4	2313	JO58	2209	
		TA5	2094	N8	2068	
		JO38	2169	TA51	2109	
		P13	1955	TA53	2092	
		JO26	2168	V17	2242	
		T12	2338	V22	2225	
		JO37	2178	N11	2031	
		T13	2338	P16	2058	
		V36	2288	T11	2341	
		PV4	2241	CV23	2050	
		TA9	1989	JO39	2171	
		ES13	2003	TA13	2032	
		ES38	2041	N5	2091	
		JO28	2175	TA16	2051	
		PV3	2318			
		Tiro llagu	2300			
Montañas de Burgos	Castro Valnera	Torca de La Grajera	1490			Puch, 1987 Martín-Merino, 2015
Montaña Palentina	Espigüete	Las Chovas	1910	ENO-101	2140	Martínez, 1977
		Ves CV-1	2045	ENO-30	2000	GEODA, 2010
		S2	–	S6	2000	UEVa, 2011,
		ENO-86	–	ENO-107	2140	Pellitero, 2012
		ENO13	2000			
Basque Mountains	Aitzgorri	Aitzgaizto 12	1280			Eraña et al., 2008

snowfall and the presence and conservation of ice in the caves (Pellitero, 2012). Finally, to the east, in the massif of Castro Valnera (1718 m), at low altitudes, la Torca de la Grajera is found at 1490 m. This is a karstic massif formed by limestones of the Cretaceous Age which, at a distance of just 30 km from the sea and with high snowfall, constitutes a hyperhumid environment. It is a vertical cave of 185 m depth, which at its base contained a mass of perennial ice. Studies are now being carried out by GE Edelweiss, the University Complutense of Madrid, and CIEMAT. It has been found that between the first exploration in October 1981, which showed 22.25 m of height, and the present day with 14.5 m, 7.1 m of the body of ice has been lost (Puch, 1987; Martín-Merino, 2015).

The smaller surface extension of the Cantabrian high mountain, with respect to the Pyrenees, may be the reason for a possibly lower number of ice caves, although, due to the sparse attention given to them, together with a considerable lack of coordination, systematic work, or consensual agreement on cataloguing of the innumerable speleological studies, this cannot be confirmed. These ice caves in the Cantabrian Mountains have received very little scientific attention. Specifically, scientific studies are very recent, initiated by speleological and morphometric studies (Fabriol, 1975; Cerdeño and Sánchez, 2000) or synthetic ones (Maire, 1977). Since 2010 specific studies on the endoclimate and the ice dynamic in the ice caves of the Picos de Europa have been initiated (Gómez-Lende et al., 2011, 2013, 2014; Gómez-Lende and Serrano, 2012a,b,c; Berenguer-Sempere et al., 2014; Gómez-Lende, 2015, 2016). Finally in the Basque Mountains, the Aitzgaizto 12 cave with perennial ice is cited in the Aitzgorri massif (Eraña et al., 2008).

28.4 STATE OF THE ART OF THE STUDIES OF THE MAIN SPANISH ICE CAVES

28.4.1 TECHNIQUES AND METHODS CARRIED OUT IN THE STUDY OF ICE CAVES OF SPAIN

Ice caves in Spain are environmental indicators in relation to the climatic and topoclimatic conditions of the high mountain, fundamental elements of the Iberian high mountain endokarst, and serve as paleoenvironmental archives. The techniques and methods used in their study are highly diverse and seek to better understand their current state, dynamic, and past evolution.

Publications where ice caves are cited were found both in the scientific bibliography and, more importantly, in speleological publications and reports. Research was carried out on the numerous available speleological reports and topographies drawn up by many national and international speleological groups. These reports provide the inventory of ice caves and the results obtained throughout their study. The large speleological clubs that have worked in the study areas, such as Spéléo Club Alpin Languedocien of Montpellier, GEK from the Institut Universitaire de Technologie of Perpignan, ERE from the Centre Excursioniste de Catalunya, Espeleo Club Gracia of Barcelona, GEP of Toulouse, GEB of Badalona and the Centro de Espeleología of Aragón for the Pyrenees, the Groups SE Geológicas, SEII, York University Cave and Pothole Club, A S Charentaise, CES Alfa of Madrid, OUCC of Oxford, GELL of Cantabria, GE Edelweiss of Burgos and GEODA, and UEVA of Valladolid in the Cantabrian Mountains are the primary sources of data on the sites, types, and conditions of ice caves.

Distribution of ice caves shows that, currently, there is a concentration of inventoried ice caves in the Cantabrian Mountains and, in a second place, in the Pyrenees (Fig. 28.6). It is perhaps explained by the influence of Atlantic domain, which is visible also in the Pyrenees, where caves are clustered in the western-central Pyrenees, or straightforwardly by the most systematical speleological exploration in the Picos de Europa. Altitude distribution is broad, both in the Pyrenees and Cantabrian Mountains, but ice caves are more frequent in the highest calcareous massifs and in the most western ones.

Studies are concentrated on a small number of ice caves, where the permanent ice occupies a significant volume inside the cave. Figs. 28.7–28.9 show, at the same scale, the size and ice volumes in the most studied ice caves.

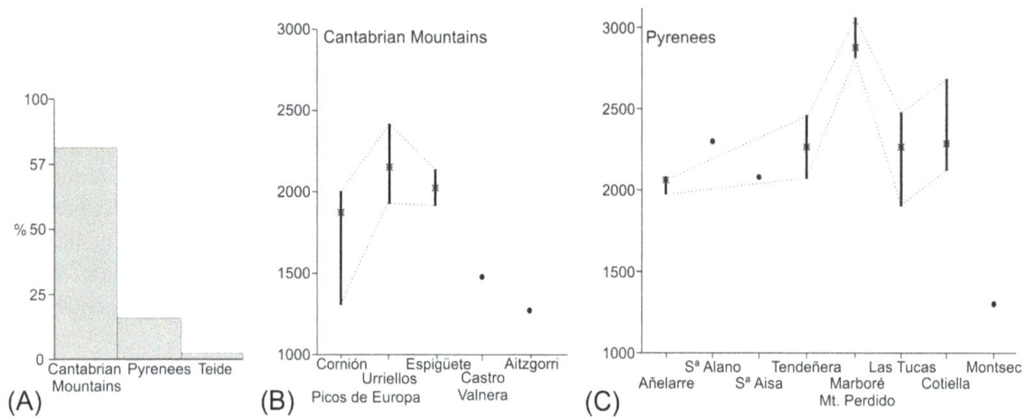

FIG. 28.6

Distribution of ice caves. (A) Percentage of ice caves in mountain system of Spain. (B) Altitudinal distribution of ice caves in the Cantabrian Mountains. (C) Altitudinal distribution of ice caves in the Pyrenees. * Average.

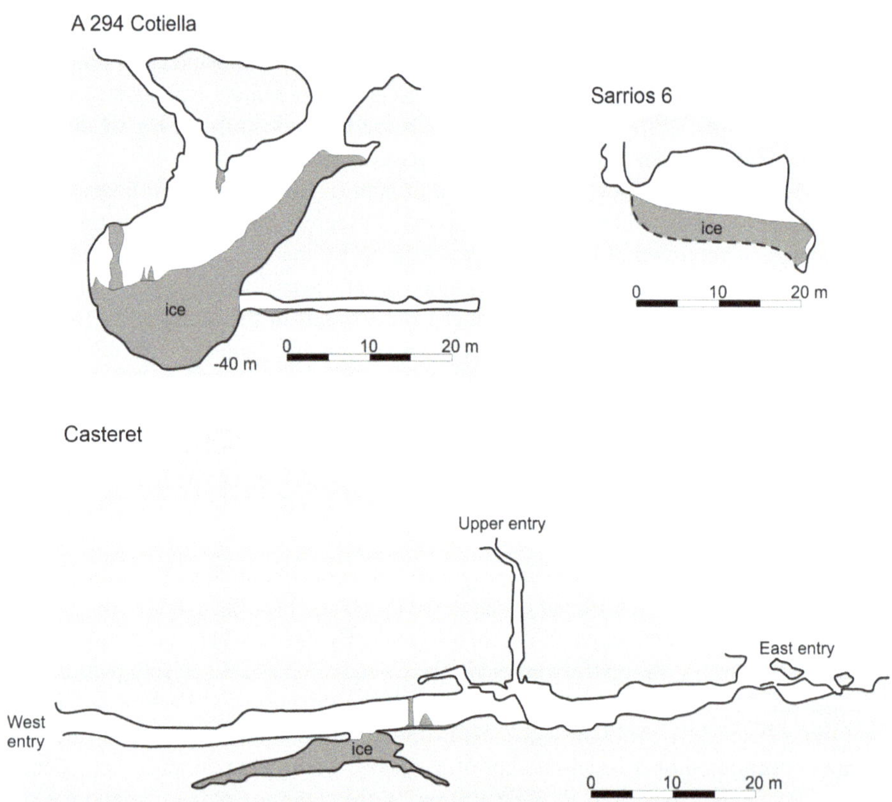

FIG. 28.7

Profiles of studied ice caves in the Pyrenees. A294 Cotiella (taken from Sancho et al., 2012), Sarrios 6 (taken from Bartolomé et al., 2016 and modified) and Casteret (taken from Leunda et al., 2015 and modified).

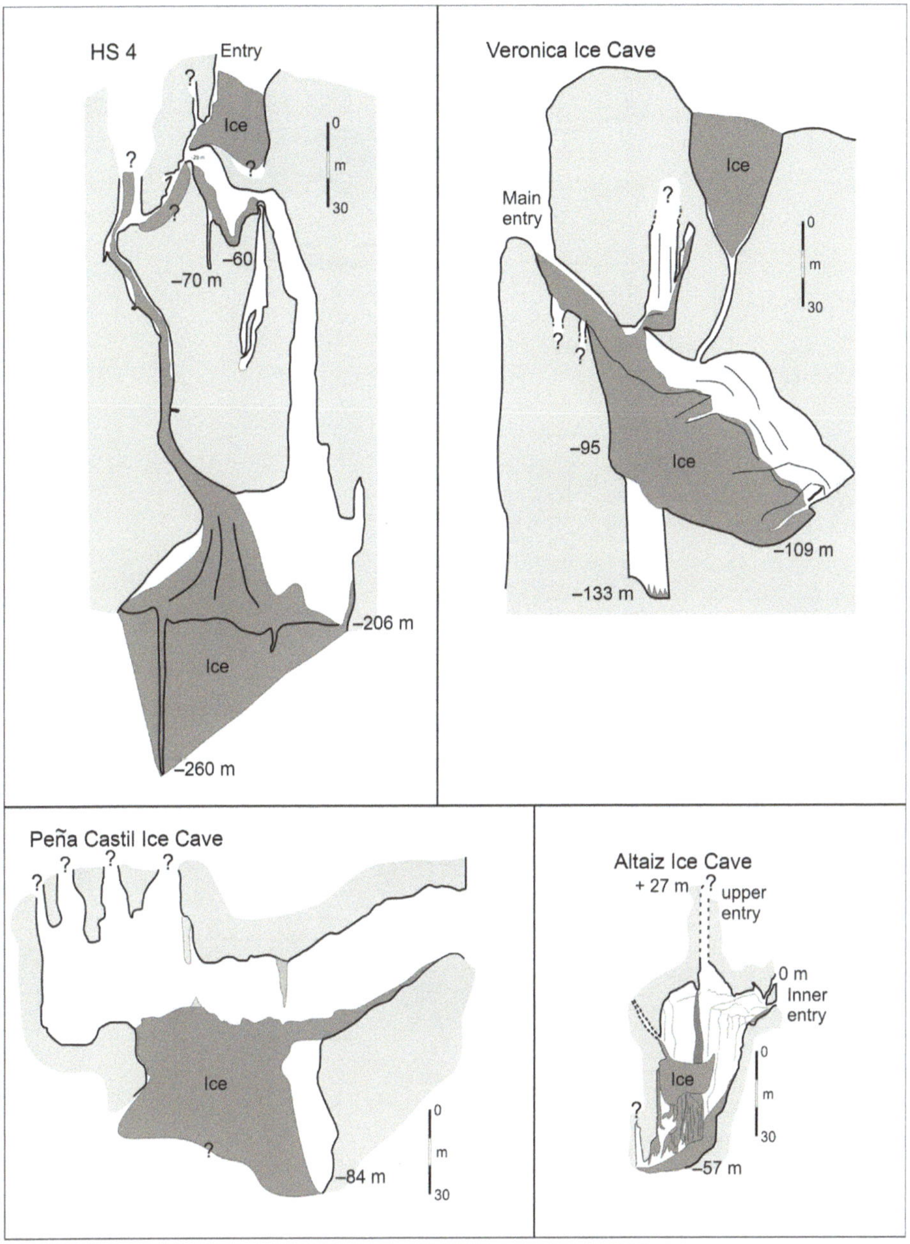

FIG. 28.8

Profiles of studied ice caves in the Picos de Europa. HS4 (taken from CES-Alfa and A.S. Charentaise, 2016, topography by B. Hivert, Y. Auffret, M. Bergueron and O. Gerbaud), Verónica, Altaiz (taken from Gómez-Lende, 2015, topography from J. Sánchez-Benítez and M. Gómez-Lende) and Peña Castil (taken from Gómez-Lende, 2015).

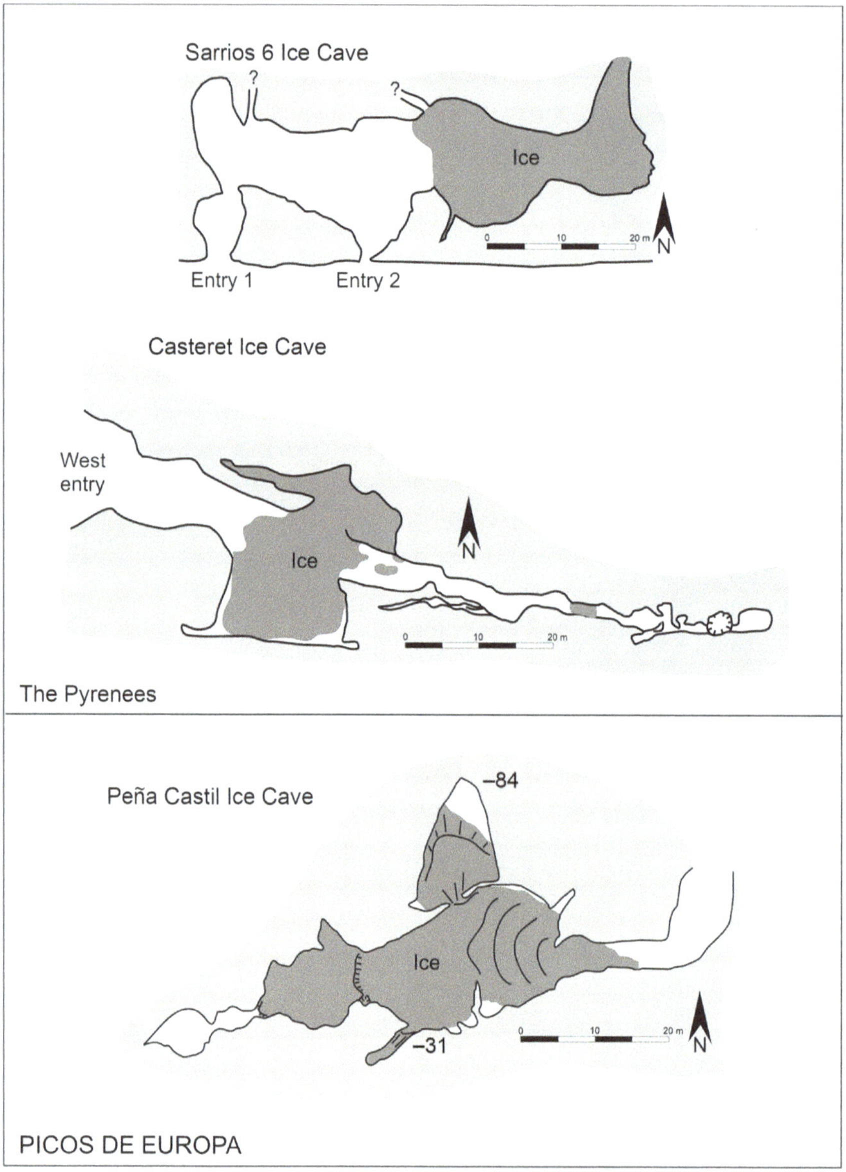

FIG. 28.9

Topography of the Sarrios 6 (taken from Bernand and Van Thienen, 1987; Bartolomé et al., 2016, modified), Casteret (taken from Leunda et al., 2015, modified) and Peña Castil (taken from Gómez-Lende, 2015, modified).

A considerable effort was made to characterize the endoclimatic characteristics of the ice caves. Knowledge of the evolution of temperature and humidity are of special interest, since we want to describe the ice dynamic, that is, the formation and melt stages in seasonal ice, and the losses or gains in permanent ice. Additionally, that information is necessary to plan cave management strategies, given that they are all in National Parks. Unfortunately, climatic conditions in ice caves are only recorded in some individual caves and not generally. Such cases are Marboré and the Forat del Gel (Montsec) (Salvayre, 1982; Cardona, 2010). More recently, the study of endoclimatic parameters is being carried out using dataloggers placed at three points in each cave in the Picos de Europa (Altaiz, Verónica and Castil since 2011 and HS4 since 2016), and in the Pyrenees (Casteret Ice Cave, since 2015 and A294; Gómez-Lende and Serrano, 2012a,b,c; Gómez-Lende et al., 2014; Belmonte, 2014; Belmonte et al., 2014; Bartolomé et al., 2015; Leunda et al., 2015; Gómez-Lende, 2015, 2016). Data was obtained daily, and seasonal variations of endoclimatic conditions were recorded, following the installation of thermal and hygrometric dataloggers at different points in these caves. In the ice cave of Castil, a thermographic camera was used to characterize and study the vertical thermal component of the air masses and thermal study in three dimensions, combining the results with the Terrestrial Laser Scanner (TLS) to generate orthothermograms (Gómez-Lende et al., 2014; Berenguer-Sempere et al., 2014). These techniques enable the detection of pockets of warm air during cold periods and the observation of heat flows within the cave. In addition, temperature maps of the caves studied were drawn up based on the data obtained using the thermal dataloggers (Gómez-Lende, 2016).

Changes in ice and mass loss were studied by comparing topographic maps, documents, and old photos with recent ones, which revealed changes in ice volume (Gómez-Lende et al., 2014; Gómez-Lende, 2015; Berenguer-Sempere et al., 2014; Bartolomé et al., 2015). In Casteret and Peña Castil the evolution of the perennial ice mass was reconstructed from the 1950s and 1980s, respectively using morphological observations, notes from different bibliographic references, and by comparing identifiable fixed points in the photos. This methodology shows that documentation is of fundamental importance if the changes in the past are to be tracked. In addition, annual variations are monitored using TLS, which facilitates the generation of high-precision three-dimensional models and orthoimages (Berenguer-Sempere et al., 2014). Through two measurements taken annually from 2011 to 2014, precise volumetric changes in the cryospeleothems and surface variations in the ice body were recorded. The reduction in size and weight of the geomatic instruments makes their use possible in any ice cave in the future. More conventional methods such as the installation of bars for the surface monitoring of ice bodies has facilitated the monitoring of interannual variations in the ice, although as we do not know the overall ice block volume, a global measurement of the ice blocks is not possible.

The ice structure has been studied in Cotiella, Sarrios, Peña Castil, and Verónica by means of the description of cryogenic structures, taking information from the deformation of perennial bodies of ice, and from past and present movements (Gómez-Lende, 2015, 2016; Bartolomé et al., 2015). The application of GPR to the study of the ice mass in Peña Castil is optimal for finding the internal structure of the ice block and determining its properties. Together with the descriptions of the structures, their dynamics has been established.

To go deeper into palaeoenvironmental studies, the first step is to date the ice bodies. Datings were made in Casteret, Sarrios-6, and Cotiella in the Pyrenees, and in Altaiz and Verónica in the

Picos de Europa. In all cases, the organic material in the ice, or levels of dung when present, were dated by radiocarbon (^{14}C). Palinological analyses were used on the levels of dung (Leunda et al., 2015; Moreno et al., 2016), and in Sarrios-6 cryogenic cave carbonates (CCC) were sampled for their micromorphologic analysis, stable isotope analyses and U-Th dating. The CCC analysis focused on rhombohedral crystals of up to 4 mm in diameter composed of a fragile core of creamy to orange colored skeletal and porous rhombohedral crystal aggregates, and an outer zone formed by brownish to orange rhombohedral crystals (Bartolomé et al., 2015). In Cotiella and Sarrios-6, the stratigraphy of the ice body profile was studied, and samples were taken using ice microcores to analyze the isotopic composition ($\delta^{18}O$ and δD) (Sancho et al., 2012; Belmonte et al., 2014; Bartolomé et al., 2015; Leunda et al. 2015; Moreno et al., 2016), which was also performed in Torca de la Grajera (Burgos) in the Cantabrian Mountains. In Casteret the cave was monitored, and measurements were taken of the environmental variables and isotopic composition of the dripping water in order to complement the analyses of the ice deposit (Leunda et al., 2015; Moreno et al., 2016). This work will improve knowledge of the ice and its relationship with infiltration waters and snow.

28.4.2 DYNAMICS AND EVOLUTION OF ICE CAVES

The calculation of the ice mass balances is one of the most common topics in the study of underground ice and is very useful for forecasting the immediate future of the ice bodies. Knowing the mass balance is especially important and urgent when the general trend is regressive and may lead to complete disappearance of the ice, which could cause the loss of its potential as a palaeoenvironmental archive (Kern and Perşoiu, 2013).

Mass balances estimated in most ice caves, independently of the cave type and its location (karstic, lava tubes, high or medium latitude, high mountain), have a negative trend, and this is also the case in Spain (Fig. 28.10). Only in a few cases are volumetric increases detected. In Spain there are no continuous quantitative records of the evolution of ice in any of the caves studied. Nevertheless, an overall negative trend was found in all of them, including in the Pyrenees, in the A-294 and Casteret ice caves (Sancho et al., 2012; Belmonte et al., 2014; Bartolomé et al., 2015), and in the Cantabrian Mountains, in Altaiz, Verónica, Castil and La Grajera (Gómez-Lende et al., 2014; Berenguer-Sempere et al., 2014; Gómez-Lende, 2015, 2016; Martín-Merino, 2015). A significant interannual variation was also found together with the regressive trend of the last decades, which allows identifying two distinct behaviors:

- *Medium-term retreat.* In all the caves studied or explored, ice volume decreased in the last decades. In Casteret cave the photographic seasonal comparison with the present day reveals a fall in the level of ice by 3.70 m in the main chamber due to melting processes, which represents an average loss of 5.5 cm a^{-1}, which is the equivalent to 42 $m^3 a^{-1}$ (Bartolomé et al., 2015). In Castil the thickness of the ice has also fallen ~2 m in the last 16 years, which is a mean of ~12.5 cm a^{-1}, and a volumetric loss of ~7862 $m^3 a^{-1}$ (Gómez-Lende et al., 2014; Gómez-Lende, 2016). In A294, in Cotiella, 12.5 $m^3 a^{-1}$ loss have been estimated (Belmonte, 2014).
  Although quantitative data for Altaiz and Verónica are not available, the fall in the mass of ice is evident, and in the Grajera a loss of ice of 7.1 m was recorded between 1985 and 2015, which is a mean loss of 36 cm a^{-1} (Martín-Merino, 2015). These estimates are consistent with the height at which the abrasion striae are located in the wall of Verónica, Altaiz, and Castil, which are all well preserved and visible to the present day. But the ablation was not continuous, as shown by Bartolomé et al. (2015) in Casteret, where there is evidence of

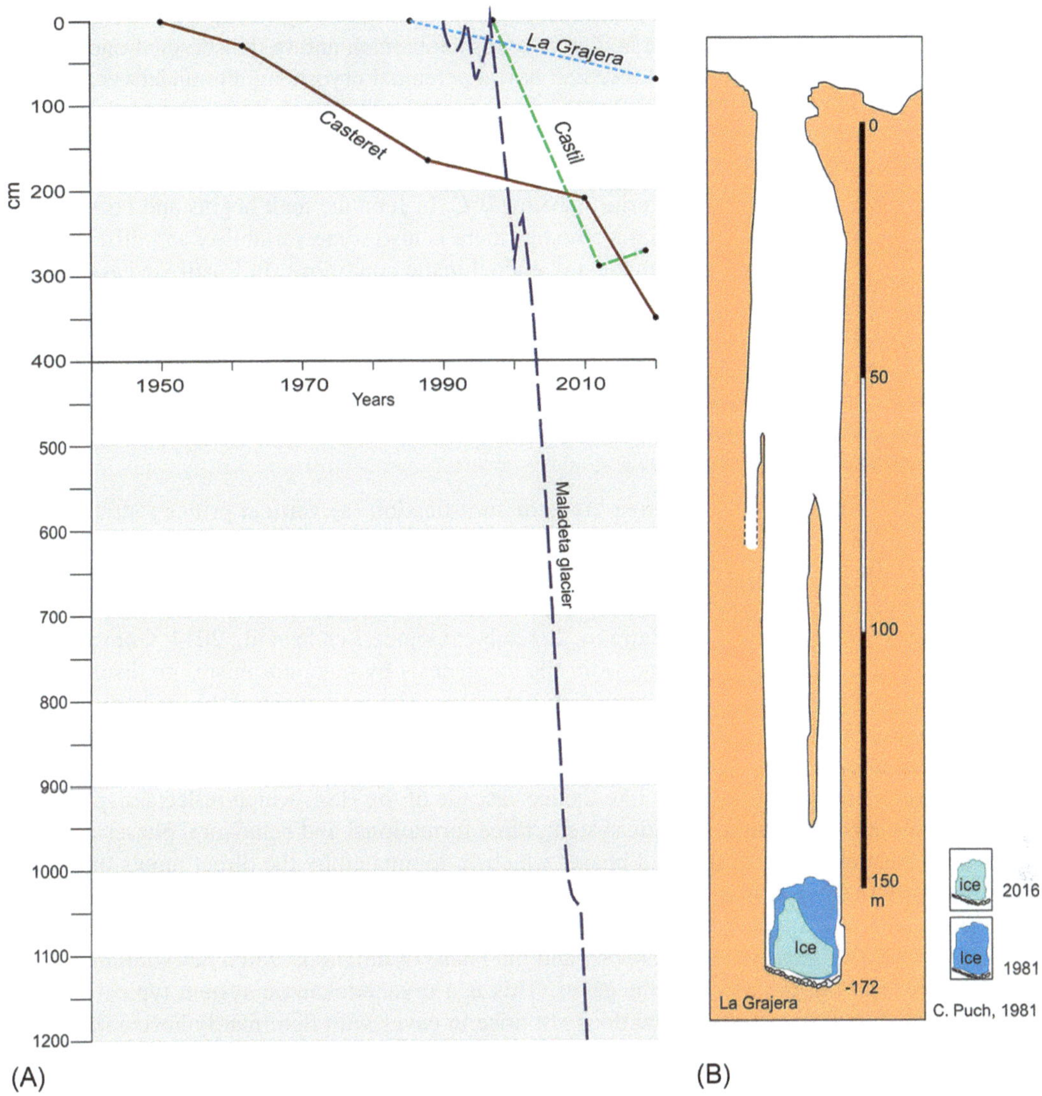

FIG. 28.10

Ice degradation in Spanish ice caves. (A) Ice lost in three ice caves: Casteret (data from Bartolomé et al., 2016), Castil (data from Gómez-Lende, 2015) and La Grajera, Valnera (data form Martín-Merino, 2015), compared to the mass lost in the Pyrenean glaciers. (B) Profile of La Grajera shaft with the ice lost between 1981 and 2016. Topography by C. Puch of 1981, data form Puch (1987) and Martín-Merino (2015).

different rhythms in the loss of volume over the last 50 years (Fig. 28.10). The ice diminished slightly from 1926 to 1950 and remained constant until the 1980s, when a drastic fall began, which has been correlated with the rhythms of glacial retreat in the Pyrenees and the disappearance of some ice columns (Bernand and Van Thienen, 1987; Bartolomé et al., 2015; Leunda et al., 2015).

- *Seasonal and interannual variations.* Annual rhythms are marked by strong fluctuations both in the perennial ice levels (ice bodies or cryospelothems), and in the seasonal ones. In Castil, Gómez-Lende (2015, 2016) revealed how a perennial cryostalagmite underwent swings of around 1–3 m^3 between 2012 and 2014, and some seasonal ice columns scanned show variations between 5 and 18 m^3. In the Pyrenees and Picos de Europa the rhythms are the same. Ice forms from the end of autumn, but mainly in spring, when the outside temperature gives rise to infiltration and the interior is below 0°C. In June the melt begins and lasts until October–November. In the perennial ice bodies there is also some variability and different behavior depending on the topoclimatic and endoclimatic conditions. In Castil ice caves have been recorded an increase in thickness of 20 cm a^{-1} by refreezing in the surface using TLS between 2011 and 2014, but in Altáiz the control bars revealed that between 2011 and 2013 the surface had lost 22 cm, an annual mean of ice melt of 11 cm a^{-1}, and 32 cm in the sides. Although the volume of the loss cannot be estimated for the whole block because its exact volume is not known, minimum surface losses of 11 $m^3 a^{-1}$ between 2011 and 2013 are cited (Gómez-Lende, 2015, 2016).

Large ice bodies can exhibit different kinds of stratification (as vertical primary stratification, uncomformities, and undulose bedded) and deformities, such as folds, compressional wedges, crack lines, and flow waves. The last ones have been observed in Castil by means of TLS, which reflects the displacement of the ice mass as a genetic system of the endokarstic ice bodies not observed in previous studies (Gómez-Lende and Serrano, 2012a,b,c; Gómez-Lende et al., 2014; Gómez-Lende, 2015, 2016). These deformations come from the ice genesis by metamorphism, its displacement, and ice motion, so that the ice body advances into the cave. The movement of the ice body has been confirmed by the existence of abrasion striae in the walls of Altaiz, Castil, and Verónica, which denote its erosive effectiveness and lead to the reconstruction of the directions of flow, as well as the breaking of perennial cryospelothems like the ice cascade of the HS4, which reflects displacement of the overlying ice body. In this genetic system, three formational and behavioral phases of the ice masses have been established: the feed phase, which is dominated by the direct inputs of snow in broad mouths with steep slopes; the transport phase, with displacement from the entrance towards the interior of the cave, including processes of metamorphism, ice genesis and deformation, as well as erosion with striae genesis and transport; and the phase of ablation, which has sublimation and refreezing processes in the depth of the caves. This is a cryoendokarstic system typical of vertical caves with abundant snow feed that does not arise in caves with dominantly horizontal karstic systems.

The Spanish caves studied, now limited to just six (Altaiz, Castil, Verónica, A294, Casteret, and La Grajera) show different endoclimatic behavior depending on the different areas of the cave, the endokarstic morphology of each cave, and the outside snow evolution. In the Picos de Europa the ice caves in which the ice bodies are housed have behaved statically during periods of research, working as cold traps in closed periods and cooled by the activation of convective air cells during open periods. The vertical morphology of the caves is fundamental to explaining this behavior, which generates mean temperatures inside the caves (2011–2013) of −0.9°C in Peña Castil, −0.4°C in Verónica, and 0°C in Altáiz (Fig. 28.11) (Gómez-Lende, 2015).

The ongoing study of the thermal regimes in the Picos de Europa has established three main periods in the interannual thermodynamic behavior: closed periods between June and October with higher

FIG. 28.11

Ice caves in the Pyrenees. (A) General aspect of casteret ice cave (summer 2016) located in Ordesa Monte Perdido Natinal Park. (B) ice body of the A294 ice cave (summer 2016), located in the Cotiella Massif.

mean daily temperatures and mean temperatures between −0.8 and +0.3°C, with isothermal behavior that has a low correlation to outside temperatures; open periods between November and May, with lower daily mean temperatures and mean temperatures between −1 and −1.7°C in the frozen chambers, with a heterothermal response and the highest correlation to the outside temperature; and periods of transition, also with heterothermal conditions, and with two sub-periods depending on whether temperatures rise or fall. From the spatial point of view, the thermal behavior is different in each chamber of the same cave, distinguishing between: the temperate of outer areas in the mouths with mean temperatures over 0°C, the cold interior frozen chambers with mean temperatures <0°C, in which traps of cold air and the influence of ice blocks determine the thermal regime, and the temperate interior areas with no thermo-regulatory influence from the ice bodies and with mean temperatures <0°C, but with greater thermal amplitude. All these results point to the need of maintaining the environmental monitoring of Spanish caves in order to know the changes in ice generation, melting periods, and changes over time associated with global warming (Fig. 28.12).

FIG. 28.12

Ice caves in the Picos de Europa. (A) Ice wall in the Verónica Ice Cave at 99m depth (summer 2012). (B) GPR works in the main room at 31m depth in the Castil Ice Cave (summer 2015). (C) Ice body bottom in the Altaiz ice cave at 57m depth (summer 2011). (D) Vertical strata in the roof of HS4 ice cave. (E) Horizontally stratified ice with scallops in the ice block of Castil Ice Cave at 40m depth (summer 2011).

28.4.3 PALAEOENVIRONMENTAL IMPLICATIONS OF ICE CAVES

The chronological range of underground ice is limited to relatively short recent periods. As a general rule, caves that have been dated, place the formation of ice between a few centuries to millennia, which is common for the Alps and Central Europe (Achleitner, 1995; May et al., 2011; Perşoiu, 2011; Luetscher et al, 2013). Only in Ledenica (Croatia) has cave ice as old as 58,000 years been found (Garašić, 2014).

The ice cave A-294 of Cotiella (Central Pyrenees) was dated back to the Holocene, between 1950 ± 35 and 5516 ± 70 cal a BP, which places the start of the ice accumulation at around 5500 years (Sancho et al., 2012; Belmonte et al., 2014). This is the oldest dating, though it coincides with an initial period of accumulation of around 5300 ± 379 a BP detected in Sarrios 1. This date corresponds to the

Neoglacial (around 5000 years ago), a cold period when the internal temperature of the cave probably did not surpass 0°C, and the water and snow coming in from the exterior accumulated in the caves, forming the ice sequences (Moreno et al., 2016).

In Casteret and Sarrios there are datings of around AD 1620, and AD 1500–1850 (Leunda et al, 2015; Moreno et al., 2016), and there are also datings of cryogenic calcite from Sarrios-6 (Bartolomé et al., 2015) which points to an age of AD 1149–1177. The formation of CCC in alpine environments is related to the winter snow melt, the infiltration of these melting waters, and their freezing in the caves, which involves the segregation of the calcite and cold conditions inside the caves. In Picos de Europa the datings obtained point to minimum ages of AD 1314–1356 in Verónica-2, of AD 1668–1682 in Verónica-1 and AD 1660–1681 in Altáiz (Gómez-Lende, 2015). Despite the small number of datings, two phases of positive balance can be seen in the Pyrenees and the Cantabrian Mountains, one related to the Medieval Climate Anomaly (MCA), and the other corresponding to the Little Ice Age (LIA).

In Sarrios and Verónica the available datings indicated that the ice formation took place from the end of the MCA onward. The severe winters and the dry summers in Europe would have been less common, at least between 1080 and 1200 (Mann, 2002; Mann et al., 2009), which may have been the reason for the melting of the ice blocks. Nevertheless, in the caves of Central Europe and the Alps, ice accumulation ratios were estimated to be relatively similar for the LIA and MCA (Perşoiu, 2011; Luetscher et al, 2013). In the Iberian Peninsula, Martín-Chivelet et al. (2011) point to the fact that during the MCA there were some relatively cold events ~1250 BP (s.VI) and ~850 BP (s.XII) ago, this may correspond to the phases of positive mass balance seen in Verónica and Sarrios (Gómez-Lende, 2015, 2016).

In Casteret, Verónica, and Altaiz there are minimum ages of the ice deposits that confirm ice formation during the LIA. The positive balances in ice caves during the LIA are common in Europe (Holmlund et al., 2005; Filipov, 2005; Kunaver, 2009; Hercman et al., 2010; Kern, 2010; Perşoiu, 2011; Perşoiu and Pazdur, 2011), such that many of the underground ice masses of the Cantabrian Mountains and the Pyrenees may be attributed to this age. But more datings in more ice caves are undoubtedly necessary to reconstruct the growth and ice melt phases in the caves.

Regarding the palaeoenvironmental potential of Spanish ice caves, the first advances are being made. The first isotopic analyses of the ice were made for the A-294 (Sancho et al., 2012), and there is work in progress in the Pyrenees, Picos de Europa, and the Mountains of Burgos, although more results are clearly needed to accomplish a paleoenvironmental reconstruction for this region. In Cotiella (A-294) the isotopic analyses show an excellent fit to the meteoric waters, with high variability and some temporary organization. The fit of ice isotopes to the local meteoric water line is interpreted as a direct response to precipitation in the form of snowfall, with values for $\delta^{18}O$ between −8.01‰ and −9.93‰ VSMOW, and for δD between −55.8‰ and −81.7‰ VSMOW. Thus, the phase with the lowest ice growth rates has the lightest values, while in the intermediate period with a high ice accumulation rate, the heavy and the light values alternate pointing to large snowfall on a regional scale. Two peaks stand out with the heaviest isotopic values, those at 4350 and 4800 cal. BP., which indicate a tendency towards more benign temperatures (Sancho et al., 2012; Belmonte et al., 2014). Pollen analyses have also begun in the layered accumulations of dung in the ice of the Casteret Ice Cave, which illustrates a great taxonomic variety (43 taxa) and balance between the proportions of arboreal (AP) and non-arboreal pollen (NAP). Although pine is the dominant taxon, the herbaceous component (Poaceae, Cichorioideae, Plantago, and ranunculaceae) is also important (Leunda et al., 2015).

28.5 CONCLUDING REMARKS

Ice caves in Spain represent a broad variety (with over 176 ice caves as seen above), from the static to the dynamic. They also have different endoclimatic regimes, and are all strongly influenced by the outer snow regimes and the presence of metamorphic ice deriving from the diagenesis of snow and congelation ice. In the more studied caves of the Picos de Europa and Pyrenees, masses of metamorphic ice dominate, which constitute the most important volumes of underground ice in Spain, although the congelation ice is important both for perennial and seasonal studies. The presence of ice caves in Spain, mainly in the mountains of the northern Iberian Peninsula and in lava caves of the Teide, indicates cold conditions for these scarcely studied environments existed until a few decades ago. Ice caves only share space with glacial phenomena in some massifs of the Pyrenees, and with inherited form the LIA ice patches in the Picos de Europa, where underground frozen masses surpass the ice patches in volume. Spanish ice caves are endokarstic periglacial phenomena and indicators of an endokarstic permafrost in the Picos de Europa (Gómez-Lende, 2015, 2016). They are considered to be indicators of permafrost when there are perennial ice bodies, as both are defined by the same thermal conditions (French, 2007), and their presence in the Pyrenees (Serrano et al., 2009) confirms the existence of a *endokarstic permafrost high mountain environment* in the calcareous mountains of the north of the Iberian Peninsula.

Ice deposits are particularly useful archives because of their sensitivity to environmental changes in the past, which provides information on both temperature and precipitation changes through the variations in ice volume, and how the isotopic composition responds to parameters associated with the climate. Finally, the ice caves of the Spanish mountains have unique heritage value, but the cave ice has a historical tendency to shrink in volume. Due to their risk of disappearance in the short or medium term, all the information contained within them should be collected, and all recent research should be continued in order to analyze the proxies they contain to fully understand these special systems and their past dynamics.

REFERENCES

Achleitner, A., 1995. Zum Alter des Höleneises in der Eisgruben-Eishöhle im Sarstein (Oberösterreich). Die Höhle 46 (1), 1–5.

ARSIP, 2009. A la recherche de Z. Massif La Pierre-San Martín. Association pour la Recherche Speleologique International á la Pierre San Martin, Lyon.

Bacallado, J.J. (Ed.), 1995. Catálogo espeleológico de Tenerife. OAMC-Museo de Ciencias Naturales, Sta. Cruz de Tenerife.

Balch, E.S., 1900. Glacières or Freezing Caverns. Allen, Lane and Scott, Philadelphia. 337 pp.

Ballesteros, D., 2016. Geomorfología y evolución geomorfológica de las cuevas alpinas de los Picos de Europa, Cordillera Cantábrica (España) (Ph.D. thesis). Universidad de Oviedo, Oviedo (Spain).

Bartolomé, M., Sancho, C., Osacar, C., Moreno, A., Leunda, M., Oliva-Urcia, B., Spötl, C., Luetscher, M., López, J., Belmonte, Á., 2015. Characteristics of cryogenic calcites in Pyrenean ice caves of the Pyrenees (Northern Spain). Geogaceta 58, 107–110.

Bartolomé, M., Sancho, C., Leunda, M., Moreno, A., Belmonte, A., Oliva, B., López-Moreno, J.I., Pierre, D., Asenjo, D., Gomollón, A., 2016. Evolución del volumen de hielo en la Cueva de Casteret en los últimos 50 años (Parque Nacional de Ordesa y Monte Perdido, Huesca). In: Proceeding XIV Reunión Nacional de Geomorfología, SEG, Málaga, pp. 713–720.

Belmonte, A., 2014. Geomorfología del macizo de Cotiella (Pirineo Oscense). cartografía, evolución paleoambiental y dinámica actual (Ph.D. thesis). Universidad de Zaragoza, Zaragoza (Spain), 582 pp.
Belmonte, A., Sancho, C., 2010. First data from a Pyrenean ice cave (A294, Cotiella massif, Spain). In: Spötl, C., Luetscher, M., Rittig, P. (Eds.), 4th International Workshop on Ice Caves. Abstracts Volume, Obertraun, Austria, p. 7.
Belmonte, A., Sancho, C., Moreno, A., López-Martínez, J., Bartolomé, M., 2014. Present-day environmental dynamics in ice cave A294, Central Pyrenees, Spain. Geogr. Fis. Din. Quat. 37, 131–140.
Berenguer-Sempere, F., Gómez-Lende, M., Serrano, E., De Sanjosé Blasco, J.J., 2014. Orthothermographies and 3D modeling as potential tools in ice caves studies: the Peña Castil Ice Cave (Picos de Europa, Northern Spain). Int. J. Speleol. 43 (1), 35–43.
Bernand, P., Van Thienen, M., 1987. Gavarnie, les grottes glacées du Marboré. Ed. Els Autors. Aubenas d'Ardeche, 136 pp.
Borreguero, M., 1986. Special Picos. Puertos de Ondón. Societé Speleologique de Suisse, Neuchatel.
Bottomley, M., 2016. Ario Caves Project Expedition – 2016. Final Report Picos de Europa, Spain. Oxford University Cave Club.
Bourguet, M.N., 2003. El mundo visto desde lo alto del Teide: Alexander von Humboldt en Tenerife. In Ciencia y romanticismo. Fundación Canaria Orotava de Historia de la Ciencia (Ed.), Tenerife, pp. 279–302.
Browne, G.F., 1865. Ice-Caves of France and Switzerland. A Narrative of Subterranean Exploration. Green and Co., London. 315 pp.
Cardona, F., 2010. El forat del Gel i la seva anomalia tèrmica. Exploracions 19, 113–126.
Casteret, N., 1933. Dix ans sous terre. Libraire académique Perrin, Paris.
Casteret, N., 1953. Dans les glaces souterraines. Les plus éleveés de Monde. Libraire Académique Perrin, Paris, p. 93.
Casteret, N., 1961. Ma vie souterraine. Flammarion, Paris.
Cedrés, R., 2015. La cueva del hielo–el Teide. Organismo Autónomo Parques Nacionales, Madrid.
Cerdeño, R., Sánchez, J., 2000. Cuevas heladas en Picos de Europa, Cantabria. Subterránea 14, 46–52.
Cerdeño, R., Sánchez, J., 2005. Picos de Europa. Macizo central o de los Urrielles. Karaitza 13, 28–37.
CES-Alfa, A.S. Charentaise, 2016. Inventario y documentación de cavidades exploradas (1971–2015) Picos de Europa, Camaleño-Macizo Central. Not published.
Citterio, M., Turri, S., Bini, A., Maggi, V., 2005. On ice and snow deposits in caves and their suitability to paleoenvironmental research. In: EGU General Assembly, Geophysical Research Abstracts, Vienna (Austria).
Collignon, B., 1985. Quelques éléments de géologie et d'hydrogéologie. Spelunca [Special Picos] 19, 7–12.
Dubois, P., 1958. Observations sur les dépots des cavités souterraines du massif du Mont Perdu (Pyrénées Centrales). In: Proceeding II Congress Internationale de Speleologie, Viena. 2, pp. 415–422.
Durán Valsero, J.J., Molina Muñoz, J.A., 1986. Sobre la existencia de formas kársticas del tipo "pozos de acumulación de nieve" en el pico La Maroma, Sierra Tejeda (Granada-Málaga). Federación Andaluza de Espeleología, pp. 103–119.
Duverneuil, F., et al., 1974/75. Plan, long profile & cross section survey of the Grotte glacée Casteret 1/500, p. 18.
Eraña, C., Arizabalaga, I., Arrue, K., Azkarate, X., Azkoaga, X., Dorado, J., Dulanto, R., Eraña, R., Esperasate, J., Expósito, A., Ezkibel, I., Galdos, A., Latantxa, J., Iglesias, G., Iglesias, J., Olalde, A., Ugarte, J., Ugarte, S., Zabaleta, P., 2008. Descripción de las Simas, pozos de nieve, formados en el cresterío de la Sierra de Aloña-Aizkorri-Aratz. Karaitza 16, 18–29.
ERE-CEC, 2003. El macizo de Alano y Peña Forca y la sima de lo Faito (-341). Huesca. Subterránea 19, 8–14.
Espeleo Club de Gràcia, 1992. La Aventura de Arañonera. Al descubrimiento del mundo subterráneo. Ed. Espeleo Club de Gràcia, Barcelona.
Estevez, J.A., 2011. Abisu l'Xelu. Subterránea 33, 8–19.
Fabriol, H., 1975. Explorations dans le massif occidental des Picos Europa (Espagne). Spelunca 2, 25–26.
FEE, 2011. Catálogo de cavidades. Simas mundiales. Federación Española de Espeleología, Madrid.

Fernández, R., Martínez, R., Ballesteros, D., Caldueño, A., Cañón, G., de Felipe, J., Rodríguez, C., Rodríguez, I., Fernández, F., de Oca, G., Sendra, G., García, A., García, J., Velilla, S., 2011. Exploraciones espeleológicas en Camino de Ario 2011 (Picos de Europa, Norte de España). Final report, Oviedo, 2011.

Fernandez-Gibert, E., Calaforra, J.M., Rossi, C., 2000. Speleogenesis in the Picos de Europa Massif, northern Spain. In: Klimchouk, A.B., Ford, D.C., Palmer, A.N., Dreybrodt, W. (Eds.), Speleogenesis, evolution of karst aquifers. UIS, pp. 352–357.

Filipov, A.G., 2005. The age of the relict firn plug in the Kremeshetskaya cave, Eastern Siberia. In: Mavlyudov, B.R. (Ed.), Glacier Caves and Glacial Karst in High Mountains and Polar Regions. Institute of Geography, Russian Academy of Sciences, Moscow, pp. 98–100.

French, H., 2007. The Periglacial Environment. Wiley, Chichester. 458 pp.

Garašić, M., 2014. New research in cave Ledenica in Bukovi Vrh on Velebit Mt in Croatian Dinaric Karst. In: Land, L., Zoltan, K., Maggi, V., Turr, S. (Eds.), Proceedings 6th International Workshop on Ice Cave, Idaho, USA, pp. 31–32.

Geoda, 2010. Memoria de actividades 2010. Technical report. http://www.geoda.net (Accessed 2011).

Gómez-Lende, M., 2015. Las cuevas heladas en Picos de Europa: clima, morfologías y dinámicas (Ph.D. thesis). Universidad de Valladolid, Valladolid (Spain), 663 pp.

Gómez-Lende, M., 2016. Cuevas heladas en el Parque Nacional Picos de Europa. Fronteras subterráneas del hielo en el Macizo Central. OAPN, Ministerio de Agricultura, Alimentación y Medio Ambiente, Madrid, p. 254.

Gómez-Lende, M., Serrano, E., 2012a. Elementos del patrimonio geomorfológico subterráneo: las cuevas heladas de Picos de Europa (Cordillera Cantábrica). In: González, A. (Ed.), Avances de la Geomorfología en España 2010-2012. Actas de la XII Reunión Nacional de Geomorfología, SEG, Santander, pp. 47–50.

Gómez-Lende, M., Serrano, E., 2012b. Morfologías, tipos de hielo y regímenes térmicos. Primeros estudios en la cueva helada de Peña Castil (Picos de Europa, Cordillera Cantábrica). In: González, A. (Ed.), Avances de la Geomorfología en España 2010–2012. Actas de la XII Reunión Nacional de Geomorfología, SEG, Santander, pp. 613–616.

Gómez-Lende, M., Serrano, E., 2012c. First thermal, morphological and ice types studies in the Peña Castil ice cave (Picos de Europa, Cantabrian Mountains, Northern Spain. In: Turri, S., Strini, A., Tomasi, F. (Eds.), 5th International Workshop on Ice Cave, Valssasina, Italy. Abstracts Volume, pp. 52–53.

Gómez-Lende, M., Serrano, E., Berenguer, F., 2011. Cuevas heladas en Picos de Europa. Primeros estudios en Verónica, Altáiz y Peña Castil. Karaitza 19, 56–61.

Gómez-Lende, M., Berenguer, F., Serrrano, E., 2014. Morphology, ice types and termal regime in a high mountain ice cave. First studies applying terrestrial laser scanner in the Peña Castil Ice Cave (Picos de Europa, Northern Spain). Geogr. Fis. Din. Quat. 37, 141–150.

Gómez-Lende, M., Serrano, E., Jordá, L., Sandoval, S., 2016. The role of GPR techniques in determining ice cave properties: Peña Castil ice cave, Picos de Europa. Earth Surf. Process. Landf. 41, 2177–2190.

González-Trueba, J.J., 2007. El paisaje natural del macizo central de los Picos de Europa. Geomorfología y sus implicaciones geoecológicas en la alta montaña cantábrica. Centro de Investigación del Medio Ambiente (CIMA), Consejería de Medio Ambiente del Gob, Cantabria, Santander, 332 pp.

Greaves, H., (Ed.), 2008. Julagua 2008. A caving and cave diving expedition to the Picos de Europa, Asturias, Spain. Expedition report. Oxford University Cave Club.

Hercman, H., Gasiorowski, M., Gradzinski, M., Kicinska, D., 2010. The first dating of cave ice from the Tatra Mountains, Poland and its implication to palaeoclimate reconstructions. Geochronometria 36, 31–38.

Holmlund, P., Onac, B.P., Hanson, M., Holmgren, K., Mörth, M., Nyman, M., Perşoiu, A., 2005. Assessing the palaeoclimate potential of cave glaciers: the example of the Scarisoara ice cave (Romania). Geogr. Ann. 87 (A), 193–201.

Humboldt, A., 1814. Le voyage aux régions equinoxiales du Nouveau Continent fait en 1799, 1800, 1801, 1802 1803 et 1804. T.I, París, 440 pp.

Jeannel, R., Racovitza, E.G., 1912. Énumération des grottes visitées, 1909–1911. Archive Zoologie Expérimentale et Générale 9 (5), 501–667.
Kemp, N., 1985. Picos de Europa 1985. Caves Caving 30, 18–20.
Kern, Z., 2010. Climate and Environmental Changes Reconstructed From Tree Rings and Cave Ice. Earth Sciences-Physical Geography, Eötvös University, Hungary. 117 pp.
Kern, Z., Perşoiu, A., 2013. Cave ice-the inminent loss of untapped mid-latitude cryospheric palaeoenvironmental archives. Quat. Sci. Rev. 67, 1–7.
Kunaver, J., 2009. The ice cave G2 in the Kanin Mts. (Slovenia), an indicator for the global warming. In: Gabrovšek, F., Mihevc, A. (Eds.), 17th Karstological School "Classical Karst", Postojna, p. 71.
Leunda, M., Bartolomé, M., Sancho, C., Moreno, A., Oliva, B., González-Sampériz, P., Gil-Romera, G., Gomollón, A., 2015. La cueva helada de Casteret (PNOMP, Huesca): primeras aportaciones del registro de hielo. In: Proceeding XIV Reunión Nacional de Cuaternario, AEQUA, Granada, pp. 78–81.
Lloyd, D.K., 1985. Expedition to the Picos de Europa 1984. Caves Caving 29, 26–28.
Lloyd, D.K., 1990. Water tracing in the Vega Huerta caves, Picos de Europa, Spain. Cave Sci., Trans. Brit. Cave Res. Assoc. 17 (3), 103–106.
López-Martínez, J., Freixes, A., 1989. El karst en los Pirineos españoles. In: Durán, J.J., López Martínez, J. (Eds.), El karst en España. vol. 4. Monografías de la SEG, Madrid, pp. 131–144.
Luetscher, M., Jeannin, P.Y., Haeberli, W., 2005. Ice caves as an indicator of winter climate evolution: a case study from the Jura Mountains. The Holocene 15 (7), 982–993.
Luetscher, M., Borreguero, M., Moseley, G.E., Spötl, C., Edwards, R.L., 2013. Alpine permafrost thawing during the Medieval Warm Period identified from cryogenic cave carbonates. Cryosphere 7 (4), 1073–1081.
Maire, R., 1977. Glaces souterraines d'altitude dans les massifs du Haut-Giffre, de l'Oberland et dans quelques régions de comparaison. Sci. Nat. Acta Geol. 54, 113–122.
Maire, R., 1990. La haute montagne calcaire. Karstologia-Mémoires, n°3. Association Française de Karstologie et Fédération Française de Spéléologie, 731 pp.
Mann, M.E., 2002. Medieval climatic optimum. In: MacCracken, M.C., Perry, J.S. (Eds.), Encyclopedia of Global Environmental Change. John Wiley and Sons, Chichester, pp. 514–516.
Mann, M.E., Zhang, Z., Rutherford, S., Bradley, R.S., Hughes, M.K., Shindell, D., Ammann, C., Faluvegi, G., Ni, F., 2009. Global signatures and dynamical origins of the Little Ice Age and Medieval Climate Anomaly. Science 326, 1256–1260.
Martínez, F., 1977. La espeleología en Palencia. Institución Tello Téllez de Meneses, Palencia.
Martínez de Pisón, E., Quirantes, F., 1981. El Teide. Estudio geográfico. Universidad de la Laguna, Tenerife. 187 pp.
Martín-Chivelet, J., Muñoz, M.B., Edwards, R.L., Turrero, M.J., Ortega, A.I., 2011. Land surface temperature changes in Northern Iberia since 4000 yr BP based in δ13C of speleothems. Global Planet. Change 77 (1–2), 1–12.
Martín-Merino, M.A., 2015. Investigación Paleoclimática en la Torca de La Grajera (Castro Valnera). GE Edelweiss, Burgos, http://grupoedelweiss.com/web/index.php/news-mainmenu-2/635-investiga-cion-paleoclimatica-en-la-torca-de-la-grajera-castro-valnera (accessed January 2016).
Martín-Moreno, R., 2006. Estudio comparativo de formas y procesos glaciares y periglaciares desde la Pequeña Edad del Hielo: altas latitudes noruegas (Spitbergen y Jotunheimen) y Altas Altitudes Españolas (Pirineos, Sistema Central y Teide) (Ph.D. thesis). Universidad Autónoma de Madrid, Madrid (Spain), 705 pp.
Martín-Moreno, R., 2010. La Pequeña Edad del Hielo en el Alto Teide (Tenerife, islas Canarias). Menciones históricas y morfogénesis periglaciar. Ería 83, 331–342.
May, B., Spötl, C., Wagenbach, D., Dublyansky, Y., Liebl, J., 2011. First investigations of an ice core from Eisriesenwelt cave (Austria). Cryosphere 5, 81–93.
Miranda, S., 2004. Los pozos de nieve de Tenerife. Estudio histórico y geográfico de la explotación de la nieve en la isla de Tenerife. Siglos XVIII y XIX. Cabildo de Tenerife, Tenerife. 183 pp.

Moreno, A., Bartolomé, M., Sancho Marcén, C., Pérez-Mejías, C., Oliva, B., Leunda, M., Sevilla, M., Belmonte, A., Stoll, H., Cacho, I., Delgado-Huertas, A., Osácar, M.C., Muñoz-Jiménez, A., Valero, B., González-Sampériz, P., 2016. Reconstrucción de cambios climáticos abruptos a partir de registros de cuevas en el Parque Nacional de Ordesa y Monte Perdido: formaciones de espeleotemas y depósitos de hielo. In: Armengol, P. (Ed.), Proyectos de investigación en Parques Nacionales: 2011-2014. OAPN, Ministerio de Agricultura, Pesca y Alimentación, Madrid, pp. 371–384.
Ortiz, I., 1990. Campaña Picos '85. Vega Huerta (Western Massif of the Picos de Europa – Leon). Garma 1, 19–23.
OUCC, 1984. The 1982 and 1983 OUCC expeditions to the Picos de Cornión, Northern Spain. In: Proceedings of the Oxford University Cave Club, vol. 11, pp. 3–49.
OUCC, 2007. Total Area Shaft Bashing Guide. Oxford University Cave Club, Oxford.
Pellitero, R., 2012. Geomorfología, paleoambiente cuaternario y geodiversidad en el macizo de Fuentes Carrionas. Montaña Palentina (Ph.D. thesis). Universidad de Valladolid, Valladolid (Spain), 1040 pp.
Perşoiu, A., 2011. Palaeoclimatic significance of perennial ice accumulations in caves: an example from Scărişoara Ice Cave, Romania (Graduate School theses and dissertations). University of South of Florida (USA), 116 pp.
Perşoiu, A., Onac, B., 2012. Ice in caves. In: White, W.B., Culver, D.C. (Eds.), Encyclopedia of Caves. Academic Press, Waltham, pp. 399–404.
Perşoiu, A., Pazdur, A., 2011. Ice genesis and its long-term balance and dynamics in Scărişoara Ice Cave, Romania. Cryosphere 5, 45–53.
Preziosi, E., Russo, N., Ayad, G., Friday, S., Girelli, L., Pesci, L., Scipioni, M., 1997. Tre anni di speleologia nel Massiccio del Cornion (Picos de Europa, Spagna). Speleologia 37, 73–80.
Puch, C., 1987. Atlas de Grandes Cavidades Españolas. Exploracions n°11. Barcelona.
Puig y Larraz, G., 1896. Cavernas y simas de España. Madrid, 392 pp.
Rogers, C., (Ed.), 2008. Canal del Montico 2008. Expedition report. Oxford University Cave Club.
Rosales, M., 1994. Historia de la espeleología en Canarias. In: Oromí, P. (Ed.), 7th International Symposium on Vulcanoespeleology, Santa Cruz de la Palma, Gran Canaria, pp. 101–108.
Rosi, C., 2004. Picos de Europa, Spain. In: Gunn, J. (Ed.), Encyclopedia of Caves and Karst Science. Fitzroy Dearborn, New York, pp. 582–585.
Salvador-Franch, F., 1977. Los fenómenos kársticos. Espeleología. In: Coll, X., Coll, M., Llobet, S. (Eds.), ERE-CEC and GE Badalona, Parque Nacional del Valle de Ordesa, Vignemale, Monte Perdido. Guía Cartográfica. Ed. Alpina, Granollers, pp. 5–10.
Salvayre, H., 1982. Aperçu sur les glaces souterraines du massif du mont Perdú-Marboré. Pyrénées 129, 19–38.
Sancho, C., Belmonte, A., Moreno, A., Bartolome, M., Calle, M., Santolaria, P., 2012. Potencial paleoclimatico de la cueva helada A294 (Macizo de Cotiella, Pirineos, Huesca). Geogaceta 52, 101–104.
Senior, K.J., 1987. Geology and speleogenesis of the M2 cave system, Western Massif, Picos de Europa, northern Spain. Cave Sci., Trans. Brit. Cave Res. Assoc. 14 (3), 93–103.
Serrano, E., González-Trueba, J.J., 2005. Assessment of geomorphosites in natural protected areas: the Picos de Europa National Park. Geomorphologie 3, 197–208.
Serrano, E., Morales, C., González-Trueba, J.J., Martín-Moreno, R., 2009. Cartografía de permafrost de montaña en los Pirineos españoles. Finisterra 87, 45–54.
Silvestru, E., 1999. Perennial ice in caves in temperate climate and its significance. Theor. Appl. Karstol. 11–12, 83–93.
Smyth, C.H.P., 1858. Report on the Teneriffe Astronomical Experiment of 1856. Richard Taylor and William Francis, London, pp. 465–577.
St. Pierre, D., 2007. Grotte Casteret. BCRA Cave Stud. Ser. 17, 48.
UEVa, 2011. Memoria de trabajos de campo 2011. Technical report, Unión Espeleológica Vallisoletana, Valladolid.
Verheyden, S., Faverjon, M., Lasborchers, N., Charavel, R., Cholin, A., Sánchez, N., Scipioni, M., Gillet, R., Delaby, S., 2007. La sima Thesaurus Fragilis. Exploraciones 2006 y 2007. Vega Huerta, macizo de Cornión, Picos de Europa. Subterránea 28 (2), 24–31.
YUCPC, 1991. Picos 91 expedition report. York University Cave and Pothole Club, 36 pp.

FURTHER READING

Belmonte, A., Sancho, C., Moreno, A., 2011. Chronology of a Pyrenean subsurface ice deposit (A294 cave, Cotiella massif, Spain). In: 8th EGU General Assembly, Geophysica Research Abstracts, vol. 13, Vienna (Austria).

Belmonte, A., Bartolomé, M., Sancho, C., Moreno, A., López-Martínez, J., 2012. Assessing the palaeoclimate potential of the A294 ice cave (Central Pyrenees, Northern Spain). In: Turri, S., Strini, A., Tomasi, F. (Eds.), 5th International Workshop on Ice Cave, Barzio, Valssasina, Italy. Abstracts Volume, p. 47.

Casteret, N., 1928. Un glaciar subterráneo en los Pirineos. Peñalara 170, 25–35.

Casteret, N., 1960. Dans les entrailles du massif calcaire. La grotte glacée Casteret. Pyrénées 44, 240–252.

Cedrés Jorge, R., 2015. La Cueva de Hielo. El Teide. Tenerife, Technical report. http://www.cedres.info/2015/08/la-cueva-del-hielo-descripciones-e.html (accessed January 2017).

CES-Alfa, 2016. Picos de Europa 2016. Resumen de las exploraciones en 2016. Camaleño-Macizo Central. CES Alfa, A.S. Charentaise, 32 pp.

Gómez-Lende, M., Berenguer, F., Serrano, E., Sanjosé, J.J., 2015. Los ortotermogramas en los estudios de las cuevas heladas. El caso de la cueva helada de Peña Castil (Picos de Europa). In: Gómez Ortiz, A., Salvador, F., Oliva, M., Salvá, M. (Eds.), Avances, métodos y técnicas en el estudio del periglaciarismo. Universitat de Barcelona, Barcelona, pp. 263–276.

Milne, D., YUCP, 2007. Explorations in the Picos de Europa. York University Cave and Pothole Club, Technical Report. http://www.picos.yucpc.org.uk/.

Sánchez-Benítez, J., 2007. Picos de Europa. Macizo central o de los Urrielles (sector cántabro de Camaleño). Karaitza 15, 52–55.

Sánchez-Benítez, J., 2009. Campaña Picos de Europa 2009. Karaitza 17, 44–49.

Sánchez-Benítez, J., Cerdeño, R., 1999. Resumen de las exploraciones de la Torca de la Nieve. Estudio del nivel freático. Subterránea 11, 36–41.

Sánchez-Benítez, J., Jordá, L., Serrano, E., Gómez-Lende, M., Hivert, B., 2011. Memoria de las exploraciones subterráneas. Macizo central de Picos de Europa, Camaleño-Cantabria. CES Alfa-AS Charentaise, Madrid. 50 pp.

UEVa, 2009. Informe sobre presencia de hielo en las cuevas del Espigüete. Technical report. Unión Espeleológica Vallisoletana, Valladolid.

CHAPTER

ICE CAVES IN SLOVAKIA

29

Pavel Bella*,†, Ján Zelinka*

*State Nature Conservancy of the Slovak Republic, Slovak Caves Administration, Liptovský Mikuláš, Slovakia**

Catholic University in Ružomberok, Ružomberok, Slovakia†

CHAPTER OUTLINE

29.1 INTRODUCTION

Ice caves present the specific underground karst phenomena (mainly by a cold microclimate and water in its solid phase, as well as psychrophilous species of fauna) that increase the geodiversity of karst landscapes. Some caves with seasonal ice fill are also in nonkarst areas (pseudokarst caves). Thick ice bodies provide important paleoclimatic evidence of the last Holocene. Show ice caves are in important tourist destinations, and ice caves opened to the public, and some other easily accessible ice caves, have a rich history. But ice caves' features are highly vulnerable because of the disruption of climate conditions and the regime of seeping meteoric waters. Therefore strict measures based on scientific principles are needed for the use and protection of ice caves. This chapter presents basic knowledge on the occurrence, geographical distribution, morphology, genesis, climate, fills, and use and protection of ice caves in Slovakia.

Ice Caves. https://doi.org/10.1016/B978-0-12-811739-2.00029-2

29.2 KARST AREAS AND CAVES IN SLOVAKIA

Karst areas of Slovakia (Central Europe) are located in the Western Carpathian Mountains (in several mountains and intermontane basins). The extent of the karst areas is more than 2700 km^2 (Fig. 29.1). The Slovakian territory within the temperate climatic zone is characterized by transitional features between oceanic and continental climate. From the climatic and orographical point of view, karst areas are divided into the high-mountain karst in the northern part of Slovakia (the highest positions of Vysoké and Belianske Tatry Mountains, Červené vrchy Mountains, Mt. Sivý vrch, Ďumbier Karst in the Nízke Tatry Mountains, Mt. Choč, and other alpine elevations above the upper boundary of forests) and the mid-mountain karst in the western, central, and eastern parts of Slovakia (Malé Karpaty Mountains, Strážovské vrchy Mountains, Veľká and Malá Fatra Mountains, Nízke Tatry Mountains, Chočské vrchy Mountains, Špis-Gemer Karst, Slovak Karst, Galmus, and other areas).

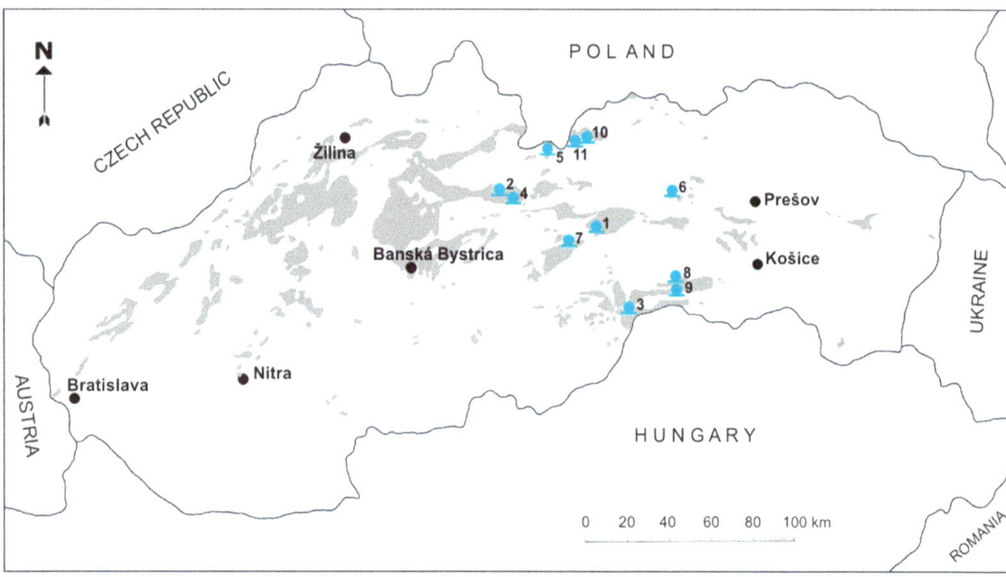

FIG. 29.1

Karst areas with a location of the most important ice caves in Slovakia: (1) Dobšinská ľadová jaskyňa Cave (Dobšiná Ice Cave) and Duča Cave, (2) Demänovská ľadová jaskyňa Cave (Demänová Ice Cave), (3) Silická ľadnica Cave (Silica Ice Cave), (4) Veľká ľadová priepasť na Ohništi Abyss (Great Ice Abyss on the Ohnište), (5) Ľadová priepasť Abyss (Ice Abyss) in Červené vrchy Mountains, (6) Ľadová jaskyňa na Dreveníku Cave (Ice Cave on the Dreveník), (7) Studňa/Ľadová jama na Muráni Cave (Well/Ice Pit on the Muráň), (8) Ľadová jaskyňa v Havranej skale Cave (Ice Cave in the Raven Rock), (9) Snežná diera Cave (Snow Hole), (10) Ľadová kamzíčia jaskyňa Cave (Ice Chamois Cave), and (11) Spišmichalova jaskyňa Cave (Spismichal's Cave).

At present, more than 7150 caves are registered in Slovakia, mostly in carbonate rocks (limestone, dolomite, travertine). Various genetic types of limestone caves are mainly in the areas of plateau karst, dissected karst of monoclinal structures and karst of massive ridges, horsts, and combined fold-fault structures. More than 50 syngenetic and epigenetic travertine caves are significant natural phenomena in some intermountain basins and valleys. But more than 180 caves originated in noncarbonate rocks (basalt, andesite, vulcaniclastic conglomerate, sandstone, granite, quartzite, quartz, or shale) as a consequence of varied geological settings in the Slovakian territory.

29.3 GEOGRAPHICAL DISTRIBUTION OF ICE CAVES

The geodiversity of several caves is accentuated by perennial or seasonal ice fill. In Slovakia, Rubín and Skřivánek (1963) mentioned 13 caves and Droppa (1973) 22 caves with ice fill. Based on existing observations, 45 perennial ice caves and 23 caves seasonally filled with ice are known in different geological settings and geographical conditions (Bella, 2008, updated; Table 29.1). The list and basic data on ice caves in Slovakia were summarized by Bella (1995). Siarzewski (1994) provided a review of ice caves in the Tatry Mountains (on both the Polish and the Slovak sides).

In Slovakia, in addition to limestone ice caves, ice fill was observed in four travertine caves and five noncarbonate caves (formed in sandstone, andesite, and basalt). From the viewpoint of cave morphology and genesis, ice fills were deposited mostly in vadose drawdown or epiphreatic leveled parts of inactive river caves (inside inclined sack-like cavities created by breakdown in original passages or closing of lower entrances by slope sediments), in crevice caves and abysses, and in corrosion or corrosion-collapsed vertical shafts (light holes).

From an altitude view point, caves with a perennial or seasonal ice fill occur in several hypsometric grades, mostly in mid-mountain and high-mountain positions (Table 29.2). The territory of Slovakia is orographically dissected from 94 to 2655 m a.s.l. (above sea level). The lowest located perennial ice cave is the Silická ľadnica (Silica Ice Cave) at 470 m a.s.l. in the Slovak Karst (Silická planina Plateau), and the highest located ones are the Priepasť v Hlúpom vrchu (Abyss in Mt. Hlúpy) at 1966 m a.s.l. in the Belianske Tatry Mountains and Ľadová priepasť (Ice Abyss) at 1938 m a.s.l. in the Červené vrchy Mountains. Compared with high-mountain and higher mid-mountain karst areas, the relatively numerous occurrences of caves with a perennial and seasonal ice fill at 700–1100 m a.s.l. (30.8%) are the result of the significant extent of karst areas in the hypsometric grade.

The most of perennial ice caves is located in the Tatry Mountains (20 caves); Spiš-Gemer Karst, which consists of Slovenský raj (Slovak Paradise) and Muránska planina Plateau (9 caves); and Nízke Tatry Mountains (6 caves).

According to the classification of climatic regions of Slovakia (Lapin et al., 2002), the caves with ice fill are located mainly in the cool region, and are less widespread in the moderately warm region. One perennial ice cave occurs even in the warm region (Table 29.3). Perennial ice caves occur mostly in the cool region (91.1%).

29.4 THE MOST SIGNIFICANT ICE CAVES

29.4.1 DOBŠINSKÁ ĽADOVÁ JASKYŇA CAVE (DOBŠINÁ ICE CAVE)

The Dobšiná Ice Cave is one of the most important ice caves in the world. It is located in the southern part of Slovenský raj (Slovak Paradise) National Park in the Spiš-Gemer Karst (north of the town of Dobšiná). The cave entrance, northwest-facing in a coniferous forest, lies on the northern slope of Mt. Duča (1141 m) at an elevation of 969 and 130 m above the bottom of the Hnilec River valley (Fig. 29.2).

Table 29.1 List and Basic Data on Perennial Ice Caves in Slovakia (Within Regional Geomorphological Units)

Regional Geomorphological Units	Cave	Entrance Altitude (m a.s.l.)	Length (m)	Depth (m)	References
Bachureň	Ľadová jaskyňa	837	82	21	Hochmuth (1995a,b)
Biele Karpaty, Vršatské bradlá, Vysoké Vršatce	Ľadnica	660		6.6	Skutil (1936) and Velič (1970)
Hornádska kotlina, Medvedie chrbty	Ľadová jaskyňa na Dreveníku	557	215	17	Wernher (1549), Ranzanus (1558), Turóczi (1768), Windisch (1780), Buchholtz (1787), Roth (1881a,b), Piovarcsy (1927), Skutil (1951), Schönviszky (1968), Vítek (1972), Prikryl (1985), Lalkovič (1993), Miháľ (2004), and others
	Malá ľadová jaskyňa	572	23	12.5	Vítek (1970, 1972)
Kremnické vrchy, Flochovský chrbát	Snežná rozsadlina	1177	91	25	Gaál et al. (2000)
Nízke Tatry, Ďumbierske Tatry, Demänovské vrchy	Demänovská ľadová jaskyňa (Demänová Ice Cave)	840	1975	57	Bel (1723, 1736), Korabinsky (1778), Windisch (1780), Hacquet (1790), Townson (1797), Bredetzky (1801, 1805), Sartori (1809), Zipser (1817), Sydow (1830), Hunfalvy (1860), Kohn (1875), Siegmeth (1880, 1898), Schwalbe (1882a,b), Szterényi (1883), Mihalik (1884a,b), Balch (1900), Vitásek (1923), Korbay (1953), Droppa (1955–56, 1957, 1964), Otruba (1958b, 1971), Halaš and Slíva (1979), Halaš (1980, 1983, 1984), Prikryl (1985), Shaw (1999), Lalkovič (2003a,b), Strug et al. (2006, 2008b), Strug and Zelinka (2008a,b,c), Strug (2011), and others
	Havrania priepasť	1372		40	Droppa (1958)
	Jelenia priepasť (Závrtová priepasť)	1487	32	18	Droppa (1958)
	Studená diera	1440	35		Bella (1985)
	Veľká ľadová priepať na Ohništi	1529 1513		125	Benický (1940, 1958), Droppa (1958), Otruba (1958a, 1958b), Hochmuth (1995c), and Holúbek (2014)
Nízke Tatry, Ďumbierske Tatry, Ďumbier	Jaskyňa studeného vetra	1714 1640	1818	129	Štéc (1998, 2001, 2013), Žák et al. (2009), and Zelinka (2011)
Slovenský kras, Horný vrch	Ľadová jaskyňa v Havranej skale	875	87	20	Scholtz (1888), Siegmeth (1891a,b), Skřivánek and Stárka (1955, 1956), and Zelinka (2007)
	Snežná diera	888	95	25	Scholtz (1888), Siegmeth (1891a,b), Skřivánek and Stárka (1955, 1956), and Psotka (2007)

Slovenský kras, Silická planina	Silická ľadnica (Silica Ice Cave)	503–470	1100	110	Bel (1723, 1739–1741), Windisch (1780), Townson (1797), Szontág (1871), Beaudant (1822, 1823), Browne (1865), Schwalbe (1882a,b), Szterényi (1883), Terlanday (1893, 1896), Siegmeth (1909), Vitásek (1930), Benický (1932), Böhm and Kunský (1938, 1941), Roth (1940a,b), Skutil and Klimeš (1957), Droppa (1962, 1974), Roda and Rajman (1971), Roda et al. (1974), Prikryl (1985), Rajman et al. (1985, 1987), Dénes (1992), Lalkovič (1992), Bárta (1995), and others
Spišsko-gemerský kras, Muránska planina	Bodolová	1080		33	Stárka (1954) and Kámen (1957a, 1958a, 1971)
	Čurička	1010	14	8	Kámen (1971)
	Ľadová jaskyňa	800	30		Kámen (1957b, 1958b, 1971)
	Studňa (Ľadová jama na Muráni)	1165	64	24	Kámen (1958a, 1971)
	Šingliarka (Ľadová jama)	1080		20	Stárka (1954) and Kámen (1964, 1971)
	Veľká Stožka	1310	120	35	Kámen (1961, 1971)
	Vrbiarka (Priepasť na Kubičkovej)	1010		17	Kámen (1957b, 1958a, 1971)
Spišsko-gemerský kras, Slovenský raj	Dobšinská ľadová jaskyňa (Dobšiná Ice Cave)	969	1483	112	Szontág (1871), Fehér (1872), Krenner (1873, 1874a,b), Pelech (1878, 1879, 1884a,b), Vitásek (1874a,b, 1900), Lowe (1879), Wünsch (1881), Schwalbe (1882a,b), Krieg (1883), Szterényi (1883), Fischer (1888a,b), Weisinger (1898), Balch (1900), Fugger (1891), Hanvai (1908), Steiner (1922a,b), Bíma (1925), Woldřich (1926), Šincl (1931), Měska (1936), Ulrich (1937), Petrovič (1952), Kvietok-Krofta (1955), Droppa (1957a, 1960, 1964), Valovič (1957), Blaha (1971), Dénes (1971), Petrovič and Šoltís (1971), Šoltís (1971), Halaš and Slíva (1979), Halaš (1980, 1984, 1985, 1989), Prikryl (1985), Tulis and Novotný (1989, 1995, 2003), Géczy and Kucharič (1995), Lalkovič (1995, 2000, 2009), Novotný and Tulis (1996, 2000, 2002, 2005), Bella (2003, 2005, 2007), Piasecki et al. (2004, 2005, 2007, 2008), Strug et al. (2004, 2008a), Clausen et al. (2007), Strug and Zelinka (2008c), Korzystka et al. (2011), Strug (2011), Gašinec et al. (2012a,b), Korzystka-Muskała et al. (2014), Gradziński et al. (2016), and others
	Duča	995	319	20	Droppa (1957a, 1960), Tulis and Novotný (1989), and Novotný and Tulis (2005)

Continued

Table 29.1 List and Basic Data on Perennial Ice Caves in Slovakia (Within Regional Geomorphological Units)—cont'd

Regional Geomorphological Units	Cave	Entrance Altitude (m a.s.l.)	Length (m)	Depth (m)	References
Strážovské vrchy, Zliechovská hornatina, Baske	Ľadová jama v Nárožnej (Ľadová diera)	600	8		Špaček (1968)
Tatry, Východné Tatry, Belianske Tatry	Jaskyňa v Škaredom žľabe (Ľadová piecka)	1510	125	18.7[a]	Audy and Šmída (1999)
	Ľadová pivnica	1433	50	18	Sekyra (1954), Droppa (1959), and Pavlarčík (1985)
	Priepasť v Hlúpom vrchu	1996		13	Sekyra (1954) and Pavlarčík (1985)
Tatry, Východné Tatry, Vysoké Tatry	A-1 (Svišťová dolinka 1)	1820		10	Pavlarčík (1984)
	A-2 (Svišťová dolinka 2)	1820		24	Pavlarčík (1984)
	A-5 (Svišťová dolinka 4)	1742	30	10	Pavlarčík (1984)
	Kostolík	1490	23		Paryski (1972) and Pavlarčík (1984)
	Ľadová jaskyňa pod Košiarom	1550	20	5	Wisniewski (1990, 1992)
	Ľadová kamzíčia jaskyňa	1700	125	20	Wisniewski (1990, 1992)
	Nižná svišťová jaskyňa (Kovalova priepasť)	1485	18	16	Paryski (1972) and Pavlarčík (1984)
	Spišmichalova jaskyňa (Ľadová jaskyňa)	1590	40		Kowalski (1960), Paryski (1972), Pavlarčík (1984), and Wisniewski (1990, 1992)
	Tichá diera (Snežná jaskyňa)	1740	9		Paryski (1972) and Pavlarčík (1984)
	Veľká ľadová puklina	1700	50	10	Wisniewski (1990, 1992)
Tatry, Západné Tatry, Červené vrchy	Harišňa	1808	7	4	Hochmuth (1982)
	Ľadová morňa	1880	45		Hochmuth (1974, 1982)
	Ľadová priepasť	1938		60	Droppa (1965), Wala (1971), and Hochmuth (1979, 1982, 1995c)
	Ľadové okno	1800	5		Hochmuth (1982)
	Snežná priepasť	1805		10	Droppa (1961) and Hochmuth (1982)
Tatry, Západné Tatry, Osobitá	Ľadová jaskyňa v Okolíku	1160	150	18	
Tatry, Západné Tatry, Sivý vrch	Košiarec	827		29	Droppa (1972b)
Veľká Fatra, Šípska Fatra	Ľadová jaskyňa na Šípe	700	29	20	Brodňanský (1969)

[a] *vertical span between entrances*

Table 29.2 The Occurrence of Caves With a Perennial and Seasonal Ice Fill Within Hypsometric Grades in the Territory of Slovakia

	Number of Caves	
Hypsometric Grades (m a.s.l.)	**With a Perennial Ice Fill**	**With a Seasonal Ice Fill**
94–299	–	–
300–699	5	4
700–1099	13	8
1100–1399	5	3
1400–1699	10	7
1700–2655	12	1

Table 29.3 The Occurrence of Caves With a Perennial and Seasonal Ice Fill Within Climatic Regions and Subregions (According to the Classification of Climatic Regions in Slovakia; Lapin et al., 2002)

		Number of Caves	
Climatic Regions	**Climatic Subregions**	**With a Perennial Ice Fill**	**With a Seasonal Ice Fill**
Warm region 50 or more summer days annually on average (with daily maximum air temperature ≥25°C)	*Moderately humid subregion with cool winter* (Jan. ≤−3°C)	1	–
Moderately warm region Less than 50 summer days annually on average (with daily maximum air temperature ≥25°C) and the Jul. mean temperature 16°C or more	*Moderately warm and moderately humid subregion of basins with cold winter* (Jan. ≤−5°C, Jul. ≥16°C)	2	1
	Moderately warm and moderately humid subregion of highlands (Jul. ≥16°C, c.500 m a.s.l.)	–	2
	Moderately warm and very humid subregion of highlands (Jul. ≥16°C, mostly above 500 m a.s.l.)	1	–
Cool region Jul. mean temperature <16°C	*Moderately cool and very humid subregion* (Jul. ≥12°C to <16°C)	14	8
	Cool mountainous and very humid subregion (Jul. ≥10°C to <12°C)	12	9
	Cold mountainous and very humid subregion (Jul. <10°C)	15	3

FIG. 29.2

Dobšiná Ice Cave: (A) entrance in the collapsed doline, (B) World Heritage table in the cave, and (C) Great Hall.

Photos: J. Zelinka (A), P. Bella (B), M. Eliáš (C).

The cave was formed by sinking palaeo-river Hnilec in the Middle Triassic Steinalm and Wetterstein limestones of Stranená Nappe. Its length is 1483 m with a vertical span of 112 m.

From a genetic point of view, the Dobšiná Ice Cave is part of the larger Stratená Cave System, which is more than 23.6 km long. In the Late Pliocene and the Early Pleistocene, underground spaces of these caves were connected. Probably in the Middle Pleistocene, they were disconnected by breakdown in connection with the origin of collapsed doline named Duča. At the same time, the drawdown, sack-like volumious cavity originated because of the collapse of the cave ceiling (with the formation of an opening to the surface) and rocky floor between (leveled?) passages (Droppa, 1957a, 1960; Jakál, 1971; Novotný and Tulis, 1996). The main part of the cave is represented by a great cavity (c.140,000 m^3) descending from the surface opening to the depth of 70 m. At present, it is mostly filled with ice that reaches the ceiling in some places and that divides the cavity into several parts: Malá sieň (Small Hall), Veľká sieň (Great Hall), Ruffínyho koridor (Ruffíny' Corridor), and Prízemie (Ground Floor). The ceiling of the biggest cavity, Great Hall, is controlled by an anticline, that is, by upwardly convex, folded structures of limestone (Novotný and Tulis, 2000, 2005). The original fluvially sculptured rocky forms have been remodeled by frost weathering.

The upper parts of Dobšiná Ice Cave at c.945 m a.s.l. relate to the largest cave level of the Stratená Cave System originating during the Late Pliocene (Jakál, 1971; Tulis and Novotný, 1989; Novotný, 1993; Novotný and Tulis, 2005; Bella et al., 2014). They also consist of nonglaciated horizontal passages and halls with original ceiling channels, flat ceilings, and preserved allochthonous fluvial sediments. Also, several carbonate speleothems (stalagmites, stalactites, flowstones, crusts, moonmilk) occur in these nonglaciated parts (Kvietok, 1949; Tulis and Novotný, 1989; Novotný and Tulis, 1996, 2002). Smaller nonglaciated parts occur also in the lower part of the cave, Suchý dóm (Dry Chamber) and Kvapľová pivnica Dripstone Cellar.

The cave ice originated as a result of cold air stagnation and freezing of meteoric waters seeping into the cavity. In addition to the previously mentioned sack-like volumious cavity, the Zrútený dóm (Collapsed Chamber) is also partially filled with ice (under the nearby Duča collapsed doline). The ice-filled part of Dobšiná Ice Cave is located from 920 to 950 m a.s.l. The surface area (in the vicinity of cave entrance) is in the moderately cool (Jul. ≥12°C to <16°C) and very humid subregion (mean annual precipitation total 900–1000 mm) (according to Lapin et al., 2002; Faško and Šťastný, 2002). The well-known ice caves (Eisriesenwelt, Dachstein-Rieseneishöhle) in the Austrian Alps are situated in high-mountain positions. The entrances to the Rieseneishöhle Cave are 1420 and 1450 m a.s.l.; the entrance to the Eisriesenwelt Cave is 1641 m a.s.l. The ice-filled part of the remarkable Scărișoara Cave in the Bihor Mountains, Eastern Carpathians (Romania), is located at 1100–1120 m a.s.l. (its vertical entrance lies at 1165 m a.s.l.).

The ice fill in the Dobšiná Ice Cave occurs as floor ice, icefalls, ice stalagmites, draperies and stalactites, ice columns, and sublimation ice crystals. In earlier literature, the ice volume was estimated at 125,000 m^3 (Pelech, 1878, 1879, 1884a,b; this data was furnished by E. Ruffíny who first measured the cave) or at 145,000 m^3 (Droppa, 1964). The ice surface was reported at 7171 (Pelech, c.f.; the data provided by E. Ruffíny) or at 11,200 m^2 (Droppa, 1964). The 1896 edition of the Baedeker guide reported that the area covered with ice was about 8000 m^2 and that the total mass of ice was estimated at 140,000 m^3 (Baedeker, 1896). Based on geophysical and geodetical measurements in 1995, the following results were specified: the ice volume was 110,132 m^3, the surface of ice fill was 9772 m^2, the maximal thickness of floor ice reached 26.5 m, and the average thickness of floor ice was 13 m (Géczy and Kucharič, 1995; Tulis and Novotný, 1995; Novotný and Tulis, 1996). A comparison of more recent conditions with early pictures of the cave when it was discovered and maps of the time, the ice surface increased after the earlier measurements, mainly in the lower part of the cave. It was found that from the beginning of the 20th century through 1925, the debris bottom of the Ground Floor was completely covered by ice, probably because of a water supply from both halls located in the upper part of the cave (Tulis and Novotný, 2003).

The available data (published measurements) have shown that the largest volume of cave ice in the world is found in the Dobšiná Ice Cave, despite the fact that it is at altitudes below 1000 m a.s.l. (only one big cavity is largely covered and filled with congelation ice). In the Scărișoara Cave, the ice volume was reported as c.100,000 m^3 (Holmlund et al., 2005; Perşoiu and Pazdur, 2011). The total ice volume in the Eisriesenwelt Cave was estimated at c.30,000 m^3 (Klappacher and Haseke-Knapczyk, 1985; Silvestru, 1999; Friedrich, 2009; Obleitner and Spötl, 2011). However, this volume was probably larger; therefore, based on laser scanning, the ice surface in the Eisriesenwelt Cave has been determined to be 27,890 m^2 (Petters et al., 2011).

Ice stratification was formed in relation to the seepage of meteoric waters over the years.

Freezing waters form new layers of floor ice on the top surfaces of ice blocks, and the ice melts on contact with the bedrock. The assumed ages of the oldest existing ice in the Dobšiná Ice Cave were 5000–7500 years BP or 4133 years BP (Droppa, 1957a, 1960, 1964), and 2700–3000 years BP (Tulis and Novotný, 2003). The skin of an undetermined bat, found 2.9 m above the ice base level in 2002, was dated at 1178–988 years BP. Based on this radiocarbon dating, the average growth of ice during the last millennium was determined to 2.16 cm per year (Clausen et al., 2007). Furthermore, it can be assumed that the exchange of ice fill takes about 1250 years. The remains of another frozen bat (*Myotis blythii, Myotis mystacinus*) were found in the lowermost part of the perennial ice block. The radiocarbon dating of the bat's soft tissues yielded ages of 1266–1074 years BP and 1173–969 years BP. The dates testify that the ice crystallized at the turn of the Dark Ages Cold Period and the Medieval Warm Period. The calculated accumulation rate of cave ice varies between 0.7 and 1.4 cm/year at that time, and is similar to the present ice accumulation rate (Gradziński et al., 2016). The glacier-like ice body is slowly moving from the cave entrance and the Small Hall and Great Hall toward the Ground Floor and Ruffíny' Corridor (Lalkovič, 1995) at a maximum rate of 2–4 cm/year (Tulis, 1997). Therefore it presents a true (flowing) ice glacier (Ford and Williams, 2007). The floor ice decreases by melting and sublimation at contact with a rock basement.

The average annual air temperature in the Great Hall is −0.4°C to −1.0°C (in Feb., −2.7°C to −3.9°C; in Aug. around +0.2°C). The air temperature in lower parts of the cave is under the freezing point all year. The relative air humidity in the ice-filled parts is mostly 75%–90%, although sometimes more than 90%. The Dobšiná Ice Cave features a different winter and summer regime of air circulation. The colder air circulates from the surface into the cave during the winter season, and reverses during the summer season (Petrovič and Šoltís, 1971; Halaš, 1989; Piasecki et al., 2004, 2005, 2008a, 2008b; Korzystka et al., 2011; and others; Fig. 29.3). The cave is classified as a statodynamic cave with congelation ice (according to Leutscher and Jeannin, 2004). The air temperature in the nonglaciated parts ranges from +0.8°C to +3.5°C, and the relative humidity ranges from 85% to 98% (Droppa, 1957a, 1960).

The Dobšiná Ice Cave is a unique and important European site in that it is host to the bat species *M. mystacinus* and *Myotis brandtii* that hibernate in an aggregation of 400–500 individuals in the nonglaciated parts of the cave. The northern bat *Eptesicus nilssonii* hibernates regularly with an approximate maximum number of 60 specimens. Among the 12 bat species registered in the cave, also found

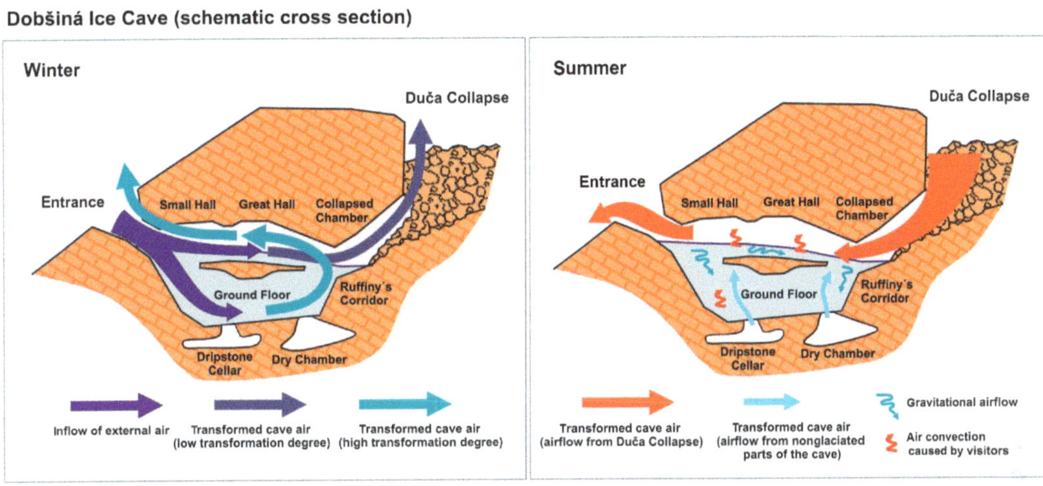

FIG. 29.3

Seasonal changes of air circulation in Dobšiná Ice Cave (Korzystka et al., 2011).

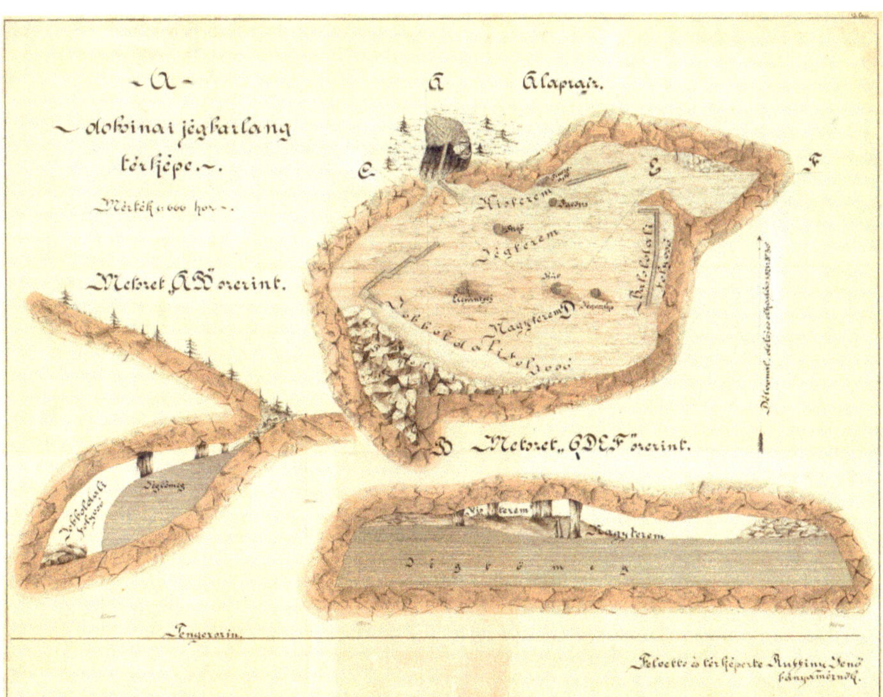

FIG. 29.4

E. Ruffíny's map of the Dobšiná Ice Cave from 1871. Archive of the Slovak Museum of Nature Protection and Speleology, Liptovský Mikuláš.

are *Myotis dasycneme* and *Myotis nattereri*, two of the rarest bat species in Slovakia (Uhrin, 1998; Bobáková, 2002; Kováč et al., 2014; and others).

The opening to the cave, called the "ice hole," has been known for a long time. However, no one descended to the underground until E. Ruffíny accompanied by G. Lang, A. Mego, and F. Fehér did in 1870. The cave was opened to the public thanks to the town of Dobšiná in 1871 (Fig. 29.4). E. Ruffíny

FIG. 29.5

Historical skating in Dobšiná Ice Cave: (A) postcard elaborated after Károly Divald's photo and (B) postcard elaborated after György Conräder's painting. Archive of the Slovak Museum of Nature Protection and Speleology, Liptovský Mikuláš.

elaborated the first plan of the cave in 1871. The first climatic observations were done by F. Fehér in 1870 and 1871. J.A. Krenner, appointed by the Royal Hungarian Natural Scientific Society, investigated the cave in 1873. E.J. Pelech published the studies on the cave in 1878, 1879 (in London), and 1884 (in Budapest). E. Hanvai observed the cave from 1882 to 1886 (results were published in 1900). M. Fischer dealt with the problematics of the cave ice in 1888. The State Meteorological Institute installed the meteorological station inside the cave (in the Great Hall) in 1911. At the time, the cave climate research was done by L. Steiner.

The Dobšiná Ice Cave was one of the first electrically illuminated show caves in the world. The first attempt to install electric lighting was made in 1881 (electric light first lit up the cave Jul. 24, 1881). K. Münnich implemented lighting using Bunsen burners in 1882. Standard electric light was installed in 1887, and electric energy was provided by a diesel engine mounted next to the cave entrance (Lalkovič, 2009).

Many important personalities and researchers of the time visited the cave (e.g., Prince Aug. von Sachsen Gotha with his suite in 1872; J.M. Petzval, a well-known scientist and professor from Vienna in 1876; F. Lesseps, developer of the Suez Canal, accompanied by several French writers in 1884; the Tsar Ferdinand I of Bulgaria; King Milan I of Serbia in 1887; American researcher E.S. Balch in 1895; and Norwegian polar explorer F. Nansen in 1900). A concert was held in the Great Hall in 1890 as a tribute to Charles Ludwig von Habsburg during his visit the Gemer Region. It was the first concert in the history of ice caves (Lalkovič, 2000; and others). M. Markó initiated and organized seven summer ice skating events in the Great Hall between 1893 and 1909 (Holzman, 1999; Székely and Horváth, 2009; Fig. 29.5).

In 1914 new illumination was installed in the cave, supplied by the local electric distribution network. The upper nonglaciated parts of the cave were discovered in 1947. From 1947 to 1952, Czechoslovakian elite figure skaters trained in the cave during the summer months. At present, after several reconstructions and innovations of the technical infrastructure of the cave, the tourist trail is 475 m long with a vertical span of 43 m. By 1970 the attendance at the cave ranged from 52,297 to 128,233 visitors per year (the cave is open from May 15 to Sep. 30). The Dobšiná Ice Cave was included into the World Heritage in 2000 (in the framework of the extension of Slovak-Hungarian site named Caves of Aggtelek and Slovak Karst).

29.4.2 DEMÄNOVSKÁ ĽADOVÁ JASKYŇA CAVE (DEMÄNOVÁ ICE CAVE)

This cave is located on the northern side of the Nízke Tatry Mountains, on the right side of the Demänovská dolina Valley. The cave entrance, west-facing in a coniferous forest, is in the rock cliff named Bašta, at the elevation of 840 m a.s.l. and about 90 m above the valley bottom. The cave was formed by an allochthonous underground flow in the Demänovka River, in the outflow part of the Demänová Cave System (Droppa, 1966, 1972a). Its length is 1975 m with a vertical span of 57 m. The cave consists of fluvially modeled passages (with ceiling and wall channels) in three developmental levels and several chambers remodeled by breakdown and frost weathering. The cave spaces descend from the entrance to a depth of 40–50 m (Vitásek, 1923; Droppa, 1955–56, 1957b).

The ice fill (namely floor ice, ice columns, stalactites, and stalagmites) occurs in the lower parts of the cave, mostly in the Kmeťov dóm (Kmeť's Chamber; Fig. 29.6). According to Droppa (1957b, 1964), the ice volume was c.880 m^3, the surface area was c.440 m^2, the average thickness ranged from 2 to 3 m, and the ice age was 400–500 years. Based on the measurements in 2005, the floor ice's area was calculated at 1407 m^2, its volume at 1042 m^3, its average thickness at 0.74 m, and its maximum thickness at 3.03 m (Strug et al., 2006). The conditions for glaciation started after lower openings

FIG. 29.6

Demänová Ice Cave, Kmeť's Chamber.

Photos: P. Bella.

were closed to the surface, which was caused by clastic sediments created from the slope processes that began after the floor valley below the cave was downcut. As a result of the changes in the cave's morphology, its microclimate conditions changed. Cold air is kept in the lower parts of the cave, and seeping meteoric waters freeze in overcooled underground spaces. In this statodynamic cave with congelation ice (according to Leutscher and Jeannin, 2004), the air temperature in the glaciated parts oscillates around 0°C and in the nonglaciated parts rises from +1.3°C to +5.7°C. The air's relative humidity ranges from 92% to 98% (Otruba, 1958a,b; Halaš, 1984; Piasecki et al., 2007; Strug and Zelinka, 2008a). The surrounding surface area is a moderately cool (Jul. ≥12°C to <16°C) and very humid subregion (mean annual precipitation total 900–1000 mm) (according to Lapin et al., 2002; Faško and Šťastný, 2002).

Original carbonate speleothems are preserved in several places in the cave (mostly stalactites, stalagmites, and flowstones); in the glaciated part, they were destroyed by frost weathering. Dripstone and flowstone formations on the surface have been colored gray to black by the soot of tar torches, oil burners, and paraffin lamps, which were used for lighting until 1924. Cryogenic cave pearls occur on the cave floor in the periglacial zone of the cave, in the Halašov dóm (Halaš's Chamber) and Čierna galléria (Black Gallery) on the edge of the Kmeť's Chamber with perennial ice fill (Žák et al., 2013; and others).

The cave is among the places long known as places where the bones of various vertebrates are found, including those of the cave bear (*Ursus spelaeus*), which were in the first half of the 18th century considered to be dragons' bones, which is why the cave was called Dragon Cave in the past. Recent but community includes representatives of six species, but only *E. nilssonii*, *M. mystacinus/brandtii*, and *Myotis mystis/blythi* hibernate regularly and in more numerous populations (Višňovská, 2007; and others). Each of these three species prefers places with different microclimatic conditions. Eight bat zones were distinguished based on the average amplitude of the monthly air temperature (Višňovská et al., 2007).

It is said that the Demänová Ice Cave has been known since time immemorial. The first mention of the openings to caves in the Demänovská dolina Valley was recorded in an Esztergom Chapter document in 1299. The first written mention of the Demänová Ice Cave related to the description of a cave not far from the town of Liptovský Mikuláš in 1672 by J.P. Hain, who was interested in cave bear bones and who considered them to be dragon bones (Lalkovič, 2003a,b). Further mentions of the Demänová Ice Cave are connected with G. Buchholtz Jr., who surveyed its underground spaces in 1719. He sent the descriptions together with a sketch of the cave to M. Bel, who published the data in 1723 and 1736. German physician and naturalist F.E. Brückmann visited the cave in 1724.

The cave has gradually became known to the public. An imperial commission, headed by a courteous mathematician J.A. Nagel, visited the cave in 1751. J.M. Korabinsky included data on the cave in the Almanac of Hungary published in 1778. K.G. Windisch mentioned the cave in a two-part geography of Hungary in 1780. The cave was also visited by B. Hacquet, a professor at the universities of Ljubljana and Lviv, who published a report on the cave in 1790. The English physician and naturalist R. Townson wrote about the cave in 1797. The Polish geoscientist, philosopher, and statesman S. Staszic, an early surveyor of the geology of Central and Eastern Europe, visited the cave in 1799. The Lviv superintendent S. Bredetzky described the cave in 1801 and 1805. F. Sartori, a royal government secretary, included the cave in a work on natural attractions of the Austrian Empire published in 1809. K.A. Zipser wrote about the cave in 1817. A year later, it was visited by the French mineralogist F.S. Beudant. In 1827 the cave was investigated by the Prussian lieutenant and geographer A.W. Sydow, who focused on the origin of the ice fill. The English naturalist J. Paget studied the cave in 1835 and the Polish geologist and paleontologist L. Zejszner in 1838. The Hungarian geographer J. Hunfalvy wrote about the cave in 1860. Among those studying ice caves, the German naturalist B. Schwalbe visited the Demänová Ice Cave in 1881. The cave was described in detail by J. Mihalik in 1884. The American geographer E.S. Balch, who dealt with cave ice, visited the cave in 1896. Numerous signatures on the cave's walls and the rich literature on the cave are evidence of early researchers, scientific groups, and the general public's great interest in this cave.

The initial opening of the cave to tourists happened around the last half of the 19th century. Interest in the Demänová Ice Cave decreased after the Demänovská jaskyňa slobody Cave opened to the public in 1924. The upper dripstone parts of the cave were discovered in 1926. The cave was reconstructed, including installation of electric lighting, and reopened to public from 1950 to 1952. The Demänovská jaskyňa mieru Cave was discovered through the Jazerná chodba (Lake Passage) of the Demänová Ice Cave in 1952. Because the temperature regime was disturbed due to the excavation of a connecting siphon between these caves, several measures had to be taken in 1953 and 1954 to renovate the ice fill. The artificial opening to the Dóm trosiek (Debris Chamber) was closed, and a stone wall dividing the glaciated and nonglaciated parts was built. The first climatic observations were done from 1953 to 1956. The tourist trail is 650 m long with a vertical span of −48 m. By 1970 attendance at the cave ranged from 51,428 to 106,900 visitors per year (the cave is open from May 15 to Sep. 30).

29.4.3 SILICKÁ ĽADNICA CAVE (SILICA ICE CAVE)

This remarkable ice-filled cave is on the Silická planina Plateau in the Slovak Karst, 2 km west of the village of Silica. Its entrance lies at an elevation of 470 m, north-facing in a leafy forest (the upper edge of the collapse doline-like depression extends to a height of 503 m a.s.l.). The surface area is in the warm, moderately humid subregion with cool winters (Jan. ≤−3°C, mean annual precipitation total 600–700 mm) (according to Lapin et al., 2002; Faško and Šťastný, 2002). The Silica Ice Cave is the lowest-lying perennial ice cave up to the latitude of 50 degrees north, in the temperate climatic zone. Among the lowest-lying ice caves is also the French Grotte de la Glacière Cave in the Jura Mountains, whose entrance is at an altitude of 525 m. Among the better known ice caves, only the Kungur Ice Cave is in the lower altitude (its entrance is at 118 m a.s.l.). However, this cave is located farther north (latitude 57 degrees north), in the continental part of temperate climatic zone (Russia, Perm Region, the western foothills of the Ural Mountains).

The Silica Ice Cave was formed in the Middle Triassic Wetterstein limestones of Silica Nappe. Morphologically, it consists of two different parts. The upper part presents a steeply inclined spacious cavity with a large opening to the surface (corrosive-collapsed abyss, *light hole*) (Roth, 1940a,b; Droppa, 1962, 1964; Fig. 29.7). The lower, mostly subhorizontal fluvially modeled passages were formed by the underground stream, Čierny potok (Black Brook). In 1931 J. Majko entered the Archeologický dóm (Archaeological Chamber) through a hole dug in the bottom of the abyss, which was filled by clastic sediments, mainly debris. The other passages and halls along the Black Brook were discovered by Czech speleodivers in 1988. The cave is 1100 m long and 110 m deep. The Black Brook flows from the Silica Ice Cave into the Gombasecká jaskyňa Cave and the Čierna vyvieračka (Black Spring).

The upper abyss of the cave (from the entrance to the depth of 50 m) is partially filled with ice (at 425–465 m a.s.l.). The drawdown spacious cavity with an accumulation of cold air was glaciated as a result of the restriction of air flow between the upper and lower parts of the cave because of debris that blocked the connecting section. The air temperature in the ice-filled part oscillates between −10.4°C and +1.0°C and in the lower Archaeological Chamber (without cave ice) between +3.0°C and +5.8°C (Rajman et al., 1987). Roda et al. (1974) reported an ice surface ranging from 710 to 970 m^2 and an ice volume ranging from 213 to 340 m^3. The surface and volume of the ground ice varies depending on the

FIG. 29.7

Silická ľadnica Cave (Silica Ice Cave): (A) entrance controlled by a steep fault, (B) cave breakdown portal, (C) ice formations in the entrance part, and (D) icefall.

Photos: P. Bella (A, B, C), P. Hipman (D).

temperature and precipitation conditions during individual years. The floor ice mass is moving downward on an inclined rock basement, but in the lower part of the abyss, ice is melting because of warmer air circulating through the debris plug (Rajman et al., 1985). An ice stalactite about 7 m long, curving inward (in the direction of warmer air flow), formed in the cave's entrance during colder periods (Kunský, 1939; Roda and Rajman, 1971). During the past several decades, only shorter ice stalactites have occurred seasonally on the cave ceiling.

Originally (before the discovery of the lower parts along the Black Brook), the Silica Ice Cave presented a static cave with congelation ice (according to Leutscher and Jeannin, 2004). Recent thermodynamic conditions for the origin of ice (during winter, transitional, and summer phases) were studied by Rajman et al. (1987). Three bat species have been recorded in the Silica Ice Cave during the winter including the rare *M. nattereri*.

The cave was settled several times before the beginning of its cave ice. Archaeological findings in the lower part of the cave (Archaeological Chamber with the Black Brook) are dated from the Neolithic Age, the Bronze Age, and the La Tène Age. Findings of cinders suggest possible Early Palaeolithic settlement (Böhm and Kunský, 1938, 1941; Bánesz, 1962; Bárta, 1995; and others). Based on archaeological findings, the formation of ice in the upper part of the cave started around 400 BC to 0 BC (Roth, 1940a; Bárta, 1995). Böhm and Kunský (1938, 1941) as well as Kunský (1943, 1950) proposed the origin of the cave ice to be c.2000 years ago.

The first information about the Silica Ice Cave was published in the first half of the 18th century. A sketch of underground spaces of the cave, elaborated by J. Buchholtz, dates to 1719. He wrote about this cave to M. Bel, who published the data on the Silica Ice Cave in 1723 and from 1739 to 1741. At that time, the Silica Ice Cave was one of the most famous caves in the Slovak Karst. The first measurement of air temperature in the cave was done by the English scientist, R. Townson, in 1793 (he described it in a travel book in 1797). French geologist F.S. Beaudant visited the Silica Ice Cave in 1818; he published a description of the cave, including an explanation of the ice's origin in 1822. E.J. Terlanday dealt with measurements of air temperature and the observation of the cave ice from 1892 to 1894. The Silica Ice Cave is a part of the site Caves of Aggtelek and Slovak Karst inscribed in the World Heritage List in 1995.

29.4.4 VEĽKÁ ĽADOVÁ PRIEPASŤ NA OHNIŠTI ABYSS (GREAT ICE ABYSS ON THE OHNIŠTE)

This cave is located on the northern side of the Nízke Tatry Mountains, on the south slope of the Ohnište Plateau (rock cliffs and ledges with belts of coniferous forest). Its upper (main) east-facing entrance lies at 1529 m a.s.l. and its lower south-facing entrance at 1513 m a.s.l. The *aven* type abyss is 125 m deep (Fig. 29.8). It was formed from the Middle Triassic Gutenstein limestones of Choč Nappe. The surface area is in the cool mountainous and very humid subregion (Jul. ≥10°C to <12°C, mean annual precipitation total 1000–1200 mm) (according to Lapin et al., 2002; Faško and Šťastný, 2002). In 1956 from Aug. 6 to 8, the air temperature decreased from 14.5°C (outside, at the mouth of the abyss) to 0.4°C at a depth of 50 m, it was constant at a depth of 50–80 m, but it increased to 0.8°C at a depth of 80 m (Otruba, 1958a,b).

In this static cave with congelation ice, the ice plug occurs at the bottom of the first abyss at a depth of 80 m (Droppa, 1958). Its volume was estimated to range from 800 to 1000 m^3 (Benický, 1958). Based on measurements taken on Jun. 3, 1973, the estimated volume of the ice plug was c.525 m^3

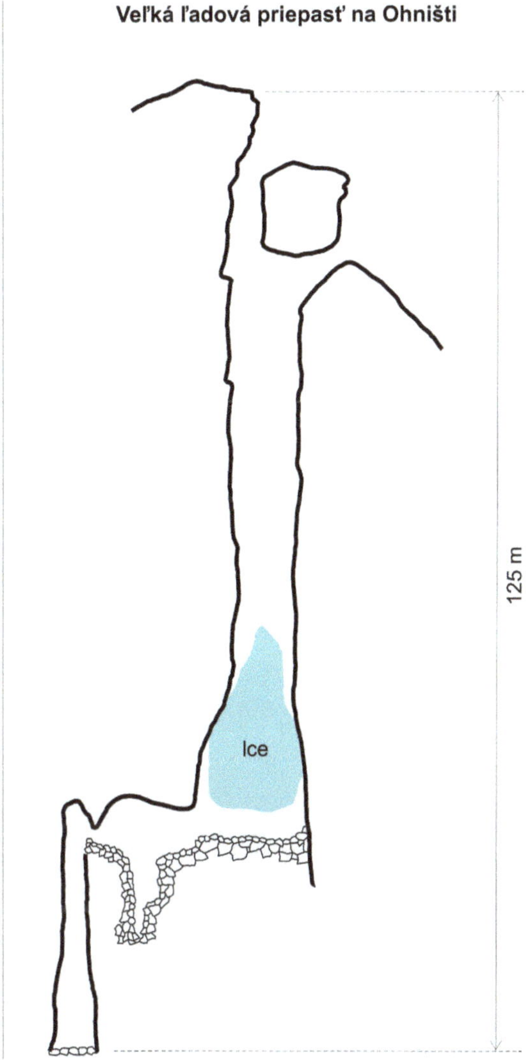

FIG. 29.8

Veľká ľadová priepasť na Ohništi Abyss (Great Ice Abyss on the Ohnište), Nízke Tatry Mountains.

From Droppa, A., 1958. Geomorfologický charakter priepastí na Ohništi [Geomorphological character of abysses on the Ohnište]. Slovenský kras 1, 14–23 (in Slovak).

(Hochmuth, 1995c). However, based on measurements taken on Jul. 24, 2011, the ice volume had reduced 150–200 m^3 during the intervening 38 years, and the morphology of the ice plug had changed markedly (Hochmuth and Holúbek, 2011). Melting of the cave ice can be attributed to climate warming, to increased precipitation in summer and reduced precipitation in winter, or to air circulation from the Silvošova diera Cave (Hochmuth and Holúbek, 2011; Holúbek, 2014).

29.4.5 ĽADOVÁ PRIEPASŤ ABYSS (ICE ABYSS)

This cave is located in the Červené vrchy Mountains (Západné Tatry Mountains, northern Slovakia), on the southern slope of Mt. Kresanica (2121 m), in the alpine zone (Fig. 29.9). Its south-facing entrance lies at 1938 m a.s.l. (Droppa, 1965), in the cold mountainous and very humid subregion (Jul. <10°C, mean annual precipitation total more than 2000 mm) (according to Lapin et al., 2002; Faško and Šťastný, 2002). The *aven* type abyss is 60 m deep. It is formed from Middle Triassic Gutenstein limestones of the cover sequence of the Tatric Unit and presents a static cave with firn and congelation ice. Two ice bodies have been described: a firn and ice plug in the upper part of the abyss at a depth ranging from 17 to 25 m (c.200 m^3) and an ice monolith in the lower part at a depth ranging from 30 to 55 m. The estimated total ice volume is c.650 m^3 (Hochmuth, 1979, 1982, 1995c). During winter the mouth of the abyss is filled with snow. In the autumn of 1968, Polish cavers descended through fissures between ice bodies and a limestone wall to a depth of 70 m (Wala, 1971).

FIG. 29.9

Ľadová priepasť Abyss (Ice Abyss), Červené vrchy Mountains.

Photo: Ľ. Kováč.

29.5 ICE CAVES DISTURBED BY HUMAN INFLUENCES

Ice caves belong to the most fragile and vulnerable caves, as well as natural geocomplexes. In Slovakia, several caves with perennial or seasonal ice fills have been disturbed by human impacts related to tourism and speleological exploration.

The decrease of ice volume in the Demänová Ice Cave was started after the excavation of new opening to the surface in the lower part of the cave during its tourist development in 1950 (an improper intention to change the tourist trail) as well as after the discovery of Demänovská jaskyňa mieru Cave through the siphon by its excavation from clastic sediments at the end of Demänová Ice Cave in 1952. To stabilize the cave's climate, these openings must be filled again. Also, a stone wall to counter the air flow was built in 1954 between the ice-filled and other parts of the cave leading to the Demänovská jaskyňa mieru Cave (Boček, 1954; Otruba, 1958a,b, 1971).

In the Dobšiná Ice Cave, the air circulation was partially changed as a result of the discovery of nonglaciated parts in 1946. About 25,000 m^3 of ice melted within 4 years. The critical places where the air circulation changed were determined based on smoke tests. Seven dry stone walls, directing warmer air to the cave opening from the surface, were installed to repair human-influenced and changed microclimatic conditions (Ondroušek, 1952). However, climatic conditions were not stabilized until after treatment of the cave entrance in 1954 (Droppa, 1960). An artificial tunnel and a trench for a tourist path dug in the ice in the 1970s, which partially disturbed air circulation in the entrance of the cave (with a decrease of ice stalactites and the ice wall in the Small Hall), had to be filled with snow and ice to restore the area to its original state (Bobro et al., 1995a,b; Zelinka, 1996). The reconstruction of a fence around the western and northern parts of the collapsed doline with a cave entrance and the cutout of several trees that had shaded the entrance area in 2010 caused climatic changes in the near-entrance area of the cave, including melting of permafrost and formation of new openings into the cave. The thermal stratification and the spatial distribution of the air temperature inside the collapsed doline was precisely determined in 2012 and compared with microclimatic conditions before the previously mentioned anthropogenic changes, based on data obtained during analogous research from 2003 to 2004 (Korzystka-Muskała et al., 2014). The project, which was aimed at stabilizing the microclimate conditions in the entrance area of the cave, was realized in 2015.

The Silica Ice Cave was used to cool beer from the brewery that existed in front of the cave portal (in the eastern part of collapsed doline) from 1863 to 1867 (Szontág, 1871). The smoke from the brewery polluted the cave (Derfiňák, 2001). After the discovery of the Archaeological Chamber in the lower part of the cave in 1931, temperature conditions were disrupted by a upward circulation of warmer air to the upper ice-filled part. This negative human impact was eliminated by filling the excavated connecting hole in 1938 (Roth, 1940a). The gate between the ice-filled part and the lower part (without ice) of the cave was reconstructed or changed several times (during 1950s, 1960s, and 1980s). The effect of the upward circulation of warmer air was enhanced by partially opening the outflow siphon in the Archaeological Chamber, through which new passages along the Black Brook were discovered by Czech speleodivers in 1988 (Hovorka, 1989). Authorities decided that the cave must be closed, and the next speleological exploration of lower parts were paused (Mitter, 1991). In 1998, when the new and more airtight gate was installed, the decrease in ice volume was reduced and more or less stopped, and climatic conditions were restored almost to the original thermodynamic regime (Zelinka, 1999). Another reason for the gradual decrease in cave ice has been the higher evapotranspiration from a non-indigenous black pine forest above the cave, thus the amount of water seeping into the underground has been gradually reduced (Rajman et al., 1987). The reduction of this nonindigenous forest to 50% canopy of all trees was realized in 2007 (Gaál and Zelinka, 2008).

A multiannual attempt to restore glaciation in the lower part of the Belianska jaskyňa Cave began in 1936. Its entrance, which in some places is seasonally filled with ice, connects with the outside atmosphere through two entrances in a vertical span of 81 m, exposed to the north. During winter, both entrances were opened and cold air flowed into the cave where the air temperature reached −25°C and in May −3°C. When the outside air temperature rose above zero, both entrances were closed, and the dynamic cave was changed to a static one. In the artificially cooled part of the cave, ice stalactites up to 1.5 m long, ice columns up to 3 m high, and ice crusts with thicknesses up to 1.5 m were occasionally formed (Lutonský, 1934; Paloncy, 1934; Kunský, 1935–36, 1937, 1939, 1943). However, these conditions were not sustainable, and therefore the failed attempt to restore glaciation in the lower part of the cave caused enhanced frost weathering of bedrock, flowstones, and dripstones (Benický, 1951; Zelinka, 1997).

29.6 ICE CAVES AND NATURE PROTECTION

All caves in Slovakia, including ice caves, are natural monuments lawfully protected by the Act on Nature and Landscape Protection. Many ice caves are situated in nature reserves and national parks. The Dobšiná Ice Cave, Silica Ice Cave, Snežná diera Cave (Snow Hole), and Great Ice Abyss on the Ohnište are declared as national natural monuments. The Demänová Ice Cave is part of the Demänová Caves National Natural Monument. The Dobšiná Ice Cave, Silica Ice Cave, and Snežná diera Cave are included in the World Heritage (within the Slovak-Hungarian site named Caves of Aggtelek and Slovak Karst). The environmental protection and climate monitoring of ice caves in Slovakia are implemented and maintained by the State Nature Conservancy of the Slovak Republic, Slovak Caves Administration, in cooperation with the Slovak Speleological Society and some domestic and foreign scientific institutes.

ACKNOWLEDGMENTS

The authors would like to thank many members of the Slovak Speleological Society for actual data on ice caves in Slovakia. The work was supported by the Project of the Scientific Grant Agency VEGA No. 2/0193/15.

REFERENCES

Audy, M., Šmída, B., 1999. Škaredý žľab v Belanských Tatrách aneb staré hříchy [The Škaredý žľab slope groove in the Belianske Tatry Mountains or old sins]. Spravodaj Slovenskej speleologickej spoločnosti 30 (4), 35–38 (in Czech).

Baedeker, K., 1896. Austria, including Hungary, Transylvania, Dalmatia, and Bosnia. Handbook for Travellers. 8th ed. Karl Baedeker Publisher, Leipsic and Dulau & Co., London. 468 p.

Balch, E.S., 1900. Glacières of Freezing Caverns. Allen, Lane and Scott, Philadelphia. 337 p.

Bánesz, L., 1962. Prieskumy v Juhoslovenskom krase pri Rožňave [Surveys in the South-Slovak Karst at Rožňava Town]. Študijné zvesti AÚ SAV 9, 237–240 (in Slovak).

Bárta, J., 1995. Pomoc archeológie pri datovaní zaľadnenia Silickej ľadnice [Archeology assistence at the dating of glaciation of the Silická ľadnica Cave]. In: Bella, P. (Ed.), Ochrana ľadových jaskýň, Proceedings, Dobšinská Ľadová Jaskyňa, 21.–22. 9. 1995. SSJ, Liptovský Mikuláš, pp. 81–84 (in Slovak).

Beaudant, F.S., 1822. Voyage minéralogique et géologique en Hongrie, pendant l'année 1818, vol. 2, Verdière Paris, 614 p.

Beaudant, F.S., 1823. Travels in Hungary in 1818. R. Phillips and Co., London. 126 p.

Bel, M., 1723. Hungariae antiquae et novae prodromus, quomodo in singulis operis partibus elaborandis, versari constituerit, auctor… Norimbergae, 202 p.

Bel, M., 1736. Notitia Hungariae Novae Historico-Geographica. Tomus Secundus, Vienna.

Bel, M., 1739–1741. Dias Antrorum mirabilis Naturae, glacialis alterius, alterius Halitus noxios eructantis, ad Regiam Societatem scientiarum Londinensem, a Matthia Belio Pannonio, R.S.S. missa. Philos. Trans. R. Soc. Lond. 41 (1), 41–56.

Bella, P., 1985. Príspevok k poznaniu krasu doliny Kamenice na severnej strane Nízkych Tatier [Contribution to the karst of Kamenica Valley on the north side of Nízke Tatry Mountains]. Spravodaj Slovenskej speleologickej spoločnosti 16 (1–2), 11–16 (in Slovak).

Bella, P., 1995. Prehľad a rozšírenie jaskýň s ľadovou výplňou na Slovensku [Review and occurrence of caves with ice fill in Slovakia]. In: Bella, P. (Ed.), Ochrana ľadových jaskýň, Proceedings, Dobšinská Ľadová Jaskyňa, 21.–22. 9. 1995. SSJ, Liptovský Mikuláš, pp. 61–73 (in Slovak).

Bella, P., 2003. Glaciálne ablačné formy v Dobšinskej ľadovej jaskyni [Glacial ablation forms in the Dobšiná Ice Cave]. Aragonit 8, 3–7 (in Slovak).

Bella, P., 2005. Rovné podlahové ľadové povrchy v zaľadnených jaskyniach (Dobšinská ľadová jaskyňa, jaskyňa Scărișoara) [Flat floor ice surfaces in ice caves (Dobšiná Ice Cave, Scărișoara Cave]. Aragonit 10, 12–16 (in Slovak).

Bella, P., 2007. Morphology of ice surface in the Dobšiná Ice Cave. In: Zelinka, J. (Ed.), Proceedings of the 2nd International Workshop on Ice Caves, Demänovská Dolina, Slovak Republic, May 8–12, 2006. SSJ, Liptovský Mikuláš, pp. 15–23.

Bella, P., 2008. Geographical distribution of ice-filled caves in Slovakia. In: Turri, S. (Ed.), Volume of Abstracts, 3rd International Workshop on Ice Caves, May 12–17, 2008. Kungur, Russia, pp. 18–19.

Bella, P., Braucher, R., Holec, J., Veselský, M., 2014. Datovanie pochovania alochtónnych fluviálnych sedimentov v hornej časti Dobšinskej ľadovej jaskyne (IV. vývojová úroveň systému Stratenskej jaskyne) pomocou kozmogénnych nuklidov [Cosmogenic nuclide dating of the burial of allochthonous fluvial sediments in the upper part of the Dobšinská Ice Cave (IVth evolution level of the Stratenská cave system), Slovakia]. Slovenský kras 52 (2), 101–110 (in Slovak).

Benický, V., 1932. Výskum v Silickej Ľadnici [Research in the Silická ľadnica Cave]. Krásy Slovenska 11 (2–3), 81–85 (in Slovak).

Benický, V., 1940. Výskum priepasti na Ohništi [Research of the abyss on the Ohnište]. Krásy Slovenska 18 (2), 43–47 (in Slovak).

Benický, V., 1951. Navštívte Belanskú jaskyňu Sokola! [Visit the Belianska jaskyňa Cave of Sokol!]. Krásy Slovenska 28 (5), 109–114 (in Slovak).

Benický, V., 1958. Výskum ľadovej priepasti na Ohništi [Research of the ice abyss on the Ohnište]. Slovenský kras 1, 5–13 (in Slovak).

Bíma, K., 1925. Dobšinská ľadová jaskyňa a jej okolie [Dobšiná Ice Cave and its surroundings]. Odbor Klubu Čs. turistov, Dobšiná. 24 p. (in Slovak).

Blaha, L., 1971. Dobšinská ľadová jaskyňa – 100 rokov od jej objavenia/Die Eishöhle von Dobšiná – 100 Jahre seit ihrer Entdeckung. Slovenský kras 9, 5–26.

Bobáková, L., 2002. Zimovanie netopierov v jaskynnom systéme Dobšinská ľadová jaskyňa – jaskyňa Duča [Wintering of bats in the Dobšiná Ice Cave – Duča Cave]. Vespertilio 6, Katalóg zimovísk netopierov Slovenskej republiky [Catalogue of the hibernacula of bats in the Slovak Republic], pp. 245–248 (in Slovak).

Bobro, M., Hančuľák, J., Zelinka, J., 1995a. Súčasné mikroklimatické pomery v Dobšinskej ladovej jaskyni [Current microclimatic conditions in the Dobšiná Ice Cave]. In: Bella, P. (Ed.), Ochrana ľadových jaskýň, Proccedings, Dobšinská Ľadová Jaskyňa, 21.–22. 9. 1995. SSJ, Liptovský Mikuláš, pp. 29–34 (in Slovak).
Bobro, M., Krepelka, P., Zelinka, J., 1995b. Rekonštrukcia vstupného priestoru v Dobšinskej ladovej jaskyni vo vzťahu k úbytku ľadu [Reconstruction of the entrance space in the Dobšiná Ice Cave in relation to ice decrease]. In: Bella, P. (Ed.), Ochrana ľadových jaskýň, Proccedings, Dobšinská Ľadová Jaskyňa 21.–22. 9. 1995. SSJ, Liptovský Mikuláš, pp. 35–36 (in Slovak).
Boček, A., 1954. Pre záchranu krás Demänovskej ľadovej jaskyne [To the saving of the beauties of Demänová Ice Cave]. Československý kras 7, 69–70 (in Slovak).
Böhm, J., Kunský, J., 1938. Silické lednice [Silická ľadnica Cave]. Sborník Československé společnosti zeměpisné 44, 129–133 (in Czech).
Böhm, J., Kunský, J., 1941. Lednice, die Eishöhle von Silice. Wierner Prähistorische Zeitschrift 28, 96–127.
Bredetzky, S., 1801. Die Höhle bey Deménfalva. In: Topografisches Taschenbuch für Ungern, auf das Jahr 1802, Oedenburg, pp. 166–181.
Bredetzky, S., 1805. Die Drachenhöhle und Fluss-Grotte nich weit von Deménfalva, einem Dorfe im Liptauer Komitate. In: Beiträge zur Topographie des Königreichs Ungern, I. Band, Wien, pp. 140–152.
Brodňanský, J., 1969. Jaskyňa v Šípe pri Zaškove [Jaskyňa v Šípe Cave at Zaškov Village]. Slovenský kras 7, 99–100 (in Slovak).
Browne, R.G., 1865. Ice-caves of France and Switzerland. Longmans, Green, and Co., London. 244 p.
Buchholtz, J., 1787. Abermalige Reise in die Karpatischen Gebirge, und die angränzenden Gespanschaften. Ungarisches Magazin 4, 251–291.
Clausen, H.B., Vrana, K., Hansen, S.B., Larsen, L.B., Baker, J., Siggaard-Andersen, M.-L., Sjolte, J., Lundholm, S.C., 2007. Continental ice body in Dobšinská Ice Cave (Slovakia) – part II. Results of chemical and isotopic study. In: Zelinka, J. (Ed.), Proceedings of the 2nd International Workshop on Ice Caves, Demänovská Dolina, Slovak Republic, May 8–12, 2006. SSJ, Liptovský Mikuláš, pp. 29–37.
Dénes, G., 1971. Klimatische Beobachtungen in der Eishöhle von Dobšiná in der ersten 50 Jahren nach der Entdeckung/Klimatické pozorovania v Dobšinskej ľadovej jaskyni v prvých 50 rokoch po objavení. Slovenský kras 9, 163–169.
Dénes, G., 1992. Darstellung der Höhle von Szilice in einem Manuskript vom Beginn des 18. Jahrhunderts. Karszt és Barlang, special issue. In: Proceedings of the ALCADI '92 International Conference on Speleo History, pp. 13–16.
Derfiňák, P., 2001. Silická ľadnica pred 135 rokmi [Silická ľadnica Cave 135 yers ago]. Spravodaj Slovenskej speleologickej spoločnosti 32 (3), 35–36 (in Slovak).
Droppa, A., 1955–56. Demänovská ľadová jaskyňa [Demänová Ice Cave]. Československý kras 8–9, 92–110 (in Slovak).
Droppa, A., 1957a. Dobšinská ľadová jaskyňa [Dobšiná Ice Cave]. Geografický časopis 9 (2), 99–118 (in Slovak).
Droppa, A., 1957b. Demänovské jaskyne – Krasové zjavy Demänovskej doliny [Demänová Caves – Karst phenomena of the Demänovská dolina Valley]. SAV, Bratislava. 289 p. (in Slovak).
Droppa, A., 1958. Geomorfologický charakter priepastí na Ohništi [Geomorphological character of abysses on the Ohnište]. Slovenský kras 1, 14–23 (in Slovak).
Droppa, A., 1960. Dobšinská ľadová jaskyňa [Dobšiná Ice Cave]. Šport, Bratislava. 115 p. (in Slovak).
Droppa, A., 1959. Belanská jaskyňa a jej kras [Belanská Cave and its karst]. Šport, Bratislava. 136 p. (in Slovak).
Droppa, A., 1961. Vysokohorský kras Červených vrchov [High-mountain karst of the Červené vrchy Mountains]. Slovenský kras 3, 21–36 (in Slovak).
Droppa, A., 1962. Gombasecká jaskyňa [Gombasecká jaskyňa Cave]. Šport, Bratislava. 80 p. (in Slovak).
Droppa, A., 1964. Ledjanye peščery Čechoslovakii [Ice caves of Czechoslovakia]. Peščery 4 (5), 85–94 (in Russian).
Droppa, A., 1965. Geomorfologický výskum priepastí v Červených vrchoch [Geomorphological research of abysses in the Červené vrchy Mountains]. Slovenský kras 5, 42–48 (in Slovak).

Droppa, A., 1966. The correlation of some horizontal caves with river terraces. Stud. Speleol. 1, 186–192.

Droppa, A., 1972a. Geomorfologické pomery Demänovskej doliny [Geomorphological settings of the Demänovská dolina Valley]. Slovenský kras 10, 9–46 (in Slovak).

Droppa, A., 1972b. Kras skupiny Sivého vrchu v Západných Tatrách [Karst of the Sivý vrch group in the Západné Tatry Mountains]. Československý kras 23, 77–98 (in Slovak).

Droppa, A., 1973. Prehľad preskúmaných jaskýň na Slovensku [Review of investigated caves in Slovakia]. Slovenský kras 11, 111–157 (in Slovak).

Faško, P., Šťastný, P., 2002. Mean annual precipitation totals. In: Landscape Atlas of the Slovak Republic. Ministry of Environment of the Slovak Republic, Bratislava/Slovak Environmental Agency, Banská Bystrica, p. 99 (Map No. 54).

Fehér, J., 1872. Dobsinai jégbarlang [Dobšiná Ice Cave]. Természettudományi Közlöny 4 (29), 10–13 (in Hungarian).

Fischer, M., 1888a. Dobsinai-jégbarlang physikai magyarázata [Physical explanation of the Dobšiná Ice Cave]. A Magyarországi Kárpát-egyesület Évkönyve 15, Igló, pp. 161–199 (in Hungarian).

Fischer, M., 1888b. Die physilakische Erklärung der Dobschauer Eishöhle. Jahrbuch des Ungarischen Karpathen-Vereines 15, Igló, pp. 152–193.

Ford, D.C., Williams, P.W., 2007. Karst Hydrogeology and Geomorphology. Wiley, Chichester. 562 p.

Friedrich, V., 2009. Eisriesenwelt die größte Eishöhle der Erde. Eisriesenwelt GmbH, Kunstverlag Peda, Passau.

Fugger, R., 1891. Eishöhlen und Windröhren. XXIV Jahresbericht der k.k. Ober-Realschule in Salzburg, Salzburg, 70 p.

Gaál, Ľ., Zelinka, J., 2008. Regulačné lesohospodárske zásahy nad Silickou ľadnicou [Regulative forest management measures above the Silická ľadnica Cave]. Aragonit 13 (2), 56 (in Slovak).

Gaál, Ľ., Grego, J., Holúbek, P., Mlejnek, R., 2000. Jaskyne pod Skalkou v Kremnických vrchoch [Caves under Mt. Skalka in the Kremnické vrchy Mountains]. Slovenský kras 38, 141–154 (in Slovak).

Gašinec, J., Bella, P., Gašincová, S., Imreczeová, A., 2012a. Digitálny výškový model podlahových ľadových povrchov v Dobšinskej ľadovej jaskyni [Digital elevation model of floor ice surfaces in the Dobšiná Ice Cave]. Slovenský kras 50 (1), 31–40 (in Slovak).

Gašinec, J., Gašincová, S., Černota, P., Staňková, H., 2012b. Uses of terrestrial laser scanning in monitoring of ground ice within Dobšinská Ice Cave. Inzynieria Mineralna 13 (2), 31–42.

Géczy, J., Kucharič, L., 1995. Stanovenie mocnosti ľadovej výplne vo vybraných miestach Dobšinskej ľadovej jaskyne [Determination of the ice thickness at selected sites oif the Dobšiná Ice Cave]. In: Bella, P. (Ed.), Ochrana ľadových jaskýň, Proceedings, Dobšinská Ľadová Jaskyňa, 21.–22. 9. 1995. SSJ, Liptovský Mikuláš, pp. 17–23 (in Slovak).

Gradziński, M., Hercman, H., Peresviet-Soltan, A., Zelinka, J., Jelonek, M., 2016. Radiocarbon dating of fossil bats from Dobšina Ice Cave (Slovakia) and potential palaeoclimatic implications. Ann. Soc. Geol. Pol. 86 (3), 341–350.

Hacquet, B., 1790. Neueste physikalisch-politische Reisen in den Jahren 1788 und 1789 durch die Dacischen und Sarmatischen oder Noerdlichen Karpathen. I. Theil, Neuernberg, 207 p.

Halaš, J., 1980. Vplyv fyzikálnych veličín ovzdušia na genézu ľadových útvarov v Dobšinskej a Demänovskej ľadovej jaskyni [Influence of the physical quantities of atmosphere on the origin of ice forms in the Dobšiná and Demänová ice caves]. Slovenský kras 18, 139–145 (in Slovak).

Halaš, J., 1983. Niektoré poznatky z merania teploty horninového plášťa v Dobšinskej ľadovej jaskyni a Demänovskej ľadovej jaskyni [Some knowledge from temperature measurement of bed rock in the Dobšiná and Demänová ice caves]. Slovenský kras 21, 79–91 (in Slovak).

Halaš, J., 1984. Demänovská ľadová jaskyňa – niektoré poznatky a výsledky z merania teploty vzduchu za obdobie 1970–1982 [Demänová Ice Cave – some knowledge and results from air temperature measurement during 1970–1982]. Slovenský kras 22, 111–129 (in Slovak).

Halaš, J., 1985. Novšie poznatky z merania teploty horninového plášťa [Newer knowledge from temperature measurement of bed rock]. Slovenský kras 23, 69–88 (in Slovak).

Halaš, J., 1989. Tepelná bilancia Dobšinskej ľadovej jaskyne [Thermic balance of the Dobšiná Ice Cave]. Slovenský kras 27, 57–71 (in Slovak).

Halaš, J., Slíva, L., 1979. Príspevok k problému merania teploty v ľadových jaskyniach [Contribution to the problematics of temperature measurement in ice caves]. Slovenský kras 17, 41–57 (in Slovak).

Hanvai, J.E., 1908. Die Dobsinauer Eishöhle und ihre Umgebung, Dobsina, 24 p.

Hochmuth, Z., 1974. Červené vrchy 73 [Červené vrchy Mountains 1973]. Spravodaj Slovenskej speleologickej spoločnosti 5 (4), 18–19 (in Slovak).

Hochmuth, Z., 1979. Ľadová priepasť v Červených vrchoch [Ice Abyss in the Červené vrchy Mountains]. Slovenský kras 17, 133–137 (in Slovak).

Hochmuth, Z., 1982. Súčasný stav výskumu jaskýň Červených vrchov [Current state of the research of caves in the Červené vrchy Mountains]. Slovenský kras 20, 19–47 (in Slovak).

Hochmuth, Z., 1995a. Pseudokrasové gravitačné jaskyne v pohorí Bachureň (Centrálnokarpatský flyš) [Pseudokarst gravity caves in the Bachureň Mountains (Central Carpathian flysh)]. In: Gaál, Ľ. (Ed.), Preserving of pseudokarst caves, Proceedings of International Working Meeting, Rimavská Sobota – Salgótarján, 1995. SAŽP, Banská Bystrica, pp. 61–67 (in Slovak).

Hochmuth, Z., 1995b. Pseudokrasové javy pri Renčišove (Centrálnokarpatský flyš, geomorfologický celok Bachureň) [Pseudokarst phenomena at Renčišov (Central Carpathian flysh, geomorphological unit Bachureň)]. Spravodaj Slovenskej speleologickej spoločnosti 26 (4), 7–11 (in Slovak).

Hochmuth, Z., 1995c. Ľadové monolity v priepastiach na Slovensku [Ice monoliths in abysses in Slovakia]. In: Bella, P. (Ed.), Ochrana ľadových jaskýň, Proceedings, Dobšinská Ľadová Jaskyňa, 21.–22. 9. 1995. SSJ, Liptovský Mikuláš, pp. 75–79 (in Slovak).

Hochmuth, Z., Holúbek, P., 2011. Zmeny ľadovej výplne vo Veľkej ľadovej priepasti na Ohništi [Changes of ice fill in the Great Ice Abyss on the Ohnište]. Aragonit 16 (1–2), 73 (in Slovak).

Holmlund, P., Onac, B.P., Hansson, M., Holmgren, K., Mörth, M., Nyman, M., Perşoiu, A., 2005. Assessing the palaeoclimate potential of cave glaciers: the example of the Scărişoara Ice Cave (Romania). Geogr. Ann. 87A (1), 193–201.

Holúbek, A., 2014. Poznámky k úbytku ľadu vo Veľkej ľadovej priepasti na Ohništi v Nízkych Tatrách [Notes to the ice decrease in the Great Ice Abyss on the Ohnište in the Nízke Tatry Mountains]. Slovenský kras 52 (2), 207–214 (in Slovak).

Holzman, H., 1999. Sommer-eislaufen in der Dobschauer Eishöhle und in weiteren Höhlen. Sowie berichte über Eisstockschiessen in Höhlen. Slovenský kras 37, 99–108.

Hovorka, J., 1989. Speleopotápěčský průzkum Černého potoka v Silicko-gombaseckém jeskynním systému [Speleological survey of the Čierny potok Brook in the Silica-Gombasek cave system]. Speleofórum ′89, 7–10 (in Czech).

Hunfalvy, J., 1860. Magyarország és Erdély eredeti képekben, II. kötet [Hungary and Transylvania original pictures], Darmstadt, 427 p. (in Hungarian).

Jakál, J., 1971. Morfológia a genéza Dobšinskej ľadovej jaskyne/Die Morphologie und Genese der Eishöhle von Dobšiná. Slovenský kras 9, 27–33.

Kámen, S., 1957a. Priepasť Bodolová [Bodolová Abyss]. Krásy Slovenska 34 (7), 259–260 (in Slovak).

Kámen, S., 1957b. Nová ľadová jaskyňa na Muráni [New ice cave at Muráň]. Krásy Slovenska 34 (12), 474–476 (in Slovak).

Kámen, S., 1958a. Muránsky a Tisovský kras [Tisovec and Muráň Karst]. Slovenský kras 1, 99–105 (in Slovak).

Kámen, S., 1958b. Ľadová jaskyňa na Muránskej planine [Ice cave on the Muránska planina Plateau]. Československý kras 11, 249–250 (in Slovak).

Kámen, S., 1961. Priepasť Veľká Stožka [Veľká Stožka Abyss]. Slovenský kras 3, 119–121 (in Slovak).

Kámen, S., 1964. Prieskum priepasti Šingliarka na Muránské plošine [Survey of the Šingliarka Abyss on the Muránska planina Plateau]. Československý kras 15, 154–155 (in Slovak).
Kámen, S., 1971. Tisovský a Muránsky kras [Tisovec and Muráň Karst]. Manuscript, Archive of the SMOPaJ Liptovský Mikuláš, 46 p. (in Slovak).
Klappacher, W., Haseke-Knapczyk, H., 1985. Eisriesenwelt Kat. Nr. 1511/24. Salzburger Höhlenbuch, Band 4. Landesverein für Höhlenkunde, Salzburg, 556 p.
Kohn, S., 1875. Die Drachenhöhle zu Demänfalva im Liptauer Comitate. Tourist 16, Wien.
Korabinsky, J.M., 1778. Almanach von Ungarn auf das Jahr 1778. Wien, Preßburg. 434 p.
Korbay, F., 1953. Príspevok k dejinám výskumu Demänovských jaskýň na Slovensku [Contribution to the history of research of caves in Slovakia]. Geografický časopis 5 (3–4), 275–279 (in Slovak).
Korzystka, M., Piasecki, J., Sawiński, T., Zelinka, J., 2011. Climatic system of the Dobšinská Ice Cave. In: Bella, P., Gažík, P. (Eds.), Proceedings of the 6th Congress International Show Caves Association, Demänovská Dolina, Slovak Republic, October 18–23, 2010. Slovak Caves Administration, Liptovský Mikuláš, pp. 85–97.
Korzystka-Muskała, M., Sawiński, T., Zelinka, J., Piasecki, J., Muskała, P., 2014. Climatic effect of environmental modifications in entrance area of the Dobšinská Ice Cave (Slovak Paradise, Slovakia) – the latest results. Slovenský kras 52 (2), 147–172.
Kováč, Ľ., Elhottová, D., Mock, A., Nováková, A., Krištůfek, V., Chroňáková, A., Lukešová, A., Mulec, J., Košel, V., Papáč, V., Ľuptáčik, P., Uhrin, M., Višňovská, Z., Hudec, I., Gaál, Ľ., Bella, P., 2014. The Cave Biota of Slovakia. State Nature Conservancy of the Slovak Republic, Slovak Caves Administration, Liptovský Mikuláš. 192 p.
Kowalski, K., 1960. Činnost polských speleologů v roce 1958 [The activity of Polish speleologists in 1958]. Československý kras 12, 245–247 (in Czech).
Krenner, A.J., 1873. A Dobsinai jégbarlang [Dobšiná Ice Cave]. Természettudományi Közlöny 5 (49), 346–353 (in Hungarian).
Krenner, A.J., 1874a. A Dobsinai jégbarlang [Dobšiná Ice Cave]. Magyar Természettudományi Társulat, Budapest, pp. 6–40 (in Hungarian).
Krenner, A.J., 1874b. Die Eishöhle von Dobsau, Budapest, 37 p.
Krieg, O., 1883. Über die Dobschauer Eishöhle und Gletscherspuren im Riesengebirge, Hirschberg in Schlesien.
Kunský, J., 1935–36. Umělé zaledňovaní Bielské jaskyně v Tatrách [Artificial glaciation of the Belianska jaskyňa Cave]. Vesmír 14, 39–42 (in Czech).
Kunský, J., 1937. Die künstiche Vereisung der Bélaer Tropsteinhöhle in der Hohen Tatra. Zentralbllat für Mineralogie, Geologie und Paläontologie, Abt. B 5, 209–215.
Kunský, J., 1939. Některé formy ledových krápníků [Some forms of ice stalactites and stalagmites]. Rozpravy II. Třídy České akademie 49 (22). 8 p. + 3 tab. (in Czech).
Kunský, J., 1943. Ledové jeskyně [Ice caves]. Širým světem 21 (2), 33–40 (in Czech).
Kunský, J., 1950. Kras a jeskyně [Karst and caves]. Přírodovědecké nakladatelství, Praha. 200 p. (in Czech).
Kvietok, L., 1949. Dobšinská ľadová jaskyňa a v nej objavené kvapľové jaskyne [Dobšiná Ice Cave and its discovered dripstone parts]. Československý kras 2, 104–111 (in Slovak).
Kvietok-Krofta, L., 1955. Dobšinská ľadová jaskyňa [Dobšiná Ice Cave]. Štátne telovýchovné nakladateľstvo, Bratislava. 54 p. (in Slovak).
Lalkovič, M., 1992. Der unbekannte Aufriss von Silická ľadnica aus der 1. Hälfte des 18. Jahrhunderts. Karszt és Barlang, special issue. In: Proceedings of the ALCADI ´92 International Conference on Speleo History, pp. 69–70.
Lalkovič, M., 1993. Príspevok k počiatkom záujmu o jaskyne na Slovensku [Contribution to the early interest on caves in Slovakia]. Slovenský kras 31, 61–74 (in Slovak).
Lalkovič, M., 1995. On the problems of the ice filling in the Dobšiná Ice Cave. Acta Carsol. 24, 313–322.
Lalkovič, M., 2000. Z histórie Dobšinskej ľadovej jaskyne (na okraj 130. výročia jej objavu) [From the history of Dobšiná Ice Cave (at the 130th anniversary of its discovery)]. Aragonit 5, 30–32 (in Slovak).

Lalkovič, M., 2003a. Najstaršia písomná zmienka o Demänovskej ľadovej jaskyni [The oldest written record on the Demänová Ice Cave]. Aragonit 8, 48–51 (in Slovak).

Lalkovič, M., 2003b. Z histórie Demänovskej ľadovej jaskyne [From the history of Demänová Ice Cave]. Slovenský kras 41, 129–164 (in Slovak).

Lalkovič, M., 2009. K začiatkom elektrického osvetlenia Dobšinskej ľadovej jaskyne [To the beginnings of electric lighting of the Dobšiná Ice Cave]. Aragonit 14 (1), 51–57 (in Slovak).

Lapin, M., Faško, P., Melo, M., Šťastný, P., Tomain, J., 2002. Climatic regions. In: Landscape Atlas of the Slovak Republic. Ministry of Environment of the Slovak Republic, Bratislava/Slovak Environmental Agency, Banská Bystrica, p. 95 (map No. 27).

Leutscher, M., Jeannin, P.-Y., 2004. A process-based classification of alpine ice caves. Theor. Appl. Karstol. 17, 5–10.

Lowe, W.B., 1879. The ice cavern of Dobschau. Nature 20 (502), 151–153.

Lutonský, A., 1934. Čo sudím o zaľadnení Bielskej jaskyne? [What do I judge on the glaciation of Belianska jaskyňa Cave?]. Krásy Slovenska 13 (4–6), 102–103 (in Slovak).

Měska, G., 1936. Ledová jeskyně Dobšinská – fenomén větraní krasových skal [Dobšiná Ice Cave – a phenomena of karst rocks weathering]. Bratislava 10, 253–258 (in Czech).

Miháľ, F., 2004. Významné rozsadlinové jaskyne Dreveníka [Important crevice caves of the Dreveník]. Spravodaj Slovenskej speleologickej společnosti 35 (2), 23–29 (in Slovak).

Mihalik, J., 1884a. Deménfalvi jég- és cseppköbarlang [Demänová ice and dripstone cave]. In: A Magyarországi Kárpát-egvesület Évkönyve 11, Igló, pp. 42–47. (in Hungarian).

Mihalik, J., 1884b. Die Deménfalver Eis- und Tropfsteinhöhle. In: Jahrbuch des Ungarischen Karpathen-Vereins, XI Jahrgang, Igló, pp. 42–48.

Mitter, P., 1991. Uzatvorenie CHPV jaskyne Silická ľadnica [Closing of the Silica Ice Cave], Jaskyniar, p. 62 (in Slovak).

Novotný, L., 1993. Treťohorné jaskynné úrovne a zarovnané povrchy v Slovenskom raji [Tertiary cave levels and planation surfaces in the Slovak Paradise]. Slovenský kras 31, 55–59 (in Slovak).

Novotný, L., Tulis, J., 1996. Výsledky najnovších výskumov v Dobšinskej ľadovej jaskyni [Results of the newest researches in the Dobšiná Ice Cave]. Slovenský kras 34, 139–147 (in Slovak).

Novotný, L., Tulis, J., 2000. Litologické a štruktúrno-tektonické pomery sprístupnenej časti Dobšinskej ľadovej jaskyne [Lithological and structural-tectonic settings of the accessible part of Dobšiná Ice Cave]. In: Bella, P. (Ed.), Výskum, využívanie a ochrana jaskýň, Proceedings of the 2nd Scientific Conference, Demänovská Dolina, 16.–19.11.1999. SSJ, Liptovský Mikuláš, pp. 59–65 (in Slovak).

Novotný, L., Tulis, J., 2002. Nové poznatky o kvapľových častiach Dobšinskej ľadovej jaskyne [New knowledge on the dripstone parts of Dobšiná Ice Cave]. In: Bella, P. (Ed.), Výskum využívanie a ochrana jaskýň, Proceedings of the 3rd Scientific Conference, Stará Lesná, 14.–16.11.2001. SSJ, Liptovský Mikuláš, pp. 36–49 (in Slovak).

Novotný, L., Tulis, J., 2005. Kras Slovenského raja [The karst of Slovak Paradise]. Správa slovenských jaskýň, Slovenská speleologická spoločnosť. Liptovský Mikuláš, Žilina. 175 p. (in Slovak).

Obleitner, F., Spötl, C., 2011. The mass and energy balance of ice within the Eisriesenwelt cave, Austria. Cryosphere 5, 245–257.

Ondroušek, O., 1952. Záchrana Dobšinské ledové jeskyně [The saving of Dobšiná Ice Cave]. Československý kras 5, 84–85 (in Czech).

Otruba, J., 1958a. Teplotné pomery v Ľadovej priepasti na Ohništi [Thermal conditions in the Ice Abyss at Ohnište]. Slovenský kras 1, 24–28 (in Slovak).

Otruba, J., 1958b. Problém mikroklímy a znovuzaľadnenia Demänovskej ľadovej jaskyne [Problem of microclimate and glaciation of the Demänová Ice Cave]. Slovenský kras 1, 36–58 (in Slovak).

Otruba, J., 1971. Meteorologické podmienky a zaľadnenie v Demänovskej ľadovej jaskyni/Meteorological conditions and glaciation in the Demänová Ice Cave. Slovenský kras 9, 193–211.

Paloncy, E., 1934. Zalednění Bielské jeskyně [Glaciation of the Belianska jaskyňa Cave]. Krásy Slovenska 13 (4–6), 97–102 (in Czech).

Paryski, W.H., 1972. Tatry Wysokie [Vysoké Tatry Mountains]. vol. 15. Przewodnik taternicki, Warszawa. 212 p. (in Polish).

Pavlarčík, S., 1984. Speleologický výskum krasových javov obalovej jednotky severnej strany Vysokých Tatier [Speleological research of karst phenomena of the cover unit of the northern side of Vysoké Tatry Mountains]. Slovenský kras 22, 41–67 (in Slovak).

Pavlarčík, S., 1985. Členenie krasových oblastí Vysokých a Belianskych Tatier [Division of karst areas of the Vysoké and Belianske Tatry Mountains]. Manuscript, archive of the SMOPaJ Liptovský Mikuláš, 56 p. (in Slovak).

Pelech, J.E., 1878. A Sztracenai völgy és a Dobsinai jégbarlang [The Stratená Valley and Dobšiná Ice Cave]. Külön-lenyomat a Magyarországi Kárpátegylet 5, Igló, 29 p. (in Hungarian).

Pelech, J.E., 1879. Valley of Stracena and the Dobschau Ice-Cavern. Trubner & Co., Ludgate Hill, London. 31 p.

Pelech, J.E., 1884a. A Sztraczenai völgy és a Dobsinai jégbarlang [The Stratená Valley and Dobšiná Ice Cave]. Franklin-társulat könyvnyomdája, Budapest. 40 p. (in Hungarian).

Pelech, J.E., 1884b. Das Stracenaer Thal und die Dobschauer Eishöhle. Buchdruckerei des Franklin-verein, Budapest. 51 p.

Perşoiu, A., Pazdur, A., 2011. Ice genesis and its long-term mass balance and dynamics in Scărişoara Ice Cave, Romania. Cryosphere 5, 45–53.

Petrovič, Š., 1952. Mikroklimatický prieskum ľadovej jaskyne u Dobšinej [Microclimatic survey of the ice cave at Dobšiná Town]. Meteorologické zprávy 5 (4), 102–106 (in Slovak).

Petrovič, Š., Šoltís, J., 1971. Stručná mikroklimatická charakteristika Dobšinskej ľadovej jaskyne/Kurzgefaßte mikroklimatische Charakteristik der Eishöhle von Dobšiná. Slovenský kras 9, 41–56.

Petters, C., Milius, J., Buchroithner, M.F., 2011. Eisriesenwelt: terrestrial laser scanning and 3D visualisation of the largest ice cave on Earth. In: Proceedings European LiDAR Mapping Forum, Salzburg, Austria. 10 p.

Piasecki, J., Zelinka, J., Pflitsch, A., Sawiński, T., 2004. Structure of air flow im the upper parts of the Dobšinská Ice Cave. In: Bella, P. (Ed.), Výskum využívanie a ochrana jaskýň, Proceedings of the 4th Scientific Conference, Tále, 5.–8. 10. 2003. SSJ, Liptovský Mikuláš, pp. 113–124.

Piasecki, J., Sawiński, T., Zelinka, J., 2005. Spatial differentiation of the air temperature in the entrance collapse of Dobšinská Ice Cave as contribution to the recognition of the problem of air exchange between cave and the surface. Slovenský kras 43, 81–96.

Piasecki, J., Sawiński, T., Strug, K., Zelinka, J., 2007. Selected characteristics of the microclimate of the Demänovská Ice Cave. In: Zelinka, J. (Ed.), Proceedings of the 2nd International Workshop on Ice Caves, Demänovská Dolina, Slovak Republic, May 8–12, 2006. SSJ, Liptovský Mikuláš, pp. 50–61.

Piasecki, J., Pflitsch, A., Sawiński, T., Zelinka, J., 2008a. Dobsinska Ice Cave thermal-circulation system in the light of the newest studies. In: Kadebskaya, O., Mavludov, B., Pyatunin, M. (Eds.), Proceedings of the 3rd International Workshop on Ice Caves, Kungur, Russia, May 12–17, 2008. Perm State University, Perm, pp. 23–25.

Piasecki, J., Sawiński, T., Zelinka, J., 2008b. The structure of air flow inside the lower part of the Dobšinská Ice Cave (the Underground Floor – Ice Cliff area). Slovenský kras 46 (1), 127–141.

Piovarcsy Jr., K., 1927. Die Höhlen am Dreveník. Turistik, Alpinismus und Wintersport 8, 9–10.

Prikryl, Ľ.V., 1985. Dejiny speleológie na Slovensku [History of the speleology in Slovakia]. Veda, Bratislava. 158 p. (in Slovak).

Psotka, J., 2007. Príspevok k názvosloviu a geológii Snežnej diery na Borčianskej planine [Contribution to the nomenclature and geology of the Snežná diera Cave on the Borčianska planina Plateau]. Spravodaj Slovenskej speleologickej spoločnosti 38 (4), 29–33 (in Slovak).

Rajman, L., Roda, Š., Ščuka, J., 1985. Výskum dynamiky ľadu v Silickej ľadnici [Research of ice dynamics in the Silická ľadnica Cave]. Slovenský kras 23, 253–260 (in Slovak).

Rajman, L., Roda, Š., Roda Jr., Š., Ščuka, J., 1987. Termodynamický režim Silickej ľadnice [Thermodynamic regime of the Silická ľadnica Cave]. Slovenský kras 25, 29–63 (in Slovak).

Ranzanus, P., 1558. Epitome Rerum Ungaricarum velut per Indices Descripta, Nyomtatásban, Viennae.

Roda, Š., Rajman, L., 1971. Referát o výskume vertikálnych kvapľových útvarov v Silickej ľadnici/Berich über die Erforschung der vertikalen Eisformationen in der Höhle Silická ľadnica. Slovenský kras 9, 249–256.

Roda, Š., Rajman, L., Erdös, M., 1974. Výskum mikroklímy a dynamiky zaľadnenia v Silickej ľadnici [Research of microclimate and dynamics of the glaciation of Silická ľadnica Cave]. Slovenský kras 12, 157–174 (in Slovak).

Roth, S., 1881a. Felsö-Magyaország néhány barlangjának leirása [Description of some caves in Upper Ungars]. In: A Magyarországi Kárpát-egyesület Évkönyve 8, Igló, pp. 367–398 (in Hungarian).

Roth, S., 1881b. Einige Höhlen Ober-Ungarns. In: Jahrbuch des Ungarischen Karpathen-Vereines 8, Igló, pp. 399–430.

Roth, Z., 1940a. Několik geomorfologických poznámek o Jihoslovenském Krasu a o Silické Lednici [Some geomorphological notes on the South-Slovak Karst and Silická ľadnica Cave]. Rozpravy II. tř. ČA 49 (8), 1–24 (in Czech).

Roth, Z., 1940b. Charakteristika a vývoj šachtovitých závrtů a propastí typu *Light holes* Cvijić a novotvořených portálů vyšších pater říčních jeskyň [Characteristics and development of shaft-like dolines and abysses of type *Light holes* Cvijić and newly formed portals of the upper storeys of river caves]. Věstník Královské české společnosti nauk 6, 1–11 (in Czech).

Rubín, J., Skřivánek, F., 1963. Československé jeskyně [Czechoslovak caves]. Sportovní a turistické nakladatelství, Praha. 106 p. (in Czech).

Sartori, F., 1809. Länder- und Völker-Merkwürdigkeiten des österreichischen Kaiserthumes. Doll, Wien.

Scholtz, A., 1888. A tornai platói és környéke [Tornai Plateau and its surroundings]. Pest, pp. 3–40 (in Hungarian).

Schönviszky, L., 1968. A Kárpát-medence legrébben ismert jégbarlangja [The oldest known ice cave of the Carpathian Basin]. Karszt és Barlang 1–2, 11–16 (in Hungarian).

Schwalbe, B., 1882a. Die drei Eishöhlen von Demanova, Dobschau und Szilicze. In: Gaea, Natur und Leben XVIII, Köln, Leipzig, pp. 626–628.

Schwalbe, B., 1882b. Die drei Eishöhlen von Demanova, Dobschau und Szilicze in Ungarn und die Eishöhle in der Frauenmauer bei Eisenerz in Steiermark. In: Verhandlungen der Gesselschaft für Erdkunde zu Berlin, Band IX, No. 4, pp. 203–207.

Sekyra, J., 1954. Velehorský kras Bělských Tater [High-mountain karst of the Belianske Tatry Mountains]. Nakladatelství ČSAV, Praha. 121 p. (in Czech).

Shaw, T., 1999. Demänovská ľadová cave in 1859. Slovenský kras 37, 187–200.

Siarzewski, W., 1994. Jaskinie lodowe w Tatrach [Ice caves in the Tatry Mountains]. In: Grodzicki, J. (Ed.), Jaskinie Tatrzańskiego Parku Narodowego. vol. 5. Polskie Towarzystwo Przyjaciół Nauk o Ziemi, Tatrzański Park Narodowy, Warszawa, pp. 11–47 (in Polish).

Siegmeth, K., 1880. Deménfalver Höhle. Nach und durch Ungarn, III. Bändchen. Von Wien; Budapest in die Hohe Tátra, Oderberg; Zürich, pp. 44–45.

Siegmeth, K., 1898. Notes sur les cavernes de Hongrie. vol. III. Mémoires de la Société de Spéléologie, Paris, pp. 147–164.

Siegmeth, K., 1891a. Az Abauj-torna-gömöri barlangvidék II [Abauj-Torna-Gömör cave area]. In: A Magyarországi Kárpát-egvesület Évkönyve 18, Igló, pp. 33–52 (in Hungarian).

Siegmeth, K., 1891b. Das Abauj-Torna-Gömörer Höhlengebiet II. In: Jahrbuch des Ungarischen Karpathen-Vereines 18, Igló, pp. 34–54.

Siegmeth, K., 1909. La grotte de glace de Szilice. Spelunca 7 (57), 46–50.

Silvestru, E., 1999. Perennial ice in caves in temperate climate and its significance. Theor. Appl. Karstol. 11–12, 83–94.

Šincl, J., 1931. Dynamika tvoření ledu v Dobšinské ledové jeskyni [Dynamics of ice origin in the Dobšiná Ice Cave]. Sborník Prírodovedného odboru Slovenského vlastivedného muzea v Bratislave 1924–1931, Bratislava, pp. 123–133 (in Czech).

Skřivánek, F., Stárka, V., 1955. Krasové zjevy Horního vrchu v Jihoslovenském krasu [Karst phenomena of the Horný vrch Plateau in the South-Slovak Karst]. Sborník Československé společnosti zeměpisné 60 (3), 221–222 (in Czech).

Skřivánek, F., Stárka, V., 1956. Krasové zjevy státní přírodní rezervace Zádielska dolina [Karst phenomena of the Zádielska dolina Valley State Nature Reserve]. Ochrana přírody 11 (10), 289–295 (in Czech).

Skutil, J., 1936. Příspěvek k soupisu slovenských jeskyň [Contribution to the inventory of Slovak caves]. Krásy Slovenska 15 (10), 146–148 (in Czech).

Skutil, J., 1951. Dreveník. Krásy Slovenska 28 (10), 249–251 (in Czech).

Skutil, J., Klimeš, F., 1957. Belova londýnska správa o Silickej ľadovej a Rybárskej plynovej jaskyni z r. 1744 [The London report of M. Bel on the Silica Ice Cave and Fishing Gas Cave from 1744]. Geografický časopis 9 (4), 240–246 (in Slovak).

Šoltís, J., 1971. Výsledky mikroklimatického prieskumu v Dobšinskej ľadovej jaskyni [Results of microclimatic survey in the Dobšiná Ice Cave]. In: Kvitkovič, J. (Ed.), Problémy geografického výskumu, Proceedings, X. jubilejný zjazd československých geografov (Prešov, 2.–5. 9. 1965). Vydavateľstvo SAV, Bratislava, pp. 185–189 (in Slovak).

Špaček, J., 1968. Správa o činnosti jaskyniarskej skupiny v Trenčianskych Tepliciach za roky 1962–1965 [Report on the activity of speleological group at Trenčianske Teplice during 1962–1965]. Slovenský kras 6, 133–135 (in Slovak).

Stárka, V., 1954. Propasti Bodolová a Šindliarka – Ľadoveň na Muránské plošině [Bodolová and Šindliarka abysses – ice mass on the Muránska planina Plateau]. Československý kras 7, 190–191 (in Czech).

Štéc, M., 1998. Novoobjavená jaskyňa Studeného vetra v Ďumbierskom vysokohorskom krase [Newly discovered Jaskyňa studeného vetra Cave (The Cave of Cold Wind) in the Ďumbier high-mountain karst]. Spravodaj Slovenskej speleologickej spoločnosti 29 (4), 7–9 (in Slovak).

Štéc, M., 2001. Jaskyňa studeného vetra – nová ľadová jaskyňa na Slovensku [Jaskyňa studeného vetra Cave (Cold Wind Cave) – a new ice cave in Slovakia]. Spravodaj Slovenskej speleologickej spoločnosti 32 (1), 36–37 (in Slovak).

Štéc, M., 2013. Jaskyňa mŕtvych netopierov [Jaskyňa mŕtvych netopierov Cave (Dead Bat Cave)]. Ďumbiersky vysokohorský kras [Ďumbier high-mountain karst], 250 p. (in Slovak).

Steiner, L., 1922a. Hőmérsékleti viszonyok a Dobsinai jégbarlangban [Thermal conditions of the Dobšiná Ice Cave]. Matematekai és Természettudományi Értesito 39, 61–75 (in Hungarian).

Steiner, L., 1922b. Die Temperaturverhältnisse in der Eishöhle von Dobšina. Meteorol. Z. 39 (7), 193–199.

Strug, K., 2011. Klimatyzne uwarunkowania rozwoju zjawisk lodowych w jaskiniach o odmiennych cechach środowiska [Climatic conditions of the development of ice phenomena in caves with different environmental features]. 18 Rozprawy Naukowe Instytutu Geografii i Rozwoju Regionalnego Uniwersytetu Wrocławskiego, Wocław. 163 p. (in Polish).

Strug, K., Zelinka, J., 2008a. The dependence between changes of range of ice forms and thermal conditions in the Demänovská Ice Cave. Slovenský kras 46 (1), 141–161.

Strug, K., Zelinka, J., 2008b. The Demänovská Ice Cave – the volume balance of the ice monolith in 2003–2007. Slovenský kras 46 (2), 369–386.

Strug, K., Zelinka, J., 2008c. The influence of extremely different external thermal conditions on the ice forms occurrence in Slovak ice caves in the 2005–2007 period. In: Kadebskaya, O., Mavludov, B., Pyatunin, M. (Eds.), Proceedings of the 3rd International Workshop on Ice Caves, Kungur, Russia, May 12–17, 2008. Perm State University, Perm, pp. 74–84.

Strug, K., Piasecki, J., Sawiński, T., Zelinka, J., 2004. The ice crystals deposit in the Dobšinská Ice Cave. In: Bella, P. (Ed.), Výskum, využívanie a ochrana jaskýň, Proceedings of the 4th scientific conference, Tále, 5.–8. 10. 2003. SSJ, Liptovský Mikuláš, pp. 125–133.

Strug, K., Piasecki, J., Szymanowski, M., Sawiński, T., Zelinka, J., 2006. Quantitative character of the bottom ice in the Demänovská Ice Cave. In: Bella, P. (Ed.), Výskum, využívanie a ochrana jaskýň, Proceedings of the 5th Scientific Conference, Demänovská Dolina, 26.–29. 9. 2005. SSJ, Liptovský Mikuláš, pp. 167–174.

Strug, K., Perşoiu, A., Zelinka, J., 2008a. Preliminary results of ice temperature measurements in the Dobšinská Ice Cave (Slovakia) and Scarisoara Ice Cave (Romania). In: Kadebskaya, O., Mavludov, B., Pyatunin, M. (Eds.), Proceedings of the 3rd International Workshop on Ice Caves, Kungur, Russia, May 12–17, 2008. Perm State University, Perm, pp. 16–22.

Strug, K., Sobik, M., Zelinka, J., 2008b. The thermal balance of the ice part in the Demänovská Ice Cave (Slovakia) between 2005 and 2007. In: Kadebskaya, O., Mavludov, B., Pyatunin, M. (Eds.), Proceedings of the 3rd International Workshop on Ice Caves, Kungur, Russia, May 12–17, 2008. Perm State University, Perm, pp. 85–92.

Sydow, A., 1830. Die Demenfalver Drachenhöhle. In: Bemerkungen auf einer Reise im Jahre 1827 durch die Beskiden über Krakau und Wieliczka nach den Central-Karpathen, Berlin, pp. 310–315.

Székely, K., Horváth, P., 2009. Korčuliarske slávnosti v Dobšinskej ľadovej jaskyni [Skating events in the Dobšiná Ice Cave]. Slovenský kras 47 (2), 315–328 (in Slovak).

Szontág, M., 1871. Gömöri jégbarlang [Gemer ice cave]. Természettudományi Közlöny 3 (19), 39–40 (in Hungarian).

Szterényi, H., 1883. A deményfalvi, dobsinai és sziliczei jégbarlangokról [Demänová, Dobšiná and Silica ice caves]. Természettudományi Közlöny 15 (166), 270–273 (in Hungarian).

Terlanday, E., 1893. A sziliczei jégbarlangokról [Silica Ice Cave]. Természettudományi Közlöny 25 (288), 404–411 (in Hungarian).

Terlanday, E., 1896. A sziliczei jégbarlangokról [Silica Ice Cave]. Természettudományi Közlöny 28 (320), 183–192 (in Hungarian).

Townson, R., 1797. Travels in Hungary With a Short Account of Vienna in the Year 1793. G.G. and J. Robinson, London. 506 p.

Tulis, J., 1997. Pohyb ľadu v Dobšinskej ľadovej jaskyni [Ice movement in the Dobšiná Ice Cave]. Aragonit 2, 6–7 (in Slovak).

Tulis, J., Novotný, L., 1989. Jaskynný systém Stratenskej jaskyne [The Cave System of Stratenská jaskyňa Cave]. Osveta, Martin. 464 p. (in Slovak).

Tulis, J., Novotný, L., 1995. Čiastková správa o morfometrických parametroch v zaľadnených častých Dobšinskej ľadovej jaskyne [Partial report on morphomertric parameters in the glaciated parts of Dobšiná Ice Cave]. In: Bella, P. (Ed.), Ochrana ľadových jaskýň, Procedings, Dobšinská Ľadová Jaskyňa, 21.–22. 9. 1995. SSJ, Liptovský Mikuláš, pp. 25–28 (in Slovak).

Tulis, J., Novotný, L., 2003. Zmeny zaľadnenia v Dobšinskej ľadovej jaskyni [Glaciation changes in the Dobšiná Ice Cave]. Aragonit 8, 7–9 (in Slovak).

Turóczi, L., 1768. Ungaria suis cum regibus compendio data. Typis Collegii Academici Societatis Jesu, Tyrnaviae. 631 p.

Uhrin, M., 1998. Prehľad poznatkov o netopieroch (Mammalia: Chiroptera) systému Dobšinská ľadová jaskyňa – Stratenská jaskyňa [Review of knowledge on bats (Mammalia: Chiroptera) of the Dobšiná Ice Cave – Stratenská jaskyňa Cave System]. Aragonit 3, 15–18 (in Slovak).

Ulrich, F., 1937. Dobšinská ledová jeskyně a příbuzné zjavy na Slovensku [Dobšiná Ice Cave and related phenomena in Slovakia]. Věda přírodní 18, 253–258 (in Czech).

Valovič, Š., 1957. Mikroklimatický prieskum v Dobšinskej ľadovej jaskyni v r. 1952–1956 [Microclimatic survey in the Dobšiná Ice Cave during 1952–1956]. Meteorologické zprávy 10 (4), 91–94 (in Slovak).

Velič, M., 1970. Z činnosti oblastnej skupiny č. 17 – Dubnica nad Váhom [From the activity of regional group No. 17 – Dubnica nad Váhom]. Spravodaj Slovenskej speleologickej spolećnosti 1 (2), 31–36 (in Slovak).

Višňovská, Z., 2007. Jaskyne Demänovskej doliny – významné zimoviská netopierov [The caves of the Demänovská dolina Valley – important bat wintering sites]. Aragonit 12, 54–61 (in Slovak).

Višňovská, Z., Zelinka, J., Strug, K., 2007. Spatial distribution of hibernating bats (Chiroptera) in relation to climatic conditions in the Demänovská Ice Cave (Slovakia). In: Zelinka, J. (Ed.), Proceedings of the 2nd International Workshop on Ice Caves, Demänovská Dolina, Slovak Republic, May 8–12, 2006. SSJ, Liptovský Mikuláš, pp. 87–97.

Vitásek, J.R., 1874a. Dobšínská ledové jeskyně [Dobšiná Ice Cave]. Komenský 22 (15), 213–215 (in Czech).
Vitásek, J.R., 1874b. Dobšínská ledové jeskyně [Dobšiná Ice Cave]. Komenský 22 (16), 231–232 (in Czech).
Vitásek, J.R., 1900. Dobšínská ledové jeskyně [Dobšiná Ice Cave]. Časopis turistů 15, 336–342 (in Czech).
Vitásek, F., 1923. Demänovská ledová jeskyně [Demänová Ice Cave]. Věda přírodní 35 (161–166), 193–200 (in Czech).
Vitásek, F., 1930. Silický kras a jeho ledová jeskyně [Silica Karst and its ice cave]. Sborník Československé společnosti zeměpisné 36 (7–8), 209–212 (in Czech).
Vítek, J., 1970. Z pekla do ráje [From hell to paradise]. Krásy Slovenska 47 (11), 506–507 (in Czech).
Vítek, J., 1972. Formy krasu v travertinech na Dreveníku u Spišského Podhradí [Karst forms in travertine on the Dreveník at Spišské Podhradie Town]. Československý kras 23, 99–114 (in Czech).
Wala, J., 1971. Lodowa studnia v Krzesanicy [Ice well in the Mt. Kresanica]. Taternik 1 (71), 34–35 (in Polish).
Weisinger, K., 1898. A jégbarlangokról [Ice caves]. Különös tekintettel a hazai barlangokra [With Specific Regard to Domestic Caves], Sümeg (in Hungarian).
Wernher, G., 1549. De admirandis Hungariae aquis hypomnematon, Basel.
Windisch, K.G., 1780. Geographie des Königreichs Ungarn: Erster Theil. Bey Anton Löwe, Preßburg. 389 p.
Wisniewski, W.W., 1990. Nowe jaskinie w Wysokich Tatrach [New caves in the Vysoké Tatry Mountains]. Eksplorancik 15 (1), 22–42 (in Polish).
Wisniewski, W.W., 1992. Po stopách hľadačov pokladov (nové jaskyne vo Vysokých Tatrách) [In the footsteps of treasure hunters (new caves in the Vysoké Tatry Mountains)]. Spravodajca Slovenskej speleologickej spoločnosti 23 (2), 9–13 (in Slovak).
Woldřich, J., 1926. O ledových jeskyních – zvláště Dobšinské – a jejich vzniku [On ice caves – especially Dobšiná – and their origin]. Příroda 19, 97–104 (in Czech).
Wünsch, J., 1881. Ledová jeskyně Dobšinská [Dobšiná Ice Cave]. Osvěta 11 (1), 79–89 (in Czech).
Žák, K., Herman, H., Orvošová, M., Jačková, I., 2009. Cryogenic cave carbonates from the Cold Wind Cave, Nízke Tatry Mountains, Slovakia: extending the age of cryogenic cave carbonate formation to the Saalian. Int. J. Speleol. 38 (2), 139–152.
Žák, K., Orvošová, M., Filippi, M., Vlček, L., Onac, B.P., Perşoiu, A., Rohovec, J., Světlík, I., 2013. Cryogenic cave pearls in the periglacial zones of ice caves. J. Sediment. Res. 83 (2), 207–220.
Zelinka, J., 1996. Rekonštrucia vstupných častí Dobšinskej ľadovej jaskyne [Reconstruction of the entrance parts of the Dobšiná Ice Cave]. Aragonit 1, 15–16 (in Slovak).
Zelinka, J., 1997. Dôsledky mrazového zvetrávania Belianskej jaskyne [Consequences of frost weathering of the Belianska jaskyňa Cave]. Slovenský kras 35, 141–146 (in Slovak).
Zelinka, J., 1999. Problematika stabilizácie speleoklimatických pomerov NPP Silická ľadnica [Problematics of the stabilisation of speleoclimatic conditions of the Silická ľadnica (Silica Ice Cave) National Nature Monument]. Aragonit 4, 18–19 (in Slovak).
Zelinka, J., 2007. Výsledky speleoklimatického monitoringu jaskyne Snežná diera v klimatickom roku 2005–2006 [Results of speleoclimatic monitoring of the Snežná diera Cave in the climatic year 2005–2006]. Aragonit 12, 43–46 (in Slovak).
Zelinka, J., 2011. Teplotný režim Jaskyne studeného vetra v Ďumbierskom krase [Temperature regime of the Jaskyňa studeného vetra Cave (Cold Wind Cave) in the Ďumbier Karst]. Aragonit 16, 74–75 (in Slovak).
Zipser, C.A., 1817. Versuch eines topographisch – mineralogischen Handbuches von Ungern. Bey carl friedrich wigand, Oedenburg. 439 p.

FURTHER READING

Pflitsch, A., Piasecki, J., Sawiński, T., Strug, K., Zelinka, J., 2007. Development and degradation of ice crystals sediment in Dobšiná Ice Cave. In: Zelinka, J. (Ed.), Proceedings of the 2nd International Workshop on Ice Caves, Demänovská Dolina, Slovak Republic, May 8–12, 2006. SSJ, Liptovský Mikuláš, pp. 38–49.

CHAPTER

ICE CAVES IN SLOVENIA

30

Andrej Mihevc
Karst Research Institute, Postojna, Slovenia

CHAPTER OUTLINE

Slovenia is a middle latitude (46°N, 14°E) southern Central European country. It is located on the north side of the Adriatic Sea, the most northerly and deeply protruding into the continent part of the Mediterranean. The total area is 20,273 km^2, and the karst with 12.148 known caves covers about 44% of the country. The karst of Slovenia belongs to Alps and Dinaric Mountains. The lowest parts of the karst are on the coast and the highest on the plateaus of to 1800 m high Dinaric Mountains and up to 2800 m high in Alps.

Alpine karst landscape belongs to Southern Calcareous Alps with the highest peak Triglav 2864 m a.s.l. Uplift and Pleistocene glaciations formed and reshaped fluvial valleys and separated karst plateaus so that up to 2 km thick vadose zone developed between them. In the valleys, large karst springs are recharging the Alpine rivers. Currently there are seven caves deeper than 1 km in the Alps.

Dinaric Karst is the major karst area of Slovenia and is the main type of relief of Dinaric Mountains. The central part of the mountains consists of rows of 1000–1700 m high karst plateaus. On the plateaus, only karst morphology is developed. The main features are large dolines and residual conical hills. Because of vertical drainage, deep vadose shafts prevail; horizontal caves are in lower elevations closer to springs. During Pleistocene glaciations, the snow line was about 1200 m a.s.l., so some of the plateaus were glaciated and slightly modified by glaciers. On the plateaus, shafts prevail.

The climate of the country is in general defined by the western circulation, which brings precipitation and intrusions of the cold air from the east, which, especially in winter, brings long periods of cold and dry weather. Most of the karst areas face continental and mountainous climate, depending on the altitude and distance from the sea.

At the coast, the mean annual temperature is 14°C. Mean July temperature is 24°C, and January temperature, 4°C. Within a short distance from the sea, the relief rises rapidly. At an altitude of 530 m and 37 km from the coast, the mean annual temperature is 8.8°C, mean July temperature 17.6°C, and January temperature −1°C. Precipitation amounts to around 1551 mm, and the snow cover lasts on

Ice Caves. https://doi.org/10.1016/B978-0-12-811739-2.00030-9

average 100 days per year. At the same distance from the coast, in the central part of the ridge of Dinaric mountains, in an altitude of 937 m the annual precipitation is 2928 mm. Mean temperature is 6.7°C, in July 15.5°C and in January −3.5°C and there are 127 days with a mean daily temperature below freezing. In the Alps, the tree line is at about 1800 m, and 0°C isoline is at an elevation of about 2400 m. At the meteorological station Kredarica located at about 2500 m, the mean annual temperature is −1.6°C. In the same altitude are the last remains of the Triglav glacier, which has only 5000 m^2 of surface left and will probably disappear in next few years.

30.1 HISTORY OF ICE CAVE EXPLORATION

The ice caves were known and used traditionally by shepherds or foresters as natural refrigerators or the only source for water on the karst mountains (Gams, 1974). The first written reports and observations of cave ice are from the 17th century (Valvasor, 1689). Later caves with ice were studied by naturalists, (Hacquet, 1784) or visited and considered curiosities by travellers and locals (Robič, 1879). Towards the end of the 19th Ccentury, many caves from Slovenia were visited or included on the lists of scientists studying the phenomena (Balch, 1900). Important contributions were made by Moser (1889) in describing the ice cave Paradana. Some ice caves were explored during World War I, when they were used for water supply on the high karst plateaus of Dinaric karst and the Alps (Kunaver, 1922). Later ice caves were studied by Michler (1950), Gams (1963, 1974), Mihevc and Gams (1979), and Mihevc (2008). Some of the ice caves were studied together with the temperature and inversion of vegetation in the entrance parts of such caves (Martinčič, 1977) or ice mass balance. Mihevc (2015) connected the strong temperature inversion in depressions connected with the cave climate and found also patchy permafrost due to the underground air circulation through karst.

30.2 ICE CAVES

In the cave register of Slovenia, permanent ice is reported to be in 551 caves (Fig. 30.1), and there are many more caves where ice or snows are persistent through a large part of the year.

The lowest cave with permanent ice was reported to be Ledena jama pri Ograji at 530 m asl (Balch, 1900), but today, ice in it melts every year towards the middle of the summer. So, at the moment the lowest cave with permanent ice appears to be Laniško brezno, with its entrance at an altitude of 645 m. The cave is a simple pit with a large opening and is 70 m deep. Mean annual temperature at that height is about 7°C.

There are 28 ice caves at altitudes below 1000 m, and another 479 between 1000 and 2400 m. Above 2400 m, where the mean annual temperature is generally below 0°C, 12 caves are known, and in 4 of them, permanent ice was reported.

The majority of the ice caves are simple shafts with large entrances. Many of them are at the bottoms of the dolines or similar depressions which collect cold air that cools the entire cave system.

Most ice caves are formed in the elevation where mean annual temperature is above 0°C; therefore, there have to be favorable cave conditions for ice formation as described by Luetscher (2005). In the Dinaric and Alpine plateaus, the conditions are favorable; less favorable are narrow ridges or steep slopes of mountains. But as the elevations are higher, and temperatures are much lower, ice is formed in caves more easily.

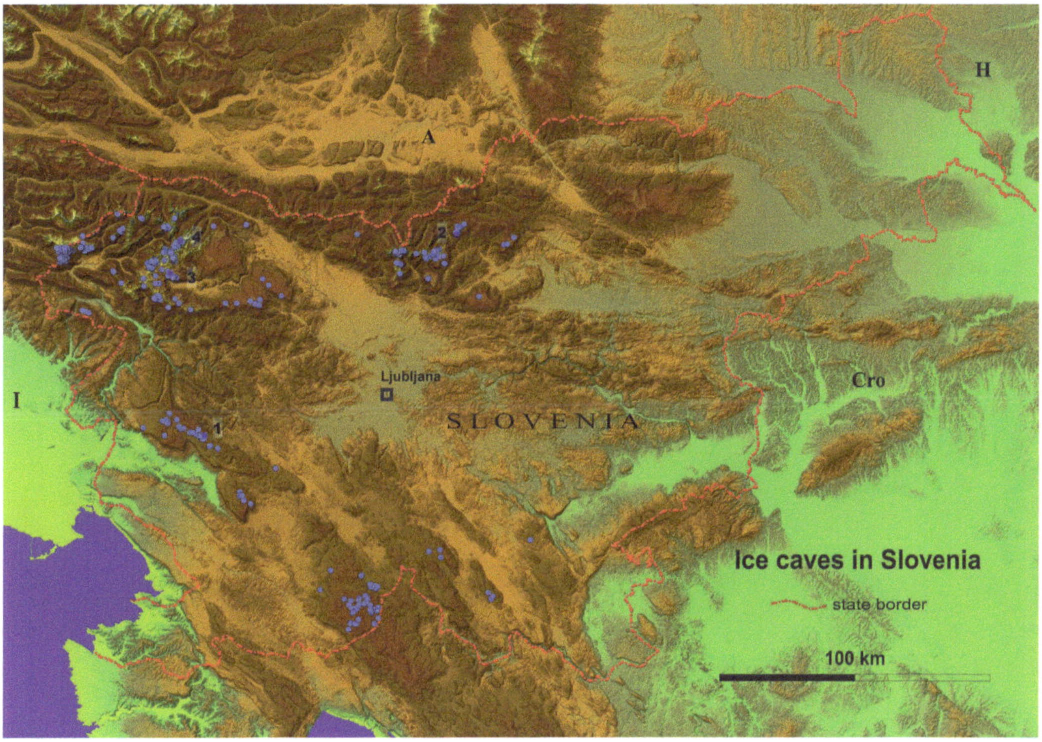

FIG. 30.1

Distribution of the ice caves in Slovenia. Ice caves are marked as *blue dots*. Letters mark important caves mentioned in text: A, Paradana; B, Snežna jama na Raduhi.

The ice is mostly formed from firn in the entrance parts and by freezing of the percolating water in the inner parts of the caves. Hoarfrost ice (or ice needles) and ice formed in sediments are negligible by quantity, but important in cave floor morphology, causing solifluction or formation of rock glaciers in the caves with large entrances.

30.3 IMPORTANT ICE CAVES

Velika ledena jama v Paradani (Great Ice cave in Paradana, Figs. 30.2 and 30.3) (45°59′19.70″N, 13°50′40.24″E), known as Paradana is probably the best-known ice cave in Slovenia. It is situated on the high Dinaric karst plateau Trnovski gozd. The cave is 6534 m long and 858 m deep. The entrance is 1130 m above sea level in the bottom of the large, 50 m deep, 500 m long, and 250 m wide doline between the Golaki peaks that are over 1400 m high. The entrance is a funnel-like depression formed by collapse that continues into three chambers and a few meters of horizontal passages. From here, several series of inner shafts and typical pitch-ramp morphology continue into the depth. Only below 700 m of depth are there some subhorizontal passages that end in collapses or too-narrow segments. Research in the cave is still going on.

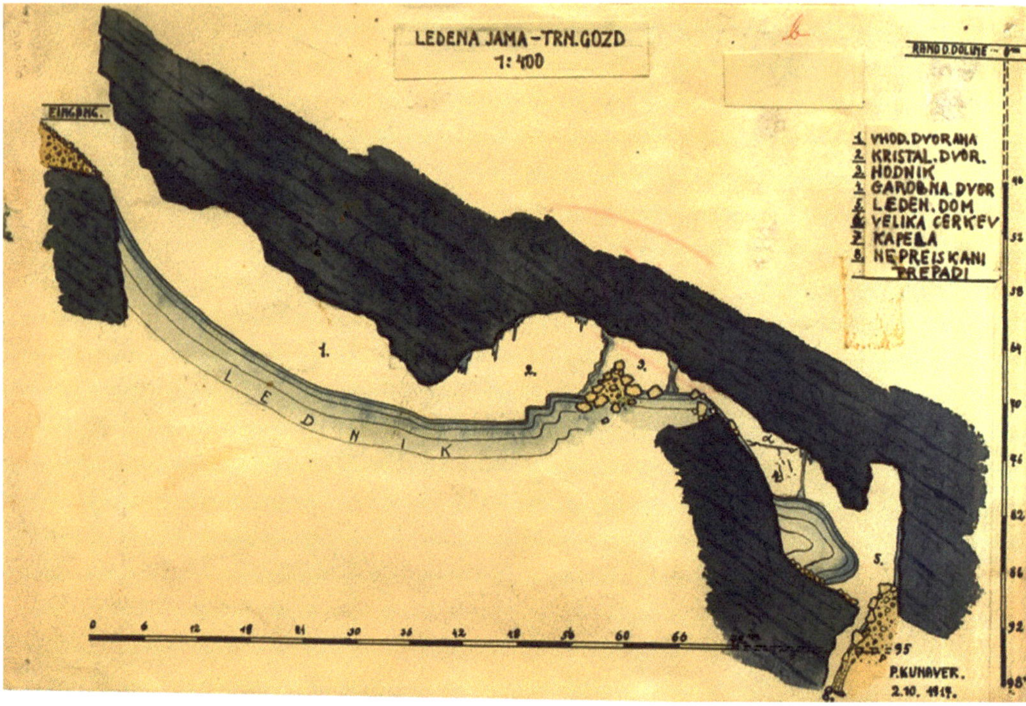

FIG. 30.2

Plan of Paradana ice cave by Pavel Kunaver from 1917. Archive of the Karst Research Institute in Postojna.

Moser in 1889 first described and measured temperatures in Paradana and other ice caves in the vicinity. Kunaver (1922) attributed the formation of ice in the cave to its position at the bottom of a doline, and the pocketlike shape of the cave, and to other factors such as exposure, the presence of a forest at the entrance, and avalanches. Later, after the discovery of the inner parts of the cave and on the basis of measurements of air movements and temperatures in deeper parts of the cave, ice formation was attributed exclusively to air circulation (Mihevc and Gams, 1979; Mihevc, 2008).

To describe air temperatures and movement in Paradana, we used data from sporadic measurements and observations since 1978, regular weekly observations during the summer of 2012, and regular monitoring of temperatures from 2008 to 2015.

The measurements showed that in the cold half of the year, when the outside temperature drops below 4°C, Paradana starts to suck in air (winter circulation). The velocity of the air entering the cave in winter can reach $2\,ms^{-1}$ in the main passage, which has a cross section of ~$4\,m^2$. Air also entered through numerous small holes and passages. We estimate the total flow of air into the cave to be $15\,m^3s^{-1}$. The air current warms up with the depth. On March 27, 1989, when the outside temperature was −1.6°C, the air entering the cave had warmed to 0°C at a depth of 140 m, to 1.1°C at 180 m, to 1.5°C at 240 m, to 3°C at 380 m, and to 4°C at 650 m.

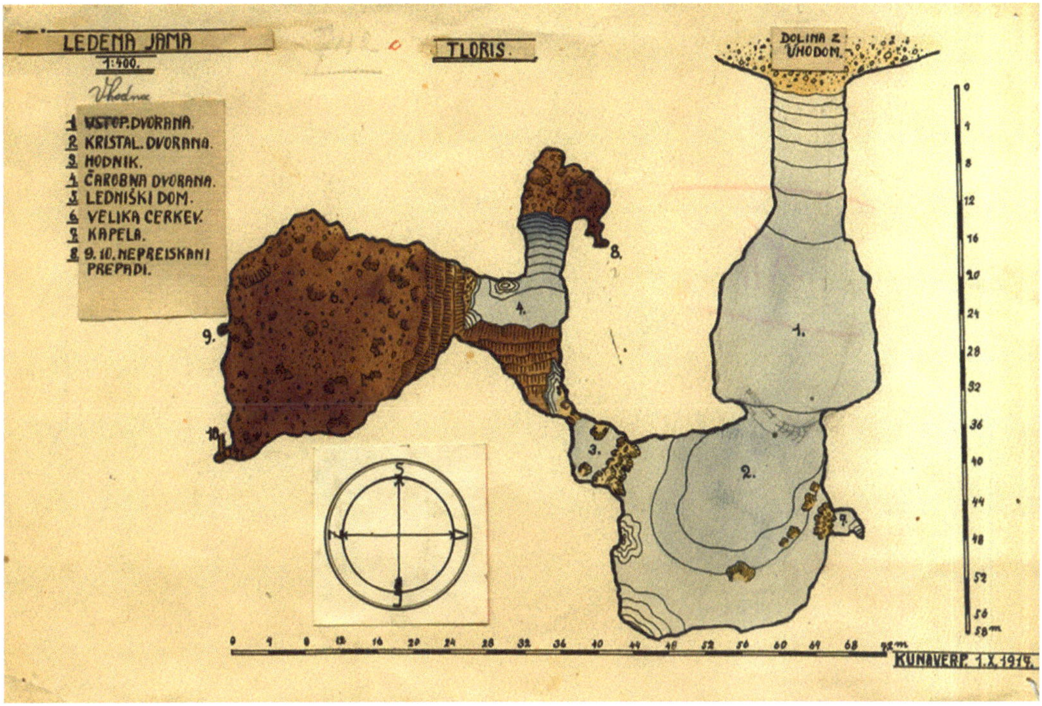

FIG. 30.3

Cross section of Paradana ice cave by Pavel Kunaver from 1917. Archive of the Karst Research Institute in Postojna.

We assume from what we know of the cave that most of the air moves through a system of connected cavities in the Golaki and exits from a system of interconnected passages to the surface at about 1400 m (Fig. 30.4), where numerous snow melting points have been observed. As the entrance to Paradana (Fig. 30.5 and Fig. 30.6) lies at 1130 m, and the height difference in the system is about 270 m, the chimney effect provides sufficient suction.

When air temperature rises above the temperature of the cave system in the warm part of the year (summer circulation), the airflow reverses its direction. Cave air in the Golaki is cooler and denser with respect to the outside air. It moves from the mountain into the lower part of Paradana and from there is pushed up shafts, crosses the ice and flows out of the cave. In August 2013, the temperature was 3.9°C at a depth of depth 680 m, 2.2°C at 250 m, and 1.1°C at 110 m, on the lower edge of permanent ice. At a depth of about 30 m, where the largest expanse of ice is, measurements were taken at about 50 cm above the ice. The temperature of the air was 1°C. Approximately 15 m^3 of air at a temperature of ~1°C flows through the entrance and maintains a stable lake of cold air in the depression at the entrance, the upper part of which shows a sharp temperature transition to the outside air temperature (Fig. 30.7).

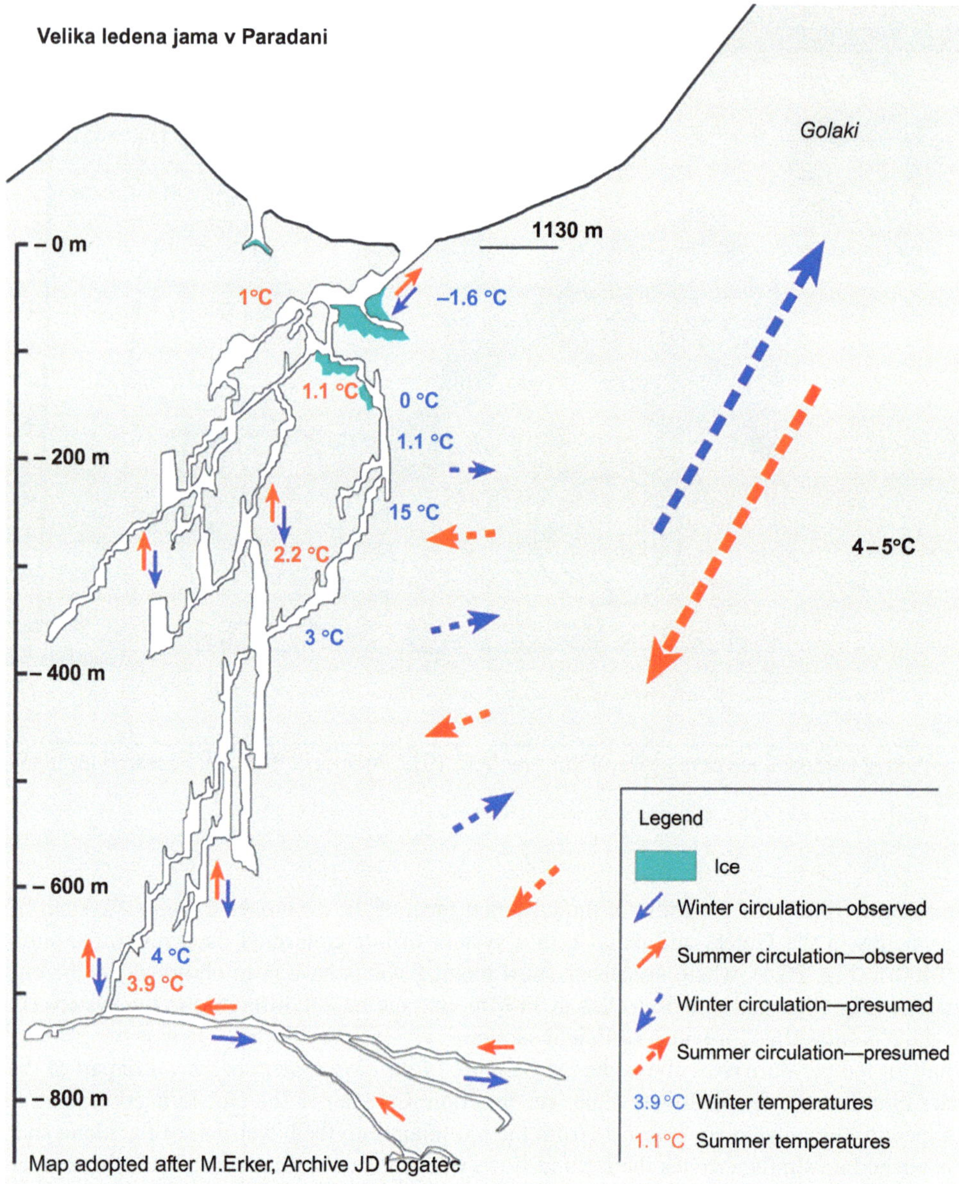

FIG. 30.4

Schematic cross section of Paradana ice cave with observed winter-summer air circulation. Temperatures measured at winter are in *blue* and *red* for one measured in summer. *Arrows* show the directions of winter (*blue*) and summer (*red*) air current.

FIG. 30.5

Entrance part of Paradana. Ice melted about 5 m since 1970 and exposed big rocks and logs.

Both the winter and summer temperature regimes and flow directions are stable and last for longer periods, but between them there are transitional periods with numerous short fluctuations in temperature and flow direction.

Ice estimated volume of about 8.000 m^3 is only in the entrance parts (Fig. 30.6), to a depth of about 100 m, but the cold air currents can freeze water as deep as 200 m. The ice was quarried from the cave already in the 19th century when the forestry roads were built. In 1889, Moser reported that in the year 1867, 800 m^3 of ice was extracted from the cave and sold to coastal city Trieste. The ice was used for the supply of the town and the port. Some ice was shipped also to Alexandria, where it was sold at a good price.

Cave ice level and volume changes can be roughly reconstructed for the past 100 years. Around 1917, ice volume was slightly larger than today. After that, in spite of ice extraction, the ice mass remained stable or even increased. From the late 1950s to 1977, the cave entrance was completely filled with ice. In 1978, the ice melted, and the cave was accessible again. Ice mass melted rapidly until 1982, after which the ice grew for a few years, but retreated again after 1986.

FIG. 30.6

Layers of snow and water ice in lower part of the entrance chamber of Paradana Ice Cave in 2015.

Around the surface of the plateau, close to Paradana ice cave, there are numerous blowholes where we observed strong air currents with similar characteristics in terms of direction and temperatures. They are part of the same cave air circulation as ice caves on the plateau. These are most common are at the bottoms of dolines and on the slopes of large karst depressions. They significantly influence the climate and vegetation. They are common on karst plateaus and also have regional importance. The best example of this phenomena is Smrekova draga (Fig. 30.8).

Smrekova draga (45°59′15.07″N, 13°52′3.97″E) is an about 1 km wide, 1.5 km long, and more than 150 m deep closed depression formed in Triassic dolomites and dolomitic limestones.

It is composed of two dolines. The deepest doline's bottom at 1105 m a.s.l., its middle is at 1130 m a.s.l., and the bottom of the southern depression is at 1125 m a.s.l. A small part of the floor and sides of Smrekova Draga is covered by slope talus and morainic material.

The floor of the southern depression is bare or covered with dwarf pine (*Pinus mugho*), and the rest of the floor is covered by coniferous forest, while higher slopes are covered by beech (*Fagus silvatica*). This vegetation inversion was initially attributed to temperature inversion in a closed karst depression. More detailed botanic research showed that the distribution of inversion is irregular, depending mostly on the cold air currents coming from blowholes, which botanists attribute to air circulation within the talus and moraine.

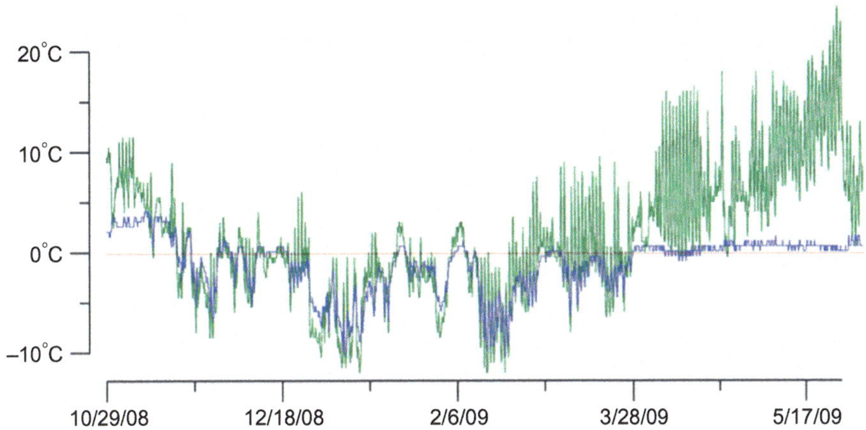

FIG. 30.7

Temperatures in Paradana at depth 50 m (*blue line*) and on the surface above cave (*green*) from October 2008 and June 2009. Temperature of air drops below 0°C because of flow of cold air entering cave. When temperatures outside are higher than ~4°C the direction of air current reverse. Air that crosses the cave ice cools down to only 1°C.

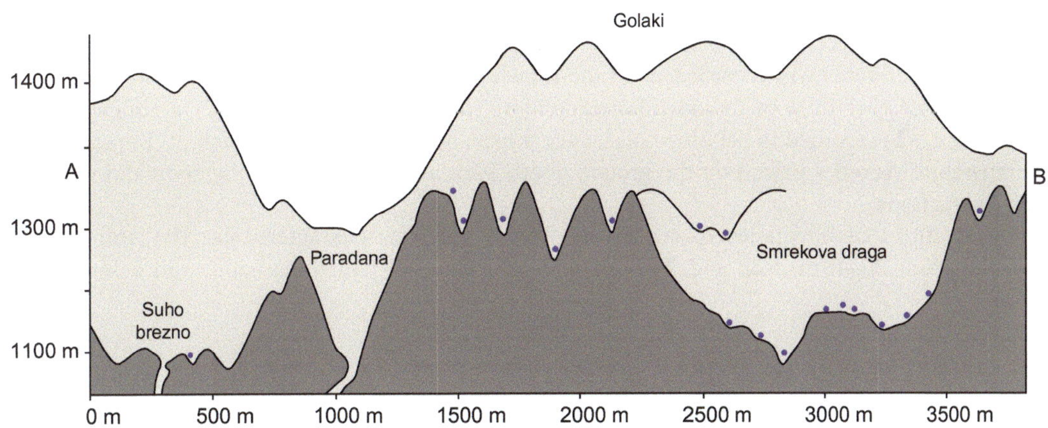

FIG. 30.8

Schematic profile across the discussed area. Important ice caves Suho brezno and Paradana are marked with *names*, and *dots* are blowholes where we measure air currents or we have installed dataloggers.

However, the distribution of the springs of cold air and low temperatures maintained all summer (about 1–3°C) indicate that the cold air of these blowholes originate from caves and not the spaces between the blocks. Only large cave system of the whole massive above the Smrekova draga can provide such a source of cold air. Smrekova draga is the lowest local relief, and the climate in it is partly a cave climate.

We observed and measured air temperature and velocity in around 40 blowholes, which are scattered or distributed in groups. The blowholes were located in the large Smrekova Draga depression, on the floors of smaller dolines, and also on slopes and level surfaces (Fig. 30.8).

Monitoring surface temperatures on the edge and floor of Smrekova Draga and some other dolines have shown longer periods of temperature inversion in the cold part of the year and diurnal temperature inversions during the summer. These are clearly the result of the position in a closed depression.

Monitoring the blowholes shows annual temperature fluctuations that match those we observed in Paradana. We observed about 40 blowholes and placed temperature recorders in four of them at a depth of around 0.5 m. Some of the blowholes were located in shallow 1–2 m deep depressions, some in karren, but most were in slope or moraine talus. They shared similar properties: air temperature was between 0.7°C and 3.7°C, and airflow velocity was 0.1–1.8 ms^{-1}. Both air temperature and air flow were clearly dependent on the degree of mixing of air in talus and the size and number of holes among the rocks. The temperature of the airflow in the blowholes was stable during the warm period of year. They did not reflect diurnal or temporary changes in the weather and did not increase towards the end of the summer. This is similar to the behavior of air in ice caves.

We estimated that on warm summer days, around 14 m^3s^{-1} of cold air was coming from the blowholes in Smrekova Draga. This air formed a thin temperature inversion. It mixed with the air in the vicinity so that at 0.5 m above the ground the temperature was approximately 10°C, while at 2.5 m from the ground it was normal (between 18°C and 22°C). The vegetation around groups of blowholes reflects extremely low summer temperatures. A layer of mosses up to 0.5 m thick, with blueberries and thick undecomposed organic matter, covers the rocks. The largest area with densely grouped blowholes is covered with dwarf pine (*Pinus mugho*)

In winter, the direction of airflow in the blowholes is reversed (Fig. 30.9). We observed this during winter visits on a tube we placed on one blowhole to avoid it being covered by snow. A similar range of temperatures was recorded in all the blowholes, including the one on which we mounted the tube. After each drop of the outside temperature (to a minimum of −26°C), a similar drop in temperature was recorded in the blowholes, even when snow cover was 1 m thick. At the same time, a temperature recorder placed on the ground about 10 m away and covered by snow did not show such fluctuations.

Observations of blowholes in six dolines showed similar characteristics. The dolines were funnel-shaped, 20–40 m deep and 100–200 m wide. They all had rocky sides and a few meters

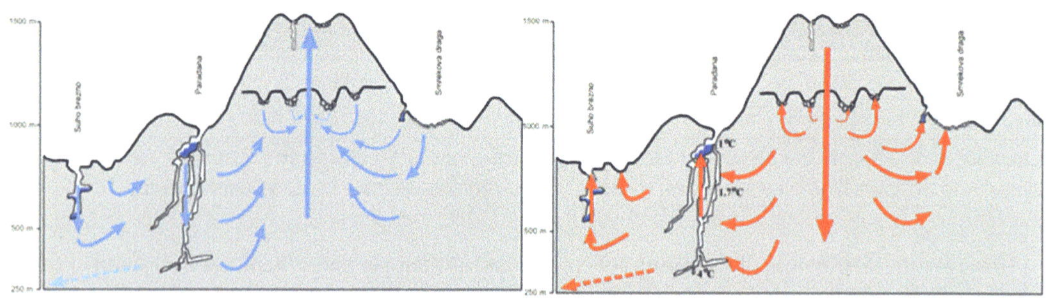

FIG. 30.9

Conceptual model of the air circulation in Paradana ice cave, Golaki, and Smrekova draga. Winter *(blue)* and summer *(red)* circulation in the karst massive Golaki creates patchy permafrost expressed as ice caves and patchy permafrost below blowholes.

of talus at the bottom, which was not enough to form cold scree anomalies. At the bottom of the dolines, there were strong blowholes with summer air temperatures of 1–3°C and air velocity of about 1 ms^{-1}. This inflow maintains the temperature inversion in the air at ground level, but at 2.5 m above the ground, the temperature was already normal. In one doline, blowholes with cold air were also located in the middle of the slope, while in the other there were blowholes on the outer rim. The doline floors around the blowholes were bare or covered with coarse humus and mosses with blueberries, while one or two meters higher, low bushes could already be observed, followed by normal vegetation.

Snežna jama cave (46°23′53.71″N, 14°44′31.52′E, Fig. 30.10) is located in the steep southern slope of Raduha Mountain (2062 m), which belongs to Southern Calcareous Alps. The whole mountain group is well karstified, and the deepest cave is over 1000 m deep. The prevailing type of caves are vadose pitch-ramp cave systems. Only in the Raduha mountain are some fragments of a large horizontal cave system preserved.

The entrance to the cave is on the slope of the mountain at 1514 m a.s.l. in a collapse dolina with permanent snow, which gave the name to the cave. Climatic conditions at the cave interpolated from nearest climatic stations are a mean air temperature of about 2.9°C, −6.1°C in January, and 11.7°C in July. The amount of precipitation is 1600 mm, and precipitation is distributed evenly throughout the year.

The main part of the cave consists of a large 1600 m long gallery with a diameter of more than 10 m and several smaller side passages. The main passage is practically horizontal at an altitude of about 1500 m a.s.l. A collapse terminates the cave close to the steep or vertical northern slopes of the mountain ridge.

The cave gallery was formed in phreatic and epiphreatic conditions. Dating of sediments and speleothems revealed that sedimentation started at before 5 Ma and ceased at about 2.6 Ma when speleothems started to grow (Häuselmann et al., 2015). Denudation of the slopes of Raduha later collapsed part of the passage and formed the present entrance.

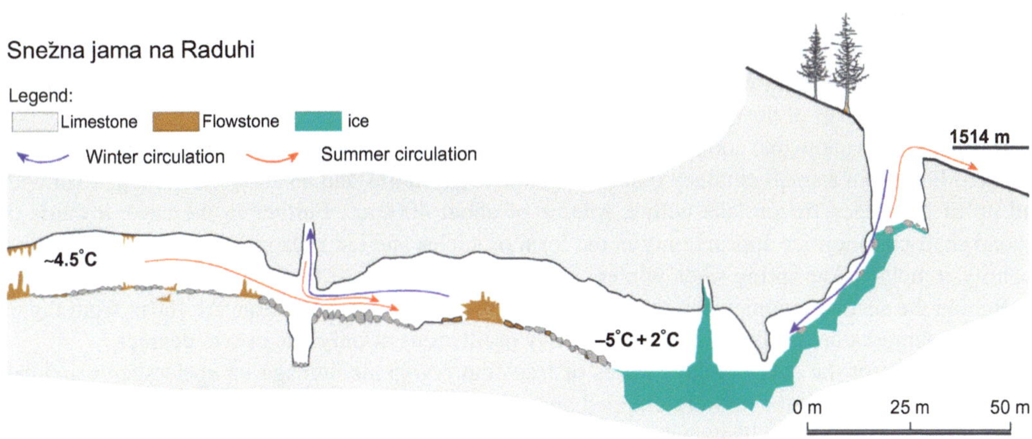

FIG. 30.10

Schematic cross section of the entrance part of Snežna jama with permanent ice. Extent of firn and ice are marked as well as the summer and winter air circulation in the cave. Freezing and ice formation in the cave occurs only between entrance and the second vadose shaft—chimney, which enables the efficient cooling during winter circulation. Inner parts of the cave have stable temperature.

FIG. 30.11

Main ice chamber in Snežna jama na Raduhi.

The main gallery is penetrated by several younger, active, vadose shafts which cut through the horizontal passage. Between the entrance and the first two shafts, a strong airflow establishes during winter, which cools down part of the cave, making conditions for perennial ice.

Ice forms from snow that collects at the entrance. The main ice body (Fig. 30.11) represents frozen vadose inflow from a small chimney that forms large ice columns and an about 25 m long, 15 m wide, and up to 10 m deep frozen lake with a volume of about 4000 m^3. Further in the cave, towards the second shaft-chimney, ice appears only in the form of icicles and ice stalagmites. Because of its small quantity, it melts in late spring when winter circulation stops (Fig. 30.10).

Behind the second chimney, the temperature is too high for ice formation. At 460 m from the entrance, the temperature is about 4.5°C, with yearly oscillations of only one or two degrees.

But this part of the cave displays traces of frost and cryogenic damage on speleothems and cave walls in the past. Speleothems were chipped away, and debris moved by cryoturbation formed a patterned ground. This is most likely a result of another open entrance in the past.

Observations of the cave ice date back to 1980, when the ice and snow at the entrance melted and allowed entrance to the cave. Since then, the volume of snow and ice at the entrance has dropped, and the level of ice in the lake dropped more than 1 m. Moreover, large ice columns that were stable at the beginning collapse every year. As the cave is managed as a touristic cave by cavers, they block summer circulation with a curtain and encourage winter circulation.

REFERENCES

Balch, E.S., 1900. Glacieres or Freezing Caverns. Allen, Lane & Scott, Philadelphia, PA. 337p.

Gams, I., 1963. Dopolnilne raziskave Triglavskega brezna leta 1962. Naše jame 4 (1–2), 21–22. Ljubljana.

Gams, I., 1974. Kras. Slovenska matica, Ljubljana.

Hacquet, B., 1784. Oryctographia Carniolica, oder Physikalische Erdbeschreibung des Herzogthums Krain, Istrien, und zum Theil der benachbarten Länder. Dritter Theil, Johann Gottlob Immanuel Breitkopf, Leipzig. 184 p.

Häuselmann, P., Mihevc, A., Pruner, P., Horáček, I., Čermák, S., Hercman, H., Bosák, P., 2015. Snežna jama (Slovenia): interdisciplinary dating of cave sediments and implication for landscape evolution. Geomorphology (Amsterdam, Netherlands) 247, 10–24.

Kunaver, P., 1922. Kraški svet in njegovi pojavi. Učiteljska tiskarna, Ljubljana. 104p.

Luetscher, M., 2005. Processes in Ice Caves and their Significance for Paleoenviromental Reconstructions. Swiss Institute for Speleology and Karst Studies (SISKA), La Chaux-de-Fonds. 154 p.

Martinčič, A., 1977. Beitrag zur Kenntnis der Ökologie der Kältelöcher in Slowenien: botanisch-ökologische Skizze. Razprave SAZU. Razred za prirodoslovne vede 20 (5), 91. Ljubljana.

Michler, I., 1950. Velika in mala ledena jama v Trnovskem gozdu. Proteus 12 (1949/50), 209–214. Ljubljana.

Mihevc, A., 2008. Distribution and characteristics of ice caves in Slovenia. In: 3rd International Workshop on Ice Caves, Kungur Ice Cave, Perm Region, Russia, May 12–17. Mining Institute of Ural Branch of RAS, Perm State University, Kungur, pp. 61–64.

Mihevc, A., 2009. Cryoturbation of the sediments at the cave entrances—case studies from Skednena jama, Potočka zijalka and Bestažovca cave. In: Hugo Obermaier Society for Quaternary Research and Archaeology of the Stone Age, 51st Annual Meeting in Ljubljana, April 14–18, pp. 26–27.

Mihevc, A., 2015. The influence of caves on the climate of the karst surface and karst permafrost. In: Čolak, I. (Ed.), Zbornik radova. Mostar, pp. 117–129.

Mihevc, A., Gams, I., 1979. Nova odkritja v veliki ledeni jami v Paradani (kat.št. 742). Naše jame 20, 7–20. Ljubljana.

Moser, C., 1889. Die Eishohlen des Tarnowaner und Birnbaumer Waldgebirges. Z. Dtsch. Oesterreich. Alpenver. 20, 351–368.

Robič, S., 1879. Ledenica na Veliki planini kamniški. Kmetijske in rokodelske novice 37 (28), 220.

Valvasor, J.V., 1689. Die Ehre dess Hertzogthums Crain. Laybach, zu finden bey Wolfgang Moritz Endter. Buchhändlern in Nürnberg, Erstes Buch. 696 p.

FURTHER READING

Kataster jam, 2017. Cave register of Karst Research Institute and Speleological Association of Slovenia.

CHAPTER

ICE CAVES IN THE USA

31

Stephen R. Higham*, Arthur N. Palmer†

Consultant, Georgetown, MA, United States State University of New York, Oneonta, NY, United States†*

CHAPTER CONTENTS

31.1 INTRODUCTION

Ice caves in the USA are not as abundant as one might expect for a country with so many mountain systems. Although altitudes in the "lower" 48 states range up to 4420m, cavernous rocks are very sparse above 2500m. The contiguous USA also lies at a relatively low latitude—the north-south extent of the 48 states ranges from north 25 degrees to north 47 degrees, which is equivalent to the range from central Egypt to the southern coast of England.

The high latitude of Alaska (55–71 degrees N) might seem to favor ice caves, but most of Alaska's karst is located at low elevations along the Pacific seacoast. No ice caves are yet documented there. A very sparse population, and difficult access, contribute to the lack of information on this subject.

Another factor in the small number of ice caves in the USA is the relatively dry condition of many of the western mountain ranges. Many parallel mountain chains of the far west receive abundant precipitation, but those farther east, in the prevailing down-wind direction, are relatively dry because of the rain-shadow effect. Note that neither latitude nor elevation alone is a good predictor of perennial ice in caves, and that any cave with abundant air circulation is likely to be ice-free for at least part of the year.

Well-known caves that contain ice in the lower 48 states of the western USA have fairly low altitudes. Examples are found in the Uinta Mountains, Utah (~2500m), Northern Rocky Mountains, Montana (~2200m), Teton Mountains, Wyoming (~2450m), Lava Beds National Monument, California

Ice Caves. https://doi.org/10.1016/B978-0-12-811739-2.00031-0

(~1370 m), and lava caves of El Malpais, New Mexico (~2200 m). The locations of major ice caves in the United States are shown on the accompanying map (Fig. 31.1).

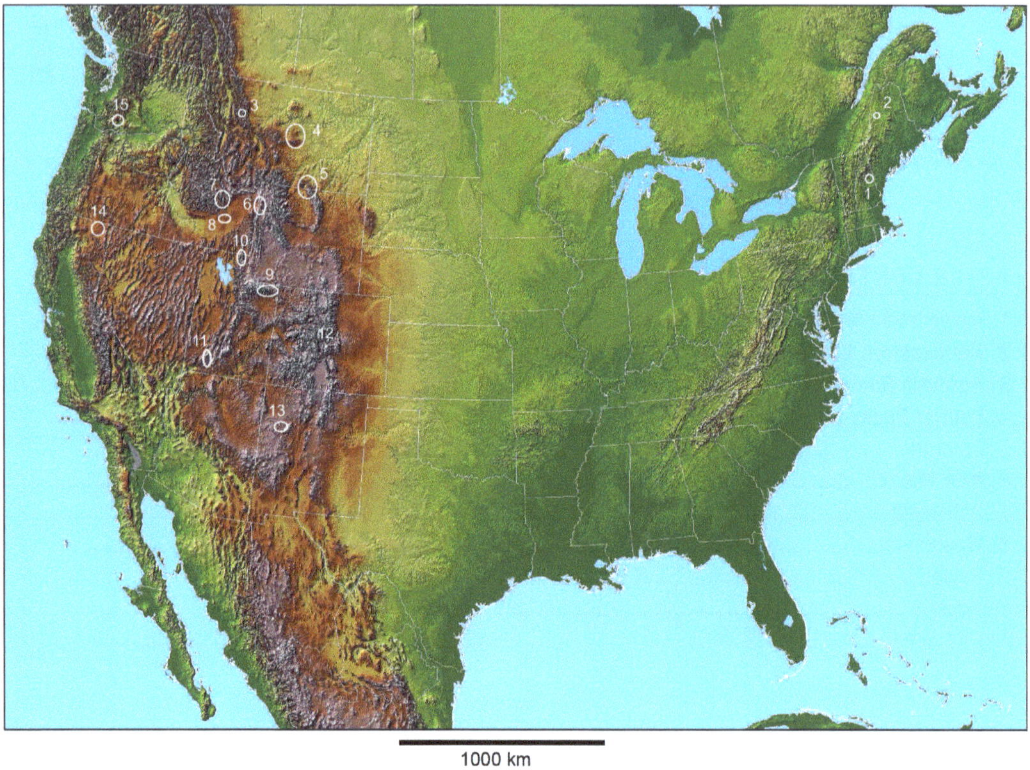

FIG. 31.1

Map of major ice caves of the USA. See text for site descriptions. **1** = Ice Gulch, New Hampshire; **2** = Debouille Lakes, Maine; **3** = Ice Cave, Silvertip Mountain, Montana; **4** = Judith Mountains, Montana; **5** = Pryor Mountains, Montana-Wyoming; **6** = Fossil Mountain Ice Cave, Wyoming; **7** = Lost River Mountains, Idaho; **8** = Craters of the Moon National Monument, Idaho; **9** = Uinta Mountains, Utah; **10** = Bear River Range, Utah; **11** = Markgunt Plateau, Utah; **12** = general region of high Rocky Mountains, Colorado; **13** = El Malpais lava beds, New Mexico; **14** = Lava Beds National Monument; **15** = Ice Cave, Washington.

31.2 NORTHEASTERN USA

The northeastern USA (New York and New England) have many so-called ice caves, but none of them now contain perennial ice. However, recent studies in two regional talus (pseudokarst) locales have found perennial ice among the blocks. Both areas contain small enterable caves, but it is not clear whether perennial ice has been documented within them. Variations in ice extent are being measured in Ice Gulch, northern New Hampshire, at elevations of 650–690 m, 44 degrees N (#1 on location map). Preliminary results are given in Holmgren and Pflitsch (2014). Changes in temperature within talus that contains perennial ice show strong temperature anomalies in comparison to surrounding areas. Ice, especially year-round ice, is not common below 700 m above sea level. Yearly mean air temperatures correlate strongly with yearly variations in ice level, showing promise that ice caves in talus and gorges can be used as climate indicators.

A second area is the Deboullie Lakes Ecological Reserve in northern Maine, at 350–450 m, 47 degrees N, where perennial subsurface ice has been found in inactive and relict rock glaciers (Putnam and Putnam, 2009; #2 on map).

31.3 NORTHERN ROCKY MOUNTAINS—MONTANA, WYOMING, IDAHO

Some ice caves of this region are of simple solutional origin (e.g., Fig. 31.2, #3 on map). In Montana the most numerous are of two types: gravity-slide and paleokarst solution breccia. In gravity slide

FIG. 31.2

Ice Cave, Silvertip Mountain, Montana (location 3 on Fig. 31.1).

Courtesy of Dave Bunnell.

caves, limestone, dolomite, or travertine overlies claystone or shale (Campbell, 2009). The shale provides a seal at the base of the fractures for ice to accumulate. Snow melt is the water source. Collapse of fossil caves in the Mississippian Madison Limestone, and infilling with a mix of breccia blocks and clay, can lead to the formation of ice caves. Cretaceous-Tertiary uplift and excavation of an old cave system can produce ice caves if the clay and limestone blocks seal off a chamber and there is a surface opening. The ice caves in the Big Snowy and Pryor Mountains are examples (#4 and #5 on map). Ice appears to collect in paleokarst caves of the early Carboniferous age because these voids are not adjusted to patterns of modern air or water flow. The through-flow of air and water in most caves in the region inhibits freezing temperatures. Notable Montana examples include Crater Ice Cave, Little Ice Cave, and Red Pryor Ice Caves.

On the western slope of the Teton Range of Wyoming, the Fossil Mountain Ice Cave was formed in Mississippian age Madison limestone (Figs. 31.3–31.5; #6 on map). This cave contains layered ice and sediments up to 13 m thick. Oxygen isotope analysis shows a temperature increase in the last 160 years (Medville, 2009; Fuller, 2006).

FIG. 31.3

Teton Shelf, near Fossil Mountain Ice Cave (Teton Mountains in background).

Courtesy of Doug Medville.

Strickler Cavern, in the Lost River Mountains of Idaho (# 7 on map), consists of a single room containing a large block of ice (personal communication, Lawrence Spangler, Salt Lake City, Utah, 2017). The ice has been dated to several hundred years.

Halliday (2009) describes many volcanic caves and open vertical volcanic conduits in the Snake River Plain, Idaho, some of which contain perennial ice. At Craters in the Moon National Monument, the opening known as Snow Cone contains a permanent ice cone and a frozen waterfall (#8 on map).

31.4 CENTRAL ROCKY MOUNTAINS—UTAH

There are many ice caves in the Uinta Mountains of northeastern Utah (personal communication, Lawrence Spangler, Salt Lake City, Utah, 2017). They are located in the vicinity of #9 on the map and include the following.

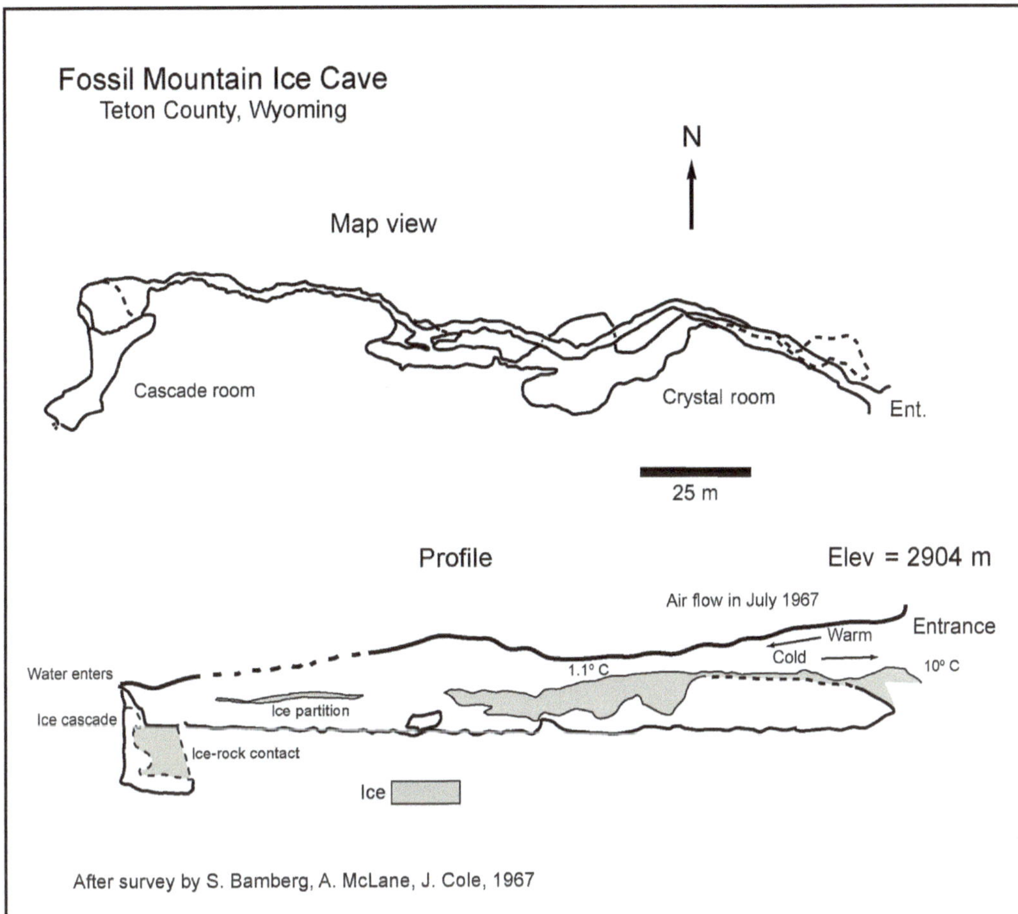

FIG. 31.4

Map and profile of Fossil Mountain Ice Cave, by Bamberg, McLane, and Cole.

Bear Ice Cave is a single room with a small persistent ice deposit. It has never been scientifically studied.

Grandview Ice Cave is a complex, strongly drafting cave with several small deposits of persistent ice. No scientific studies have been made there, other than a few temperature and wind-speed measurements.

Winter Wonderland Cave is a recently discovered linear cave with strong air flow and abundant, thick, persistent ice deposits. A few wind and temperature measurements have been made, along with several age estimates of organic debris within the ice. A temperature monitoring study is currently underway.

Jim Peck Ice Cave is a large and complex cave located at high altitude. It is a natural cold-air trap with a 100-meter-deep entrance shaft, which contains a massive block of perennial ice. However,

FIG. 31.5

Interior of Fossil Mountain Ice Cave.

Courtesy of Dave Bunnell.

the ice is essentially restricted to the entrance pit and a single large room directly below the pit. This cave has the potential to provide a long-term climatological record.

Big Brush Creek Cave is a large, complex, floodwater maze (Fig. 31.6). It contains perennial ice deposits in a side-passage complex near the entrance (Fig. 31.7). A large ice pillar also forms in the entrance room of the cave, which can be persistent from year to year but melts away every few years or decades during warmer periods, or when large floods enter the cave (Green, 2009). The entrance areas stay very cold year-round, because they form a cold-trap where winter air temperatures <0°C collect in a large ponor with an entrance about 50 m wide. Because the stream passage is sumped at the downstream end, there is no air flow to a lower entrance, so the cold air persists year-round in the entrance areas.

Toothbrush Cave contained no ice when it was washed open and first explored in 2005. However, the lower sections of the cave have become colder from year to year and are beginning to accumulate persistent ice. Extreme seasonal flooding will likely destroy the ice deposits every

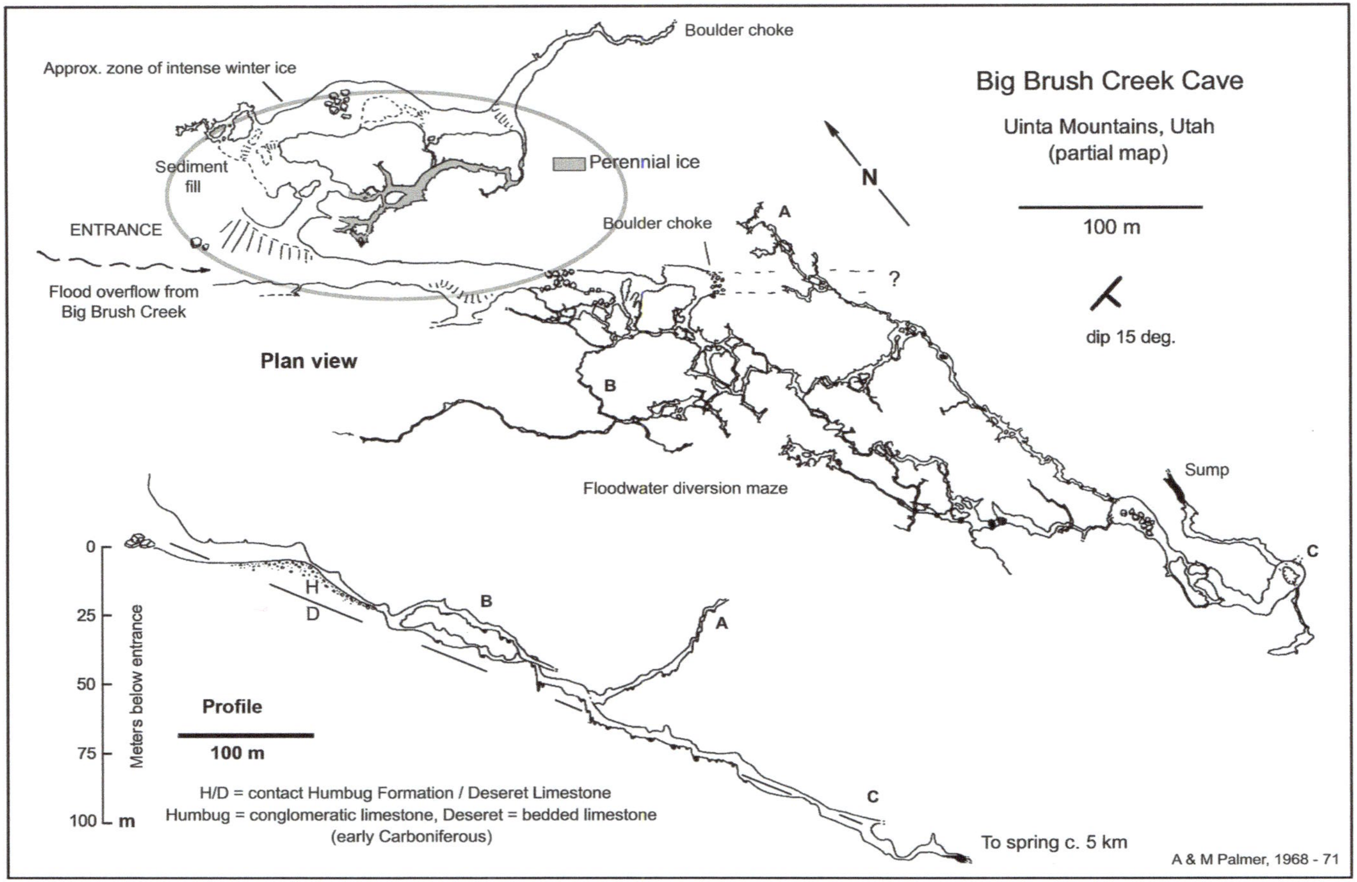

FIG. 31.6

Partial map and profile of Big Brush Creek Cave, Uinta Mountains, Utah, from Palmer (1975). Cold air is trapped in the huge ponor entrance of the cave at the sink point of Big Brush Creek, which drains a large area near the crest of the anticlinal mountain system. During the winter, water freezes in the upper few hundred metres of the cave (see Fig. 31.7), while temperatures persist at a few degrees above zero in the steeply descending lower sections of the cave. The entrance ice usually melts entirely during the summer, but ice persists year-round in the northeastern part of the cave, which is nearly blocked from the main passage by a high ridge of bedrock, sediment, and collapse material. That part of the cave terminates in breakdown at its southeastern end, which blocks most air flow to lower parts of the cave. Ice in that area blocks some passages entirely and forms extensive sheets and stalagmites on the floor.

FIG. 31.7

Ice choke in northeastern maze of Big Brush Creek Cave, Utah, August 1970. Note tree branch encased in ice, suitable for radiocarbon dating, and which suggests periodic melting and refreezing in the remote past.

Courtesy of A. Palmer.

few years or decades. A few temperature measurements have been made (down to 25F), but no significant studies have been made.

The following caves are located in the Bear River Range of Utah (#10 on map; personal communication, Lawrence Spangler, Salt Lake City, Utah, 2007).

Nielsons Cave contains occasionally persistent ice, which accumulates in a large room directly below the 100-meter-deep vertical entrance. This room had small persistent ice and snow deposits in the 1980s and 1990s, but no significant studies have been made, and the ice has retreated or melted entirely in recent years.

Main Drain Cave is Utah's deepest cave. A large cone of persistent snow occupies the floor of its 76 m vertical entrance. No real studies have been conducted here.

Several other caves in the Bear River Range (e.g., Hunters Cave, Glacier Cave, and likely others) also contain large cones or snow accumulations below vertical pit entrances, as well as ice build-ups on passage floors, which can persist for years. As is the case elsewhere in the area, there have been no significant studies or documentation.

A number of caves in southwestern Utah are located in the Markagunt Plateau (~2700 m), #11 on map. They are mainly lava tubes, and although some contain local areas of perennial ice, we are not aware of any ice-related studies in the region.

31.5 COLORADO

The greatest accumulation of subterranean ice in the USA is almost certainly in abandoned mines (lead-zinc, etc.) in the Rocky Mountains of Colorado. According to employees of the US Geological Society in Denver, nearly all such mines are choked with ice. These are located at high elevations, generally >3000 m. Their general location is in the vicinity of #12 on the map. Typically, entrances to these mines are their highest points, so they act as cold-air traps. They appear to behave like the paleokarst caves mentioned in Montana by Campbell (2009), and in fact much of the Pb-Zn ore is located in paleokarst features.

31.6 NEW MEXICO

Thick lava flows cover most of the landscape in northwest-central New Mexico (#13 on map). This is the Malpais region, which is covered by lava flows dating mainly from several thousand years ago. The average elevation is about 1300 m, but rises northward to as much as 2400 m in the vicinity of the Candelaria Ice Cave. The ice deposit is estimated to be about 3000 years old but is currently shrinking (Dickfoss et al., 1997). A few other lava-tube caves in the region contain year-round ice where there are no low entrances to prevent cold air from accumulating (Fig. 31.8).

FIG. 31.8

Typical lava tube in El Malpais National Monument, New Mexico. This cave contains no ice, but some with limited air circulation do.

Courtesy of A. Palmer.

31.7 CALIFORNIA

There are hundreds of known caves in Lava Beds National Monument (Bruce Rogers, Fremont, California, personal communication, 2017). The Monument is located at #14 on the map. Perennial ice is present in several of the caves, and changes in ice volume are being monitored by the National Park Service staff and others. Simple ice-level monitoring has been performed in 16 of the 35 known ice caves since 1990, supplemented with varying amounts of photo monitoring (Fig. 31.9). Although this monitoring reveals changes in the level of many ice floors, it does not detect changes in ice volume, nor does it detect differential changes across an ice floor (Thomas, 2010).

Late autumn ice levels are generally used as the standard for comparison. Records of ice levels in the Merrill Ice Cave have been analyzed from when records began in 1956 by stratigraphic and photographic methods. In general, there has been a great decline in ice volume (Fuhrmann, 2007). However,

FIG. 31.9

Crystal Ice Cave, in Lava Beds National Monument, is limited to ranger-guided visits to limit degradation of ice by body heat.

Courtesy of Dave Bunnell.

in Skull Ice Cave, seasonal variations in ice volume within individual years show no seasonal cycles. Of 18 ice-cave records examined by the group, this is the only one in the northern hemisphere in which the ice mass is increasing over multiple decades.

In other caves, perennial melting of previously stable ice floors is increasing, with some caves experiencing total ice loss where deposits were once greater than 2 m thick. To increase the quality of ice monitoring, the Park Service staff are refining a combination of surface-area and ice-level measurements to estimate changes in ice volume in the Monument's five most significant ice caves. This new approach is being established in accordance with the National Park Service Klamath Inventory and Monitoring Network's Integrated Cave Entrance Community and Cave Ecosystem Long-term Monitoring Protocol (Krejca et al., 2011). The goal of this long-term monitoring strategy is to document changes in cave environments using several different parameters, including ice.

31.8 WASHINGTON

Ice Cave, near Trout Lake, Washington, is a lava tube that contains some perennial ice in isolated areas (#15 on the map). This cave was studied for seven months in 1986 and 1987, with data gathered on temperature, relative humidity, wind speed, and ice speleothem melt rate (Martin and Quinn, 1990). Ice Cave is a segmented lava tube that traps cold air during winter, permitting the development of ice speleothems. Four separate daily visits from January 1986 through July 1987 allowed weather data collection in this cave. Winter visits in 1986 and 1987 revealed cool temperatures of −4°C and +1°C, respectively, in the stable cave interior in response to differing weather regimes. Spring and summer observations in 1987 indicated a varied temperature gradient from cave entrances to the deep interior. Stability in temperature and relative humidity increased when going deeper into the cave. Small temporal variations in temperature (0.5–1°C) and relative humidity were detected at all stations, indicating turbulent processes occasionally extended throughout the cave. Growth and decay of ice speleothems varied strongly both seasonally and spatially. Their spring and summer measurements agree roughly with profiles generated by classic heat-flow dynamics described by Wigley and Brown (1976).

31.9 HAWAII

Hawaii contains many lava tubes, but most are fairly warm and contain no ice. The main island is at a latitude of 19 degrees N. However, the upper slopes of Mauna Loa, a still-active volcano, are at a high enough elevation (~3350 m) that at least two lava tubes containing ice are known on its upper slopes. However, the ice is rapidly retreating, according to members of the Hawaii Speleological Society (Pflitsch et al., 2016). Although the lava tubes were formed by downward-flowing lava, their lower ends are blocked by solidified lava, so the caves act as cold traps. The largest ice cave contains a frozen lake with a surface area of about 250 m^2.

REFERENCES

Campbell, N.P., 2009. Northern Rockies of Montana. In: Palmer, A., Palmer, M. (Eds.), Caves and Karst of the USA. National Speleological Society, Huntsville, AL, pp. 220–228.

Dickfoss, P., Betancourt, J., Thompson, L., Turner, R., Thornstrom, S., 1997. History of ice at Candelaria Ice Cave. In: Mabery, K. (Ed.), Natural history of El Malpais Monument. New Mexico Bureau of Mines & Mineral Resources Bulletin 156, pp. 91–112.

Fuhrmann, K., 2007. Monitoring the disappearance of a perennial ice deposit in Merrill Cave. J. Cave Karst Stud. 69, 256–265.

Fuller, N., 2006, A reconstruction of Late Holocene climate in the Teton area from layers of sediment contained in ice from Fossil Mountain Ice Cave, Teton County, Wyoming: Alpine Karst, vol. 2, p. 1–14.

Green, D., 2009. Central Rockies of northeastern Utah. In: Palmer, A., Palmer, M. (Eds.), Caves and Karst of the USA. National Speleological Society, Huntsville, AL, pp. 237–238.

Halliday, W.R., 2009. Lava plateaus of Idaho. In: Palmer, A., Palmer, M. (Eds.), Caves and Karst of the USA. National Speleological Society, Huntsville, AL, p. 302.

Holmgren, D., Pflitsch, A., 2014. Analysis of selected climatological observations of talus and gorge caves in New England. In: Sixth International Workshop on Ice Caves, National Cave and Karst Institute Symposium, vol. 4, pp. 67–71.

Krejca, J., Myers, G., Mohren, S., Sarr, D., 2011. Integrated cave entrance community and cave environment long-term monitoring protocol: Natural Resource Report NPS/KLM/NRR-2011/XXX. National Park Service, Fort Collins, CO.

Martin, K., Quinn, R., 1990. Meteorological observations of Ice Cave. NSS Bulletin, vol. 5(1). Trout Lake, WA: National Speleological Society, pp. 45–51.

Medville, D., 2009. Teton Mountains, Northwestern Wyoming. In: Palmer, A., Palmer, M. (Eds.), Caves and Karst of the USA. National Speleological Society, Huntsville, AL, pp. 234–236.

Palmer, A., 1975. The origin of maze caves. National Speleological Society, NSS Bulletin, vol. 37(3), pp. 56–76.

Pflitsch, A., Schörghofer, N., Smith, S., Holmgren, D., 2016. Massive ice loss from the Mauna Loa ice cave, Hawaii. Arctic Antarctic Alpine Res. 48 (1), 33–43.

Putnam, A., Putnam, D., 2009. Inactive and relict rock glaciers of the Deboullie Lakes Ecological Reserve, northern Maine, USA. J. Q. Sci. 24 (7), 773–784.

Thomas, S.C., 2010. Monitoring cave entrance communities and cave environments in the Klamath Network, 2010 Pilot Study Results. Natural Resource Data Series NPS/KLMN/NRDS—2010. National Park Service, Fort Collins, CO.

Wigley, T.M.L., Brown, M.C., 1976. The physics of caves. In: Ford, T.D., Cullingford, C.H.D. (Eds.), The Science of Speleology. Academic Press, London, pp. 329–358.

FURTHER READING

Despain, J., 2009. California. In: Palmer, A., Palmer, P. (Eds.), Caves and Karst of the USA. National Speleological Society, Huntsville, AL, pp. 291–297.

Hill, C., Sutherland, W., Tierney, L., 1976. Caves of Wyoming. Geological Survey of Wyoming, Bulletin 59, 230 p.

Smith, K.J., 2014. Ice cave monitoring at Lava Beds National Monument. In: International Workshop on Ice Caves VI: National Cave and Karst Research Institute, Symposium, vol. 4, pp. 88–93.

Sowers, J., Devereaux, B., 2000. The ice cavity at Merrill Cave: Cave Research Foundation. Q. Newslett. 28 (2), 4–6.

Index

Note: Page numbers followed by *f* indicate figures and *t* indicate tables.

D

H

I

P

Q

R

FSC
www.fsc.org
MIX
Papier | Fördert gute Waldnutzung
FSC® C083411

Kolibri 360 GmbH
Ferdinand-Jühlke-Straße 7
99095 Erfurt, Deutschland
produktsicherheit@kolibri360.de